T0191525

Lecture Notes in Artificial Intelligence   10191

Subseries of Lecture Notes in Computer Science

More information about this series at http://www.springer.com/series/1244

Ngoc Thanh Nguyen · Satoshi Tojo
Le Minh Nguyen · Bogdan Trawiński (Eds.)

Intelligent Information and Database Systems

9th Asian Conference, ACIIDS 2017
Kanazawa, Japan, April 3–5, 2017
Proceedings, Part I

 Springer

Editors

Ngoc Thanh Nguyen
Wrocław University of Science
 and Technology
Wrocław
Poland

Satoshi Tojo
Japan Advanced Institute of Science
 and Technology
Nomi
Japan

Le Minh Nguyen
Japan Advanced Institute of Science
 and Technology
Nomi
Japan

Bogdan Trawiński
Wrocław University of Science
 and Technology
Wrocław
Poland

ISSN 0302-9743   ISSN 1611-3349   (electronic)
Lecture Notes in Artificial Intelligence
ISBN 978-3-319-54471-7   ISBN 978-3-319-54472-4   (eBook)
DOI 10.1007/978-3-319-54472-4

Library of Congress Control Number: 2017932640

LNCS Sublibrary: SL7 – Artificial Intelligence

Printed on acid-free paper

This Springer imprint is published by Springer Nature
The registered company is Springer International Publishing AG
The registered company address is: Gewerbestrasse 11, 6330 Cham, Switzerland

Preface

ACIIDS 2017 was the ninth event in a series of international scientific conferences on research and applications in the field of intelligent information and database systems. The aim of ACIIDS 2017 was to provide an international forum for scientific research in the technologies and applications of intelligent information and database systems. ACIIDS 2017 was co-organized by Japan Advanced Institute of Science and Technology (Japan) and Wrocław University of Science and Technology (Poland) in co-operation with IEEE SMC Technical Committee on Computational Collective Intelligence, Quang Binh University, Vietnam, Yeungnam University, South Korea, Bina Nusantara University, Indonesia, Universiti Teknologi Malaysia, and the University of Newcastle, Australia. It took place in Kanazawa (Japan) during April 3–5, 2017.

The conference series ACIIDS is well established. The first two events, ACIIDS 2009 and ACIIDS 2010, took place in Dong Hoi City and Hue City in Vietnam, respectively. The third event, ACIIDS 2011, took place in Daegu (Korea), followed by the fourth event, ACIIDS 2012, which took place in Kaohsiung (Taiwan). The fifth event, ACIIDS 2013, was held in Kuala Lumpur in Malaysia while the sixth event, ACIIDS 2014, was held in Bangkok in Thailand. The seventh event, ACIIDS 2015, took place in Bali (Indonesia). The last event, ACIIDS 2016 was held in Da Nang (Vietnam).

We received more than 400 papers from 42 countries all over the world. Each paper was peer reviewed by at least two members of the international Program Committee and international reviewer board. Only 154 papers with the highest quality were selected for oral presentation and publication in the two volumes of ACIIDS 2017 proceedings.

Papers included in the proceedings cover the following topics: knowledge engineering and Semantic Web, social networks and recommender systems, text processing and information retrieval, intelligent database systems, intelligent information systems, decision support and control systems, machine learning and data mining, computer vision techniques, advanced data mining techniques and applications, intelligent and context systems, multiple model approach to machine learning, applications of data science, artificial intelligence applications for e-services, automated reasoning and proving techniques with applications in intelligent systems, collective intelligence for service innovation, technology opportunity, e-learning and fuzzy intelligent systems, intelligent computer vision systems and applications, intelligent data analysis, applications and technologies for the Internet of Things, intelligent algorithms and brain functions, intelligent systems and algorithms in information sciences, IT in biomedicine, intelligent technologies in smart cities in the twenty-first century, analysis of image, video and motion data in life sciences, modern applications of machine learning for actionable knowledge extraction, mathematics of decision sciences and information science, scalable data analysis in bioinformatics and biomedical informatics, and technological perspectives of agile transformation in IT organizations.

The accepted and presented papers highlight new trends and challenges of intelligent information and database systems. The presenters showed how new research could lead to novel and innovative applications. We hope you will find these results useful and inspiring for your future research.

We would like to extend our heartfelt thanks to Jarosław Gowin, Deputy Prime Minister of the Republic of Poland and Minister of Science and Higher Education, for his support and honorary patronage of the conference.

We would like to express our sincere thanks to the honorary chairs, Prof. Testsuo Asano (President of JAIST, Japan) and Prof. Cezary Madryas (Rector of Wrocław University of Science and Technology, Poland), for their support.

Our special thanks go to the program chairs, special session chairs, organizing chairs, publicity chairs, liaison chairs, and local Organizing Committee for their work for the conference. We sincerely thank all the members of the international Program Committee for their valuable efforts in the review process, which helped us to guarantee the highest quality of the selected papers for the conference. We cordially thank the organizers and chairs of special sessions who contributed to the success of the conference.

We also would like to express our thanks to the keynote speakers (Prof. Tu-Bao Ho, Prof. Bernhard Pfahringer, Prof. Edward Szczerbicki, Prof. Hideyuki Takagi) for their interesting and informative talks of world-class standard.

We cordially thank our main sponsors, Japan Advanced Institute of Science and Technology (Japan), Wrocław University of Science and Technology (Poland), IEEE SMC Technical Committee on Computational Collective Intelligence, Quang Binh University (Vietnam), Yeungnam University (South Korea), Bina Nusantara University (Indonesia), Universiti Teknologi Malaysia (Malaysia), and the University of Newcastle (Australia). Our special thanks are due also to Springer for publishing the proceedings, and to all the other sponsors for their kind support.

We wish to thank the members of the Organizing Committee for their very significant work and the members of the local Organizing Committee for their excellent work.

We cordially thank all the authors, for their valuable contributions, and the other participants of this conference. The conference would not have been possible without their support.

Thanks are also due to many experts who contributed to making the event a success.

April 2017

Ngoc Thanh Nguyen
Satoshi Tojo
Le Minh Nguyen
Bogdan Trawiński

Organization

Honorary Chairs

Testsuo Asano   President of Japan Advanced Institute of Science
and Technology, Japan

Cezary Madryas   Rector of Wrocław University of Science
and Technology, Poland

General Chairs

Satoshi Tojo   Japan Advanced Institute of Science and Technology,
Japan

Ngoc Thanh Nguyen   Wrocław University of Science and Technology, Poland

Program Chairs

Tzung-Pei Hong   National University of Kaohsiung, Taiwan

Le Minh Nguyen   Japan Advanced Institute of Science and Technology,
Japan

Bogdan Trawiński   Wrocław University of Science and Technology, Poland

Steering Committee

Ngoc Thanh Nguyen   Wrocław University of Science and Technology, Poland
(Chair)

Longbing Cao   University of Science and Technology Sydney, Australia

Suphamit Chittayasothorn   King Mongkut's Institute of Technology Ladkrabang,
Thailand

Ford Lumban Gaol   Bina Nusantara University, Indonesia

Tu Bao Ho   Japan Advanced Institute of Science and Technology,
Japan

Tzung-Pei Hong   National University of Kaohsiung, Taiwan

Dosam Hwang   Yeungnam University, Korea

Lakhmi C. Jain   University of South Australia, Australia

Geun-Sik Jo   Inha University, Korea

Hoai An Le-Thi   University Paul Verlaine, Metz, France

Toyoaki Nishida   Kyoto University, Japan

Leszek Rutkowski   Technical University of Czestochowa, Poland

Ali Selamat   Universiti Teknologi Malaysia, Malaysia

Special Session Chairs

Dariusz Król   Wrocław University of Science and Technology, Poland
Kiyoaki Shirai   Japan Advanced Institute of Science and Technology,
                                    Japan

Liaison Chairs

Ford Lumban Gaol   Bina Nusantara University, Indonesia
Bao Hung Hoang   Viethanit, Vietnam
Mong-Fong Horng   National Kaohsiung University of Applied Sciences,
                                    Taiwan
Dosam Hwang   Yeungnam University, Korea
Ali Selamat   Universiti Teknologi Malaysia, Malaysia

Organizing Chairs

Atsuo Yoshitaka   Japan Advanced Institute of Science and Technology,
                                    Japan
Adrianna Kozierkiewicz-   Wrocław University of Science and Technology, Poland
    Hetmańska

Publicity Chairs

Danilo S. Carvalho   Japan Advanced Institute of Science and Technology,
                                    Japan
Maciej Huk   Wrocław University of Science and Technology, Poland
Bernadetta Maleszka   Wrocław University of Science and Technology, Poland

Publication Chair

Marcin Maleszka   Wrocław University of Science and Technology, Poland

Webmaster

Marek Kopel   Wrocław University of Science and Technology, Poland

Keynote Speakers

Tu-Bao Ho   Japan Advanced Institute of Science and Technology,
                                    Japan
Bernhard Pfahringer   University of Waikato, New Zealand
Edward Szczerbicki   The University of Newcastle, Australia
Hideyuki Takagi   Kyushu University, Japan

Special Sessions Organizers

1. Special Session on Advanced Data Mining Techniques and Applications (ADMTA 2017)

Tzung-Pei Hong	National University of Kaohsiung, Taiwan
Bac Le	University of Science, VNU-HCM, Vietnam
Tran Minh Quang	Ho Chi Minh City University of Technology, Vietnam
Bay Vo	Ho Chi Minh City University of Technology, Vietnam

2. Special Session on Intelligent and Contextual Systems (ICxS 2017)

Maciej Huk	Wrocław University of Science and Technology, Poland
Goutam Chakraborty	Iwate Prefectural University, Japan
Basabi Chakraborty	Iwate Prefectural University, Japan
Qiangfu Zhao	University of Aizu, Japan

3. Multiple Model Approach to Machine Learning (MMAML 2017)

Tomasz Kajdanowicz	Wrocław University of Science and Technology, Poland
Edwin Lughofer	Johannes Kepler University Linz, Austria
Bogdan Trawiński	Wrocław University of Science and Technology, Poland

4. Special Session on Applications of Data Science (ADS 2017)

Fulufhelo Nelwamondo	Council for Scientific and Industrial Research (CSIR), South Africa
Vukosi Marivate	Council for Scientific and Industrial Research (CSIR), South Africa

5. Special Session on Artificial Intelligence Applications for E-services (AIAE 2017)

Chen-Shu Wang	National Taipei University of Technology, Taiwan
Deng-Yiv Chiu	Chung Hua University, Taiwan

6. Special Session on Automated Reasoning and Proving Techniques with Applications in Intelligent Systems (ARPTA 2017)

Jingde Cheng	Saitama University, Japan

7. Special Session on Collective Intelligence for Service Innovation, Technology Opportunity, E-Learning and Fuzzy Intelligent Systems (CISTEF 2017)

Chao-Fu Hong	Aletheia University, Taiwan
Kuo-Sui Lin	Aletheia University, Taiwan

8. Special Session on Intelligent Computer Vision Systems and Applications (ICVSA 2017)

Dariusz Frejlichowski	West Pomeranian University of Technology, Szczecin, Poland
Leszek J. Chmielewski	Warsaw University of Life Sciences, Poland
Piotr Czapiewski	West Pomeranian University of Technology, Szczecin, Poland

9. Special Session on Intelligent Data Analysis, Applications and Technologies for Internet of Things (IDAIoT 2017)

Shunzhi Zhu	University of Technology Xiamen, PR China
Rung Ching Chen	Chaoyang University of Technology, Taiwan
Yung-Fa Huang	Chaoyang University of Technology, Taiwan

10. Special Session on Intelligent Algorithms and Brain Functions (InBRAIN 2017)

Andrzej Przybyszewski	Polish-Japanese Academy of Information Technology, Poland
Tomasz Rutkowski	University of Tokyo, Japan

11. Special Session on Intelligent Systems and Algorithms in Information Sciences (ISAIS 2017)

Martin Kotyrba	University of Ostrava, Czech Republic
Eva Volna	University of Ostrava, Czech Republic
Ivan Zelinka	VŠB, Technical University of Ostrava, Czech Republic

12. Special Session on IT in Biomedicine (ITiB 2017)

Ondrej Krejcar	University of Hradec Kralove, Czech Republic
Ali Selamat	Universiti Teknologi Malaysia, Malaysia
Kamil Kuca	University of Hradec Kralove, Czech Republic
Dawit Assefa Haile	Addis Ababa University, Ethiopia
Tanos C.C. Franca	Military Institute of Engineering, Brazil

13. Special Session on Intelligent Technologies in Smart Cities in the 21st Century (ITSC 2017)

Cezary Orłowski	WSB University Gdańsk, Poland
Artur Ziółkowski	WSB University Gdańsk, Poland
Aleksander Orłowski	Gdansk University of Technology, Poland
Katarzyna Ossowska	Gdansk University of Technology, Poland
Arkadiusz Sarzyński	Gdansk University of Technology, Poland

14. Special Session on Analysis of Image, Video and Motion Data in Life Sciences (IVMLS 2017)

Kondrad Wojciechowski	Polish-Japanese Academy of Information Technology, Poland
Marek Kulbacki	Polish-Japanese Academy of Information Technology, Poland
Jakub Segen	Polish-Japanese Academy of Information Technology, Poland
Andrzej Polański	Silesian University of Technology, Poland

15. Special Session on Modern Applications of Machine Learning for Actionable Knowledge Extraction (MAMLAKE 2017)

Waseem Ahmad	Waiariki Institute of Technology, New Zealand
Paul Leong	Auckland University of Technology, New Zealand
Muhammad Usman	Shaheed Zulfiqar Ali Bhutto Institute of Science and Technology, Pakistan

16. Special Session on Mathematics of Decision Sciences and Information Science (MDSIS 2017)

Takashi Matsuhisa	Ibaraki Christian University, Japan
Vladimir Mazalov	Karelia Research Centre Russian Academy of Sciences, Russia
Pu-Yan Nie	Guangdong University of Finance and Economics, PR China

17. Special Session on Scalable Data Analysis in Bioinformatics and Biomedical Informatics (SDABBI 2017)

Dariusz Mrozek	Silesian University of Technology, Poland
Stanisław Kozielski	Silesian University of Technology, Poland
Bożena Małysiak-Mrozek	Silesian University of Technology, Poland

18. Special Session on Technological Perspective of Agile Transformation in IT organizations (TPATIT 2017)

Cezary Orłowski	WSB University Gdańsk, Poland
Artur Ziółkowski	WSB University Gdańsk, Poland
Miłosz Kurzawski	Blue Media Corporation, Poland
Tomasz Deręgowski	ACXIOM Corporation, Poland
Włodzimierz Wysocki	University of Technology Koszalin, Poland

Program Committee

Salim Abdulazeez	College of Engineering, Trivandrum, India
Ajith Abraham	Machine Intelligence Research Labs, USA

Kazimierz Choroś	Wrocław University of Science and Technology, Poland
Kun-Ta Chuang	National Cheng Kung University, Taiwan
Piotr Chynał	Wrocław University of Science and Technology, Poland
Robert Cierniak	Czestochowa University of Technology, Poland
Dorian Cojocaru	University of Craiova, Romania
Phan Cong-Vinh	Nguyen Tat Thanh University, Vietnam
Jose Alfredo Ferreira Costa	UFRN, Universidade Federal do Rio Grande do Norte, Brazil
Keeley Crockett	Manchester Metropolitan University, UK
Bogusław Cyganek	AGH University of Science and Technology, Poland
Piotr Czapiewski	West Pomeranian University of Technology, Szczecin, Poland
Ireneusz Czarnowski	Gdynia Maritime University, Poland
Piotr Czekalski	Silesian University of Technology, Poland
Paul Davidsson	Malmö University, Sweden
Mauricio C. de Souza	Universidade Federal de Minas Gerais, Brazil
Roberto De Virgilio	Università degli Studi Roma Tre, Italy
Tien V. Do	Budapest University of Technology and Economics, Hungary
Grzegorz Dobrowolski	AGH University of Science and Technology, Poland
Habiba Drias	University of Science and Technology Houari Boumediene, Algeria
Maciej Drwal	Wrocław University of Science and Technology, Poland
Ewa Dudek-Dyduch	AGH University of Science and Technology, Poland
El-Sayed M. El-Alfy	King Fahd University of Petroleum and Minerals, Saudi Arabia
Nadia Essoussi	University of Carthage, Tunisia
Rim Faiz	University of Carthage, Tunisia
Victor Felea	Alexandru Ioan Cuza University of Iasi, Romania
Thomas Fober	University of Marburg, Germany
Simon Fong	University of Macau, SAR China
Tanos C.C. Franca	Military Institute of Engineering, Brazil
Dariusz Frejlichowski	West Pomeranian University of Technology, Szczecin, Poland
Hamido Fujita	Iwate Prefectural University, Japan
Mohamed Gaber	Robert Gordon University, UK
Ford Lumban Gaol	Bina Nusantara University, Indonesia
Dariusz Gasior	Wrocław University of Science and Technology, Poland
Janusz Getta	University of Wollongong, Australia
Daniela Gifu	Romanian Academy, Iasi Branch, Romania
Dejan Gjorgjevikj	Ss. Cyril and Methodius University in Skopje, Macedonia
Daniela Godoy	ISISTAN Research Institute, Argentina
Gergő Gombos	Eötvös Loránd University, Hungary
Adam Gonczarek	Wrocław University of Science and Technology, Poland
Antonio Gonzalez-Pardo	Universidad Autonoma de Madrid, Spain

Manuel Graña	San Sebastián University, Spain
Janis Grundspenkis	Riga Technical University, Latvia
Quang-Thuy Ha	Vietnam National University, Hanoi (VNU), Vietnam
Sung Ho Ha	Kyungpook National University, Korea
Dawit Assefa Haile	Addis Ababa University, Ethiopia
Pei-Yi Hao	National Kaohsiung University of Applied Sciences, Taiwan
Ctibor Hatar	Constantine the Philosopher University, Slovakia
Marcin Hernes	Wrocław University of Economics, Poland
Bogumila Hnatkowska	Wrocław University of Science and Technology, Poland
Huu Hanh Hoang	Hue University, Vietnam
Quang Hoang	Hue University, Vietnam
Jaakko Hollmén	Aalto University School of Science, Finland
Chao-Fu Hong	Aletheia University, Taiwan
Tzung-Pei Hong	National University of Kaohsiung, Taiwan
Mong-Fong Horng	National Kaohsiung University of Applied Sciences, Taiwan
Jen-Wei Huang	National Cheng Kung University, Taiwan
Yung-Fa Huang	Chaoyang University of Technology, Taiwan
Maciej Huk	Wrocław University of Science and Technology, Poland
Zbigniew Huzar	Wrocław University of Science and Technology, Poland
Dosam Hwang	Yeungnam University, Korea
Roliana Ibrahim	Universiti Teknologi Malaysia, Malaysia
Dmitry Ignatov	National Research University Higher School of Economics, Russia
Lazaros Iliadis	Democritus University of Thrace, Greece
Hazra Imran	Athabasca University, Canada
Agnieszka Indyka-Piasecka	Wrocław University of Science and Technology, Poland
Mirjana Ivanovic	University of Novi Sad, Serbia
Sanjay Jain	National University of Singapore, Singapore
Jarosław Jankowski	West Pomeranian University of Technology, Szczecin, Poland
Chuleerat Jaruskulchai	Kasetsart University, Thailand
Khalid Jebari	LCS Rabat, Morocco
Joanna Jedrzejowicz	University of Gdansk, Poland
Piotr Jedrzejowicz	Gdynia Maritime University, Poland
Janusz Jezewski	Institute of Medical Technology and Equipment ITAM, Poland
Geun Sik Jo	Inha University, Korea
Kang-Hyun Jo	University of Ulsan, Korea
Jason J. Jung	Chung-Ang University, Korea
Janusz Kacprzyk	Systems Research Institute, Polish Academy of Sciences, Poland
Tomasz Kajdanowicz	Wrocław University of Science and Technology, Poland
Nadjet Kamel	Ferhat Abbas University of Setif, Algeria

Mehmet Karaata	Kuwait University, Kuwait
Ioannis Karydis	Ionian University, Greece
Nikola Kasabov	Auckland University of Technology, New Zealand
Arkadiusz Kawa	Poznan University of Economics, Poland
Rafal Kern	Wrocław University of Science and Technology, Poland
Chonggun Kim	Yeungnam University, Korea
Pan-Koo Kim	Chosun University, Korea
Attila Kiss	Eötvös Loránd University, Hungary
Jerzy Klamka	Silesian University of Technology, Poland
Goran Klepac	Raiffeisen Bank, Croatia
Blanka Klimova	University of Hradec Kralove, Czech Republic
Shinya Kobayashi	Ehime University, Japan
Joanna Kolodziej	Cracow University of Technology, Poland
Marek Kopel	Wrocław University of Science and Technology, Poland
Jozef Korbicz	University of Zielona Gora, Poland
Jacek Koronacki	Institute of Computer Science, Polish Academy of Sciences, Poland
Raymondus Kosala	Bina Nusantara University, Indonesia
Leszek Koszalka	Wrocław University of Science and Technology, Poland
Malgorzata Kotulska	Wrocław University of Science and Technology, Poland
Martin Kotyrba	University of Ostrava, Czech Republic
Zdzisław Kowalczuk	Gdańsk University of Technology, Poland
Jan Kozak	University of Silesia, Poland
Stanisław Kozielski	Silesian University of Technology, Poland
Adrianna Kozierkiewicz-Hetmańska	Wrocław University of Science and Technology, Poland
Bartosz Krawczyk	Wrocław University of Science and Technology, Poland
Ondrej Krejcar	University of Hradec Kralove, Czech Republic
Dalia Kriksciuniene	Vilnius University, Lithuania
Dariusz Krol	Wrocław University of Science and Technology, Poland
Marzena Kryszkiewicz	Warsaw University of Technology, Poland
Adam Krzyzak	Concordia University, Canada
Tetsuji Kuboyama	Gakushuin University, Japan
Kamil Kuca	University of Hradec Kralove, Czech Republic
Elżbieta Kukla	Wrocław University of Science and Technology, Poland
Marek Kulbacki	Polish-Japanese Academy of Information Technology, Poland
Kazuhiro Kuwabara	Ritsumeikan University, Japan
Halina Kwasnicka	Wrocław University of Science and Technology, Poland
Mark Last	Ben-Gurion University of the Negev, Israel
Annabel Latham	Manchester Metropolitan University, UK
Bac Le	University of Science, VNU-HCM, Vietnam
Hoai An Le Thi	Université de Lorraine, France
Kun Chang Lee	Sungkyunkwan University, Korea
Yue-Shi Lee	Ming Chuan University, Taiwan
Paul Leong	Auckland University of Technology, New Zealand

Pu-Yan Nie	Guangdong University of Finance and Economics, China
Yusuke Nojima	Osaka Prefecture University, Japan
Mariusz Nowostawski	University of Otago, New Zealand
Alberto Núñez	Universidad Complutense de Madrid, Spain
Manuel Núñez	Universidad Complutense de Madrid, Spain
Mariusz Ochla	IBM Center for Advances Studies, Poland
Richard Jayadi Oentaryo	Singapore Management University, Singapore
Kouzou Ohara	Aoyama Gakuin University, Japan
Tomasz Orczyk	University of Silesia, Poland
Cezary Orłowski	WSB University Gdańsk, Poland
Shingo Otsuka	Kanagawa Institute of Technology, Japan
Marcin Paprzycki	Systems Research Institute, Polish Academy of Sciences, Poland
Jakub Peksinski	West Pomeranian University of Technology, Szczecin, Poland
Danilo Pelusi	University of Teramo, Italy
Xuan Hau Pham	Quang Binh University, Vietnam
Tao Pham Dinh	National Institute for Applied Sciences, France
Maciej Piasecki	Wrocław University of Science and Technology, Poland
Bartłomiej Pierański	Poznan University of Economics and Business, Poland
Dariusz Pierzchala	Military University of Technology, Poland
Marcin Pietranik	Wrocław University of Science and Technology, Poland
Piotr Pietrzak	IBM Poland, Poland
Elias Pimenidis	University of the West of England, UK
Andrzej Polanski	Silesian University of Technology, Poland
Elvira Popescu	University of Craiova, Romania
Piotr Porwik	University of Silesia, Poland
Petra Poulova	University of Hradec Kralove, Czech Republic
Bhanu Prasad	Florida A&M University, USA
Andrzej Przybyszewski	University of Massachusetts Medical School, USA
Tran Minh Quang	Ho Chi Minh City University of Technology, Vietnam
Paulo Quaresma	Universidade de Evora, Portugal
Ngoc Quoc Ly	University of Science Ho Chi Minh City, Vietnam
Mohammad Rashedur Rahman	North South University, Bangladesh
Ewa Ratajczak-Ropel	Gdynia Maritime University, Poland
Patricia Riddle	University of Auckland, New Zealand
Manuel Roveri	Politecnico di Milano, Italy
Przemysław Różewski	West Pomeranian University of Technology, Szczecin, Poland
Leszek Rutkowski	Czestochowa University of Technology, Poland
Tomasz Rutkowski	University of Tokyo, Japan
Tiia Ruutmann	Tallinn University of Technology, Estonia
Alexander Ryjov	Lomonosov Moscow State University, Russia
Virgilijus Sakalauskas	Vilnius University, Lithuania

Daniel Sanchez	University of Granada, Spain
Cesar Sanin	University of Newcastle, Australia
Minoru Sasaki	Gifu University, Japan
Moamar Sayed-Mouchaweh	Ecole des Mines de Douai, France
Juergen Schmidhuber	Swiss AI Lab IDSIA, Switzerland
Björn Schuller	University of Passau, Germany
Jakub Segen	Gest3D, USA
Ali Selamat	Universiti Teknologi Malaysia, Malaysia
S.M.N. Arosha Senanayake	Universiti Brunei Darussalam, Brunei Darussalam
Natalya Shakhovska	Lviv Polytechnic National University, Ukraine
Andrzej Sieminski	Wrocław University of Science and Technology, Poland
Dragan Simic	University of Novi Sad, Serbia
Ivana Simonova	University of Hradec Kralove, Czech Republic
Bharat Singh	Universiti Teknology PETRONAS, Malaysia
Andrzej Skowron	Warsaw University, Poland
Leszek Sliwko	University of Westminster, UK
Adam Slowik	Koszalin University of Technology, Poland
Vladimir Sobeslav	University of Hradec Kralove, Czech Republic
Kulwadee Somboonviwat	King Mongkut's Institute of Technology Ladkrabang, Thailand
Zenon A. Sosnowski	Bialystok University of Technology, Poland
Jerzy Stefanowski	Poznan University of Technology, Poland
Serge Stinckwich	University of Caen-Lower Normandy, Vietnam
Ja-Hwung Su	Kainan University, Taiwan
Andrzej Swierniak	Silesian University of Technology, Poland
Edward Szczerbicki	University of Newcastle, Australia
Julian Szymanski	Gdansk University of Technology, Poland
Yasufumi Takama	Tokyo Metropolitan University, Japan
Zbigniew Telec	Wrocław University of Science and Technology, Poland
Krzysztof Tokarz	Silesian University of Technology, Poland
Jakub Tomczak	Wrocław University of Science and Technology, Poland
Diana Trandabat	Alexandru Ioan Cuza University of Iasi, Romania
Bogdan Trawinski	Wrocław University of Science and Technology, Poland
Hong-Linh Truong	Vienna University of Technology, Austria
Ualsher Tukeyev	al-Farabi Kazakh National University, Kazakhstan
Olgierd Unold	Wrocław University of Science and Technology, Poland
Muhammad Usman	Shaheed Zulfiqar Ali Bhutto Institute of Science and Technology, Pakistan
Pandian Vasant	Universiti Teknologi PETRONAS, Malaysia
Jorgen Villadsen	Technical University of Denmark, Denmark
Bay Vo	Ho Chi Minh City University of Technology, Vietnam
Ngoc Chau Vo Thi	Ho Chi Minh University of Technology, Vietnam
Eva Volna	University of Ostrava, Czech Republic
Gottfried Vossen	ERCIS Münster, Germany

Chen-Shu Wang	National Taipei University of Technology, Taiwan
Lipo Wang	Nanyang Technological University, Singapore
Xiaodong Wang	Fujian University of Technology, China
Yongkun Wang	University of Tokyo, Japan
Junzo Watada	Waseda University, Japan
Izabela Wierzbowska	Gdynia Maritime University, Poland
Konrad Wojciechowski	Silesian University of Technology, Poland
Michal Wozniak	Wrocław University of Science and Technology, Poland
Krzysztof Wrobel	University of Silesia, Poland
Tsu-Yang Wu	Harbin Institute of Technology Shenzhen Graduate School, China
Marian Wysocki	Rzeszow University of Technology, Poland
Farouk Yalaoui	University of Technology of Troyes, France
Xin-She Yang	Middlesex University, UK
Lina Yao	University of Adelaide, Australia
Slawomir Zadrozny	Systems Research Institute, Polish Academy of Sciences, Poland
Drago Žagar	University of Osijek, Croatia
Danuta Zakrzewska	Lodz University of Technology, Poland
Constantin-Bala Zamfirescu	Lucian Blaga University of Sibiu, Romania
Katerina Zdravkova	St. Cyril and Methodius University, Macedonia
Ivan Zelinka	VŠB, Technical University of Ostrava, Czech Republic
Vesna Zeljkovic	Lincoln University, USA
Aleksander Zgrzywa	Wrocław University of Science and Technology, Poland
Qiang Zhang	Dalian University, China
Zhongwei Zhang	University of Southern Queensland, Australia
Qiangfu Zhao	University of Aizu, Japan
Dongsheng Zhou	Dalian University, China
Zhi-Hua Zhou	Nanjing University, China
Shunzhi Zhu	University of Technology Xiamen, China
Maciej Zieba	Wrocław University of Science and Technology, Poland
Artur Ziółkowski	WSB University Gdańsk, Poland
Marta Zorrilla	University of Cantabria, Spain

Program Committees of Special Sessions

Advanced Data Mining Techniques and Applications (ADMTA 2017)

Tzung-Pei Hong	National University of Kaohsiung, Taiwan
Tran Minh Quang	Ho Chi Minh City University of Technology, Vietnam
Bac Le	University of Science, VNU-HCM, Vietnam
Bay Vo	Ho Chi Minh City University of Technology, Vietnam
Chun-Hao Chen	Tamkang University, Taiwan
Chun-Wei Lin	Harbin Institute of Technology Shenzhen Graduate School, China

Wen-Yang Lin	National University of Kaohsiung, Taiwan
Yeong-Chyi Lee	Cheng Shiu University, Taiwan
Le Hoang Son	University of Science, Ha Noi, Vietnam
Vo Thi Ngoc Chau	Ho Chi Minh City University of Technology, Ho Chi Minh City, Vietnam
Van Vo	Ho Chi Minh University of Industry, Ho Chi Minh City, Vietnam
Ja-Hwung Su	Cheng Shiu University, Taiwan
Ming-Tai Wu	University of Nevada, Las Vegas, USA
Kawuu W. Lin	National Kaohsiung University of Applied Sciences, Taiwan
Tho Le	Ho Chi Minh City University of Technology, Vietnam
Dang Nguyen	Deakin University, Geelong, Australia
Hau Le	Thuyloi University, Hanoi, Vietnam
Thien-Hoang Van	Ho Chi Minh City University of Technology, Vietnam
Tho Quan	Hochiminh City University of Technology, Vietnam
Ham Nguyen	University of People's Security Hochiminh City, Vietnam
Thiet Pham	Ho Chi Minh University of Industry, Vietnam

Intelligent and Contextual Systems (ICxS 2017)

Basabi Chakraborty	Iwate Prefectural University, Japan
Goutam Chakraborty	Iwate Prefectural University, Japan
Hideyuki Takahashi	RIEC, Tohoku University, Japan
Jerzy Świątek	Wrocław University of Science and Technology, Poland
Józef Korbicz	University of Zielona Gora, Poland
Keun Ho Ryu	Chungbuk National University, South Korea
Maciej Huk	Wrocław University of Science and Technology, Poland
Masafumi Matsuhara	Iwate Prefectural University, Japan
Michael Spratling	University of London, UK
Qiangfu Zhao	University of Aizu, Japan
Tetsuji Kubojama	Gakushuin University, Japan
Tetsuo Kinoshita	RIEC, Tohoku University, Japan
Thai-Nghe Nguyen	Can Tho University, Vietnam
Zhenni Li	University of Aizu, Japan

Multiple Model Approach to Machine Learning (MMAML 2017)

Emili Balaguer-Ballester	Bournemouth University, UK
Urszula Boryczka	University of Silesia, Poland
Abdelhamid Bouchachia	Bournemouth University, UK
Robert Burduk	Wrocław University of Science and Technology, Poland
Oscar Castillo	Tijuana Institute of Technology, Mexico
Rung-Ching Chen	Chaoyang University of Technology, Taiwan
Suphamit Chittayasothorn	King Mongkut's Institute of Technology Ladkrabang, Thailand

José Alfredo F. Costa	Federal University (UFRN), Brazil
Bogusław Cyganek	AGH University of Science and Technology, Poland
Ireneusz Czarnowski	Gdynia Maritime University, Poland
Patrick Gallinari	Pierre et Marie Curie University, France
Fernando Gomide	State University of Campinas, Brazil
Francisco Herrera	University of Granada, Spain
Tzung-Pei Hong	National University of Kaohsiung, Taiwan
Agnieszka Indyka-Piasecka	Wrocław University of Science and Technology, Poland
Konrad Jackowski	Wrocław University of Science and Technology, Poland
Piotr Jędrzejowicz	Gdynia Maritime University, Poland
Tomasz Kajdanowicz	Wrocław University of Science and Technology, Poland
Yong Seog Kim	Utah State University, USA
Bartosz Krawczyk	Wrocław University of Science and Technology, Poland
Kun Chang Lee	Sungkyunkwan University, Korea
Edwin Lughofer	Johannes Kepler University Linz, Austria
Bernadetta Maleszka	Wrocław University of Science and Technology, Poland
Hector Quintian	University of Salamanca, Spain
Andrzej Sieminski	Wrocław University of Science and Technology, Poland
Dragan Simic	University of Novi Sad, Serbia
Adam Słowik	Koszalin University of Technology, Poland
Zbigniew Telec	Wrocław University of Science and Technology, Poland
Bogdan Trawiński	Wrocław University of Science and Technology, Poland
Olgierd Unold	Wrocław University of Science and Technology, Poland
Pandian Vasant	University Technology Petronas, Malaysia
Michał Woźniak	Wrocław University of Science and Technology, Poland
Zhongwei Zhang	University of Southern Queensland, Australia
Zhi-Hua Zhou	Nanjing University, China

Special Session on Applications of Data Science (ADS 2017)

Partha Talukdar	Indian Institute of Science, India
Jp de Villiers	Council for Scientific and Industrial Research (CSIR), South Africa
George Anderson	University of Botswana, Botswana
Vukosi Marivate	Council for Scientific and Industrial Research (CSIR), South Africa
Bo Xing	University of Johannesburg, South Africa
Benjamin Rosman	Council for Scientific and Industrial Research (CSIR), South Africa
Fulufhelo Nelwamondo	Council for Scientific and Industrial Research (CSIR), South Africa

Special Session on Artificial Intelligence Applications for E-services (AIAE 2017)

| Chi-Chung Lee | Chung Hua University, Taiwan |
| Mei-Yu Wu | Chung Hua University, Taiwan |

Yuan-Chu	Hwang, National United University, Taiwan
Ming-Hsiung	Ying, Chung Hua University, Taiwan
Wei-Lun Chang	Tamkang University, Taiwan
Hsien Ting	National University of Kaohsiung, Taiwan
Duen-Ren Liu	National Chiao Tung University, Taiwan
Chih-Kun Ke	National Taichung University of Science and Technology, Taiwan

Special Session on Automated Reasoning and Proving Techniques with Applications in Intelligent Systems (ARPTA 2017)

Shoichi Morimoto	Senshu University, Japan
Yuichi Goto	Saitama University, Japan
Hongbiao Gao	Saitama University, Japan
Shinsuke Nara	Muraoka Design Laboratory, Japan
Kai Shi	Northeastern University, China
Kazunori Wagatsuma	CIJ solutions, Japan

Special Session on Collective Intelligence for Service Innovation, Technology Opportunity, E-Learning and Fuzzy Intelligent Systems (CISTEF 2017)

Albim Y. Cabatingan	University of the Visayas, Philippines
Teh-Yuan Chang	Aletheia University, Taiwan
Chi-Min Chen	Aletheia University, Taiwan
Chih-Chung Chiu	Aletheia University, Taiwan
Wen-Min Chou	Aletheia University, Taiwan
Chao-Fu Hong	Aletheia University, Taiwan
Chia-Lin Hsieh	Aletheia University, Taiwan
Chia-Ling Hsu	Tamkang University, Taiwan
Chi-Cheng Huang	Aletheia University, Taiwan
Rahat Iqbal	Coventry University, UK
Huan-Ting Lin	The University of Tokyo, Japan
Kuo-Sui Lin	Aletheia University, Taiwan
Min-Huei Lin	Aletheia University, Taiwan
Yuh-Chang Lin	Aletheia University, Taiwan
Shin-Li Lu	Aletheia University, Taiwan
Janet Argot Pontevedra	University of San Carlos, Philippines
Shu-Chin Su	Aletheia University, Taiwan
Pen-Choug Sun	Aletheia University, Taiwan
Chen-Fang Tsai	Aletheia University, Taiwan
Ai-Ling Wang	Tamkang University, Taiwan
Chia-Chen Wang	Aletheia University, Taiwan
Leuo-Hong Wang	Aletheia University, Taiwan
Hung-Ming Wu	Aletheia University, Taiwan
Feng-Sueng Yang	Aletheia University, Taiwan
Hsiao-Fang Yang	National Chengchi University, Taiwan
Sadayuki Yoshitomi	Toshiba Corporation, Japan

*Special Session on Intelligent Computer Vision Systems
and Applications (ICVSA 2017)*

Ferran Reverter Comes	University of Barcelona, Spain
Michael Cree	University of Waikato, New Zealand
Piotr Dziurzański	University of York, UK
Paweł Forczmański	West Pomeranian University of Technology, Szczecin, Poland
Marcin Iwanowski	Warsaw University of Technology, Poland
Heikki Kälviäinen	Lappeenranta University of Technology, Finland
Tomasz Marciniak	UTP University of Science and Technology, Poland
Adam Nowosielski	West Pomeranian University of Technology, Szczecin, Poland
Krzysztof Okarma	West Pomeranian University of Technology, Szczecin, Poland
Arkadiusz Orłowski	Warsaw University of Life Sciences, Poland
Edward Półrolniczak	West Pomeranian University of Technology, Szczecin, Poland
Pilar Rosado Rodrigo	University of Barcelona, Spain
Khalid Saeed	AGH University of Science and Technology Cracow, Poland
Rafael Saracchini	Technological Institute of Castilla y León (ITCL), Spain
Samuel Silva	University of Aveiro, Portugal
Gregory Slabaugh	City University London, UK
Egon L. van den Broek	Utrecht University, Utrecht, The Netherlands
Ventzeslav Valev	Bulgarian Academy of Sciences, Bulgaria

*Special Session on Intelligent Data Analysis, Applications and Technologies
for Internet of Things (IDAIoT 2017)*

Goutam Chakraborty	Iwate Prefectural University, Japan
Bin Dai	University of Technology Xiamen, China
Qiangfu Zhao	University of Aizu, Japan
David C. Chou	Eastern Michigan University, USA
Chin-Feng Lee	Chaoyang University of Technology, Taiwan
Lijuan Liu	University of Technology Xiamen, China
Kien A. Hua	Central Florida University, USA
Long-Sheng Chen	Chaoyang University of Technology, Taiwan
Xin Zhu	University of Aizu, Japan
David Wei	Fordham University, USA
Qun Jin	Waseda University, Japan
Jacek M. Zurada	University of Louisville, USA
Tsung-Chih Hsiao	Huaoiao University, China
Hsien-Wen Tseng	Chaoyang University of Technology, Taiwan

Nitasha Hasteer	Amity University Uttar Pradesh, India
Chuan-Bi Lin	Chaoyang University of Technology, Taiwan
Cliff Zou	Central Florida University, USA

Special Session on Intelligent Algorithms and Brain Functions (InBRAIN 2017)

Zbigniew Struzik	RIKEN Brain Science Institute, Japan
Zbigniew Ras	University of North Carolina at Charlotte, USA
Konrad Ciecierski	Warsaw University of Technology, Poland
Piotr Habela	Polish-Japanese Academy of Information Technology, Warsaw, Poland
Peter Novak	Brigham and Women's Hospital, Boston, USA
Wieslaw Nowinski	Cardinal Stefan Wyszynski University, Warsaw, Poland
Andrei Barborica	Research & Compliance and Engineering, FHC, Inc., Bowdoin, USA
Alicja Wieczorkowska	Polish-Japanese Academy of Information Technology, Warsaw, Poland
Majaz Moonis	UMass Medical School, Worcester, USA
Krzysztof Marasek	Polish-Japanese Academy of Information Technology, Warsaw, Poland
Mark Kon	Boston University, Boston, USA
Rafal Zdunek	Wrocław University of Science and Technology, Poland
Lech Polkowski	Polish-Japanese Academy of Information Technology, Warsaw, Poland
Andrzej Skowron	Computer Science and Mechanics, Warsaw University, Poland
Ryszard Gubrynowicz	Polish-Japanese Academy of Information Technology, Warsaw, Poland
Takeshi Okada	The University of Tokyo, Japan
Dominik Slezak	Warsaw University, Poland
Radoslaw Nielek	Polish-Japanese Academy of Information Technology, Warsaw, Poland

Special Session on Intelligent Systems and Algorithms in Information Sciences (ISAIS 2017)

Martin Kotyrba	University of Ostrava, Czech Republic
Eva Volna	University of Ostrava, Czech Republic
Ivan Zelinka	VŠB-Technical University of Ostrava, Czech Republic
Hashim Habiballa	Institute for Research and Applications of Fuzzy Modeling, Czech Republic
Alexej Kolcun	Institute of Geonics, AS CR, Czech Republic
Roman Senkerik	Tomas Bata University in Zlin, Czech Republic
Zuzana Kominkova Oplatkova	Tomas Bata University in Zlin, Czech Republic
Katerina Kostolanyova	University of Ostrava, Czech Republic
Antonin Jancarik	Charles University in Prague, Czech Republic

Igor Kostal	The University of Economics in Bratislava, Slovakia
Eva Kurekova	Slovak University of Technology in Bratislava, Slovakia
Leszek Cedro	Kielce University of Technology, Poland
Dagmar Janacova	Tomas Bata University in Zlin, Czech Republic
Martin Halaj	Slovak University of Technology in Bratislava, Slovakia
Radomil Matousek	Brno University of Technology, Czech Republic
Roman Jasek	Tomas Bata University in Zlin, Czech Republic
Petr Dostal	Brno University of Technology, Czech Republic
Jiri Pospichal	The University of Ss. Cyril and Methodius (UCM), Slovakia
Vladimir Bradac	University of Ostrava, Czech Republic
Roman Jasek	Tomas Bata University in Zlin, Czech Republic
Vaclav Skala	University of West Bohemia, Czech Republic

Special Session on IT in Biomedicine (ITiB 2017)

Golnoush Abae	Universiti Teknologi Malaysia (UTM), UTM Johor Bahru, Malaysia
Orcan Alpar	University of Hradec Kralove, Czech Republic
Dawit Assafa Haile	Addis Ababa University, Ethiopia
Branko Babusiak	University of Zilina, Slovakia
Pavel Blazek	University of Defense, Hradec Kralove, Czech Republic
Peter Brida	University of Zilina, Slovakia
Petr Cermak	Silesian University Opava, Czech Republic
Martin Cerny	VSB, Technical University of Ostrava, Czech Republic
Richard Cimler	University of Hradec Kralove, Czech Republic
Rafael Dolezal	University of Hradec Kralove, Czech Republic
Ricardo J. Ferrari	Federal University of Sao Carlos, Brazil
Tanos C.C. Franca	Military Institute of Engineering, Praça, Brazil
Michal Gala	University of Zilina, Slovakia
Jan Honegr	University Hospital Hradec Kralove, Czech Republic
Radovan Hudak	Technical University of Kosice, Slovakia
Roliana Ibrahim	Universiti Teknologi Malaysia (UTM), UTM Johor Bahru, Malaysia
Marek Kukucka	Slovak University of Technology in Bratislava, Slovakia
David Korpas	Silesian University Opava, Czech Republic
Ondrej Krejcar	University of Hradec Kralove, Czech Republic
Kamil Kuca	University of Hradec Kralove, Czech Republic
Juraj Machaj	University of Zilina, Slovakia
Jaroslav Majerník	Pavol Josef Safarik University in Kosice, Slovakia
Petra Maresova	University of Hradec Kralove, Czech Republic
Reza Masinchi	Universiti Teknologi Malaysia (UTM), UTM Johor Bahru, Malaysia
Marek Penhaker	VSB Technical University of Ostrava, Czech Republic
Jan Plavka	Technical University of Kosice, Slovakia
Teodorico C. Ramalho	Federal University of Lavras (UFLA), Brazil

Martin Rozanek	Czech Technical University in Prague, Czech Republic
Saber Salehi	Universiti Teknologi Malaysia (UTM), UTM Johor Bahru, Malaysia
Ali Selamat	Universiti Teknologi Malaysia (UTM), UTM Johor Bahru, Malaysia

Special Session on Intelligent Technologies in Smart Cities in the 21st Century (ITSC 2017)

Cezary Orłowski	WSB University Gdansk, Poland
Piotr Oskar Czechowski	Gdynia Maritime University, Poland
Ewa Glińska	Bialystok University of Technology, Poland
Joanna Godlewska	Bialystok University of Technology, Poland
Jarosław Hryszko	Wrocław University of Technology, Poland
Dariusz Kralewski	University of Gdansk, Poland
Kostas Karatzas	Aristotle University of Thessaloniki, Greece
Lech Madeyski	Wrocław University of Technology, Poland
Maciej Nowak	Jagiellonian University in Kraków, Poland
Cezary Orłowski	WSB University Gdansk, Poland
Helena Szczerbicka	Leibniz University Hannover, Germany
Paweł Węgrzyn	Jagiellonian University in Kraków, Poland
Artur Ziółkowski	WSB University Gdansk, Poland

Special Session on Analysis of Image, Video and Motion Data in Life Sciences (IVMLS 2017)

Artur Bąk	Polish-Japanese Academy of Information Technology, Poland
Leszek Chmielewski	Warsaw University of Life Sciences, Poland
Aldona Barbara Drabik	Polish-Japanese Academy of Information Technology, Poland
Marcin Fojcik	Sogn og Fjordane University College, Norway
Adam Gudyś	Silesian University of Technology, Poland
Celina Imielińska	Vesalius Technologies LLC, USA
Henryk Josiński	Silesian University of Technology, Poland
Ryszard Klempous	Wrocław University of Science and Technology, Poland
Ryszard Kozera	The University of Life Sciences, SGGW, Poland
Julita Kulbacka	Wrocław Medical University, Poland
Marek Kulbacki	Polish-Japanese Academy of Information Technology, Poland
Aleksander Nawrat	Silesian University of Technology, Poland
Jerzy Paweł Nowacki	Polish-Japanese Academy of Information Technology, Poland
Eric Petajan	LiveClips LLC, USA
Andrzej Polański	Silesian University of Technology, Poland
Joanna Rossowska	Polish Academy of Sciences, Institute of Immunology and Experimental Therapy, Poland

Jakub Segen	Gest3D LLC, USA
Aleksander Sieroń	Medical University of Silesia, Poland
Michał Staniszewski	Polish-Japanese Academy of Information Technology, Poland
Adam Świtoński	Silesian University of Technology, Poland
Agnieszka Szczęsna	Silesian University of Technology, Poland
Kamil Wereszczyński	Polish-Japanese Academy of Information Technology, Poland
Konrad Wojciechowski	Polish-Japanese Academy of Information Technology, Poland
Sławomir Wojciechowski	Polish-Japanese Academy of Information Technology, Poland

Special Session on Modern Applications of Machine Learning for Actionable Knowledge Extraction (MAMLAKE 2017)

Ajit Narayanan	AUT University, New Zealand
Simon Fong	University of Macau, SAR China
Parma Nand	AUT University, New Zealand
Muhammad Asif Naeem	AUT University, New Zealand
Philip Bright	Waiariki Institute of Technology, New Zealand
Akhtar Zaman	Waiariki Institute of Technology, New Zealand

Special Session on Mathematics of Decision Sciences and Information Science (MDSIS 2017)

Hakim Bendjenna	University of Tebessa, Algeria
Masahiro Hachimori	University of Tsukuba, Japan
Ryuichiro Ishikawa	Waseda University, Japan
Masami Ito	Kyoto Sangyo University, Japan
Yoshihiro Hoshino	Kagawa University, Japan
Evgeny Ivashko	IAMR KarRC RAS, Russia
Diang-yu Jiang	Huaihai Institute of Technology, PR China
Yuji Kobayashi	Toho University, Japan
Hidetoshi Komiya	Keio University, Japan
Michiro Kondo	Tokyo Denki University, Japan
Ridda Laouar	University of Tebessa, Algeria
Tieju Ma	CEEEM, East China University of Science and Technology, PR China
Mikio Nakayama	Keio University, Japan
Hiroyuki Ozaki	Keio University, Japan
Rohit Parikh	CUNY, USA
Leon A. Petrosjan	St. Petersburg University, Russia
Vincenzo Scalzo	University of Naples Federico II, Italy
Kunitaka Shyoji	Shimane University, Japan
Krzysztof Szajowski	Wrocław University of Science and Technology, Poland
Wataru Takahashi	Tokyo Institute of Technology, Japan

Stefano Vannucci	University of Siena, Italy
Alexander Vasin	Moscow State University, Russia
Hongbin Yan	CEEEM, East China University of Science and Technology, PR China
Jun Zhang	University of Kentucky, USA
Xingzhou Zhang	Dalian University of Technology, PR China

Special Session on Scalable Data Analysis in Bioinformatics and Biomedical Informatics (SDABBI 2017)

Hesham H. Ali	University of Nebraska, Omaha, USA
José P. Cerón-Carrasco	Universidad Católica San Antonio de Murcia (UCAM), Spain
Po-Yuan Chen	China Medical University, Taichung, Taiwan
Rudolf Fleischer	German University of Technology, Oman
Che-Lun Hung	Providence University, Taichung, Taiwan
Sergio Lifschitz	Pontificia Universidade Catolica do Rio de Janeiro, Brazil
Stanisław Kozielski	Silesian University of Technology, Poland
Xun Lan	Stanford University, USA
Jung-Hsin Lin	Academia Sinica, Taipei, Taiwan
Pradipta Maji	Indian Statistical Institute, Kolkata, India
Bożena Małysiak-Mrozek	Silesian University of Technology, Poland
Dariusz Mrozek	Silesian University of Technology, Poland
Alessandro S. Nascimento	IFSC, University of Sao Paulo, Brazil
Karin Verspoor	University of Melbourne, Australia
Quan Zou	Tianjin University, PR China

Special Session on Technological Perspective of Agile Transformation in IT organizations (TPATIT 2017)

Jakub Chabik	EBIT Company, Poland
Ireneusz Czarnowski	Gdynia Maritime University, Poland
Bogdan Franczyk	University of Leipzig, Germany
Anna Kosieradzka	Warsaw University of Technology, Poland
Dariusz Kralewski	University of Gdansk, Poland
Leszek Maciaszek	Macquarie University Sydney, Australia
Cezary Orłowski	WSB University Gdansk, Poland
Edward Szczerbicki	University of Newcastle, Australia
Artur Ziółkowski	WSB University Gdansk, Poland

Contents – Part I

Text Processing and Information Retrieval

Intelligent Database Systems

Intelligent Information Systems

Decision Support Control Systems

Machine Learning and Data Mining

Computer Vision Techniques

Advanced Data Mining Techniques and Applications

Intelligent and Context Systems

Multiple Model Approach to Machine Learning

Contents – Part II

Automated Reasoning and Proving Techniques with Applications in Intelligent Systems

Collective Intelligence for Service Innovation, Technology Opportunity, E-Learning and Fuzzy Intelligent Systems

Intelligent Computer Vision Systems and Applications

**Intelligent Data Analysis, Applications and Technologies for Internet
of Things**

Intelligent Algorithms and Brain Functions

Intelligent Systems and Algorithms in Information Sciences

IT in Biomedicine

Intelligent Technologies in the Smart Cities in the 21st Century

Analysis of Image, Video and Motion Data in Life Sciences

Modern Applications of Machine Learning for Actionable Knowledge Extraction

Mathematics of Decision Sciences and Information Science

Scalable Data Analysis in Bioinformatics and Biomedical Informatics

Technological Perspective of Agile Transformation in IT organizations

Knowledge Engineering and Semantic Web

The Knowledge Increase Estimation Framework for Ontology Integration on the Instance Level

Adrianna Kozierkiewicz-Hetmańska[✉], Marcin Pietranik,
and Bogumiła Hnatkowska

Faculty of Computer Science and Management,
Wroclaw University of Science and Technology,
Wybrzeze Wyspianskiego 27, 50-370 Wroclaw, Poland
{adrianna.kozierkiewicz,marcin.pietranik,
bogumila.hnatkowska}@pwr.edu.pl

Abstract. Integration of big collections of data is time and cost consuming process. It can be beneficial when the knowledge is distributed in different sources, or not – when different sources contain redundant information or, what worse, inconsistent information. Authors propose a formula to estimate knowledge increase after the process of ontology integration on the instance level. The validity of the formula was checked by a questionnaire and confirmed by a statistical study. The formula allows to estimate knowledge increase in objective manner.

1 Introduction

Nowadays, each company, office or factory store and process a large set of information about their management procedures, products' planning, service delivery, marketing and sales, shipping and payment, human resources etc. Processing such collection of information is very time- and cost consuming, especially when several of such collections are expected to be merged into one. In other words – one of the most important task that refers to knowledge management concerns its integration. This task can be described as a process of combining several, independent knowledge bases into a single, unified structure containing all of the knowledge extracted from the input sources and with an exclusion or resolution of any potential conflicts or inconsistencies that may occur.

Assuming that the aforementioned knowledge bases are expressed using ontologies, the task of their integration can be defined as follows: *for given n ontologies $O_1, O_2, ..., O_n$ one should determine an ontology O^* which is the best representation of given input ontologies.*

The structure of the ontology (according to [12]) can be expressed using an ontology stack consisting of levels of concepts, relations and instances. Therefore, the integration problem should also be considered on these three levels. Obviously, the process of obtaining such unified representation is even more time- and cost consuming. Therefore, a method of assessing the profitability of performing the integration is naturally desired.

© Springer International Publishing AG 2017
N.T. Nguyen et al. (Eds.): ACIIDS 2017, Part I, LNAI 10191, pp. 3–12, 2017.
DOI: 10.1007/978-3-319-54472-4_1

This paper is devoted to an easy method for assessing the increase of knowledge that can be gained after the integration of two or more ontologies. We propose a novel measure which allows to estimate such potential growth, however, our method can be used before any level of the integration is actually performed. In other words, our measure allows to answer an important question: assuming some set of ontologies, is it worth to integrate them or not? In our previous research [8] we have defined such measure for the concept level and due to the limited space available for this paper we will consider only the instance level.

The remaining part of this article is organised as follows. In the next section a brief summary of related works is given. Section 3 contains an introduction to ontologies and basic notions used throughout our research. In Sect. 4 we describe the aforementioned measure which allows to estimate the increase of knowledge after the ontology integration on the instance level. Section 5 presents the results of an experiment which was conducted to demonstrate the effectiveness of our ideas. The last section sheds some light on our upcoming research plans and eventually concludes the paper.

2   Related Works

It is possible to find some measures which allow to estimate the effectiveness of integration process [15], however most of them are difficult to calculate and the integration needs to be performed beforehand. Some papers like [6] considered ontology quality from a philosophical point of view – author presented only definition of data quality and called it "fitness for use".

In [7] authors presented a flexible, holistic ontology-based data quality framework. In this work authors used the well known measures like completeness, accuracy, consistency and confidence. In a similar way, data quality were measured in [3]. Authors defined tools like completeness, relevance, reliability etc.

In [2] information quality analysis of schemas in data integration environments was presented. Authors defined entity and relationship redundancy degree and based on those the measures for the schema minimality were proposed. All of these measures, described by Batista [2], are very useful, despite the fact that they require to conduct the integration process beforehand.

The last two papers [2,3] were not devoted directly to ontologies. In [4] authors criticised the approach where "fitness" of a dataset is computed by a set of defined quality metrics due to its time requirements. Authors introduced *Dataset Quality Ontology* called *daQ*. It is a lightweight ontology that allows a dataset to be "stamped" with its quality measures and expressing concrete, tangible values that represent the quality of the data. The *daQ* is an extensible vocabulary for attaching the results of quality benchmarking of a linked open dataset to that dataset. The main defect of this solution is the initial phase of creating the proposed ontology.

A more interesting solution was introduced in [14] where authors presented a model *OntoQA* which analyses ontology schemas along with their populations and described them through a defined set of metrics. Authors defined the whole set of instance metrics which were grouped into two categories: knowledge based metrics and class based metrics. By using them it was possible to estimate plethora of indicators, among others the following: *class richness* (how instances are distributed across classes), *average population* (the average distribution of instances in classes), *cohesion* (the number of separate connected components in the instances) etc. The main disadvantage of the proposed metrics is their simplicity and the lack of connection between concepts, relations, and instances. All of the described metrics are calculated in separation and none of the metric can give a big picture of the quality of performed integration.

Maedche and Staab [9] proposed similarity measures between ontologies on two levels: lexical and conceptual. Mülligann and others [10] created a semantic similarity measures on concept and instance levels. However, none of these measures are able to estimate the knowledge increase.

An interesting research was presented in [13]. Authors defined a quality of knowledge ontology and computed the Degree of Quality. The quality of knowledge were estimated based on some features like: cognitive adequacy, context (which are quantifiable features) and veracity, complexity, practical use and specificity (which are quantifiable features). The Degree of the quality is calculated for each ontology as the weighted sum for each attribute. Despite the mathematical representation of the ontology quality, the proposed measure is focused on the description of ontologies and their usefulness. It is not a generic solution because of their relation with domain of ontologies. Additionally, the proposed degree does not consider the integration of ontologies.

In the literature, there are some publications devoted to similarity measures which are used during ontology integration and alignment [5,11,12]. None of these works allow to estimate the knowledge increase and to decide about the profitability of the eventual integration. In this paper we propose a measure which not suffers the flaws described above. Due to the limitation of this paper only measure for instance level will be considered. The concepts level have been elaborated in [8].

3   Basic Notions

We assume the existence of a pair *(A, V)* (referenced as "a real world") where A is a finite set containing all possible attributes and V is a finite set of their valuations. Formally: $V = \bigcup_{a \in A} V_a$ where V_a is a domain of an attribute $a \in A$.

Ontology is defined as a triple:

$$O = (C, R, I) \tag{1}$$

where C is a set of concepts, R is a set of relations and I is a set of identifiers of instances belonging to concepts from the set C.

Every concept from the set C is defined as follows:

$$c = (idc, A^c, V^c, I^c) \tag{2}$$

where idc is an identifier of a particular concept c, A^c is a set of assigned attributes, V^c is a set of domains of attributes from A^c (formally: $V^c = \bigcup_{a \in A^c} V_a$) and I^c is a set of instances of the concept c.

Every ontology that meets following criteria:

1. $\forall_{c \in C} A^c \subseteq A$
2. $\forall_{c \in C} V^c \subseteq V$

will be called *(A, V)-based*. For short we can write $a \in c$ which denotes the fact that the attribute a belongs the concept's c set of attributes A^c.

Instances from the set I^c are defined as a pair:

$$i = (id^i, v_c^i) \tag{3}$$

where id^i is its identifier and v_c^i is a function with a signature:

$$v_c^i : A^c \rightarrow V^c \tag{4}$$

Instances can be interpreted as a physical materialisation of a concept c, which can be treated as a template for objects from an assumed universe of discourse. Instances can belong to many different concepts, therefore, the main set of instances I from Eq. 1 holds only theirs' identifiers. Formally it can defined as:

$$I = \bigcup_{c \in C} \{id^i | (id^i, v_c^i) \in I^c\} \tag{5}$$

Obviously the condition $card(I) \leq \sum_{c \in C} card(I^c)$ is always true. For short we can write $i \in c$ which denotes the fact that the instance i belongs the the concept c.

We define a helper function Ins^{-1} that returns set of concepts to which an instance with a given identifier belongs. It has the signature $Ins^{-1} : I \rightarrow 2^C$ and can be defined as:

$$Ins^{-1}(i) = \{c | c \in C \land i \in c\} \tag{6}$$

We assume the existence of a set $\widetilde{O}$ that contains n ontologies that will be integrated and which share the same pool of instance identifiers. The result of such integration is an ontology (defined according to the Eq. 1) in the further parts of this article denoted and referred to as $O^* = (C^*, R^*, I^*)$.

In our previous work [8] we have developed a function σ_C^{-1} with a signature $\sigma_C^{-1} : C^* \rightarrow 2^{C_1 \cup C_2 \cup ... \cup C_n}$ that takes a concept from the final ontology O^* and returns concepts from source ontologies $O_1 \cup O_2 \cup ... \cup O_n$ from the set $\widetilde{O}$ that have been merged to create it. In other words – it "unpacks" the concept that is a result of the integration of elements taken from source ontologies.

4   The Quantity of Knowledge on the Ontology Instance Level

In this paper we want to focus on the instance level of ontologies. Therefore, we define a function σ_I which takes instances of selected concepts along with their particular valuations and integrates them using some selected algorithm (e.g. taken from [11]). For given set $\tilde{O} = \{O_1, ..., O_n\}$ containing n ontologies this function has a signature $\sigma_I : \bigcup_{c \in C_1} I^c \times ... \times \bigcup_{c \in C_n} I^c \to I^*$, where $C_1, ...C_n$ are sets of concepts from respective ontologies.

By analogy we define a function σ_I^{-1} which "unpacks" a selected instance of a particular concept taken from a result of the integration. This reverse function is defined below:

$$\sigma_I^{-1}(i, c^*) = \{(id^i, c) | c \in \sigma_C^{-1}(c^*) \wedge i \in c\} \tag{7}$$

It can be understood as a set (with repetitions) containing instances of concepts and their structures from ontologies from $\tilde{O}$ that have been integrated in order to create the particular concept c^* and a selected instance i^*. Having these two element we can easily designate a particular valuations v_c^i of attributes of acquired instances (according to Eqs. 3 and 4).

The main goal of this article is developing a framework that allows to estimate the amount of knowledge that has been gained due to the performed integration of instances. Therefore, we introduce a notion Δ_I which denotes a function that calculates the knowledge that can be gained thanks to the integration of instances of selected concept c^* sharing the same identifier. This function has a signature $\Delta_I : C^* \times I^* \to [-1, 1]$. The function may appear slightly counterintuitive because it is not a metric. The reason of introducing negative values is a remark that when ontologies are integrated on the instance level the knowledge can be gained but also can be lost due to potential inconsistencies between source ontologies. We cover this issue in the next section of the article where an illustrative example is analysed.

We define an auxiliary notion aux_a that for particular attribute a taken from the set of attributes A^{c^*} of some concept c^* from the integrated ontology denotes a set $aux_a = \{(id^i, c) \in \sigma_I^{-1}(i, c^*) | a \in c\}$. This set is a subset of a set of instances that has been obtained using a function σ_I^{-1} and contains only concepts that include given attribute a in their structures. In consequence we also introduce two following sets eq_aux_a and neq_aux_a satisfying the criteria:

- $eq\_aux_a \subseteq aux_a$, $neq\_aux_a \subseteq aux_a$
- $eq\_aux_a \cap neq\_aux_a = \phi$
- $neq\_aux_a = aux_a \setminus eq\_aux_a$
- $\underset{((id^i, c_1), (id^i,, c_2)) \in eq\_aux_a \times eq\_aux_a}{\forall} v_{c_1}^i = v_{c_2}^i$

The first three criteria are straightforward and describe a mutual exclusion of considered sets. The last criterion is an actual definition of the set eq_aux_a. It contains only these instances and concepts from the source ontologies for which a particular attribute's valuations are equal, therefore no knowledge can be gained when their integration is performed. The function Δ_I is eventually calculated using an algorithm below:

Let us illustrate by simple examples how the knowledge increasing degree is calculated. Figure 1 presents three cases of ontology integration at instance level.

Algorithm 1. Knowledge gain after instance integration

Input : the concept c^* and the instance identifier id^i
Output: the value of $\Delta_I \in [-1, 1]$

1 **begin**
2 $\quad$ $\Delta = 0$;
3 $\quad$ **for** $a \in A^{c^*}$ **do**
4 $\quad\quad$ generate set aux_a;
5 $\quad\quad$ **if** $card(aux_a) = 1$ **then**
6 $\quad\quad\quad$ $\Delta = \Delta + 1$;
7 $\quad\quad$ **end**
8 $\quad\quad$ **else if** $card(aux_a) > 1$ **then**
9 $\quad\quad\quad$ generate all possible sets eq_aux_a;
10 $\quad\quad\quad$ select eq_aux_a with the largest cardinality;
11 $\quad\quad\quad$ $\Delta = \Delta - (1 - \frac{card(eq\_aux_a)}{card(aux_a)})$;
12 $\quad\quad$ **end**
13 $\quad$ **end**
14 $\quad$ $\Delta = \frac{\Delta}{card(A^c)}$;
15 $\quad$ **return** Δ ;
16 **end**

In all of them we consider the consequences of integration of two ontologies containing only one instance. The instance is identified by the same id number (1).

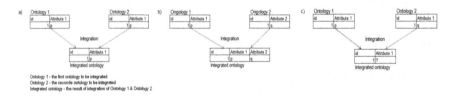

Fig. 1. Examples of ontologies integration at instance level

In the first case (see Fig. 1, a) the ontologies to be integrated are exactly the same. The set $aux_{Attribute1} = \{(1, p, Ontology1), (1, p, Ontology2)\}$. There is only one set $eq\_aux_{Attribute1} = aux_{Attribute1}$. So the integration process does not increase nor decrease the knowledge about that instance ($\Delta_I = 0$).

In the second case (see Fig. 1, b) each of integrated ontologies describes separate properties of the instance (*Ontology 1 – Attribute1*, and *Ontology 2 – Attribute 2*). The set $aux_{Attribute1} = \{(1, p, Ontology1)\}$, and $aux_{Attribute2} = \{(1, q, Ontology2)\}$. After the *for* loop $\Delta_I = 2$, but finally it is calculated as 1 (we doubled the knowledge we have).

In the last case (see Fig. 1, c) the information about *Attribute 1* is inconsistent in both ontologies. In the consequence, after the integration, no one can be sure about the exact value of this attribute (what is marked with '?'). We lose the whole knowledge we believe to have before integration ($\Delta_I = -1$).

When we have many ontologies to be integrated the final value of each attribute is established on the base of the attribute value in majority of ontologies. If in all entry ontologies the attribute value is different, it is replaced with '?' in the resulting ontology.

In order to extend the presented measure to calculate the knowledge increase coming from the integration of not only one instance of selected concepts but for whole resulting ontology $O^* = (C^*, R^*, I^*)$ we use the function $\widetilde{\Delta_I}$ with a signature $\widetilde{\Delta_I} : C^* \to [-1, 1]$. This function 'crawls' through the set of integrated concepts and calculates a mean knowledge increase for its instances. Then it calculates an overall mean value for every concept. Formally it can be defined as follows:

$$\widetilde{\Delta_I} = \frac{1}{|C^*|} \sum_{c \in C^*} \left(\frac{1}{|I^c|} \sum_{i \in I^c} \Delta_I(c, i) \right) \tag{8}$$

5   Evaluation of the Proposed Formula

This section presents how the proposed formula calculating Δ_I was evaluated. Authors address the main objective with the use of statistical study. There are several types of statistical studies used to collect data, i.e. surveys, experimental studies and observations, among which a survey was selected for the stated purpose.

Authors prepared a questionnaire consisting of 27 questions, presenting results of integration of 2 or 3 ontologies at an instance level. The questionnaire contained only symbolic data without any specific meaning (e.g. Attribute 1, x, Attribute 2, y). An ontology is a formal representation of knowledge. Using it entails using semantical content expressed within it and therefore, it can't be generated in a random way. In our experiments symbolic data can be easily replaced by exemplary data with explicit meaning (e.g. *Movie title* in place of *Attribute 1*, *The Invisible* in place of *x*, *Director* in place of *Attribute 2* etc.). However, such solution requires to known the specific ontology domain by responders. Moreover, using symbolic data allows us to ensure that the concept attributes are treated equally (some of them are not perceived by volunteers as more important than the others). Additionally, we cannot use the real ontology because the provided datasets are too big and it is very hard to process them manually by an expert. Some examples of prepared ontologies are presented in Fig. 1.

The respondents were asked to rate the level of knowledge change with the use of the scale range from -100 to 100, where -100 means that you loose the whole knowledge, 0 - the knowledge remains the same, and 100 – the increase of knowledge is maximal. The questionnaire was offered to staff members of Faculty of Computer Science and Management and to authors' acquaintances somehow connected with an IT industry. The population was self-selected, and biased (as the participants weren't randomly selected). We decided to use that type of sampling because of specific needs the participants shout satisfy. We expected them to be familiar with the concept of ontologies or at least the concept of databases, and to know a role of entities' identifiers. The questionnaire was filled by 21 people, both women and men, of different age (from 21 to 62), and different education status (5 with a Bachelor or similar degree, 8 with an MSc or equivalent degree, 8 with a PhD degree).

In the first step of the analysis for each question the consensus was determined based on collected experts' answers. The experts in the experiments were the volunteers who took part in our research and filled the questionnaire described in the previous section of the article.

According to the literature [11] it is possible to determine consensus satisfying a *1-* or *2-optimality* criterion. The *1-optimality* postulate requires the consensus to be as near as possible to elements of the profile (input set describing elements that are integrated). The consensus generated by a function satisfying this postulate plays a very important role because it can be understood as the best representation of the profile. The *2-optimality* postulate requires that the sum of the squared distances between a consensus and the profile elements is minimal. For our aim, the consensus satisfying the *1-optimality* criterion has reflected the better experts' opinions. The median of experts' answers for each question has been calculated as the final estimation of the knowledge increase of the ontology integration cases presented in the questionnaire.

For the analysis we had two samples. The first sample contained 27 elements, each of which was determined as the consensus of expert's opinion referring to the knowledge increase for the cases presented in the questionnaire. The second sample was processed based on the formula presented in Algorithm 1. Gathered results are presented on Fig. 2. The whole analysis was made with a significance level $\alpha = 0.05$.

Before selecting a proper test we had analysed the distribution of both samples using Shapiro-Wilk test. For the first sample we obtained the value of statistical test equal to 0.945923 and the p-value equal to 0.1707. For the second sample W the statistic was equal to 0.91379 and the p-value equal to 0.076464. Therefore, we couldn't reject the null hypothesis (both samples come from a normal distribution).

For the further analysis we selected the Intraclass Correlation Coefficient test (ICC) [1]. This test measures the strength of inter-judge reliability – the degree of their's assessments concordance. We had two samples for which we have tested the consistency and the absolute agreement. We obtained the statistically significant concordance of knowledge increase for presented cases estimated by the group of experts and the calculated value based on the formula described in this paper.

The obtained values therefore entail a high-variance between answers for questions and a low variance between both methods of knowledge increase estimation. For both tests (the absolute agreement and the consistency) the ICC was around 0.975 and the p-value less then 1×10^{-6}. We claim that the proposed method corresponds with a natural way people estimate their increase

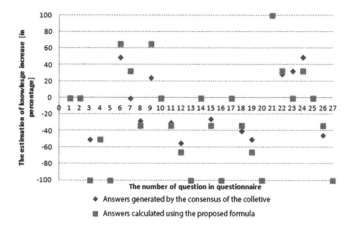

Fig. 2. The results of experiment

of knowledge making it interesting direction of upcoming research. Moreover, it can be easily adapted to other elements of ontologies.

6 Summary and Future Works

This paper is a continuation of another work [8], which addresses the problem of ontology integration on the concept level. This paper is devoted to the problem of ontology integration on the instance level. Authors have proposed how to estimate the knowledge increase resulting from integration process in objective manner. The proposed formula has been statistically verified. The conducted research demonstrated how the developed measure reflect the way people evaluate knowledge increase. The interclass correlation coefficient test has been around 0.975 what allows to state that the proposed formula evaluate knowledge increase in intuitive for human experts way.

The solution proposed in this paper has wide and practical applications. In a real situation, we need to make a decision based on experts' opinions. Let us imagine situation, where each ontology represents some expert's knowledge. It could be obtained, for example, during tests, surveys or interviews. It is obvious, that more experts give us more reliable opinions, the more robust and well grounded the final decision can be. However, more experts entail higher total costs. In case of creating a working team, we would like to hire as small group of people as possible and at the same time, preserving the highest possible level of competence of such team. The evaluation of the increase of knowledge resulting from adding an expert to the final group, but done before doing so, could allow to decide whether or not it is worth to employ a new team member.

In the future, authors plan to extend the framework of elements referencing to ontology integration on the relation level. The integration process is not symmetric. Sometimes it is easier and more beneficial if we join one ontology to another then in opposite way. Therefore, the development of methods which allow to estimate the potential and subjective (from a point of view of the integrated ontology) growth of knowledge after the integration of a set of ontologies is planned.

References

1. Bartko, J.J.: The intraclass correlation coefficient as a measure of reliability. Psychol. Rep. **19**(1), 3–11 (1966). doi:10.2466/pr0.1966.19.1.3
2. Batista, M.D.C.M., Salgado, A.C.: Information quality measurement in data integration schemas. In: QDB, pp. 61–72 (2007)
3. Bobrowski, M., Marré, M., Yankelevich, D.: Measuring data quality. Universidad de Buenos Aires. Report. 1999:99–002 (1999)
4. Debattista, J., Lange, C., Auer, S.: daQ, an ontology for dataset quality information. In: LDOW (2014)
5. Euzenat, J., Petko, V.: An integrative proximity measure for ontology alignment. In: Proceedings of the ISWC-2003 Workshop on Semantic Information Integration, No commercial editor (2003)
6. Frank, A.U.: Data quality ontology: an ontology for imperfect knowledge. In: Winter, S., Duckham, M., Kulik, L., Kuipers, B. (eds.) COSIT 2007. LNCS, vol. 4736, pp. 406–420. Springer, Heidelberg (2007). doi:10.1007/978-3-540-74788-8_25
7. Geisler, S., Weber, S., Quix, C.: An ontology-based data quality framework for data stream applications. In: 16th International Conference on Information Quality, pp. 145–159 (2011)
8. Kozierkiewicz-Hetmanska, A., Pietranik, M.: The knowledge increase estimation framework for ontology integration on the concept level. J. Intell. Fuzzy Syst. **32**(2), 1161–1172 (2017)
9. Maedche, A., Staab, S.: Measuring similarity between ontologies. In: Gómez-Pérez, A., Benjamins, V.R. (eds.) EKAW 2002. LNCS (LNAI), vol. 2473, pp. 251–263. Springer, Heidelberg (2002). doi:10.1007/3-540-45810-7_24
10. Mülligann, Ch., Trame, J., Janowicz, K.: Introducing the new SIM-DL A semantic similarity measurement plug-in for the Protégé ontology editor. In: Proceedings of the 1st ACM SIGSPATIAL International Workshop on Spatial Semantics and Ontologies. ACM (2011)
11. Nguyen, N.T.: Advanced Methods for Inconsistent Knowledge Management. Springer, London (2008)
12. Pietranik, M., Nguyen, N.T.: A multi-atrribute based framework for ontology aligning. Neurocomputing **146**, 276–290 (2014). doi:10.1016/j.neucom.2014.03.067
13. Supekar, K., Patel, C., Lee, Y.: Characterizing quality of knowledge on semantic web. In: FLAIRS Conference, pp. 472–478 (2004)
14. Tartir, S., Arpinar, I.B., Moore, M., Sheth, A.P., Aleman-Meza, B.: OntoQA: metric-based ontology quality analysis (2005). http://lsdis.cs.uga.edu/library/download/OntoQA.pdf
15. Zhu, L., Yang, Q., Chen, W.: Research on ontology integration combined with machine learning. In: Second International Conference on ICICTA 2009, vol. 1. IEEE (2009). 10.1109/ICICTA.2009.119

Online Integration of Fragmented XML Documents

Handoko[1(✉)] and Janusz R. Getta[2]

[1] Electronic and Computer Engineering Department,
Satya Wacana Christian University, Salatiga, Indonesia
handoko@staff.uksw.edu
[2] School of Computer Science and Software Engineering,
University of Wollongong, Wollongong, NSW, Australia
jrg@uow.edu.au

Abstract. Online data integration of large XML documents provides the most up-to-date results from the processing of user requests issued at a central site of heterogeneous multi-database system. The fragments of large XML documents received from the remote sites are continuously combined with the most current state of integrated documents. Online integration of fragmented XML documents has a positive impact on performance of entire online data integration system.

This paper presents the online integration procedures for the fragments of large XML documents. We propose a new model of data for fragmented XML documents and we define a set of operations to manipulate the fragments. A new optimisation procedure presented in the paper finds the smallest core of each new fragment that can be integrated with the documents available at a central site. We show that processing of the smallest cores of XML fragments significantly reduces overall processing time.

Keywords: Data integration · Online algorithm · Fragmented XML documents · Semistructured data

1 Introduction

The recent growth of wide area networks allows for data integration across many diverse systems with various data formats at the remote endpoints. In a global data model approach, a central site in a data integration system breaks down user requests and sends a number of sub-requests to the remote sites to get the results. Then, such results are processed by the data integration system accordingly to the earlier prepared data integration plans.

Online integration is a continuous consolidation of data transmitted over a network with data already available at a central site. It provides a user with the most up-to-date results of a query being processed by the system. Online integration applies online processing algorithms where a smallest unit of data increment is instantly processed without having entire set of data available [5].

© Springer International Publishing AG 2017
N.T. Nguyen et al. (Eds.): ACIIDS 2017, Part I, LNAI 10191, pp. 13–23, 2017.
DOI: 10.1007/978-3-319-54472-4_2

In a distributed and heterogeneous multi-database system, the large chunks of data retrieved at the remote sites are decomposed and sent to a central site as XML fragments. Then, the incoming XML fragments are combined at a central site to form the original documents for further processing. In fact, integration of the fragments can be started immediately if the fragments available at a central site have enough data to complete the integration process. Hence, it is important to identify the minimum requirements for processing of XML fragments to allow processing of a smaller cores of fragments rather than the complete ones. Then, a modified version of the algorithms described in [5] allows for more efficient processing of the fragments of XML documents.

The structure of this paper is the following. Section 2 covers the previous work in an area of integration of fragmented XML documents. The principles of processing XML fragments are described in Sect. 3. Section 4 covers fragmentation of XML documents, and Sect. 5 describes XML algebraic operations on fragmented XML documents. In Sect. 6 we describe an online integration algorithm for fragmented XML documents, and Sect. 7 concludes the paper.

2   Previous Work

XML fragmentation improves the performance of query processing through the decompositions of queries into smaller sub-queries operating in a parallel mode on the fragments of XML documents. It is achieved as either *ad-hoc* or *structured* fragmentation [7]. Hole-filler is the most popular *ad-hoc* fragmentation technique. In this approach, every XML fragment has a unique *filler* ID and a set of *holes* which represent empty places where other fragments could be connected to. A structure of original document and its fragmentation schema is presented in a simple DTD named *Tag Structure* [2]. On the other hand, *structured* fragmentation follows the rules of the relational model fragmentation [3,6].

XML data stream is a theoretically infinite sequence of XML documents. Processing of XML data stream is a challenging process because it requires very efficient algorithms to integrate the most recently received increments with the already completed results of processing. Most of query processing techniques on XML data streams are based on XQuery and XPath streaming evaluation [9]. Some of XML stream processing techniques based on XQuery evaluation refer to XML fragments in the concept of *hole-filler* [1,2,8]. In addition to *hole-filler* model, Bose [2] proposed a query algebra for XQuery on XML stream data.

Fegaras [4] proposed an incremental query processing for a large-scale database called MRQL Streaming. A streaming query is expressed as $q(\overline{S})$ where $S_i \in \overline{S}$ and $S_i : i = 0, \ldots, n$ is a streaming data source. S_i contains an initial dataset and followed by a continuous incremental stream ΔS_i in a time interval Δt. Incremental processing is performed by combining the result query at time t with results of query on ΔS_i, such that $h(\overline{S \uplus \Delta S}) = h(\overline{S}) \bigotimes h(\overline{\Delta S})$. $\bigotimes$ is a merge function which is implemented as a partitioned *join*. Online integration system proposed in [5] performs continuous integration where an incoming and complete XML document triggers the computations of a data integration expression.

3   Principles and Assumptions

We consider a model of data integration where the data increments are transmitted to a central site as the fragments of XML documents. We found in an earlier work [5] that it is possible to increase performance of an online integration system when the processing is performed on the smaller cores of XML fragments. Integration of semistructured data described in this work is based on the following assumptions. (1) The remote sites have the ability to disassemble XML documents into XML fragments with the characteristics described in the next section. (2) XML fragments retrieved at the remote sites are received by a central site in a random order. (3) Due to a high level of autonomy of the remote sites, a central site has no impact on the priorities with which the fragments are retrieved at and sent by the remote sites.

Based on these assumptions, we adopt the following more specific principles for *online data integration* process.

1. A fragmented XML document is a set of XML fragments.
2. Every data container located at a central site contains fragmented XML documents.
3. At a pre-processing stage described in [5], we generate online integration plans for every user request received by the central site. First, we transform a user query into a *global query expression* $(f(q_1, \ldots, q_k))$. Then, we transform it into a *data integration expression* $(f(D_1, \ldots, D_k))$ by systematic replacement of symbol $q_1, \ldots, q_k$ with the data container $D_1, \ldots, D_k$. In the next step, we generate the *increment expressions* which allow for instant computation of new data increments of the particular data containers at a central site. We use the increment expressions obtained earlier to generate the online integration plans for each data container. The main difference between an approach described in [5] and this work is online integration performed on XML fragments instead of complete XML documents.
4. The attributes used in the filter expressions related to data containers are stored in the adequate lists to find when a set of XML fragments is ready for processing.

4   Fragmented XML Document

The operations on XML fragments allow us to process incomplete XML documents and to append their missing parts at the end of integration. Based on an assumption that every node in an XML document can be identified by a unique path (i.e. path and index), we define an XML fragment in the following way.

Definition 1. *An XML fragment is a tuple $\langle x_i(m_i), o, p, H \rangle$ where $x_i(m_i)$ is an XML document that represents a body of the XML fragment. A component o is the identity of parent XML document the fragment comes from, and p (hook) is a unique path where the root of XML fragment is located in the parent XML document. A component H is a set of paths that represents the missing XML fragments (holes) in $x_i(m_i)$.*

An XML fragment ($\langle x_i(m_i), o, p, H \rangle$) has the following characteristics. (1) An XML fragment body ($x_i(m_i)$) is a well-formed XML document. (2) It has a component (o) to store id of its parent XML document. (3) It has a *hook* (p) and a *hole* component (H) that allow for reconstruction of XML fragments into the parent XML document. (4) A *hook* component (p) is represented by a path, and determines a location of XML fragment in its parent XML document. (5) There is exactly one XML fragment that includes a root node of the parent XML document, i.e. its *hook*="xml". (6) A *hole* (H) component is a set of paths to the roots of missing fragments. If $H = \emptyset$ then a fragment is complete.

Definition 2. *A fragmented XML document is a set of XML fragments* $\{\langle x_i(m_i), o_i, p_i, H_i \rangle : i = 1, \ldots, n\}$.

Note, that o_i does not need to be the same for all fragments in a fragmented XML document.

Definition 3. *Let* $x(m) = \{\langle x_i(m_i), o_i, p_i, H_i \rangle : i = 1, \ldots, n\}$ *be a fragmented XML document. A complete XML document is defined as a fragmented XML document where* $\bigcup_i p_i - \{\text{"xml"}\} = \bigcup_i H_i$.

In the other words, a set of XML fragments is a *complete XML document* if for every *hole* the set includes a respective fragment with a matching *hook* and the set also includes a *root fragment* with a path "xml" where a virtual node "xml" plays a role of a root node of every XML document.

Example 1. Let $x(m)$ be a fragmented XML document disassembled into six XML fragments visualised in Fig. 1. The details of the fragments are listed in Table 1.

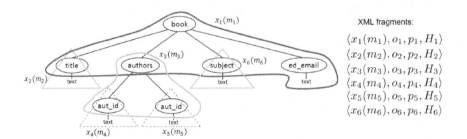

Fig. 1. Fragmented XML document.

5   Operations of XML Algebra

The operations on XML algebra belong to one of the following three types depending on an abstraction level of an operation: (1) operations on XML fragments, (2) operations on fragmented XML documents, and (3) operations on data containers with fragmented XML documents. A *fusion* of fragmented XML

Table 1. The components of XML fragments in Fig. 1

XML fragment	Origin id	Hook	Holes
(a)	o_1="1"	p_1="xml"	H_1={"xml/book/title","xml/book/authors", "xml/book/subject"}
(b)	o_2="1"	p_2="xml/book/title"	H_2={}
(c)	o_3="1"	p_3="xml/book/authors"	H_3={"xml/book/authors/aut_id[1]", "xml/book/authors/aut_id[2]"}
(d)	o_4="1"	p_4="xml/book/authors/aut_id[1]"	H_4={}
(e)	o_5="1"	p_5="xml/book/authors/aut_id[2]"	H_5={}
(f)	o_6="1"	p_6="xml/book/subject"	H_6={}

documents is needed to combine two XML documents where one of them is a part of the other, see Definition 4 below. The definition assumes that a structure of each XML document is represented by an *Extended Tree Grammar (ETG)*[5].

Definition 4. *Let* $G = (N_g, T_g, A_g, S_g, P_g)$ *and* $H = (N_h, T_h, A_h, S_h, P_h)$ *be ETGs,* $N_g \cap N_h \neq \emptyset$, *Y be a non terminal symbol, and* $y \in (N_g \cap N_h)$. *Let* S→xml[id](Y) *be a production rule for start symbol in H. Let* $p_g \in P_g$ *and* $p_h \in P_h$ *be production rules for a non terminal symbol Y in both ETGs. A* *fusion operation on two ETGs is denoted as* $F = G \oplus H$ *and is an operation that combines G and H, such that* $F = (N, T, A, S, P)$ *is an ETG where* $N = N_g \cup N_h$, $T = T_g \cup T_h$, $A = A_g \cup A_h$, *and* $P = P_g \cup P_h - \{p_g\}$.

A *hook* operation combines two XML fragments which have matching *hook* and *hole* components to form a more complete XML fragment. The operation creates a new XML fragment where one of the *holes* in the first argument is filled with the second argument, see Definition 5 below.

Definition 5. *Let* $\langle x_i(m_i), o_i, p_i, H_i \rangle$, $\langle x_j(m_j), o_j, p_j, H_j \rangle$ *be XML fragments with ETG G and H respectively. Let* $o_i = o_j$ *and* $p_j \in H_i$. *A* *hook operation on two XML fragments is defined as* $\langle x_i(m_i), o_i, p_i, H_i \rangle \hookleftarrow \langle x_j(m_j), o_j, p_j, H_j \rangle = \langle x_r(m_r), o_r, p_r, H_r \rangle$. $x_r(m_r)$ *is an XML fragment result after a hook operation,* $o_r = o_i = o_j$, *and* $p_r = p_i$. $H_r = H_i \cup H_j - \{p_j\}$ *is a set of holes after a hook operation. The XML fragment result has an ETG* $F = G \oplus H$.

A *union* of two sets of XML fragments is possible when the sets do not contain the fragments with the same *hook* and it is defined in a usual way as a theoretical set union. At some points of processing, we might want to apply *defragmentation* procedure that first uses *union* operation to group all relevant fragments in one set and later on it uses *hook* operation to combine the XML fragments which have the matching values of *hook* and *hole*.

To speed up defragmentation, we sort the XML fragments by their o and p components. The sorted XML fragments shows that the position of XML fragments in a fragmented XML document represents their location at the origin XML document. If the XML fragments are sorted, then for two XML fragments

$\langle x_i(m_i), o_i, p_i, H_i \rangle$ and $\langle x_j(m_j), o_j, p_j, H_j \rangle$ where i, j are the element indexes and $i < j$, we can perform a *hook* operation $\langle x_i(m_i), o_i, p_i, H_i \rangle \hookleftarrow \langle x_j(m_j), o_j, p_j, H_j \rangle$ but not the opposite.

Next, we need an efficient algorithm that processes the incoming XML fragments. The algorithm must determine where to place an incoming fragment, when to combine a fragment into any of existing fragmented documents in the data containers on input to data integration and materializations containing the intermediate results of data integration, and how to perform defragmentation process on the fragmented XML documents.

A *minion* operation integrates two data containers with fragmented XML documents, see Definition 6 below.

Definition 6. *Let* $D(\mathcal{G}), D(\mathcal{H})$ *be data containers of fragmented XML documents. Let* $x(m) \in D(\mathcal{G})$, $x(m) = \{\langle x_i(m_i), o_i, p_i, H_i \rangle : i = 1, \dots, s\}$ *and* $y(n) \in D(\mathcal{H})$, $y(n) = \{\langle y_j(n_j), o_j, p_j, H_j \rangle : j = 1, \dots, t\}$. *Let* $O_i = \{o_i : \exists \langle x_i(m_i), o_i, p_i, H_i \rangle \in x(m) : i = 1, \dots, s\}$, $O_j = \{o_j : j = 1, \dots, t\}$ *and* $O_i \cap O_j \neq \emptyset$. *A* <u>*minion*</u> *(merge-union) operator is defined as* $D(\mathcal{G}) \uplus D(\mathcal{H}) = \{z(l) : \exists x(m) \in D(\mathcal{G}), y(n) \in D(\mathcal{H}), z(l) = x(m) \cup y(n)\}$.

We organise the input data containers and processing of incoming fragments in the following way.

1. Every data container (D) is divided into a *bounded* data container (D^b) and a *rover* data container (D^r) $(D = D^b \uplus D^r)$. D^b is used to store fragmented XML documents which are ready for processing. Meanwhile, not processed fragmented XML documents are placed in a *rover* data container (D^r).
2. A new incoming XML fragment is placed as an element of a fragmented XML document in a data container D^r according to its original XML document.
3. When a fragmented XML document in D^r satisfies the minimal requirements for processing, it is transferred to a *bounded* data container D^b, and its processing is started.
4. We need an operation on two data containers of fragmented XML document to combine the fragmented XML documents which have the same identity in both data containers. For example, such operation is needed at the end of processing in order to add the unprocessed XML fragments to fragmented XML document results.

Since we use a concept of fragmented XML document to replace a complete XML document, most of the XML algebra operators described in [5] are applicable. Nevertheless, XML algebraic operations have to be re-defined. Some XML algebraic operators require examination of a condition expression (φ) on the path expressions. For complete XML documents, path expressions refer to navigation paths from the root node of XML documents. Meanwhile, path evaluation on a fragmented XML document requires a procedure to discover a particular node in XML fragments which may not have its root element. Below, we redefine the concepts of *selection, join,* and *antijoin* for fragmented XML documents.

Definition 7. *Let $D(\mathcal{G})$ be a data container of fragmented XML documents, $x(m) \in D(\mathcal{G})$, $x(m) = \{\langle x_i(m_i), o, p_i, H_i \rangle : i = 1, \ldots, n\}$, and $x_f = \langle x_i(m_i), o_i, p_i, H_i \rangle$. Selection on $D(\mathcal{G})$ is a unary operator denoted by $\sigma_\varphi(D(\mathcal{G})) = \{x(m) : \exists x_f \in x(m) \ (f(x_f, \varphi) = true)\}$, where $f(x_f, \varphi) \in \{true, false\}$, φ is a condition expression.*

Definition 8. *Let $x(m) = \{\langle x_i(m_i), o_i, p_i, H_i \rangle : i = 1, \ldots, n\}$ and $y(n) = \{\langle y_j(n_j), o_j, p_j, H_j \rangle : j = 1, \ldots, m\}$ be fragmented XML documents. Let $x_f = \langle x_i(m_i), o_i, p_i, H_i \rangle$ and $y_f = \langle y_j(n_j), o_j, p_j, H_j \rangle$. Join operation on fragmented XML documents is defined as $x(m) \bullet_\varphi y(n) = x(m) \cup y(n) : \exists x_f \in x(m) \exists y_f \in y(n)$ and $f(x_f, y_f, \varphi) = true$. φ is a condition expression and f is an evaluation function such that $f(x_f, y_f, \varphi) \in \{true, false\}$.*

Definition 9. *Let $D(\mathcal{G}), D(\mathcal{H})$ be data containers of fragmented XML documents. Join operation is defined as $D(\mathcal{G}) \bowtie_\varphi D(\mathcal{H}) = \{z(o) : \exists x(m) \in D(\mathcal{G}), y(n) \in D(\mathcal{H}), z(o) = x(m) \bullet_\varphi y(n)\}$.*

Definition 10. *Let $D(\mathcal{G}), D(\mathcal{H})$ be data containers of fragmented XML documents. Antijoin operator is defined as $D(\mathcal{G}) \sim_\varphi D(\mathcal{H}) = \{x(m) : x(m) \in D(\mathcal{G})$ and $\overline{\forall y(n)} \in D(\mathcal{H}) \neg \exists (x(m) \bullet_\varphi y(n))\}$, where φ is a condition expression.*

Let D_i, D_j, D_k be data containers for fragmented XML documents. A *minion* (*merge-union*) operation has the following properties.

1. A *minion* operation is commutative, i.e. $(D_i \uplus D_j) = (D_j \uplus D_i)$.
2. A *minion* operation is associative, i.e. $D_i \uplus (D_j \uplus D_k) = (D_i \uplus D_j) \uplus D_k$.
3. If a condition φ can be evaluated in D_i, then *minion* operation is distributive over *selection* operation, i.e. $\sigma_\varphi(D_i \uplus D_j) = \sigma_\varphi(D_i) \uplus D_j$.
4. If a property for *join* operation condition φ exists in D_j, then *minion* operation is distributive over *join* operation: $D_i \bowtie_\varphi (D_j \uplus D_k) = (D_i \bowtie_\varphi D_j) \uplus D_k$.
5. A *minion* operation is left and right distributive over *union* operation, i.e. $D_i \cup (D_j \uplus D_k) = (D_i \cup D_j) \uplus D_k$ and $(D_i \uplus D_j) \cup D_k = (D_i \cup D_k) \uplus D_j$.
6. If a condition φ can be evaluated in D_i, then *minion* operation is right distributive over *antijoin* operation, i.e. $(D_i \uplus D_j) \sim_\varphi D_k = (D_i \sim_\varphi D_k) \uplus D_j$.
7. A *minion* operation can reduce the *antijoin* operation, i.e. $D_i \sim_\varphi (D_j \uplus D_k) = (D_i \sim_\varphi D_j)$, if XML fragment D_j contains elements in operation condition (φ) $D_i \sim_\varphi (D_j \uplus D_k) = (D_i \sim_\varphi D_k)$ and if XML fragment D_k contains elements in operation condition φ.

6   Online Integration of XML Fragments

Initially, online integration of XML fragments is similar to integration described in [5]. An incoming XML fragment is placed in a *rover* data container (D^r) and it is added to a corresponding fragmented XML document accordingly to its identity. If an incoming XML fragment does not match to any existing fragmented XML document, we create a new fragmented XML document and place the incoming XML fragment into it. If a fragmented XML document in the *rover* data container has enough properties for processing then it is transferred to a corresponding *bounded data container*. Incoming data at the *bounded* data containers triggers the following processing of a data integration expression.

The online integration system adjusts the pre-processing phase to deal with XML fragments. The central site is responsible to determine all condition (φ) attributes for all operations, all elements involved in the conditions for every data container, and all available elements in the fragmented XML documents. Then, the existing nodes/elements are applied to decide whether a fragmented XML document has enough properties for further processing.

It may happen that a data container is used several times in a data integration expression. Therefore, a list of adequate properties is applied.

Example 2. Let $f(D_1, D_2, D_3) = (D_1 \bowtie D_2) \cup (D_3 \sim D_1)$ be a data integration expression. A *join* operation $D_1 \bowtie_{(\varphi_1 \vee \varphi_2)} D_2$ has two condition expressions as follows: $\varphi_1 = $ `xml/book/authors/aut_id[1]=//aut_id` and $\varphi_2 = $ `xml/book/authors/aut_id[2]=//aut_id`.

XML paths `xml/book/authors/aut_id[1]` and `xml/book/authors/aut_id[2]` are unique locations of node elements of a fragmented XML document. Meanwhile, `//aut_id` is a path of a node element of a fragmented XML document in the data container D_2. Hence, we generate an adequate list as in the Table 2.

Table 2. An adequate lists of data containers for a data integration expression $f(D_1, D_2, D_3) = (D_1 \bowtie D_2) \cup (D_3 \sim D_1)$

Data container	Operation	Path	Opr	Path or value
D_1	Operation 1	`xml/book/authors/aut_id[1]`	=	`//aut_id`
D_1	Operation 1	`xml/book/authors/aut_id[2]`	=	`//aut_id`
D_2	Operation 1	`//aut_id`	=	`xml/book/authors/aut_id`

Processing of fragmented XML documents may create fragmented XML documents at data containers, materializations (M_j), where the fragmented XML documents have been computed, and remove lists (L_d), where they do not meet criteria for evaluation of the required condition (φ).

6.1   Data Integration Expression

We use a *bounded* D_i^b and *rover* D_i^r input data containers to save the incoming fragmented XML documents. As a consequence, a data integration expression should be transformed in the following way.

1. We replace all data containers D_i in a data integration expression with $D_i^b \uplus D_i^r$, for $i = 1 \ldots k$.

2. Next, we transform the data integration expression by multiple applications of *minion* properties such that all *rover* data containers are moved to the end of computation process.

Example 3. A data integration expression for XML fragments.

Consider the following input data containers D_i represented as *minion* of respective *bounded* and *rover* containers $D_i^b \uplus D_i^r$, for $i = 1 \ldots 6$. A data integration expression $(D_1 \bowtie (D_2 \sim D_3)) \cup ((D_4 \bowtie D_5) \sim D_6)$ is transformed through systematic replacement of the containers D_i with the expressions $D_i^b \uplus D_i^r$ and distribution of *minion* operation over *join*, *antijoin* and *union* operations.

$((D_1^b \uplus D_1^r) \bowtie (D_2 \sim D_3)) \cup ((D_4 \bowtie D_5) \sim D_6)$; D_1 is replaced with $D_1^b \uplus D_1^r$
$((D_1^b \bowtie (D_2 \sim D_3)) \uplus D_1^r) \cup ((D_4 \bowtie D_5) \sim D_6)$
$((D_1^b \bowtie (D_2 \sim D_3)) \cup ((D_4 \bowtie D_5) \sim D_6)) \uplus D_1^r$; D_6 is replaced with $D_6^b \uplus D_6^r$
$((D_1^b \bowtie (D_2 \sim D_3)) \cup ((D_4 \bowtie D_5) \sim (D_6^b \uplus D_6^r))) \uplus D_1^r$
$((D_1^b \bowtie (D_2 \sim D_3)) \cup (((D_4 \bowtie D_5) \sim D_6^b) \uplus D_6^r)) \uplus D_1^r$
$(((D_1^b \bowtie (D_2 \sim D_3)) \cup ((D_4 \bowtie D_5) \sim D_6^b)) \uplus D_1^r) \uplus D_6^r$
and so on.

At the end we obtain an expression given below.
$((((((D_1^b \bowtie (D_2^b \sim D_3^b)) \cup (D_4^b \bowtie D_5^b) \sim D_6^b) \uplus D_1^r) \uplus D_2^r) \uplus D_4^r) \uplus D_5^r) \uplus D_6^r$

6.2 Increment Expression

In the next step, we transform a data integration expression into an increment expression for every data container by applications of XML algebra rules as described in [5]. A *defragmentation* process can be performed at the end of processing because the *rover* data containers have been moved to the end of data integration expression.

Example 4. An increment expression generation.

Let δ_1 be an increment data at a *bounded* data container D_1^b. The data integration expression

$$f(D_1, \ldots, D_6) = ((((((D_1^b \bowtie (D_2^b \sim D_3^b)) \cup ((D_4^b \bowtie D_5^b) \sim D_6^b)) \uplus D_1^r) \uplus D_2^r) \uplus D_4^r) \uplus D_5^r$$

can be transformed as follows:

$((((((D_1 \cup \delta_1) \bowtie (D_2^b \sim D_3^b)) \cup ((D_4^b \bowtie D_5^b) \sim D_6^b)) \uplus D_1^r) \uplus D_2^r) \uplus D_4^r) \uplus D_5^r$
$(((((((D_1^b \bowtie (D_2^b \sim D_3^b)) \cup (\delta_1 \bowtie (D_2^b \sim D_3^b))) \cup ((D_4^b \bowtie D_5^b) \sim D_6^b)) \uplus D_1^r) \uplus D_2^r) \uplus D_4^r) \uplus D_5^r$
$(((((((D_1^b \bowtie (D_2^b \sim D_3^b)) \cup ((D_4^b \bowtie D_5^b) \sim D_6^b)) \cup (\delta_1 \bowtie (D_2^b \sim D_3^b))) \uplus D_1^r) \uplus D_2^r) \uplus D_4^r) \uplus D_5^r$
$(((((f(D_1^b, \ldots, D_6^b) \cup (\delta_1 \bowtie (D_2^b \sim D_3^b))) \uplus D_1^r) \uplus D_2^r) \uplus D_4^r) \uplus D_5^r$
$(((((f(D_1^b, \ldots, D_6^b) \cup (\delta_1 \bowtie M_1)) \uplus D_1^r) \uplus D_2^r) \uplus D_4^r) \uplus D_5^r$ $M_1 = ((D_2^b \sim D_3^b) \uplus D_2^r)$

Using the same transformation procedures, we obtain a set of increment expressions for the rest of data containers.

$\delta_1 : (((((f(D_1^b, \ldots, D_6^b) \cup (\delta_1 \bowtie M_1)) \uplus D_1^r) \uplus D_2^r) \uplus D_4^r) \uplus D_5^r$

$\delta_2 : ((((f(D_1^b, \ldots, D_6^b) \cup (D_1 \bowtie (\delta_2 \sim D_3))) \uplus D_1^r) \uplus D_2^r) \uplus D_4^r) \uplus D_5^r$

$\delta_3 : ((((f(D_1^b, \ldots, D_6^b) \sim (\delta_3 \sim M_4)) \uplus D_1^r) \uplus D_2^r) \uplus D_4^r) \uplus D_5^r$

$\delta_4 : ((((f(D_1^b, \ldots, D_6^b) \cup ((\delta_4 \bowtie D_5^b) \sim D_6^b)) \uplus D_1^r) \uplus D_2^r) \uplus D_4^r) \uplus D_5^r$

$\delta_5 : ((((f(D_1^b, \ldots, D_6^b) \cup ((D_4^b \bowtie \delta_5) \sim D_6^b)) \uplus D_1^r) \uplus D_2^r) \uplus D_4^r) \uplus D_5^r$

$\delta_6 : ((((f(D_1^b, \ldots, D_6^b) \sim (\delta_6 \sim M_1))) \uplus D_1^r) \uplus D_2^r) \uplus D_4^r) \uplus D_5^r$

where $M_1 = ((D_2^b \sim D_3^b) \uplus D_2^r)$; $M_2 = (((D_1^b \bowtie (D_2^b \sim D_3^b)) \uplus D_1^r) \uplus D_2^r)$; $M_3 = (((D_4^b \bowtie D_5^b) \uplus D_4^r) \uplus D_5^r)$; $M_4 = ((((D_4^b \bowtie D_5^b) \sim D_6^b) \uplus D_4^r) \uplus D_5^r)$

Finally, the increment expressions are the following.

$g_1 = (\delta_1 \bowtie M_1)$; $g_2 = (D_1 \bowtie (\delta_2 \sim D_3))$; $g_3 = (\delta_3 \sim M_4)$; $g_4 = ((\delta_4 \bowtie D_5) \sim D_6)$; $g_5 = ((D_4 \bowtie \delta_5) \sim D_6)$; $g_6 = (\delta_6 \sim M_1)$

6.3   Online Integration Plans for XML Fragments

The increment expressions generated from a data integration expression on fragmented XML documents are the extensions of processing on complete XML documents. Therefore, we can utilize the algorithms for online integration plan and scheduling described in [5]. However, since fragmented XML documents in the *rover* data containers have no effect to the rest of computation, we apply *minion* operations at the very end of computation process. Fragmented XML documents in the *rover* data containers are excluded from computation of data integration expression until the results are ready to send.

Example 5. Let $g_2 = (D_1 \bowtie (\delta_2 \sim D_3))$ be an increment expression generated in Example 4. We consider a data increment (δ_2) arrives at a data container D_2. Transformation of g_2 into an online integration plan d_2 is performed as follows.
(1) In the first step, we map an expression $(\delta_2 \sim D_3)$ into a step $\Delta_1 = (\delta_2 \sim D_3)$ and an expression $(D_1 \bowtie \Delta_1)$ into a step $\Delta_2 = (D_1 \bowtie \Delta_1)$.
(2) Then, we append $M_e = (M_e \sim \Delta_2)$ to combine the computation results with the previous final materialization.
(3) Next, we update a data container $D_2 : D_2 = (D_2 \cup \delta_2)$.
(4) An intermediate materialization M_1 is identified to be affected to update. M_1 is a computation result of a data integration expression $h_1(D_2, D_3) = (D_2 \sim D_3)$. Therefore $h_1(D_1, D_2)$ is transformed into an increment expression $g_{M1} = (\delta_2 \sim D_3)$. A plan to update M_1 is generated as follows: $d_{M1} : \Delta_{M1} = (\delta_2 \sim D_3)$; $M_1 = (M_1 \cup \Delta_{M1})$. These steps are appended to the steps produced earlier.

The complete online integration plan for increment expression g_2 is a sequence of operations $p_1 : \Delta_1 = (\delta_2 \sim D_3)$; $p_2 : \Delta_2 = D_1 \bowtie \Delta_1$; $p_3 : M_e = (M_e \sim \Delta_3)$; $p_4 : D_2 = (D_2 \cup \delta_2)$; $p_5 : \Delta_{M1} = (\delta_2 \sim D_3)$; $p_6 : M_1 = (M_1 \cup \Delta_{M1})$.

Then, we perform *minion* operations to combine *rover* data containers with the final materialization before we send the results to users. A sequence of operations includes $M_e = (M_e \uplus D_1^r)$; $M_e = (M_e \uplus D_2^r)$; $M_e = (M_e \uplus D_4^r)$; $M_e = (M_e \uplus D_5^r)$. Operations to combine *rover* data containers to the intermediate materializations are not necessary since we have reached the end of computation process.

To increase performance of the online integration system, we apply *defragmentation* procedure at the very end of computation or only if it is needed. At this stage, the scheduling algorithms for online integration plans described in [5] are applicable.

7 Summary and Future Work

Online integration system proposed in this paper optimizes the integration of large size XML documents by processing the fragments of data increments. Our approach allows for the dynamic identification of a core within every data increment and for processing of each increment in a moment when a core is complete even if an increment itself is not complete yet. It reduces time an increment is waiting for the processing. Meanwhile, the defragmentation procedures are performed at the end of processing to reduce overall online data integration time. We provide the formal backgrounds to show that the online data integration system proposed in the paper is implementable.

A number of interesting research problems remain to be investigated. Online integration system proposed in this paper is applicable for data integration in Internet of Things environment, by associating sensor nodes to the data containers. The first problem is the data structure alignment when the sensors use their own internal data structures. The second problem is related to the large number of sensors involved. The last problem involves properties of usually small and frequently changed data.

References

1. Bose, S., Fegaras, L.: Data stream management for historical XML data. SIGMOD **99**(3), 403–422 (2004)
2. Bose, S., Fegaras, L., Levine, D., Chaluvadi, V.: A query algebra for fragmented XML stream data. In: Lausen, G., Suciu, D. (eds.) DBPL 2003. LNCS, vol. 2921, pp. 195–215. Springer, Heidelberg (2004). doi:10.1007/978-3-540-24607-7_13
3. Braganholo, V., Mattoso, M.: A survey on XML fragmentation. SIGMOD Rec. **43**(3), 24–35 (2014)
4. Fegaras, L.: Incremental query processing on Big Data streams. CoRR, abs/1511.07846 (2015)
5. Handoko, Getta, J.R.: Dynamic query scheduling for online integration of semistructured data. In: 2015 IEEE 39th Annual Computer Software and Applications Conference (COMPSAC), vol. 3, pp. 375–380, July 2015
6. Ma, H., Schewe, K.-D.: Fragmentation of XML documents. J. Inf. Data Manage. **1**(1), 21–33 (2010)
7. Özsu, T.M., Valduriez, P.: Principles of Distributed Database Systems, 3rd edn. Springer, Heidelberg (2011)
8. Wang, G., Huo, H., Han, D., Hui, X.: Query processing and optimization techniques over streamed fragmented XML. World Wide Web **11**(3), 339–359 (2008)
9. Wu, X., Theodoratos, D.: A survey on XML streaming evaluation techniques. VLDB J. **22**(2), 177–202 (2013)

Merging Possibilistic Belief Bases by Argumentation

Thi Hong Khanh Nguyen[1]([⊠]), Trong Hieu Tran[2], Tran Van Nguyen[2], and Thi Thanh Luu Le[2]

[1] Electricity Power University of Vietnam, Hanoi, Vietnam
khanhnth@epu.edu.vn
[2] Vietnam National University, Hanoi, Vietnam
hieutt@vnu.edu.vn, tranvan.vtty@gmail.com, lt.thanhluu@gmail.com

Abstract. Belief merging is one of active research fields with a large range of applications in Artificial Intelligence. Most of the work in this research field is in the centralized approach, however, it is difficult to apply to interactive systems such as multi-agent systems. In this paper, we introduce a new argumentation framework for belief merging. To this end, a constructive model to merge possiblistic belief bases built based on the famous general argumentation framework is proposed. An axiomatic model, including a set of rational and intuitive postulates to characterize the merging result is introduced and several logical properties are mentioned and discussed.

Keywords: Belief merging · Argumentation

1 Introduction

In recent years, Belief Merging, a research field study on the integration of the knowledge bases, has become an attractive research area in Artificial Intelligence. It is applied in a large range of areas such as Information Systems, Multi-agent Systems, Data Retrieval, and Distributed Systems. The advantage of belief merging approach is the richness of information we can obtain but its trade-off is the inconsistencies that we have to solve. In literature, there are two main approaches to deal with the inconsistencies arisen when we combine multiple information sources.

In the first approach, we will try to adopt the inconsistency in obtaining information source by improving the classical reasoning methods. One of typical instances of this idea is the family of paraconsistent logics [7, 8, 15]. This approach needs a simple operation to collect and store information from source, but it requires a highly computational complexity reasoning operation. Unfortunately, the reasoning operation is more frequently used than another, thus this approach is only suitable for a specific class of applications.

In the second approach is using belief merging in which we try to build a consistent information system from multiple information sources. Precisely, from

© Springer International Publishing AG 2017
N.T. Nguyen et al. (Eds.): ACIIDS 2017, Part I, LNAI 10191, pp. 24–34, 2017.
DOI: 10.1007/978-3-319-54472-4_3

the given belief bases $\{K_1, \ldots, K_n\}$ we build a consistent belief base K^* which best represents for these belief bases. There are two settings in this approach, the centralized and distributed ones. In centralized setting, belief merging is considered as an arbitration in which all belief bases are submitted to a mediator, and this mediator will decide which is the common belief base. This is the main trend in belief merging with a large range of works such as [17–19,22,23]. Obviously, in this setting the merging result is depend on the mediator, the participating agents have to expose all their own beliefs and they are omitted in merging process. Therefore, it is difficult to apply to high interactive systems such as multiagent systems.

In the second setting, each merging process is considered as a game in which participants step by step give their proposals until an agreement is reached. The first direction in this setting is belief merging by negotiation with some typical works are as follows: a family of game-based merging operators [16], a two-stage belief merging process [9,10], a bargaining game solution [29] and a game model for merging stratified belief bases [21,24,25,27,28].

The second direction is belief merging by argumentation in which merging process is organized as a debate and participants uses their own beliefs and manipulates argumentation skills to reach the agreement. Typically, an argumentation framework for merging weighted belief bases [14] and other framework for merging the belief bases in possibilistic logic by Amgoud et al. [3]. In [26], a general framework for merging belief bases by argumentation is introduced, however, the semantics of argumentation extensions are not mentioned and discussed.

In this work, we propose a new argumentation framework for merging possibilistic logic bases. The contribution of this paper is two-fold. First, we introduce a framework to merge possibilistic belief bases in which a general argumentation framework is applied in possibilistic belief bases to obtain meaningful results in comparison to other belief merging techniques for belief merging in possibilistic logic such as [3,4,6,20]. Second, an axiomatic model including rational and intuitive postulates for merging results is introduced and several logical properties are discussed.

The rest of this paper is organized as follows: We review about possibilistic logic in Sect. 2. Belief merging for prioritized belief bases by possibilistic logic framework is presented in Sect. 3. Sections IV and V introduce a general argumentation framework and a model to merge belief bases by this framework. Postulates for belief merging by argumentation and logical properties is introduced an discussed in Sect. 4. Some conclusions and future work are presented in Sect. 5.

For the sake of representation, we consider the following example:

Example 1. A terrible environmental crisis, which cause mass fish deaths (a), happened in the seabed in Middle of Vietnam. There are some opinions ordered in time series as follows:

The public and scientists: The mass fish deaths (a) are caused by the toxic spill disaster of a steel factory (b): ($b \rightarrow a$).

Steel factory: We have a modern waste water treatment system (c), thus the water was cleaned before it was discharged: ($c, c \rightarrow \neg b$).

Communication agencies: The steel factory has imported hundreds of tons of chemical toxic (d) and its underground tube is put at wrong position (f): (d, f).

The public and scientists: A diver has died (g) with the symptom caused by toxin from water (b): ($g, b \rightarrow g$).

Steel factory: We have imported chemical material to detergent our tubes (d) however the water was cleaned before it was discharged: ($\neg(d \rightarrow b)$). Our underground tube is in the right position and it has not been completed, thus it can discharge now: ($\neg f$).

Official Authorities: There are two causes of mass fish deaths. It may be from chemical toxin (d) or it may cause by algae bloom phenomenon (e): ($d \rightarrow a$) $\vee$ ($e \rightarrow a$). We have not yet had any clue about the relation between mass fish deaths and the discharge of steel factory.

The public and scientists: The mass fish deaths cannot cause by the algae bloom phenomenon because there is no body of algae, water did not change color and fishes died at the bottom: ($\neg e$).

From the progression of events as above, we have the sets of beliefs as follows:

$K_1 : \{b \rightarrow a, g, b \rightarrow g, \neg e\}$,
$K_2 : \{c, c \rightarrow \neg b, \neg(d \rightarrow b), \neg f\}$,
$K_3 : \{d, f\}$,
$K_4 : \{(d \rightarrow a) \vee (e \rightarrow a)\}$.

2 Possibilistic Logic

In this work, we consider a propositional language $\mathcal{L}$ built on a finite alphabet $\mathcal{P}$ and common logic connectives including $\neg, \wedge, \vee$, and $\rightarrow$. The classical consequence relation is $\vdash$. We use Ω to denote a finite set of interpretations of $\mathcal{L}$. Given $\omega \in \Omega$, $\omega \models \psi$ represents that ω is a model of the formula ψ.

A *possibilistic formula* (ψ, α) Fformula ψ and a weight $\alpha \in [0, 1]$. A *possibilistic belief base* is a finite set of possibilistic formulas $K = (\psi_i, \alpha_i)|i = 1, \ldots, n$. We denote K^* an associated belief base w.r.t K defined as follows: $K^* = \{\psi_i|(\psi_i, \alpha_i) \in K\}$. Obviously, a possibilistic belief base K is consistent if K^* is consistent and vice verse. We also denote $\mathbb{K}$ and $\mathbb{K}^*$ set of all possibilistic belief bases and their associated belief bases, respectively.

For each possibilistic belief base K, the possibility distribution of K, denoted by π_K as follows:

Definition 1 [12]. $\forall \omega \in \Omega$

$$\pi_K(\omega) = \begin{cases} 1 & if \ \forall (\psi_i, \alpha_i) \in K, \omega \models \psi_i \\ 1 - max\{\alpha_i : (\psi_i, \alpha_i) \in K \ and \ \omega \nvDash \psi_i\} & otherwise \end{cases} \quad (1)$$

Example 2. Continuing Example 1. Suppose that $K = \{(a, 0.8), (\neg c; 0.7), (b \rightarrow a, 0.6), (c; 0.5), (c \rightarrow \neg b; 0.4)\}$. According to Definition 1, we can determine the possibility distribution for K as follows: $\pi_K(a \neg b \neg c) = 1, \pi_K(abc) = 0.6, \pi_K(ab \neg c) = 0.5, \pi_K(a \neg bc) = 0.3$, and $\pi_K(\neg abc) = \pi_K(\neg ab \neg c) = \pi_K(\neg a \neg bc) = \pi_K(\neg a \neg b \neg c) = 0.2$

Definition 2. *Given a possibilistic belief base K and $\alpha \in [0, 1]$, the $\alpha - cut$ of K is denoted by $K_{\geq \alpha}$ and defined as follows: $(K_{\geq \alpha} = \{\psi \in K^* | (\psi, \beta) \in K, \beta \geq \alpha\})$. Similarly, a strict $\alpha - cut$ of K is denoted by $K_{> \alpha}$ and defined as follows: $(K_{> \alpha} = \{\psi \in K^* | (\psi, \beta) \in K, \beta > \alpha\})$.*

Definition 3. *Possibilistic belief base K_1 is equivalent to possibilistic belief base K_2, written as $K_1 \equiv K_2$ if and only if $\pi_{K_1} = \pi_{K_2}$.*

It is easy to prove that $K_1 \equiv K_2$ iff for all $\alpha \in [0, 1]$ $(K_1)_{\geq \alpha} \equiv (K_2)_{\geq \alpha}$

2.1 Possibilistic Inference

Definition 4. *The inconsistency degree of possibilistic belief base K is as follows:*

$$Inc(K) = max\{\alpha_i : K_{\geq \alpha_i} \text{ is inconsistent}\} \qquad (2)$$

The inconsistency degree of possibilistic belief base K is the maximal value α_i such that the $\alpha_i - cut$ of K is inconsistent. Conventionaly, if K is consistent, then $Inc(K) = 0$.

Definition 5. *Given a possibilistic belief base K and $(\psi, \alpha) \in K$, (ψ, α) is a subsumption in K if:*

$$(K \setminus \{(\psi, \alpha)\})_{\geq \alpha} \vdash \psi \qquad (3)$$

Respectively, (ψ, α) is a strict subsumption in K if $K_{> \alpha} \vdash \psi$.

We have the following lemma [5]:

Lemma 1. *If (ψ, α) is a subsumption in K then $K \equiv (K \setminus \{(\psi, \alpha)\})$.*

Definition 6. *Given a possibilistic belief base K, formula ψ is a plausible consequence of K if:*

$$K_{> Inc(K)} \vdash \psi \qquad (4)$$

Definition 7. *Given a possibilistic belief base K, formula (ψ, α) is a possibilistic consequence of K, denoted $K \vdash_\pi (\psi, \alpha)$, if:*

- $K_{> Inc(K)} \vdash \psi$
- $\alpha > Inc(K)$ and $\forall \beta > \alpha, K_{> \beta} \nvdash \psi$

In any inconsistent possibilistic belief base K, all formulas with certainty degrees smaller than or equal to $Inc(K)$ will be omitted in the inference process.

Example 3. Continuing Example 2, obviously K is equivalent to $K' = \{(a, 0.8), (\neg c, 0.7), (b \rightarrow a, 0.6), (c, 0.5)\}$. Formula $(c \rightarrow \neg b, 0.4)$ is omitted because of $Inc(K) = 0.5$. We have:

- Plausible inferences of K are $\neg a, c \rightarrow a, b \rightarrow a, \ldots$
- Possibilistic consequences of K are $(c \rightarrow a, 0.7), (b \rightarrow a, 0.6), \ldots$.

3   Belief Merging by Argumentation in Possibilistic Logic

In this section, we consider an implementation of general framework above in order to solve the inconsistencies occur when we combine belief bases $(K_1, \ldots, K_n)$. Let us start with the concept of argument.

Definition 8. *Each argument is presented as a double* $\langle S, s \rangle$, *where s is a formula and S is set of formulas such that:*

(1) $S \subseteq \mathcal{K}^*$,
(2) $S \vdash s$,
(3) *S is consistent and S is minimal w.r.t. set inclusion.*

S is the support and s is the conclusion of this argument. We denote $\mathcal{A}(\mathcal{K})$ *the set of all arguments built from* $\mathcal{K}$.

We recall an argumentation framework in [2], it is extended from the famous one proposed by Dung in [13].

Definition 9. *An argumentation framework is a triple* $\langle \mathcal{A}, \mathcal{R}, \succeq \rangle$ *in which* $\mathcal{A}$ *is a finite set of arguments,* $\mathcal{R}$ *is a binary relation represented the relationship among the arguments in* $\mathcal{A}$, *and* $\succeq$ *is a preorder on* $\mathcal{A} \times \mathcal{A}$. *We also use* $\succ$ *to represent the strict order w.r.t* $\succeq$.

Definition 10. *Let* X, Y *be two arguments in* $\mathcal{X}$.

– *Y attacks X if* $Y \succeq X$ *and* $Y \mathcal{R} X$.
– *If* $Y \mathcal{R} X$ *but* $X \succ Y$ *then X can defend itself .*
– *X set of arguments* $\mathcal{A}$ *defends X if Y attacks X then there always exists* $Z \in \mathcal{A}$ *and Z attacks Y .*

Definition 11. *A set of arguments* $\mathcal{A}$ *is conflict-free if* $\nexists X, Y \in \mathcal{A}$ *such that* $X \mathcal{R} Y$

The attack relations among arguments include undercut and rebut. They are defined as follows:

Definition 12. *Let* $\langle S, s \rangle$ *and* $\langle S', s' \rangle$ *be arguments of* $\mathcal{A}(\mathcal{K})$. $\langle S, s \rangle$ *undercuts* $\langle S', s' \rangle$ *if there exists* $p \in S'$ *such that* $s \equiv \neg p$.

Namely, an argument is under undercut attack if there exists at least one argument in its support is attacked.

Definition 13. *Let* $\langle S, s \rangle$ *and* $\langle S', s' \rangle$ *be arguments of* $\mathcal{A}(\mathcal{K})$. $\langle S, s \rangle$ *rebuts* $\langle S', s' \rangle$ *if* $s \equiv \neg s'$.

Informally, two arguments rebut each other if their conclusions are conflict.

In [1], the authors argued that each argument has a degree of influence. It allows us to compare arguments to choose the best one. When the priorities of arguments are explicit, the higher certain beliefs support, the stronger the argument is. The strength of the argument is defined as follows:

Definition 14. *The force of an argument $A = \langle S, s \rangle$, denoted by $force(A)$ is determined as follows:*

$$force(A) = min\{\alpha_i : \psi_i \in S \text{ and } (\psi_i, \alpha_i) \in \mathcal{K}\}. \tag{5}$$

We consider any aggregation operator $\oplus$ satisfied the following properties:

(1) $\oplus(0, \ldots 0) = 0$,
(2) If $\alpha \geq \beta$ then for all $i = 1, \ldots, n$, then
$\oplus(x_1, \ldots, x_{i-1}, \alpha, x_{i+1}, \ldots, x_n) \geq \oplus(x_1, \ldots, x_{i-1}, \beta, x_{i+1}, \ldots, x_n)$.

Several common aggregation operators considered in literature are maximum (Max), sum (Σ) and lexicographical order($GMax$).

Proposition 1. *Let $\mathcal{K} = \{K_1, \ldots, K_n\}$ be a set of n possibilistic belief bases and $A = \langle S, s \rangle$ be an argument in $\mathcal{A}(\mathcal{K})$, then*

- $\forall \psi_i \in S, K_i \vdash (\psi_j, a_{ji}), i = 1, ..., n$.
- $force(A) = min\{\oplus(a_{j1}, \ldots, a_{jn})\}$.

By the force of argument, we can compare arguments as follows:

Definition 15. *Argument X is preferred to argument Y, denoted by $X \succ Y$ if $force(X) > force(Y)$.*

Example 4. Given $K = \{(\neg b \vee a, 0.9), (b, 0.7), (\neg d \vee a, 0.6), (d, 0.5)\}$, we have: $K = \{(\neg b \vee a, 0.9), (b, 0.7), (\neg d \vee a, 0.6), (d, 0.5)\}$. We have two arguments related to a :

- $A_1 = < \{\neg b \vee a, b\}, a >$,
- $A_2 = < \{\neg d \vee a, d\}, a >$.

However, A_1 is preferred to A_2 because $force(A_1) = 0.7$ and $force(A_2) = 0.5$.

The inconsistence of a possibilistic belief base K_i can be calculated from the force of inconsistent arguments as follows:

Definition 16. *Let K be a possibilistic belief base and $\langle \mathcal{A}(\mathcal{K}), Undercut, \succ \rangle$ be an argumentation framework.*

$$Inc^{att}(K) = max\{min(force(X), force(Y)) \mid \alpha_i \; att \; A_j\}. \tag{6}$$

where $att \in \{undercut, rebut\}$.

Example 5. Let $K_1 = \{(a \vee \neg b; 0.9), (f; 0.9), (g; 0.8), (\neg d \vee \neg e; 0.5), (\neg e; 0.5), (d; 0.5), (a \vee \neg d; 0.4), (\neg b \vee g; 0.3), (a \vee \neg e; 0.3), (a; 0.2), (a \vee \neg d \vee \neg e; 0.1)\}$, $K_2 = \{(c; 0.8), (\neg f; 0.8), (\neg b \vee \neg c; 0.2), (\neg b \wedge d; 0.3)\}$, and $\oplus$ be an aggregation function defined as follows: $\oplus(\alpha, \beta) = \alpha + \beta - \alpha.\beta$. We have: $\mathcal{K}_\oplus = \{(a \vee \neg b \vee c; 0.98), (c \vee f; 0.98), (a \vee \neg b \vee \neg f; 0.98), (c \vee g; 0.96), (\neg f \vee g; 0.96), ((a \vee \neg b) \wedge (a \vee \neg b \vee d); 0.93), ((\neg b \vee f) \wedge (d \vee f); 0.93), (a \vee \neg b \vee \neg c; 0.92), (\neg b \vee \neg c \vee f; 0.92), (a \vee \neg b; 0.9), (f; 0.9), (c \vee \neg d \vee \neg e; 0.9), (c \vee \neg e; 0.9), (c \vee d; 0.9), (\neg d \vee \neg e \vee \neg f; 0.9), (\neg e \vee \neg f; 0.9), (d \vee \neg f; 0.9), (a \vee c \vee \neg d; 0.88), (\neg b \vee c \vee g; 0.88), (a \vee \neg d \vee$

$\neg f; 0.88), (\neg b \lor \neg f \lor g; 0.88), (a \lor c \lor \neg e; 0.86), ((g \lor \neg b) \land (g \lor d); 0.86), (a \lor \neg e \lor \neg f; 0.86), (a \lor c; 0.84), (\neg b \lor \neg c \lor g; 0.84), (a \lor \neg f; 0.84), (a \lor c \lor \neg d \lor \neg e; 0.82), (a \lor \neg d \lor \neg e \lor \neg f; 0.82), (g; 0.8), (c; 0.8), (\neg f; 0.8), ((\neg b \lor \neg e) \land (d \lor \neg e); 0.65), ((\neg b \lor d) \land (d); 0.65), (\neg b \lor \neg c \lor \neg d \lor \neg e; 0.6), (\neg b \lor \neg c \lor \neg e; 0.6), (\neg b \lor \neg c \lor d; 0.6), (a \lor \neg b \lor \neg c \lor \neg d; 0.52), ((\neg b \lor g) \land (\neg b \lor g \lor d); 0.51), ((a \lor \neg b \lor \neg e) \land (a \lor d \lor \neg e); 0.51), (\neg d \lor \neg e; 0.5), (\neg e; 0.5), (d; 0.5), (\neg b \lor \neg c \lor g; 0.44), (a \lor \neg b \lor \neg c \lor \neg e; 0.44), ((a \lor \neg b) \land (a \lor d); 0.44), (a \lor \neg d; 0.4), (\neg b \lor g; 0.3), (a \lor \neg e; 0.3), (\neg b \land d; 0.3), (a \lor \neg b \lor \neg c \lor \neg d \lor \neg e; 0.28), (a; 0.2), (\neg b \lor \neg c; 0.2), (a \lor \neg d \lor \neg e; 0.1)\}$.

Table 1 is the set of arguments built from $\mathcal{K}_\oplus$ an their force. We have:
$Undercut = (A_{11}, A_{32}), (A_{11}, A_{33}), (A_{32}, A_{11}), (A_{32}, A_{12}), (A_{32}, A_{16}), (A_{32}, A_{17}), (A_{32}, A_{18}), (A_{32}, A_{19}), (A_{32}, A_{21}), (A_{32}, A_{22}), (A_{32}, A_{25}), (A_{32}, A_{28}), (A_{32}, A_{29}), (A_{32}, A_{30})$.

We have:
$Inc^{undercut}(\mathcal{K}_\oplus) = max\{min(0.9, 0.8), min(0.9, 0.8), min(0.8, 0.9), min(0.8, 0.9), min(0.8, 0.9), min(0.8, 0.9), min(0.8, 0.9), min(0.8, 0.88), min(0.8, 0.88), min(0.8, 0.88), min(0.8, 0.86), min(0.8, 0.84), min(0.8, 0.82), min(0.8, 0.82)\} = 0.8$. Therefore, the inconsistency degree of $\mathcal{K}_\oplus$ is 0.8.

Now, we can define the belief merging by argumentation as follows:

Definition 17. Let $\mathcal{K} = \{K_1, \ldots, K_n\}$ be a set of possibilistic belief bases. Belief merging operator is defined as follows:
$\Delta_\oplus^{att}(\mathcal{K}) = \{\psi | (\psi, a) \in \mathcal{K}_\oplus, a > Inc^{att}(\mathcal{K}_\oplus)\}$ where $att \in \{indercut, rebut\}$.

We call $\Delta_\oplus^{att}$ the family of BMA (*Belief Merging by Argumentation*) operators.

Example 6. Continuing Example 5, with $att = undercut$ and $\oplus(\alpha, \beta) = \alpha + \beta - \alpha.\beta$ we have:
$\Delta_\oplus^{att}(\mathcal{K}) = \{\{(a \lor \neg b \lor c), (c \lor f), (a \lor \neg b \lor \neg f), (c \lor g), (\neg f \lor g), ((a \lor \neg b) \land (a \lor \neg b \lor d)), ((\neg b \lor f) \land (d \lor f)), (a \lor \neg b \lor \neg c), (\neg b \lor \neg c \lor f), (a \lor \neg b), (f), (c \lor \neg d \lor \neg e), (c \lor \neg e), (c \lor d), (\neg d \lor \neg e \lor \neg f), (\neg e \lor \neg f), (d \lor \neg f), (a \lor c \lor \neg d), (\neg b \lor c \lor g), (a \lor \neg d \lor \neg f), (\neg b \lor \neg f \lor g), (a \lor c \lor \neg e), ((g \lor \neg b) \land (g \lor d)), (a \lor \neg e \lor \neg f), (a \lor c), (\neg b \lor \neg c \lor g), (a \lor \neg f), (a \lor c \lor \neg d \lor \neg e), (a \lor \neg d \lor \neg e \lor \neg f)\}$.

4   Postulates and Logical Properties

We recall that $\mathcal{K} = \{K_1, \ldots, K_n\}$ is a finite set of possibilistic belief bases, AF_s is an argumentation framework is determined from $\mathcal{K}$. Aggregation function $\mathcal{K}_\oplus$ is defined as follows: $\mathcal{K}_\oplus : \mathbb{K}^n \to \mathbb{K}^*$. The set of postulates is introduced as follows:

(SYM) $\mathcal{K}_\oplus(\{K_1, \ldots, K_n\}) = \mathcal{K}_\oplus(\{K_{\pi(1)}, \ldots, K_{\pi(n)}\})$, where π is a permutation in $\{1, \ldots, n\}$.
Postulate (SYM), sometimes called (ANON)[11], ensures the equity of participants. It states that the result of an argumentation process should reflect the arguments of the participants rather than their identity.

Table 1. Forces of arguments.

Argument	Force
$A_1 =< \{a \vee \neg b \vee c\}, a \vee \neg b \vee c >$	0.98
$A_2 =< \{c \vee f\}, c \vee f >$	0.98
$A_3 =< \{a \vee \neg b \vee \neg f\}, a \vee \neg b \vee \neg f >$	0.98
$A_4 =< \{c \vee g\}, c \vee g >$	0.96
$A_5 =< \{\neg f \vee g\}, \neg f \vee g >$	0.96
$A_6 =< \{(a \vee \neg b) \wedge (a \vee \neg b \vee d)\}, (a \vee \neg b) \wedge (a \vee \neg b \vee d) >$	0.93
$A_7 =< \{((\neg b \vee f) \wedge (d \vee f)\}, (\neg b \vee f) \wedge (d \vee f) >$	0.93
$A_8 =< \{a \vee \neg b \vee \neg c\}, a \vee \neg b \vee \neg c >$	0.92
$A_9 =< \{\neg b \vee \neg c \vee f\}, \neg b \vee \neg c \vee f >$	0.92
$A_{10} =< \{a \vee \neg b\}, a \vee \neg b >$	0.9
$A_{11} =< \{f\}, f >$	0.9
$A_{12} =< \{f, \neg f \vee g\}, g >$	0.9
$A_{13} =< \{c \vee \neg d \vee \neg e\}, c \vee \neg d \vee \neg e >$	0.9
$A_{14} =< \{c \vee \neg e\}, c \vee \neg e >$	0.9
$A_{15} =< \{c \vee d\}, c \vee d >$	0.9
$A_{16} =< \{\neg d \vee \neg e \vee \neg f, f\}, \neg d \vee \neg e >$	0.9
$A_{17} =< \{\neg e \vee \neg f, f\}, \neg e >$	0.9
$A_{18} =< \{d \vee \neg f, f\}, d >$	0.9
$A_{19} =< \{a \vee c \vee \neg d, d \vee \neg f, f\}, a \vee c >$	0.88
$A_{20} =< \{\neg b \vee c \vee g\}, \neg b \vee c \vee g >$	0.88
$A_{21} =< \{a \vee \neg d \vee \neg f, f\}, a \vee \neg d >$	0.88
$A_{22} =< \{\neg b \vee \neg f \vee g, f\}, \neg b \vee g >$	0.88
$A_{23} =< \{a \vee c \vee \neg e\}, a \vee c \vee \neg e >$	0.86
$A_{24} =< \{(g \vee \neg b) \wedge (g \vee d)\}, (g \vee \neg b) \wedge (g \vee d) >$	0.86
$A_{25} =< \{a \vee \neg e \vee \neg f, f\}, a \vee \neg e >$	0.86
$A_{26} =< \{a \vee c\}, a \vee c >$	0.84
$A_{27} =< \{\neg b \vee \neg c \vee g\}, \neg b \vee \neg c \vee g >$	0.84
$A_{28} =< \{a \vee \neg f, f\}, a >$	0.84
$A_{29} =< \{a \vee \neg d \vee \neg e \vee \neg f, f\}, a \vee \neg d \vee \neg e >$	0.82
$A_{30} =< \{a \vee \neg d \vee \neg e \vee \neg f, f, d \vee \neg f\}, a \vee \neg d \vee \neg f >$	0.82v
$A_{31} =< \{c\}, c >$	0.8
$A_{32} =< \{\neg f\}, \neg f >$	0.8
$A_{33} =< \{\neg b \vee \neg c \vee f, c, \neg f\}, \neg b >$	0.8
$A_{34} =< \{\neg b \vee \neg c \vee g, c\}, \neg b \vee g >$	0.84
$A_{35} =< \{(\neg b \vee \neg e) \wedge (d \vee \neg e)\}, (\neg b \vee \neg e) \wedge d \vee \neg e) >$	0.65
$A_{36} =< \{(\neg b \vee d) \wedge (d)\}, (\neg b \vee d) \wedge (d) >$	0.65
$A_{37} =< \{\neg b \vee \neg c \vee \neg e, c\}, \neg b \vee \neg e >$	0.6
$A_{38} =< \{\neg b \vee \neg c \vee d, c\}, \neg b \vee d >$	0.6
$A_{39} =< \{(\neg b \vee g) \wedge (\neg b \vee g \vee d)\}, (\neg b \vee g) \wedge (\neg b \vee g \vee d) >$	0.51
$A_{40} =< \{(a \vee \neg b \vee \neg e) \wedge (a \vee d \vee \neg e)\}, (a \vee \neg b \vee \neg e) \wedge (a \vee d \vee \neg e) >$	0.51
$A_{41} =< \{a \vee \neg b \vee \neg c \vee \neg e, c\}, a \vee \neg b \vee \neg e >$	0.44
$A_{42} =< \{(a \vee \neg b) \wedge (a \vee d)\}, (a \vee \neg b) \wedge (a \vee d) >$	0.44

(CON) $\not\exists \psi \in \mathcal{L}(\mathcal{K}_\oplus(\{K_1,\ldots,K_n\}) \vdash \psi) \wedge (\mathcal{K}_\oplus(\{K_1,\ldots,K_n\}) \vdash \neg\psi)$

Postulate (CON) states that belief merging by argumentation should return a consistent result.

(UNA) if $K_1^* \equiv \ldots \equiv K_n^*$ then $\mathcal{K}_\oplus(\{K_1,\ldots,K_n\}) \equiv K_1^*$.

Postulate (UNA) presents the assumption of unanimity. It states that if all participants possess the same set of beliefs, then this set of belief should be the result of argumentation process. Clearly, Postulate (UNA) is more general than postulate (IDN) and it also implies (IDN) which is defined as follows:

(IDN) $\mathcal{K}_\oplus(\{K_i,\ldots,K_i\}) \equiv K_i^*$

It states that if all participants have the same possibilistic belief base, then after the argumentation process, we should have the result as its associated belief base.

(CLO) $\cup_{i=1}^n B_i^* \vdash \mathcal{K}_\oplus(\{K_i,\ldots,K_i\})$

Postulate (CLO) requires the closure of the result of argumentation process. It states that any belief in argumentation result should be in at least some input belief base.

(MAJ) if $|\{K_i^* \vdash \psi, i = 1\ldots n\}| > \frac{n}{2}$ then $\mathcal{K}_\oplus(\{K_i,\ldots,K_i\}) \vdash \psi$.

Postulate (MAJ) states that if a belief is supported by the majority group of participants, it should be in the result of argumentation process.

(COO) if $K_i^* \vdash \psi, i = 1\ldots n$ then $\mathcal{K}_\oplus(\{K_i,\ldots,K_i\}) \vdash \psi$.

Postulate (COO) states that if a belief is supported by all participants, it should be in the result of argumentation process.

We have the following lemma:

Lemma 2. *It holds that:*

- *(UNA) implies (IDN);*
- *(MAJ) implies (COO).*

Investigate the properties of belief merging operator defined in the previous section we have:

Theorem 1. *Family of BMA operators satisfies the following postulates* (SYM), (CON), (UNA), *and* (CLO). *It does not satisfy* (MAJ).

5   Conclusion

In this paper, a framework for merging possibilistic belief bases by argumentation is introduced and discussed. The key idea in this work is using the inconsistent degree as a measure together with the notion of undercut to construct an argumentation framework for belief merging. A set of postulates is introduced and logical properties are mentioned and discussed. They assure that the proposed model is sound and complete. The deeper analysis on the set of postulates and logical properties, and the evaluation of computational complexities of belief merging operators in this framework are reserved as future work.

Acknowledgment. The authors would like to thank Professor Quang Thuy Ha and Knowledge Technology Lab, Faculty of Information Technology, VNU University of Engineering and Technology for expertise support.

References

1. Amgoud, L., Cayrol, C.: Inferring from inconsistency in preference-based argumentation frameworks. Int. J. Approximate Reasoning **29**, 125–169 (2002)
2. Amgoud, L., Cayrol, C.: A reasoning model based on the production of acceptable arguments. Ann. Math. Artif. Intell. **34**, 197–216 (2002)
3. Amgoud, L., Kaci, S.: An argumentation framework for merging conflicting knowledge bases. Int. J. Approximate Reasoning **45**(2), 321–340 (2007). Eighth European Conference on Symbolic and Quantitative Approaches to Reasoning with Uncertainty (ECSQARU 2005)
4. Benferhat, S., Dubois, D., Kaci, S., Prade, H.: Possibilistic merging and distance-based fusion of propositional information. Ann. Math. Artif. Intell. **34**(1–3), 217–252 (2002)
5. Benferhat, S., Dubois, D., Prade, H., Williams, M.-A.: A practical approach to fusing prioritized knowledge bases. In: Barahona, P., Alferes, J.J. (eds.) EPIA 1999. LNCS (LNAI), vol. 1695, pp. 222–236. Springer, Heidelberg (1999). doi:10. 1007/3-540-48159-1_16
6. Benferhat, S., Kaci, S.: Fusion of possibilistic knowledge bases from a postulate point of view. Int. J. Approximate Reasoning **33**, 255–285 (2003)
7. Béziau, J.: Paraconsistent logic from a modal viewpoint. J. Appl. Logic **3**(1), 7–14 (2005)
8. Blair, H.A., Subrahmanian, V.: Special issue paraconsistent logic programming. Theoret. Comput. Sci. **68**(2), 135–154 (1989)
9. Booth, R.: A negotiation-style framework for non-prioritised revision. In: Proceedings of the 8th Conference on Theoretical Aspects of Rationality and Knowledge, TARK 2001, pp. 137–150. Morgan Kaufmann Publishers Inc. (2001)
10. Booth, R.: Social contraction and belief negotiation. Inf. Fusion **7**, 19–34 (2006)
11. Delobelle, J., Haret, A., Konieczny, S., Mailly, J., Rossit, J., Woltran, S.: Merging of abstract argumentation frameworks. In: Baral, C., Delgrande, J.P., Wolter, F. (eds.) Principles of Knowledge Representation and Reasoning: Proceedings of the Fifteenth International Conference, KR 2016, Cape Town, South Africa, April 25–29, 2016, pp. 33–42. AAAI Press (2016)
12. Dubois, D., Lang, J., Prade, H.: Possibilistic logic. In: Gabbay, D., et al. (eds.) Handbook of Logic in Artificial Intelligence and Logic Programming, pp. 439–513 (1994)
13. Dung, P.M.: On the acceptability of arguments and its fundamental role in non-monotonic reasoning, logic programming and n-person games. Artif. Intell. **77**, 321–357 (1995)
14. Gabbay, D., Rodrigues, O.: A numerical approach to the merging of argumentation networks. In: Fisher, M., Torre, L., Dastani, M., Governatori, G. (eds.) CLIMA 2012. LNCS (LNAI), vol. 7486, pp. 195–212. Springer, Heidelberg (2012). doi:10. 1007/978-3-642-32897-8_14
15. Hunter, A.: Paraconsistent Logics. Springer, Dordrecht (1998)

16. Konieczny, S.: Belief base merging as a game. J. Appl. Non-Classical Logics **14**(3), 275–294 (2004)
17. Konieczny, S., Perez, R.P.: On the logic of merging. In: Proceedings of the Sixth International Conference on Principles of Knowledge Representation and Reasoning (KR98), Trento, pp. 488–498 (1998)
18. Lin, J.: Integration of weighted knowledge bases. Artif. Intell. **83**, 363–378 (1996)
19. Lin, J., Mendelzon, A.: Merging databases under constraints. Int. J. Coop. Inf. Syst. **7**(1), 55–76 (1998)
20. Qi, G., Du, J., Liu, W., Bell, D.A.: Merging knowledge bases in possibilistic logic by lexicographic aggregation. In: Grünwald, P., Spirtes, P. (eds.) UAI 2010, Proceedings of the Twenty-Sixth Conference on Uncertainty in Artificial Intelligence, Catalina Island, CA, USA, July 8–11, 2010, pp. 458–465. AUAI Press (2010)
21. Qi, G., Liu, W., Bell, D.A.: Combining multiple prioritized knowledge bases by negotiation. Fuzzy Sets Syst. **158**(23), 2535–2551 (2007)
22. Rescher, N., Manor, R.: On inference from inconsistent premises. Theor. Decis. **1**, 179–219 (1970)
23. Revesz, P.Z.: On the semantics of theory change: arbitration between old and new information. In: 12th ACM SIGACT-SIGMOD-SIGART Symposium on Principles of Databases, pp. 71–92 (1993)
24. Tran, T.H., Vo, Q.B.: An axiomatic model for merging stratified belief bases by negotiation. In: Nguyen, N.-T., Hoang, K., Jędrzejowicz, P. (eds.) ICCCI 2012. LNCS (LNAI), vol. 7653, pp. 174–184. Springer, Heidelberg (2012). doi:10.1007/978-3-642-34630-9_18
25. Tran, T.H., Nguyen, N.T., Vo, Q.B.: Axiomatic characterization of belief merging by negotiation. Multimedia Tools Appl. 1–27, June 2012
26. Tran, T.H., Nguyen, T.H.K., Ha, Q.T., Vu, N.T.: Argumentation framework for merging stratified belief bases. In: Nguyen, N.T., Trawiński, B., Fujita, H., Hong, T.-P. (eds.) ACIIDS 2016. LNCS (LNAI), vol. 9621, pp. 43–53. Springer, Heidelberg (2016). doi:10.1007/978-3-662-49381-6_5
27. Tran, T.H., Vo, Q.B., Kowalczyk, R.: Merging belief bases by negotiation. In: König, A., Dengel, A., Hinkelmann, K., Kise, K., Howlett, R.J., Jain, L.C. (eds.) KES 2011. LNCS (LNAI), vol. 6881, pp. 200–209. Springer, Heidelberg (2011). doi:10.1007/978-3-642-23851-2_21
28. Tran, T.H., Vo, Q.B., Nguyen, T.H.K.: On the belief merging by negotiation. In: 18th International Conference in Knowledge Based and Intelligent Information and Engineering Systems, KES 2014, Gdynia, Poland, 15–17, 2014, pp. 147–155, September 2014
29. Zhang, D.: A logic-based axiomatic model of bargaining. Artif. Intell. **174**, 1307–1322 (2010)

Towards Common Vocabulary for IoT Ecosystems—preliminary Considerations

Maria Ganzha[1,3], Marcin Paprzycki[1,4], Wiesław Pawłowski[2(✉)], Paweł Szmeja[1], and Katarzyna Wasielewska[1]

[1] Systems Research Institute, Polish Academy of Sciences, Warsaw, Poland
{maria.ganzha,marcin.paprzycki,pawel.szmeja,
katarzyna.wasielewska}@ibspan.waw.pl
[2] Faculty of Mathematics, Physics, and Informatics,
University of Gdańsk, Gdańsk, Poland
wieslaw.pawlowski@inf.ug.edu.pl
[3] Department of Mathematics and Information Technologies,
Warsaw University of Technology, Warsaw, Poland
[4] Department of Management, Warsaw Management Academy, Warsaw, Poland

Abstract. The INTER-IoT project aims at delivering a comprehensive solution to the problem of interoperability of Internet of Things platforms. Henceforth, semantic interoperability also has to be addressed. This should involve a hierarchy of ontologies, starting from an upper ontology, through core and domain ontologies. As a starting point, we have analyzed ontological models of the concepts of *thing, device, observation* and *deployment*, as occurring in the IoT domain. We have chosen five popular ontologies: SSN, SAREF, oneM2M Base Ontology, IoT-Lite, and OpenIoT, as candidates for a central INTER-IoT ontology.

Keywords: Internet of Things · Semantic interoperability · IoT ontology

1 Introduction

Lack of interoperability between Internet of Things (IoT) platforms/systems/applications is recognized as an important issue that prevents faster development of IoT ecosystems. Therefore, the European Commission has funded seven research projects, to find a comprehensive solution. Among them, the INTER-IoT project will use semantic technologies to deal with meta-level interoperability. Specifically, the semantic interoperability will be established through the use of a modular *central ontology*, ontology alignments, and semantic transformations. Therefore, one of key questions becomes: what should the central ontology be based on? Therefore, we took an initial look at the state-of-the-art IoT ontologies and analyzed how they conceptualize *thing, device, observation* and *deployment*.

In what follows, we report our findings. We start by briefly outlining the INTER-IoT approach to semantic interoperability. This allows us to discuss the

© Springer International Publishing AG 2017
N.T. Nguyen et al. (Eds.): ACIIDS 2017, Part I, LNAI 10191, pp. 35–45, 2017.
DOI: 10.1007/978-3-319-54472-4_4

proposed structure and role of the central ontology. Next, we describe key technical details of selected ontologies. Finally, we present a general analysis of the ontologies and their applicability in the INTER-IoT. Here, let us note that for the purpose of this contribution, we will use the term "IoT artifact" to denote any entity that can join an IoT ecosystem, e.g. platforms, systems, applications, etc.

2   Semantic Interoperability—the INTER-IoT Way

The goal of the INTER-IoT project [2] is to facilitate interoperability across the hardware-software stack. However, here, we are solely interested in semantic interoperability. In [10], we have outlined the state-of-the-art in ontologies of the Internet of Things (as well as these related to project's main use cases). As expected, we have found that there is no single, comprehensive, *all agreed* ontology of the IoT. Taking this into account, in [11], we have proposed the following approach to reaching semantic interoperability. Here, for simplicity, only the core assumptions and the basic flow of information is described.

1. We assume that multiple IoT artifacts are to be joined into an ecosystem (we try to avoid conceptual traps of a scenario where only 2 artifacts are considered). This process involves human developers, who will establish the necessary data flows. We assume that bringing about interoperability should force only minimal changes to the joining artifacts (ideally, none).
2. For each artifact, its semantics is lifted to an OWL-based representation (see, [13]). If the original semantics was not OWL-based, bi-directional converters (named *producer* and *consumer*) are created, to allow communication in "own language". Specifically, original semantics and data format is mapped onto the OWL ontology in the RDF format (and a similar mapping is created for communication "back").
3. A central modular ontology is instantiated. Its modules capture necessary aspects of the IoT, as well as domain specific concepts. Here, the key assumption is *modularity*. For instance, if in a given IoT ecosystem it is not necessary to provide geospatial information (e.g. when all sensing devices are placed in stationary locations), the geospatial module will not be included.
4. Ontologies representing each joining artifact are *aligned* with appropriate modules of the central ontology (see, [12]). The resulting alignments are persisted and form the basis for translation between communicating artifacts.
5. Communication, in addition to conversions performed by producers and consumers, involves semantic translations (using alignments) from semantics of a source artifact to the common semantics and then, to the semantics of the target artifact. Obviously, process is repeated "on the way back".

Clearly, construction of the central ontology, plays key role in the proposed approach to semantic interoperability, and thus, is the focus of this paper.

3   Comparing IoT-Related Ontologies

The space of ontologies is fragmented, regardless of the domain of interest. The richer an ontology is, the larger area it spans. Hence, uniqueness and intersections with other ontologies become more intricate and complex. Internet of Things spans enormous number of domains, and expands with the growing popularity of "smart devices". Use of ontologies in the IoT mimics this expansiveness. There are many ontologies that represent models relevant to the IoT, including, but not limited to, devices, units of measurement, data streams, data processing, geolocation, data provenance, computer hardware, methods of communication, etc. We assume that the centerpiece of the IoT is a smart device capable of communication. From this perspective, we have identified ontologies that capture the idea of a device, and are well established in the IoT space: SSN, SAREF, oneM2M Base Ontology, IoT-Lite, and OpenIoT. Each of them takes a different approach to modeling the IoT space but, despite the differences in conceptualization, they cover intersecting fragments of the IoT landscape. Below, we discuss divergence, contrariness and overlaps between these ontologies.

SSN, or "Semantic Sensor Network" [4,8] is an ontology centered around sensors and observations. It is a de-facto extension of the SensorML language. SSN focuses on measurements and observations, disregarding hardware information about the device. Specifically, it describes sensors in terms of capabilities, performance, usage conditions, observations, measurement processes, and deployments. It is highly modular and extendable. In fact, it depends on other ontologies in key areas (e.g. time, location, units) and, for all practical purposes, needs to be extended before actual implementation of an SSN-based IoT system. SSN, formulated on top of DUL[1], is an ontological basis for the IoT, as it tries to cover any application of sensors in the IoT.

SAREF [9], or "The Smart Appliances REFerence" ontology covers the area of smart devices in houses, offices, public places, etc. It does not focus on any industrial or scientific implementation. The devices are characterized predominantly by the function(s) they perform, commands they accept, and states they can be in. Those three categories serve as building blocks of the semantic description in SAREF. Elements from each can be combined to produce complex descriptions of multi-functional devices. The description is complemented by device services that offer functions. A noteworthy module of SAREF is the energy and power profile that received considerable attention, shortly after its inception[2]. SAREF uses WGS84 for geolocation and defines its own measurement units.

oneM2M Base Ontology (oneM2M BO; [3,6]) is a recently created ontology, with first non-draft release in August 2016. It is relatively small, prepared for the release 2.0 of oneM2M specifications, and designed with the intention of providing a shared ontological base, to which other ontologies would align. It is similar to the SSN, since any concrete system necessarily needs to extend it

---

[1] http://www.ontologydesignpatterns.org/ont/dul/DUL.owl.
[2] https://goo.gl/1OXTJb, https://goo.gl/ZaGjCJ.

before implementation. It describes devices in a very broad scope, enabling (in a very general sense) specification of device functionality, networking properties, operation and services. The philosophy behind this approach was to enable discovery of semantically demarcated resources using a minimal set of concepts. It is a base ontology, as it does not extend any other base models (such as DUL or Dublin Core). However, alignments to other ontologies are known [5].

IoT-Lite [7] is an "instantiation" of the SSN, i.e. a direct extension of some of its modules. It is a minimal ontology, to which most of the caveats of the SSN apply. Specifically: focus on sensors and observations, reliance on other ontologies (e.g. time or units ontologies), high modularity and extendability. The idea behind the IoT-Lite was to create a small/light semantic model that would be less taxing (than other, more verbose and broader models) on devices that process it. At the same time, it needed to cover enough concepts to be useful. The ontology describes devices, objects, systems and services. The main extension of the SSN, in the IoT-Lite, lies in addition of actuators (to complement sensors, as a device type) and a coverage property. It explicitly uses concepts from a geolocation ontology [1] to demarcate device coverage and deployment location.

OpenIoT [15,16] ontology was developed within the OpenIoT project. However, here, we use the term "OpenIoT" to refer to the ontology. It is a comparatively big model that (re)uses and combines other ontologies. Those include all modules of the SSN (the main basis for the OpenIoT), SPITFIRE (including sensor networks), Event Model-F, PROV-O, LinkedGeoData, WGS84, Cloud-Domain, SIOC, Association Ontology and others, including smaller ontologies developed at the DERI (currently, Insight Centre). It also makes use of ontologies that provide basis for those enumerated earlier, e.g. DUL. Other than concepts from the SSN, OpenIoT, uses a large number of SPITFIRE concepts, e.g. network and sensor network descriptions. While some mentioned ontologies are not imported by the OpenIoT explicitly, they appear in all examples, documentation, and project deliverables. Therefore, one can treat OpenIoT as a combination of parts of all of those. Similarly to the SSN, OpenIoT does not define its own location concepts and does not explicitly import geolocation ontologies. It relies on other ontologies for that but, in contrast to the SSN, it clearly indicates Linked-GeoData and WGS84 as sources of geolocation descriptions. It defines a limited set of units of measure (e.g. temperature, wind speed), but only when they were relevant to the OpenIoT project pilot implementation.

The rich suite of used ontologies means that OpenIoT provides very rich description of devices, their functionalities, capabilities, provenance, measurements, deployments and position, energy, relevant events, users and many others. Interestingly enough, it does not explicitly describe actuators or actuating properties/functions. It can be observed that the broad scope of the ontology makes it rather complicated. This is also because, it is not documented well-enough, i.e. the detail level and ease-of-access of the documentation do not match the range of coverage of concepts in the model. Moreover, it is not clearly and explicitly modularized, despite being an extension of the SSN.

Let us note that, while there are other IoT models of potential interest (such as OGC Sensor Things, UniversAAL ontologies, FAN FPAI, IoT Ontology[3], M3 Vocabulary), we will not consider them here. This is because of (a) space limitation, and (b) the fact that they have generated much less "general interest". Nevertheless, we plan to include these ontologies in subsequent work.

Let us now compare the selected ontologies side-by-side. To do this, we have selected key aspects, or categories, *directly pertaining to the IoT*; placed the first column of Table 1. However, because of intricacies and disparate philosophies behind compared ontologies (see, above), each category needs to be further investigated. In other words, proposed categorization is a tentative way of visualizing and analyzing similarities and differences between ontologies of choice. Here, we follow an approach proposed by Raúl García-Castro during June 2016 European Platform Initiative (IOT EPI[4]) meeting.

Before proceeding it should be noted that there are numerous approaches to ontology evaluation, e.g. [14,17]. We have, however, found that applying them would not help in the context of specific, project-related, problem. Specifically, we were more interested in capturing and comparing *details* of each area that the selected ontologies cover, rather than their *overall evaluation* by some standard.

Table 1. IoT ontologies comparison

Category (Subdomain)	SSN	SAREF	oneM2M BO	IoT-Lite[†]	OpenIoT[†]
Thing	X	X	X	X	X
Device	X	X	X	X	X
Device deployment	X^α	X	$X^{\oslash\alpha}$	X	X
Device properties & capabilities	X				X
Device energy	X	X^ϵ			X
Function & service		X	X	X^S	
Sensing & sensor properties	X	X^β		$X^\oslash$	X
Observation	X^α	X	X		X
Actuating & actuator properties		X^β		$X^\oslash$	
Conditionals	X				

† Extends modules of SSN
α No time or location
β Implicit, implied by device functions
ϵ Rich energy model
S Service only
$\oslash$ Only small or provisional description, or a stub

In what follows, we discuss selected categories from Table 1. While, due to space limitation, we had to pick only some categories, this discussion should be valuable to anyone interested in use of semantic technologies in the IoT.

---

[3] http://ai-group.ds.unipi.gr/kotis/ontologies/IoT-ontology.
[4] http://iot-epi.eu/.

Thing. This category describes the general approach and provision of properties to any class of an ontology. All considered ontologies are, understandably, generic in this regard. Each contains only a handful of relevant properties that pertain to the very generic concepts. SSN's Things can have FeatureOfInterest (an abstraction of a real world phenomena, such as person, event or, literally, anything) and display Properties (a specification of DUL Quality; needs to be observable and inseparable from the SSN thing). SAREF defines a, similarly general, Property (specifying anything that can be sensed, measured or controlled). IoT-Lite extends the SSN with an Object (any physical entity) and its Attribute (any property exhibited by the Object that can be exposed by a Service). OpenIoT does not provide independent extensions or departures from the approach taken by the SSN. Instead, it provides subclasses for the SSN Property, mostly to describe entities needed in pilots of the project (e.g. WindSpeed, AtmospherePressure).

OneM2M BO is unique in its description of things, because the entire ontology is very general. It defines its own Thing class that captures, quite literally, any entity identifiable in a oneM2M system. OneM2M BO does not extend any upper ontologies, and its Thing is a direct subclass of owl:Thing. Here, a Thing can have ThingProperty (which has a self-explanatory, all-encompassing definition). In this way, oneM2M BO displays characteristics of an upper ontology.

Device. Devices are at the core of the IoT. This is reflected in all ontologies. OneM2M BO proposes the simplest structure of a Device class that uses a written description, instead of rich ontological relations. Device has a single subclass of InterworkedDevice (one that does not directly implement oneM2M interfaces). A Device can consistOf a number of other Devices.

In the SSN, the central taxonomy subtree consists of Device, Sensor, and SensingDevice subsuming both previous classes. An SSN System can represent any part of an infrastructure of devices connected in some way. In particular, it can be any Device in the System. Any System is comprised of subsystems (also of class System). IoT-Lite expands this structure with the addition of an ActuatingDevice and a (passive) TagDevice. Strangely, there is no definition of an Actuator. OpenIoT does not expand the basic structure of the SSN.

SAREF borrows from both, oneM2M and SSN. SAREF Device consistsOf any number of Devices, and has a DeviceCategory that, in turn, has its own subclass structure (which starts with FunctionRelated, EnergyRelated and BuildingRelated categories). It is meant to represent a given perspective (point of view) on a device (e.g. of user, administrator, manufacturer, etc.). On top of that, the ontology defines a couple of subclasses of the Device class, which range from general, such as a Sensor, to quite specific, like a WashingMachine (with classes, such as Switch, in between). Interestingly, Sensor and Actuator are not "neighbors" (the first being a subclass of a Device, and the latter of a DeviceFunction).

Observation. The second crucial element of any IoT ontology is the way that observations are modeled. They are fundamental data items, and their description very strongly affects possible use of a model and functionality of a concrete systems. In oneM2M BO, observations revolve around three general classes:

Variable, Aspect and Metadata. Variable class encompasses input and output variables, as well as a ThingProperty, that pertains to any entity and can have additional Metadata. The latter class is a catch-all way of annotating observations (e.g. with units or precision), which lacks specification, i.e. any property structure is permissible under the BO Metadata. Aspects describe functionality as well as input or output Variables. This simplistic, high-level model of observations allows for great flexibility. On the other hand, there are no examples, and the intended use is very tersely explained. Lack of documentation, combined with elasticity of interpretation, may lead to systems being barely interoperable, despite using the same base ontology.

SSN proceeds differently, by extending the general model proposed by DUL. It introduces the Observation class. Each Observation results in a SensorOutput, a class with relations with other relevant information, such as ObservationValue, or the Sensor that made the Observation. Observations have FeatureOfInterest that describes their characteristics, e.g. precision, latency, range, response time, etc. In general, the SSN Observation is a record of an occurrence of measurement, along with structured meta-data about the observation value, its properties, as well as the process leading to the Observation. Since the SSN lacks explicit units or time definitions, it needs to be complemented with relevant ontologies.

IoT-Lite does not extend the SSN Observation related modules. Instead, it proposes a vast simplification by introducing a Metadata class, similarly to the oneM2M BO. It is a generic class, intended to model any entity that does not fit the Unit or QuantityKind classes (a separate ontology is needed to describe the actual quantities). Observed values are not stored in the structure of the IoT-Lite. Instead, sensors are described in terms of types/kinds of observations made by them. For instance, one can construct a full description of a temperature sensor with meta-data of precision, unit, etc. However, within IoT-Lite, a series of concrete observations cannot be described.

OpenIoT extends the SSN Observation model by providing a Context, however, because of lack of documentation, the intended usage of this class is not clear. Nevertheless, it preserves the SSN Observation structure.

Finally, SAREF observations are described in terms of device Functions (in particular, SensingFunction and MeteringFunction). While lacking an explicit observation class, Functions have a number of properties that pertain to concrete values of measurements. Every relevant Function has a time value (e.g. hasMeterReadingTime) and an "observation" value (e.g. hasMeterReadingValue). These values are described in terms of Properties, which have concrete values alongside the UnitsOfMeasure. SAREF proposes its own taxonomy of units of measurements (currency, power, temperature, etc.). Other than the values of concrete measurements, Functions have "reading types" (e.g. gas, pressure, energy, etc.), which are implied to be relatively constant, vis-a-vis, for instance, meter readings of time and value. Compared to the SSN, the observation model in SAREF is simpler, and more focused on devices and their functions. It does not treat observations as pieces of data with their own structure and place in the system, which enables advanced data processing, e.g. analysis of historical data (within the structure given by the ontology). Instead, the SAREF model presents observations as tentative "outputs" of a function.

Device Deployment. A deployment description is a very important information in any system with multiple distributed devices. OneM2M BO interprets this category as a basic information about a network environment (AreaNetwork), but only if the device is proxied (InterworkedDevice). There is no standard way to model deployment information for any oneM2M BO Device.

SSN describes device deployment in terms of Platform(s) a Device is on, and System(s) it is part of. Even though the SSN itself does not define time or location properties, it is strongly implied that Devices, Systems and Platforms should be annotated with such information (no specific ontology to fulfill that function is suggested). SSN also defines a Deployment, a process with subprocesses (DeploymentRelatedProcess) that lead to the device becoming deployed. IoT-Lite extends the deployment aspect of the SSN by explicit use of geolocation from the WGS84 model. OpenIoT, on the other hand, provides a very peculiar extension of the SSN, namely it adds an OperatingProperty of Device, named EaseOfDeployment. No further description or explanation of its usage is provided.

In SAREF, deployment is understood in terms of physical space, in which a device is deployed, i.e. BuildingSpace, annotated with geolocation data from the WGS84. This is an interesting design decision, as it restricts SAREF Devices to be deployed only in buildings. It seems to contradict the design-time assumption that SAREF devices, i.e. smart appliances, can be located also in public spaces.

4   Summary of Key Findings

Each of considered ontologies proposes a different approach to modeling the IoT space. The biggest differences are in the details. **(a)** OneM2M BO proposes a small base ontology, similar to upper ontologies that provides only a minimal set of highly abstract entities. This allows for a very broad set of domain ontologies to be easily aligned with it. It also means that the BO itself is not enough to model any concrete problem (or solution) in the IoT. Furthermore, it does not capture some aspects that are very common in other ontologies. **(b)** OpenIoT contrasts this philosophy by providing a detailed model for a specific problem (i.e. pilot implementations from the OpenIoT project) that can be also applied in a more general case, or in other solutions. Its heavy usage of external ontologies provides high semantic interoperability by design. **(c)** SSN is a developed model of the IoT in general, but with strong focus on sensors. It is based on DUL, and is clearly modularized, which makes it a good candidate for extensions into concrete systems and implementations. This is evidenced by the fact that other ontologies, evaluated here, make good use of it. When it comes to specificity, it places itself in the middle between oneM2M BO and OpenIoT. **(d)** IoT-Lite is an extension of selected SSN modules, mainly to include actuators. Rather than focusing on providing a detailed description of a delimited problem space within the IoT, it approaches the modeling problem from the perspective of an implementation device. It aims to deliver a small, but complete, model in order to simplify processing of semantic information. This is also its distinctive characteristics. **(e)** SAREF is a model with a strong focus on its own area—of smart appliances. Even though mappings to other standards exist, SAREF was developed from scratch to represent a specific area of application of the IoT. In this area, it delivers a strong and detailed base, that is also clear and easy to understand. At the same time, it is general enough to be used when extended to other domains, or solutions. Interestingly, all these ontologies almost completely disregard hardware specifications. It seems that the "place" of a device in an IoT system is much more important to ontology engineers than its hardware specification and resulting capabilities.

5   Concluding Remarks

The aim of this work was to compare how selected (most popular) ontologies capture and formally represent key aspects of "the world of IoT". The results of this investigation are important in the context of the INTER-IoT project, where the question: which ontology should be used (if any) to provide foundation of the central ontology, is of utmost importance. Moreover, the results of our comparison can be of use to the Semantic Interoperability Working Group of the IoT EPI initiative.

Results of our preliminary investigations show how different can be existing conceptualizations of the same domain, depending on the context of the approach, and the applied ontology engineering methodology. Separately, we conclude that, while each considered ontology has its uses and caveats, two of them stand out in the context of potential use in the INTER-IoT project. These are SSN and SAREF. The first presents a model focused on sensors, but still robust enough, and with strong ontological basis. Those features make it a good choice in terms of interoperability (which is the focus of the project). In addition, the SSN is modular, extendable, and has been actually implemented and extended in other systems and ontologies (e.g. IoT-Lite and OpenIoT). SAREF, on the other hand, is a thoroughly modern ontology with many recommendations and relatively large scope, despite targeting only smart appliances. It already has alignments with other models, thus improving its interoperability.

In the immediate future, we will complete work reported here, by including less popular ontologies and extending the list of concepts. This will allow us to choose the base device ontology for the INTER-IoT project. Next, we will extend it (or align with other ontologies) to create a modular core ontology for interoperability in the IoT.

Acknowledgments. This research was partially supported by the European Union's "Horizon 2020" research and innovation programme as part of the "Interoperability of Heterogeneous IoT Platforms" (INTER-IoT) project under Grant Agreement No. 687283.

References

1. Basic Geo (WGS84 lat/long) vocabulary. https://www.w3.org/2003/01/geo/
2. INTER-IoT Project. http://www.inter-iot-project.eu
3. oneM2M–standards for M2M and the Internet of Things. http://www.onem2m. org/
4. Semantic Sensor Network XG final report (2011). http://www.w3.org/2005/ Incubator/ssn/XGR-ssn-20110628/
5. SmartM2M: Smart Appliances; Reference Ontology and oneM2M mapping. Technical specification 103 264, European Telecommunications Standards Institute (2015)
6. Ben Alaya, M., Medjiah, S., Monteil, T., Drira, K.: Towards semantic data interoperability in oneM2M standard. IEEE Commun. Mag. **53**(12), 35–41 (2015). https://hal.archives-ouvertes.fr/hal-01228327
7. Bermudez-Edo, M., Elsaleh, T., Barnaghi, P., Taylo, K.: IoT-Lite: a lightweight semantic model for the Internet of Things. In: Proceedings of the IEEE Conferences on Ubiquitous Intelligence & Computing, Toulouse, France, July 2016
8. Compton, M., Barnaghi, P., Bermudez, L., Garcia-Castro, R., Corcho, O., Cox, S., Graybeal, J., Hauswirth, M., Henson, C., Herzog, A., Huang, V., Janowicz, K., Kelsey, W.D., Phuoc, D.L., Lefort, L., Leggieri, M., Neuhaus, H., Nikolov, A., Page, K., Passant, A., Sheth, A., Taylor, K.: The SSN ontology of the W3C semantic sensor network incubator group. Web Semant. Sci. Serv. Agents World Wide Web **17**, 25–32 (2012). http://www.websemanticsjournal.org/index.php/ps/article/view/312
9. Daniele, L., Hartog, F., Roes, J.: Created in close interaction with the industry: the smart appliances reference (SAREF) ontology. In: Cuel, R., Young, R. (eds.) FOMI 2015. LNBIP, vol. 225, pp. 100–112. Springer, Heidelberg (2015). doi:10. 1007/978-3-319-21545-7_9
10. Ganzha, M., Paprzycki, M., Pawłowski, W., Szmeja, P., Wasielewska, K.: Semantic interoperability in the Internet of Things: an overview from the INTER-IoT perspective. J. Netw. Comput. Appl. (in press, 2016)
11. Ganzha, M., Paprzycki, M., Pawłowski, W., Szmeja, P., Wasielewska, K.: Towards semantic interoperability between Internet of Things platforms (submitted for publication). Springer (2016)
12. Ganzha, M., Paprzycki, M., Pawłowski, W., Szmeja, P., Wasielewska, K., Fortino, G.: Tools for ontology matching—practical considerations from INTER-IoT perspective. In: Li, W., Ali, S., Lodewijks, G., Fortino, G., Di Fatta, G., Yin, Z., Pathan, M., Guerrieri, A., Wang, Q. (eds.) IDCS 2016. LNCS, vol. 9864, pp. 296–307. Springer, Cham (2016). doi:10.1007/978-3-319-45940-0_27
13. Ganzha, M., Paprzycki, M., Pawłowski, W., Szmeja, P., Wasielewska, K., Palau, C.E.: From implicit semantics towards ontologies–practical considerations from the INTER-IoT perspective (submitted for publication). In: Proceedings of 1st edn. of Globe-IoT 2017: Towards Global Interoperability among IoT Systems (2017)
14. Gruber, T.R.: Toward principles for the design of ontologies used for knowledge sharing? Int. J. Hum. Comput. Stud. **43**(5), 907–928 (1995)

15. Jayaraman, P.P., Calbimonte, J.P., Quoc, H.N.M.: The schema editor of OpenIoT for semantic sensor networks. In: Kyzirakos, K., Henson, C.A., Perry, M., Varanka, D., Grütter, R., Calbimonte, J.P., Celino, I., Valle, E.D., Dell'Aglio, D., Krötzsch, M., Schlobach, S. (eds.) Proceedings of the 1st Joint International Workshop on Semantic Sensor Networks and Terra Cognita (SSN-TC 2015) and the 4th International Workshop on Ordering and Reasoning (OrdRing 2015) co-located with the 14th International Semantic Web Conference (ISWC 2015), Bethlehem, PA, United States, October 11–12th, 2015. CEUR Workshop Proceedings, vol. 1488, pp. 25–30. CEUR-WS.org (2015)

16. Soldatos, J., Kefalakis, N., Hauswirth, M., Serrano, M., Calbimonte, J.-P., Riahi, M., Aberer, K., Jayaraman, P.P., Zaslavsky, A., Žarko, I.P., Skorin-Kapov, L., Herzog, R.: OpenIoT: open source Internet-of-Things in the cloud. In: Podnar Žarko, I., Pripužić, K., Serrano, M. (eds.) Interoperability and Open-Source Solutions for the Internet of Things. LNCS, vol. 9001, pp. 13–25. Springer, Heidelberg (2015). doi:10.1007/978-3-319-16546-2_3

17. Vrandečić, D.: Ontology Evaluation, pp. 293–313. Springer, Heidelberg (2009). doi:10.1007/978-3-540-92673-3_13

Graphical Interface for Ontology Mapping with Application to Access Control

Michał Drozdowicz[1]([✉]), Motasem Alwazir[1], Maria Ganzha[1,2], and Marcin Paprzycki[1]

[1] Systems Research Institute, Polish Academy of Sciences, Warsaw, Poland
{michal.drozdowicz,motasem.alwazir,maria.ganzha, marcin.paprzycki}@ibspan.waw.pl
[2] Department of Mathematics and Information Sciences, Warsaw University of Technology, Warsaw, Poland

Abstract. Proliferation of smart, connected devices brings new challenges to data access and privacy control. Fine grained access control policies are typically complex, hard to maintain and tightly bound to the internal structure of the processed information. We thus discuss how semantic inference can be used together with an intuitive ontology management tool to ease the management of Attribute Based Access Control policies, even by users not experienced with semantic technologies.

1 Introduction

Rise of the Internet of Things (IoT), results in growing interest in data access control. Obviously, assuring privacy and security of data is of utmost importance. However, creation of complex ecosystems results in need of establishing, which data is going to be exposed, to which stakeholder, why, when, for how long, etc. Recently (see, [2,3]), we have proposed an Attribute Based Access Control system utilizing semantic reasoning to enrich available information, when making access control decisions (SXACML). However, even IT professionals have limited knowledge of semantic technologies. Therefore, we consider how an ontology non-expert can effectively define and/or manage an ontology within the Policy Administration Point.

To this effect, Sect. 2 introduces the SXACML system, and provides a use case scenario. In Sect. 3, we outline the state-of-the-art in ontology modeling tools. Section 4, describes OntoPlay, a module that provides needed ontology management capabilities. Next, in Sect. 5, we outline how OntoPlay has been integrated with the SXACML.

2 SXACML

The eXtensible Access Control Markup Language (XACML; [1]) allows implementation of the Attribute Based Access Control (ABAC; [7]) mechanisms. In the XACML, attributes are grouped into four categories:

© Springer International Publishing AG 2017
N.T. Nguyen et al. (Eds.): ACIIDS 2017, Part I, LNAI 10191, pp. 46–55, 2017.
DOI: 10.1007/978-3-319-54472-4_5

- *Subject* – the entity (possibly a person) requesting access,
- *Resource* – the entity, access to which is under control,
- *Action* – that the *Subject* requests to be performed on the *Resource*,
- *Environment* – other attributes that bring additional context.

The XACML specification defines also a reference architecture, comprised of:

- *Policy Enforcement Point* (PEP) – responsible for actual enabling or preventing access to the resource. It also coordinates the execution of, so called, *Obligations* – additional operations that should be performed when a decision has been made (e.g. logging the request for auditing purposes).
- *Policy Information Point* (PIP) – a source of values of attributes. Commonly handled by data stores, such as relational databases or LDAP directories.
- *Context Handler* – converts requests and responses between native formats and the XACML canonical representation and coordinates, with the PIP, gathering of the required attribute values.
- *Policy Decision Point* (PDP) – evaluates policies and issues the final authorization decision.
- *Policy Administration Point* (PAP) – defines, stores and manages policies.

In this context, in [2,3], we introduced a semantics-driven implementation of the Policy Information Point. There, request attributes, combined with information stored in an ontology, allowed inferring additional data, necessary for access control decision. Advantages of such approach include, but are not limited to:

1. Simplified policies – information common to multiple policies can be extracted into the ontology.
2. Better support for Role Based Access Control.
3. More flexibility in defining relationships between concepts in policies.
4. Possibility of greater interoperability, by semantic mapping of disparate concepts in the request and in policies.

However, note that, while the first two points can be solved using only an ontology describing the domain of the organization and, perhaps, a simple static mapping of XACML terms into ontology concepts; the last points require a more robust and dynamic solution, allowing administrators to manage the mapping of concepts. Hence, we will now focus on defining additional relationships between domain concepts, and providing interoperability in the access control context.

Use case scenario. To put our work in a real-world context, consider a somewhat simplified example originating from the INTER-IoT project[1]. Let us consider controlling access to facilities of a cargo port, and assume that policies, stating which persons and vehicles may access the port premises, have been defined using a Policy Administration Point. One of them states that drivers (and trucks) employed by Globex Corporation can access the area. We assume

---

[1] http://www.inter-iot-project.eu/.

that authentication mechanisms are in place, to verify that information provided by the drivers/trucks is correct and valid. Now, consider that one day transport of goods is handled by a subcontractor of Globex Corp – Stark Transport. Here, subcontracted drivers/trucks should also be granted access. This scenario involves two aspects that necessitate additional processing of request attributes in order to make the authorization decision:

1. Relationship between Stark Transport and Globex Corp cannot be stored in access policies (as Globex hired Stark "incidentally" to deal with shortage of trucks, and *only then* they should be granted access).
2. Stark Transport uses a slightly different terminology than Globex Corp and thus concepts from the XACML access request have to be mapped to these used in the port's policies.

3 Existing Ontology Modelling Tools

Let us now assume that semantic technologies are to be used in this scenario. Hence, the system should facilitate defining relationships between concepts related to access control. Moreover, this function should be accessible to users unfamiliar with ontology modeling. Existing tools for defining concepts and relations between them, manipulation and searching of data represented as an ontology; can be divided into two groups: (i) ontology editors, and (ii) SPARQL query editors.

Ontology editors. Ontology editors are integrated development environment (IDE) for creating new and managing existing ontologies. Their GUI is provided as a desktop, or a web, application. The World Wide Web Consortium (W3C)[2], lists 12 Ontology Editors[3]. In addition to ontology creation and modification, they may provide extra capabilities. For instance, *Protégé*[4], *WebProtege*[5], *Cognitum*[6] and *OWLGrEd*[7] display a graphical representation of concepts and relations between them (as defined in the ontology). They also support defining SWRL Rules, use of reasoners, etc. Those editors (or others, see [5,8]) are very useful for ontology engineers or developers during the stage of designing, updating and utilizing an ontology but they are not suitable for non ontology experts, even when it comes to simple tasks, such as creating an instance of a class. For example, in *Protégé*, which, according to its website, is trusted by more than 300,000 users, such simple tasks cannot be performed without knowing the elements of an ontology. It also requires the user to understand the structure of the loaded ontology to be able to add consistent taxonomies. In other words,

---

[2] https://www.w3.org.
[3] https://www.w3.org/wiki/Ontology_editors.
[4] http://protege.stanford.edu/.
[5] http://protege.stanford.edu/products.php#web-protege.
[6] http://www.cognitum.eu/semantics/FluentEditor/.
[7] http://owlgred.lumii.lv/.

Protégé does not facilitate easy management and mapping of XACML concepts in the ontology, by administrators with minimal knowledge of ontologies.

SPARQL editors. Tools included in this category, e.g. *SPARQL Editor*[8], *YASQE*[9] and *Virtuoso*[10] allow writing and executing SPARQL queries, to search and retrieve data from an ontology. They provide a simple user interface, for user to select the ontology she wants to query, and a text area to write the queries. A more user friendly way to write SPARQL queries is a GUI designed to support building them; found in tools like: Visual *SPARQL BUILDER*[11] and *Gruff*[12]. Nevertheless, no simple way exists to use such editors by persons who do not know SPARQL. A work around would be to prepare queries, which user might need for mapping XACML concepts. However, this would mean that the code would have to be changed whenever the ontology is modified. Furthermore, available queries would work only with a given ontology (could not be reused).

Summarizing, a number of user interfaces to semantically demarcated information exists. However, none of them could serve as a lightweight front-end, allowing creation of nested descriptions of OWL individuals and class expressions. Moreover, all of them require the user to have understanding of ontologies.

4 OntoPlay

Let us start by observing that users of, for example, database-centered applications do not need to know about databases, tables and relations. The same is needed for ontology-driven systems. Here, front-end/GUI should be straightforward and should *"hide"* use of ontologies. It should guide "ontology-illiterate" users to define individuals and/or classes expressions to query the ontology. Moreover, in an "Open World assumption", one might need to change the ontology, to model more concepts, or add new properties. To avoid changing the system whenever vocabularies are modified, the front-end should be ontology-agnostic (i.e. ontology change should result in a minimum changes to its code). The GUI should be rendered dynamically, based on the underlying ontology. For example, consider a system that uses the Pizza ontology[13]. The GUI should allow people, who do not understand semantics, to add a new instance of the class Pizza. It should also make it easy to find a pizza from Italy.

The OntoPlay (see, [4]) is a web-based front-end plugin satisfying these requirements. Its GUI contains the *condition builder*, which allows "any user" to create class conditions, or individuals, translated to OWL expressions, and merge them with the underlying ontology. In [4], we showed how it helps dealing with various scenarios involving ontologies. Let us stress that the OntoPlay operates with any OWL ontology, if it is syntactically and semantically correct.

---

[8] http://sparql.carsten.io.
[9] http://yasqe.yasgui.org/.
[10] https://dbpedia.org/sparql.
[11] http://leipert.github.io/vsb.
[12] http://franz.com/agraph/gruff.
[13] http://protege.stanford.edu/ontologies/pizza/pizza.owl.

Recently, the level of OWL expressiveness, in the OntoPlay, had been increased through handling relations, and individuals using *annotation* properties[14] (OWL 2.0 entities). Here, user is to be able to use annotations "without knowing it". Additional interfaces were put in the OntoPlay as well.

The main components of the new OntoPlay are the same as before: Client, Server and Gateway [4]. The Gateway was not changed. On the client side, the AngularJS[15] was used to allow data binding between HTML elements and models defined using JavaScript. This was needed in the condition builder interface, for example, when user changes data property to object property, or creates nested class expressions. On the server side, using the functionality and the routing of the Play framework, web services were defined, to read the structure (or the data) in the ontology and return it in JavaScript Object Notation (JSON). Combining Angular with web services made the user interface more dynamic and efficient from a programming point of view. Examples of defined services are: get properties of a class, get range type of a property, get individuals in the range of a property, etc. In addition, the structure of the JSON build has been changed to include the *annotation* properties for the new created entity, and relations defined within it. Let us now enumerate new interfaces available in the OntoPlay and the functionality of each one of them.

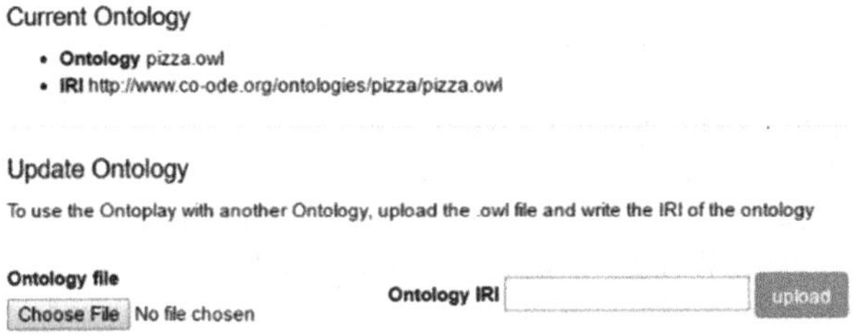

Current Ontology

- **Ontology** pizza.owl
- **IRI** http://www.co-ode.org/ontologies/pizza/pizza.owl

Update Ontology

To use the Ontoplay with another Ontology, upload the .owl file and write the IRI of the ontology

Ontology file
Choose File No file chosen

Ontology IRI [   ]   upload

Fig. 1. Administrating the ontology connected to the OntoPlay

Managing ontology. A simple interface which allows the "system administrator" to manage the current ontology. The importance of this interface is that the Java code does not have to be changed when a different ontology is to be used. When using this interface (see, Fig. 1), the administrator can see the current ontology connected to the OntoPlay and update/replace it, if needed, by uploading the OWL file with the new ontology and writing its *IRI*.

Class instances. A web page displaying all individuals of a class, in the underlying ontology. Just like a system displaying data from the database, it allows users to perform operations: view, add, edit and delete data, defined in the ontology knowledge base. Figure 2 shows this view for the Country class defined in the Pizza Ontology.

---

[14] https://www.w3.org/TR/owl2-syntax/#Annotation_Properties.

[15] https://angularjs.org/.

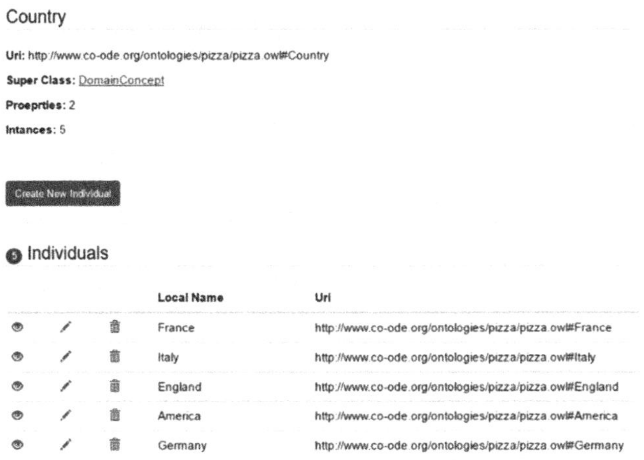

Country

Url: http://www.co-ode.org/ontologies/pizza/pizza.owl#Country

Super Class: DomainConcept

Proeprties: 2

Intances: 5

Create New Individual.

● Individuals

			Local Name	Uri
◉	✎	🗑	France	http://www.co-ode.org/ontologies/pizza/pizza.owl#France
◉	✎	🗑	Italy	http://www.co-ode.org/ontologies/pizza/pizza.owl#Italy
◉	✎	🗑	England	http://www.co-ode.org/ontologies/pizza/pizza.owl#England
◉	✎	🗑	America	http://www.co-ode.org/ontologies/pizza/pizza.owl#America
◉	✎	🗑	Germany	http://www.co-ode.org/ontologies/pizza/pizza.owl#Germany

Fig. 2. Class interface displaying Countries defined in the ontology

Annotation properties. Observe that OWL does not put any constraints on the domain (it can be any IRI) and the range (any IRI or data literal) of the *Annotation Properties*. Some annotation properties are predefined by OWL (e.g. owl:versionInfo, redfs:label, rdfs:comment, redfs:seeAlso and redfs:isDefinedBy), while custom ones can be added to an ontology. They can be added for several reasons. One of them is to define N-ary relations[16]. For example, one wants to establish the "amount" of a pizza topping. For example "a little" (or "a lot") of black olive topping). To satisfy this, *withAmount* can be introduced as an annotation, to describe the relation hasTopping.

Predefined, and custom, annotation properties can be used to describe classes, data and object properties, individuals and even annotation properties. An ontology engineer may be able to use annotation properties in the context they were defined, meant for, without having constraints on the domain and range. On the other hand, user assumed here, would not know how to do it "the right way". That is why we have introduced a mechanism to limit the annotation properties and make them available only where they are meant to be. To do that, a new admin panel (Fig. 3) has been developed to allow "system administrator" to restrict, which annotation property users can use to describe an individual, or a relation within an individual. Currently, OntoPlay supports only a data literal, as a value for any annotation property (custom or predefined). In the condition builder/individual creator interface, a new dialog is opened when the user chooses to annotate a property, or an expression. This dialog allows user to choose one of annotations pre-assigned to that property or class. Figure 4 illustrates how the user can specify that she wants a little fruit topping on her Pizza.

---

[16] https://www.w3.org/TR/swbp-n-aryRelations/.

Here, **More** will allow her to choose one of default annotation properties and assign a value to it.

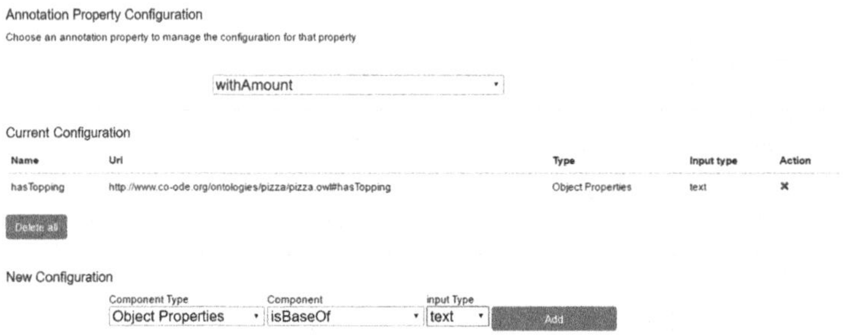

Annotation Property Configuration
Choose an annotation property to manage the configuration for that property

withAmount

Current Configuration

Name	Url	Type	Input type	Action
hasTopping	http://www.co-ode.org/ontologies/pizza/pizza.owl#hasTopping	Object Properties	text	✕

Delete all

New Configuration

Component Type	Component	Input Type	
Object Properties	isBaseOf	text	Add

Fig. 3. Managing *withAmount* annotation property in the admin panel

5   Integration of Redesigned OntoPlay into the SXACML

Let us now discuss how the OntoPlay can be used in our use case scenario. As assumed, the semantic PIP uses OWL ontologies as the basis for inferring values of attributes that were not contained in the incoming access request, but required for policy evaluation. Typically, three ontologies are used by the system.

1. Request ontology – common for all scenarios where the SXACML is used. It contains vocabulary related to the XACML specification, such as Subject, Resource and Action.
2. Domain ontology – containing structured knowledge related to a particular organization or business. We assume it is modeled by an expert and filled with axioms from an external data source (database). This ontology needs not be directly related to the context of access control.
3. Mapping ontology – links the generic Request ontology with the Domain ontology. At the very least, it needs to define what, in domain terms, is the Subject and the Resource. It may also contain axioms defining relationships; e.g. hierarchies of resources and clustering of subjects into groups or roles.

While the first two ontologies are static or, at least, need not be modified together with the access control policies, the mapping ontology needs to be easy to update (also for an administrator unfamiliar with ontology modeling). To facilitate this, OntoPlay is configured to work with the Mapping Ontology, which in the beginning is empty, aside from the import statements referring to the Domain Ontology and the Request Ontology. This makes it possible to use any terms, from these ontologies, when creating class expressions in the UI. From the SXACML administration page it is then possible to define *equivalentClass* expressions, for the following classes declared in the RequestOntology.

- urn:oasis:names:tc:xacml:1.0:subject-category:access-subject
- urn:oasis:names:tc:xacml:3.0:attribute-category:resource
- urn:oasis:names:tc:xacml:3.0:attribute-category:action
- urn:oasis:names:tc:xacml:3.0:attribute-category:environment

To cover the scenario introduced in Sect. 2, we have modelled a simplistic domain ontology – Logistics Ontology – that represents knowledge concerning operations of the port. Here, note that in real-life, instead of creating a simple ontology, we would consider use of (possibly somewhat modified) existing transport and logistics ontologies (see, [6]). Integration of a more sophisticated ontology is planned, as future work, within the INTER-IoT project. The sample ontology consists of the following entities (we omit namespaces for brevity):

Fig. 4. Specifying the desired amount of the fruit topping using annotation property

- Person – an employee of the port or other institution.
- Company – a company employing or contracting persons.
- ContractedHaulier – a subtype of Company. A transport organization that has a fixed contract with the port authority.
- Location – an area of the port.
- Role – role of a person, e.g. a driver.
- Yard – the area within the port premises where truck and container movements occur. A subtype of Location.
- isHiredBy – a property describing the relationship between a Person and a Company.
- isContractedBy – a property describing the relationship between a Company and another Company. This property is transitive, therefore, a company contracted by company A that is contracted by company B, will also be inferred as contracted by company B. The ontology also contains a statement that Stark is subcontracted by Globex.

The policy that we focus on has the following structure:

- It specifies conditions for a Permit result
- The Subject needs to be of type HiredDriver
- The Resource needs to be Yard
- The Action needs to be Entry

Lastly, the request from the Stark Transport driver to access the facilities contains the following attributes (and their values):

- Subject category has two attributes: id = cristiano.cosio@starktransport.com, isHiredBy = Stark Transport
- Resource category has attribute id = InternalParking
- Action category has attribute id = Entry

We can see that the request does not use the same attribute set as the one used in the policy. Specifically, it does not refer to type HiredDriver and uses InternalParking instead of Yard. Here, note that the default XACML engine would not be able to deduct that Cristiano is an instance of type HiredDriver, therefore the permission would not be granted and our requirement of permission delegation would not be satisfied.

Let us demonstrate how we can employ semantic reasoning for this purpose. First of all, we need to make sure we map the generic subject category to the Person class from the Logistics Ontology. We do so by creating a class expression specifying that Subject is an equivalent class of Person. Subsequently, we should define another simple expression, tying together the concepts of Yard and InternalParking. This would be done in the same way – as an equivalentClass expression. Finally, we would like to specify the meaning of HiredDriver used in the policy, but not existing in the ontology itself. To this effect we create a class expression built of the following conditions:

1. HiredDriver is a subtype of Subject
2. it hasRole of Driver
3. it isHiredBy a ContractedHaulier OR it is hired by a Company that is subcontracted by a ContractedHaulier

The associated expression as defined in OntoPlay is shown in Fig. 5.

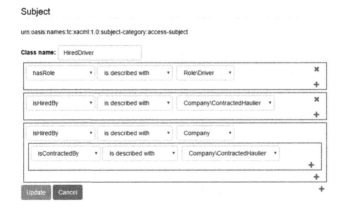

Fig. 5. Class expression defining the class HiredDriver

Based on this knowledge, the semantic Policy Information Point can infer that Cristiano matches the Subject conditions of the policy and that InternalParking matches the Yard requirement, therefore permitting the request.

6   Conclusions and Future Work

We have discussed how the redesigned OntoPlay, a simple and flexible ontology management UI component, can be utilized within an access control module to provide ontology mapping capabilities. The main improvement of this solution, when compared to other ontology editors, is that it is assumed that the user possesses minimal/no prior knowledge or experience with semantic technologies. We have illustrated how the combination of OntoPlay and semantic inference can simplify access control policies and improve the interoperability of the system. In the next steps, we plan to use transport and logistics ontologies listed in [6] and verify applicability of this approach to real-world use cases defined in the INTER-IoT project. We will also work on extending possibilities to easily map not only classes but also properties, thus enabling flexible attribute mapping.

Acknowledgments. Research presented in this paper has been partially supported by EU-H2020-ICT grant INTER-IoT 687283.

References

1. eXtensible Access Control Markup Language (XACML) version 3.0 (2013). http://docs.oasis-open.org/xacml/3.0/xacml-3.0-core-spec-os-en.html
2. Drozdowicz, M., Ganzha, M., Paprzycki, M.: Semantic policy information point – preliminary considerations. In: Loshkovska, S., Koceski, S. (eds.) ICT Innovations 2015. AISC, vol. 399, pp. 11–19. Springer, Cham (2016). doi:10.1007/978-3-319-25733-4_2
3. Drozdowicz, M., Ganzha, M., Paprzycki, M.: Semantically enriched data access policies in eHealth. J. Med. Syst. **40**(11), 238 (2016)
4. Drozdowicz, M., Ganzha, M., Paprzycki, M., Szmeja, P., Wasielewska, K.: OntoPlay - a flexible user-interface for ontology-based systems. In: AT, pp. 86–100 (2012)
5. Falco, R., Gangemi, A., Peroni, S., Shotton, D., Vitali, F.: Modelling OWL ontologies with graffoo. In: Presutti, V., Blomqvist, E., Troncy, R., Sack, H., Papadakis, I., Tordai, A. (eds.) ESWC 2014. LNCS, vol. 8798, pp. 320–325. Springer, Cham (2014). doi:10.1007/978-3-319-11955-7_42
6. Ganzha, M., Paprzycki, M., Pawłowski, W., Szmeja, P., Wasielewska, K.: Semantic interoperability in the Internet of Things: an overview from the INTER-IoT perspective. J. Netw. Comput. Appl. **81**, 111–124 (2016)
7. Hu, V.C., Ferraiolo, D., Kuhn, R., Schnitzer, A., Sandlin, K., Miller, R., Scarfone, K.: Guide to attribute based access control (ABAC) definition and considerations. NIST Spec. Publ. **800**, 162 (2014)
8. Petersen, N., Lange, C., et al.: TurtleEditor: an ontology-aware web-editor for collaborative ontology development. In: 2016 IEEE Tenth International Conference on Semantic Computing (ICSC), pp. 183–186. IEEE (2016)

Influence of Group Characteristics on Agent Voting

Marcin Maleszka[(✉)]

Faculty of Computer Science and Management,
Wroclaw University of Science and Technology,
Wyb. Wyspianskiego 27, 50-370 Wroclaw, Poland
marcin.maleszka@pwr.edu.pl

Abstract. A collective of identical agents in a multi-agent system often works together towards the common goal. In situations where no supervisor agents are present to make decisions for the group, these agents must achieve some consensus via negotiations and other types of communications. We have previously shown that the structure of the group and the priority of communication has a high influence on the group decision if consensus theory methods are used. In this paper, we explore the influence of preferential communication channels in asynchronous group communication in situations, where majority vote and dominant value are used. We also show how this relates to consensus approach in such groups and how to use a combination of both approaches to improve performance of real-life multi-agent systems.

Keywords: Knowledge integration · Multiagent system · Collective knowledge · Consensus theory · Agent voting

1 Introduction

Knowledge management and different related tasks are becoming more and more important in modern information society. Different methods of decision making, information retrieval and knowledge integration are being used in various applications, often requiring no input from the user. Over the years, multiple methods have been developed to solve the problem that sometimes occur during these tasks, including consensus theory to handle one cause of the problems in knowledge integration, that is knowledge inconsistency. Inconsistency is a feature of knowledge which may be characterized by the lack of possibility for inference processes, therefore solving it is an essential task in many cases of knowledge management [12]. In our overall research we focus on time-related aspects of this problem and in this paper we focus on applying the lessons learned in decentralized multi-agent systems using consensus theory to similar systems using choice theory [5].

In our previous papers [8,10] we have focused mainly on various approaches to collective knowledge integration based on consensus theory [12] and tested

© Springer International Publishing AG 2017
N.T. Nguyen et al. (Eds.): ACIIDS 2017, Part I, LNAI 10191, pp. 56–64, 2017.
DOI: 10.1007/978-3-319-54472-4_6

them both in simulation environment and a prototype practical application. We used asynchronous communication and preferred communication channels to better represent several real-world situations. We researched some previously unexplored aspects of agent knowledge state change when using this type of integration. Here we focus on the area that was already researched more thoroughly by others (e.g. [5]). As our first runs of the prototype have shown, a using dominant value of the group as a result of integration may give better results in this type of applications. In this paper we make some observations of agent collective behaviour when this method of knowledge integration is used, exploring it both in simulated environment and in real world application.

This paper is organized as follows: in Sect. 2 we provide a detailed description of research most relevant to ours, mostly focusing on currently described research, but also including our previous work, upon which we based this paper; Sect. 3 presents a short overview of a multi-agent system with decentralized voting used in a simulated environment, as well as our observations of its various runs; Sect. 4 describes a prototype weather prediction system using results from the simulated environment and built upon our previous iterations of this practical application; in Sect. 5 we provide some concluding results, detailing possible further applications of our research.

2   Related Works

In our overall research we have considered multiple areas, including multi-agent systems, collective knowledge integration, decentralized systems, asynchronous communication protocols and more. In this section we will provide short overview of various papers most relevant to the research presented in this paper. As far as we know, some parts of our previous research and methodology have not been researched yet. In this paper however, we use methods mostly known in literature in a similar way as our previous approach, leading to some new and interesting results.

Our research stems from the consensus theory and the possibility of consensus changing over time. This is somewhat similar to continuous-time consensus in multi-agent systems, e.g. autonomous robots or network systems [16], where it is used for attitude alignment, flocking, formation control, negotiations, etc. One important aspect is finite-time consensus – if agents will reach consensus in finite time. Another is the stability of multi-agent systems [1] – if all agents knowledge states converge to the same value. This was considered both in centralized and decentralized agent systems [6]. Somewhat similar situation is determining the optimal number of experts in group decision making [14] for example by determining when knowledge added by a new agent is lower than some threshold.

Another aspect of our research is the centralized and decentralized approach to communication in multi-agent systems. There are the traffic control systems [3], where each basic agent controls a single subsystem or functionality (e.g. one crossroads) and the supervisor agent supervises the traffic flow over the whole area. Similarly in decision support systems using multi-agent approach basic agents may communicate between themselves, sharing some basic data, but the overall decision is made by some main facilitator agent [13]. There are also fully decentralized multi-agent systems with no observing agents, e.g. in [5] the authors show that a group of diverse random agents deciding by majority vote gives better results in the game of Go that a set of uniform agents. A decentralized system may be also used for surveillance [15], where each agent has its own knowledge but shares it to influence other agents.

Our approach to decentralized multi-agent system is based on asynchronous communication and some additional structure similar to a social network of agents, which we use for determining additional, preferential communication channels. With possibility of using supervisor agents, we were able to explore both centralized and decentralized systems, e.g. employees in a company and social networks with *friend* relation. We base this on research done in [9], where it was proposed as a method for improving the teaching process and other knowledge dissemination cases. A similar approach was used in [2] to show that only strong ties in a social network lead to improvement in groupwork results.

In this paper we use the same approach, but substitute consensus theory [12] by voting rules. A similar approach may be found in [5], but there the authors work with a small group of agents (up to 25) with diverse knowledge states. The model proposed by authors is universal for finite systems and was tested on a discretized Go game with both uniform and diverse agents. As the theoretical background of this approach is thoroughly explained in literature, we will not be redefining it for purposes of this paper.

3 Decentralized Multi-agent System in a Simulated Environment

In our overall research we are using a previously developed simulation environment based on JADE agent framework [4]. We previously used it to simulate increasing groups of agents trying to achieve consensus [7] and to observe the influence of the mode of communication on the process of knowledge integration [8,10]. The centralized nature of JADE framework is used for the purpose of observing the behaviour of a collective of implemented agents and is independent of the communicating group. The collective itself may communicate either in centralized or decentralized manner. As centralized voting was studied multiple times by other researchers, we focused on the decentralized approach.

By the decentralized dominant value voting we understand the following, expanded from [10]:

- Each agent has a number of preferred other agents (*friends*).
- Each agent starts communication in irregular intervals, sending its knowledge to a random other agent, with preference for its *friends*.
- Agent communication is unidirectional - after agent sends a message, he does not expect a reply.

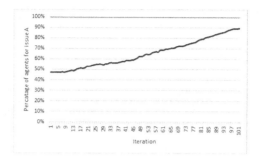

Fig. 1. Example simulation run for gathering 3 votes and no preferred communications.

- Receiver agent gathers incoming knowledge. Once a given number of other agents has contacted it with their knowledge, it integrates it and changes his own knowledge state. The voting approach to integration is based on determining the most often occurring knowledge state and changing own knowledge to this one in the next step.

In the framework we use identical *SocialAgents* representing members of the collective and a supervisor gathering the observations, tied to the centralized architecture of JADE. The collective itself operates in a decentralized manner. On initialization the social agents generate some random knowledge (a basic example: a single integer values between 0 and k). In irregular intervals (each tick T each agent has a random chance to initiate communication) these agents will then communicate with each other, sending their own current knowledge state to a random other agent (with probability p to an agent from the whole population, chosen with uniform distribution; and with probability $1-p$ to a single agent from a subgroup of *preferred* agents, chosen with uniform distribution). The receiving agent gathers knowledge from several others and integrates it, changing its own knowledge state. Meanwhile, the centralized observer gathers current knowledge states from all agents and prepares information about the collective as a whole (e.g. it may show the result of a centralized integration from the whole group of agents).

In our previous research we used consensus theory [12] as a basis for integration, but our practical application has also shown interesting results when using dominant value [11]. Following this, we use dominant value as a basis for integration in this series of experiments. Each agent gathers up to v knowledge states from himself and other agents, then changes its own knowledge state to the dominant value of this set. The preferred agent votes are not weighted differently, but their is a higher chance of their knowledge state being included in the voting set.

We conducted several series of experiments, changing parameters such as type of knowledge, total number of agents, number of preferred agents, probability of communication to preferred agents, number of votes to gather before integration and others. The key observations we gathered may be distilled to the case of a

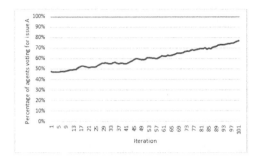

Fig. 2. Example simulation run for gathering 3 votes, 20 preferred agents and 40% chance of communication to preferred agents.

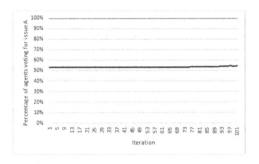

Fig. 3. Example simulation run for gathering 20 votes, 10 preferred agents and 20% chance of communication to preferred agents.

binary choice made by a large group of agents (here: 500). Some interesting runs are presented in Figs. 1, 2, 3. We show the *winning* decision as the increasing one in the chart.

Our observations of the simulation runs may be outlined as follows:

The smaller the number of different opinions gathered, the higher the (long term) chance that the agent will change its own knowledge state. This corresponds to earlier research in [2,5]. The first paper shows, that for small number of diverse agents in the group, the results are more often biased towards one side (in case of the game of Go this leads to losing strategy). The latter describes the case of only strong ties in a social network actually influencing the groupwork – here this corresponds to stronger influence of a single relation if the number of relations (votes) is smaller.

For larger number of gathered outside opinions, the chance of the agent changing its knowledge state is significantly smaller, as the opinions will correspond more closely to the division of knowledge in the overall group.

The larger the chance of selecting preferential communication channels, the higher the chance that the agents in *friend* relation will influence the opinion of an agent (i.e. the agents opinion will be closer to the subgroup opinion).

This leads to certain subgroups of the collective becoming more distinct from others, but uniform among their members.

On the other hand, a large number of diverse agents in the preferred group lowers the chance of the agent changing its opinion. As a large group reflects the whole collective more closely than a smaller one, the chance that the group is biased towards some opinion is lower. This observation is especially important for our practical application, as it allows fine tuning of the integration process – the number of gathered votes determines the overall trend and the size of *friend* group regulates the speed of the occurring change.

4   Application: Weather Forecast System

During our research of the influence of agent communication on knowledge integration, we are using a prototype of a weather forecasting multi-agent system using our theoretical and simulation results. In our early research [11] we have observed that using dominant value in integration, instead of consensus theory approach, gave slightly better results. This spurned the research presented in this paper. After conducting the simulation runs, we were able to further fine tune our prototype application and improve the predictions of the whole system.

The architecture of the prototype system consists of three distinct layers:

- Source layer - a collective of various agents (corresponding to *SocialAgents* in the simulation environment) with two functions: creating their own predictions (based on historical data, copied from internet sources, etc.) and using the decentralized voting approach to modify their predictions based on other agents forecasts. Different integration methods may be used for modifying agents knowledge, both based on consensus theory and dominant value. Some agents may drop out when their sources are unavailable, but this does not influence overall system.
- Integration layer - after the time for individual communication between agents ends, a single supervisor agent gathers results from all agents and integrates them to a single forecast, using different approaches (consensus or dominant value).
- Interaction layer - in a practical system this is the interface layer, presenting the final forecast to the user. In experimental setup this is the observation agent, similar to the one used in the simulation environment. It gathers both the final forecast and any partial results and preprocesses them for later analysis.

We are conducting evaluations of the weather prediction system using real world data in Wroclaw, Poland. For each iteration of the system (with added different integration methods), we have conducted experiments on all its variants (basic dominant value and centralized consensus in April-May 2015, centralized and decentralized consensus in October 2015, centralized consensus, decentralized consensus and decentralized consensus with *friend* relation in April-May 2016). Based on simulation environment results, we decided to test various methods of integration in source layer combined with using dominant value for integration layer. As previously we have run the other variants of the weather prediction system in parallel, to compare their effectiveness. The results for all time periods and methods are shown in Table 1.

Table 1. All observed runs of the weather prediction system in all variants: basic dominant value (B. Dominant), centralized consensus (Cen. Cons.), decentralized consensus (Dec. Cons.), decentralized consensus with *friend* relation (D.-S. Cons.) and two new variants of dominant value approach: full decentralized voting in source layer (Dom. Dec.) and voting interchangeable with consensus in source layer (Dom. Mix.)

System-Run	MAE	Comp. w/Best Src.	Comp. w/Worst Src.	Comp. w/Avg. Src.
B. Dominant (IV-V '15)	1,857	89%	17% better	3% better
Cen. Cons. (IV-V '15)	2,018	82%	7% better	95%
Cen. Cons. (X '15)	2,132	75%	2% better	90%
Dec. Cons. (X '15)	1,984	83%	9% better	97%
Cen. Cons. (IV '16)	1,991	85%	6% better	93%
Dec. Cons. (IV '16)	1,994	85%	6% better	93%
D.-S. Cons.(IV-V '16)	1,989	85%	6% better	93%
B. Dominant (X '16)	1,193	89%	15% better	99%
Cen. Cons. (X '16)	1,956	87%	12% better	97%
Dec. Cons. (X '16)	1,931	88%	14% better	98%
D.-S. Cons. (X '16)	1,933	88%	14% better	98%
Dom. Dec. (X '16)	1,898	90%	16% better	*Equal*
Dom. Mix. (X '16)	1,892	90%	16% better	1% better

Our previous research focused on the consensus based approach to integrate knowledge, but for this practical application repeated experiments have shown that various approaches based on dominant value give better results. The three tested variants of dominant value approach may be described as follows:

- Basic Dominant Value – this most basic approach does not require any additional communication in the source layer. The basic agents calculate their own forecast and the supervisor agent in the integration layer selects the most often occurring value as the final prediction.
- Dominant Value in Decentralized System – in this approach the agents in the source layer have some time to communicate and change their predictions based on other agents information, before the supervisor agent selects the most often occurring value as the final prediction. The source layer agents communicate as in the simulation environment with tuned parameters: gathering 3 votes, 15 preferred agents and 40% chance of communication to preferred agents.

– Dominant Value mixed with Consensus Integration – this approach is similar to the second one, but there is additional 50% chance that the knowledge change in the source layer will be based on consensus theory (best results as presented in [10]) instead of dominant value.

Our overall experiments with the prototype weather prediction system have shown that the best results were determined using the Dominant Value mixed with Consensus Integration approach. In each tested situation the various dominant value approaches had the smallest MAE and were the closest to the single most accurate prediction.

5   Conclusions

This paper finalizes our research into observing the changes of knowledge states of agents in a decentralized collective during integration process. Previously we have studied centralized systems and various types of decentralized systems with integration accomplished by means of consensus theory. In this paper we used the voting methods described in literature and applied them to the same type of system. We have observed the behaviour of such system in a simulated environment, gathering some guidelines for fine tuning the integration process. We then applied these observations to a prototype weather prediction system we have been developing parallel to our research. Here we have shown that a specific mix of voting and consensus, used in a hybrid centralized-decentralized system provides best results for this specific practical applications.

Our future research also includes developing our own agent platform, independent on frameworks such as JADE. We will use it both as a basis for the next version of our simulation environment and for future version of the prototype weather forecast application. Disassociating our research from platform limitations should allow for larger scale experiments and platform independent applications. This will allow us to implement our other prototype applications, such as economic and traffic prediction systems. In particular for a traffic system we consider single agents observing single drivers in different road situations and monitoring them continuously. After some finite time a single agent should determine if the driver is a good or a bad one (this is more related to our research in [7], which would especially important to insurance companies. Additionally, monitoring individual drivers on a local (via decentralized integration) and global (centralized) level would allow improvements to the traffic flow. We also consider the same approach to determine important, but not explicit topics discussed in web commentaries.

Acknowledgment. This research was co-financed by Polish Ministry of Science and Higher Education grant.

References

1. Bhat, S.P., Bernstein, D.S.: Finite-time stability of continuous autonomous systems. Siam J. Control Optim. **38**(3), 751–766 (2000)
2. De Montjoye, Y.-A., Stopczynski, A., Shmueli, E., Pentland, A., Lehmann, S.: The strength of the strongest ties in collaborative problem solving. Sci. Rep. **4**, 5277 (2014). Nature Publishing Group
3. Iscaro, G., Nakamiti, G.: A supervisor agent for urban traffic monitoring. In: IEEE International Multi-Disciplinary Conference on Cognitive Methods in Situation Awareness and Decision Support (CogSIMA), pp. 167–170. IEEE (2013)
4. JADE, Java Agent Development Framework. http://jade.tilab.com/
5. Jiang, A., Marcolino, L.S., Procaccia, A.D., Sandholm, T., Shah, N., Tambe, M.: Diverse randomized agents vote to win. In: Advances in Neural Information Processing Systems, pp. 2573–2581 (2014)
6. Li, S., Dua, H., Lin, X.: Finite-time consensus algorithm for multi-agent systems with double-integrator dynamics. Automatica **47**, 1706–1712 (2011)
7. Maleszka, M.: Consensus with expanding conflict profile. In: Barbucha, D., Nguyen, N.T., Batubara, J. (eds.) New Trends in Intelligent Information and Database Systems. SCI, vol. 598, pp. 291–299. Springer, Cham (2015). doi:10.1007/978-3-319-16211-9_30
8. Maleszka, M.: Knowledge in asynchronous social group communication. In: Nguyen, N.T., Trawiński, B., Fujita, H., Hong, T.-P. (eds.) ACIIDS 2016. LNCS (LNAI), vol. 9621, pp. 364–373. Springer, Heidelberg (2016). doi:10.1007/978-3-662-49381-6_35
9. Maleszka, M., Nguyen, N.T., Urbanek, A., Wawrzak-Chodaczek, M.: Building educational and marketing models of diffusion in knowledge and opinion transmission. In: Hwang, D., Jung, J.J., Nguyen, N.-T. (eds.) ICCCI 2014. LNCS (LNAI), vol. 8733, pp. 164–174. Springer, Cham (2014). doi:10.1007/978-3-319-11289-3_17
10. Maleszka, M.: Local and global consensus in asynchronous group communication. In: 2016 IEEE International Conference on Systems, Man, and Cybernetics, SMC 2016, 9–12 October 2016, Budapest, Hungary, pp. 3071–3076. IEEE (2016)
11. Mercik, J., Tolkacz, O., Wojciechowska, J., Maleszka, M.: Wykorzystanie integracji wiedzy do zwiekszenia efektywnosci prognozowania w warunkach niepewnosci. In: Porebska-Miac, T. (ed.) Systemy Wspomagania Organizacji 2015, Wydawnictwo Uniwersytetu Ekonomicznego w Katowicach, Katowice (2015)
12. Nguyen, N.T.: Advanced Methods for Inconsistent Knowledge Management. Springer, Heidelberg (2007)
13. Nagata, T., Sasaki, H.: A multi-agent approach to power system restoration. IEEE Trans. Power Syst. **17**(2), 457–462 (2002). IEEE
14. Nguyen, V.D., Nguyen, N.T.: An influence analysis of the inconsistency degree on the quality of collective knowledge for objective case. In: Nguyen, N.T., Trawiński, B., Fujita, H., Hong, T.-P. (eds.) ACIIDS 2016. LNCS (LNAI), vol. 9621, pp. 23–32. Springer, Heidelberg (2016). doi:10.1007/978-3-662-49381-6_3
15. Peterson, C.K., Newman, A.J., Spall, J.C.: Simulation-based examination of the limits of performance for decentralized multi-agent surveillance and tracking of undersea targets. In: SPIE Defense+ Security, p. 90910F. International Society for Optics and Photonics (2014)
16. Ren, W., Beard, R.W., Atkins, E.M.: A survey of consensus problems in multi-agent coordination. In: Proceedings of the 2005 American Control Conference, pp. 1859–1864. IEEE (2005)

Collective Knowledge: An Enhanced Analysis of the Impact of Collective Cardinality

Van Du Nguyen[✉]

Department of Information Systems, Faculty of Computer Science and Management,
Wrocław University of Science and Technology, Wrocław, Poland
van.du.nguyen@pwr.edu.pl

Abstract. In this paper, we present an enhanced analysis of the impact of collective cardinality on the quality of collective knowledge. Collective knowledge is a knowledge state determined on the basis of collective members' knowledge states on the same subject (matter) in the real world. The collective members (which are often autonomous units) have their own knowledge bases, thus their knowledge states can be different from each other. The quality is based on the difference between the collective knowledge and the real state of the subject. For this aim, we introduce a new factor named *diam* presenting the maximal difference between a knowledge state in the collective and the real state. The simulation experiments reveal that the quality is not only dependent on the collective cardinality but also dependent on the diam value. Additionally, the number of collective members needed to achieve a difference level between the collective knowledge and the real state is also investigated.

Keywords: Collective intelligence · Collective knowledge · Consensus methodology

1 Introduction

The term *Collective Intelligence* has been widely used and attracted the attention of many researchers in recent years, but maybe it has existed for a very long time in forms of a collective of individuals doing things at least sometimes intelligent [15]. In [6] Newell has stated that an intelligent collective is a social system in which a small group or a large formal organization, that is capable to act (even approximately) as a single, rational agent. Collective Intelligence is considered as an intelligence that emerges from the collaboration and competition of many individuals [4, 12]. According to these works, Collective Intelligence is the result of the process of consensus decision-making from various forms of the collective knowledge state [8].

In this paper, we assume that a set of autonomous units, which have their own knowledge bases, are asked for being given their views or opinions about some common subjects (matters) in the real world. These views or opinions are called *knowledge states*. These knowledge states to some degree reflect the real states of the subjects because of incompleteness and uncertainty. Of course, the real state of a subject exists but it is not known when the autonomous units are asked for being given their knowledge

© Springer International Publishing AG 2017
N.T. Nguyen et al. (Eds.): ACIIDS 2017, Part I, LNAI 10191, pp. 65–74, 2017.
DOI: 10.1007/978-3-319-54472-4_7

states. By a collective, we consider a set of knowledge states of the autonomous units (collective members) on the same subject. These knowledge states can be different from each other. The collective knowledge is a knowledge state determined on the basis of the knowledge states in the collective and considered as a representative for the collective as a whole. In [7] the author has developed many consensus-based algorithms for determining the knowledge of a collective for different knowledge representations such as logic expressions, relational structures, ontology. In the process of collective knowledge determination, there exist two cases may take place: *objective case* and *subjective case* [9]. The classification of these cases is based on the relationship between the real state and the knowledge of a collective. The collective knowledge of a collective in objective case reflects the real state to some degree because of incompleteness and uncertainty. For example in the problem of weather forecast for a particular region on a future day, the collective forecast, which is determined on the basis of forecasts given by meteorological stations, reflects the real weather state to some degree. Conversely, with subjective case they are identical.

In [3] the author has investigated the problem of consensus susceptibility for some structures of knowledge presentations such as real number, binary vector, ordered partitions, or ordered coverings. The main aim of the paper is to answer the question "*Whether the collective knowledge of a collective is good enough?*" by taking into account the collective cardinality. However, the author did not take into account the existing of the real state of the subject that the knowledge states in a collective reflects. In the previous work [10, 11] we have concluded that the large collective cardinality has positive impact on the quality of its knowledge. In this paper, we present an enhanced analysis of the impact of collective cardinality on the quality of collective knowledge for objective case by introducing a new factor named *diam*. The value of diam presents the maximal difference between the knowledge states in a collective and the real state of the subject that they reflect. Meanwhile, the collective cardinality presents the number of members in a collective. The quality of collective knowledge is based on the difference between the real state and the collective knowledge. For this aim, we simulate collectives with different scenarios of collective cardinalities and diam values to determine their impact on the quality of collective knowledge. We consider the representation of collective members' knowledge states as numbers and multi-dimensional vectors. By means of simulation experiments, the quality of collective knowledge is not only dependent on the collective cardinality but also dependent on the diam value. Moreover, the number of collective members needed to achieve a level difference between the collective knowledge and the real state is also investigated.

The remainder of the paper is structured as follows. In the next section, we will briefly introduce some related works. Some basic notions related to collective, knowledge of a collective and the quality of collective knowledge are mentioned in Sect. 3. The simulation experiments and their evaluation are presented in Sect. 4. Some conclusions and future works are pointed out in the last section.

2   Related Works

In this section, we briefly introduce some approaches related to the impact of collective cardinality on the collective opinion. The term *Collective Intelligence* has widely appeared after the publication of the influential book of Surowiecki titled *The Wisdom of Crowds*. By means of statistical analysis, the author has shown that a collective is better than a single member in making some common judgment tasks [13]. For example, in the game show *"Who wants to be a millionaires?"* the answers given by the audience were 91% on the target while only 65% those of experts were on target. The main reason of the phenomenon is each member can be a good specialist in the field in which he/she is professional, whereas problems in the real world come from many fields. Then the author has put forward a hypothesis *"A collective is more intelligent than an individual"* which has been confirmed by Surowiecki in [14]. In that work, the author has described several experiments, which served for proving the hypothesis. One of the most influential experiments is about guessing the number of beans in a jar given by a number of members. The collective guess is only 2.5% difference from the actual number of beans in the jar. In [17] the authors have invited a collective of 500 members for solving some common problems such as forecast temperature, guessing the weights of specific amounts of coffee, milk, gasoline, air, and gold. Through experimental analysis with different collective cardinalities such as 10, 20, 50, 100, and 200, the authors have concluded that the large collective has positive influence on the collective judgment accuracy. In [2], by using prediction market, the authors have conducted experiments with collective of 18 members and 8 members. The results reveal that the large collective cardinality has positive impact on the quality of collective forecast. In [1], the author has also concluded that the large collective cardinality has positive impact on the accuracy of decision. That is *the higher the number of decision makers, the more accurate the decision made*. In [5] the author investigated the problem of collective knowledge determination by expanding collective cardinality with different structures of knowledge representations. Through simulation, the temporary knowledge of small collectives is convergent to the knowledge of the large collective. In the previous works [9–11], we have preliminarily investigated the impact of some factors such as collective cardinality, inconsistency knowledge on the quality of collective knowledge. In [10, 11], through simulation, the large collective cardinality has positive impact on the quality of its knowledge. In [9], the authors have investigated the impact of inconsistency knowledge on the quality of collective knowledge. The inconsistency knowledge, which is presented the difference level between members in a collective, is measured by means of inconsistency degree functions. With some restrictions, the hypothesis *"the higher the inconsistency degree, the better the quality of collective knowledge"* has been formally proved.

In this work, we will investigate the impact of not only collective cardinality but also diam value on the quality of collective knowledge. Additionally, the number of members needed to achieve a difference level between the collective knowledge and the real state is also investigated.

3 Preliminaries

3.1 Collective

Let U be a set of objects representing the potential elements of knowledge referring to a concrete real-world subject. By symbol $\prod (U)$ we denote a set of all non-empty finite subsets with repetitions of set U. A set $X \in \prod (U)$ is called a collective in which each member $x \in X$ represents the knowledge state of a collective member. In this paper a collective is described as follows:

$$X = \{x_1, x_2, \ldots, x_n\}$$

where n is the cardinality of collective representing the number of members in collective X.

3.2 Collective Knowledge

As aforementioned, collective knowledge is a knowledge state determined on the basis of members' knowledge states in a collective. It is often considered as the representative for the collective as a whole. In [7], many consensus-based algorithms are proposed to determine collective knowledge with different structures of knowledge representations such as logic expressions, ontology. Generally, collective knowledge is determined by:

- criterion O_1 if the sum of differences between x^* and each knowledge state in collective X to be minimal. That is $d(x^*, X) = \min_{y \in U} d(y, X)$.
- criterion O_2 if the sum of squared difference between x^* and each knowledge state in collective X to be minimal. That is, $d^2(x^*, X) = \min_{y \in U} d^2(y, X)$.

where x^* presents the collective knowledge of collective X, $d(x^*, X) = \sum_{i=1}^{n} d\left(x^*, x_i\right)$ and $d^2(x^*, X) = \sum_{i=1}^{n} d^2\left(x^*, x_i\right)$.

3.3 Quality of Collective Knowledge

The quality of collective knowledge is defined by taking into account the difference between the real state and collective knowledge [7, 8]. The highest quality of collective knowledge is the closest the collective knowledge to the real state.

$$Q_X(x^*) = 1 - d(r, x^*)$$

where x^* represents the knowledge of collective X, r represents the real state of the subject in which the knowledge states in collective X reflect.

According to the previous work [10, 11], we propose some postulates related to the impact of increasing/decreasing the number of collective members on the quality of collective knowledge as follows:

P1. If $Y = X \dot\cup \{x^*\}$, then $Q_X(x^*) = Q_Y(y^*)$

P2a. If $(Y = X \dot\cup \{y\}) \wedge (d(r, y) \leq d(r, x^*))$, then $Q_Y(y^*) \leq Q_X(x^*)$

P2b. If $(Y = X \dot\cup \{y\}) \wedge (d(r, y) \geq d(r, x^*))$, then $Q_Y(y^*) \geq Q_X(x^*)$

P3a. If $(Y = X \backslash \{y\}) \wedge (d(r, y) \leq d(r, x^*))$, then $Q_Y(y^*) \geq Q_X(x^*)$

P3b. If $(Y = X \backslash \{y\}) \wedge (d(r, y) \geq d(r, x^*))$, then $Q_Y(y^*) \leq Q_X(x^*)$

Some commentaries for above postulates: Postulate 1 states that if adding a member that is the collective knowledge of a collective to that collective, then the quality is unchanged. Intuitively, based on consensus choice, the collective knowledge of new collective (after added member x^*) is identical with that of the original collective. From this fact, this kind of member can be useful for improving the inconsistency degree of a collective. In other words, it is helpful in improving the consistency degree between members in a collective. According to Postulate 2, if adding a member that is closer to the real state than the collective knowledge, then the quality will be better. Otherwise, the quality will be worse. In case of removing members, Postulate 3 states that the quality will be better if a member added is closer to the real state than the collective knowledge. Otherwise, the quality will be worse.

4   Simulation Experiments and Their Evaluation

4.1   Simulation Design

In this paper we simulate collectives with different collective cardinalitys and diam values. After that, we will determine their impact on the quality of collective knowledge. Concretely, for each diam value (here we called a *scenario*), collective cardinality will be increased from 3 to 1000 (each step by one member). For each simulation experiment, the number of repetitions is 50. The detail parameters of these approaches are as follows:

- Single-attribute collectives: U is a set of integers in which the maximal difference to the real state does not exceed the diam value. For example, with diam of 1000, a collective of n members will involve n numbers belonging to interval $[0, 2000]$. These knowledge states are randomly generated. The values of diam are described in Table 1. The Manhattan distance is used for measuring the difference between knowledge states in a collective. We use the same real state for all scenarios.

Table 1.   Paramenters of single-attribute collectives

	Real	Min	Max	Diam
Scenario 1	1000	0	2000	1000
Scenario 2	1000	500	1500	500
Scenario 3	1000	750	1250	250

- Multi-attribute collectives: The number of attributes is three. In this case, U is a set of points in the sphere with the same center (100, 100, 100) and a predefined radius. For example, with diam of 10, a collective of n members (which are randomly generated) will involve n numbers belonging to the circle with center (100, 100, 100) and radius 10. These knowledge states are randomly generated. The diam values are described in Table 2. The Euclidean distance is used for measuring the difference between knowledge states in a collective. Notice that the real state is identical for all scenarios.

Table 2. Parameters of multi-attribute collectives

	Real	Diam
Scenario 1	(100, 100, 100)	50
Scenario 2	(100, 100, 100)	20
Scenario 3	(100, 100, 100)	10

4.2 Simulation Results

In Subsect. 3.3, the closer the collective knowledge to the real state is, the higher the quality of collective knowledge will be. The difference between two knowledge states in a collective is normalized into [0, 1]. In this work, however, the diam values used are too large, which can cause the impact of increasing the collective cardinality on the quality seems to be insignificant. Thus, we do not normalize the difference between two members in a collective into [0, 1].

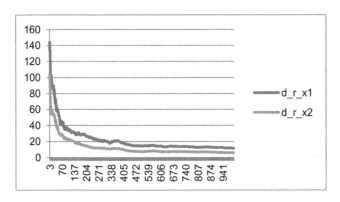

Fig. 1. Single-attribute collectives with diam 500

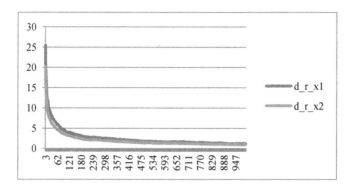

Fig. 2. Multi-attribute collectives with diam 50

In Figs. 1 and 2, d_r_x1 and d_r_x2 present the differences between the real state and the collective knowledge determined by using criterion O_1 and O_2, respectively. From these figures, we can state that the large collective cardinality plays an important role in reducing the difference between the real state and the collective knowledge. In other words, the hypothesis *"The higher the collective cardinality, the closer the collective knowledge of a collective to the real state"* is true. However, when the collective cardinality is large enough, its increase does not cause the difference between the collective knowledge and the real state to be significant. The relationship between the log of collective cardinality and the difference between the collective knowledge and the real state is described in Tables 3 and 4.

Table 3. Statistical test for single-attribute collectives

	Intercept coefficient	Log_Size coefficient	R^2	p-value
Scenario 1	278.00	−89.50	0.87	0.00
Scenario 2	118.14	−37.47	0.87	0.00
Scenario 3	64.82	−20.59	0.89	0.00

Table 4. Statistical test for multi-attribute collectives

	Intercept coefficient	Log_Size coefficient	R^2	p-value
Scenario 1	15.13	−4.50	0.84	0.00
Scenario 2	6.29	−2.01	0.83	0.00
Scenario 3	3.63	−1.27	0.86	0.00

According to Tables 3 and 4, the linear relationships between collective cardinality and the difference between the collective knowledge and the real state for different scenarios of diam values are described as follows:

For single-attribute collectives:

- Scenario 1: $d(r, x_1^*) = 278 - 89.5 \log_{10}(size)$
- Scenario 2: $d(r, x_2^*) = 118.14 - 37.47 \log_{10}(size)$
- Scenario 3: $d(r, x_3^*) = 64.82 - 20.59 \log_{10}(size)$

In this case, we can see that the number of members needed to achieve 5% difference between the collective knowledge and the real state is 355 (in case of scenario 1), 66 (in case of scenario 2), 5 (in case of scenario 3).

For multi-attribute collectives:

- Scenario 1: $d(r, x_1^*) = 15.13 - 4.5 \log_{10}(size)$
- Scenario 2: $d(r, x_2^*) = 6.29 - 2.01 \log_{10}(size)$
- Scenario 3: $d(r, x_3^*) = 3.63 - 1.27 \log_{10}(size)$

Similarly, the number of members needed to achieve 5% difference between the collective knowledge and the real state is 640 (in case of scenario 1), 428 (in case of scenario 2), 291 (in case of scenario 3). From this fact, we can state that with different diam values, the number of collective members needed to achieve a difference level between the collective knowledge and the real state is different. In practice, however, choosing a proper value of diam for a concrete subject in the real world is a difficult task.

Figure 3 presents the difference between the real state and the collective knowledge with different collective cardinalities (3, 100, 500, 1000) and diam values (250, 500, 1000) in case of single-attribute collectives. Notice that, in this figure, $A250\_1/A250\_2$ is the case in which diam value of 250 and criterion O_1/O_2 is used to determine the collective knowledge of a collective (similar to other diam values).

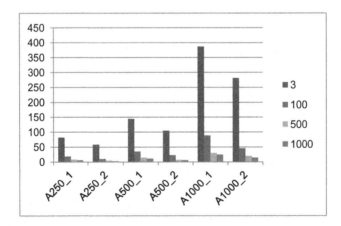

Fig. 3. The Impact of diam value in case of single-attribute collectives

Figure 4 presents the difference between the real state and the collective knowledge of each collective with different collective cardinality (3, 100, 500, 1000) and diam values (10, 20, 50) in case of multi-attribute collectives. In this figure, $R10\_1/R10\_2$ is the case in which the diam value is 10 and criterion O_1/O_2 is used to determine the collective knowledge of a collective (similar to other diam values).

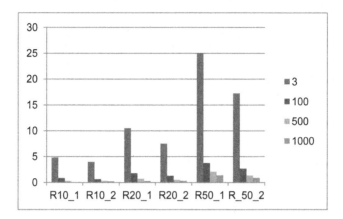

Fig. 4. The impact of diam value in case of multi-attribute collectives

From above figures, we can conclude that the large collective cardinality has positive impact on the quality of collective knowledge besides diam value. Additionally, determining a number of collective members needed to achieve a difference level between the real state and the collective knowledge is also dependent on diam value. However, this research problem should be investigated intensively in the future because of its highly practical impact.

5   Conclusions and Future Works

In this paper the previous research related to the impact of collective cardinality on the quality of collective knowledge for objective case is expanded by introducing a new factor named *diam*. It presents the maximal difference between a knowledge state in a collective and the real state. For this aim, we have simulated collectives with different collective cardinalities and diam values to determine the impact of these factors on the quality of collective knowledge. By means of simulation experiments, the quality is not only dependent on the collective cardinality but also dependent on the diam value. Additionally, the number of collective members needed to achieve a difference level between the collective knowledge and the real state is also determined.

In the future work, we will investigate intensively the relationship between collective cardinality and diam value on the quality of collective knowledge. A mathematical model will be proposed to prove the impact of these factors on the quality. Additionally, the problem of creating a collective consisting of data or knowledge states referred from a large number of sources on the Internet such as social networks, forums, blogs where more and more data gets dramatically produced is also taken into consideration [16].

References

1. Conradt, L.: Collective behaviour: when it pays to share decisions. Nature **471**, 40–41 (2011)
2. Cui, R., Gallino, S., Moreno, A., Zhang, D.J.: The Operational Value of Social Media Information (2015). http://dx.doi.org/10.2139/ssrn.2702151
3. Kozierkiewicz-Hetmanska, A.: Analysis of susceptibility to the consensus for a few representations of collective knowledge. Int. J. Softw. Eng. Knowl. Eng. **24**, 759–775 (2014)
4. Levy, P.: Collective Intelligence: Mankind's Emerging World in Cyberspace. Perseus Books, Cambridge (1997)
5. Maleszka, M.: Consensus with expanding conflict profile. In: Barbucha, D., Nguyen, N.T., Batubara, J. (eds.) ACIIDS 2015, Studies in Computational Intelligence, vol. 598, pp. 291–299. Springer, Heidelberg (2015)
6. Newell, A.: Unified Theories of Cognition. Harvard University Press, Cambridge (1990)
7. Nguyen, N.T.: Advanced Methods for Inconsistent Knowledge Management. Springer, London (2008)
8. Nguyen, N.T.: Inconsistency of knowledge and collective intelligence. Cybern. Syst. **39**, 542–562 (2008)
9. Nguyen, V.D., Nguyen, N.T., Truong, H.B.: A preliminary analysis of the influence of the inconsistency degree on the quality of collective knowledge. Cybern. Syst. **47**(1–2), 69–87 (2016)
10. Nguyen, V.D., Nguyen, N.T.: A method for improving the quality of collective knowledge. In: Nguyen, N.T., Trawiński, B., Kosala, R. (eds.) ACIIDS 2015. LNCS (LNAI), vol. 9011, pp. 75–84. Springer, Heidelberg (2015). doi:10.1007/978-3-319-15702-3_8
11. Nguyen, V.D., Nguyen, N.T.: An influence analysis of the inconsistency degree on the quality of collective knowledge for objective case. In: Nguyen, N.T., Trawiński, B., Fujita, H., Hong, T.-P. (eds.) ACIIDS 2016. LNCS (LNAI), vol. 9621, pp. 23–32. Springer, Heidelberg (2016). doi:10.1007/978-3-662-49381-6_3
12. Russell, P.: The Global Brain Awakens: Our Next Evolutionary Leap, 2nd edn. Global Brain Inc., Palo Alto (1995)
13. Shermer, M.: The Science of Good and Evil. Henry Holt, New York (2004)
14. Surowiecki, J.: The Wisdom of Crowds. Anchor, New York (2005)
15. Tovey, M.: Collective Intelligence Creating a Prosperous World at Peace. Earth Intelligence Network, Oakton (2008)
16. Vossen, G.: Big data as the new enabler in business and other intelligence. Vietnam J. Comput. Sci. **1**(1), 1–12 (2013)
17. Wagner, C., Suh, A.: The wisdom of crowds: impact of collective size and expertise transfer on collective performance. In: 47th Hawaii International Conference on System Sciences, pp. 594–603 (2014)

A New Ontology-Based Approach
for Construction of Domain Model

Bogumiła Hnatkowska, Zbigniew Huzar, Lech Tuzinkiewicz,
and Iwona Dubielewicz[✉]

Wrocław University of Science and Technology, Wyb. Wyspiańskiego 27,
50-370 Wrocław, Poland
{Bogumila.Hnatkowska,Zbigniew.Huzar,
Lech.Tuzinkiewicz,Iwona.Dubielewicz}@pwr.edu.pl

Abstract. Domain model is one of the most important artefacts in software engineering. It can be built with the use of domain ontologies. The objective of the authors' research is to elaborate an effective approach to domain model construction based on knowledge extraction from existing ontologies. A significant element of the approach is knowledge extraction algorithm. In this paper, a modified, more flexible version of the extraction algorithm is presented. A comparison of the new algorithm with the old one is conducted based on a case study. Both algorithms produce similar results regarding quality measures. In contrast to the old algorithm, the new is parameterized and therefore can be applied in an incremental way what is a valuable feature.

Keywords: Ontology · Knowledge acquisition · Conceptual modeling

1 Introduction

In the series of previous publications [2, 3, 7] we presented an approach to domain model construction based on SUMO-like ontologies [8–10]. In this paper, we present a new version of the approach, which is more general than the previous one. It may be applied for any, not only SUMO-like, kind of ontology. Additionally, the approach is parametrized and therefore more flexible and convenient for users.

The Fig. 1 outlines the entire process of domain model construction. The starting point is a set of initial domain notions *DNS* extracted from textual requirements specification. After identification which nouns have equivalent ontology notions, in the contrast to the previous approach, we take into account only these that are mapped into ontology classes. They form the input for the new algorithm.

The second input for the algorithm is a consistent domain ontology with an ontology set of notions *ONT*. The *DSN* set is mapped into a subset of ontology notions *ONT*. For each domain notion, a domain expert defines, if possible, an equivalent ontology notion. So, a usually partial function $EQ : DNS \rightarrow 2^{ONT}$ is defined. The range of the function *EQ*, *ran(EQ)*, is a subset of ontology notions, and contains key ontology notions related to the initial domain notions. Next, the set $KNO = ran(EQ)$ is extended with other related ontology notions. The extension may be seen as a result of some knowledge extension function $KE : 2^{ONT} \rightarrow 2^{ONT}$. Taking the set *KNO* as an argument

© Springer International Publishing AG 2017
N.T. Nguyen et al. (Eds.): ACIIDS 2017, Part I, LNAI 10191, pp. 75–85, 2017.
DOI: 10.1007/978-3-319-54472-4_8

the function *KE* delivers $ONT' = KE(KNO)$. The transformation of the set ONT' ONT' into the domain model is defined automatically by a function $TR : ONT' \rightarrow DM$ [6].

The domain model *DM* is checked for completeness and redundancy. The subset DM_{irr} represents these elements of *DM* that are irrelevant for the domain problem. The set DM_{mis} is disjoint with *DM* and represents the elements that are missing from the domain model.

The quality of the domain model is assessed by two standard measures – precision and recall, defined as follows:

$$precision = \frac{\#(DM) - \#(DM_{irr})}{\#(DM)}$$

and

$$recall = \frac{\#(DM) - \#(DM_{irr})}{\#(DM) - \#(DM_{irr}) + \#(DM_{mis})}$$

where notation #(*X*) means the number of elements of a set *X*. The harmonic mean of precision and recall gives the balanced F-score measure defined as:

$$F = 2 \cdot \frac{precision \cdot recall}{precision + recall}$$

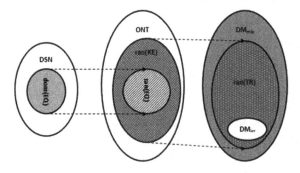

Fig. 1. Schema of transformations from a requirements specification to a domain model [1].

2   Core Ontology

A definition of an ontology, commonly used in information systems, comes from [4]:

An ontology is an explicit, formal specification of a shared conceptualization of a domain of interest.

Explicit means that the types of concepts used and the constraints on their use are explicitly defined; *formal* means that the ontology should be machine-readable; *shared* means that the ontology captures consensual knowledge; *the domain of interest* means that it reflects a model relevant to a given task not to the whole world.

A separate comment relates to the notion of conceptualization as an abstract model of phenomena in the observed fragment of reality. According to [4], conceptualization was understood as an extensional relational structure consisting of a set of the universe of discourse D and a set of extensional relations defined on the universe R, i.e. it is the pair <D, R>. Let us note that an extensional definition enumerates everything that falls under that definition. It means that the conceptual model is considered as equivalent to a class diagram expressed in UML.

Elements of the universe D may represent sets or classes. Remember that a set is a collection of distinct objects, considered as an object in its own right. In knowledge representation, a class is also a collection of individuals or objects. Objects of sets and objects of classes are called their instances. The main difference between set's and class's instance is that class's instances have own identity.

In [5] it was proved a significant drawback of this extensional approach and proposed more refined one. Let W be a set of possible worlds (situations, states, circumstances, etc.), D – the universe, and R – a set of intentional (conceptual) relations. An intentional definition gives the meaning of a term by specifying necessary and sufficient conditions for when the term should be used. In the case of nouns, this is equivalent to specifying the properties that an object needs to have to be counted as a referent of the term. For each possible world $w \in W$ a subset of R is assigned. In further we will not reference to the set W. Now, the conceptualization C is defined as the pair

$$C = <D, R> .$$

In this approach, the conceptual model may be considered as equivalent to a class diagram expressed in UML.

Each n-ary relation $r \in R$ has a signature $r \subseteq D_1 \times \ldots \times D_n$, where $D_1, \ldots, D_n \in D$. For the relation r an auxiliary function $dom(r, i) = D_i$, for $i = 1, \ldots, n$, is defined.

The core ontology is defined as a structure:

$$ONT = <D, \preccurlyeq_D, R, \preccurlyeq_R >$$

where $\preccurlyeq_D, \preccurlyeq_R$ are subsumption relations on the universe and the set of relations R, respectively. If $D_1, D_2 \in D$, the subsumption relation $D_1 \preccurlyeq_D D_2$ means that $D_1 \subseteq D_2$. D_1 is called a subconcept of the superconcept D_2. If $D_1 \preccurlyeq_D D_2$ and there is no D_3 such that $D_1 \preccurlyeq_D D_3$ and $D_3 \preccurlyeq_D D_2$ then D_1 is called a direct subconcept of the superconcept D_2.

In a similar way, the subsumption relation is defined for relations. The subsumption relation is also called "hyponym-hypernym relation".

The notion of the core ontology is useful because it may be derived from any concrete domain ontology, and thus a common algorithm for knowledge extraction may be applied.

3 Algorithm

The algorithm bases on the following assumptions:

- Given a set of initial domain notions *DNS* excerpted in some way from an informal, natural language description of o requirements specification. The considered notions are nouns only.
- Given a domain ontology $ONT = <D, \preccurlyeq_D, R, \preccurlyeq_R>$ related with the requirements specification. It is assumed that the ontology is consistent, but maybe incomplete.

Step 1. An equivalence function *EQ* between domain and ontology notions is established. Its signature is of the form: $EQ : DNS \rightarrow 2^D$, where D means a set of all ontology concepts. The function must be defined by a domain and ontology experts. Usually, it is a partial function which means that not all initial domain notions have equivalences in the domain ontology. It may be result of incompleteness of the domain ontology or may effected from improper description of the requirements specification. The set $D_0 = EQ(DNS)$ represents ontology classes equivalent to some of domain notions.

Step 2. A set of additional ontology classes that are related by a generalization for the set of classes $EQ(DNS)$ is defined. Considered are the ontology classes that are superclasses at most at m level above, and subclasses at most at n level below the classes from the set $EQ(DNS)$. Let introduce notation: $sup^1(C, C')$ means $C \preccurlyeq_D C'$ and $sub^1(C, C')$ means $C' \preccurlyeq_D C$.

Further, $sup^k(C, C')$ means that there are classes $C_1, \ldots, C_k$ such that $sup^1(C, C_1), \ldots, sup^1(C_{k-1}, C_k)$ and $C_k = C'$. Similarly $sub^k(C, C')$ is defined.

Now, the set of the additional classes is defined as follows:

$$Sup^{1..m}(C) = \left\{ C' \in D \mid \bigcup_{C \in D_0} \left(sup^1\left(C, C'\right) \cup \ldots \cup sup^m\left(C, C'\right) \right) \right\}$$

$$Sub^{1..n}(C) = \left\{ C' \in D \mid \bigcup_{C \in D_0} \left(sub^1\left(C, C'\right) \cup \ldots \cup sub^n\left(C, C'\right) \right) \right\}$$

$$Gen^{m,n}(C) = Sup^{1..m}(C) \cup Sub^{1..n}(C)$$

$$Gen^{m,n}(D_0) = \bigcup_{C \in D_0} Gen^{m,n}(C)$$

Step 3. A set of associations that are directly related to the set of classes

$$D_1^{m,n} = D_0 \cup Gen^{m,n}(D_0)$$

is defined as follows:

$$R_0 = \{r \in R | r {\subseteq} C_1 \times \ldots \times C_n \text{ where } C_1, \ldots, C_n \in D_1^{m,n}\}$$

Now, the associations that are related indirectly by composition of other k associations to the classes D_1 are defined.

$$R_1 = R_0 \cup \{r \in R | r {\subseteq} C_1 \times \ldots \times C_n \text{ and } C_1, \ldots, C_{i-1}, C_{i+1}, \ldots, C_n \in D_1^{m,n} \text{ and } C_i \notin D_1^{m,n}\}$$

$$D_2^{m,n} = D_1^{m,n} \cup \{C_i \in D | \exists r \in R_1 \cdot r {\subseteq} C_1 \times \ldots \times C_n \text{ and}$$
$$C_1, \ldots, C_{i-1}, C_{i+1}, \ldots, C_n \in D_1^{m,n} \text{ and } C_i \notin D_1^{m,n}\}$$

$$\ldots \ldots \ldots$$

$$R_j = R_{j-1} \cup \{r \in R | r {\subseteq} C_1 \times \ldots \times C_n \text{ and } C_1, \ldots, C_{i-1}, C_{i+1}, \ldots, C_n \in D_j^{m,n} \text{ and } C_i \notin D_j^{m,n}\}$$

$$D_{j+1}^{m,n} = D_j \cup \{C_i \in D | \exists r \in R_j \cdot r {\subseteq} C_1 \times \ldots \times C_n \text{ and } C_1, \ldots, C_{i-1}, C_{i+1}, \ldots, C_n \in D_j^{m,n} \text{ and } C_i \notin D_j^{m,n}\}$$

For $j = 1, 2, \ldots, k$

Step 4. Finally, the sub-ontology generated by D_0 and parameterized by m, n, k is defined as pair: $<D_k^{m,n}, R_{k+1}>$.

4 Comparison of Algorithms - Case Study

4.1 General Remarks

The old [1] and new versions (see Sect. 2) of algorithms will be compared with the use of quality metrics, i.e. precision, recall, and F-score. To be able to calculate the quality metrics for comparison purposes we need to reference to an "ideal" model that can be potentially retrieved from existing ontologies, assuming that the ontology is consistent and complete. "Ideal" means a model extracted by hand by a domain expert, without any irrelevant notions, flexible enough to address all the needs expressed by requirement specification.

To evaluate the quality of the proposed algorithm the following procedure was established:

1. Selection of domain of interest.
2. Preparation of user-stories being an initial input for algorithm and extraction of the domain notions.
3. Preparation of an "ideal" domain model for algorithm verification purposes.
4. Domain knowledge extraction with the usage of the old version of the algorithm. Quality metrics calculation.
5. Domain knowledge extraction with the usage of the new version of the algorithm. Quality metrics calculation. Data analysis and conclusions.

The presentation of evaluation procedure reuses the same domain, user-stories and test-cases as in the paper [1].

4.2   Evaluation Procedure – Steps 1–2

The evaluation process is as follows:

Step 1. The selected domain of interest is the 'hotel domain" included in Hotel.kif ontology [12].

Step 2. The prepared user-stories presented in the form [11]:

As a <type of user>, I want <some function> so that <some reason> are as follows:

User Story 1: As a potential customer, I want to see information about *hotel, hotel rooms, rooms' amenities* and *prices* so that I can decide whether to become a customer.

User Story 2: As a potential customer, I want to check availability of selected *room type (room availability)* in a given *reservation period* so that to be able to decide if to make *reservation* or not.

The domain notions used in User Stories are written in italics, and they form the *DNS* set:

DNS = {*hotel, hotel room, room's amenity, room price, room type, reservation period, reservation*}

These notions were mapped to ontology notions in the following way:

EQ = {*hotel → HotelBuilding, hotel room → HotelRoom, room's amenity → roomAmenity, room price → not translated, there is a price but it relates to a reservation, room type → HotelRoomAttribute, reservation period → {reservationStart, reservationEnd}, reservation → HotelReservation*}

Note, that the mapping is a relation not a function – "reservation period" was mapped to two different notions: "reservationStart", and "reservationEnd".

The mappings of notions which refer to the relationships (not to classes), are excluded from consideration. Therefore the set D_0 contains only {*HotelBuilding, HotelRoom, HotelRoomAttribute, HotelReservation*}.

4.3   Evaluation Procedure – Steps 3–5

The aim of step 3 is to prepare an "ideal" domain model for algorithms verification. "Ideal" means a model extracted by hand by domain expert, without any irrelevant notions, flexible enough to address all the needs expressed not only by user-stories but also by their acceptance criteria (test-cases). For both user-stories, presented in the previous subsection, test cases in the format *Given-When-Then* were defined [4]. An example of such a test-case is presented below.

Test Cases for User Story 1

Scenario1: Basic information about hotel

Given that a hotel is defined with: *<hotel name>*, *<hotel postal address>*, and *<hotel category>*

When a customer navigates to the main hotel page

Then he/she should be informed about hotel *<hotel name>*, *<hotel postal address>*, and *<hotel category>*

Scenario2: List of hotel rooms

Given that a hotel rents room types defined with: *<room type>*, *<capacity>*, *<price per night>*, *<amenities>*

When a customer navigates from main hotel page to the list of room types page

Than he/she should be informed about hotel *<room type>* s

The full list of considered test cases was defined in [1].

The "ideal" model was created manually by a domain expert from existing SUMO ontologies included in: Merge.kif, Mid-level-ontology.kif, Hotel.kif [12].

The domain expert has prepared the model on the basis of IDN_R set which is defined as:

$$IDN_R = IDN \cup IDN_{TC}$$

where IDN_{TC} contains the notions used only in test-cases (and not in user-stories- see [1]) i.e.

IDN_{TC} = {*room type, hotel name, hotel postal address, hotel category, room capacity*}

Only "structural" knowledge was extracted, e.g. axioms (implications, equivalence sentences); in the consequence some relations described in axioms are not presented on the resulting "ideal" model (e.g. HotelBuilding is not connected with HotelUnits).

The expert had to define mapping of all notions included in IDN_{R} regardless they are mapped to classes or associations. A subset of these mappings for IDN_{TC} is presented below:

$EQ_{TC}(IDN_{TC})$ = {*room type* → *HotelRoomAttribute, hotel name* → *names, hotel postal address* → *postAddressText, hotel category* → *HotelRating*}

Only necessary elements were extracted, i.e. these resulting from mapping of elements from IDN_R to ontology notions (classes, relationships); generalization relationships are incomplete (e.g. in SUMO each class is a subclass on Entity) and shown only between classes which inherit interesting properties.

The result of manual knowledge extraction – the "ideal" model – is presented in Fig. 2. The model diagram consists of 27 classes, 18 generalizations, and 9 named relations. We have also 3 unnamed relations between a power type and its general classifier. The unnamed relations result from the fact that in SUMO a relation domain can be defined as subclass domain, e.g. one of domains of relation *reservedRoom* is defined as a set of subclasses of *HotelUnit* class. This power type is represented on the diagram as *HotelUnitType*, and it is connected with unnamed association with *HotelUnit* (its general classifier). The unnamed relations result from translation to the UML

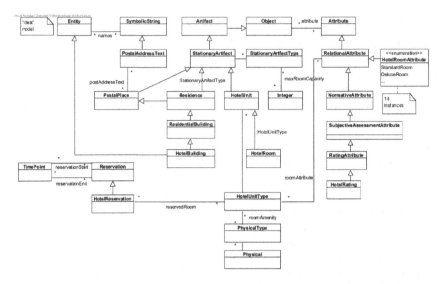

Fig. 2. The "ideal" model manually extracted from existing domain ontologies

following SUMO relations: between *HotelUnit*, and *HotelUnitType* (relations: *reservedRoom*, *roomAttribute*), between *Physical* and *PhysicalType* (relation: *roomAmenity*), between *StationaryArtifact* and *StationaryArtifactType* (relation: *maxRoomCapacity*). In consequence, the number of total relevant elements on the domain diagram is equal to 57 (27 + 18 + 9 + 3).

4.4   Quality Evaluation of the Old Version of Algorithm

Both, the knowledge extraction process (step 2), and the calculation of quality metrics were fully automatized for the old version of algorithm. The algorithm has no input parameters so it produces only one domain model in result (see [1]).

Below a selected data describing the results of the extraction process are presented:

- Generalizations (Total number: 2, Relevant number: 2)
- Classes (Total number: 11, Relevant number: 9)
- Relations (Total number: 5, Relevant number: 4)
- Unnamed relations (Total number: 1, Relevant number: 1)

On the basis of the data presented above the quality metrics were calculated:
Precision = (2 + 9 + 4 + 1)/(2 + 11 + 5 + 1) = 16 /19 = 84%
Recall = 16 /57 = 28%
F = 0,42%

Table 1. Calculation of precision and recall dependently on input parameters

Value of	m=0 k=0	m=0 k=1	m=0 k=2	m=1 k=0	m=1 k=1	m=1 k=2	m=2 k=0	m=2 k=1	m=2 k=2
relevant elements	4	9	18	12	22	32	15	30	35
retrieved elements	4	11	32	14	46	230	26	81	475
precision	1	0,82	0,56	0,86	0,48	0,14	0,58	0,37	0,07
recall	0,07	0,17	0,33	0,22	0,41	0,59	0,28	0,56	0,64
F	0,13	0,28	0,42	0,35	0,44	0,23	0,38	0,45	0,13

4.5   Quality Evaluation of the New Version of Algorithm

Similarly to the previous case, the knowledge extraction process as well as the calculation of quality metrics were fully automatized. It was especially useful because the new algorithm takes three parameters on input (m, n, k) which meaning was described in Sect. 2.

We conducted several experiments to analyse the produced domain models – see Table 1. It should be noted that the initial set of notions the algorithm started with consisted of only four classes:

$EQ(IDN) = \{HotelBuilding, HotelRoom, HotelReservation, HotelRoomAttribute\}$

We manipulated the values of m and k parameters. The third parameter n didn't influence the resulting model in our example as input classes have no subclasses. We changed the scope of the rest of parameters from 0 to 2, what gave us 9 possible combinations – see Table 1 for details.

The worst results (regarding F-score) were obtained for the extreme values of parameters (n, m) equal to $(0, 0)$ and $(2,2)$. It is not surprising that in the former case, the domain model consists of only four classes without any relationships (precision = 1.0). The result in the latter case is astonishing – the model consists of many irrelevant elements (475), its precision is very small (0,07), and it cannot be accepted, despite the fact that the recall measure is gaining satisfactory value (0,64). The best results of F-score were obtained for two pairs of parameters (n, m): (1,1), (2,1), and it was about 0,45.

The previous version of algorithm extracts essential elements from ontology with the precision equal to 0,84 but also ignores many required elements of the model (recall = 0,28), and the F-score = 0,42.

Summarizing results of the experiments we can conclude that k parameter should be set to 1 at the beginning of the exploration process. Increasing the value of this parameter can have an adverse influence on the extraction of redundant elements. The m and n parameters should be set to 1 and 0 respectively.

5 Conclusions

Presented considerations seem to confirm that the proposed approach is useful to domain model construction. At the same time, the approach enables an iterative extension of previously elaborated models. An iteration brings incremental extraction of knowledge from a domain ontology, the gradual extension of the domain glossary, and possibility of assessment of completeness of this ontology. The approach resembles an agile approach.

The proposed approach creates new opportunities for the exploration of knowledge represented in the form of ontology. Parameterizing of the algorithm gives the possibility of:

- dynamic analysis and extraction of needed knowledge in the context of defined requirements,
- testing and analysing of a domain ontology,
- evaluation of completeness of knowledge which is available as an ontology,
- configure the algorithm according to the particular solution problem (depending on the scope and the form of a problem specification).

Mention above aspects are the purpose of the future works and expected results allow us to develop the extended version of the algorithm including also analysis of relationships. The target aim is to elaborate a practical approach of the business process modeling basing on ontologies.

References

1. Dubielewicz, I., Hnatkowska, B., Huzar, Z., Tuzinkiewicz, L.: Domain modeling based on requirements specification and ontology. In: Madeyski, L., Śmiałek, M., Hnatkowska, B., Huzar, Z. (eds.) Software Engineering: Challenges and Solutions. AISC, vol. 504, pp. 31–45. Springer, Heidelberg (2017). doi:10.1007/978-3-319-43606-7_3
2. Hnatkowska, B., Huzar, Z., Dubielewicz, I., Tuzinkiewicz, L.: Development of domain model based on SUMO ontology. In: Zamojski, W., Mazurkiewicz, J., Sugier, J., Walkowiak, T., Kacprzyk, J. (eds.) Theory and Engineering of Complex Systems and Dependability. AISC, vol. 365, pp. 163–173. Springer, Heidelberg (2015). doi:10.1007/978-3-319-19216-1_16
3. Hnatkowska, B., Huzar, Z., Dubielewicz, I., Tuzinkiewicz, L.: Problems of SUMO-Like ontology usage in domain modelling. In: Nguyen, N.T., Attachoo, B., Trawiński, B., Somboonviwat, K. (eds.) ACIIDS 2014. LNCS (LNAI), vol. 8397, pp. 352–363. Springer, Heidelberg (2014). doi:10.1007/978-3-319-05476-6_36
4. Gruber, T.: A translation approach to portable ontology specifications. Knowl. Acquis. 5(2), 199–220 (1993)
5. Guarino, N., Giaretta, P.: Ontologies and knowledge bases towards a terminological clarification. https://csee.umbc.edu/courses/771/papers/KBKS95
6. Hnatkowska, B.: Towards automatic SUMO to UML translation. In: Kościuczenko P., Śmiałek M. (eds.) From Requirements to Software, Research and Practice, pp. 87–100, Polskie Towarzystwo Informatyczne (2015)

7. Bogumiła, H., Zbigniew, H., Lech, T., Iwona, D.: Conceptual modeling using knowledge of domain ontology. In: Nguyen, N.T., Trawiński, B., Fujita, H., Hong, T.-P. (eds.) ACIIDS 2016. LNCS (LNAI), vol. 9622, pp. 554–564. Springer, Heidelberg (2016). doi:10.1007/978-3-662-49390-8_54

8. Niles, I., Pease, A.: Linking lexicons and ontologies: mapping wordnet to the suggested upper merged ontology. In: Proceedings of the IEEE International Conference on Information and Knowledge Engineering, pp. 412–416 (2003)

9. Niles, I., Pease, A.: Toward a standard upper ontology. In: Welty, C., Smith, B. (eds.) Proceedings of the 2nd International Conference on Formal Ontology in Information Systems (FOIS-2001), pp. 2–9 (2001)

10. Pease, A.: Ontology: A Practical Guide. Articulate Software Press, Angwin (2011)

11. User stories. https://www.agilealliance.org/glossary/user-stories/

12. Wikipedia. Suggested upper merged ontology. https://github.com/ontologyportal/sumo/blob/master/Hotel.kif

Social Networks and Recommender Systems

A Power-Graph Analysis of Non-fast Information Transmission

Jacek Mercik[(✉)]

WSB University in Wroclaw, Wroclaw, Poland
`jacek.mercik@wsb.wroclaw.pl`

Abstract. Specific types of information (e.g. knowledge, intellectual capital, conversation) require different models in the analysis of transmission, called non-fast transmission model. This paper introduces graphs with logical structure and proposes a method for evaluating the components of such graphs of transmission of information. The method may allow for optimization of transmission. This involves the maximization of the dissemination of non-fast information, and analysis of the impact of the information or its speed to reach various participants of the process of transmission.

Keywords: Slow information · Graph · Power

1 Non-fast vs. Rapid Transmission of Information

Assume we are analysing a system for transmission of information. Such a system consists of a set of elements (participants), who can be both senders and recipients of information, and the relationships between the participants indicating to whom and from whom the information is sent, and the manner in which it is done.

For example, in computer networks the participants are different computers, and their relationships are the data transmission lines. Information, in this case, is data sets. In the telephone network, on the other hand, the participants are both elements of technical infrastructure and people. Information is data sets, transmitted both between people and technical elements of the telephone network. Transmission lines are of "physical" (telephony) and "non-physical" character (e.g. a social conversation between two personal assistants). In such understanding of the systems of transmission of information, a conversation during the birthday party can be described as a system of transmitting information between people without a "physical" transmission line.

It seems that everything being transmitted between the participants of such a system in the form of data is information. However, for the purpose of this work the concept of information will also encompass the "physical" transmission, assuming that each physical transfer can be presented as equivalent information transfer. In other words, by transmission of information we sometimes understand pure information, and occasionally (depending on context) the medium (e.g. a letter sent by post).

It is currently assumed that the efficiency of the information system is determined by the speed and accuracy of information transfer between its various elements.

© Springer International Publishing AG 2017
N.T. Nguyen et al. (Eds.): ACIIDS 2017, Part I, LNAI 10191, pp. 89–99, 2017.
DOI: 10.1007/978-3-319-54472-4_9

Most research is dedicated to improving the speed of transmission, selection of appropriate algorithms to improve transmission and ensuring its integrity and reliability.

Individual papers relate to another aspect of the information transmission, i.e. to dissemination or reception of the information by its audience. In such situations, it seems, the speed of data transmission is of secondary importance, and in some cases completely negligible. This work is dedicated to such information transfer, and will be hereinafter referred to as non-fast transmission.

Note that in non-fast transmission the part of the system which slows down transmission is not necessarily a human. For example, in the analysis of the flow of intellectual capital (which, for convenience, can be called knowledge), the flow of information between two academic institutions, within the meaning of the work, is non-fast transmission, although it involves elements of high speed transmission (e.g. e-mail, fax, phone calls, publications).

Moreover, different transmission methods suggest that there are several transmission channels (relationships) directly connecting the same elements of the system of transmission of information. The two academic centres mentioned above can easily combine e-mail, fax or teleconferences, which means there is a variety of different relationships connecting them varying in terms of the speed of transmission of information, reliability, complexity and intensity. This means that, when modelling the relationships linking two selected components of the system, we should take into account the simultaneous multiple binary relations. Although these relationships exist between the same elements, they are different as a matter of principle.

2   The Logical Structure of the Network (Graph)

Consider first a part of the graph with one vertex (Fig. 1). The flow of information through the vertex may be dependent on logical conditions imposed on both the input and output of the node (vertex). For example, the conjunction condition on the entry to the node means that for the information to be further transmitted, all sources of information which enter the node (represented by arrows in the drawing) must finish the transmission of information. A good example of this situation is collecting, by a standardization company, all the information prior to developing a relevant standard.

Fig. 1. Fragment of a relationship graph.

The alternative condition imposed on the input means that at least one of the sources must terminate the information transmission. Analogously, a condition of disjunction means that only one source triggers further transmission of information, and the other sources are ignored. Note that the classic "gossip protocol" operates according to the entering procedure with a logical condition alternative.

Meeting a logical condition on entry to the node initiates the logical condition for the output: conjunction means that the information is sent farther across all transmitting channels simultaneously; alternative means that at least one transmission channel is active; disjunction means that only one of the transmission channels is active.

Moreover, in addition to the transferred information the nodes themselves may also generate new information, and be the source of information.

It seems that no vertex may be the absorbing node (where no transmission of information starts[1]) or an isolated one, where no information enters or comes out. Similarly, no vertex can be only the source of information, which means that for each node there is at least one source of information entering it.

Depending on where new information appears in the system, an undirected graph is transformed into a directed graph in one of the possible situations: (a) a starting node, (b), an ending node (c) starting and ending node. Figure 2 shows a fragment of a sample directed graph. Consequently, numbering of the vertices of a graph of transmission and directions of transmission (represented in the figure by arrows) must be adjusted in accordance with the principle that the node of a lesser number may not be the recipient of information from a node with a higher number. Also, the graph eliminates all transmission loops considering that from the point of view of new information, a loop acts as a "black hole" absorbing information [2].

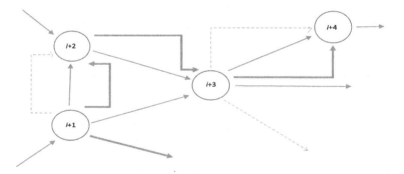

Fig. 2. A fragment of a directed graph of transmission of new information with various connections between objects represented by the different kinds of arrows.

3   Undirected and Directed Communication Graphs

A flat graph G is an ordered pair $G = \langle N, U \rangle$ wherein each u edge corresponds to at least one pair of ordered vertices, $\langle x, y \rangle \in N \times N$ such that $\langle x, u, y \rangle \in N \times U \times N$. We assume that U is not an empty set, which means that the N set is not empty either. We also assume that in a graph G there are no so-called loops, $\langle x, u, x \rangle \notin N \times U \times N$.

---

[1] In extreme cases, it may be a single output describing the impact of the environment on the node.

In an analytic form the structure of communication between nodes will be represented by a graph. Any graph with non-empty sets of nodes (N) and edges (U), where all nodes and edges are adequately numbered may be uniquely defined by incidence matrix $A(G) = [a_{ij}]_{nxm}$, where: $n = |N|$ represents cardinality of a set of nodes N, $m = |U|$ represents cardinality of a set of edges U, $a_{ij} = 1$ if there is relation u such that $\langle i, u, j \rangle \in N \times U \times N$ (where for simplicity i and j denote x_i and y_j respectively). An incidence matrix maybe generalized by assuming $a_{ij} > 0$ *instead of* $a_{ij} = 1$ if there is relation u such that $\langle i, u, j \rangle \in N \times U \times N$ – this is the reason why we presented an incidence matrix in an expanded form by showing all connections connecting every pair of nodes.

If in a given graph G there is at least one pair of relations $\{\langle i, u, j \rangle, \langle j, u, i \rangle\} \in N \times U \times N$, such a graph is called an undirected graph. Note that in an undirected graph there may be two vertices i and j for which there is only one of the two relations $\{\langle i, u, j \rangle, \langle j, u, i \rangle\} \subset N \times U \times N$.

Determination of the starting or ending node (vertex) or both converts an undirected graph into a directed graph. This means that at any given time for any pair of relations $\{\langle i, u, j \rangle, \langle j, u, i \rangle\} \in N \times U \times N$ there can be only one relation from $\{\langle i, u, j \rangle, \langle j, u, i \rangle\} \in N \times U \times N$. Its kind depends on whether a starting vertex, an ending vertex or both are chosen, and what logical conditions are imposed on inputs and outputs of the vertices. Note that the direction applies only to symmetrical relationship and means that if an undirected relation occurs between two selected vertices, $\langle i, u, j \rangle \in N \times U \times N \Rightarrow \neg \langle j, u, i \rangle \in N \times U \times N$ during the conversion of the undirected graph into a directed graph the relationship $\langle i, u, j \rangle\rangle$ is maintained[2].

The model proposed in this paper assumes the only thing known is a set of nodes with the structure of the connections between them, G (graph of communication), the weighting attributed to each node[3], w_i, logical conditions of enter and exit for every node, and particular weight of a given connection between nodes i and j, $p_{i,j}^r$, where index r refers to every consecutive relation connecting these nodes[4]. First, we define the character of the undirected graph and analyse the relationships occurring in it. Then, prior to transforming the undirected graph into a directed one, we determine the starting, ending, or both nodes.

We assume that the assessment, $t_{i,j}$, of a connection between two vertices i and j of a graph G equals

$$t_{i,j}^r = \alpha_{i,j} w_{i,j} p_{i,j}^r, \tag{1}$$

where:

---

[2] It is also possible that in the process of orientation a communication graph can turn into two disjoint sub-graphs. In this case, it is impossible to transfer information from one sub-graph into another sub-graph in a direct way. This paper will not discuss such cases.

[3] We assume that the weight describes the "importance" of particular i node in the context of transmitted information.

[4] Remember that in a graph G many different relationships connecting two adjacent vertices are acceptable.

$\alpha_{i,j}$ – arbitrary factor assigned for the connection $(i\,j)$ on the basis of historical data to achieve a better match (we assume it is initially equal to 1), r - subsequent number of the connection between the vertices i and j, $r = 1, 2, ...$; $w_{i,j}$ – weight assigned to any combination of a vertex i with a vertex j resulting from the assigned weighting of the vertices. This weight is determined in the proposed process of attributing weights to connections.

Definition. Path $s_{i,j}^r$ is a sequence of edges and vertices joining vertices i and j, where each node and each edge can occur only once, i.e. $s_{i,j}^r = \{i, u_{ik}, \dots, u_{lj}, j\}$ for $k, l \neq i, j, k \neq l$, $r = 1, 2, \dots$.

Much like Rosenthal [5], we attribute certain weight, $w_{i,j}$, to connections without giving them any additional interpretation. These weights are designed to render validity of specific connections resulting only from the importance of the nodes, described in turn, by their weights (consistently with the second axiom of Myerson [4][5]). The "weighing" of connections will be achieved through marking of the G graph.

Following the idea presented in Mercik [3], we'll now mark the edges. We assume that the weights of vertices are non-negative, which means that the marking also applies to the vertices with a weight equal 0.

Definition. Let i and j be two neighbouring vertices, i.e. vertices connected at least by a relationship $\{u_{i,j}^r, r = 1, 2, \dots\}$. If their respective weights are w_i and w_j, the relationship between nodes i and j weights $w_{ij} = w_i + w_j$. If i and j vertices are not adjacent, yet connected by at least one path[6], $\{s_{i,j}^r, r = 1, 2, \dots\}$, the weight of the relationship equals[7] for $k, l \in s_{ij}, k \neq i, l \neq j, w_{ij} = w_{ik} + \sum_{\substack{(k, l) \in s_{ij} \\ k \neq l}} w_{kl} + w_{lj} - \sum_{\substack{k \in s_{ij} \\ k \neq i,j}} w_k$. Moreover, the weight of each path will meet the inequality $w_{ij} \geq w_i + w_j$ and can be different, for each path $s_{i,j}^r$. Note that for every $r = 1, 2, \dots$ each path $s_{i,j}^r$ has the same weight $w_{i,j}$.

Definition. If for the path $s_{i,j}^r$ its weight w_{ij} meets the inequality $w_{ij} \geq q$, the path $s_{i,j}^r$ is referred to as a q-path, $^q s_{i,j}^r$. Depending on the application of the proposed approach, the value q and, consequently, the q-path will change. Such a description allows for the analysis of e.g. information propagation in networks.

Definition. The G graph exclusively specifying the relationships between vertices may therefore be modified to graph G_w where any non-zero value a_{ij} that occurs in the matrix of incidence $A(G)$ is replaced by the corresponding value w_{ij}, thus forming a weighted incidence matrix $A_w(G)$.

---

[5] The second axiom refers to the belief that the relationship between two nodes has equal value for either of them.

[6] There can be a number of paths connecting the vertices i and j. In this case, each of these paths will be distinguished, s_{ij}^r, for $r = 1, 2, \dots$.

[7] This means that the assessment of the path is the sum of assessment of individual edges.

Note that if the weight of one of the vertices is zero, the weight of the edgeh connecting it directly to another vertex with a non-zero weight will remain non-zero, thus the incidence matrix $A_w(G)$ retains relationships between vertices. We interpret it in such a way that sometimes, for maintaining a route connecting two selected nodes, a node with zero weight must be enabled.

If the graph describes a structure in which there is no start and end nodes (non-oriented graph), in such a graph we consider all the paths from such node. If in the analysis of a specific situation the start or end of a path node must be present, all the analysed paths must contain it.

Relative Measure for Weighting Individual Components of the Directed Graph.

During the implementation the basic problem with the analysis of systems behaving in a stochastic is the impossibility to predict which of possibilities will be implemented. Moreover, there is usually no data (or the data is statistically unreliable) allowing to assume a particular distribution of possibilities. Hence, we assume that every possibility is equally likely (Laplace's criterion) and if some characteristics are calculated, it is assumed they will be the expected values. A consequence of this approach is the so-called a priori analysis, which provides the answer to the question what can occur rather than what will occur. In this way we will investigate the importance of individual components of the communication graph for transmitting non-fast information.

The proposed a priori measure of individual elements of the graph of information transmission will be modelled on the so-called power indexes, known from the theory of cooperative games. In the literature there are two dominating power indices: the Shapley-Shubik [6] power index and the Banzhaf [1] power index. Both indices are based on the Shapley value concept and belong to so called family of a priori power indices.

Let s_{ii} be a path consisting of only one element[8] (excluding the possibility of loops in the graph describing the structure). Its length, by assumption, is equal to the weight value of the node, i.e. w_i.

Let S_i be the set of all paths s_i (including q-paths) passing through the i-th element (an edge or a node). Let S_i^q be the set of all paths s_i^q passing through the i-th element of the graph (an edge or a node) whose weight is not less than q, $S_i^q \subseteq S_i$.

Let $S_{i.}$ be the set of all paths $s_{i.}$ (including q-paths) extending from the i-th node. Let $S_{i.}^q$ be the set of all paths $s_{i.}^q$ extending from the i-th node, whose weight is not less than q, $S_{i.}^q \subseteq S_{i.}$.

$S = \bigcup_{i=1}^{N} S_{i.}$ is the set of all paths in the graph described by the incidence matrix $A_w(G)$ and $S^q = \bigcup_{i=1}^{N} S_{i.}^q$ is the set of all paths with length not less than q and starting with the i = 1, 2, ..., N nodes.

Note that in a similar way we can define $S_{.i}$ i.e. a set of all paths ending in an i node. From the symmetricity of the definition $S^q = \bigcup_{i=1}^{N} S_{i.}^q = \bigcup_{i=1}^{N} S_{.i}^q$.

---

[8] For simplicity of marking, we drop the index r, considering however, that there can be multiple connections for a given pair of vertices i and j.

A power index of graph elements is a mapping $\varphi:S \rightarrow R^m$. For each $i \in N$ and S, the ith coordinate of φ, $\varphi(i)$, is interpreted as the voting power of element i in the graph G, m denotes cardinality of a set being a sum of nodes and edges of the graph G.

Given the fact that relations between nodes are described by the set graph structure, we can now determine the power index of graph G element as follows.

Thus, for the i-th node of a graph structure described by G graph an absolute power index is defined as follows[9]:

$$\varphi_i^p = \frac{|S_i^q|}{|S|} = \frac{|S_i^q|}{\left|\bigcup_{i=1}^{N} S_{i.}\right|}, \tag{2}$$

where φ_i^p denotes power index of a node of graph G. By analogy, you can define an a priori power index for edges (connections) φ_i^c, where c stands for communication (and an edge between nodes for a given path)[10].

Note that the proposed power index does not match the Shapley-Shubik power index, as every node present on a particular path is a pivotal node (their absence eliminates the path as a connection between the start and end nodes, or it may shorten the path so that it is no longer a type q-path for a given q value). Moreover, in axiomatics of both classical indices, i.e. the Shapley-Shubik power index and the Banzhaf power index there is a postulate (an axiom) of an empty player (node): an empty player has no power. Hence the proposed indices φ_i^c and φ_i^p as not meeting the postulate cannot be considered equal to Shapley-Shubik or Banzhaf indexes. The axiomatics of the proposed indices will be dealt with in subsequent papers.

The Algorithm for Determining a Priori Power Index for the Elements of the Communication Graph.

Regardless of whether we are dealing with a directed or undirected graph, it is necessary to find a set S, i.e. the set of all paths in the graph G described by the incidence matrix $A_w(G)$. Such algorithm may involve reviewing all possibilities. Running time of such algorithm for finding all the paths in such graphs is dependent on two parameters: the size of the graph defined by the number of vertices N and the value q in q-paths. It can reasonably be assumed that the value N in the analysis of transmission of non-fast information will not be unrealistically high, which allows to perform calculations in real time. A full search of all paths is dependent on the number of connections to a degree $r_{max}N2^N$ where r_{max} it means the maximum amount of direct relations between any two vertices in a graph G.

Step 1. On the Basis of the Matrix $A_w(G)$ We Find a Set S Consisting of:

- In a graph with only the start vertex all paths extending from the start vertex,
- In a graph with only the end vertex all paths terminating at the end vertex,

---

[9] We use the | . | operator to denote the cardinality of a finite set.

[10] The analysed paths also contain linked affiliate connections, resulting from logical conditions relating to the input and output of a given vertex. The definition of these connections is provided hereinafter.

- In a graph with the end and start vertices all paths extending from the start vertex and ending at the end vertex,
- In an undirected graph G we find all paths starting successively at each vertex of the graph G.

Step 2. After finding a set S appropriate for the given graph G, we find for every i in the set S all paths belonging to the set S_i.

Step 3. Using the formula (2) we calculate the value φ_i^p or φ_i^c for every i respectively from either the set of vertices or connections between the vertices.

The information about the a priori power of individual edges can thus be added to a description of the graph and so-called incidence-power matrix, $A_p(G)$, can be created. A priori incidence-power matrix contains not only local values associated with the connection between two given vertices but also describes the nature of this connection against the other relationships in the communication graph. Thus, we have modified the values for the relationship u so it describes not only the connection between given nodes, but also evaluates this connection globally for the entire communication graph, $u = [\varphi_i^c]$ for edges between nodes. We also believe that the method of marking edges in graph G introduces, by incorporating the weights of players, information on the power of individual nodes to the features of their connecting edges.

Assessment of the Paths Connecting Two Given Vertices.

Each of the paths of graph G can be assessed using weights relating to each individual connection (formula 1). If the logical conditions of entry and exit of vertices are on the path are of OR|OR type, an assessment of this path is performed by assessing only the elements of the path. If on a given path, in relation to the input of vertices of this path, exist logical conditions other than OR, the connection occurring in these logic conditions should participate in the evaluation of the path, despite not being on the same path. These connections will be called associated conditions, $S_{i,j}^L$ - a collection of associated connections to the path $s_{i,j}$. Note that $S_{i,j}^L$ may be an empty set. We believe that the assessment of a path should be reduced by the value of assessment of all the connections conditioning the execution of the path (associated connections). We therefore propose the following method to assess the paths.

Step 1. For each two designated vertices i, j of graph G find all the paths connecting them.

Step 2. Find all associated connections for each of the paths found.

Step 3. Calculate the value which is an assessment of each of the paths found, taking into account the associated connections.

$$w_{i,j}^L = t_{i,j} - \sum_{(k,p) \in S_{i,j}^L} w_{k,p}$$

Note that the evaluation of a specific path can return negative values. Such assessments are not ascribed any interpretation beyond the use for the ranking of paths.

Example. Consider the example of a non-fast data transmission shown in Fig. 2. Note that there are four vertices in this graph: $i + 1, i + 2, i + 3, i + 4$. There are also three types of relationships: the relationship marked by the dashed line (type a), solid line (type b), and the solid bold line (type c). Although all visible relationships between the vertices are directed, it cannot be ruled out that the graph in Fig. 2 represents a fragment of an undirected graph. However, the moment the vertex i + 1 is considered the input and i + 4 the output, the graph in Fig. 2 becomes a directed one. Assume that the weights of vertices, w_i, are respectively: 2, 2, 4, 3.

We assume that the vertices $i + 2, i + 3, i + 4$ have logic input designed so that the connection marked with a dotted line (type a) is not sufficient to trigger further transmission of information from the vertex such connection enters. Triggering it requires the entry of at least one other active connection (type a or type b).

According to the formula (1) weights of direct connections between each pair of vertices are attributed (Table 1). For simplicity the correction factor value was assumed $\alpha_{i,j} = 1$.

Table 1. Assessment of connections between vertices in the analysed example.

Connection number	Vertex - exit	Vertex - entry	Particular weight of a given connection $p_{i,j}^r$	Weight of edge $w_{i,j}$	Overall assessment of connection $t_{i,j}$
1	$i + 1$	$i + 2$	1	4	4
2	$i + 1$	$i + 2$	2	4	8
3	$i + 1$	$i + 2$	3	4	12
4	$i + 2$	$i + 3$	2	6	12
5	$i + 2$	$i + 3$	2	6	12
6	$i + 1$	$i + 3$	2	6	12
7	$i + 3$	$i + 4$	1	7	7
8	$i + 3$	$i + 4$	2	7	14
9	$i + 3$	$i + 4$	3	7	21

The results of calculations presented in Tables 1 and 2 enable the evaluation of elements of the directed graph shown in Fig. 2 (i + 1 start vertex, i + 4 end vertex) with a predetermined logical structure:

(1) Power index of vertices i + 2 and i + 3 is (30/34, 34/34) respectively, which means that i + 3 is more important than the vertex i + 3 in the transmission of information between the vertices i + 1 and i + 4.
(2) Power index of connections:

Connection	1	2	3	4	5	6	7	8	9
Index φ_i^c	14/34	17/34	8/34	16/34	14/34	4/34	16/34	16/34	16/34

(3) Assessments of paths allow their ranking in order of importance: 3–4–9, 3–5–9, 2–5–9, 2–5–8, 2–4–8, and so on. Analysis of the example shows that the path $i + 1, i + 2, i + 3, i + 4$ is the path of most importance in the transmission of information between the vertices $i + 1$ and $i + 4$.

(4) Note that q-paths, i.e. paths with a predetermined weight of information of at least q, eliminate some of the paths. These paths can change the a priori weight assessment of particular vertices without changing the ranking of the remaining paths.

Table 2. Assessment of paths between vertices $i + 1$ and $i + 4$ in the analysed example.

Basic path	Additional conditions for connections	Assessment of given path	Basic path	Additional conditions for connections	Assessment of given path
1–4–7	2, 8	$23 - 22 = 1$	2–4–9	–	33
1–4–7	2, 9	$23 - 29 = -6$	2–5–7	8	$27 - 14 = 13$
1–4–7	2, 8	$23 - 22 = 1$	2–5–7	9	$27 - 21 = 6$
1–4–7	2, 9	$23 - 29 = -6$	2–5–8	–	34
1–4–8	2	$30 - 8 = 22$	2–5–9	–	41
1–4–8	3	$30 - 12 = 18$	3–4–7	8	$31 - 14 = 17$
1–4–9	2	$37 - 8 = 29$	3–4–7	9	$31 - 21 = 10$
1–4–9	3	$37 - 12 = 25$	3–4–8	–	38
1–5–7	2	$23 - 8 = 15$	3–4–9	–	45
1–5–7	3	$23 - 12 = 11$	3–5–7	8	$31 - 14 = 17$
1–5–8	2	$30 - 8 = 22$	3–5–7	9	$31 - 21 = 10$
1–5–8	3	$30 - 12 = 18$	3–5–8	–	38
1–5–9	2	$37 - 8 = 29$	3–5–9	–	45
1–5–9	3	$37 - 12 = 25$	6–7	8	$19 - 14 = 5$
2–4–7	8	$27 - 14 = 13$	6–7	9	$19 - 21 = -2$
2–4–7	9	$27 - 21 = 6$	6–8	–	26
2–4–8	–	34	6–9	–	33

4 Summary and Conclusions

The occurrence of the non-fast information transmission is a fact. Existing methods of analysis of the transmission of such information may prove to be inadequate due to the nature of the information. The proposed method of modelling non-fast information transmission allows for quantitative assessment of the transmission, identification of key paths and evaluation of the weight of each vertex in the system processing this type of information.

It appears that the proposed a priori evaluation of the elements of the transmission of information may also be used for any other kind of information transmission, in particular, due to its indeterminate logical structure, for all conditional transmissions.

References

1. Banzhaf III, J.F.: Weighted voting doesn't work: a mathematical analysis. Rutgers Law Rev. **19**, 317–343 (1965)
2. Burt, R.S.: Structural holes and good ideas. Am. J. Soc. **110**(2), 349–399 (2004)
3. Mercik, J.: Formal a priori power analysis of elements of a communication graph. In: Nguyen, N.T., Trawiński, B., Fujita, H., Hong, T.-P. (eds.) ACIIDS 2016. LNCS (LNAI), vol. 9621, pp. 410–419. Springer, Heidelberg (2016). doi:10.1007/978-3-662-49381-6_39
4. Myerson, R.B.: Graphs and cooperation in games. Math. Oper. Res. **2**, 225–229 (1977)
5. Rosenthal, E.C.: Communication and its costs in graph-restricted games. Soc. Netw. **10**, 275–286 (1988)
6. Shapley, L.S., Shubik, M.: A method of evaluating the distribution of power in a committee system. Am. Polit. Sci. Rev. **48**(3), 787–792 (1954)

Group Recommendation Based on the Analysis of Group Influence and Review Content

Chin-Hui Lai[1][(✉)] and Pei-Ru Hong[2]

[1] Department of Information Management, Chung Yuan Christian University, Taoyuan, Taiwan
chlai@cycu.edu.tw
[2] Institute of Information Management, National Chiao Tung University, Hsinchu, Taiwan
x79110@hotmail.com

Abstract. With the development of internet, users not only receive information passively but also share their own opinions on the social networking websites. Accordingly, users' preferences for items may be affected by others through opinion sharing and social interactions. Moreover, users with similar preferences usually form a group to share related information with others. Users' preferences may be affected by group members. Existing researches often focus on analyzing personal preferences and group recommendation approaches without user influence. In this work, we propose a novel group recommendation approach which combines the group influence, rating-based score and profile similarity to predict group preference. The group influence is composed of group member influences, review influence and recommendation influence. The profile similarity is derived from the analysis of item descriptions and review content. The experimental results show that considering the group influence and content information in group recommendation approach can effectively improve the recommendation performance.

Keywords: Latent dirichlet allocation (LDA) · Group recommendation · Social influence · Information retrieval · Collaborative filtering

1 Introduction

Due to the development of social network, people can share information, interact with others, follow others' posts and so on. Users' behavior can impact on other users' thinking, behavior and decision making through the social networks. Users are influenced with each other. This phenomenon is also called social influence. In addition, users usually have the information load problem because they have difficulty in finding the needed information on a social network. The recommender systems can resolve this problem by providing suitable information or items based on users' preferences or the contents of items. Some researches utilized users' social influence in recommender systems to improve the recommendation quality. However, they did not analyze the social groups on social networks. Most of people like to join interested social groups to share their opinions and obtain others' information. However, identifying the preference of groups is more difficult than that of individuals. The previous researches on group

N.T. Nguyen et al. (Eds.): ACIIDS 2017, Part I, LNAI 10191, pp. 100–109, 2017.
DOI: 10.1007/978-3-319-54472-4_10

recommendation method merge individual preference as group preference. They did not analyze users' social influence or social interactions based on users' behavior on a social network.

Therefore, in this work, a novel group recommendation method is proposed on a book social networking website, which allows users to share books and reading comments. The proposed method first analyze group member influence, review influence and recommendation influence based on users' behavior in a group. The group member influence means that one member in a group may follow the likes of another member's to read books. The review influence means that a user is influenced by others' reviews to read a book which may be interesting to him/her. In addition, the recommendation influence means that a user in a group recommends his/her unread books to other members and then they read and like these recommended books. These three types of influences are combined as a user's group influence to represent a user's weighting in a group.

Besides, group members' ratings on books, review content and book descriptions are also analyzed in the proposed method. Ratings is used to represent users' preferences, but not for groups'. Thus, these ratings given by group members and the group influence of each member are integrated into a rating-based score for measuring the preference of a group on a book. In addition, the book descriptions and reviews are analyzed by using *tf-idf* [1] and Latent Dirichlet Allocation (LDA) [2] methods to create group and book profiles. The similarities among group and book profiles are evaluated. Finally, the proposed hybrid method merges the rating-based score and profile similarity to make predictions for groups. The top-N books with high predicted scores are recommended to target groups.

The remainder of the paper is organized as follows. Section 2 introduces the related works including social influence in recommender systems and group recommender systems. In Sect. 3, we introduce the proposed recommendation methods based on social influence. Section 4 shows the experimental results and evaluations. The conclusion and future works are discussed in Sect. 5.

2   Related Works

Social influence means that a user's behavior is affected by his/her friends. The recommender systems based on social influence make recommendations to the active users by analyzing their behavior with their friends and followers. However, most of the research estimates the social influence in recommender systems by taking a single perspective instead of the perspectives of both the influenced and influential user. For example, the impact of a Twitter user on his/her followers is determined by the relative numbers of content the follower received from the user [3]. The influence of a user is measured by the number of his/her friends who click the shared web post; or by the ratio of the number of items a target user received to the number of items a recommender recommended. In addition, the mathematical programming models [4] are applied to the influence networking sites for identifying user groups with large joint influential power. Generally, information can be disseminated quickly through those influential users, so that the items recommended by those influential users are easily accepted by others.

Group recommender systems are used for music, movies, television programs, tourist attractions and so on. For making group recommendation, some systems [5–7] aggregate the profiles/preferences of various users to form a group profile/preference. There are two most common methods that aggregate members' preference to be groups'. The one is average scores from group members and the other is least misery which means to find the lowest score as group's score. Thus, such systems have high probability in discovering valuable recommendations that will satisfy the majority of the group's members. Additionally, some systems merge individual recommendation lists to form a group recommendation list [8]. Such systems give users more information when they need to make decisions, and the recommendation results are relatively easy to explain. The limitation of group recommendation is that it is very time consuming if the group is large. Group's information is larger than individual's because it contains some activities and interaction records. Data sparsity is the other limitation. It is difficult to analyze the preference of a target group because the group's members are not active and the number of ratings is only a few.

3 Group Recommendation Method

3.1 Overview

In this work, a novel group recommendation method, which combines group influence analysis and group profiling, is proposed on a social networking website for sharing

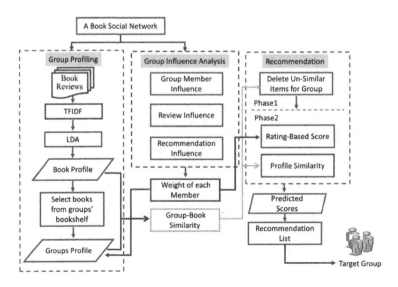

Fig. 1. Overview of the proposed group recommendation method

book reviews, i.e. *Goodreads*[1]. Figure 1 shows an overview of our group recommendation method. The proposed method is composed of three phases: group influence analysis, group profiling and recommendation. According to the implicit relations among users, the group member influence, review influence and recommendation influence are defined and investigated in the phase of group influence analysis. Because users' behavior may impact other users to give items ratings, these three kinds of influences are combined to represent a user's weight in a group. In the phase of group profiling, book reviews are analyzed by using information retrieval and Latent Dirichlet Allocation (LDA) [2] methods for creating book profiles and group profiles. The last phase is the recommendation. Based on the profile similarity between book and group profiles, unsimilar items are filtered and then the predicted scores of books are calculated by combing the rating-based and content-based scores. Finally, the books with high predicted scores will be recommended to target groups.

3.2   Group Influence Analysis

In this section, we will introduce the method of group influence analysis, composed of group influence, review influence and recommendation influence. For each type of influence, the member-to-member influences for all pairs of members in a group are evaluated. Then, these influences will be combined and regarded as a user's weight in a group.

Group Member Influence. Group member influence means that group members are influenced by a specific member to rate items, which have been rated by the member. The time factor is used to measure the influence strength of influenced users on an accepted item. Let $tf_{v,i}$ be a time factor of user v on book i, defined in Eq. (1) where $t_{v,i}$ represents the time that user v gives a rating to book i, and τ is a parameter for evaluating the impact of time factor.

$$tf_{v,i} = e^{-\tau\left(t_{now}-t_{v,i}\right)} \tag{1}$$

Let users u and user v be group g's members. U_g is a set of members in group g. Equation (2) defines the whole membership influence of user u in group g from the perspective of user u. It summarizes the influence scores of user u on other members in group g as a group member influence.

$$MIG_u^g = \sum_{v\in U_g, v\neq u}\left(\frac{\sum_{i\in Ir_{u\to v}^g} tf_{v,i}}{|Ir_u^g|}\right) \tag{2}$$

$Ir_u^g = \left\{i\middle| i \in I_u \cap I_g \text{ and } r_{u,i} \neq 0\right\}$ is a set of books that user u have read and rated from group g's bookshelf. $Ir_{u\to v}^g = \left\{i\middle| i \in Ir_u^g \cap Ir_v \text{ and } tf_{v,i} > tf_{u,i}\right\}$ is a set of books in member v's bookshelf. These books were read and rated by the member v after the

---

[1] https://www.goodreads.com/.

member u has rated them. The ratings of book i given by both member u and member v should be greater than 3 because the positive effect between members is considered.

Review Influence. Besides users' rating behavior, users' reviews are analyzed to obtain the review influence. The review influences of user u to group g, as shown in Eq. (3). All pairs of the review influence between user u and the other member (i.e. user v) in group g are measured first. Then, for the group g, all pairs of user u's review influence are summarized as RIG_u^g.

$$RIG_u^g = \sum_{v \in U_g, v \neq u} \left(\frac{\sum_{i \in IR_{u \to v}^g} tf_{v,i}}{|IR_u^g|} \right) \tag{3}$$

Let IR_u be a set of books that user u has written the reviews. $IR_u^g = \left\{ i \middle| i \in IR_u \cap I_g \text{ and } r_{u,i} \neq 0 \right\}$ is a set of books in group g's bookshelf, which user u have written the reviews and given the ratings. $IR_{u \to v}^g = \left\{ i \middle| i \in IR_u^g \cap Ir_v \text{ and } tf_{v,i} > tf_{u,i} \right\}$ is a set of books in group g's bookshelf, which user v read and rated these books after user u have written reviews and given ratings. The books, whose ratings given by user u and user v are greater than 3, are considered in this computation. $tf_{v,i}$ is a time factor defined in Eq. (1).

Recommendation Influence. The recommendation influence means that a group member can recommend books in which group members may be interested to the group's bookshelf. On the website of *Goodreads*, a reading group is generated according to the tags in book categories. Thus, the group members generally have certain preference or similar preference. Any member can recommend books which are popular or suitable for the group's reading preference to the group even though he does not have read these books. The user u's recommendation influence is defined in Eq. (4), which summarizes all pairs of recommendation influences between user u and other members (i.e. user v).

$$SIG_u^g = \sum_{v \in U_g, v \neq u} \left(\frac{\sum_{i \in IS_{u \to v}^g} tf_{v,i}}{|IS_u^g|} \right) \tag{4}$$

IS_u^g is a set of books which have been recommended to group g by user u. $IS_{u \to v}^g = \left\{ i \middle| i \in IS_u^g \text{ and } tf_{v,i} > tf_{u,i} \right\}$ is a set of books in group g's bookshelf, and member v read and rated these books after member u have recommended. The ratings of books given by user v are greater than 3. $tf_{v,i}$ is a time factor which user u recommends book i to the group (i.e. Eq. (1)).

Group Influence. For a user in a group, the group member influence, review influence and recommendation influences are combined as a group influence, as defined in Eq. (5). GI_u^g is the group influence of user u in group g, which summarized these three influence of member u (i.e. MIG_u^g, RIG_u^g and SIG_u^g). Because not all users write the reviews or recommend books to the group, some of them may not have the review influence or the recommendation influence except for group member influence. Thus, to avoid getting 0 in the group influence, these two kinds of influence are plus 1 respectively before all influences are multiplied together.

$$GI_u^g = MIG_u^g \times \left(1 + RIG_u^g\right) \times \left(1 + SIG_u^g\right) \tag{5}$$

3.3 Group Profiling

The book profiles and group profiles are generated by analyzing the content of books.

Book Profile. A book profile is represented as an n-dimensional topic vector, which comprises the key terms derived from the reviews of the book and their respective weights derived by the normalized *tf-idf* approach. Based on the term weights, terms with high values are selected as discriminative terms to describe the features of a book. Then, Latent Dirichlet Allocation (LDA) [2] method is applied to these terms for generating k topics. According to the distribution of terms in each book, we can conduct a k-dimensional topic vector to each book. Thus, the book profile is represented as a topic vector, i.e. $BP_x = \langle bt_{1,x}:btw_{1,x}, bt_{2,x}:btw_{2,x}, \dots, bt_{k,x}:btw_{k,x}\rangle$, where $bt_{i,x}$ is a topic term i and $btw_{i,x}$ is a weight of the topic term i (i.e. topic probability) in book x.

Group Profile. Similar to the book profile, a group profile is also represented as an n-dimensional topic vector, extracted from books in a group's bookshelf. Let the group topic vector be $GP_g = \langle gt_{1,g}:gtw_{1,g}, gt_{2,g}:gtw_{2,g}, \dots, gt_{k,g}:gtw_{k,g}\rangle$, where $gt_{i,g}$ is a topic term i and $gtw_{i,g}$ is a weight of the topic term i in group g's bookshelf. $gtw_{i,g}$ is derived from the integration of time factor, the weights of topic features derived from the books in group g's bookshelf and users' group influence in group g, as defined in Eq. (6).

$$gtw_{i,g} = \frac{\sum_{z \in I_g} btw_{i,z} \times tf_{g,z} \times \sum_{u \in U_g \cap Ur_z} GI_u^g}{\sum_{z \in I_g} tf_{g,z} \times \sum_{u \in U_g \cap Ur_z} GI_u^g} \tag{6}$$

where I_g is a set of books in group g's bookshelf and $btw_{i,z}$ is a weight of topic i in book z. $tf_{g,z}$ is a time factor when book z is added into group g's bookshelf. $U_g \cap Ur_z$ is a set of users who have rated book z in group g. GI_u^g is a group influence of user u who have rated books in group g's bookshelf, defined in Eq. (5).

3.4 Recommendation

In this section, we propose a two-phase model to make recommendations for a target group. In the first phase, the *cosine* similarities between a group profile and book profiles are evaluated. If the values of similarities are lower than a specific threshold, it means that these books are not related to the preference of the group. Therefore, these un-similar books are filtered out for creating a candidate list for a group.

In the second phase, the hybrid method which combines the rating-based score (i.e. RS) and profile similarity between groups and books (i.e. PS) is proposed. It is used to predict the ratings for the books in the candidate list derived from phase 1 for a target group. The rating-based score represents a predicted rating of a group on a book. It is measured by users' group influence, including group member influence, review influence and recommendation influence, and the ratings of books given by group members in the candidate list. Equation (7), i.e. IR_i^g, defines the formula for measuring the rating-based score of book i in group g.

$$IR_i^g = \frac{\sum_{u \in U_g \cap Ur_i} \left(r_{u,i} \times MIG_u^g \times \left(1 + z_{u,i} \times RIG_u^g \right) \times \left(1 + SIG_u^g \right) \right)}{\sum_{u \in U_g \cap Ur_i} \left(MIG_u^g \times \left(1 + z_{u,i} \times RIG_u^g \right) \times \left(1 + SIG_u^g \right) \right)} \tag{7}$$

Ur_i is a set of users who have rated book i. $U_g \cap Ur_i$ is a set of users in group g who also have rated book i. $r_{u,i}$ is a rating of book i given by user u. MIG_u^g, RIG_u^g and SIG_u^g are user u's group member influence, review influence and recommendation influence respectively. $Z_{u,i}$ is 1 if user u has written a review to book i (i.e. user u has review influence on book i); otherwise it is 0.

Then, the rating-based score and the profile similarity between group and book profiles are aggregated to predict a rating of a book i for a target group g, as defined in Eq. (8). Such hybrid method considers both users' rating information and book content.

$$R_i^g = \alpha \times N(IR_i^g) + (1 - \alpha) \times N(sim(g, i)) \tag{8}$$

α is an adjust parameter between the rating-based score and profile similarity, which ranges from 0 to 1. Because the rating-based score and profile similarity have different scale (i.e. the rating-based scores range from 0 to 5 and the profile similarity ranges from 0 to 1), normalization for these two scores is necessary. Thus, $N(IR_i^g)$ indicates a normalized rating-base score, while $N(sim(g, i))$ indicates a normalized profile similarity. Books with high predicted ratings are selected into a recommendation list. Then, top-N books are recommended to a target group. The proposed hybrid method overcomes the problem of data sparsity, because it not only analyzes the rating information but also content features of groups and books discussed in Sect. 3.3. In addition, the hybrid method takes the group influence into account, so that the predictions may be more accurate.

4   Experiments and Evaluations

4.1   Experiment Setup

In our experiment, we collected a data set from an international social networking website, i.e. *Goodreads*. Users of *Goodreads* can create his/her own bookshelf, write reviews for books to share personal opinions, give ratings to books, and make friends or follow someone's bookshelf. The dataset consists of books, review, and user information from Jan 1, 2011 to Mar 31, 2014. For the data set, we first select 7 groups and their members are regarded as root users. Each member in a group has at least rated two books in a group bookshelf. Each book in a group's bookshelf is rated by at least one member. Therefore, the collected dataset is composed of 361 users, 600000 reviews and over 25000 books. There are 1731 books with 8093 ratings in 7 target groups' bookshelf. On average, each book was read by 3 people, while each user may have read about 70 books. The dataset is divided into 70% for training and 30% for testing. The training set is used to implement the proposed method and then generate recommendation lists. The testing set is used to verify the performance of these recommendations.

Evaluation Metric. To evaluate the recommendation performance of the proposed method, *Precision* and *recall* metrics [1, 9] are employed in this work, which are widely used in recommender systems. *Precision* is the fraction of recommended books which are in target groups' book lists. *Recall* is the fraction of a target group's books that are recommended by the proposed method. These measures are simple to calculate, but they are in conflict sometimes. For example, increasing the size of the recommendation set may improve recall, but may reduce precision. Therefore, to balance the trade-off between precision and recall, the *F1*-metric, which combine the precision and recall by assigning equal weights to both metrics, is used in the experiments for evaluation. The F1-metric is shown in Eq. (9).

$$F1 = \frac{2 \times recall \times precision}{recall + precision} \tag{9}$$

4.2 Experimental Results

In this experiment, different variations of the proposed methods are compared. GI-H is the proposed hybrid method which combines the group influence, rating-based score and profile similarity (i.e. Eq. (8)). GI-RS is the method which combines the group influence and rating-based score (i.e. Eq. (7)), while the GI-PS is the method which combines the group influence and profile similarity. H, which combines rating-based scores and profile similarities, is the hybrid method without any influence. UCF is the traditional user-based collaborative filtering method. It only uses similar neighbors' ratings based on *Pearson* similarity to predict group members' ratings on books. Then, for a specific book, all members' predicted ratings in the same group are averaged as a predicted rating of the group.

Figure 2 shows the comparison results from top-20 to top-120 when α = 0.8. GI-H performs better than other methods. These three influences are complements to each other. When recommending top-60 books, H method performs slightly better than GI-H. GI-RS performs worse than other methods. UCF performs better than MRS-RS when recommending more than top-20 books. Generally, GI-H, GI-PS and H methods are better than UCF. Thus, analyzing the content information can effectively improve the recommendation performance comparing to the traditional UCF. In addition, recommendation based on the combination of group influence, rating-based score and profile similarity overcomes the sparse problem and leads a better performance.

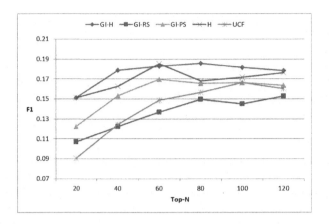

Fig. 2. Comparison of the proposed methods under different top-*N*

For the hybrid method, Fig. 3 shows the comparison results of the methods under different combination of influences. GI represents the group influence which combines the group member influence (i.e. M), review influence (i.e. R) and recommendation influence (i.e. S). For example, MS-H method is the hybrid method with the combination of group member influence and recommendation influence. For each compared method, the recommendation result is obtained by averaging the F1 values under various top-N (i.e. top-20, top-40, top-60, top-80, top-100 and top-120) recommendations. In Fig. 3, R-H performs better than S-H and M-H, And S-H performs better than M-H. For the hybrid methods with one influence, their performance are R-H>S-H>M-H. That is, using the review influence in making recommendations leads better performance than other influence. H method also performs better than other methods which combines one or two influence. Generally, GI-H still has the best performance than other compared methods. Therefore, the proposed hybrid method can effectively improve the recommendation quality.

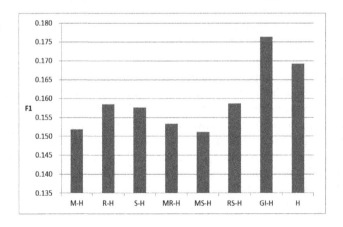

Fig. 3. Comparison of different variations of the proposed methods

5 Conclusions

Generally, traditional group recommendation systems did not analyze the group influence based on group members' behavior. In our work, we proposed a novel recommendation method, which combines group influence, rating-based scores and profile similarity, for recommending books to groups. The group influence consists of group member influence, review influence and recommendation influence. These influences are measured by analyzing users' reading behavior in a group. Then, the rating-based scores and profile similarity are integrated into a hybrid recommendation model to make predictions for groups. The rating-based score represents users' and their groups' preference to books, while the profile similarity represents the preference of content features among groups and books. The experimental results show that the proposed recommendation method performs better than the methods with considering one or two influence. It also outperforms both the method without influences and traditional user-based CF. Therefore, analyzing the group influence and content features in a group recommendation method can indeed enhance the recommendation performance.

Acknowledgements. This research was supported by the Ministry of Science and Technology of Taiwan under the grant MOST 104-2410-H-033-027.

References

1. Salton, G., Buckley, C.: Term-weighting approaches in automatic text retrieval. Inf. Process. Manag. **24**, 513–523 (1988)
2. Blei, D.M., Ng, A.Y., Jordan, M.I.: Latent dirichlet allocation. J. Mach. Learn. Res. **3**, 993–1022 (2003)
3. Weng, J., Lim, E.P., Jiang, J., He, Q.: Twitterrank: finding topic-sensitive influential twitterers. In: Proceedings of the Third ACM International Conference on Web Search and Data Mining, pp. 261–270. ACM, New York (2010)
4. Xu, K., Guo, X., Li, J., Lau, R.Y.K., Liao, S.S.Y.: Discovering target groups in social networking sites: An effective method for maximizing joint influential power. Electron. Commer. Res. Appl. **11**, 318–334 (2012)
5. Choonsung, S., Woontack, W.: Socially aware TV program recommender for multiple viewers. IEEE Trans. Consum. Electron. **55**, 927–932 (2009)
6. Kim, J.K., Kim, H.K., Oh, H.Y., Ryu, Y.U.: A group recommendation system for online communities. Int. J. Inf. Manag. **30**, 212–219 (2010)
7. Masthoff, J.: Group adaptation and group modelling. In: Virvou, M., Jain, L. (eds.) Intelligent Interactive Systems in Knowledge-Based Environments, vol. 104, pp. 157–173. Springer, Berlin Heidelberg (2008)
8. O'Connor, M., Cosley, D., Konstan, J.A., Riedl, J.: PolyLens: a recommender system for groups of users. In: Proceedings of the Seventh Conference on European Conference on Computer Supported Cooperative Work, pp. 199–218. Kluwer Academic Publishers, Bonn (2001)
9. Sarwar, B., Karypis, G., Konstan, J., Riedl, J.: Analysis of recommendation algorithms for e-commerce. In: Proceedings of the 2nd ACM Conference on Electronic Commerce, pp. 158–167. ACM, Minneapolis, Minnesota, United States (2000)

On Rumor Source Detection and Its Experimental Verification on Twitter

Dariusz Król[1]([✉]) and Karolina Wiśniewska[2]

[1] Department of Information Systems,
Wrocław University of Science and Technology, Wroclaw, Poland
Dariusz.Krol@pwr.edu.pl
[2] Faculty of Computer Science and Management,
Wrocław University of Science and Technology, Wroclaw, Poland

Abstract. This paper analysis the rumor source detection on three Twitter networks of different sizes: 1K, 10K and 100K tweets. At first step, an algorithm was designed, that selects from all users a set of potential rumormongers, who initiated the fake content tweet. The next step was based on tracking of propagation trails by (1) randomly distributed, (2) maximum, (3) minimum, and (4) median weight of node in the retweet trees. Given these postulates, the study describes an empirical investigation of finding the position of the rumor-teller, calculating the length of propagation path and using statistical methods to interpret and then report basic results. The results showed that we are not able to separate the initial rumor users from the most influential spreaders in the small networks. However, in the big network - 100K - those classifications are expected to bring a satisfactory result.

Keywords: Information diffusion · Spreading phenomenon · Cascading behaviour · Rumor propagation · Real-world social networks

1 Introduction

Rumors are part of our everyday life, and its spread has a significant impact on human lives. Propagation of the rumor starts when some people spread one information to their acquaintances whose follow their predecessors [7]. The rapid growth of online social media has made it possible for rumors to spread in a much more efficient manner than before. Therefore, rumor detection become one of the critical research topics of information credibility and trust evaluation on social media.

Twitter has gained a reputation over the years as a prominent news source, often disseminating information in a very short time, often not matched by traditional media. Detecting credible or trustworthy information in the network, especially during important events, can be very valuable and may have practical applications for news consumers, financial markets, journalists, and emergency services, and more generally to help minimize the impact of mis- and

© Springer International Publishing AG 2017
N.T. Nguyen et al. (Eds.): ACIIDS 2017, Part I, LNAI 10191, pp. 110–119, 2017.
DOI: 10.1007/978-3-319-54472-4_11

false information. This is exacerbated by the fact that (i) anyone can publish (incorrect) information and (ii) it is hard to tell who the original source of the information is [16].

Due to the dynamic nature of Twitter, fake news or rumors spread quickly. Hence, the evaluation of tweets must be done also very fast to hinder the propagation of non-credible content.

On Twitter, a rumor is a collection of tweets, all asserting the same or slightly changed unverified statement, propagating through the network, in a multitude of cascades. The maximum length of the single tweet can be 140 characters. Each post is characterized by two main components: the tweet and the user (source) who posted the tweet. The process takes place along the links between ignorants, spreaders and stiflers like in SIR model [3]. The ignorants, similar to *Susceptible*, have never heard about the rumors. The spreaders, similar to *Infective*, are sending the rumor to others. The stiflers, similar to *Removed*, who knows the rumor, do not spread it [19].

As tweets are retweeted, it can be difficult to identify who is being addressed and cited. Differences in approach can lead to inaccuracies as messages are retweeted. When one interprets attribution as citation, credit may go to the wrong person [1]. Although rumors are usually initiated by a small number of people [9], but still, it is hard to tell who is the original source of the information. The main question we are studying in this paper is as follows: if a rumor is initiated by a single source in a social network, how the actual source can be identified, what is the converse of this presented in [11]. The aim of this research work to build the solution for mitigating misinformation spread was two-fold: (1) to develop an algorithm, that selects the top suspect spreaders, who initiated the rumor tweets, and (2) to evaluate the use of automated methods for tracking its very probable propagation trails.

The rest of this paper is organized as follows. Section 2 review related work for finding rumor sources. Section 3 discusses the simulations and results. Section 5 concludes with ongoing and future work and contributions.

2   Using a Methodology for Finding Rumor Source

In this section, we discuss how we can today investigate rumor propagation, in particular finding rumor source in social network like Twitter.

A number of methods and tools trace rumor spreading in the system. For example, Shah et al. studied the problem of rumor source finding [17,18]. They model rumor spreading with a variant of an SIR model and define rumor centrality as a source estimator, to evaluate the likelihood that each node is the actual rumor source. The major events in which misinformation or rumors were studied on Twitter include: the 2010 earthquake in Chile, Hurricane Sandy in 2012 [5] and the Boston Marathon blasts in 2013 [8]. Mendoza et al. [12] studied tweets about 2010 earthquake in Chile and found that rumors and non-rumors are retweeted (commented) in a different way. Castillo et al. [2] used a machine leaning technique that makes use of text in tweets, user characteristics and tweet propagation pattern to classify rumors and non-rumors. Ratkiewicz et al. [15] developed a tool that visualizes tweet propagation which can be used to detect abusive behaviors. Gupta et al. [6] focused on the credibility of events instead of individual tweets by an event graph-based optimization.

More recently, specific web-based systems have been proposed to analyse rumors. These include: (1) *TwitterTrails.com* [13], a tool, that allows users to investigate the propagation characteristics of a rumor and its refutation, if any, (2) *Snopes.com* [14], a popular website documenting memes and urban legends, (3) *TweedCred* [4], a real-time system to assess credibility of content on Twitter, and (4) *Emergent.info* [10], a real-time rumor tracker that focuses on emerging stories on the web and examines their truthfulness. The fact-checking capabilities of these systems that confirm or debunk the veracity of latest rumors on the Internet range from completely automatic to semi-automatic. However, they do not monitor the spreading process and do not identify all potential source(s).

In the next sections we use many different concepts which need explanation. For example, the rumor, the potential rumor source, the spreader.

3   Simulation Analysis and Results

Automatic detection and verification of rumors in Twitter are very difficult tasks. Therefore, we focused on dividing the dataset on groups, which could help as to separate the group of potential rumor-tellers and spreaders.

We performed our experiments on a snapshot of the Twitter network that was crawled in January 2009. It consists of 100,000 records and 15,808,402 dispatched messages. More details are given in Table 1.

Figures 1, 2, 3, and 4 depict the results of our simulation study. In addition to visually depicting the trend in the data with a regression line, you can also calculate the equation of the regression line. This equation can either be seen in a dialogue box and/or shown on your graph. How well this equation describes the data (the 'fit'), is expressed as a correlation coefficient, R^2 (R-squared). The closer R^2 is to 1.00, the better the fit. This too can be calculated and displayed in the graph.

Table 1. Summary statistics for the studied datasets

	Network 1	Network 2	Network 3
Number of tweets (PR)	1000	10000	100000
Number of potential rumor sources (PB)	245	2646	16590
Number of stiflers (S)	617	4036	17073
Max number of retweets (MxSM)	42	344	15764
Min number of retweets (MnSM)	1	1	1
Median number of retweets (MedSM)	1	1	2

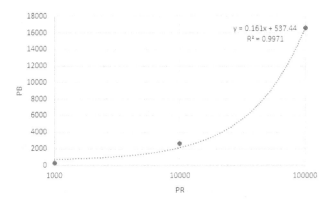

(a) Relationship between the number of tweets (PR) and the number of potential rumor sources (PB).

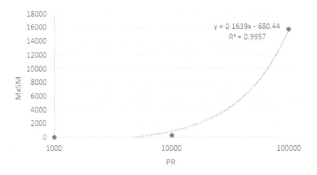

(b) Relationship between the number of tweets (PR) and the maximum number of retweets (MxSM).

Fig. 1. The $MxSM$–PR–PB relationship for random propagation trail.

In this section, we apply our two phase identification method in a social network to find the initial spreader (source) of a rumor. To apply our algorithm to a real social network, we extracted a graph from Twitter. We conduct extensive experiments on three real-world data sets ranging from small to large scale: 1K, 10K and 100K tweets. The experiments are run on a Microsoft Windows machine with a 3.2 GHz quad-core Intel Core i7 and 16 GB memory.

4   Discussion

In this paper, we proposed an approach for finding the source of the rumor. The algorithm shows good potential to help users in identifying rumors and their sources. The empirical results show that our methods can trace the potential sources of misinformation with up to 50%–80% accuracy. Experiments illustrate

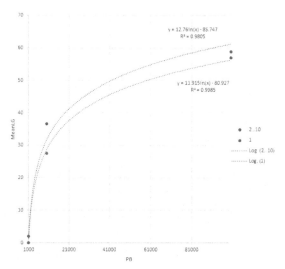

(a) Relationship between mean length of rumor (MeanLG) and the number of potential rumor sources (PB). The fitting of the suggested function varies between 0.98 and 0.99.

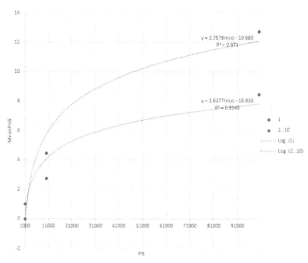

(b) Relationship between mean position of potential rumor sources (MeanPoB) and the number of potential rumor sources (PB). The fitting of the first suggested function is hardly moderate.

Fig. 2. The *MeanLG–PB–MeanPoB* relationship for random propagation trail.

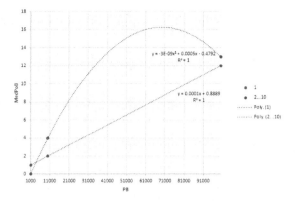

(a) Median position of potential rumor sources (PB).

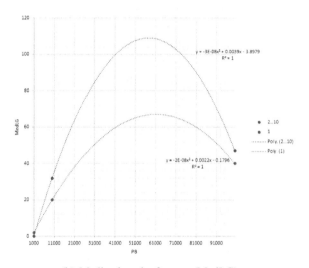

(b) Median length of rumor (MedLG).

Fig. 3. The *MeanPoB–PB–MeanLG* relationship for random propagation trail.

that our rumor detection method obtain significant improvement, compared with the state-of-the-art approaches. The bigger network the better we detach the rumor-tellers from other users.

The initial solution of finding the potential rumor source and separate them from spreaders has not produced satisfactory results. When we analysed first results, we see that the number of potential rumors rise in the linear way with the good fitting rate circa 0.99. The same situation meets the figure with the maximum number of send message in the function of network size and also has a good fitting. The other situation presents the figure with variance of sent messages in the function of network size. The fitting of this plot has the shape of second-degree polynomial. The variance rises faster than other values that

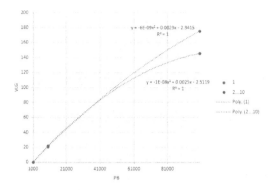

(a) Relationship between variance length of rumor and the number of potential rumor sources (PB).

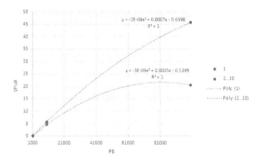

(b) Relationship between variance position of potential rumor sources and the number of potential rumor sources (PB).

Fig. 4. The *VLG–PB–VPoB* relationship for random propagation trail.

means the set of numbers of sent messages change rapidly in comparison with maximum number of sent messages. Comparing the results we could tell that the length of rumor in the function of the number of potential source is a logarithmic function with fitting 0.99–0.98 from two sets of data (1 sent message and 2–10 sent messages). The mean position of rumor-teller has the same profile like the mean length of rumor that means it shows the nature of logarithmic with the fitting 0.97–0.91. The median length of the rumor has a polynomial of the second degree with ideal fitting but with a negative coefficient. The similar situation we could observe in the figures which present the variance of length of the rumor and variance of the position of the source.

We collected the results from finding the propagation trail to four cases: random trail, maximum, minimum, median weight node propagation trail. We observed that in the first network (1 000 records) the smallest number of propagation is included. When the size is small, we cannot find the starts of the rumor propagation what causes the minimal number of investigated potential sources.

In the small network case, we cannot see a result which help us to separate the sources from others spreaders.

When we would like to characterized the result of the biggest set of data we could see that there are many parts of investigated data. We could see a similar situation to that in the case of the first network. The position of the potential rumor sources do not change in the last part of data and has a number between 2–2.33. This could indicate that this collection is most similar, which in turns implies that we are dealing with a set of sources. We could be confident that in this collection, most if not all users are sources. We see that the length of the gossip varies between the number 57.33–59 and the maximum length is for median weight propagation trail. In this case are the greatest variations in the length of the propagation, it can definitely due to the length of the dataset. We could tell that when the dataset size rise the variation of the length of the trail change more.

5   Conclusion

In this work we implemented two algorithms of finding the propagation trail for the rumor. There are algorithms that are inspired by the nature or epidemic spreading. There are harvested measures that are used to describe the spreading of the rumour propagation. In the experiments we did not concentrate on the rumour semantic that means that content of the tweet is not examined. That caused the difficulties of finding the dataset which could be used in the research of finding the source, because in Internet there do not exist the servers with data which are perfectly relevant to use here. The semantic is not investigated because the time for preparing this paper was limited and the time used for the experiments was to long because of big data processing. The first part of the experiment study took very long using two dataset (10 000 records and 100 000 records), the time of finding the potential rumor source took over two weeks and several times the experiment was interrupted. The university server on which the survey was calculated, was restarted many times. The result from the first network, which consist of only 1 000 records, is supposed that when we have small data the result do not change because the number of all investigated trails is small and in this case took 3.18% and it is not a good result. We could say that on the basis of these results, we are not able to determine the course and behaviour of rumours in small datasets. With the second set of data we can probably separate the first set with only one send message and suggest that there are many rumor-tellers as indicated by the median of the set of position of potential busybody. When we analyse the same results we see the minimum weight propagation trail. In the last case, the biggest set of data, we could say that the last one tested partition is only one. Previous statement can confirm by observing that position of potential rumor source has not much changed and take second position in the propagation trail. The confirmation of that statement we could find in the table where is a comparison of spreader and rumor-teller. Based on the results we confirmed that when size of the dataset rises we can find more sources and the variation of length of the propagation trail rises. But if we have a big set of data and investigate more than 100 000 records we can even better separate the group of potential rumor sources. The last and the biggest set of data helps us recognize that in the last part of the dataset (part with [51,100] sent messages) a group of rumor-tellers are masking and it corresponds with the number of sent messages, and the position on the propagation trail in most cases is on the second place.

From the study we know that mean length of rumor and mean position of source in function of dataset size has a logarithmic distribution. We conclude that variance position of rumor source and variance length of the rumor have a polynomial distribution with an exponent of 2. In the future work to improve a division of group of potential sources is to treat each user as a separate element of the set, and in this way to note the differences between them. This solution could take more time but could bring better result of dataset separation. To speed up the processing time of the algorithm we think that we could use methods of statistical distribution of data or data separating methods, for example, clustering, because anyone measure the propagation of the rumour without analysing the semantic and we do not know which of those method could be the best. Perhaps starting with the semantic analyses of the rumour could bring us better results in separating the dataset.

This is preliminary work in a larger research effort on understanding the propagation of rumors through social media. We eventually would like to develop methods for automatically identifying misinformation by detecting the corrections. In the immediate future, we intend to analyze a larger set of rumors related to multiple real-world crisis events. We hope to identify patterns or common types of rumors, possibly using signatures or characteristic patterns of misinformation and corrections over time.

Acknowledgments. This research received financial support from the statutory funds at the Wrocław University of Science and Technology, Poland.

References

1. Boyd, D., Golder, S., Lotan, G.: Tweet, tweet, retweet: conversational aspects of retweeting on twitter. In: Proceedings of the 2010 43rd Hawaii International Conference on System Sciences, HICSS 2010, pp. 1–10. IEEE Computer Society, Washington, DC (2010)
2. Castillo, C., Mendoza, M., Poblete, B.: Information credibility on twitter. In: Proceedings of the 20th International Conference on World Wide Web, WWW 2011, pp. 675–684. ACM, New York (2011)
3. Cheng, J.J., Liu, Y., Shen, B., Yuan, W.G.: An epidemic model of rumor diffusion in online social networks. Eur. Phys. J. B **86**(1), 29 (2013)
4. Gupta, A., Kumaraguru, P., Castillo, C., Meier, P.: TweetCred: real-time credibility assessment of content on twitter. In: Aiello, L.M., McFarland, D. (eds.) SocInfo 2014. LNCS, vol. 8851, pp. 228–243. Springer, Cham (2014). doi:10.1007/978-3-319-13734-6_16

5. Gupta, A., Lamba, H., Kumaraguru, P., Joshi, A.: Faking sandy: characterizing and identifying fake images on twitter during hurricane sandy. In: Proceedings of the 22nd International Conference on World Wide Web, WWW 2013, pp. 729–736. ACM, New York (2013)
6. Gupta, M., Zhao, P., Han, J.: Evaluating event credibility on twitter, pp. 153–164
7. Król, D.: How to measure the information diffusion process in large social networks? In: Nguyen, N.T., Trawiński, B., Kosala, R. (eds.) ACIIDS 2015. LNCS (LNAI), vol. 9011, pp. 66–74. Springer, Cham (2015). doi:10.1007/978-3-319-15702-3_7
8. Lee, J., Agrawal, M., Rao, H.R.: Message diffusion through social network service: the case of rumor and non-rumor related tweets during boston bombing 2013. Inf. Syst. Front. **17**(5), 997–1005 (2015)
9. Liang, G., Yang, J., Xu, C.: Automatic rumors identification on sina weibo. In: 2016 12th International Conference on Natural Computation, Fuzzy Systems and Knowledge Discovery (ICNC-FSKD), pp. 1523–1531, August 2016
10. Liu, X., Nourbakhsh, A., Li, Q., Fang, R., Shah, S.: Real-time rumor debunking on twitter. In: Proceedings of the 24th ACM International on Conference on Information and Knowledge Management, CIKM 2015, pp. 1867–1870. ACM, New York (2015)
11. Luo, Z., Osborne, M., Tang, J., Wang, T.: Who will retweet me?: finding retweeters in twitter. In: Proceedings of the 36th International ACM SIGIR Conference on Research and Development in Information Retrieval, SIGIR 2013, pp. 869–872. ACM, New York (2013)
12. Mendoza, M., Poblete, B., Castillo, C.: Twitter under crisis: can we trust what we RT? In: Proceedings of the First Workshop on Social Media Analytics, SOMA 2010, pp. 71–79. ACM, New York (2010)
13. Metaxas, P.T., Finn, S., Mustafaraj, E.: Using twittertrails.com to investigate rumor propagation. In: Proceedings of the 18th ACM Conference Companion on Computer Supported Cooperative Work & Social Computing, CSCW 2015 Companion, pp. 69–72. ACM, New York (2015)
14. Nourbakhsh, A., Liu, X., Shah, S., Fang, R., Ghassemi, M.M., Li, Q.: Newsworthy rumor events: a case study of twitter. In: 2015 IEEE International Conference on Data Mining Workshop (ICDMW), pp. 27–32, November 2015
15. Ratkiewicz, J., Conover, M., Meiss, M., Gonçalves, B., Patil, S., Flammini, A., Menczer, F.: Truthy: mapping the spread of astroturf in microblog streams. In: Proceedings of the 20th International Conference Companion on World Wide Web, WWW 2011, pp. 249–252. ACM, New York (2011)
16. Seo, E.S.: Failure diagnosis in distributed systems. Ph.D. dissertation, University of Illinois at Urbana-Champaign (2012)
17. Shah, D., Zaman, T.: Rumors in a network: who's the culprit? IEEE Trans. Inf. Theory **57**(8), 5163–5181 (2011)
18. Shah, D., Zaman, T.: Finding rumor sources on random trees. Oper. Res. **64**(3), 736–755 (2016)
19. Zhao, L., Wang, J., Chen, Y., Wang, Q., Cheng, J., Cui, H.: SIHR rumor spreading model in social networks. Phys. A Stat. Mech. Appl. **391**(7), 2444–2453 (2012)

Topic Preference-based Random Walk Approach for Link Prediction in Social Networks

Thiamthep Khamket, Arnon Rungsawang, and Bundit Manaskasemsak$^{(\boxtimes)}$

Massive Information and Knowledge Engineering Laboratory,
Department of Computer Engineering, Faculty of Engineering,
Kasetsart University, Bangkok 10900, Thailand
thiamthep.k@ku.th, {arnon,un}@mikelab.net

Abstract. Link prediction is a challenging problem in complex graph, but has many impacts on various fields, such as detecting missed linkages in general graphs, collaborative filtering in co-authorship networks, and predicting protein-protein interactions in Bioinformatics. In this paper, we present a new link prediction method applied for friend suggestion in social networks. The benefits not only help social players easily find new friends but also enhance their loyalties to the social sites. Unlike existing methods that commonly employ statistical attributes of vertices (e.g., in- and out-degree) and topological structures (e.g., distance and path), we contribute to exploit topical information extracted from users' posted messages. We also introduce a new similarity measure that takes into account users' topic preferences and popularity in topical domains for effective ranking associated friends. Experimental results conducted on a real Twitter data show that the proposed approach outperforms other three state-of-the-art methods in all the cases.

Keywords: Topic preference · Random walk · Personalized PageRank · Link prediction · Social network

1 Introduction

During the past decade, online social media have become an essential part of human life. They virtually connect worldwide players together by some purposes such as communication and information sharing. In scientific modeling, those virtual communities can be visualized as graphs, where a vertex represents an individual, and an edge connects two vertices with some relations. The edge can be either directed (e.g., in case of followership of Twitter) or undirected (e.g., in case of the friendship of Facebook). In general, social graphs are very dynamic. Vertices and edges are created and deleted over time. Understanding characteristics and mechanisms of the evolution of the graphs is a great challenge, and yields many benefits such as helping for resource manipulation and service improvement. What factors and how will they affect any two persons to be associated in the near future? are some specific question instances related to the *link prediction* problem [5,7,14] that we concentrate on in this paper.

© Springer International Publishing AG 2017
N.T. Nguyen et al. (Eds.): ACIIDS 2017, Part I, LNAI 10191, pp. 120–129, 2017.
DOI: 10.1007/978-3-319-54472-4_12

Link prediction is the task of estimating the likelihood of the existence of an unobserved link between two vertices, based on attributes of both vertices, their neighbors, and other observed links. Link prediction has been studied in various kinds of graphs and also applied in various domains. In social networks, a friend recommendation service is an important application of link prediction. It aims to detect hidden relationships between users that have not been established yet or missed during social network evolution. The service not only helps users in finding new friends but also enhances their loyalties to the social sites. Similarly, in co-authorship networks among scientists, link prediction can help to find researchers with compatible expertise and suggest collaborations [17,19]. In Bioinformatics, it has been used in protein-protein interaction prediction [2].

In this paper, we study link prediction for friend recommendation systems in online social networks. Based on Twitter data as a case study, we hypothesize that two Twitter users are more likely to be connected (i.e., one follows the other) if they are interested in the same topics and the followee is also popular in those topical domains. To that end, we first consider users' preferential topics by analyzing their posted tweets. We then adopt the personalized PageRank algorithm [10] to identify topic-specific popular users. Finally, a new prediction metric is presented to suggest top-k potential linkages. Experimental results show that our method performs significantly better than the state-of-the-art ones.

The remainder of this paper is organized as follows. Section 2 mentions studies related to ours. Section 3 details the proposed topic preference-based random walk method. Section 4 reports performance evaluation. Finally, Sect. 5 concludes the paper.

2 Background and Related Work

2.1 Problem Formulation

Based upon the definition in [14], the link prediction problem can be formulated as follows. Given a social network $\mathcal{G}(\mathcal{V}, \mathcal{E})$ where $\mathcal{V}$ represents the set of individuals and an edge $e(u, v) \in \mathcal{E}$ represents some form of interactions—followings on Twitter in our case—between individuals at a particular time t. For time $t_i < t_i'$, we assume that $\mathcal{G}_{[t_i, t_i']}(\mathcal{V}_i, \mathcal{E}_i)$ denotes the subgraph of $\mathcal{G}$ in which edges are restricted within the timestamps between t_i and t_i'. In a common supervised learning process, we may choose a training data $\mathcal{G}_{[t_i, t_i']}(\mathcal{V}_i, \mathcal{E}_i)$ and a test one $\mathcal{G}_{[t_j, t_j']}(\mathcal{V}_j, \mathcal{E}_j)$, where $\mathcal{V}_i = \mathcal{V}_j$ and $t_i' < t_j$. So that, there are all $|\mathcal{V}_j| \cdot (|\mathcal{V}_j| - 1)$ possible edges. The task of link prediction is thus to find out these edges that did not exist in $\mathcal{G}_{[t_i, t_i']}$ but possibly appeared in $\mathcal{G}_{[t_j, t_j']}$, and to rank them with a similarity score s_{xy} in descending order, where $x, y \in \mathcal{V}_j$.

2.2 Existing Methods for Link Prediction

In particular, the problem of link prediction has been investigated based on the assumption that two vertices are more likely to be connected if they are more similar. A vertex similarity can be defined by several manners. The vertex-neighborhood methods, such as *common neighbors* [16], *Jaccard coefficient* [11], *Adamic-Adar* [1], and *preferential attachment* [3], use information about degrees and nearest neighbors of vertices; while the path-ensemble *Katz* method [12] and other methods using random walk, such as *hitting & commute times* [6,8] and *rooted PageRank* [4], require global knowledge of the network topology. Recently, the *local random walk* method [15] has been claimed to use less information with only few-step random walk on the network.

Our approach can be classified as one of the random walk-based methods. However, the key difference is that all those existing methods consider statistical properties of vertices and topological structure from the network only, whereas ours also takes into account the topic model as well as vertex popularity in domains. In addition, for the comparison reason, we then mention to some of them in detail here.

(i) *Common neighbors* (*CN*). For a vertex x, we let $\Gamma(x)$ denote the set of outlinked neighbors of x (i.e., persons whom are followed by x, or x's followees). Intuitively, two vertices x and y are more likely to be connected if they have many common neighbors. The simplest measure of this neighborhood overlap is the directed count:

$$s_{xy}^{CN} = |\Gamma(x) \cap \Gamma(y)|. \tag{1}$$

(ii) *Random walk with restart* (*RWR*). This method is a directed application of the personalized PageRank algorithm [4,10] (i.e., rooted PageRank). Consider a random walker starting from a vertex x, who will iteratively move to a random neighbor with probability ϵ and then return to x with probability $1 - \epsilon$. We let $\boldsymbol{r}_x$ be the stationary distribution of landing starting from x, so that r_{xy} is the probability that the walker locates at vertex y in the steady state. Thus, we have

$$\boldsymbol{r}_x = \epsilon P^T \boldsymbol{r}_x + (1 - \epsilon)\boldsymbol{e}_x, \tag{2}$$

where P is the transition probability matrix with $p_{xy} = \frac{1}{|\Gamma(x)|}$ if x links to y; otherwise, $p_{xy} = 0$. Suppose $|\mathcal{V}| = n$, the $\boldsymbol{e}_x$ is an n-dimensional column vector with the xth element equal to 1 and all others equal to 0. This equation can be solved by the power iteration method [9].

Since in our case the network is a directed graph, the similarity score of the edge $e(x, y)$ is thus defined as:

$$s_{xy}^{RWR} = r_{xy}. \tag{3}$$

(iii) *Local random walk* (*LRW*). Similar to *RWR*, given a random walker who starts from a vertex x. We let $r_{xy}^{(t)}$ be the probability that the walker locates at vertex y after t steps. Then, we have

$$\boldsymbol{r}_x^{(t)} = P^T \boldsymbol{r}_x^{(t-1)}, \tag{4}$$

where $\boldsymbol{r}_x^{(0)}$ is an initialized n-dimensional column vector with the xth element equal to 1 and remainders equal to 0.

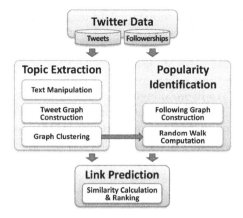

Fig. 1. The workflow of topic preference-based random walk for link prediction

For the directed graph, we adapt the traditional LRW and thus define the similarity score of the edge $e(x, y)$ as:

$$s_{xy}^{LRW} = \frac{\Gamma(x)}{m} \cdot r_{xy}^{(t)}, \tag{5}$$

where m is the number of edges in the network.

3 Topic Preference-based Random Walk

This section presents in detail our proposed Topic Preference-based Random Walk ($TPRW$) approach for link prediction, using Twitter data as a case study. The data were downloaded and stored into two buckets separately: users' posted tweets and followerships. The system workflow, depicted in Fig. 1, is composed of three main modules. First, the *topic extraction* aims to identify individuals' preferential topics by analyzing their posted tweets. Second, the *popularity identification* employs the random walk computation to find out persons who are popular in specific topics. Last, the *link prediction* incorporates both topic preference and popularity to calculate similarity scores of all possible nonexistent links, and ranks them.

3.1 Topic Extraction

In order to effectively recommend friends with similar interests, our effort in this section is to extract what topics an individual Twitter user is interested in, based on his/her posted tweets. To that end, we introduce here three processing steps as follows.

Text Manipulation. We aim to manipulate textual contents of users' posted tweets and construct their term-vector representations. Since tweets in fact are short messages (i.e., 140-character limit), understanding their semantics is a great challenge. To avoid noise of meaningless and polluted messages, we select tweets having a label *retweet* only since a tweet with high retweet counts is likely to have semantics (i.e., to be meaningful) in itself and also has more impacts on readers [13,21]. In addition, since Twitter provides the *hashtag* service to allow users to easily follow topics they are interested in, we then also restrict to tweets having at least one hashtag. Afterwards, the textual contents of those chosen tweets need to be processed by the following four basic tasks as suggested in most information retrieval (IR) systems.

Word segmentation: This task converts stream of characters to stream of words. Hence, we simply tokenize terms by using whitespace characters as a delimiter.

Stopword removal: Stopwords are words that tend to have less meaning or low discrimination power. In natural language, they usually refer to article, preposition, conjunction, etc. Removing stopwords can help the system not only reduce noise but also potentially organize computational resources.

URL removal: Since Twitter allows users to write a tweet with limited characters, for a long URL of an embedded link, it needs to be shorten before posting. So that, it is usually meaningless and should be removed.

Stemming and lemmatization: In this task, we group together the different inflected forms of a word as a single unit, in order to reduce the number of terms to be further processed.

Finally, we represent each tweet as a vector of n distinct terms, which each element is assigned with the *tf-idf* value [18].

Tweet Graph Construction. Since our objective is to group together tweets that are semantically addressed the same topic, we thus need to know how they are related to (or similar in our case). Therefore, in this step we first define relationships of tweets, called a *tweet graph*, and further describe how to group them in the next step. The graph represents tweets as a set of vertices. Any two vertices are symmetrically connected with a weight defined by their similarity value.

Let $G(V, E, W)$ be a weighted undirected graph representing tweet relations, and $\boldsymbol{\tau}_i \in V$ be an n-dimensional vector $\{\tau_{i1}, \tau_{i2}, \ldots, \tau_{in}\}$ representing the ith tweet. Then, a weight $w_{ij} \in W$ assigned to the edge $e(\boldsymbol{\tau}_i, \boldsymbol{\tau}_j) \in E$ is defined as:

$$w_{ij} = sim(\boldsymbol{\tau}_i, \boldsymbol{\tau}_j) = \frac{\sum_{k=1}^{n}(\tau_{ik} \times \tau_{jk})}{\sqrt{\sum_{k=1}^{n} \tau_{ik}^2} \times \sqrt{\sum_{k=1}^{n} \tau_{jk}^2}}. \tag{6}$$

That is, the weight refers to the cosine similarity used in classical IR system [18].

Graph Clustering. Given a tweet graph G, we attempt to group related tweets together. As in most IR systems, words contained in such a document (i.e., tweet) are assumed to semantically describe itself. A topical cluster is thus defined as a group of tweets that share highly common words.

Fortunately, there exist several graph algorithms, such as graph clustering and community detection, that can help to accomplish this task. We therefore employ the Markov clustering (MCL) [20] since it divides the graph without requiring the number of clusters as an input parameter. The algorithm assumes that there exist clusters in a graph, and takes a random walk approach to proceed clustering. That is, a random walk through the graph will result in longer time spent walking within a cluster, and less time spent traveling along edges joining two different clusters. Thus, MCL uses such intuition and groups vertices whose random walker stops at the same vertex.

We now inspect such a topic and estimate how much probability an individual is interested in. For each topical cluster c_i obtained from MCL, given a Twitter user x, we let $\mathcal{P}(c_i|x)$ be the probability that x is interested in c_i, defined as:

$$\mathcal{P}(c_i|x) = \frac{\eta_{c_i \cap x}}{\eta_x}, \tag{7}$$

where $\eta_{c_i \cap x}$ is the number of x's posted tweets that are clustered in c_i, and η_x is the total number of x's tweets. In other words, the equation denotes the proportion of x's tweets only addressed to a topic c_i to the entire ones.

3.2   Popularity Identification

Motivated by a random walk approach to estimate popularity of vertices on a graph, in this section we adopt the personalized PageRank algorithm [10] to identify popular Twitter users for a given topic. We design two processing steps detailed as follows.

Following Graph Construction. We construct a social graph $\mathcal{G}(\mathcal{V}, \mathcal{E})$ represented Twitter users with their followerships. An edge $e(u, v) \in \mathcal{E}$ is defined if the user $u \in \mathcal{V}$ has followed the user $v \in \mathcal{V}$.

Random Walk Computation. Based upon the definition of personalized PageRank, the stationary distribution of a random walk is calculated as follows. At each step, with probability ϵ a random walker follows a randomly chosen outgoing edge from the current vertex, and with a certain probability $1 - \epsilon$ jumps to another vertex non-uniformly. In this study, all the jumps are made in accordance with users' topic preferences. Let $r_{c_i}(y)$ be the probability that the walker locates at vertex y in the steady state for a given topic c_i. Thus, we have

$$\boldsymbol{r}_{c_i} = \epsilon P^T \boldsymbol{r}_{c_i} + (1 - \epsilon) \boldsymbol{e}_{c_i}, \tag{8}$$

where P is the transition probability matrix with $p_{xy} = \frac{1}{|\Gamma(x)|}$ if x follows y and $\Gamma(x)$ is a set of x's outlinked neighbors; otherwise, $p_{xy} = 0$. The $\boldsymbol{e}_{c_i}$ is a non-uniform vector, which each xth element is defined as the probability that the user x is interested in the topic c_i:

$$e_{c_i}(x) = \mathcal{P}(c_i|x). \tag{9}$$

3.3   Link Prediction

In the last step, the proposed *TPRW* defines a new similarity score of a predicted nonexistent edge $e(x, y)$ as the aggregation of x's preference and y's popularity of all topical clusters:

$$s_{xy}^{TPRW} = \sum_{i=1}^{|\mathcal{C}|} \mathcal{P}(c_i|x) \cdot r_{c_i}(y), \qquad (10)$$

where $|\mathcal{C}|$ is the total number of topical clusters outputted from MCL. A list of friends suggested to a user x is thus provided from sorting those scores of all non-followees y in descending order.

4   Performance Evaluation

4.1   Dataset and Experimental Setup

We conducted experiments on English-written Twitter data downloaded via the Twitter REST APIs.[1] The data were collected during two periods: from June 1st to 7th, 2016 for the training process; and from July 8th to 15th, 2016 for the test one. Within both periods, we consider only 97,235 users in common, and 29,092,546 and 33,506,897 following linkages of the former and the latter, respectively (i.e., on average increasing 45.4 followings per user). Since our *TPRW* approach takes into account users' topics, we then also downloaded their tweets during the first periods for the analysis. After the steps of text manipulation and graph clustering, as described in Sect. 3.1, we have average 130 tweets per user and 35 topical clusters, respectively.

We compared the performance in predicting potential links/friends with three state-of-the-art methods: *CN*, *RWR*, and *LRW*. For the parameter setting of the training process, we assigned the parameter ϵ to 0.85 [4] (i.e., in Eqs. (2) and (8)) of both *RWR* and *TPRW*. For *LRW*, we set t to 2 iteration runs (i.e., in Eq. (4)) since we focus on predicting friends within 2 hops away from a given user during the testing process of all experiments.

We use the standard *precision* metric to indicate how many predicted links of a method are correct. Consider a set of source users $\mathcal{V}$. For each user x, we aim to predict his/her top-k friends. We let $\mathcal{L}(x)$ be the set of the first k predictions based on the training data, and let $\Gamma(x)$ be the set of correct answers given from the test data. Thus, the average precision at k is calculated by:

$$precision = \frac{1}{|\mathcal{V}|} \sum_{x \in \mathcal{V}} \frac{|\mathcal{L}(x) \cap \Gamma(x)|}{k}. \qquad (11)$$

---

[1] https://dev.twitter.com/rest/public.

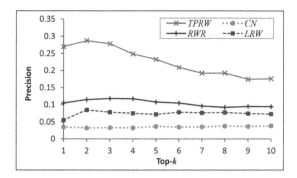

Fig. 2. Top-k comparison under average precision

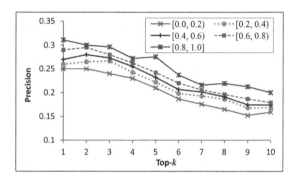

Fig. 3. Impact of degrees of users' topic preference in *TPRW*

4.2   Results and Discussions

To compare performances of the methods, we randomly select 200 source users as a test set. Figure 2 shows the average precisions examined for the first 10 predicted links. As it can be seen in the figure, all the methods yield not so high average precisions. The reason is that, as in fact, the characteristic of real-world dataset is extremely *imbalance*: the number of edges known to be present is often significantly less than the number of edges known to be absent, as described in Sect. 2.1. However, our proposed *TPRW* significantly outperforms the other three state-of-the-art methods in all the cases, indicating that the users' topic preference can contribute to an effectiveness of the link prediction. Moreover, *TPRW* can also suggest friends more correctly in higher orders of the ranking.

   We further investigate how topics have an impact on the prediction. Consider a topic given by MCL as described in Sect. 3.1, we first classify the entire users into five groups based on their degree of preference in that topic (i.e., Eq. (7)): in ranges $[0, 0.2)$, $[0.2, 0.4)$, $[0.4, 0.6)$, $[0.6, 0.8)$, and $[0.8, 1]$. Then, for each range, we randomly select 100 source users in total from all topics as a test set. Figure 3 shows the average precisions contributed from different degrees of users' topic preference. As it can be seen, the higher degree is, the more accurate of prediction we get. The finding also supports that the users' topic preference indeed affects on the prediction.

5 Conclusion

In this paper, we concentrate on the problem of link prediction for social networks and investigate how content topics have an impact on friend suggestion. We compare our *TPRW* which takes into account the users' topic preferences with other three state-of-the-art methods which only employ statistical properties. Experimental results reveal that topical information has indeed affected the prediction performance. In other words, any two users with more similar preferential topics tend to highly become friends in the near future.

For future work, we anticipate extending our experimental study to other social data such as Facebook. Moreover, there still exist several factors such as hot trend, temporal aspect, etc., that we are also interested to explore.

References

1. Adamic, L.A., Adar, E.: Friends and neighbors on the web. Soc. Netw. **25**(3), 211–230 (2003)
2. Airoldi, E.M., Blei, D.M., Fienberg, S.E., Xing, E.P., Jaakkola, T.: Mixed membership stochastic block models for relational data with application to protein-protein interactions. In: Proceedings of the International Biometrics Society Annual Meeting (2006)
3. Barabási, A.L., Albert, R.: Emergence of scaling in random networks. Science **286**(5439), 509–512 (1999)
4. Brin, S., Page, L.: The anatomy of a large-scale hypertextual web search engine. Comput. Netw. ISDN Syst. **30**(1–7), 107–117 (1998)
5. Clauset, A., Moore, C., Newman, M.E.J.: Hierarchical structure and the prediction of missing links in networks. Nature **453**, 98–101 (2008)
6. Fouss, F., Pirotte, A., Renders, J.M., Saerens, M.: Random-walk computation of similarities between nodes of a graph with application to collaborative recommendation. IEEE Trans. Knowl. Data Eng. **19**(3), 355–369 (2007)
7. Getoor, L., Diehl, C.P.: Link mining: a survey. ACM SIGKDD Explor. Newslett. **7**(2), 3–12 (2005)
8. Göbel, F., Jagers, A.A.: Random walks on graphs. Stochast. Process. Appl. **2**(4), 311–336 (2005)
9. Golub, G.H., Van Loan, C.F.: Matrix Computations. Johns Hopkins University Press, Baltimore, London (1996)
10. Haveliwala, T.H.: Topic-sensitive pagerank: a context-sensitive ranking algorithm for web search. IEEE Trans. Knowl. Data Eng. **15**(4), 784–796 (2003)
11. Jaccard, P.: Étude comparative de la distribution florale dans une portion des alpes et du jura. Bulletin de la Société Vaudoise des Sciences Naturelles **37**(142), 547–579 (1901)
12. Katz, L.: A new status index derived from sociometric analysis. Psychometrika **18**(1), 39–43 (1953)

13. Kwak, H., Lee, C., Park, H., Moon, S.: What is Twitter, a social network or a news media? In: Proceedings of the 19th International Conference on World Wide Web, pp. 591–600 (2010)
14. Liben-Nowell, D., Kleinberg, J.: The link-prediction problem for social networks. J. Am. Soc. Inf. Sci. Technol. **58**(7), 1019–1031 (2007)
15. Liu, W., Lü, L.: Link prediction based on local random walk. EPL (Europhys. Lett.) **89**(5), 58007 (2010)
16. Lorrain, F., White, H.C.: Structural equivalence of individuals in social networks. J. Math. Sociol. **1**(1), 49–80 (1971)
17. Pavlov, M., Ichise, R.: Finding experts by link prediction in co-authorship networks. In: Proceedings of the 2nd International Conference on Finding Experts on the Web with Semantics, pp. 42–55 (2007)
18. Salton, G., Buckley, C.: Term-weighting approaches in automatic text retrieval. Inform. Process. Manage. Int. J. **24**(5), 513–523 (1988)
19. Sun, Y., Barber, R., Gupta, M., Aggarwal, C.C., Han, J.: Co-author relationship prediction in heterogeneous bibliographic networks. In: Proceedings of the International Conference on Advances in Social Networks Analysis and Mining, pp. 121–128 (2011)
20. Van Dongen, S.: Graph clustering via a discrete uncoupling process. SIAM J. Matrix Anal. Appl. **30**(1), 121–141 (2008)
21. Welch, M.J., Schonfeld, U., He, D., Cho, J.: Topical semantics of twitter links. In: Proceedings of the 4th ACM International Conference on Web Search and Data Mining, pp. 327–336 (2011)

Level of Education and Previous Experience in Acquiring ICT/Smart Technologies by the Elderly People

Ivana Simonova[✉] and Petra Poulova

Faculty of Informatics and Management, University of Hradec Králové,
Hradec Králové, Czech Republic
{ivana.simonova,petra.poulova}@uhk.cz

Abstract. The article introduces the results of research conducted within the elderly people of Municipality of Hradec Kralove, Czech Republic. The main research objective is to discover whether (1) the level of education and (2) exploitation of ICT/smart technologies on work positions in the productive age correlate to the education and training in this field in the post-productive age. The method of questionnaire applied to the research sample of 432 respondents to provide answers to five questions and to verify two hypotheses. The collected data were processed by the IBM SPSS Statistics software. The findings, correlating to similar studies conducted in other countries, proved the necessity to develop methodologies on acquiring new technologies by the elderly people in the future.

Keywords: ICT · Smart technology · Information and communication technology · Research · Elderly · Retired · Post-productive period

1 Introduction

Either we accept the fact, or not, the world' population is ageing, and the same finding was discovered in the Czech Republic. Having a more detailed view, differences were detected between male and female population, as displayed in Fig. 1.

This situation is reflected in the research results presented below.

Ageing brings changes to the lifecourse, particularly in the health state, family life and professional career. From the view of last criterion, a child is developing to a mature person who step by step becomes older. In other words, the pre-productive age is steadily marching to the productive period, reaching the substantial change one day – the retirement age. Despite being different in various countries, a person leaves the job and ceases to work because s/he has reached a particular age [2], s/he stops employment completely and is withdrawn from occupation [3].

The retired person is often called the elderly one been defined as being past middle age and approaching old age; [4]. In the Czech Republic the retirement law defining the retirement age is still under the development, currently providing the opportunity to leave from work at the age of 53, under some rather unfavourable financial conditions, or even earlier in selected professions. Behind this point, life seems not to bring new challenges, however, the reverse is true. The loss of some social contacts developed within the

© Springer International Publishing AG 2017
N.T. Nguyen et al. (Eds.): ACIIDS 2017, Part I, LNAI 10191, pp. 130–139, 2017.
DOI: 10.1007/978-3-319-54472-4_13

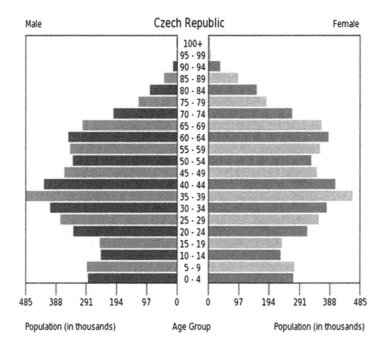

Fig. 1. Age structure of Czech population [1]

professional life, it is highly appreciated the elderly person found/developed new ones so as to face the new social situation. Meanwhile, the situation has been changing, particularly the knowledge and skills necessary for succeeding in the e-society.

Reflecting the above mentioned, the main aim of this article is to introduce the results of research focusing on the elderly people and their attitudes to (1) ICT/smart technologies and devices and (2) towards reaching the required skills.

2   Research Objectives and Questions

The main objective of this research is to discover whether (1) the level of education and (2) previous experience in exploitation of ICT/smart technologies on work positions in the productive age correlate to the further education and training in this field in the post-productive age. Two research questions were defined:

1. Do the elderly people with higher level of professional education acquire new knowledge and skills in exploitation latest ICT/smart technologies through attending IT courses in the post-productive age? In other words, did these people attend IT courses, after they retired?
2. Do the elderly people who worked with ICT/smart technologies during their productive age (and had reached a certain level of knowledge and skills) continue in acquiring new skills in this field in the post-productive age? In other words, did these people attend IT courses, after they retired?

The criterion of IT courses attendance was selected as a form of institutional education, and other criteria as the support from family (children, grandchildren) and friends in acquiring ICT/smart technologies skills were not considered within this research question.

2.1 Hypotheses

Reflecting the research questions, following two hypotheses were defined.

H1: Respondents who reached a higher level of professional education attended IT courses in their post-productive age more often compared to those with lower level of education.

H2: Respondents who worked with ICT/smart technologies in the productive age, attended IT courses in the post-productive age more often compared to those who did not work with ICT/smart technologies at work.

2.2 Methodology and Tool

The data were collected by the method of questionnaire. It contained 12 questions, both describing the research sample (questions 1–3) and respondents' attitudes questions 4–12). Answers to the questions were in the multiple-choice format with one, four or all choices, in Yes/No format or in the open format. Following five questions are under the focus in this article:

1. Did you work with PC in your last work position?
2. What is the highest level of education you reached?
3. What mobile devices do you possess now?
4. What mobile devices are you able to exploit now?
5. How did you learn to exploit the devices and technologies?

The questionnaire was disseminated face-to-face in the printed form and collected immediately after been filled in.

2.3 Research Sample

Questionnaires were distributed to 437 respondents, participants of University of the Third Age courses (U3 V) and IT courses organized by the Municipality of Hradec Kralove (MHK). It means that only those elderly people who are interested in self-education (either in the non-ICT field at U3V, or in the ICT field in IT courses of MHK), were included in the research sample. Reflecting this fact, it means this is a convenient, not representative sample. As such, it is limited to respondents interested in their further education in the post-productive age and dwelling in Hradec Kralove region. The authors have been aware of these limitations. Therefore, the research results are going to be exploited for the purpose of the pilot study in the field of further education of the elderly people in the Czech Republic.

Following the demographic curve of the Czech population for the group of elderly people (i.e. in the post-productive age) [5, 6], the research sample consisted of 432 respondents: 317 female (73%) and 11 male respondents (17%). Most respondents (32%) were born in the period of 1945–49, the oldest ones were 87+ years, i.e. they were 67–71 years old.

From the view of gender, there were 15 respondents 85+ years old: 12 female (F) and 3 male ones (M). Most respondents were females in the 1942–51 age group (188), i.e. 65–74 years old (Fig. 2).

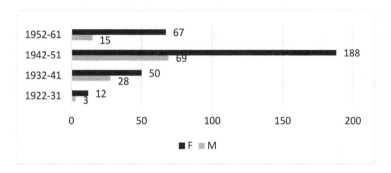

Fig. 2. Research sample: age/gender structure of the whole sample (n)

Level of education reached by the respondents was structured into three groups:

– vocational (i.e. three-year upper level secondary education without school leaving examination, mostly required for working class and crafts professions;
– upper secondary (i.e. four-year education graduating with school leaving examination),
– tertiary (i.e. university, higher) education.

The research sample consisted of 53 respondents who had vocational level of education (13%), 287 respondents reached the upper secondary level of education (66%) and the tertiary level of education was detected with 91 respondents (21%) (Fig. 3).

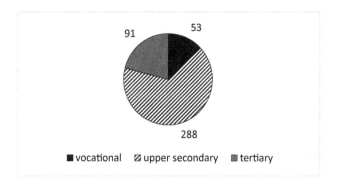

Fig. 3. Research sample: level of education of the whole sample (n)

3   Research Results: Descriptive Statistics

The data collected by the questionnaire were processed by the IBM SPSS Statistics software. The results follow the structure of five questions, as described in Sect. 2.2. Data are displayed in relative frequencies (n).

3.1   Question 1: Did You Work with PC in Your Last Work Position?

Within the total amount of 432 respondents 255 of them (59%) declared they had worked with ICT/smart technologies and devices in their last work position, if they had been available at that time. They were 198 female (78%) and 57 male respondents (22%). Most of them were born in the 1945–49 period (102) and 1950–54 period (77). It was rather surprising that the "youngest" respondents born in 1955–59 period did not belong to the largest group (Fig. 4).

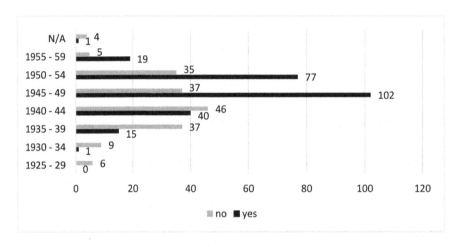

Fig. 4.   Research sample: age and gender structure of ICT/smart technologies users (n)

3.2   Question 2: What Is the Highest Level of Education You Reached?

Within the amount of 255 respondents – ICT/smart technologies users on their last work position called the 'with PC' group in following figures, most of them reached 174, i.e. 68%, upper secondary education. This level is usually required from administrative and middle medical staff (probably female) and middle technical staff in building, electro-industry (probably male respondents, but fact was not detected by the questionnaire). The tertiary level of education was reached by 76 respondents (30%), and only five respondents with the vocational level of education worked with ICT/smart technologies within their professions (Fig. 5).

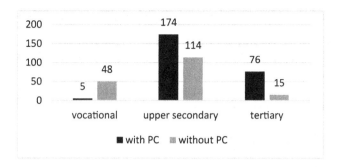

Fig. 5.   Research sample: education structure of ICT/smart technologies users (n)

3.3   Question 3: What Mobile Devices Do You Possess Now?

Most of those respondents who worked with ICT/smart technologies on their last work position use mobile phones (223 respondents, 87%) followed by users of PC (153 respondents, 60%) and notebooks (122 respondents, 48%). Belonging to devices which widely spread in last few years, the possession of smart phones was rather low, only 49 respondents declared the exploitation (19%) (Fig. 6).

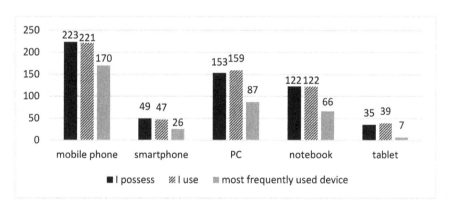

Fig. 6.   Possession, exploitation and the most frequent use of devices within the ICT/smart technologies users (n)

3.4   Question 4: What Mobile Devices Are You able to Exploit Now?

As possessing a device and the ability of its real exploitation are two completely different matters, respondents' answers might differ substantially. However, they do not (Fig. 6), only the number of mobile phone exploiters is two respondents lower, i.e. two of them possess but do not use their mobile phones. The same difference was detected with smartphones. The amount of PC users is even higher that PC owners. The amounts with notebooks are identical.

3.5   Question 5: How Did You Learn to Exploit the Devices and Technologies?

This is the core question of this research. Besides 135 respondents who consider themselves to be the self-learners (53%), 95 were supported by children (37%), 31 by grandchildren (12%) and 90 by friends (35%). Totally, 127 respondents (49.8%) attended IT
courses (Fig. 7). However, they also might be included in the self-learner group but this
criterion was not intentionally distinguished.

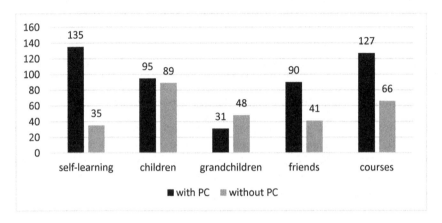

Fig. 7. Support to learning how to exploit devices and technologies within the ICT/smart
technologies users (n)

The IT courses the respondents attended aimed at the first steps in the work with
devices and technologies – Introductory courses, but they were also for advanced users,
those interested in digital photography or communication (e-mailing, skyping, blog
writing etc.). Their length varied from several hours to several months.

4   Research Results: Verification of Hypotheses

Both hypotheses were verified by the Pearson chí-square test of independence of variables. In Table 1, Counts (observed frequencies) and Expected (expected frequencies)
are displayed. The former show how many records were detected with the combinations
of the researched variables (education and IT courses attendance in post-productive age),
the latter reflect the amount of expected records containing this combination, if the variables are statistically independent. If asymptotic significance <0.05, the variables are
dependent.

The statistic results for verification of H1 are displayed in Table 1.

As the asymptotic significance 0.000 is lower than 0.05, this result means the variables are strongly dependent on the significance level $\alpha = 0.05$.

Table 1. Chí-square test for H1: education/attendance of IT courses

	Value	df	Asymp. Sig. (2-sided)
Pearson chi-square	81.053[a]	2	.000
Likelihood ratio	88.606	2	.000
N of valid cases	432		

[a] 0 cells (0,0%) have expected count less than 5. The minimum expected count is 2.53.

The statistic results for verification of H2 are displayed in Table 2.

Table 2. Chí-Square test for H2: work with ICT/smart technologies in the last work positions in productive age/attendance to IT courses in post-productive age

	Value	df	Asymp. Sig. (2-sided)
Pearson chi-square	6.622[a]	1	.010
Continuity correction[b]	6.125	1	.013
N of valid cases	432		

[a] 0 cells (0,0%) have expected count less than 5. The minimum expected count is 79,08.

[b] Computed only for a 2 × 2 table

As the asymptotic significance 0.007 is lower than 0.05, this result means the variables are strongly dependent on the significance level $\alpha = 0.05$.

5   Conclusions, Discussions

It can be clearly seen from the rom the above presented results, the elderly people have shown through the attendance to IT courses that they are interested in further education and development of knowledge and skills they either partially reached during their productive age, and/or their higher level of professional education led them this aim. This research also detected particular activities they were interested in, e.g. e-mailing (73%), skyping (38%), e-banking (40%) and e-shopping (34%), digital photography (38%), writing documents (38%) and reading newspaper (25%) etc. However, information on health was under the focus of one third of them (33%). Compared to the study conducted by Bujnowska-Fedak and Pigorowicz [7], who also studied this field within a wider topic focusing on the needs and preferences of the group of nearly 300 patients of 60+ age group. Despite the results been also limited, they found out the respondents (84%) expressed their favourable attitudes, and even the desire, to receive simple medical recommendations via mobile/smartphones or computers. Moreover, they would also appreciate receiving results of health tests and reminders of scheduled visits or prescribed medications via these devices in the form of short message service or e-mails (60%) and nearly half of the respondents (47%) would request doctor's appointments online.

To reflect new social reality and make such services into effect, knowledge and skills are required. Therefore, permanent education for elderly people was researched by

Modad et al. [8]. They applied both the quantitative and qualitative methods on the sample of 96 persons. Establishing the Human Sciences Faculty, the Course of Digital Literacy for Older Adults was conducted there, working also as a means of social inclusion of the elderly people. The similar topic was under the interest of Agudo et al. [9] who also focused on the relation of the elderly to information and communication technologies. They used the method of discussion groups in the sample of more than 200 elderly people and found out that computers with the Internet access were the most frequently exploited device, and no differences were detected in availability to this device and technology from the view of respondents' age and gender.

Learning and exploitation of new technologies were understood as a demand for integration of the elderly people to society and a factor of active ageing by Gonzales et al. [10]. Compared to Agudo et al., they monitored attitudes of 240 participants to the ICT course and particularly focused on their training in IT skills development. Data collected by the questionnaire showed respondents' appreciation and involvement in new technologies which resulted in their positive attitude and interest in working with them, as well as in lifelong learning. These findings correlate to those introduced in the above presented research.

Other experience in training the elderly in using computers and the Internet were described by Garcia et al. [11]. The focus on the detection and analysis of the most common difficulties which the elderly people face during the training and basic recommendations on how to succeed in this process are provided to them.

As resulting from the selected studies and researches discussed above, these days the Internet and computers are an inevitable part of lives of the elderly people. In some cases special approaches caused by limited health and age conditions should be applied to the process of acquiring the new knowledge and skills in this field. However, the findings clearly prove the interest of the elderly people, motivated both by their desire for socializing (both within the IT courses and consequently by applying the developed skills). Reflecting the demographic curve, which shows the increase in amount of the elderly people in the future decades, the problem of how to learn them to acquire new technologies is going to be highly topical. Not only from this reason, it desires researchers' attention.

Acknowledgements. This work and the contribution were supported by project of Students Grant Agency – FIM, University of Hradec Kralove, Czech Republic (under ID: UHK-FIM-SP-2017). In addition, the authors thank Monika Borkovcova for her help with the survey.

References

1. CIA World Factbook, 02 October 2016. https://www.cia.gov/library/publications/the-world-factbook/geos/ez.html
2. Oxford Living Dictionaries, 02 October 2016. https://en.oxforddictionaries.com/definition/retire
3. American Heritage Dictionary, 02 October 2016. http://www.yourdictionary.com/retirement#americanheritage

4. American Heritage. Dictionary of the English Language. 5th edn. Houghton Mifflin Harcourt Publishing Company (2016)
5. Czech Statistical Office, Seniors, 16 September 2016. https://www.czso.cz/csu/czso/seniori
6. Czech Statistical Office, Population by one-year age and sex, 18 September 2016. https://vdb.czso.cz/vdbvo2/faces/en/index.jsf;jsessionid=HEB1rMEUbIS8yjTGoJjJroN2bnt U67fmRNQ7RcR64jPqYXCMROCx!1897923919?page=vystup-objekt&z=G&f=GRAFI CKY_OBJEKT&pvo=DEMD001&c=v3~2__RP2015MP12DP31&str=v4&rouska=true& clsp=null
7. Bujnowska-Fedak, M.M., Pigorowicz, I.: Support for e-health services among elderly primary care patients. Telemedicine e-Health **20**(8), 696–704 (2014)
8. Modad, R.G.H., Encinas, K.L.P., Arriaga, L.G.L.: Permanent education for older adult. In: EDULEARN15: 7th international conference on education and new learning technologies. In: Gomez Chova, L., Lopez Martinez, A., Candel Torres, I. (eds.) EDULEARN proceedings, pp. 5866–5871 (2015)
9. Agudo, S., Pascula, M.A., Fombona, J.: Uses of digital tools among the elderly. Comunicar **39**, 193–201 (2012)
10. Gonzales, A., Ramirez, M.P., Viadel, V.: Attitudes of the elderly toward information and communication technologies. Educ. Gerontol. **38**(9), 585–594 (2012)
11. Garcia, E., Giret, A., Salido, M.A.: Experiences training the elderly to use computers and internet. INTED **2012**, 928–934 (2012)

The Effect of Presentation in Online Advertising on Perceived Intrusiveness and Annoyance in Different Emotional States

Kaveh Bakhtiyari[1,2(✉)], Jürgen Ziegler[1], and Hafizah Husain[2]

[1] Interactive Systems, Department of Computer and Cognitive Science,
Faculty of Engineering, University of Duisburg-Essen, 47057 Duisburg, Germany
{kaveh.bakhtiyari,juergen.ziegler}@uni-due.de
[2] Department of Electrical, Electronics, and System Engineering, Universiti Kebangsaan
Malaysia (The National University of Malaysia), 43600 Bangi, Selangor Darul Ehsan, Malaysia
hafizahh@ukm.edu.my

Abstract. Online advertising is a rapidly growing area with high commercial relevance. This paper investigates the effect of different types of ad presentation, varying in frame size, position and animation level on visual intrusiveness and annoyance as perceived by users. Furthermore, we investigate the influence of users' emotional states on perceived intrusiveness and annoyance. This research has been carried out through a survey study. The analysis of the data shows a linear correlation between the visual attention of the ads and its features. Also, a positive influence of emotion has been found on various types of ad presentations. In addition, the participants with emotions of positive valence and low arousal showed more tolerance to the same ad as the users with a different emotional state. This research proposes a new aspect in computational advertising to adapt the recommendations based on the user's emotional state and the parameters of the online advertisements.

Keywords: Online advertising · Visual salience · Annoyance · Emotional influence · Computational advertising · Visual intrusiveness

1 Introduction

Online advertising is an important source of revenue for internet companies. There are many forms of online ads; S. Rodgers and Thorson [1] identified few types of online advertisements: banners, interstitials, pop-ups, pop-unders, sponsorships, hypertext links and websites. Among them, banners seem to be more prevalent [2], so that they are making 16% of the revenues in online-businesses, and about 54% on mobile devices. In 2015, Search Engine Marketing (SEM) [3] was the most profitable form of internet advertising. Yet, there are many open questions in this business, such as: What are the dependent factors of profitable ads? How can an effective form of ad be recognized and presented?

There are two aspects on delivering an effective (more profitable) ad. The first aspect is about "content", which is not in the scope of this paper, and the second aspect is about

© Springer International Publishing AG 2017
N.T. Nguyen et al. (Eds.): ACIIDS 2017, Part I, LNAI 10191, pp. 140–149, 2017.
DOI: 10.1007/978-3-319-54472-4_14

"form of presentation" that is our target. Each form of the presentation is different in intrusiveness and visual saliency. Some ads force the users to watch, however, some other ads may not be even recognized.

This research studies the effect of emotion on visual saliency and the user's perception on different types of ad presentations. It aims to deliver a more appropriate form of ad's presentation to the users in order to increase the user's satisfaction and to improve the user's experience in online advertising. To the best of our knowledge, there is no much empirical study about the intrusiveness and its correlation with the emotions on various types of advertisements [4, 5].

At the end, this study would help the computational advertising systems to recommend more tolerance-friendly advertisements to the users to prevent or decrease the annoyance attitudes. This research contributes by answering the following research questions:

- RQ1: How do the different types of ad's presentation attract the visual attention?
- RQ2: What is the extent of perceived annoyance for different types of ad's presentations?
- RQ3: Is there any significant difference between male and female users' perceived attention and annoyance against the ads?
- RQ4: Do the emotions effect the user's attention and perception on online advertisements?

This paper is organized as follows: a short background is provided in Sect. 2. Then the research methodology is elaborated. At the end, the collected data is analyzed and every research question is answered and discussed.

2 Background

Advertisement effectiveness depends on many factors such as the ad's features, and user's cognitive and emotional state. The ad's features such as logo design, content, presentation, color and images affect the brand awareness and branding memory [6, 7]. Also the user's cognitive and emotional state affect the user's perception and tolerance. Based on a well-established literature [8, 9], there is a dual mode process Elaboration Likelihood Model of information processing. This model proposes two processes based on the user's levels of involvement. High involvement situations follow a "central route" processing. It means that they take a cognitive effort to evaluate statements. In this process, nonessential stimuli, such as color and/or sound are "secondary" elements and they are not being considered heavily. On the other hand, in the low-involvement situations, users use "peripheral route" processing. In this process, the user's subconscious is more involved, and affective elements are ruling the process, and they are highly affected through peripheral cues such as music and visual components.

From the marketing perspective, ad content and design can be classified into two groups of cognitive and affective. In cognitive contents, the users are more likely to click on an ad which offers an incentive for action [10], such as promotional offers or discounts. A research in 1999 [11] showed that 66% of users read and look at the banners to find a free offer before they click to another page. In an affective advertisement, the content gains the user's attention by using an emotional appeal [12]. This mode of content expects to attract more attention. In general, most of the consumer level advertisements are more effective and less factual [13].

The role of emotion in marketing and consumer satisfaction has been investigated and evaluated by many researchers. Emotion is an influencing factor in information processing and responding to persuasive appeals. From a marketing point of view, the key is to develop an advertising concept that speaks to the target consumers directly. This can be achieved by taking into account the emotions that the target consumers are likely to be feeling at this point. The visual and textual content of the advertisement should work together to initiate a message in line with the emotional appeal the marketer hopes to make [7]. Also the ad's content itself can have an influence on human emotions, for more details we refer the readers to [14].

Users perceived feelings toward an ad may lead to their judgment about its provider credibility. Based on the literature on online advertisements, more visually intrusive ads are more disturbing [15] and annoying [16]. A website with annoying ads can easily jeopardize the reputation of the service provider and the promoted brand [17]. Therefore, it is very important to deliver an advertisement with a balance of visual saliency and annoyance in order to gain more profit without losing the reputation and the users' attention. Visual saliency is the perceptual quality or state which makes an item stands out from the neighbors and it can attract the visual attention immediately. But the visual saliency and annoyance of a single form of presentation may not be the same all the time. Users may experience different states of visual attention and tolerance on different types of presentations during a course of a time.

3   Methodology

This study was carried out by analyzing and considering the ad's features, user's emotions and their level of visual attention and annoyance for each form of ads throughout a survey. Figure 1 shows the whole concept of this research.

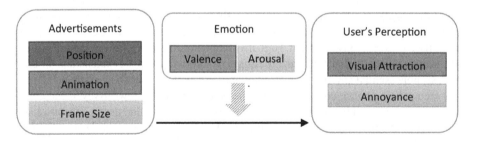

Fig. 1. The research concept and the role of emotion on advertisements' presentation

To investigate the answers of the research questions, various types of presenting online advertisements have been explored. Online ads brokers are offering different specific forms of ads presentation, and most of them are common among many brokers.

These brokers have issued some rules and regulations for the host clients to provide ethical experience of online advertisements for the users. We have judged the ethical and unethical types of advertisements based on the rules published by Google Inc., and unethical types of ads such as pop-up and pop-under windows are ignored [4, 16, 18].

3.1 Advertisements

Twelve forms of common and ethical online advertisements have been selected in this survey that cover the majority of ads' styles. Each form has three main presentation features including position, frame size [3] and animation [19]. The position of an ad can be inline or outline based on the location of the main content of the web page and the advertisement. The animation level can be considered as low or high. The last feature is the frame size, which is either small or big. These features are defined as binomial parameters due to the complexity of the analysis. The different forms of ads and their characteristics are listed in Table 1, and each one is assigned a specific case number. The model of skyscraper is called to those vertical ads which are usually located at the left or right sides of the page.

Table 1. List of 12 types of selected online advertisements

Type	In/Out-line	Position	Animation	Size	Case
Horizontal	Outline	Top – Bottom	Low	Small	1
	Outline	Top	Low	Big	5
	Outline	Bottom (fixed)	Low	Small	6
Vertical (Skyscraper)	Outline	Right – Left	Low	Big	3
	Outline	Right – Left	Med. (fly-in)	Big	8
	Outline	Right – Left	High (fly-in)	Big	9
Square	Outline	Right – Left	Low	Big	2
	Inline	Middle of text	Low	Small	4
	Inline	Middle of text	High – Muted	Small	10
	Inline	Middle of text	High – Sound	Small	12
Half/Full	Outline	Middle/Center	Low	Half	7
Screen	Inline	Interstitial	Low/Medium	Full	11

In this survey, 45 banners were selected with a variety of animation levels and sizes. These are selected from 16 different categories (e.g. computer, shopping, charity, etc.) targeting mostly both females and males. These cases are presented in a random order for each participant.

3.2 Emotions

To capture the user's current emotional state, the participant is asked to report his/her own emotion and its strength at two stages: Firstly, at the beginning, and secondly, at the end of the questionnaire. Self-reporting at two stages helps us to detect inaccuracies and anomalies in self-reported emotions. This survey and its contents are designed to be emotionally neutral in order to prevent any emotional moderation. Therefore, if the first and the last reported emotions are statistically and psychologically different, the reported record is considered as unreliable and removed from further analysis.

Emotion reporting is presented based on the Pick-A-Mood (PAM) model on 9 emotions [20]. These emotions are also presented in a mannequin-like style introduced in the PAM model with separately designed icons for male and female for each emotion. PAM model eases the selection of user's current emotional state [20]. PAM has clustered eight emotions (except Neutral) from the two dimensions of Valence – Arousal [20]. The valence dimension represents an emotion, either positive (attractiveness) or negative (aversiveness). The arousal dimension defines the energy of an emotion. The neutral emotion is separately categorized as EG0. These clustering and the assigned abbreviations are shown in Table 2.

Table 2. Emotional clustering measured by PAM

	Pleasant	Unpleasant
Energized	*EG1*	*EG4*
	Excited – Lively	Tense – Nervous
	Cheerful – Happy	Irritated – Annoyed
Calm	*EG2*	*EG3*
	Calm – Serene	Sad – Gloomy
	Relaxed – Carefree	Bored – Weary

To report the strength of the selected emotion, a scale of 1 to 5 is provided to be chosen [21, 22]. The participant has to choose a pair of emotion and its strength to report how strong they feel the selected emotion.

3.3   Questions

After selecting the emotions, the 12 cases were being presented in a random order. Each case was followed by 4 questions to cover RQ1 and RQ2. The two questions of VI1 and VI2 were presented to cover RQ1 about visual saliency (attraction), and the questions of AN1 and AN2 were presented to cover RQ2 about the user's perception and annoyance on each form of advertisement. These four questions are provided below:

- (VI1) How much does the presented advertisement attract your attention?
- (AN1) How much does the presented type of advertisement annoy you?
- (VI2) How much does the presented type of advertisement distract your concentration (focus) on the main content?
- (AN2) If this type of advertisement is being shown frequently, how much does it make a **negative** perception/attitude towards the product/service/brand?

For each question, the participant has to provide an answer in scale of 1 to 5 interpreting 1: Very low; 3: Medium; and 5: Very high.

4   Results

The survey was carried out online by 180 participants (90 Males – 90 Females) from various countries to ensure comprehensiveness of culture as it may effects on emotions [23]. After the initial analysis, 20 records (8 Males – 12 Females) were considered as outlier, and were removed due to the significantly different emotional states reported at the beginning and the end of the survey. The participants were from different geographical regions with mostly 48% from Europe and 23% from the Asia and Australia.

The reported emotions of Tense, Irritated and Sad, which are all negative emotions have the lowest number of entries in the collected data. We left EG4 out from analysis, because they were only few cases (only 5 records). The other emotional groups of EG0 to EG3 have 27, 31, 72 and 25 participants respectively.

The overall average of the answers given to the four questions on 12 cases are shown in Fig. 2. This figure shows that the cases 1 and 12 are respectively the worst and the best in user's visual attention; and the cases 1 and 5 has the least and the cases 11, 12 and 10 have the most annoyance, in the same order.

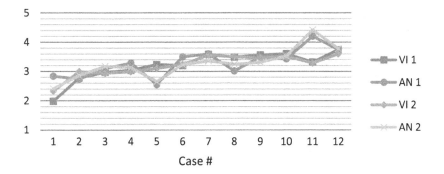

Fig. 2. The overall average answers of the four questions

As it was planned, the similar questions (Questions 1, 3: VI1, VI2; and Questions 2, 4: AN1, AN2) haves the highest correlations as expected (as shown in Table 3). Also, this table provides evidence that the participants could understand the questions, and answers were not given randomly. Therefore, VI2 and AN2 can also be ignored, because they are highly correlated with the VI1 and AN1. In the rest of research VI1 and AN1 are called VI and AN respectively.

Table 3. The correlation of the answers among four questions

	VI1	AN1	VI2	AN2
VI1	1			
AN1	0.533025	1		
VI2	0.979866	0.53969	1	
AN2	0.652699	0.91148	0.650396	1

To answer RQ3, about the possibility of significant difference between the male and female users, mean and variance values are calculated. The t-Test results between male and female responses with the alpha value of 0.05 (VI: $t(17) = 1.027$, $p = 0.321$; AN: $t(22) = 0.069$, $p = 0.945$) show that there is no significant difference between the genders, even though the mean values of the male group shows a higher attraction to the online advertisements.

4.1   The Effects of the Ads Features

The mean values, standard deviation (SD), and the effect of various emotions on each case have been studied. Figure 3 presents the mean and error values of ad's features of the experiment.

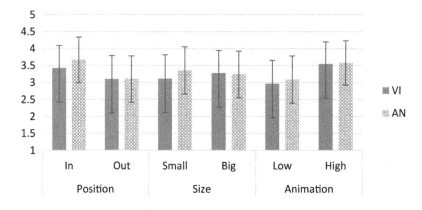

Fig. 3. The means and error bars of the ad's features

The effect of each feature on intrusiveness (VI) and annoyance (AN) is investigated by ANOVA. The results of ANOVA is tabulated in Table 4 with $F(1,1918)$ and their Eta-squared values. Based on the reported results for VI and AN, regardless of the user's emotions, a ANOVA analysis and Eta-squared values show the significant effect and effect size of the ads features for visual attention and annoyance.

Table 4. ANOVA with Eta-squared analysis on the effects of features on VI and AN

	VI			AN		
	F	p	Eta-squared	F	p	Eta-squared
Position	24.6102	< 0.00	0.0126	73.1269	< 0.00	0.0368
Frame Size	6.4398	0.011	0.0033	60.2119	0.0483	0.0057
Animation	85.5155	< 0.00	0.0428	3.9068	< 0.00	0.0302

* *df (Between groups):* 1; *df (Within groups):* 1918; *F-crit:* 3.8463

The results also show that the animation level of the ad plays a more important role in visual saliency and intrusiveness, and position has more effect on the user's annoyance. For both VI and AN, frame size has less effect than the other features. These results are limited to the participants in this survey, and it can be extended by being tested in practical and laboratory experiments.

4.2   Emotional Analysis

In this section, Mean, SD and ANOVA results with Eta-squared values have been presented. The ANOVA analysis with Bonferroni correction for each case separately will not be reported in this paper due to the space limit. Table 5 shows that there is a difference among the different emotional groups for both VI and AN. To measure the significance of this difference, ANOVA test has been employed among these groups with the alpha value of 0.05.

Table 5. The mean and standard deviation values of VI and AN

	VI	Mean	SD	AN	Mean	SD
EG0		1.37037	0.629294		2.703704	1.409168
EG1		2.032258	0.948116		2.774194	1.453953
EG2		2.180556	1.039046		2.930556	1.356532
EG3		2.08	0.862168		2.88	1.129897
EG4		1.6	0.894427		2.4	1.516575

The ANOVA test on emotional groups (Table 6) shows that emotion has a significant effect on VI and AN. Also, the emotional groups were compared using a post-hoc analysis of Bonferroni correction, and it showed significant differences among themselves.

Table 6. ANOVA with Eta-squared values on the emotions for VI and AN

	VI			AN		
	F	p	Eta-squared	F	p	Eta-squared
Emotion	10.30881	< 0.00	0.021	11.91755	< 0.00	0.0242

* df (Between groups): 4; df (Within groups): 1915; F-crit: 2.376574

5   Discussion and Conclusion

The analysis of the collected data showed that those advertisements with highly animated content, and animated position are, in average, more annoying and distracting the user's focus. This study also took the users' gender into the account. Although men showed slightly more attracted to the advertisements, the level of annoyance was similar in both genders. Also the result showed that despite a slight difference in attraction, there is no significant difference between the attraction and tolerance of male and female users. Previously, a research showed that there is a significant difference in the positive attitudes on online ads [24].

Also, ANOVA analysis shows a significant effect of the discussed ad's features on visual saliency, intrusiveness and annoyance. Among the features, animation and position have more effect on visual saliency and annoyance respectively.

The ANOVA results showed a significant effect of emotion on the participants' perception on different advertisement presentations, also the post-hoc analysis showed that the users' emotions with positive valence and/or low arousal are more attracted and tolerant for intrusive advertisements (Table 7). Therefore, it is recommended to only show less intrusive ads to the users with EG2, otherwise they get annoyed easily, and do not focus on the ad.

Table 7. Intrusiveness recommendation based on the emotional dimensions

Dimensions	Intrusive ads recommended	Non-intrusive ads recommended
Valence	Positive (+)	Negative (−)
Arousal	Low ↓	High ↑

6 Future Work

In the next step, we would like to check and detect the user's attention and tolerance by measuring the interaction features such as the keyboard keystroke dynamics, mouse movements and eye movements. It is expected to find a model between these features. Therefore, the user's perception and emotion can be predicted while he/she is interacting with a PC without being explicitly questioned [25, 26].

The results of this research can be integrated as a Decision Support System (DSS) into the computational advertising systems to recommend the online ads in more effective forms of presentation. Ad's features and user's affective states are the independent variables of this research, in order to provide a balance between the ad's intrusiveness and user's annoyance and tolerance toward an ad.

Acknowledgement. We express our gratitude and appreciation to Mrs. Mona Taghavi for her comments and reviews on this manuscript. This research is supported by Research Grant for Binationally Supervised Doctoral Degrees through Deutscher Akademischer Austauschdienst (DAAD), Germany.

References

1. Rodgers, S., Thorson, E.: The interactive advertising model: how users perceive and process online ads. J. Interact. Advert. **1**, 41–60 (2000)
2. Hoffman, D., Novak, T.: Advertising pricing models for the World Wide Web. In: Hurley, D., Kahn, B., Varian, H., Cambridge, E. (eds.) Internet Publishing and Beyond: The Economics of Digital Information and Intellectual Property, p. 2. MIT Press, Cambridge (2000)
3. Agarwal, A., Hosanagar, K., Smith, M.D.: Location, location, location: an analysis of profitability of position in online advertising markets. J. Market. Res. **48**, 1057–1073 (2011)

4. McCoy, S., Everard, A., Polak, P., Galletta, D.F.: An experimental study of antecedents and consequences of online ad intrusiveness. Int. J. Hum. Comput. Interact. **24**, 672–699 (2008)
5. Lin, Y.-L., Chen, Y.-W.: Effects of ad types, positions, animation lengths, and exposure times on the click-through rate of animated online advertisings. Comput. Ind. Eng. **57**, 580–591 (2009)
6. Lohtia, R., Donthu, N., Hershberger, E.K.: The impact of content and design elements on banner advertising click-through rates. J. Advert. Res. **43**, 410–418 (2003)
7. Petrovici, I.: Aspects of symbolic communications in online advertising. Proc. Soc. Behav. Sci. **149**, 719–723 (2014)
8. Meyers-Levy, J., Malaviya, P.: Consumers' processing of persuasive advertisements: an integrative framework of persuasion theories. J. Market. **63**, 45–60 (1999)
9. Petty, R.E., Cacioppo, J.T.: Communication and Persuasion: Central and Peripheral Routes to Attitude Change. Springer, New York (1986)
10. Krishnamurthy, S.: Deciphering the internet advertising puzzle. Market. Manage. **9**, 35–39 (2000)
11. Mullaney, T.J.: Online Marketing is Clicking. Business Week, New York (1999)
12. Holbrook, M.B., Batra, R.: Assessing the role of emotions as mediators of consumer responses to advertising. J. Consum. Res. **14**, 404–420 (1987)
13. Lambert, D.R., Morris, M.H., Pitt, L.F.: Has industrial advertising become consumerized? a longitudinal perspective from the USA. Int. J. Advert. **14**, 349–364 (1995)
14. Brown, S.P., Stayman, D.M.: Antecedents and consequences of attitude toward the ad, a meta-analysis. J. Consum. Res. **19**, 34–51 (1992)
15. Reed, M.: Going beyond the banner ad. Marketing **29**, 25–26 (1999)
16. McCoy, S., Everard, A., Polak, P., Galletta, D.F.: The effects of online advertising. Commun. ACM **50**, 84–88 (2007)
17. Danaher, P.J., Mullarkey, G.W.: Factors affecting online advertising recall: a study of students. J. Advert. Res. **43**, 252–267 (2003)
18. Rettie, R.: An exploration of flow during Internet use. Internet Res. Electron. Netw. Appl. Policy **11**, 103–113 (2001)
19. Kuisma, J., Simola, J., Uusitalo, L., Öörni, A.: The effects of animation and format on the perception and memory of online advertising. J. Interact. Mark. **24**, 269–282 (2010)
20. Desmet, P., Vastenburg, M., Van Bel, D., Romero Herrera, N.: Pick-A-Mood; development and application of a pictorial mood-reporting instrument. In: Out of Control: Proceedings of the 8th International Conference on Design and Emotion, London, UK, 11–14 September 2012 (2012)
21. Hartel, C.E.J., Ashkanasy, N.M., Zerbe, W.J.: Functionality, Intentionality and Morality. Elsevier, New York (2007)
22. Watson, D., Clark, L.A., Tellegen, A.: Development and validation of brief measures of positive and negative affect: the PANAS scales. J. Person. Soc. Psychol. **54**, 1063–1070 (1988)
23. Bakhtiyari, K., Husain, H.: Fuzzy model of dominance emotions in affective computing. J. Neural Comput. Appl. **25**, 1467–1477 (2014)
24. Wolin, L.D., Korgaonkar, P.: Web advertising: gender differences in beliefs, attitudes and behavior. Internet Res. **13**, 375–385 (2003)
25. Bakhtiyari, K., Taghavi, M., Husain, H.: Hybrid affective computing—keyboard, mouse and touch screen: from review to experiment. J. Neural Comput. Appl. **26**, 1277–1296 (2015)
26. Bakhtiyari, K., Taghavi, M., Husain, H.: Implementation of Emotional-Aware Computer Systems Using Typical Input Devices. In: Nguyen, N.T., Attachoo, B., Trawiński, B., Somboonviwat, K. (eds.) ACIIDS 2014. LNCS (LNAI), vol. 8397, pp. 364–374. Springer, Heidelberg (2014). doi:10.1007/978-3-319-05476-6_37

Runtime Verification and Quality Assessment for Checking Agent Integrity in Social Commerce System

Najwa Abu Bakar[1], Mohd Hafiz Selamat[1], and Ali Selamat[1,2(✉)]

[1] UTM-IRDA Digital Media CentrUTM-IRDA Digital Media Centre
and Software Engineering Department, Faculty of Computing, Universiti Teknologi Malaysia,
81310 Skudai, Johor Darul Takzim, Malaysia
najwa.abakar@gmail.com, {mhafiz,aselamat}@utm.my
[2] FIM, Center for Basic and Applied Research, University of Hradec Kralove,
Rokitanskeho 62, 500 03 Hradec Kralove, Czech Republic

Abstract. In social commerce systems, integrity is an important quality factor to ensure trust and reputation. Buyer agents need to trust Seller agents before performing online purchases. Since the interaction and transaction are performed online, there are certain conditions due to the dynamic of Seller agent activities and preferences that affect integrity. Hence, in this research, a solution is proposed to detect Seller agent integrity violations during runtime. Identifiability, interaction availability, trustability and information accessibility are some of the identified requirements that contribute to agent integrity. The proposed solution includes the definition of integrity as Seller agent quality properties and the implementation of Runtime Verification and Quality Assessment process. The effectiveness of the proposed solution is evaluated by implementing the checking and assessment process within social commerce system model.

Keywords: Agent integrity · Runtime verification · Quality assessment · Agent quality · Identifiability · Availability · Trustability · Accessibility · Agent systems · Social commerce system

1 Introduction

The integration of e-commerce systems with social media such as Facebook [1], Instagram [2] and Pinterest [3] has formed the next generation of e-commerce systems called social commerce [4]. Social commerce is an agent system that facilitates social shopping activities [5]. It is a new form of electronic commerce where the commerce activities are performed via social networking systems. Social commerce systems provide virtual environments where a group of agents who are the potential Buyer and Seller agents interact with each other. Agents do not only interact and share information with each other in an online community, but also perform buying and selling activities. Agents can create friendship links, join interest groups and share contents related to their businesses. Seller agents from time to time advertise items for sale and interested buyers can create virtual interest link relationship to access information. Potential buyers gather information and make purchasing decisions via their personal social networks [6, 7].

© Springer International Publishing AG 2017
N.T. Nguyen et al. (Eds.): ACIIDS 2017, Part I, LNAI 10191, pp. 150–159, 2017.
DOI: 10.1007/978-3-319-54472-4_15

Similar to social networking systems and traditional e-commerce that can be modeled as agent systems [8], Social Commerce Systems (SCS) also can be modeled as open dynamic multi-agent systems. It is an open system as agents can freely join and leave the system and it is dynamic because every agent has its own decisions and tasks that keep changing throughout their active period. While performing activities in open system, agent personal information, availability for interaction, relationship links with other agents and shared contents may be restricted due to several reasons such as privacy and data confidentiality protection [9, 10]. These factors are continuously changing during runtime depending on the activities executed within the system. As a result, Seller agent integrity and trust during commerce activities are affected as the Seller agent's identity at certain time may be unknown and agents may not be available for interaction. Seller and Buyer agents also may not trust each other as they are not connected via friendships or interest groups in social network. Finally, Seller agent contents also may be restricted and not accessible for public.

Several works that study social commerce systems have been focusing on integrity quality properties [4, 11–13]. However, further research needs to be done especially for checking agent integrity during runtime. Thus, a proper solution to detect integrity violations during Seller agent's active period is needed. This can be achieved via runtime verification and quality assessment by considering Seller agent's contextual information i.e. privacy preferences of profile, contents and interaction, as well as relationship links with the buyers. In this study, the identifiability, availability, trustability and accessibility are assessed to determine integrity status of the Seller agents.

This paper is organized as follows: Sect. 2 provides the background and related works. In Sect. 3, we describe in detail the implementation of the proposed Runtime Verification and Quality Assessment (RVQA) process. Section 4 states the experimentation and results. Finally, Sect. 5 provides the conclusion and future work.

2 Related Works

This paper focuses towards integrity properties of electronic and social commerce systems. There are several existing works that have been focusing on integrity issues in electronic and social commerce systems. Alberti [11] defined Social Integrity Constraints to ensure trust between users and their representative agents with other agents in the systems to perform interactions during auction activities in e-commerce systems. Yao [12] addressed integrity as one of the vulnerabilities in e-commerce systems and suggested the defense mechanisms as well as discussed the solutions to be implemented at the application level. Magentix2 [13], the proposed privacy-enhancing agent platform, considers integrity during communication between its agents. Finally, Lee [4] defined social commerce as the next generation of e-commerce. Trust is the main quality issue in social commerce in which integrity also plays its part to ensure trusted transactions. Lee [4] also includes identifiability as the necessary integrity element to enhance accountability and reliability during interaction in social commerce systems.

Table 1 shows the comparison matrix that maps e-commerce and social commerce systems literatures with the discussed quality properties. From the comparison matrix, it can be seen that trust is the quality property that is commonly studied by many researchers while other properties including integrity are also important and need to be further researched.

Table 1. The mapping of electronic and social commerce literatures with the quality properties.

No	Electronic and social commerce systems literatures	Quality properties					
		Trust	Reputation	Availability	Privacy	Confidentiality	Integrity
1	Korba [14]				√		
2	Sierra [15]	√	√				
3	Alberti et al. [11]	√					√
4	Huynh et al. [16]	√	√		√		
5	Osman [17]	√	√				
6	Krupa [18]	√			√		
7	Yao et al. [12]	√	√	√	√	√	√
8	Such et al. [13]	√	√		√	√	√
9	Balakrishnan and Majd [19]	√	√				
10	Majd and Balakrishnan [20]	√					
11	Lee [4]	√					√
12	Lu et al. [21]	√	√				

3   Implementation

This work is a continuation of our previous works [22, 23]. In this research, the Runtime Verification and Quality Assessment (RVQA) process is implemented within Social Commerce System (SCS) model. Figure 1 shows the SCS architecture that incorporates the RVQA component. The RVQA process is implemented within SCS architecture to detect conditions that may violate integrity of Seller agents. The RVQA component includes the checking of critical states reachability and the assessing of integrity statuses. These processes are performed based on the defined integrity requirements and violations conditions that consider Seller Agent's profile privacy, interaction availability, social connections and contents accessibility preferences. Creation of Seller agents, social connections and dynamic changes of preferences are modeled and simulated so that the violations that may occur during execution can be detected by the RVQA component. Finally, the integrity violations.

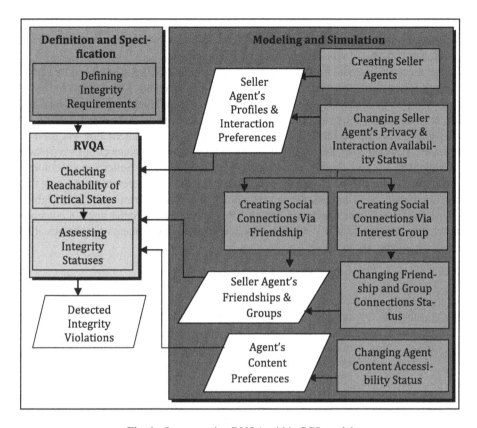

Fig. 1. Incorporating RVQA within SCS model

3.1 Definition and Specification of Agent Integrity

Since integrity is the main concern in every electronic commerce transactions, social commerce user requirements are in contrast of online social networking personal user requirements in terms of identifiability, availability, trustability and accessibility [4]. For online social networking, user identity and privacy need to be protected [23]. On the other hand, for social commerce, identifiability, availability, trustability and accessibility of the agents are very important to ensure integrity during buying and selling activities. Active Seller agents can choose to make its profile identifiable by setting the privacy preference of the profile to be exposed to public users. The Seller agent availability to communicate via messages also can be set to be available for public so that public users can contact Seller agents. Social connections also can be established between Seller agents and Buyers agents via friendship links and interest groups to improve trust. Seller agent contents also should be shared to public to increase information accessibility and trust of its business. It is important for Seller agents to satisfy the integrity requirements in order to maintain trust and reputation with the potential Buyer agents. In short, to fulfill integrity quality, Seller agents have to make their identity identifiable, be available for interaction, provide contents, perform social activities and join interest groups. Agents that are not being identifiable, available, trustable or inaccessible are considered violating the integrity requirements.

The agent quality goal for this case study is to monitor Seller agent integrity status in order to improve trust between Seller agents and Buyer agents. There are three levels of agent integrity requirements that are requirements at individual level, social level and content level. Agent integrity criteria for individual level include the identifiability of Seller agent's profile and the availability of Seller agents for interaction via private messages. At social level, the requirements include agents social connections via friendship and interest group. Finally, at content level, it is required for Seller agent's content to be accessible for public. Consequently, integrity violations are determined during the conditions when the Seller agent's identity is not identifiable, interaction is not available, Seller agents and Buyer agents are not socially connected and Seller agent's contents are not shared to public. The definition of agent integrity requirements and the possible violations are enlisted in Table 2.

Table 2. Agent integrity requirements

Integrity Levels	Requirements	Violations
Individual Profile	When a Seller agent is added (Seller agent is active), it reaches its critical state. The Seller agent is identifiable when its profile privacy preference is set to be opened for public.	The violation of this requirement is when a Seller agent is active and its privacy preference is restricted as private profile.
Individual Interaction	When a Seller agent is added (Seller agent is active), it reaches its critical state. The Seller agent is available when its private message privacy preference is set to be opened for communication with public.	The violation of this requirement is when a Seller agent is active and its private message privacy preference is restricted from public.
Social Friendship	When a Seller agent is added (Seller agent is active), it reaches its critical state. The Seller agent is trustable when it is connected with the Buyer agent via friendship link.	The violation of this requirement is when the Seller and Buyer are not socially connected via friendship.
Social Group	When a Seller agent is added (Seller agent is active), it reaches its critical state. The Seller agent is trustable when it is connected with the Buyer agent via interest group.	The violation of this requirement is when the Seller and Buyer are not socially connected via interest group.
Content	When a Seller agent is added (Seller agent is active), it reaches its critical state. The Seller agent's information is accessible when it publishes content for public.	The violation of this requirement is when Seller agent's content is not accessible to the public.

3.2 Modeling and Execution in SCS

During execution, user agents in SCS perform social commerce and trading activities with other agents. These activities are modeled using Anylogic multi-method modeling tool. The modeling and execution of the SCS extends and modifies our previous work that model and execute social networking system [23].

First, User agent population is created at certain specified rate in the main SCS environment. State chart is used to represent behavior of each individual in the User agent population. The UserStatechart, as shown in Fig. 2, is composed of the hierarchy of states and the transitions between states. When a User agent is created, it is in the ProfileActive state. Based on the defined integrity requirements (refer to Table 2), Runtime Verifier identifies that ProfileActive is a critical state. In ProfileActive state, there are two possible states for User agents i.e. ProfilePublic and ProfilePrivate states. As soon as the profile is active (agent enters ProfileActive state), the User agents are by default in ProfilePublic state. Later, User agents can choose to stay in the ProfilePublic state or to move to ProfilePrivate state by changing the privacy preference setting. Therefore, the rest of the agent activities such as becoming Seller agents, establishing social connections and sharing contents are performed by agents while staying in ProfilePublic or ProfilePrivate states.

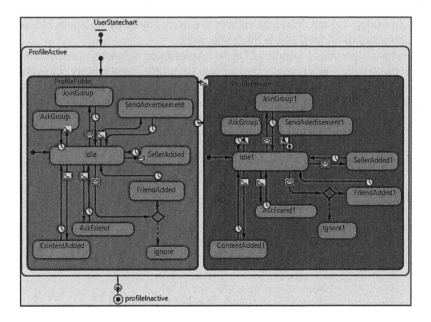

Fig. 2. State chart to model user agent activities in social commerce system

Second, it is followed by the creation of Friend and GroupMember agent population as a result of interaction among User agents to request and to accept the formation of friendship and group membership links, respectively. All active User agents in SCS environment are by default the potential Buyer agents. Every potential Buyer agents can choose other agents in the SCS environment to become its active Seller agents, with whom the social commerce and trading activities can be performed. Simultaneously, the chosen Seller agents can share contents in order to publish business information.

3.3   RVQA Process in SCS

The Runtime Verification process is implemented to identify when the defined critical states for each Seller agent are reached. The critical states are the states where the conditions that can violate Seller agent integrity are possible to occur. At the critical states, the Quality Assessment processes are performed. RVQA algorithms assess Seller agent integrity at individual, social and content levels by considering the Seller agent's dynamic contextual information. The contextual information i.e. Seller agent's privacy preferences, social activities and status of the published contents determine agent identifiability, availability, trustability, and accessibility status (refer to Table 2). Finally, the detected integrity violations are classified and reported. The number of integrity violations increases every time the agent integrity violations are detected. This verification and assessment processes are performed continuously for all levels whenever the defined critical states are reached. The processes continue to track the number of integrity violations detected in order to measure the Seller agent integrity level during runtime. Increased number of detected integrity violations means that the integrity violation level is high (low integrity) and decreased number of detected integrity violations means that the integrity violations level is low (high integrity).

Algorithm 1 presents a sample of Quality Assessment algorithms to determine the status of the Seller agent's integrity requirement i.e. profile identifiability. As explained in the definition section (Sect. 3.1), Seller agent's active status determines that the Seller agent is in critical state. At critical state, Seller agent's decisions to make its profile identifiable determine the Seller agent's individual integrity status.

Algorithm 1. Non-Identifiable Profile Detection

```
   Input: agent index: x, current time: t
   Output: noNonIdentifiableProfile, noIdentifiableProfile
1  begin
2  if SellerActive(x,t) then
3     Critical(x,t) ← true
4     if ProfileNotPublic(x,t) then
5        NonIdentifiableProfile(x,t)
6        increase(x,noNonIdentifiableProfile)
7     end
8     if ProfilePublic(x,t) then
9        IdentifiableProfile(x,t)
10       increase(x,noIdentifiableProfile)
11    end
12 end
13 return noNonIdentifiableProfile, noIdentifiableProfile
```

For Seller agent, x at time, t, the dynamic profile privacy preference set to be either public (ProfilePublic(x,t)) or non-public (ProfileNotPublic(x,t)) determines whether the Seller agent's identity is identifiable (`IdentifiablePro-file(x,t)`) or non-identifiable (`NonIdentifiableProfile(x,t)`), respectively. The number of non-identifiable profile (`noNonIdentifiableProfile`) is increased whenever non-identifiable profile is detected and the number of identifiable profile (`noIdentifiableProfile`) is increased whenever identifiable profile is detected. The number of non-identifiable profile is classified as the detected integrity violations that occur at that particular time.

4   Experimentation and Results

The objective of this experiment is to detect conditions that can violate Seller agent integrity during SCS execution. The RVQA implementation within SCS agent model was executed and the monitoring processes were performed to check the reachability of critical states at individual, social and content levels of Seller agent activities. When the critical states were reached, the integrity statuses of Seller agent identifiability, availability, trustability, and accessibility status of each Seller agent were assessed so that the integrity violations can be classified and detected.

The integrity violations detection results are presented in the form of time graphs for the first 100 days of the SCS execution. Figure 3 presents the number of Seller agents that are not identifiable and not available for interaction throughout the execution period. The time graph represents the agent integrity violations detection result at individual level. At the end of 100 days of execution, RVQA managed to detect 54% of Seller agent profiles that are non-identifiable and 50% of Seller agents that are unavailable for interaction as compared to the number of active Seller agent profiles.

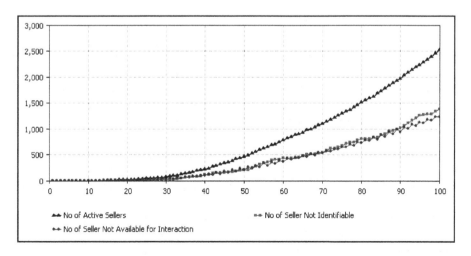

Fig. 3. Number of integrity violations detected at individual level vs time (days)

Figure 4 shows the number of classified Seller agents that are not socially connected to Buyer agents via friendship and interest group throughout the execution period.

The time graph represents the agent integrity violations detection results at social level. At the end of 100 days of execution, RVQA managed to detect 63% of Seller agents that are non-trustable via friendship and 25% of Seller agents that are non-trustable via interest group as compared to the number of active Seller agents.

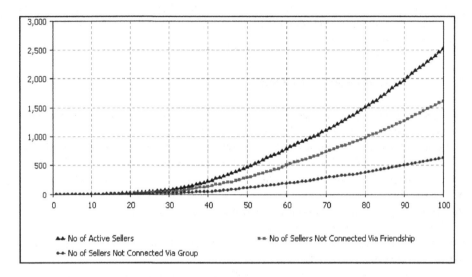

Fig. 4. Number of integrity violations detected at social level vs time (days)

5   Conclusion and Future Work

The significant contribution of this research is the detection of agent integrity violation conditions that occur during SCS execution. This paper discusses the issue of agent integrity in social commerce, proposes new RVQA process model and defines agent integrity requirements for SCS. The SCS architecture is designed and modeled to incorporate the RVQA component. The model was executed and the detected integrity violations conditions were classified. The results manage to highlight end users that integrity violations do occur during the execution of social commerce activities.

This work complements the existing electronic and social commerce works. The detected agent integrity violations can be used by end users to improve the agent integrity during the execution of social commerce activities. Seller agents can improve their integrity level while Buyer agents can use the information to choose trusted Seller agents. In the future, the detected agent integrity information can be analyzed further using advanced classification and measurement techniques. More integrity requirements also can be defined and the existing requirements can be improved according to the agent integrity verification and assessment goal.

Acknowledgements. The Universiti Teknologi Malaysia (UTM) and Ministry of Higher Education Malaysia under research Fundamental Research Grant Scheme 4F904 are hereby acknowledged for some of the facilities that were utilized during the course of this research work.

References

1. Facebook: Facebook (2016). https://www.facebook.com/
2. Instagram: https://instagram.com/ (2016). https://instagram.com/. Accessed: 17 Sep 2015
3. Pinterest: Pinterest: Discover and Save Creative Ideas (2016). https://www.pinterest.com/. Accessed: 17 Sep 2015
4. Lee, J.Y.: Trust and social commerce. Univ. Pittsburgh Law Rev. **77**(2), 137–146 (2015)
5. Johnson, L.: What Marketers Need to Know About 5 New Types of Social Commerce. Adweek (2015)
6. Beisel, D.: (The Beginnings of) Social Commerce (2005). http://genuinevc.com/2005/12/06/the-beginnings-of-social-commerce/
7. Guo, S., Wang, M., Leskovec, J.: The role of social networks in online shopping: information passing, price of trust, and consumer choice. In: ACM Conference on Electronic Commerce, pp. 157–166 (2011)
8. Yu, H., Shen, Z., Leung, C., Miao, C., Lesser, V.R.: A survey of multi-agent trust management systems. IEEE Access **1**, 35–50 (2013)
9. Butler, E.: Privacy setting awareness on facebook and its effect on user-posted content. Hum. Commun. **14**(1), 39–55 (2011)
10. Debatin, B., Lovejoy, J.P., Horn, A.K., Hughes, B.N.: Facebook and online privacy: attitudes, behaviors, and unintended consequences. J. Comput. Commun. **15**(1), 83–108 (2009)
11. Alberti, M., Chesani, F., Guerri, A., Gavanelli, M., Lamma, E., Mello, P., Milano, M., Torroni, P.: Expressing interaction in combinatorial auction through social integrity constraints. In: Procëedings of W(C)LP, vol. 2005–1, pp. 53–64 (2005)
12. Yao, Y., Ruohomaa, S., Xu, F.: Addressing common vulnerabilities of reputation systems for electronic commerce. J. Theor. Appl. Electron. Commer. Res. **7**(1), 3–4 (2012)
13. Such, J.M., Garc, A., Bellver, J.: Magentix2: a Privacy-enhancing Agent Platform (2013)
14. Korba, L.: Privacy in distributed electronic commerce. In: Proceedings of the Annual Hawaii International Conference on System Sciences, vol. 2002–January, pp. 4017–4026 (2002)
15. Sierra, C.: Agent-mediated electronic commerce. Auton. Agent. Multi. Agent. Syst. **9**, 285–301 (2004)
16. Huynh, T.D., Jennings, N.R., Shadbolt, N.R.: An integrated trust and reputation model for open multi-agent systems. Auton. Agent. Multi. Agent. Syst. **13**(2), 119–154 (2006)
17. Osman, N.Z.: Runtime verification of deontic and trust models in multiagent interactions. University of Edinburgh (2008)
18. Krupa, Y.: Privacy as Contextual Integrity in Decentralized Multi-Agent Systems (2012)
19. Balakrishnan, V., Majd, E.: A comparative analysis of trust models for multi-agent systems. Lect. Notes Softw. Eng. **1**(2), 183–185 (2013)
20. Majd, E., Balakrishnan, V.: Analysis of trust models to improve E-commerce multi-agent systems. In: First International Conference on Advanced Data and Information Engineering (DaEng), vol. 285, pp. 615–622 (2014)
21. Lu, B., Fan, W., Zhou, M.: Social presence, trust, and social commerce purchase intention: an empirical research. Comput. Hum. Behav. **56**, 225–237 (2016)
22. Bakar, N.A., Selamat, A.: Runtime verification of multi-agent systems interaction quality. In: Selamat, A., Nguyen, N.T., Haron, H. (eds.) ACIIDS 2013. LNCS (LNAI), vol. 7802, pp. 435–444. Springer, Heidelberg (2013). doi:10.1007/978-3-642-36546-1_45
23. Bakar, N.A., Selamat, A.: Runtime verification and quality assessment for privacy violations detection in social networking system. Front. Artif. Intell. Appl. **286**, 346–357 (2016)

Evaluation of Tensor-Based Algorithms for Real-Time Bidding Optimization

Andrzej Szwabe[✉], Paweł Misiorek, and Michał Ciesielczyk

Institute of Control and Information Engineering,
Poznan University of Technology, Piotrowo 3a, 60-965 Poznan, Poland
{andrzej.szwabe,pawel.misiorek,michal.ciesielczyk}@put.poznan.pl

Abstract. In this paper we evaluate tensor-based approaches to the Real-Time Bidding (RTB) Click-Through Rate (CTR) estimation problem. We propose two new tensor-based CTR prediction algorithms. We analyze the evaluation results collected from several papers – obtained with the use of the iPinYou contest dataset and the Area Underneath the ROC curve measure. We accompany these results with analogical results of our experiments – conducted with the use of our implementations of tensor-based algorithms and approaches based on the logistic regression. In contrast to the results of other authors, we show that biases – in particular those being low-order expectation value estimates – are at least as useful as outcomes of high-order components' processing. Moreover, on the basis of Average Precision results, we postulate that ROC curve should not be the only characteristic used to evaluate RTB CTR estimation performance.

Keywords: Big Data · Context-aware recommendation · Tensor decomposition · Logistic regression · Click-Through Rate prediction · WWW · Display advertising · Real-Time Bidding · Demand-Side Platform

1 Introduction

In the recent years, some of the tensor-based data processing algorithms have been identified as addressing Big Data processing challenges, including those widely referred to as 'the four Vs of Big Data' [7]. Although the area of the research on tensor-based Big Data processing has emerged quite recently [3], the results that have been already achieved, indicate that, at least in online advertising application scenarios, tensor-based data processing is able to outperform many alternative approaches, including those based on the matrix factorization techniques and deep learning algorithms [15,18].

Real-Time Bidding (RTB), with its very high data volume, heterogeneity, extreme sparsity and the need for the high-velocity stream processing and the low-delay querying, is one of the best examples of widely-used Big Data technologies [12]. As confirmed by many authors, the research on RTB algorithms involves facing many challenges that are typical for Big Data. In particular,

© Springer International Publishing AG 2017
N.T. Nguyen et al. (Eds.): ACIIDS 2017, Part I, LNAI 10191, pp. 160–169, 2017.
DOI: 10.1007/978-3-319-54472-4_16

RTB algorithms must be capable to process highly heterogeneous data streams of the volume having the order of terabytes rather than gigabytes [2]. Moreover, to by applicable in a real-world environment, an RTB optimization algorithm must be able to provide its results in tens of milliseconds [12].

In the most typical scenario of an offline Click-Through Rate (CTR) estimation system evaluation, the system is assumed to provide 'binary decisions' to queries representing potential impressions, rather than providing real-valued estimates of the CTR expectation values. These 'decisions' may be seen as estimates of logical values indicating whether the system estimates the corresponding proposition (i.e. the proposition that the given user will click the given creative in response to the given impression) as being true. Therefore, at least in the context of the most widely used methodology of the CTR estimation evaluation, the purpose of a CTR estimation system is to discriminate between true and false propositions of this kind [18]. Such an approximate reasoning capability may be unambiguously evaluated by means of trivial comparisons of the system operation outcomes with the appropriate values from the test set. As agreed by many authors investigating the CTR prediction problem, the most reasonable way to evaluate a CTR prediction system is to use the Area Underneath the ROC curve (AUROC) measure [2,18], which is discrimination-centric and copes well with the imbalance between the number of positives (i.e. clicks) and negatives (i.e. non-clicks) [4,6,16].

1.1   Real-Time Bidding Click-Through Rate Estimation Problem

As noticed by many researchers – from both the industry [10,20] and the academia [15,19] – there is a strong and still growing demand for advances in the research on optimization of RTB automatic decision-making algorithms. RTB optimization algorithms include those referred to as solutions to the optimal RTB problem and solutions to the CTR estimation problem [19]. These optimization algorithms became the key source of the competitive advantage of RTB Demand Side Platforms (DSPs) [10,12,17,20] and a very interesting topic of the research on recommender systems [10,15].

Within the area of the research on RTB optimization solutions, the CTR estimation is probably the most intensively investigated problem [2,15,20]. Many authors even assume that a CTR estimation system is an obligatory component of any RTB DSP optimization system architecture [15,18].

1.2   CTR Estimation as a Recommendation Problem

Authors of several papers presenting RTB optimization algorithms consider a CTR estimation system as a special case of a recommender system [15,18]. Indeed, the main purpose of a CTR estimation system that operates within an RTB DSP is to 'recommend', or to 'select', ad impressions that are likely to be valuable to users, e.g., ad impressions that lead to clicks or conversions [15,18].

The convention of treating a CTR estimation system as a recommender system is also followed in the way that CTR estimation systems are evaluated.

As explained in Sect. 4, an offline evaluation of a CTR estimation system is typically performed in accordance with one of the most popular methodologies of recommendation systems evaluation [15,16,18].

2   Related Work

In the context of the RTB DSP optimization scenario, it is very important to make the recommendation algorithm (the one that is used to perform CTR estimation) both highly contextual and capable to exploit various data augmentations [15,19]. In the papers presenting CTR estimation algorithms these two requirements are sometimes integrated into the single, more general requirement. Specifically, recommender systems are required to be able to model the heterogeneous data attributes explicitly from multiple alternative and complementary 'aspects' [15]. The idea of such 'multi-aspect' data modeling is very familiar to researchers working on tensor-based data representations or data processing methods [1,9]. Moreover, the need for 'multi-aspect' data modeling has been also recognized by the authors of tensor-based RTB CTR estimation systems [15] and by the authors of advanced classification systems theoretically-grounded on the rough set theory [8]. All these types of systems perform some type of 'multi-aspect' data modeling by using, in one way or another, nonlinear combinations of multiple 'interacting' features [2,15].

AUROC is the most popular measure used by researchers working on RTB CTR estimation algorithms, in particular those using the iPinYou dataset [13,15,18,20] (see Table 1 for the result comparison). The recent results of [15] indicate that it is possible to achieve a high performance of CTR estimation using tensor factorization such as Tucker Decomposition (TD), Canonical Decomposition (CD) or the newly proposed model referred to as Fully-Coupled interaction Tensor Factorization (FCTF). However, in the same paper, the authors have shown that a comparable (or even higher) performance may be achieved using the approach based on Logistic Regression (LR) which does not take into account the higher-order interactions between features. High performance of the LR approach is additionally confirmed by the results reported in [13,20]. In [20], the authors show that their implementation of the LR algorithm outperforms the method based on Gradient Boosting Regression Tree (GBRT). The CTR estimation solution presented in [13] is, in fact, the gradient algorithm based on modifications of the LR model (for scenarios of risk and utility modeling).

LR models have already been widely applied to CTR estimation systems [2,14,15], partly due to their ability to handle large-scale problems. In contrast, high AUROC results have not been obtained by means of the deep learning approach introduced in [18] – mainly due to the fact that the relatively complicated neural model is not able to use many features. The authors report the results obtained for two Sampling-based Neural Networks variants, Sampling-Based Denoising Auto-Encoder (SNN-DAE) and Sampling-based Restricted Boltzmann Machine (SNN-RBM), as well as for Factorization-machine supported Neural Networks (FNN).

Table 1. The comparison of AUROC (%) results for each advertiser in the iPinYou dataset; the best results in each column are highlighted by bold font setting.

Method	iPinYou campaign								
	1458	3358	3386	3427	3476	2259	2261	2821	2997
LR [15]	97.93	96.80	78.48	97.17	92.04	72.34	**65.21**	67.02	53.30
TD [15]	97.73	96.88	75.50	96.18	93.82	69.16	61.26	66.79	56.20
CD [15]	97.94	97.63	74.79	96.15	92.69	71.80	63.81	66.85	52.29
FCTF [15]	98.18	**98.31**	76.22	96.54	94.29	**72.61**	64.99	**68.63**	53.02
LR [20]	**98.82**	97.53	79.08	**97.35**	**96.25**	68.65	62.38	63.25	60.39
GBRT [20]	97.07	97.22	76.86	93.42	94.22	67.91	57.39	58.20	59.79
FNN [18]	70.52	-	**80.55**	-	-	69.74	62.99	-	61.41
SNN-DAE [18]	70.46	-	79.22	-	-	68.08	63.72	-	**61.58**
SNN-RBM [18]	70.49	-	80.07	-	-	68.34	63.72	-	61.45
Gradient [13]	97.7	98.0	77.8	96.0	95.0	69.1	61.9	63.9	60.8

The AUROC results presented in [15] confirm that higher-order tensors may be successfully applied to estimate CTR. However, the deeper analysis of these results indicates that the high performance of the presented algorithms relies heavily on the application of biases based on CTR statistics for individual features.

3   CTR Estimation Algorithms

Motivated by the recent research on tensor-based approaches to the RTB CTR estimation problem [15] and by more theoretical works on the so-called tensor centering [1], we propose two new RTB CTR estimation algorithms:

- Bias Modeling based on 1st-order Tensors (BMT) – the algorithm investigating deviations of observed CTR for events' features,
- Centered Tucker Decomposition (CTD) – the algorithm based on the decomposition of a centered tensor modeling the interactions between the features.

3.1   BMT Algorithm

Similarly as in [15], we model each bid request as a triple of the form of (u, p, a), where $u \in U$, $p \in P$, and $a \in A$ represent the user, the publisher context, and the ad, respectively. Let $C(u)$, $C(p)$, and $C(a)$ denote the sets of single-value attributes (such as region, domain, etc.) for u, p, and a correspondingly. Additionally, let $T(u)$ denote a set of tags (the only multi-value attribute in the iPinYou dataset) for the user u.

As pointed out in [15], much of the observed variation in click events is caused by biases which are independent of any interactions. To evaluate the impact of these biases on the performance of the tensor-based methods, we have proposed a simplified approach based solely on average CTR values observed in the training data. From the perspective of tensor modeling, the bias values are equal to the entries of centered first-order tensors build for each event attribute separately. As a result, for a given feature defined in the domain of a given attribute, the bias is modeled as the deviation of observed CTR for all events having this feature from the overall CTR observed in the whole training set. Specifically, the approximation of the first-order bias related to the tuple (u, p, a) is equal to:

$$b_{u,p,a} = \frac{\sum_{t \in T(u)} b_t}{|T(u)|} + \sum_{c \in C(u) \cup C(p) \cup C(A)} b_c \,, \tag{1}$$

where b_t indicates the observed deviations of the tag t, and b_c indicates the observed deviations of the feature c that the user u, publisher p or ad a possess.

The final estimation of the click probability obtained by means of the BMT algorithm corresponds to the value $b_{u,p,a}$. Thus, the algorithm may be used to evaluate the impact of the bias modeling to the performance of tensor-based methods that use it as one of the estimation outcome components, i.e., the CTD algorithm as well as all tensor-based algorithms proposed in [15].

3.2   CTD Algorithm

In the feature-based CTD model, each user, publisher and ad is represented as a sum of vectors corresponding to the features defined in the domains of single-value or multiple-value attributes, i.e.:

$$u = \sqrt{|T(u)|} \sum_{i \in T(u)} t_i + \sum_{i \in C(u)} c_i \,, \qquad p = \sum_{i \in C(p)} c_i \,, \qquad a = \sum_{i \in C(a)} c_i \,.$$

Note, that as in [15], we normalize the sum of tags. The 3-mode tensor $\mathcal{T} \in \mathbb{R}^{|U| \times |P| \times |A|}$ is modeled as $\sum_{u \in U} \sum_{p \in P} \sum_{a \in A} u \otimes p \otimes a$, where $\otimes$ denotes the tensor product.

In contrast to [15], we additionally center the data, as such an operation can significantly improve the performance of the algorithm. Each *click* event is represented as 1 and each *non-click* event as $-\frac{number\ of\ clicks}{number\ of\ non-clicks}$. Next, tensor $\mathcal{T}$ is centered across the tensor slices corresponding to the elements of each mode. The centering operation is performed by subtracting the mean from values of the cells of a given tensor slice [1].

We use the Tucker decomposition approach [9], by reducing the rank of tensor $\mathcal{T}$, to minimize the impact of the noise on the underlying population. Specifically, $\mathcal{T}$ is factored into $\mathcal{C} \times_U \mathcal{U} \times_P \mathcal{P} \times_A \mathcal{A}$, in which $\mathcal{C} \in \mathbb{R}^{k \times k \times k}$, $\mathcal{U} \in \mathbb{R}^{|U| \times k}$, $\mathcal{P} \in \mathbb{R}^{|P| \times k}$, $\mathcal{A} \in \mathbb{R}^{|A| \times k}$, and k is the number of latent factors. The estimate corresponding to any triple may be obtained as follows:

$$t_{u,p,a} = \sum_{l}^{k} \sum_{m}^{k} \sum_{n}^{k} \mathcal{C}_{l,m,n} \cdot \mathcal{U}_{u,l} \cdot \mathcal{P}_{p,m} \cdot \mathcal{A}_{a,n} \,. \tag{2}$$

Finally, we incorporate the biases related to each tuple (u, p, a), presented in Eq. 1, into the CTD model. Thus, the estimation of the click probability takes into account the bias values and is equal to $t_{u,p,a} + b_{u,p,a}$.

4   Evaluation Methodology

The goal of a CTR estimation system is to predict whether a specific user, in a given context, will click on an ad creative. Thus, it may be considered as a binary classification problem, in which the system has to decide whether an impression will result in a *click* event. In such a scenario, one of the most widely used metrics are AUROC [2,10,15,18,20] and Root Mean Squared Error (RMSE) [14,15,20]. However, in cases of heavy class imbalance (like in the case of the number of *click* and *non-click* events in RTB) the RMSE metric is typically not appropriate – as it has been reported in [18]. For instance, a very trivial, and by all means, useless classifier predicting all the events as *non-clicks* would perfectly fit most of the observations. Therefore, in this paper, we use the AUROC metric as a classification-oriented quality measure that, according to [6], enables one to directly evaluate the systems' ability to distinguish between accurate and inaccurate predictions.

Nonetheless, AUROC, despite being very useful for recommendation or link prediction systems [4,16], may not provide a full insight into the RTB CTR estimation problem. Therefore, complementarily to the AUROC results, we show the Average Precision (AP) results (equivalent to the area underneath the precision-recall curve). While both metrics measure the proportion of preferred items that are actually recommended (i.e., true positive rate), AP emphasizes the proportion of recommended items that are preferred (i.e. precision) while AUROC emphasizes the proportion of items that are not preferred but end up being recommended (i.e., false positive rate) [4,16]. Such a difference of how the true negatives are treated is especially evident when the number of negative observations (*non-clicks*) is a few orders of magnitude higher than the positive ones (*clicks*). The ROC curves scale up with class imbalance [6], while in the case of a precision-recall curve, the true negatives do not bias the score. Specifically, when the number of false positives substantially increases, it is reflected in the AP and is missed in the AUROC analysis, as illustrated in [5].

4.1   Dataset

To evaluate the proposed methods we used the first publicly available large-scale real-world RTB dataset released in 2014 by iPinYou Information Technologies Co., Ltd [11]. iPinYou is currently one of the largest DSP in China [20]. The major dataset statistics are shown in Table 2. The dataset contains impression, click, and conversion logs collected from several advertisers during various days and is divided into a training set and a test set.

Generally, each record contains four types of information: (i) user features, (ii) ad features, (iii) publisher features, and (iv) other features regarding the

Table 2. Dataset statistics.

Season	Dataset	Impressions	Users	Ad slots	URLs	Clicks	CTR
2	Training set	$12,190,438$	$10,107,786$	$141,515$	$2,362,123$	$8,838$	0.073%
	Test set	$2,521,627$	$2,310,300$	$48,458$	$663,216$	$1,873$	0.074%
3	Training set	$3,147,801$	$2,816,492$	$53,571$	$963,764$	$2,700$	0.086%
	Test set	$1,579,071$	$1,490,721$	$43,660$	$567,044$	$1,135$	0.072%

RTB auction. The average CTR – denoting the positive user feedback ratio in the dataset – is less than 0.1%, far less than in the most popular datasets used to evaluate recommender systems, such as Movielens or Netflix. Moreover, depending on the selection of the particular advertiser represented in the iPinYou contest dataset, the user behavior – directly influencing the performance of bidding algorithms – significantly differs. More detailed information on the dataset may be found in [15, 20].

4.2   Experimental Setup

The accuracy of the algorithms proposed in Sect. 3 have been compared against our implementations of CTR prediction methods proposed by other authors:

- Logistic Regression (LR) [20],
- regularized Logistic Regression (rLR),
- feature-based Tucker Decomposition (TD) [15],

We implemented the LR as specified in [20]. We used stochastic gradient descent to learn the model parameters, the cross entropy between the predicted click probability and the ground-truth result as the loss (in both LR and rLR), and L2 regularization (in rLR). The learning rate in all the experiments was set to 0.01 and the tolerance for the stopping criterion was set to 0.0001. For rLR, the regularization strength (ten values in a logarithmic scale between 10^{-4} and 10^4) was optimized using cross-validation on the training set. In our experiments, we have $801,889$ and $589,871$ binary features for season 2 and 3, correspondingly.

In the case of TD, as defined in [15], each *click* entry (u, p, a) is modeled as equal to 1 whereas each *non-click* entry is modeled as 0, and each triple probability is estimated according to Eq. 2. In our experiments, we set k equal to 32 for both TD and CTD, since such a number of latent factors enables to achieve stable prediction quality on the iPinYou dataset [15].

5   Evaluation Results

The results of comparison of the proposed algorithms with our implementations of state-of-the-art algorithms, performed using AUROC and AP measures and

Table 3. CTR estimation performance in terms of AUROC and AP(%); the best and second best results are correspondingly highlighted by bold and slanted font setting.

Season	Advertiser	AUROC (%)					AP (%)				
		LR	rLR	BMT	TD	CTD	LR	rLR	BMT	TD	CTD
2	1458	*98.37*	**98.69**	97.93	74.82	98.43	*69.781*	**71.075**	5.692	0.937	65.353
	3358	**97.58**	91.35	94.06	74.14	*95.38*	**42.560**	9.421	1.889	0.840	*22.745*
	3386	**77.52**	68.99	*74.16*	61.62	68.87	**2.105**	*1.150*	0.421	0.128	0.821
	3427	**97.51**	*97.31*	82.36	66.63	87.15	**48.276**	*47.533*	0.312	0.641	0.665
	3476	**96.47**	*83.68*	70.81	63.43	65.75	**15.243**	*0.388*	0.135	0.116	0.113
	Total	**92.57**	*90.82*	84.23	65.03	80.15	6.977	**15.501**	0.029	0.002	*10.277*
3	2259	66.53	65.64	**68.31**	64.57	*67.73*	*0.083*	0.079	0.077	0.068	**0.088**
	2261	60.18	**61.54**	*61.18*	60.38	59.39	*0.065*	0.049	**0.066**	0.062	0.046
	2821	**63.20**	57.62	*60.05*	51.44	58.87	**0.095**	0.081	*0.092*	0.066	0.083
	2997	*57.34*	**60.68**	56.88	49.95	54.89	*0.470*	**0.556**	0.427	0.335	0.429
	Total	*75.81*	75.70	74.58	73.22	**76.37**	**0.335**	0.252	0.002	0.002	*0.253*

the iPinYou dataset, are presented in Table 3. The application of our own implementations of LR, rLR and TD methods was necessary to perform evaluations comparable in terms of both the AUROC measure and AP measure. The results have confirmed the high performance of LR methods. These methods achieve the highest scores in most of the scenarios in terms of both the measures.

The approach based on the modeling of CTR deviations (i.e., BMT) enables achieving AUROC results similar to those obtained when using tensor-based modeling of higher-order feature interactions as an additional component of the estimation outcome. This observation indicates an interesting direction of the future work on an efficient integration of the averages modeling and the feature interactions modeling. Moreover, the importance of the tensor centering for the reliable modeling of higher-order interactions may be observed when comparing the result of TD and CTD methods.

6   Conclusions

Some of the authors of the most recent works on tensor-based RTB CTR estimation recognize the theoretical importance and the practical value of low-order expectation value estimates – the so-called biases [15]. However, this finding has not been followed by any further investigation yet – at least in the area of the research on the RTB CTR estimation problem. In this paper we show that even a naive modeling of low-order expectation value estimates only, such as the one applied to the algorithms proposed in this paper (i.e., BMT and CTD), may be valuable. Our results indicate that the modeling of low-order expectation values may be even more useful than an application of a complex, computationally much more demanding model of high-order tensor components based on the widely-used tensor decomposition methods which neglect the impact of biases and the tensor centering.

As we experimentally confirm in this paper, methods based on LR enable to achieve higher RTB CTR estimation performance than the state-of-the-art methods based on multi-linear modeling. Such a finding is especially evident in the case of the AP results. While many authors of the tensor-based approaches focus on representing hundreds of thousands of features using vectors of significantly reduced dimensionality (even to just a few dimensions) and on adding proper biases to the model, they neglect the problem of training data under-fitting. Although such an under-fitting may remain unnoticed while using measures such as AUROC – due to heavy class imbalance – it is clearly visible (as we show in Sect. 5) when the precision metric is used.

Obviously, any claims based on results of laboratory experimentation must refer to a precisely specified evaluation methodology including a definition of a widely-used dataset and a set of appropriately selected evaluation measures [16]. Our claim about the superiority of the algorithms based on logistic regression over the evaluated tensor-based algorithms (observed in the reported experimental cases), remains true independently from the selection of the particular evaluation measure. On the other hand, in most of the cases of the tensor-based algorithms evaluation shown in Sect. 5, the method that achieves the best performance in terms of AUROC measure, does not achieve the highest AP score. While analyzing this ambiguity, one should take into account that AUROC reflects the algorithm's ability to distinguish any *click* from *non-click* event in the test set whereas AP measure depends mainly on the algorithm ability to estimate high click probabilities [5]. As the bidder's capability of increasing the CTR level is one of the most important properties of any RTB DSP, one may conclude that AUROC should not be the only measure used for the evaluation of RTB CTR estimation performance.

Acknowledgments. This work is supported by the Polish National Science Centre, grant DEC-2011/01/D/ST6/06788.

References

1. Bro, R., Smilde, A.K.: Centering and scaling in component analysis. J. Chemometr. **17**(1), 16–33 (2003)
2. Chapelle, O., Manavoglu, E., Rosales, R.: Simple and scalable response prediction for display advertising. ACM Trans. Intell. Syst. Technol. **5**(4), 61:1–61:34 (2014). http://doi.acm.org/10.1145/2532128
3. Cichocki, A.: Era of Big Data processing: a new approach via tensor networks and tensor decompositions. CoRR abs/1403.2048 (2014). http://arxiv.org/abs/1403.2048
4. Ciesielczyk, M., Szwabe, A., Morzy, M., Misiorek, P.: Progressive random indexing: dimensionality reduction preserving local network dependencies. ACM Trans. Internet Technol. **17**(2) (2017). doi:10.1145/2996185
5. Davis, J., Goadrich, M.: The relationship between precision-recall and ROC curves. In: Proceedings of the 23rd International Conference on Machine Learning, ICML 2006, NY, USA, pp. 233–240 (2006). http://doi.acm.org/10.1145/1143844.1143874

6. Fawcett, T.: An introduction to ROC analysis. Pattern Recogn. Lett. **27**(8), 861–874 (2006). http://dx.doi.org/10.1016/j.patrec.2005.10.010
7. Japkowicz, N., Stefanowski, J.: Big Data Analysis: New Algorithms for a New Society. Studies in Big Data. Springer International Publishing, Heidelberg (2015)
8. Kruczyk, M., Baltzer, N., Mieczkowski, J., Draminski, M., Koronacki, J., Komorowski, J.: Random reducts: a Monte Carlo rough set-based method for feature selection in large datasets. Fundam. Inform. **127**(1–4), 273–288 (2013)
9. Lathauwer, L.D., Moor, B.D., Vandewalle, J.: A multilinear singular value decomposition. SIAM J. Matrix Anal. Appl. **21**, 1253–1278 (2000)
10. Lee, K., Orten, B., Dasdan, A., Li, W.: Estimating conversion rate in display advertising from past performance data. In: Proceedings of the 18th ACM SIGKDD International Conference on Knowledge Discovery and Data Mining, KDD 2012, NY, USA, pp. 768–776 (2012). http://doi.acm.org/10.1145/2339530.2339651
11. Liao, H., Peng, L., Liu, Z., Shen, X.: iPinYou global RTB bidding algorithm competition dataset. In: Proceedings of the Eighth International Workshop on Data Mining for Online Advertising, ADKDD 2014, NY, USA, pp. 6:1–6:6 (2014). http://doi.acm.org/10.1145/2648584.2648590
12. Provost, F., Fawcett, T.: Data science and its relationship to big data and data-driven decision making. Big Data **1**(1), 51–59 (2013)
13. Ren, K., Zhang, W., Rong, Y., Zhang, H., Yu, Y., Wang, J.: User response learning for directly optimizing campaign performance in display advertising. In: Proceedings of the 25th ACM International on Conference on Information and Knowledge Management, CIKM 2016, NY, USA, pp. 679–688 (2016). http://doi.acm.org/10.1145/2983323.2983347
14. Richardson, M., Dominowska, E., Ragno, R.: Predicting clicks: estimating the click-through rate for new ads. In: Proceedings of the 16th International Conference on World Wide Web, WWW 2007, NY, USA, pp. 521–530 (2007). http://doi.acm.org/10.1145/1242572.1242643
15. Shan, L., Lin, L., Sun, C., Wang, X.: Predicting ad click-through rates via feature-based fully coupled interaction tensor factorization. Electron. Commer. Res. Appl. **16**, 30–42 (2016). http://www.sciencedirect.com/science/article/pii/S1567422316000144
16. Shani, G., Gunawardana, A.: Evaluating recommendation systems. In: Ricci, F., Rokac, L., Shapira, B., Kantor, P.B. (eds.) Recommender Systems Handbook, pp. 257–297. Springer, Heidelberg (2011). http://link.springer.com/chapter/10.1007/978-0-387-85820-3_8
17. Wang, J., Yuan, S.: Real-time bidding: a new frontier of computational advertising research. In: Proceedings of the Eighth ACM International Conference on Web Search and Data Mining, WSDM 2015, NY, USA, pp. 415–416 (2015). http://doi.acm.org/10.1145/2684822.2697041
18. Zhang, W., Du, T., Wang, J.: Deep learning over multi-field categorical data. In: Ferro, N., Crestani, F., Moens, M.-F., Mothe, J., Silvestri, F., Nunzio, G.M., Hauff, C., Silvello, G. (eds.) ECIR 2016. LNCS, vol. 9626, pp. 45–57. Springer, Cham (2016). doi:10.1007/978-3-319-30671-1_4
19. Zhang, W., Yuan, S., Wang, J.: Optimal real-time bidding for display advertising. In: Proceedings of the 20th ACM SIGKDD International Conference on Knowledge Discovery and Data Mining, KDD 2014, NY, USA, pp. 1077–1086 (2014). http://doi.acm.org/10.1145/2623330.2623633
20. Zhang, W., Yuan, S., Wang, J.: Real-time bidding benchmarking with iPinYou dataset. CoRR abs/1407.7 (2014). http://arxiv.org/abs/1407.7073

A Consensus-Based Method to Enhance a Recommendation System for Research Collaboration

Dinh Tuyen Hoang[1], Van Cuong Tran[1], Tuong Tri Nguyen[2],
Ngoc Thanh Nguyen[3], and Dosam Hwang[1(✉)]

[1] Department of Computer Engineering, Yeungnam University,
Gyeongsan, South Korea
hoangdinhtuyen@gmail.com, vancuongqbuni@gmail.com,
dosamhwang@gmail.com
[2] Hue University's College of Education, Hue, Vietnam
tuongtringuyen@gmail.com
[3] Faculty of Computer Science and Management,
Wrocław University of Science and Technology, Wrocław, Poland
Ngoc-Thanh.Nguyen@pwr.edu.pl

Abstract. With the development of scientific societies, research problems are increasingly complex, requiring scientists to collaborate to solve them. The quality of collaboration between researchers is a major factor in determining their achievements. This study proposes a collaboration recommendation method that takes into account previous research collaboration and research similarities. Research collaboration is measured by combining the collaboration time and the number of co-authors who already collaborated with an author. Research similarity is based on authors' previous publications and academic events they attended. In addition, a consensus-based algorithm is proposed to integrate bibliography data from different sources, such as the *DBLP Computer Science Bibliography*, *ResearchGate*, *CiteSeer*, and *Google Scholar*. The experimental results show that this proposal improves the accuracy of the recommendation systems, in comparison with other methods.

Keywords: Recommendation system · Collaboration · Consensus · Integration

1 Introduction

Recommendation systems have been used in many fields of industry, such as e-commerce, Amazon product recommendations, movie recommendations, and e-learning. In the academic domain, research collaborator recommendation has attracted a lot of attention in recent years [1,4,6]. Normally, research collaboration is based on working together in the research process to achieve the common goal of finding new scientific knowledge. Thus finding collaborators is one of the

© Springer International Publishing AG 2017
N.T. Nguyen et al. (Eds.): ACIIDS 2017, Part I, LNAI 10191, pp. 170–180, 2017.
DOI: 10.1007/978-3-319-54472-4_17

key factors that determine the quality of scientific results. The problem is how to find collaborators who are suitable for a particular research purpose. Especially, scientists who are students or young researchers will have big difficulty to find suitable collaborators because of lacking knowledge and information about the topic of interest. In addition, previous studies have confirmed that scientists or groups with well-connected collaboration networks tend to yield more research [1,6]. Therefore, it is necessary and important for scientists to make the acquaintance of new, worthy collaborators in academic social networks. To solve this problem, some methods have been proposed to recommend new and likely connections [1,4]. An academic social network can be considered as a weighted directed graph of nodes (such as authors or groups) that have certain relationships (for example, friendship and co-authorship, as shown in Fig. 1). The relationships represent some academic background (research interests, co-author information, and academic events attended, etc.). Since scholarly big data, recommending collaborators in academic social networks is thus an increasingly important topic. Some research collaboration methods have been investigated, and some bibliographic systems have been developed, such as the *DBLP Computer Science Bibliography*[1], *CiteSeerX*[2], *ResearchGate*[3]. However, the problem is that most of them works separately, and they do not satisfactorily mine the entire academic network. This leads to lack of a unified method to efficiently model the academic network. In previous studies [1,4], different types of information were modeled individually, and thus, the relationships between them did not consider obtaining the correct datasets.

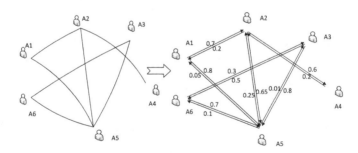

Fig. 1. An academic social network

In this study, a new consensus-based recommendation system (CRS) is proposed by taking into account co-authorship, academic event attendance records, and authors' publications. The dataset has been integrated from different sources, such as *DBLP, CiteSeerX, Google Scholar, ResearchGate*, and *Microsoft Academic Search*. Our method extracts research collaboration, publication similarity, and content from attended conferences in order to measure

---

[1] http://dblp.uni-trier.de/.

[2] http://citeseerx.ist.psu.edu/.

[3] https://www.researchgate.net/.

relationship strengths between authors. In addition, the relationship between authors who never collaborated before is also taken into account to improve the accuracy of the recommendation process.

The rest of this paper is organized as follows. In the next section, related work on recommendation systems concentrating on research collaboration is investigated briefly. In Sect. 3, the proposed method is presented. The experimental results and evaluations are shown in Sect. 4. Lastly, conclusions and future work are presented in Sect. 5.

2   Related Works

Research problems are increasingly sophisticated and complex. Thus, scientists need to cooperate with each other to solve them. The quality of collaboration between scientists has an important impact on the quality of results. Recommending collaborators is helpful in increasing group's connections and improving the quality of scientific research. Some approaches have been worked out for recommending collaboration between researchers. For instance, Chen et al. [1] proposed a relation strength similarity method (called RSS). They built a co-authorship network from authors who have common publications. They have defined an asymmetric vertex similarity measure that can be applied in weighted networks. Newman [7] used statistical properties of scientific collaboration. The results showed that the numbers of collaborators were different if researchers had different regulations. Li et al. [6] took into account how to find scholars' collaborations in a coauthors network. Several metric were considered through mutual paper coauthoring information, which helped to determine the link importance. These authors analyzed the DBLP dataset to measure the cooperation strengths.

Most of the existing methods are based only on one data source, such as *DBLP* or *CiteSeer*, etc. This is an essential limitation for efficient and effective collaborator recommendation process. In this paper, we assume that datasets are integrated from different resources, such as *DBLP*, *CiteSeer*, and *ResearchGate*. We will try to show that the quality of datasets is an important factor for improving the recommendation systems.

Data integration involves combining data residing in different sources and giving users a unified view of these data [9]. This process becomes necessary in a diversity of situations such as commercial and scicentific domains. Data integration appears with growing frequency both the volume and the appropriate to share existing data explodes which become the focus of extensive theoretical work, and a lot of open problems remain unsolved [8]. For example, Udin M.N. et al. [12] proposed a method for finding semantic relationships among tags. They took into account the pairwise relationships between tags, resources, and users, and the relationships among them. Duong et al. [3] proposed a method for building the collaborative ontology. The quality of consensus is used to reach consensus among participants in the collaborative group. The Vietnamese WordNet was built by using the consensus quality. In the academic context, Do et al. [2] proposed a framework to integrate bibliographic data in computer science publications from heterogeneous digital libraries. The framework consists of three

---

Algorithm 1. Integration algorithm

Input: - a set of authors $A = \{a_1, a_2, ..., a_n\}$;
        - a set of data sources $S = \{s_1, s_2, ..., s_m\}$;
Output: Homogeneous dataset;

 1: Import DBLP data;
 2: **for** each author $a_i \in$ A **do**
 3:    P_i; // the set of publications of author a_i from DBLP data
 4:    **for** each source $s_j \in$ S **do**
 5:      **if** $P_{ij} \notin P_i$ **then**
 6:        $P_i = P_i \cup P_{ij}$ // P_{ij} is the set of publications of author a_i in source s_j
 7:      **end if**;
 8:    **end for**;
      $P_i = integrate(P_i)$
 9: **end for**;
10: **return** Homogeneous dataset;

---

key components: A publication collector, a bibliographic parser, and a duplication checker. Schallehn et al. [10] proposed a generic adapter that can be used in highly distributed scenarios of XML and associated technology to transfer and homogenize data. Global citation linking to integrate digital libraries was developed as an application scenario.

3 Consensus-Based Recommendation Method

This section discusses how to build a research collaboration recommendation system based on consensus. The method focuses on finding a set of collaborators to recommend to the target author. Let $A = \{a_1, a_2, ..., a_N\}$ be a set of authors, P_{a_i} be a set of all publications of author a_i. The system architecture of the proposed CRS is shown in Fig. 2 and includes two stages: Integration identification and recommendation. Integration identification aims at integrating bibliography data from different sources like *DBLP*, *CiteSeer*, *ResearchGate*, etc. In the recommendation stage, collaboration strength is computed and ranked to get a set of collaborators who should be recommended to the target author. The details of the proposed method are presented in the following subsections.

3.1 Integration Identification Stage

In this subsection, a consensus methodology is used to integrate the data from the different bibliography sources like *DBLP*, *Academic.edu*, *Google Scholar*, etc. [2]. We do not consider the problem of author disambiguation assuming that each author has an independent name. Concretely, the integration stage is shown in Algorithm 1.

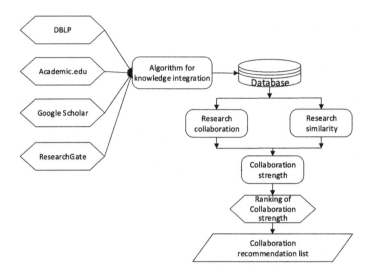

Fig. 2. The workflow of the integration method

Algorithm 2. Recommendation algorithm

Input: - Homogeneous data;

   - Target author a_m;

Output: A set of potential collaborators;

1: Preprocessing data;
2: Creating a weighted directed graph of target author a_m;
3: **for** each author a_i in list co-authors **do**
4:     Calculating $D(a_m, a_i)$;
5:     Getting K authors with largest collaboration strength for target author a_m;
6: **end for**;
7: **return** A set of potential collaborators to recommend to target author a_m;

3.2   Recommendation Method

Intuitively, an academic network is considered as a weighted directed graph, as shown in Fig. 1. A weighted directed graph is defined as $G = (A, E, W)$, where A is a set of vertices, which is a set of authors. We denote a set of directed edges by E, which represents the connections among authors and W represents the set of values for collaboration strength between authors.

In this study, collaboration strength is taken into account by combining research collaboration and research similarity. These factors are helpful in giving a more accurate collaborator recommendation to a target author.

Research Collaboration: Research collaboration is defined by taking into account collaboration time, the co-authors, and the conferences attended by the authors.

Definition 1. *The research collaboration between authors a_m and a_n is defined as follows:*

$$C(a_m, a_n) = \frac{1}{2}\left(\frac{|P_{a_m} \cap P_{a_n}|}{|P_{a_m}|} + \frac{|C_{a_m} \cap C_{a_n}|}{|C_{a_m}|}\right) \times \sum_{p_i \in (P_{a_n} \cap P_{a_m})} \frac{t_{p_i} - t_{p_1} + 1}{t_c - t_{p_1} + 1} \quad (1)$$

where $C(a_m, a_n)$ represents the research collaboration between authors a_m and a_n. $|P_{a_m}|$, $|P_{a_n}|$ are the numbers of all publications of authors a_m and a_n, while $|C_{a_m}|$ and $|C_{a_n}|$ represent the number of all attended academic events by these authors, respectively. We denote the publishing date of publication p_i by t_{p_i} and t_{p_1} is the publishing date of the first publications by authors a_m and a_n, while t_c is the current date (i.e., 2016 in this paper). Simply, $C(a_m, a_n)$ is the normalized frequency of the co-author and academic events attended within the range from 0 to 1. The value of $C(a_m, a_n)$ is greater for correspondingly more intensive collaboration between author a_n to author a_m in recent times.

Research Similarity: Co-authors can reveal the strength collaboration among themselves. However, it is worthy of note that if collaborators have never cooperated with the target author then recommending new collaborators is more useful and practical in the reality of academia. To tackle this problem, research similarity between authors is measured by taking into account their publications and the content of attended academic events. In this case, a deep learning **Doc2Vec** model, which achieves state-of-the-art results in document similarity [5], is used to compute the content similarity among many publications and many academic events.

Definition 2. *The research similarity between a_m and a_n is defined as follows:*

$$R(a_m, a_n) = \sum_{p_i \in P_{a_m}, p_j \in P_{a_n}} \frac{sim(p_i, p_j)}{(|P_{a_m}| \times |P_{a_n}|)} \times \sum_{c_i \in C_{a_m} c_j \in C_{a_n}} \frac{sim(c_i, c_j)}{(|C_{a_m}| \times |C_{a_n}|)} \quad (2)$$

where $R(a_m, a_n)$ is the research similarity between authors a_m and a_n; $|P_{a_m}|$, $|P_{a_n}|$ indicates the number of all publications of authors a_m and a_n. Function $sim(p_i, p_j)$ returns the content similarity between the abstracts of publications p_i and p_j, while $|C_{a_m}|$ and $|C_{a_n}|$ represent the number of all attended conferences for authors a_m and a_n. Function $sim(c_i, c_j)$ is the value of the content similarity between conferences c_i and c_j, while $|C_{a_m}|$ and $|C_{a_n}|$ represent the number of all attended conferences of authors a_m and a_n. Textual descriptions and topics of interest from conferences in WikiCFP data are extracted to calculate $sim(c_i, c_j)$. The value of $R(a_m, a_n)$ is greater for correspondingly closer the topic research among authors.

Collaboration Strength: The strength of interaction between authors is calculated by aggregating research collaboration and research similarity.

Definition 3. *The collaboration strength between authors a_m and a_n is defined as follows:*

$$T(a_m, a_n) = (1 - \alpha) \times C(a_m, a_n) + \alpha \times R(a_m, a_n) \tag{3}$$

where $0 \leq T(a_m, a_n) \leq 1$, which indicates the strength of the interaction between authors a_m and a_n. Parameter α is a constant ($\alpha \in [0, 1]$), which controls the rates of reflecting two important values on research collaboration and research similarity.

Normally, the collaboration strength between authors who have been connected in the past can compute as follows:

$$D_1(a_m, a_n) = \begin{cases} T(a_m, a_n), & \text{if } a_m \text{ direct connection to } a_n. \\ 0, & \text{otherwise.} \end{cases} \tag{4}$$

However, many potential collaborators have no direct connections, although they may have indirect connections [1]. Assume that $a_i = \{a_1, a_2...a_k\}$ is a set of mediate authors that connect author a_m and author a_n. Let p_j be a set of paths from author a_m to author a_n. The collaboration strength between authors a_m and a_n for indirect connections is considered as follows:

$$D_2(a_m, a_n) = \begin{cases} Max_{j=1..p}\left(\prod_{i=1}^{k-1} T(a_i, a_{i+1}) \right), & \text{if } a_m \text{ indirect connection to } a_n. \\ 0, & \text{otherwise.} \end{cases} \tag{5}$$

If the number of mediate authors is large, then the value of $D_2(a_m, a_n)$ will be too small and time-consuming to be calculated. Therefore, the number of mediate authors should be less than a threshold value ($k \leq h$).

Recommendation score is created by combining Eqs. (4) and (5) as follows:

$$D(a_m, a_n) = (1 - \beta) \times D_1(a_m, a_n) + \beta \times D_2(a_m, a_n) \tag{6}$$

where $\beta \in [0, 1]$ and controls the weight of the collaboration strength between $D_1(a_m, a_n)$ and $D_2(a_m, a_n)$. The value of $D(a_m, a_n)$ is within the range [0,1]. A set of collaborators that is generated based on ranking the recommendation scores and that is recommended for target author a_m is returned by Eq. (6).

4   Experiments

4.1   Datasets

We chose DBLP to start collecting the dataset, mainly for two reasons: (i) with more than 3.4 million publications by over 1.7 million authors and about 4896 conferences, DBLP is a good site to find authors and publications; and (ii) all bibliographic records are stored in the file *dblp.xml*[4]. The data was downloaded in May 2016 and contains these attributes: title, authors, year, type, names

---

[4] http://dblp.uni-trier.de/xml/.

of conferences, and DOI links. In addition, Algorithm 1 was used to integrate the bibliographic data from several online digital libraries like Google Scholar, ResearchGate, etc. The abstracts of the publications were retrieved with Elsevier API. The academic events data was crawled from WikiCFP [5,6,7] that contains information about 9871 conferences (from 2008 to 2011). We divided the dataset into two parts: the training data from 2008 to 2011 and the testing data from 2012 to 2014.

4.2 Evaluation

In this study, the most popular method from the recommendation system literature was used for evaluating the performance of the proposed method, which are *Precision, Recall*, and *F_measure* [11]. Normally, comparing the collaborator list in the testing dataset, there are four sets: (i) recommended and collaborated (TP), (ii) recommended but not collaborated (TN), (iii) not recommended but collaborated (FP), (iv) not collaborated and not recommended (FN). The values of *Precision, Recall*, and *F_measure* were computed as follows:

$$Precision = \frac{TP}{TP + FP} \tag{7}$$

$$Recall = \frac{TP}{TP + FN} \tag{8}$$

$$F\_measure = 2 \times \frac{Precision \times Recall}{Precision + Recall} \tag{9}$$

A co-author network is generated when authors collaborated on at least one publication. There are some authors who have not worked as co-authors with other researchers. Intuitively, they have no impact on other authors. Thus, we do not consider them in this study. To evaluate the performance of this proposed method, a set of 50 target authors was randomly selected. Then Algorithm 2 was used to create a set of recommended collaborators for each target author. For objective reasons, two scenarios were considered. First, we ranked the most valuable collaborators from recommendation systems and compared them with the test dataset. Second, we only considered new collaborators who have never collaborated with the target author. Values for *Precision, Recall*, and *F_measure* were calculated by averaging the 50 corresponding values.

4.3 Results and Discussions

The proposed method was compared with another recommendation method called RSS, which was proposed by Chen et al. [1]. Correctly, the RSS method

---

[5] http://wikicfp.com/examples/wikicfp.v1.2008.xml.gz.

[6] http://wikicfp.com/examples/wikicfp.v1.2009.xml.gz.

[7] http://wikicfp.com/examples/wikicfp.v1.2010.xml.gz.

uses the number of coauthored papers as the weight of each edge. Therefore, the collaboration strength between authors a_m and a_n was calculated as follows:

$$T(a_m, a_n) = \frac{|P_{a_m} \cap P_{a_n}|}{|P_{a_m}|} \qquad (10)$$

where P_{a_m}, P_{a_n} represent the numbers of all publications for authors a_m, a_n. Then, values of the collaborators from the recommendation system are ranked from highest to lowest.

Table 1. Comparison recommending collaborator methods

Method	Precision	Recall	F_measure
CRS	0.63	0.62	0.62
RSS	0.55	0.53	0.54

Table 2. Comparison recommending new collaborator methods

Method	Precision	Recall	F_measure
CRS	0.15	0.16	0.15
RSS	0.1	0.12	0.11

As can be seen from Tables 1 and 2, the proposed CRS reached better results than the RRS method in both cases. Simply, the CRS method combines research similarity and research collaboration to compute collaboration strength, which is used to represent collaboration between authors. The dataset is enriched by using the integration algorithm, and the publications are analyzed to produce the authors' features. Moreover, the CRS method takes into account the collaboration time and attended academic events. Therefore, this method has distinct benefits (e.g., greater accuracy to represent collaboration strength and rich information) for recommending collaborators, in comparison with the RSS method.

5   Conclusion and Future Work

This work proposed a new recommendation method for collaborator recommendation. First, the proposed method integrated data from many resources to improve the quality of the dataset. Secondly, the value of collaboration strength is calculated by means of the previous research collaboration and research similarity. Lastly, a weighted directed graph is built to compute the collaboration strength for potential authors who have not connected in the past. Collaboration time and the number of co-authors are considered to increase the accuracy in research collaboration. In addition, authors' publications and academic events attended are analyzed to obtain the research similarity. Deep learning algorithms are applied to improve the accuracy of the recommendation system. The experimental results show that authors have a tendency toward cooperation with other authors who have more mutual interactions. That is also true of authors who tend to collaborate with recent co-authors. Close research topics are also an important factor driving collaboration between authors. From the experiment analysis, the proposed method gives higher accuracy, compared with another method.

Since topic research can change over time, in future work, publications will be analyzed to detect their research topics. This is an important factor for improving the accuracy of the proposed method.

Acknowledgment. This paper is supported by the BK21+ program of the National Research Foundation (NRF) of Korea.

References

1. Chen, H.H., Gou, L., Zhang, X.L., Giles, C.L.: Discovering missing links in networks using vertex similarity measures. In: Proceedings of the 27th Annual ACM Symposium on Applied Computing, pp. 138–143. ACM (2012)
2. Do, T., Lam, D., Huynh, T.: A framework for integrating bibliographical data of computer science publications. In: 2014 International Conference on Proceeding of ComManTel, pp. 245–250. IEEE (2014)
3. Duong, T.H., Tran, M.Q., Nguyen, T.P.T.: Collaborative vietnamese wordnet building using consensus quality. Vietnam J. Comput. Sci., 1–12 (2016). doi:10.1007/s40595-016-0077-x
4. Klamma, R., Cuong, P.M., Cao, Y.: You never walk alone: recommending academic events based on social network analysis. In: Zhou, J. (ed.) Complex 2009. LNICSSITE, vol. 4, pp. 657–670. Springer, Heidelberg (2009). doi:10.1007/978-3-642-02466-5_64
5. Le, Q.V., Mikolov, T.: Distributed representations of sentences and documents. arXiv preprint (2014). arxiv:1405.4053
6. Li, J., Xia, F., Wang, W., Chen, Z., Asabere, N.Y., Jiang, H.: Acrec: a coauthorship based random walk model for academic collaboration recommendation. In: Proceedings of the Companion Publication of the 23rd International Conference on World Wide Web Companion, pp. 1209–1214. ACM (2014)
7. Newman, M.E.: Scientific collaboration networks. i. network construction and fundamental results. Phys. Rev. E **64**(1), 016131 (2001)
8. Nguyen, N.T.: Using consensus methods for solving conflicts of data in distributed systems. In: Hlaváč, V., Jeffery, K.G., Wiedermann, J. (eds.) SOFSEM 2000. LNCS, vol. 1963, pp. 411–419. Springer, Heidelberg (2000). doi:10.1007/3-540-44411-4_30
9. Nguyen, N.T.: Advanced Methods for Inconsistent Knowledge Management. Advanced Information and Knowledge Processing. Springer, London (2008)
10. Schallehn, E., Endig, M., Sattler, K.U.: Integrating bibliographical data from heterogeneous digital libraries. In: Proceeding of ADBIS-DASFAA Symposium, pp. 161–170 (2000)

11. Tran, V.C., Hwang, D., Jung, J.J.: Twisner: semi-supervised method for named entity recognition from text streams on Twitter. J. Univ. Comput. Sci. **22**(6), 782–801 (2016)
12. Uddin, M.N., Duong, T.H., Nguyen, N.T., Qi, X.M., Jo, G.S.: Semantic similarity measures for enhancing information retrieval in folksonomies. Expert Syst. Appl. **40**(5), 1645–1653 (2013)

International Business Matching
Using Word Embedding

Didier Gohourou[1(✉)], Daiki Kurita[1], Kazuhiro Kuwabara[2],
and Hung-Hsuan Huang[2]

[1] Graduate School of Information Science and Engineering,
Ritsumeikan University, Kusatsu, Shiga 525-8577, Japan
`gr0259sh@ed.ritsumei.ac.jp`
[2] College of Information Science and Engineering,
Ritsumeikan University, Kusatsu, Shiga 525-8577, Japan

Abstract. Recommender systems, which help users discover information or knowledge they might need without requiring them to have specific previous knowledge, are gaining popularity in our age of information overload. In addition, natural language processing techniques like word embedding offer new possibilities for extracting information from a massive amount of text data. This work explores the possibility of applying word embedding as the foundation for a recommender system to help international businesses identify appropriate counterparts for their activities. In this paper, we describe our system and report preliminary experiments using Wikipedia as a corpus. Our experiments attempt to provide answers to support business decision makers when they are considering entering a relatively unknown market and are seeking better understanding or appropriate partners. Our experiment shows promising results that will pave the way for future research.

Keywords: Recommender system · Word embedding · Word2vec · International business

1 Introduction

In an age where data are produced exponentially, people are frequently confronted by information overload, which is a term that describes the inability to filter out relevant from irrelevant information based on individual needs. One main reason why recommendation systems are gaining popularity is because they help discover what someone might need without requiring that person to have specific previous knowledge about what is being looked for. Recommendation systems are generally based on one of three approaches: collaborative filtering, content-based filtering, or a combination of the two called hybrid recommender system [1]. Due to their efficiency and popularity, recommender systems have been applied to such areas as e-commerce, social networks, and entertainment. Other domains exist where they can also support knowledge discovery.

© Springer International Publishing AG 2017
N.T. Nguyen et al. (Eds.): ACIIDS 2017, Part I, LNAI 10191, pp. 181–190, 2017.
DOI: 10.1007/978-3-319-54472-4_18

In the business area, information overload is a prominent problem in business organizations due to the very dynamic nature of the field. Business decision makers (who demand growth) often seek good business matches for either partnering in activities, merging with, or acquiring them. Information retrieval systems like search engines provide more knowledge about things we already know. However, most of the time, we are unaware of the range of options or cannot identify the most appropriate knowledge. This problem is common among business leaders who have difficulty using information technologies to find accurate business counterparts. Thus, they have to rely mainly on word-of-mouth to extend their business overseas. Recommender systems can alleviate this limitation.

Regarding the number and the diversity of worldwide business structures, it can be tedious and inaccurate to define a single model for businesses and apply traditional recommendation mechanisms. Another limitation of classic recommendation algorithms for this task is the lack of updated data for accurate recommendations. Gamified crowdsourcing might support data gathering and knowledge refinement [3,5]. Applying this approach to collect data on businesses requires a large number of crowdworkers, which may not be feasible for the business industry. A better knowledge base for this task is probably the text data continuously produced on the web by news articles related to businesses.

The emergence of language modeling techniques from the natural language processing area, such as word embedding, are opening new possibilities to extract information from massive amounts of data. Word embedding can overcome the lack of data often faced by classic recommendation algorithms, because it is based on unsupervised learning techniques. Word embedding models can be trained using news articles relevant to business domains. Popular software packages such as word2vec [6] or GloVe [10] are frequently used for this kind of task. Previous attempts successfully used word embedding for recommending items including software in application stores [9] or locations to visit [2].

Our work aims to build a content-based recommendation system based on word embedding for international business matching. Here, we present the design of our system and experimental results using Wikipedia article dumps as corpora for training word2vec models to determine whether word embedding is a suitable approach for business matching.

This paper begins by reviewing the related works. We then describe our recommendation methodology for international business matching, followed by an experimental evaluation and a discussion of the obtained results. Finally, we enumerate the challenges of subsequent steps and describe how we will address them.

2   Related Work

Word embedding is a set of natural language processing techniques that map words from the vocabulary of a corpus to vectors of real numbers in a continuous space. This technique has been used in many natural language processing (NLP) tasks, such as syntactic parsing, sentiment analysis, machine translation, etc.

Word2vec, which was recently developed by Mikolov et al. [6] is a popular word embedding package. In this section we review recommending tasks that exploit word embedding techniques, especially with word2vec.

The need for business to business recommender systems has been discussed for electronic commerce [12] in a work that proposed an experimental methodology that used a collaborative filtering recommendation approach. Even though the study targeted only e-commerce businesses, three different recommendation processes have been specified according to which party the recommendation was made: buyers, suppliers or intermediaries. Their proposed data acquisition method requires input from all three parties and must be connected to the internal database of the intermediary part. Even though this methodology initially gives quick results, scaling it might be hard since it involves the maintenance of three different recommendation processes.

Word embedding is getting a lot of attention as a recommendation mechanism, due to the unsupervised learning property of this technique. Musto et al. evaluated three widespread word embedding mechanisms as content-based filtering recommender system [7]: latent semantic indexing, random indexing, and word2vec as word embedding on MovieLens and DBbook datasets. This study, whose results are promising, was a prelude for future research in word embedding as a content-based recommender system mechanism. The work was later extended with a semi-automatic mechanism to map Wikipedia pages to items in the MovieLens and DBbook datasets [8]. Word embedding algorithms were evaluated again as content-based filtering recommender systems and proved to be comparable to state-of-the-art approaches based on collaborative filtering.

Word embedding, especially word2vec, has been applied to practical recommendation tasks. An example is the e-commerce platform, Rakuten, where word2vec was applied to item to item recommendations to find relevant items (given an item) and user to item recommendations to find relevant items (given a user's behavior) [11]. The method outperformed collaborative filtering when using click rate as a performance indicator. Another practical application is a location recommendation system for a social network called Foursquare [9], where a recommendation for the next place to visit is based on previous places visited by the user.

Some applications of word2vec to recommendation tasks led to the creation of word2vec extensions like Item2Vec [2]. Item2Vec is an item-based collaborative filtering, which slightly modifies the word2vec objective function.

These works mainly focused on mainstream recommendation tasks such as those on social networks or e-commerce; in contrast, in this paper, we focus on international business matching. Since we target international businesses, we also plan to consider the multilingual aspects of the underlying text corpora.

3   Methodology

3.1   Business Recommendations

The following is a description of a use case for our proposed system. A typical scenario is when a manufacturing business, for example, is planning to expand its activities to a new country, where unfortunately, most of its key partners such as its suppliers, are not yet implanted. The expanding business must find suitable local companies in the target country as potential business partners. Our proposed system is intended to recommend companies in the target country that satisfy the needs of the prospective business. Recommendations are made in a way that the recommended companies will be very similar to the current partners of the business where it is currently operating. The system uses corpora that contain information related to worldwide companies to produce recommendations. The system is then operational even when no appropriate businesses directory exists in the target country.

3.2   System Design

Our proposed system consists of three main modules: a crawler, a recommendation mechanism, and a syntactic parser. Figure 1 illustrates how they interact with each other. The following is a brief description of the interaction among the system mechanisms:

1. The crawler retrieves text data from articles of business news websites. The text data are preprocessed to remove unnecessary characters, such as non-alphanumeric, and stored. The stored text data are used as a corpus to train the model.
2. Word2vec is used to train a model using the stored text data as a corpus. Tools such as the Gensim python package[1] can continuously train the model as more text data become available from the crawler.
3. The front-end of the system uses a syntactic parser to convert user queries entered as natural language into vector operations.
4. These operations will be run against the trained model by the recommending mechanism.
5. The recommending mechanism uses the trained model to run the vector calculations to find similarities among the companies.
6. The recommended items (companies) are refined by user selection.

4   Experimental Evaluation

4.1   Recommendation Using Word2vec

For a better understanding of the system design choices and the experiment, this section first provides a brief description of word embedding with word2vec and

---

[1] https://radimrehurek.com/gensim/.

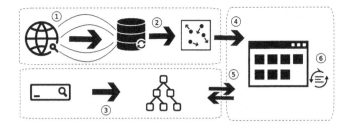

Fig. 1. System architecture consisting of a model trained from data coming from a web crawler, a recommendation mechanism, and a syntactic parser

defines some use cases of the proposed recommender system before reporting our experiment and its results.

Word2vec is a group of related two-layer neural networks for word embedding that takes a large corpus as input and produces a high-dimensional space where each word of the corpus is matched to a corresponding vector in the space. To train models of distributed representation of words, word2vec can use two types of architecture. One is called continuous bag of words (CBOW), where the model predicts a word from the surrounding context words. The other one is skip-gram (SG), where from a word the model predicts the surrounding context words.

Various vector calculations, essentially based on the cosine similarity of the vectors, can be run on the resulting model trained by word2vec. Some of these operations find the most similar words to one word by identifying which words have the closest vector representation to the defined one, or find a word that match another word in order to be similar to a given pair of words. The second operation can be better described by the following. Let the statement $king - queen = man - woman$ express that the relation between the vectors representing of the word $king$ and the word $queen$ is identical to the relation between the vectors representing the word man and the word $woman$. Then we have the popular equation from [6] $woman + king - man = queen$, which can be read as, $queen$ is for $woman$, as $king$ is for man. Then a statement like: $woman + king - man = ?$ request X from the model such as X is for woman as king is for man. That is, which word should be paired with $woman$ to obtain a similar relation between the words $king$ and man.

Following our business recommendation scenario, the following is one potential use case for our system: let B_1 be a business in location L_1. B_1 is using supplier S_1 in L_1. B_1 would like to extend its activities into another location L_2 where S_1 is not present. B_1 business leader of might be interested in finding supplier S_2 in L_2, which resembles S_1 and can deliver the same service in L_2. Here is another use case: by identifying a (good) partnering relation between two companies, a business leader might get advice about whom to partner with in order to get the same type of relationship. Referring to the previous overview of word2vec, we can answer the above use cases if they are properly formulated as vector calculations.

4.2 Experimental Setup

Since this paper's purpose is assessing the usefulness of word embedding as a foundation of a recommender system for international business matching, we used the English, French, and Japanese versions of Wikipedia dump files to train different word2vec models. The original implementation in the C language of the word2vec[2] package was used for training the models because the training time can be reduced significantly compared to other implementations like the python version (without cython). We used the skip-gram training model, since it tends to produce better accuracy [6]; the size of word vector dimension was set to 200, and the maximum skip length between words was set to 5.

The training took around 740 min for the English Wikipedia and 200 min each for the French and the Japanese Wikipedia dumps. The three models were trained on a laptop with a core i5 processor, 20 GB of memory, and an SSD hard-drive. We used the Gensim python package has been used to load and run calculation on the models from the training.

Considering some business industries, sample vector calculations reflecting business matching problems were run against the trained models. Some example targeted questions include *how similar is the relation between the car manufacturer and their countries of origin?*; *what is the equivalent for one company given the relation between two others?*; *how can we know the industry of a given company?*. The obtained answers can be discussed by either comparing them to real world business environment facts or pointing out how they can be exploited to support decision making.

4.3 Results and Discussion

In this section we qualitatively evaluate the results obtained from the viewpoint of our purpose, which is to determine whether word embedding can be the basis for an effective recommender system for business matching.

We start by interpreting the visual representation of the relation between car manufacturers and their countries of origin. We used principal component analysis (PCA) as a technique of dimensionality reduction for a 2-dimensional representation. Figures 2a, b, and c show the resulting graphs for the three models. The manufacturers and their nationalities are clearly separated as clusters, and many of the vectors representing manufacturer-nationality relations tend to have the same direction. This means the relationships are more or less similar.

Second, Tables 1a, b, and c show the top five results of the two queries we ran that targeted major businesses. For the first query results from the corpora of different languages, at least one automotive company is mentioned. Their would have been the one recommended by the system. The results from the English corpus are more significant since in the real business environment, Renault and Nissan have a successful partnership, while Isuzu was affiliated with General

---

[2] https://code.google.com/p/word2vec/.

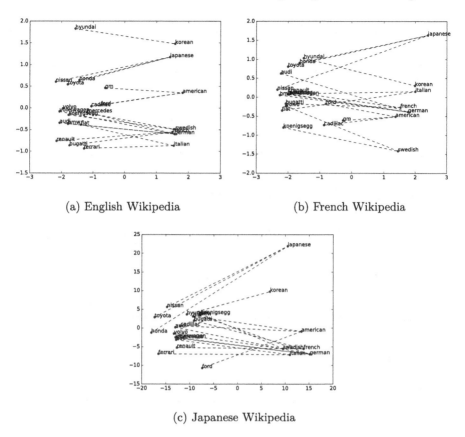

(a) English Wikipedia   (b) French Wikipedia

(c) Japanese Wikipedia

Fig. 2. Cars manufacturers and their countries of origin

Motors (GM). For the second query, the results from different corpora contain electronic/electric companies. Once again, the results from the English corpus are more significant since they are names of Japanese electronics companies.

A third observation was made by plotting top companies with a different representation by industry. Table 2 shows the industries the companies used. In Figs. 3a, b, and c, the companies are clustered by industry. Such visualization might help discover a company's industry on which we have very little information by just plotting a vector that represents the name of the given company. The English and French Wikipedia versions have better clustering than the Japanese one, partly due to the different representations some companies have in Japanese Wikipedia. For example, IBM is represented as either 'IBM' or 'アイビーエム'.

Additionally, notice the overall more or less significant differences in the quality of the results obtained from the different corpora (English, French, and Japanese versions of Wikipedia). The results extracted from the English version of Wikipedia tend to be more meaningful. This might reflect the fact that English

Table 1. Top 5 most similar words for two word2vec queries with cosine similarities

(a) English Wikipedia

gm + nissan - renault = ?		japan + samsung - korea = ?	
motors	0.698	toshiba	0.649
powertrain	0.640	panasonic	0.607
isuzu	0.628	hitachi	0.604
daewoo	0.626	fujitsu	0.578
ioniq	0.623	sanyo	0.573

(b) French Wikipedia

gm + nissan - renault = ?		japan + samsung - korea = ?	
fairlady	0.480	fujitsu	0.622
muzychuk	0.475	toshiba	0.567
calsonic	0.475	phablette	0.560
sx	0.443	panasonic	0.543
ferronato	0.438	reneas	0.519

(c) Japanese Wikipedia. Note that corresponding Japanese words were used during the experiment. (for example 'nissan' was represented by 日産自動車)

gm + nissan - renault = ?		japan + samsung - korea = ?	
toyota	0.581	nec	0.563
mitsubishi	0.548	casio	0.512
kawasaki	0.545	ricoh	0.511
honda	0.517	hamamatsu	0.504
mitsubishi corp	0.517	seiko	0.498

Table 2. Companies and industry

Motor vehicles	Information technologies	Oil and gas	Banking	Retail
Volkswagen	Apple	Exxon	Barclays	Walmart
Toyota	Google	Petrochina	Jpmorgan	Costco
Chevrolet	Microsoft	Chevron	Hsbc	Walgreens
Ferrari	Oracle	Total	Ubs	Tesco
Peugeot	Ibm	Gazprom	Bnp	Carrefour

Wikipedia has over five million articles, compared to the French and Japanese versions with fewer than two million articles each. These differences show the importance of the choice of a corpus for a particular task. In our case, since Wikipedia content may be too generic, we plan to address this in future works.

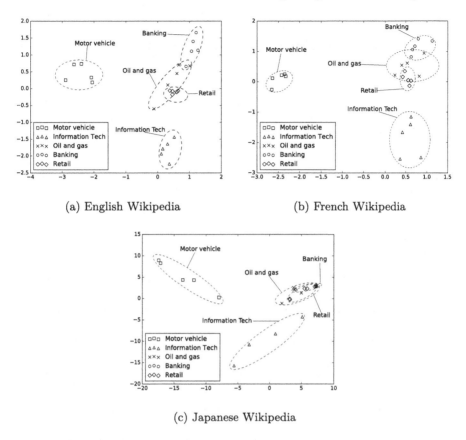

(a) English Wikipedia   (b) French Wikipedia

(c) Japanese Wikipedia

Fig. 3. Companies scattered by industry

5   Conclusion and Future Work

In most cases, our experiment results, which aimed at using word embedding as a recommender system for international business, showed meaningful interpretations in the real business world. This motivates us to expand our proposed research direction. However, we sometimes encountered results that were difficult to match to the real business world, perhaps due to the generic nature of Wikipedia's content as corpus. In the future, we plan to use specific business content as a corpus, which will be collected by running a crawler on business magazine websites, as defined in our system design.

So far we split the results by models trained on corpora of different languages. To address the multilingual aspects of international business more simply for the user, we plan to use a previous approach [4] that consists of constraining the model training to keep the representations of similar words in different languages close to each other. This step will help us extract single results from multilingual corpora.

Discovering similarities and other types of relationships between companies and using them for automatic recommendations will support business decision makers who want to find the best partner to grow their activities globally. We also plan to implement our prototype and evaluate its effectiveness in business settings.

Acknowledgment. This work was partially supported by JSPS KAKENHI Grant Number 15K00324.

References

1. Adomavicius, G., Tuzhilin, A.: Toward the next generation of recommender systems: a survey of the state-of-the-art and possible extensions. IEEE Trans. Knowl. Data Eng. **17**(6), 734–749 (2005)
2. Barkan, O., Koenigstein, N.: Item2vec: neural item embedding for collaborative filtering. CoRR abs/1603.04259 (2016)
3. Eickhoff, C., Harris, C.G., de Vries, A.P., Srinivasan, P.: Quality through flow and immersion: gamifying crowdsourced relevance assessments. In: Proceedings of the 35th International ACM SIGIR Conference on Research and Development in Information Retrieval, SIGIR 2012, pp. 871–880. ACM, New York (2012)
4. Green, S., Andrews, N., Gormley, M.R., Dredze, M., Manning, C.D.: Entity clustering across languages. In: Proceedings of the 2012 Conference of the North American Chapter of the Association for Computational Linguistics: Human Language Technologies, NAACL HLT 2012, pp. 60–69. Association for Computational Linguistics, Stroudsburg (2012)
5. Kurita, D., Roengsamut, B., Kuwabara, K., Huang, H.-H.: Knowledge base refinement with gamified crowdsourcing. In: Nguyen, N.T., Trawiński, B., Fujita, H., Hong, T.-P. (eds.) ACIIDS 2016, Part I. LNCS (LNAI), vol. 9621, pp. 33–42. Springer, Heidelberg (2016). doi:10.1007/978-3-662-49381-6_4
6. Mikolov, T., Chen, K., Corrado, G., Dean, J.: Efficient estimation of word representations in vector space. In: Proceedings of Workshop at ICLR (2013)
7. Musto, C., Semeraro, G., de Gemmis, M., Lops, P.: Word embedding techniques for content-based recommender systems: an empirical evaluation. In: RecSys 2015 Poster Proceedings, September 2015
8. Musto, C., Semeraro, G., de Gemmis, M., Lops, P.: Learning word embeddings from wikipedia for content-based recommender systems. In: Ferro, N., Crestani, F., Moens, M.-F., Mothe, J., Silvestri, F., Nunzio, G.M., Hauff, C., Silvello, G. (eds.) ECIR 2016. LNCS, vol. 9626, pp. 729–734. Springer, Heidelberg (2016). doi:10.1007/978-3-319-30671-1_60
9. Ozsoy, M.G.: From word embeddings to item recommendation. CoRR abs/1601.01356 (2016)
10. Pennington, J., Socher, R., Manning, C.D.: Glove: global vectors for word representation. In: Empirical Methods in Natural Language Processing (EMNLP), pp. 1532–1543 (2014)
11. Phi, V.T., Chen, L., Hirate, Y.: Distributed representation based recommender systems in e-commerce. In: DEIM Forum 2016, C8-1 (2016)
12. Zhang, X., Wang, H.: Study on recommender systems for business to business electronic commerce. Commun. IIMA **5**(4), 53–62 (2005)

Mixture Seeding for Sustainable Information Spreading in Complex Networks

Jarosław Jankowski[✉]

West Pomeranian University of Technology, ul. Żołnierska 49, 71-210 Szczecin, Poland
jjankowski@wi.zut.edu.pl

Abstract. A high intensity of online advertising often elicits a negative response from web users. Marketing companies are looking for more sustainable solutions, especially in the area of visual advertising. However, the research efforts related to information spreading processes and viral marketing are focused mainly on the maximization of coverage. This paper presents a sustainable seed selection solution based on mixtures of seeds with different characteristics. The proposed solution makes it possible for the information spreading processes to reach more diverse audiences. Mixture seeding avoids overrepresentation of nodes with similar characteristics, and thus decreases campaign intensity while maintaining acceptable coverage.

1 Introduction

In recent years, online social networks have become one of the most important channels for marketers for spreading information about new products, ideas and opinions. Research in this field is related to modelling of information diffusion processes [23], identification of factors affecting their dynamics [16], mechanisms of social influence [4], and initial seed selection strategies [15]. The main focus is placed on reaching nodes with high centrality measures and on increasing coverage based on the number of activated nodes within the network [6]. Selection of initial nodes with the use of centrality measures such as degree delivers a seed set with overrepresentation of nodes with high degree and underrepresentation of other nodes [20]. As a result of intensive viral marketing, a large amount of content is passed to the most influential nodes, with possible negative impact on target users [18]. Degraded user experience due to intensive marketing creates the need for more sustainable solutions [11]. However, these are mainly focused on visual content [12, 13]. To address this issue in social networks and viral marketing, a sustainable approach, using a mixture of seeds rather than homogeneous seed sets, is proposed. Different proportions of seeds in mixtures, with less top nodes, affects the characteristics of the nodes reached, and increases their diversity. The remainder of the paper is organized as follows: Sect. 2 presents a review of the literature, and Sect. 3 the conceptual framework and assumptions for the proposed approach. Empirical results are presented in Sect. 4, followed by a summary in Sect. 5.

© Springer International Publishing AG 2017
N.T. Nguyen et al. (Eds.): ACIIDS 2017, Part I, LNAI 10191, pp. 191–201, 2017.
DOI: 10.1007/978-3-319-54472-4_19

2   Literature Review

Online social networks, which are the media for information spreading processes, have played key roles in major social and political change, and are used for viral marketing and the spreading of rumours. Results from companies using word-of-mouth techniques show that properly designed campaigns can give better results than traditional marketing [25]. With similarities to the spread of infectious diseases, related research is focused on the use of epidemic models to predict and understand phenomena within social networks [14]. Other modelling techniques used are derived from the field of innovation diffusion such as the Linear Threshold Model (LT) and, more specific to social networks, the Independent Cascades Model (IC) defined by Kempe et al. [15]. The performance of information spreading processes is dependent on many factors, including network structures and social relations [1]. The selection of initial nodes for starting the information flow plays a crucial role in determining final coverage. The research problem defined by Kempe [15] as optimal seed set selection is widely discussed in the literature from the perspective of areas such as combinatorial optimization, physics, sociology and marketing [10].

Several methods are used to select seeds in such a way as to initiate information cascades and deliver high coverage within the network [7]. Proposed solutions are usually based on heuristics and selection of nodes with specific characteristics such as a high degree [5] and computationally expensive greedy selection [15] and its extensions [26]. Comparison of various seeding strategies shows the high performance of seeding based on centrality measures [10]. While early research was based mainly on static network structures, recent research moves towards temporal networks with higher focus on the role of network changes in the information spreading process [19]. Seed selection is usually based on single stage seeding, and computed initial network characteristics are used at all stages of the process. Recently, new adaptive approaches have been presented with evolving strategies that take into account changes of networks and process parameters [24].

Earlier research approaches are mainly focused on increasing the coverage represented by the proportion of activated nodes within the whole network rather the characteristics of the infected nodes [2]. Most studies are based on seed sets selected using a given criterion and the ranking of nodes. Reaching nodes with high centrality measures does not always give positive results: even though they have many connections within a network, they are not necessarily willing to pass on content. Among the many connections within a network, only a small proportion are based on strong social relations. For high degree nodes engaged in spreading information, performance and the influence on the recipients can be low due to weak ties and high volumes of unsolicited messages [18]. To change the proportion of nodes with similar values of high centrality measures in the target set, the sustainable approach presented in this paper assumes the reaching of nodes with different characteristics and a more diverse target group.

3   Conceptual Framework for Mixture Seeding

There are various seed selection strategies (denoted S_1, S_2, ..., S_n) for initiating information spreading processes within complex networks. Options include selection of nodes based on centrality measures such as degree, closeness or betweenness. For each strategy i ($i = 1, 2, ..., n$), a rank i of nodes is generated, according to their potential for spreading process based on the value of a centrality measure. Rank i takes the form of an ordered set of seed nodes $N_j = (N_1, N_2, ..., N_m)$, where for each element $Rank_j(N_i) > Rank_j(N_i + 1)$. A number m of top nodes from these rankings is selected as seeds, where m denotes the number of seeds, usually based on a given percentage of seed nodes within the network (SP). In typical seeding only one strategy is used and top nodes from one ranking are selected. After the seeding, according to the Independent Cascades (IC) model used in this paper, all seeds contact all of their neighbors and activate them with propagation probability PP. The process continues for all newly activated nodes until the information spreading process stops and no more activations are observed.

Selection of seeds by a single heterogeneous strategy may result in overrepresentation of nodes with a high value of one property and the potential of other nodes in network for spreading information not being used. For example, in degree based seeding, selection of additional seeds with high betweenness could improve communication between different segments of the network. The proposed mixture seeding overcomes this problem and uses all rankings $Rank_1$, $Rank_2$, ..., $Rank_n$ simultaneously. From each ranking, seeds are taken in proportion to the parameters of the mixture. In this experiment, k mixtures (M_1, M_2, ..., M_k) are created, where each mixture M_l ($l = 1, 2, ..., k$) uses different proportions of seeds from each ranking. The number of mixtures is dependent on the number of factors (strategies in this case) and the number of available proportion levels for each factor. Methods of mixture selection and evaluation are based on the initial concept described by H. Scheffe [23] and extended in later research [16]. Figure 1 illustrates examples of mixture based seeding for three mixtures of three node selection strategies. Green denotes nodes selected with the use of strategy S_1, blue nodes are selected according to strategy S_2, and red nodes are selected according to strategy S_3. Mixtures are generated using the approach presented in [16], with the sum of all proportions equal to one. Mixture M_1 uses the same proportion of seeds of each type (1/3, 1/3, 1/3). In mixture M_2, 2/3 of seeds are selected according to strategy S_2 and 1/3 of seeds are selected according to strategy S_3, with no S_1 based seeds. Mixture M_3 comprises 1/6 of S_1 seeds and the same proportion of S_2 based seeds while 2/3 of the seeds are selected according to S_3 strategy.

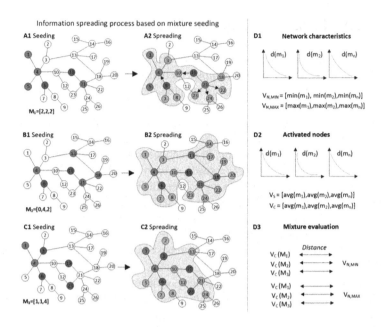

Fig. 1. Example showing seeding based on three types of mixtures M_1, M_2, M_3 with different proportions of seeds of each type, and mixture evaluation procedure

Mixture seeding assumes the ability to cover networks with a higher diversity of parameters observed in activated nodes. After the seed selection, the information spreading process starts; infected nodes (marked in orange) are shown for each mixture in diagrams A2, B2 and C2 of Fig. 1, with the resulting infected segment of network for each mixture. While a single strategy delivers seeds and final activations with high centrality measures according to the strategy used, the mixture seeding makes it possible to obtain more diverse activations and avoid overrepresentations of seeds with similar characteristics. Early research related to optimal mixtures assumes quadratic representation [23] but following [16], because of bounded regions of mixtures, cubic models are more appropriate. For the analysis of results in terms of coverage achieved for each mixture, the cubic model in the form presented in Eq. (1) can be used:

$$y = \sum_{i=1}^{q} \beta_i x_i + \sum_{i=1}^{q-1} \sum_{j=i+1}^{q} \beta_{ij} x_i x_j + \sum_{i=1}^{q-2} \sum_{j=i+1}^{q-1} \sum_{k=j+1}^{q} \beta_{ijk} x_i x_j x_k + \varepsilon \qquad (1)$$

with q mixture components, where coefficient β_i represents the expected response for $x_i = 1.0$, β_{ij} indicates the amount of quadratic curvature along the edges of the simplex region consisting of binary mixtures of x_i and x_j, β_{ijk} is the equivalent for the full cubic model, and ε represents the estimated error. The fitted parameters will indicate the influence of each selection strategy on the final response. Different mixtures can produce different outcomes not only in terms of coverage and the number of activations but also in terms of the characteristics of the nodes reached. Final results from each mixture can be evaluated in terms of the proportion of activated nodes and their relation to distributions of network measures. Distributions of each structural measure, including degree, closeness and others (represented by $m_1, m_2,..., m_n$), are computed (Fig. 1 D1). Two vectors are used: minima $V_{N,MIN} = [min(m_1), min(m_2), ..., min(m_n)]$ and maxima $V_{N,MAX} = [max(m_1), max(m_2), ..., max(m_n)]$. In the next stage (Fig. 1 D2), average values are computed for activated nodes: a vector of seeds $V_S = [avg(m_1), avg(m_2), ..., avg(m_n)]$ and a vector related to coverage based on activated nodes $V_C = [avg(m_1), avg(m_2),..., avg(m_n)]$. The Euclidean distances between the whole network and vectors from each mixture are computed, and used for the evaluation of the response to the mixture in terms of diversification.

4   Results

For the verification of the proposed approach, three seed selection strategies were used for mixtures, including random selection (R), degree based selection (D) and eigenvector based selection (E). For mixture experiment design simplex-lattice designs was used [23]. For three factors of seeds, a share in the mixture and three levels assumed the design with $i = 10$ mixtures M_i were created. The share of nodes from ranks Rank(R), Rank(D), Rank(E) in mixtures was designed as follows: $M_1 = [1, 0, 0]$, $M_2 = [2/3, 1/3, 0]$, $M_3 = [1/3, 2/3, 0]$, $M_4 = [0, 1, 0]$, $M_5 = [2/3, 0, 1/3]$, $M_6 = [1/3, 1/3, 1/3]$, $M_7 = [0, 2/3, 1/3]$, $M_8 = [1/3, 0, 2/3]$, $M_9 = [0, 1/3, 2/3]$, $M_{10} = [0,0,1]$ to cover full factorial experiments, with all combinations taken into an account. Mixtures M_1, M_4 and M_{10} containing only one type of seed are the equivalent of the conventional seeding for random, degree and eigenvector based seeding. Experiments were conducted within five real social networks, including N1 - High-Energy Theory Collaborations Network [21], N2 - General Relativity and Quantum Cosmology Collaboration Network [17], N3 - Hamsterster friendships within social network [9], N4 - Bible network [3] and N5 - Email communication network [8]. Experiments were conducted for the number of seeds based on five seeding percentages (SP), which were equal to 1%, 2%, 3%, 4% and 5% of all nodes within the network. A seeding percentage lower than 1% resulted in a seed set, which was too small to create mixtures, while values bigger than 5% were not realistic for a typical viral marketing campaign. For information flow, an independent cascade model (IC) and agent based stimulation was used [15]. In an IC model, each activated node has a single attempt to contact all connected nodes and activate them with propagation probability (PP). It is repeated in the next steps for all newly activated nodes, and the process continues, until no new activations are registered. As a propagation probability parameter, five distinct values were used for PP equal to 0.01, 0.02, 0.03, 0.04 and 0.05. Earlier experiments showed that propagation probabilities lower than 0.01 are resulting processes with very low dynamics; on the other hand, for values greater than 0.05, most strategies deliver a high coverage. Following the above settings, experimental space is a result of the Cartesian Product of N x SP x PP = 125 configurations and a simulation was performed for each mixture M_i $(i = 1, 2,...,10)$ and each configuration. Due to the probabilistic nature of the process, results were averaged from 100 runs. Each simulation delivered information about final coverage from each simulation represented by a number of activated nodes within the network. As a comparison between all configurations, results were normalized and are presented in the Table 1 from each mixture as a normalized coverage (NC). Overall results show that the best results were achieved for a typical degree based seeding (NC = 0.82), whose with homogenous seed consisted of high degree nodes, and this confirms many other studies showing performance of degree based seeding. However, other mixtures with 0.66 share of high degree nodes (M_3 and M_7) achieved a high performance with NC = 0.77 and NC = 0.71 respectively.

Results for all networks separately showed that a homogenous degree based selection only delivered best results for two networks (N4 and N5). The best results for network N1 were achieved for mixture M_7 with a share of high degree nodes DS = 2/3 and eigenvector share ES = 1/3. For network N2, the best results were achieved for the seed set with M_2 and only 1/3 share of high degree nodes with a high representation of random nodes. Network N3 achieved the best results for M_3. For network N4 and N5, typical degree based seeding delivered the best results. Analyses related to seeding percentage showed the best performance for degree based seeding, apart from low percentage SP = 1%. Role or high degree nodes decrease, along with the growth of propagation probability. For the lowest propagation probability (PP = 0.01) NC = 0.89 for a typical degree based selection, while NC = 0.82 and NC = 0.81 for .66 degree based seeds were mixed with eigenvector and ransom selection respectively. A similar situation takes place for PP = 0.02 and PP = 0.03 while, for higher propagation, the result will probably be better than for a degree based selection; these are observed for M_3 and M_7 with a 2/3 share of high degree nodes.

In the next stage, simulation cases were analysed with regards to the relation between mixture-based seeding, as shown in Fig. 2, and a mixture of seeding coverage denoted by the red line. For all the cases above, a red line was used as a reference, and this shows better results in terms of the number of activated nodes, while for cases below coverage was smaller than for mixture seeding. The proportion between conventional and mixture seeding is shown in Fig. 3, with values > 1 for cases with mixture seeding better than a conventional approach. Normalised coverage from all configurations shows the best performance for the conventional degree seeding (M_{10}); however, M_3 (2/3 share of degree and 1/3 share of random) and M_7 (2/3 share of degree and 1/3 share of eigenvector), results were at high levels as well, while lowest coverage was observed for conventional random seeding (M_1) (Fig. 4). Figure 5 shows simulation results for single cases with mixtures containing a degree based selection for networks N3, PP = 0.05 and SP = 0.05, while M_2 (0.66, 0.33, 0) showed better results.

Table 1. Normalized coverage from experimental results

M_i	Mixture share			All	Network					Seeding percentage					Propagation probability				
	RS	DS	ES		N1	N2	N3	N4	N5	1%	2%	3%	4%	5%	0.01	0.02	0.03	0.04	0.05
M_1	1	0	0	0.25	0.01	0.61	0.41	0.19	0.01	0.18	0.23	0.25	0.26	0.30	0.05	0.08	0.16	0.38	0.56
M_2	2/3	1/3	0	0.66	0.56	**0.85**	0.65	0.74	0.50	0.69	0.64	0.65	0.63	0.70	0.66	0.72	0.64	0.65	0.64
M_3	1/3	2/3	0	0.77	0.78	0.68	**0.77**	0.78	0.83	**0.80**	0.76	0.81	0.73	0.74	0.81	0.84	0.81	0.69	**0.69**
M_4	0.00	1	0	**0.82**	0.92	0.75	0.69	**0.81**	**0.91**	0.74	**0.81**	**0.85**	**0.89**	**0.8**	**0.89**	**0.94**	**0.86**	0.71	0.68
M_5	2/3	0	1/3	0.45	0.10	0.51	0.56	0.66	0.44	0.52	0.47	0.43	0.43	0.42	0.46	0.45	0.42	0.48	0.46
M_6	1/3	1/3	1/3	0.65	0.74	0.58	0.59	0.73	0.63	0.70	0.62	0.71	0.62	0.61	0.70	0.71	0.64	0.61	0.60
M_7	0.00	2/3	1/3	0.71	**0.96**	0.45	0.63	0.73	0.8	0.74	0.72	0.74	0.65	0.71	0.82	0.79	0.73	**0.72**	0.50
M_8	1/3	0	2/3	0.44	0.24	0.30	0.43	0.69	0.55	0.52	0.46	0.42	0.40	0.40	0.54	0.52	0.46	0.43	0.24
M_9	0	1/3	2/3	0.61	0.86	0.34	0.57	0.65	0.65	0.67	0.58	0.66	0.60	0.56	0.72	0.69	0.67	0.57	0.42
M_{10}	0	0	1	0.4	0.27	0.02	0.48	0.65	0.57	0.51	0.36	0.44	0.36	0.32	0.50	0.49	0.40	0.33	0.26

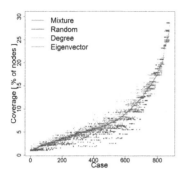

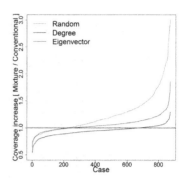

Fig. 2. Relation of coverage from conventional seeding to the cases from mixture seeding.

Fig. 3. Number of mixture based cases with higher coverage than conventional seeding

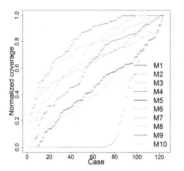

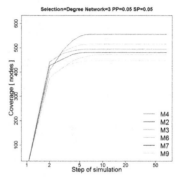

Fig. 4. Normalised coverage for all mixtures M_1-M_{10}.

Fig. 5. Example diffusion process for mixtures with different proportion of seeds with high degree.

In the next stage, a statistical model was built to fit the response, using the R package with the *mixexp* library and method presented in [16] and coefficients computed for the model (1). For the random component (R) value 0.27 for coefficient was achieved. The highest value of coefficient (0.80) was achieved for the degree, and the same value for mixtures, in which the degree and random nodes are represented. The model shows that the usage of all high degree nodes is not necessary. If the proportion of degree nodes is at a high level, i.e. 0.66, other used seeds can be selected randomly. The combination of high degree nodes with high eigenvector nodes did not bring positive results and a low model coefficient was achieved (0.27). The model response is showed in Fig. 6, whose visualisation is based on curves proposed by [22] for mixtures with three components. The effect plot in Fig. 7 shows the direction for three used factors, and the proportion of random (R), degree (D) and eigenvector share (E).

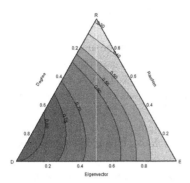

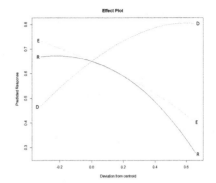

Fig. 6. Response from mixture model   **Fig. 7.** Effect plot based on Piepel direction

During the next step, an analysis is performed, and it includes the characteristics of all infected nodes for each used configuration of the network, propagation probability (PP) and seeding percentage for each mixture M_1-M_{10}. For each seed, five measures were retrieved: degree, second-level degree as a sum of degree and degrees of neighbours, closeness, eigenvector and Page Rank, and assigned to each infected node in the form of vector V_S = [avg(D), avg(D2), avg(CL), avg(EV), avg(PR)]. The Euclidean distance between V_S and minimal values from the whole network $V_{N,MIN}$ = [min(m_1), min(m_2), min(m_n),] as well as maximal values $V_{N,MAX}$ = [max(m_1), max(m_2), max(m_n)], were computed and showed in Fig. 8.

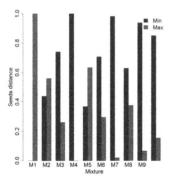

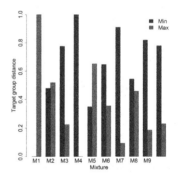

Fig. 8. Distance from vector V_S representing seeds to whole network *min* and *max* vectors   **Fig. 9.** Distance from vector V_c for reached nodes to whole network *min* and *max* vectors

The degree-based selection (M_4) gives the shortest distance to maximal vector for most cases, and a random selection gives the shortest distance to the minimal vector. The opposite is observed for random selection (M_1). Mixture M_2 [0.66, 0.33, 0] shows the most diversified results with a similar distance from minimal and maximal values. Figure 9 shows a similar relation between distance of reached nodes represented by the coverage vector V_C, the whole network's properties ($V_{N,MIN}$, $V_{N,MAX}$) and same sustainable results were achieved for mixture M_2.

5   Summary

Recent trends in digital marketing show the need for searching for sustainable solutions to limit the negative impact on target users. While this is discussed usually in relation to visual advertising content, information diffusion and viral marketing studies are usually focussed on reaching a high coverage within the network. From the performance point of view it is not only coverage that is important, but also the characteristics of the nodes reached. The approach presented here is an attempt to search for sustainable solutions that not only reach high degree nodes but achieve balance in nodes reached. The results show that the mixture-based approach can be used to search for sustainable solutions and helps to avoid overrepresentation of nodes with specific characteristics among reached nodes, and that the proportion of seeds in the mixture affects the target group. In many cases it was possible to obtain better results with mixture seeding than with conventional seed selection. The results showed that it was possible to decrease the proportion of high degree nodes. These can be expensive to reach and can have many connections with weak ties, which can decrease performance. The experiment presented revealed several future directions. Firstly, more networks can be verified with detailed analysis of factors affecting their performance, such as distributions of network measures and other structural characteristics. Other propagation probabilities can be tested to observe mixture based seeding, which will improve performance. Another direction could be the use of more mixture components and other structural characteristics; however, this will greatly increase the experimental space.

Acknowledgements. This work was partially supported by the Polish National Science Centre, grant no. 2016/21/B/HS4/01562.

References

1. Bampo, M., Ewing, M.T., Mather, D.R., Stewart, D., Wallace, M.: The effects of the social structure of digital networks on viral marketing performance. Inf. Syst. Res. **19**(3), 273–290 (2008)
2. Bhagat, S., Goyal, A., Lakshmanan, L.V.: Maximizing product adoption in social networks. In: Proceedings of the Fifth ACM International Conference on Web Search and Data Mining, pp. 603–612. ACM (2012)
3. Bible network dataset – KONECT, October 2016
4. Chen, W., Wang, C., Wang, Y.: Scalable influence maximization for prevalent viral marketing in large-scale social networks. In: Proceedings of the 16th ACM SIGKDD International Conference on Knowledge Discovery and Data Mining, pp. 1029–1038. ACM (2010)
5. Chen, W., Wang, Y., Yang, S.: Efficient influence maximization in social networks. In: Proceedings of the 15th ACM SIGKDD International Conference on Knowledge Discovery and Data Mining, pp. 199–208. ACM (2009)
6. Dobele, A., Toleman, D., Beverland, M.: Controlled infection! Spreading the brand message through viral marketing. Bus. Horiz. **48**(2), 143–149 (2005)
7. Fan, X., Li, V.O.: The probabilistic maximum coverage problem in social networks. In: Global Telecommunications Conference, 2011 IEEE, pp. 1–5 (2011)

8. Guimera, R., Danon, L., Díaz-Guilera, A., Giralt, F., Arenas, A.: Self-similar community structure in a network of human interactions. Phys. Rev. E **68**(6), 065103 (2003)
9. Hamsterster friendships network dataset – KONECT, October 2016
10. Hinz, O., Skiera, B., Barrot, C., Becker, J.U.: Seeding strategies for viral marketing: an empirical comparison. J. Mark. **75**(6), 55–71 (2011)
11. Jankowski, J., Kazienko, P., Wątróbski, J., Lewandowska, A., Ziemba, P., Zioło, M.: Fuzzy multi-objective modeling of effectiveness and user experience in online advertising. Expert Syst. Appl. **65**, 315–331 (2016)
12. Jankowski, J., Ziemba, P., Wątróbski, J., Kazienko, P.: Towards the tradeoff between online marketing resources exploitation and the user experience with the use of eye tracking. In: Nguyen, N.T., Trawiński, B., Fujita, H., Hong, T.-P. (eds.) ACIIDS 2016. LNCS (LNAI), vol. 9621, pp. 330–343. Springer, Heidelberg (2016). doi:10.1007/978-3-662-49381-6_32
13. Jankowski, J., Wątróbski, J., Witkowska, K., Wolski, W.: The use of a fuzzy cognitive maps and eye tracking in exploitation of online advertising resources. In: ICEIS, pp. 467–472 (2016)
14. Kandhway, K., Kuri, J.: How to run a campaign: optimal control of SIS and SIR information epidemics. Appl. Math. Computat. **231**, 79–92 (2014)
15. Kempe, D., Kleinberg, J., Tardos, É.: Maximizing the spread of influence through a social network. In: Proceedings of the Ninth ACM SIGKDD International Conference on Knowledge Discovery and Data Mining, pp. 137–146. ACM (2003)
16. Lawson, J., Willden, C.: Mixture experiments in R using mixexp. J. Stat. Softw. Code Snippets **72**(2), 1–20 (2016)
17. Leskovec, J., Kleinberg, J., Faloutsos, C.: Graph evolution: densification and shrinking diameters. ACM Trans. Knowl. Discovery Data (ACM TKDD) **1**(1), 2 (2007)
18. Michalski, R., Jankowski, J., Kazienko, P.: Negative effects of incentivised viral campaigns for activity in social networks. In: 2012 Second International Conference on Cloud and Green Computing (CGC), IEEE, pp. 391–398 (2012)
19. Michalski, R., Kajdanowicz, T., Bródka, P., Kazienko, P.: Seed selection for spread of influence in social networks: temporal vs. static approach. New Gener. Comput. **32**(3–4), 213–235 (2014)
20. Mislove, A., Marcon, M., Gummadi, K. P., Druschel, P., Bhattacharjee, B.: Measurement and analysis of online social networks. In: Proceedings of the 7th ACM SIGCOMM Conference on Internet Measurement, pp. 29–42. ACM (2007)
21. Newman, M.E.J.: The structure of scientific collaboration networks. Proc. Natl. Acad. Sci. U.S.A. **98**, 404–409 (2001)
22. Piepel, G.F.: Measuring component effects in constrained mixture experi-ments. Technometrics **24**(1), 29–39 (1982)
23. Scheffé, H.: Experiments with Mixtures. J. Roy. Stat. Soc. B **20**, 344–360 (1958)
24. Stonedahl, F., Rand, W., Wilensky, U.: Evolving viral marketing strategies. In: Proceedings of the 12th Annual Conference on Genetic and Evolutionary Computation, pp. 1195–1202. ACM (2010)
25. Trusov, M., Bucklin, R.E., Pauwels, K.: Effects of word-of-mouth versus traditional marketing: findings from an internet social networking site. J. Mark. **73**(5), 90–102 (2009)
26. Wang, Y., Cong, G., Song, G., Xie, K.: Community-based greedy algorithm for mining top-k influential nodes in mobile social networks. In: Proceedings of the 16th ACM SIGKDD International Conference on Knowledge Discovery and Data Mining, pp. 1039–1048. ACM (2010)

Complex Networks in the Epidemic Modelling

Tomasz Biegus and Halina Kwasnicka[⊠]

Department of Computational Intelligence,
Wroclaw University of Science and Technology, Wroclaw, Poland
halina.kwasnicka@pwr.edu.pl
http://www.ii.pwr.edu.pl/~kwasnicka

Abstract. The spread of infectious diseases is analyzed in the paper using social networks. Study of this phenomenon is very important due to the safety of all of us. After the analysis of properties of generated networks using known algorithms a novel method for generating scale-free complex networks was proposed. Features of the epidemic course were studied using computer experiments. The proposed method was compared with three known strategies in terms of influence of vaccination strategies on the tempo and mode of epidemic spread and the number of infected individuals.

Keywords: Epidemic spread · Social networks · Vaccination strategy

1 Introduction

Infectious diseases have deprived people's lives since centuries. Humanity has suffered various epidemics. The risk associated with viruses and bacteria causing infections is still very high, therefore, the determination of the appropriate strategies, allowing to minimize the number of cases, is a serious problem. Currently, we observe the intensive interdisciplinary research in this area, involving specialists from various fields as physics, mathematics, sociology, biology, and medicine. The great achievement of modern science is vaccination. If an effective vaccine against a given disease exists, it becomes a powerful weapon in the fight against it. It helps to prevent the emergence of infectious diseases, and as the population gets such disease, reduce the effects. The ideal situation would be to immunize the whole population of people. But it is impossible due to a number of reasons, e.g.: economical reason, concerns about potential side effects, etc. The choice of the right group of people to be vaccinated can be crucial for avoiding or diminishing epidemic effects. Intensive researches are conducted on the strategies of selection of vaccination [1]. Some researchers propose the use of a sociological phenomenon known as the *friendship paradox* [2]: *our friends have more friends than we have*. The group of people selected for vaccination according to some 'good' strategy helps to suppress epidemic. It is also possible to use certain groups of individuals as 'sensors'; their health status can be monitored for early detection of epidemic explosion [3,4].

© Springer International Publishing AG 2017
N.T. Nguyen et al. (Eds.): ACIIDS 2017, Part I, LNAI 10191, pp. 202–213, 2017.
DOI: 10.1007/978-3-319-54472-4_20

The aim of the presented study is to develop a model that accurately reflects the phenomenon of the spread of the pathogen in a population, and use it to assess the impact of the structure of the network of contacts, and other factors, on the spread of the epidemic. The SIR (**S**usceptible, **I**nfected, **R**emoved) model is used in the presented research. Complex networks are used as a model of networking between people. We use four methods of complex artificial networks generation; three are customized classic solutions, the fourth is the authors' algorithm that generates *scale-free* networks, widely used in the field of epidemic modeling. A series of experiments examining the characteristics of the epidemic spreading in these networks have been conducted.

The paper is organized as follows. The next section presents very shortly the epidemiological models. The third section gives the overview of complex networks. Section fourth describes the networks and algorithms used for their generation. Section five reports conducted experiments and their results. Finally, the summary and future works are given.

2   Epidemiological Models

The model SI (**S**usceptible, **I**nfected) is the oldest one; it was proposed in 1927 [5] as a trial of understanding the damn epidemic in Mumbai in 1905–1906. SI models the dynamics of the epidemic by dividing the population into two groups using differential equations. Other models differ in the number of groups and the way of determining changes in cardinality of particular groups [6]. The most popular are: (1) SIR (**S**usceptible, **I**nfected, **R**emoved/**R**ecovered); (2) SIS (**S**usceptible, **I**nfected, **S**usceptible); (3) SI (**S**usceptible, **I**nfected).

Model SIR assumes that the individuals can be in the state **S** what means that they are healthy and susceptible for infection, or in the state **I** – they are infected and can infect others, or in the state **R** – they passed the infection and acquired immune resistance to it. A scheme of this model is a simple directed graph: $S \Rightarrow I \Rightarrow R$. Arrows indicate the possible transitions between groups. This model assumes a constant number of individuals, nobody dies, and everyone after infection is completely resistant to this pathogen. SIR is described by three differential Eqs. 1(a), (b) and (c):

$$\frac{d|S|}{dt} = -\beta |S||I|, \ (a) \quad \frac{d|I|}{dt} = \beta |S||I| - \gamma |I|, \ (b) \quad \frac{d|R|}{dt} = \gamma |I|, \ (c) \qquad (1)$$

where $|X|$ is a size of set X, β and γ are coefficients. Equation 1(a) describes a reduction in the number of healthy individuals, it is proportional to the product of the number of healthy and infected in the population. Equation 1(b) says that a number of sick people $|I|$ increases as much as the group S diminishes, and diminishes as group R increases. Equation 1(c) describes the recovery process depends entirely on the number of infected and a type of pathogen. The model assumes that passing the disease gives the total resistance for the pathogen (true for mumps, rubella, and some others).

Model SIS considers two groups of individuals: S – individuals healthy and susceptible to disease and I – ill (infected). An adequate scheme of this model is a graph: $S \Rightarrow I \Rightarrow S$. Here every individual can change the state from S to I and again to the state S; it means that the infection does not give the resistance for the pathogen, Eq. 2.

$$\frac{d|S|}{dt} = -\beta|S||I| + \gamma|I|, \quad \frac{d|I|}{dt} = \beta|S||I| - \gamma|I| \tag{2}$$

The oldest model SI assumes that individuals after infection can be healthy but still they can infect others (e.g., HIV). The schema of SI is: $S \Rightarrow I$. It is described by Eq. 3:

$$\frac{d|S|}{dt} = -\beta|S||I|, \quad \frac{d|I|}{dt} = \beta|S||I \tag{3}$$

3   Complex Networks

All means of infectious diseases spread require a contact (direct or indirect) between a healthy person and infected. An intrusive manner to represent a complex system, consisting of individuals and the connections between them are the graphs.

From the epidemiology point of view, graphs reflecting the structure of contacts associated with exposure to transfer the pathogen are the most interesting. The oldest approach to complex networks modeling is the use of *random* graphs, proposed by Erdös-Rényi, shortly called ER [7]. It assumes that the probability of the edge between every pair of vertices is the same. This assumption is false for social networks, as well as for many other complex networks, occurring in nature. Common characteristics of complex networks found in nature, including social networks, differ greatly from the random graphs. Two well-known features of complex networks found in nature are *scale-free* (without scalability), and the *small-world* [8]. The *small world* means that between any pair of vertices in the graph is the relatively short path [9], the algorithm of creation such networks is called **Watts-Strogatz**, we refer to it shortly, WS. Observation of real networks (energy networks, actors networks – connection exists between actors playing in the same film, and neural connections in the body of nematode C. Elegans) allowed to see the need of construction the networks with the *small-world* property.

Authors of [10] found that for natural networks, the probability $P(k)$ that a vertex has connections with k other vertices, for big values of k, is approximately proportional to the power function with negative exponent γ, $P(k) \sim k^{-\gamma}$. The power function has *long tail* resulting in a relatively high probability of the presence in the graph vertices, with a very large number of connections with others. In the context of social networks, such vertices are called *people hubs*. The *people hubs* in society significantly influence on the spread of infection. The authors propose a new algorithm for complex networks generation, known as Barabási – Albert algorithm (BA in short) [10]. A common feature of the

ER (*random*) and WS (*small-world*) models is that the probability of the new vertex which has a lot of connections (large k), decreases exponentially with increasing k. For this reason, ER and WS algorithms generate the networks in which vertices with very many connections practically do not appear. In contrast, the *tail power-law* distribution analyzed by Barabási and Albert causes relatively high probability $P(k)$ for large k, and vertices with high degree exist.

Currently, the use of complex networks in combination with models SIR, SIS, SI is widespread in the epidemic study. The potential benefits of such research on the prevention of the spread of HIV are presented in [11]. The analysis was carried out using a real network of real contacts reconstructed on the basis of the virus samples taken from different individuals. Pellis with coauthors [12] list eight current challenges in the field of modeling of epidemics using complex networks, among them is'explain the impact of the network properties on the outbreak of an epidemic' which is one of the research interest in our study. The authors of [11] also indicate the importance of distribution of degrees of vertices in the network. Networks *scale-free* with the *long-tail* characteristics are popular among researchers of various processes, in particular epidemiological studies. Authors of [13] investigated the spread of the epidemic in the model SIQRS (**S**usceptible, **I**nfected, **Q**uarantined, **R**emoved) and using *scale-free* networks generated by BA algorithm. Complex networks are useful in many diverse fields. In [14,15] the problem of complex networks and the spread of computer viruses is analyzed. Models of the networks changing over time are also studied [12,16,17].

4   Generation the Complex Networks Used in the Research

In the presented research, we use complex networks, namely generated by algorithms Erdös-Rényi ER (*random*), Watt-Strogatz WS (*small-world*), Barabási-Albert BA (*scale-free*), and the fourth is the authors' algorithm named *GammaScaleFree* – it generates *scale-free* networks.

In our study, we have interested in the influence of the average degree of vertices on the epidemic development. ER, WS, and BA algorithms do not ensure the required (assumed) value of average degree of vertices. Therefore, we modified them, for modified version we use names ERm, WSm, and BAm respectively. The ERm algorithm takes two values as input parameters: $|V|$ – a number of vertices, and $|E|$ – the number of edges. The method starts from the $|V|$ not connected vertices. Next, two vertices are randomly chosen, and if they are not connected yet, the edge between them is added. The procedure stops when $|E|$ edges is added. Here, value of $|E|$ can be: $0 \leq |E| \leq |V|(|V| - 1)/2$.

Algorithm WSm takes as the input three parameters: $|V|$, $|E|$, and $p_{reconnect}$ – a probability of edge reconnection. The algorithm consists of two steps: (1) A ring of $|V|$ vertices is created, and each vertex is connected to its neighbors, assuring that the number of edges is exactly $|E|$. (2) The reconnection phase: we choose a vertex and the edge to its nearest clockwise neighbor. With probability $p_{reconnect}$, we reconnect this edge to a vertex chosen uniformly at random over the entire ring, with duplicate edges forbidden. Otherwise, we leave the edge without change. This process is repeated by moving clockwise around the ring, considering each vertex in turn until one lap is completed.

Algorithm BAm, a modified version of BA, takes two input parameters: $|V|$ and $|E|$. First step is creation of m_0 not connected vertices, where $m_0 = \lfloor 0.05\,|V| \rfloor$. For the next step the m parameter is required, it is set as $m = \lceil |E|/(|V| - m_0) \rceil$. The second step goes as follows: Repeat $|V| - m_0$ times the following: add a new vertex $v_i, (i = m_0, m_0 + 1, ..., |V|)$; add m vertices in the following manner: choice randomly one vertex $v_j, j \in \{v_1, v_2, ..., v_{i-1}\}$ according to the probability $p_{connect}(v_j) = \frac{k_j+1}{\sum_l^{|V|}(k_l+1)}$, where k_j is a degree of vertex v_j, and add the edge between v_i and v_j; repeat it m times to assure that m edges will be added to every added vertex v_i. The whole procedure stops when the graph consists of $|V|$ vertices. Due to the ceiling operation in the calculation of m, we can obtain more than $|E|$ edges. In such a case the redundant edges are removed with equal probability. Finally, the resulting graph consists of $|V|$ vertices and $|E|$ edges.

One thing should be commented here. According to the authors of BA, this algorithm produces *scale-free* graphs with the *power-law tail* distribution of degree of vertices; for large k the negative exponent of the power function is $\gamma = 2.9 \pm 0.1$. A question arises: does BAm has the same property? There is no place here to present all results of our experiments. The conclusion of our study is that BA and BAm have a significant limitation, namely, no way to get any value of γ . Regardless of the number of edges and the number of vertices, the tail of the distribution is always proportional to a power function $f(k) = k^{-2.9}$.

The most important observation in [10] is that the probability $P(k)$ that the vertex has a connection with k other vertices, for a certain interval of k, is approximately proportional to a power function with negative exponent $P(k) \propto k^{-\gamma}$. BA and BAm can generate this type of networks, however, we cannot freely choose the value of γ. Both produce *scale-free* networks for which the γ value is 2.9 ± 0.1. Meanwhile, the same work [10] indicates that γ for different real networks has different values: actors network: $\gamma = 2.3 \pm 0.1$; www: $\gamma = 2.1 \pm 0.1$; energy network in the west of the USA: $\gamma \approx 4$; scientific citation: $\gamma = 3$. The above shows that the use of graphs any assumed value of γ would allow generating networks represent different types of contacts. This fact was the motivation to propose our algorithm generating a network of *scale free* – an algorithm allowing to select γ as a parameter.

The proposed *GammaScaleFree* algorithm takes three parameters as input data: $|V|$, $|E|$, and the negative exponent γ (see Pseudocode 1). Its idea is simple. The number of connected edges to the vertex is a random value according to the power distribution (Eq. 4), therefore the generated graph has the distribution of the degree of vertices similar to that given by the power function with assumed value of γ.

$$p_{degree}(s, \gamma, k) = \frac{k^{-\gamma}}{\sum_{i=s}^{|V|-1} i^{-\gamma}} \tag{4}$$

where $p_{degree}(s, \gamma, k)$ is a probability that a vertex will have desired degree k, $k \in s, s+1, ..., |V|-1$, s is responsible for 'shift' of the distribution, (see Fig. 1). For $k \notin \{s, s+1, ..., |V|-1\}$, $p_{degree}(s, \gamma, k) = 0$. This distribution is used to select randomly the value of desired degree $k_{desired}(v_i)$ for every $v_i \in V$. Practically, we require the resulting distribution proportional to the assumed power-law distribution. For undirected graph without feedback, consisting of $|V|$ vertices, k (i.e., degree of a vertex) can take a value from the set: $\forall_{v \in V} k(v) \in \{0, 1, ..., |V|-1\}$. The probability of adding an edge to a given vertex v_i is given by Eq. 5

$$p_{adding}(s, \gamma, v_i) = \frac{k_{desired}(v_i)}{\sum_{j=1}^{|V|} k_{desired}(v_j)} \tag{5}$$

Pseudcode 1 presents the general idea of the *GammaScaleFree* algorithm. The procedure *CALCULATE-DESIRED* is explained in Pseudocode 2. Pseudocode 3 indicates how to calculate value of s.

Pseudocode 1. The *GammaScaleFree* algorithm

1. Input $|V|$, $|E|$, γ
2. FOR EACH $v_i \in V$ DO
    2.1. DesiredDegree(v_i) = *CALCULATE-DESIRED*
    2.2. SumOfDesiredDegree = SumOfDesiredDegree + DesiredDegree(v_i)
3. FOR EACH $v_i \in V$ DO
    3.1. ProbabilityOfAddingAnEdge(v_i) = DesiredDegree(v_i)/ SumOfDesiredDegree
4. WHILE NumberOfAddedEdges< $|E|$ DO
    4.1. Choice randomly two different, not connected, vertices v_1 and v_2 according to ProbabilityOfAddingAnEdge(v_i)
    4.2. Add the edge between v_1 and v_2
5. RETURN resulting graph

Pseudocode 2. The *CALCULATE-DESIRED* procedure

1. Shift s = *CALCULATE-SHIFT*
2. DesiredDegree(v_i) = random value $\propto$ Eq. 4 with given s and γ
3. RETURN DesiredDegree(v_i)

In Pseudocode 3 we need to calculate $\overline{k}_{real}$ – an average degree of vertices resulting from given values $|V|$ and $|E|$, and the average degree of vertices given by the power-law distribution – $\overline{k}_{dist}$. The algorithm compares these two quantities and returns such value of s, at which these values are most similar to each other.

$$\overline{k}_{real} = 2|E|/|V|, \quad \overline{k}_{dist} = \sum_{i=s}^{|V|-1} P_{degree}(s, \gamma, i)i \tag{6}$$

Pseudocode 3. *CALCULATE-SHIFT* procedure

1. Calculate *DegreeFromData* as $\overline{k}_{real}$
2. $s = 1$ 2. Calculate *PreviousDegree* as $\overline{k}_{dist}$ for $p_{degree}(s, \gamma, |V|-1)$
3. Calculate *NextDegree* as $\overline{k}_{dist}$ for $p_{degree}(s+1, \gamma, |V|-1)$

3. $shift = 1$

4. WHILE $|NextDegree - DegreeFromData| < |PreviousDegree - DegreeFromData|$ DO

    4.1. $s = s + 1$ 4.2. Calculate $PreviousDegree$ as $\overline{k}_{dist}$ for $p_{degree}(s, \gamma, |V| - 1)$

    4.3. Calculate $NextDegree$ as $\overline{k}_{dist}$ for $p_{degree}(s + 1, \gamma, |V| - 1)$

5. RETURN s

Figure 1 presents the probability distributions used to draw the desired degree of each vertex, obtained with different values of s and constant values of the parameters $|V| = 11$ and $gamma = 2$. This figure also shows the role of parameter s in the distribution. We have tested the proposed method regarding the characteristics of received distributions of degree of vertices in the generated graphs. The conclusion from these test is that the resulting distributions are proportional to the power function, they are closer to the natural networks especially for $\gamma = 4$, for $\gamma = 2$ our algorithm produces graphs a bit worse than these produced by BA.

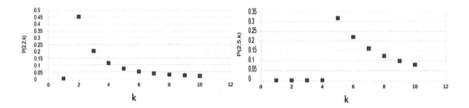

Fig. 1. The probability distribution, P_{degree}, for $s = 2$ and $s = 5$ with $gamma = 2$

5   The Spread of the Epidemic – Experimental Studies

Numerous experiments have been conducted, but due to the limited length of this paper, only a few of them are presented here. We focus on the problems: the influence of size and type of the network on the epidemic, how the different susceptibility to infection in the population influence the epidemic course, and does the vaccination strategies influence the number of infected individuals?

The SIR model is adopted in the presented research. An individual assigned to a vertex can be in one of the states: *Susceptible* (S), *Infected* (I) or *Recovered* (R). The group R contains immune individuals as a result of the disease. An assumed number of iterations in the experiment is given as parameter $MaxIter$. An initial number of infected individuals is Inf_0, the infected individuals are randomly selected with uniform probability. Individuals vaccinated are assigned to state R, their number is given as parameter $Vacc$. Four strategies of vaccination were tested, they are described further. A pathogen has associated parameter *infectivity of the pathogen* z – the ability of the pathogen to move between individuals, $z \in [0, 1]$. Every individual v_i has assigned a parameter which specifies the susceptibility to infection $su(v_i)$. It has a real value from the range of $[0, 1]$ and is randomized with the Normal distribution (given the mean and standard deviation as parameters). Each iteration includes checking all the neighbors (connections) of each vertex v_j that represents susceptible individual ($v_j \in S$). If any of the neighbor is in state I (i.e., is infected) than the considered vertex (individual) changes its state from S to I with the probability $p(v_j \in I) = z\,su(v_j)$. The preset parameter, $Iter_{inf}$, the same for all individuals, determines how many iterations it will take to change the status of the infected individual from I to R. In each iteration, the values of numbers of individuals in particular states S, I, R, and a number of newly infected individuals, are collected during this experiment. In some experiments, the average values and standard deviations of measured quantities are presented.

The impact of the size of the network (graph) on the course of the spread of the epidemic. A size of the graph means a number of individuals considered in the experiment. All considered types of networks were tested with the assumed other parameters. To ensure comparability of results, increase the number of vertices results in a proportional increase in the number of edges and the initial number of infected individuals. In this way, the average degree of vertices in the graphs and the percentage of infected individuals remained unchanged. The following parameters are constant during the experiment: number of iterations $MaxIter = 100$; an average degree of vertices $= 10$; the infectivity factor of the pathogen $z = 0.06$; the coefficients of susceptibility to infection $su(v_j) = 0.5$ for all v_j; duration of the illness $Iter_{inf} = 5$ iterations. Experiment was conducted using all types of graphs, ERm, WSm, BAs, and *GammaScale-Free*, each of them with size value of: $1\,000$, $10\,000$, $100\,000$, (100 runs for the same parameters) and $1\,000\,000 - 10$ runs was done for this size.

Observations: In the *Random* graphs (ERm) an average value of fractions of infected individuals for the three biggest sizes are similar, ≈ 0.68, while for the smallest graph it is ≈ 0.49. The difference is in standard deviations. It is $\approx 0.306; 0.01; 0.003; 0.0008$ respectively for graphs from the smallest one to the biggest. The similar results were observed for the *small-world* WSm graphs, here the respective values of average fractions of infected individuals is $\approx 0.61; 0.77; 0.77; 0.77$, and standard deviations changes $\approx 0.32; 0.008; 0.003; 0.001$, respectively. The same situation was observed for *Scale-free* (BAm) and *GammaScaleFree* graphs. The difference was observed only for the smallest graphs. It could suggest that the results obtained using quite small networks, about $10\,000$ nodes, approximate these obtained using much larger networks.

The impact of the type of the network on the epidemic. The goal of this experiment was to examine the relationship between the type of network and the spread of the pathogen, for different values of the infectivity of pathogen z. Firstly we compared the results given by ERm; WSm with $p_{reconnect} = 0.3$, and BAm algorithms. Next we present the results obtained only for the *scale-free* networks generated by: BAm, *GammaScaleFree* with $\gamma = 2.5$; *GammaScaleFree* with $\gamma = 2$; *GammaScaleFree* with $\gamma = 3$; *GammaScaleFree* with $\gamma = 5.5$. We are interested in an average fraction of individuals infected during the assumed number of iterations. Input data in this experiment are following: $|V| = 10\,000$; $|E| = 50\,000$; $Inf_0 = 10$, random vaccination; number of vaccinated: 20, $su(v_j)$ is Normal distribution with *average value* $= 0.5$ and *standard deviation* $= 0.1$,

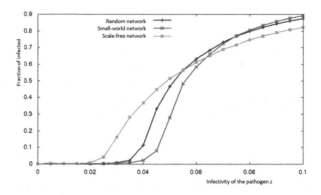

Fig. 2. A part of infected individuals in ERm, WSm and BAm networks

$MaxIter = 100$, $Iter_{inf} = 5$ iterations. The results with ERm, WSm, and BAm are collected in Fig. 2. The same dependency for networks generated by the authors' method is shown in Fig. 3.

Observations: If a network is a *scale-free*, an important part of the society may be infected when the infectivity of the pathogen is so small that in the random and *small-world* networks almost nobody would be infected. The *scale-free* networks are the most sensitive in opposite to the *small-world* networks. An interesting observation is when infectivity of the pathogen receives value around 0.06. The situation is reversed, the reason can lie in the distribution of degree of vertices. Distribution of degree of vertices in the *scale-free* network has *long tail*, there is a small but significant number of vertices with a very large number of connections with others, in social networks they are called '*people hub*'. The presence of such vertices facilitates the spread of the epidemic, even at low probabilities of infection. On the other hand, when the average degree of vertices is similar in the scale-free network to the others, it means that having tails (high degrees nodes) such network has to have a large number of vertices with the relatively low degree. This is responsible for the reversal of the situation with an increase in the likelihood of infection. The *scale-free* networks with a large number of vertices of a small number of connections result in the presence of relatively isolated areas of the graph, where the infection does not reach even with a significant probability of infection during contact.

Results presented in Fig. 3 show a high sensitivity of scale-free network on pathogens with small infectivity factor z and the relative resistance to pathogens with a high value of this ratio. In such networks, it is easy to induce small epidemic (less than 50% population), but more difficult is to induce the bigger epidemic. This effect intensifies with decreasing γ. Figure 3 shows clearly this trend.

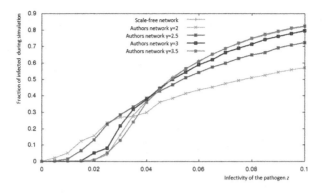

Fig. 3. A part of infected individuals in authors' networks

Distribution of susceptibility to infection in the population. Human beings differ in resistance to a pathogen, some people have a reduced resistance to some diseases. In this experiment we assumed that the distribution of susceptibility to the pathogen is the truncated normal distribution [18]. We have used only distribution with the mean= 0.5 and standard deviation much lower than 0.5.

Values of other parameters are as before, except infectivity of the pathogen $z = 0.06$. The research question is the influence of the standard deviation of susceptibility to infection on the average number of infected individuals during assumed number of iterations, for different kind of networks – *small-world* and *scale-free*.

Observations: What seems surprising is that changes in standard deviation from *zero* to 0.4 does not influence on the result. The increase of standard deviation reduces the total number of infected individuals, and it does not depend on the type of network. This observation is important from the practical point of view. According to the above result, forecasts on the base of models with constant resistance can be useful because they give excessive results in relation to the reality.

Four vaccination strategies. As we know, it is not possible to cover vaccination of the entire population. In the literature two main types of vaccination are mentioned: *random immunization* and *target immunization* – vaccinated are the individuals with degree (a number of contacts) grater than an assumed threshold. The second strategy makes use a priory information about the degree of vertices [2–4]. Based on literature, we studied following strategies: (1) random; (2) vaccination of individuals with greatest number of contacts; (3) vaccination of random friends of randomly chosen individuals; (4) authors' strategy: vaccination of friends with the largest number of contacts among friends of randomly chosen individuals. This experiment was conducted with the ERm, WSm, and BAm, all parameter values the same as in previous experiments, and the standard deviation of the distribution of susceptibility to the pathogen is equal to 0.1.

Observations: The results agree with the intuition. The best strategy is the last one, but, the differences are not very important. The biggest difference was observed for *scale-free* network: a fraction of infected individuals is from 0.61 (ER network) to 0.51 for the vaccination of friends with the largest number of contacts. The general results are shown in Fig. 4.

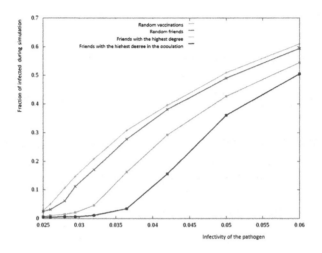

Fig. 4. A part of infected individuals in authors' networks

We can shortly summarize the results that the biggest differences in use of different vaccination strategies can be seen at the values of the infectivity of the pathogen from 0.035 to 0.045. Thought out choise of the vaccination can reduce the total number of infected especially in the *scale-free networks*.

6   Summary

We present a part of our research on the spread of epidemic modeling together with methods of construction of complex networks. We proposed an original algorithm for generating complex networks of *scale-free*, in which the γ parameter allows to create complex networks with more diverse characteristics than the BA algorithm. The ability to choose any value of γ allows creating complex networks closer to these present in nature. The proposed algorithm is universal and can potentially be used in different areas using complex networks modeling.

Experiments confirm that large complex networks, regarding the pathogen propagation, can be successfully modeled using smaller networks while maintaining the appropriate ratio. The impact of model parameters on the course of the epidemic was also examined. Our experiments demonstrate that the introducing the differentiated resistance to infection into the model results in a reduction of the total number of infected individuals. Four vaccination strategies, including one proposed by authors, were tested. The simulation results confirm that the proper selection of individuals for vaccination is important because of the ease of formation and spread of the epidemic.

Future works should concern on modeling another than SIR model of an epidemic. Another interesting issue is introducing the gradual decay of individual's resistance after passing the disease. It is also possible to introduce other factors affecting the likelihood of infection. Concerning the method of network generating, it is worth to introduce a limit on the maximal degree of a node in the graph, which would bring the generated network closer to the network existing in nature.

References

1. Wu, Q., Lou, Y.: Local immunization program for susceptible-infected-recovered network epidemic model. Chaos **26**(2), 1054–1500 (2016)
2. Feld, S.L.: Why your friends have more friends than you do. Am. J. Sociol. **6**, 1464–1477 (1991)
3. Christakis, N.A., Fowler, J.H.: Social network sensors for early detection of contagious outbreaks. PLoS One **5**(9): e12948, 1–8 (2010)
4. Shao, H., et al.: Forecasting the Flu: designing social network sensors for epidemics. CoRR. abs/1602.06866 (2016)
5. Kermack, W.O., McKendrick, A.G.: A contribution to the mathematical theory of epidemics. Proc. Roy. Soc. Lond. A Math. Phys. Eng. Sci. **115**(772), 700–721 (1927)
6. Xun, C.H., Minaya, V.: An extension of the Kermack-McKendrick model for AIDS epidemic. J. Franklin Inst. **342**(4), 341–351 (2005)
7. Erdös, P., Rényi, A.: On the evolution of random graphs. Publ. Math. Inst. Hungary. Acad. Sci. **5**, 17–61 (1960)
8. Xiao, F.W., Guanrong, C.: Complex networks: small-world, scale-free and beyond. IEEE Circuits and Syst. Mag. **3**(1), 6–20 (2003)
9. Watts, D.J., Strogatz, S.H.: Nature (6684), 440–442 (1998)
10. Barabási, A.L., Albert, R.: Emergence of scaling in random networks. Science **286**(5439), 509–512 (1999)
11. Wertheim, J.O., et al.: Using HIV transmission networks to investigate community effects in HIV prevention trials. PLOS ONE **6**(11), 1–7 (2011)
12. Pellis, L., et al.: Eight challenges for network epidemic models. Epidemics **10**, 58–62 (2015)
13. Li, T., Wang, Y., Guan, Z.-H.: Spreading dynamics of a SIQRS epidemic model on scale-free networks. Commun. Nonlinear Sci. Numer. Simul. **19**, 686–692 (2014)
14. Yang, L.-X., et al.: Epidemics of computer viruses: a complex-network approach. Appl. Math. Comput. **219**(16), 8705–8717 (2013)
15. Yang, L.-X., Yang, X.: The spread of computer viruses over a reduced scale-free network. Physica A **396**, 173–184 (2014)
16. Rizzo, A., Pedalino, B., Porfiri, M.: A network model for Ebola spreading. J. Theor. Biol. **394**, 212–222 (2016)
17. Pastor-Satorras, R., et al.: Epidemic processes in complex networks. Rev. Mod. Phys. **87**(3), 925–979 (2015)
18. Robert, C.P.: Simulation of truncated normal variables. Stat. Comput. **5**(2), 121–125 (1995)

Text Processing and Information Retrieval

Identification of Biomedical Articles with Highly Related Core Contents

Rey-Long Liu[✉]

Department of Medical Informatics, Tzu Chi University, Hualien, Taiwan
`rlliutcu@mail.tcu.edu.tw`

Abstract. Given a biomedical article *a*, identification of those articles with similar *core contents* (including research goals, backgrounds, and conclusions) as *a* is essential for the survey and cross-validation of the *highly related* biomedical evidence presented in *a*. We thus present a technique CCSE (Core Content Similarity Estimation) that retrieves these highly related articles by estimating and integrating three kinds of inter-article similarity: *goal similarity*, *background similarity*, and *conclusion similarity*. CCSE works on titles and abstracts of biomedical articles, which are publicly available. Experimental results show that CCSE performs better than PubMed (a popular biomedical search engine) and typical techniques in identifying those scholarly articles that are judged (by biomedical experts) to be the ones whose core contents focus on the same gene-disease associations. The contribution is essential for the retrieval, clustering, mining, and validation of the biomedical evidence in literature.

Keywords: Biomedical article · Highly-Related evidence · Core content · Inter-Article similarity estimation

1 Introduction

Identification of scholarly articles is a fundamental task routinely conducted by biomedical researchers. As this identification task is often motivated by specific research issues (e.g., associations among genes, diseases, chemicals, and proteins), the researchers often strive to find multiple articles that are highly related to the issues. Therefore, given an article *a*, several search engines have provided the service of retrieving those articles that are related to *a* (e.g., Google Scholar[1] and PubMed[2]), and many techniques have been developed to estimate the similarity between scholarly articles [1, 3, 6, 7, 12]. These techniques often worked on titles and abstracts of articles, as well as other text-based information (e.g., main text of each article) and link-based information, including out-link references (i.e., how an article cites others) and in-link citations (i.e., how an article is cited by others).

In this paper, we present a novel technique CCSE (Core Content Similarity Estimation) to estimate the similarity between biomedical articles. When compared with the

---

[1] Google Scholar is available at https://scholar.google.com.
[2] PubMed is available at http://www.ncbi.nlm.nih.gov/pubmed.

© Springer International Publishing AG 2017
N.T. Nguyen et al. (Eds.): ACIIDS 2017, Part I, LNAI 10191, pp. 217–226, 2017.
DOI: 10.1007/978-3-319-54472-4_21

existing search engines and techniques, CCSE has two interesting features: (1) it works on article titles and abstracts only, which are publicly available on the Internet (other parts of the articles, such as the main text, out-link references, and in-link citations, may not be publicly obtainable); and (2) it improves inter-article similarity estimation by considering the *core contents* of the articles. Core contents of a biomedical article include the research goal, background (problem description), and conclusion of the article. Two articles that share similar core contents can provide *highly related* evidence for further research and validation. CCSE can thus be used to improve biomedical search engines in retrieving the highly related articles, even when only titles and abstracts of the articles are available.

Development of CCSE is challenging, as core contents (research goal, background, and conclusion) of a biomedical article may be briefly expressed in the title and scattered in the abstract. CCSE is a novel technique that separately estimates three kinds of similarity (*goal similarity*, *background similarity*, and *conclusion similarity*), and integrates them to produce the similarity between two articles. Empirical evaluation shows that CCSE performs significantly better than a popular biomedical search engine and several typical techniques in retrieving the scholarly articles that are judged (by domain experts) to be the ones whose core contents focus on the same research issues. The contribution is essential for the retrieval, clustering, mining, validation, and curation of the research findings published in scientific literature.

2   Related Work

Previous studies have developed many techniques to estimate similarity between scholarly articles. These techniques may be *link-based* techniques, which worked on citation relationships among the articles (i.e., out-link references and in-link citations of each article). They may also be *text-based* techniques, which worked on textual contents of the articles (i.e., title, abstract, keywords, and main text of each article).

Linked-based techniques employed two kinds of citation links: *in-links* and *out-links*. For an article a, in-link citations are those articles that cite a, while out-link references are those articles that a cites. *Co-citation* is a representative technique that considers in-links of scholarly articles [15]. Two articles may be related to each other if they are co-cited by other articles. *Bibliographic coupling* (BC) is a representative technique that considers out-links [9]. It is defined in Eq. 1, where $a1$ and $a2$ are two articles, O_{a1} and O_{a2} are the sets of articles that $a1$ and $a2$ cite respectively [5, 6]. Two articles may be related to each other if they co-cite several articles. The link-based techniques may also be integrated with text-based techniques that consider textual contents of the articles [1, 4, 6, 8, 11, 12]. However, applicability of these techniques is limited, as many scholarly articles have very few (or even no) in-link citations, and out-link citations are not always publicly available either. CCSE works on titles and abstracts of scholarly articles, which are publicly available.

$$Similarity_{BC}(a1, a2) = \frac{|O_{a1} \cap O_{a2}|}{|O_{a1} \cup O_{a2}|} \tag{1}$$

Text-based techniques worked on textual contents of the articles, which can be titles, abstracts, keywords, and main bodies of the articles. Similarity between two articles are often dominated by those terms (in the two articles) that have higher weights. Latent Semantic Analysis (LSA) is a typical technique that employs singular value decomposition to estimate similarity. However, LSA did not perform well for scholarly articles [3].

Among many other text-based techniques, *PubMed* and *BM25* were identified as the ones that performed best for scholarly articles [3]. PubMed is a popular search engine that provides the "Related Citations" service for biomedical researchers. Many factors are considered by PubMed to estimate the similarity between two scholarly articles [10, 13], including (1) stemming of the terms in the articles, (2) lengths of the articles (i.e., number of terms in the articles), (3) positions of the terms in the articles (e.g., terms in titles of the articles), (4) key terms of the articles in domain-specific thesauri, and (5) weights of the terms in the articles (weighted by the term frequency in an article and how rarely the term appears in a collection of articles). On the other hand, BM25 employs Eq. 2 to estimate the score (similarity) of an article a_2 with respect to a given article a_1 [14]. In Eq. 2, k_1 and b are two parameters, $|a|$ is the number of terms in article a (i.e., length of a), *avgal* is the average number of terms in an article (i.e., average length of articles), $TF(t, a)$ is the frequency of term t appearing in article a, and $IDF(t)$ is the inverse document frequency of term t, which measures how rarely t appears in a large collection of articles

$$Similarity_{BM25}(a_1, a_2) = \sum_{t \in a_1 \cap a_2} \frac{TF(t, a_2)(k_1 + 1)}{TF(t, a_2) + k_1(1 - b + b\frac{|a_2|}{avgal})} Log_2 IDF(t) \qquad (2)$$

However, all these text-based techniques did not consider the similarity between core contents of scholarly articles. Core contents of a scholarly article mainly consist of the research goal, background, and conclusion of the article. They may be briefly expressed in the title and scattered in the abstract of the article. Our technique CCSE is proposed to recognize the core contents from the title and the abstract of the article, and based on the core contents recognized, estimate the similarity between scholarly articles. We will show that, by estimating inter-article similarity based on the core contents, CCSE can perform significantly better than PubMed and BM25 in identifying those biomedical articles that share similar core contents.

3   Core Content Similarity Estimation

Technical challenges of CCSE include (1) recognition of the core contents of an article and (2) estimation of the inter-article similarity based on the core contents recognized. We tackle the first challenge by the hypothesis that a term t in an article a may have different degrees of relatedness to the goal, the background, and the conclusion of a, depending on the *positions* where t appears in the title and the abstract of a. For example, the research goal of a tends to appear at the title of a, as well as the beginning and the end of the abstract of a, because the research goal may describe both the research problem and the main findings of a. The background mainly defines the research problem, which tends to appear at the beginning of the abstract; and conversely the conclusion mainly describes the main results, which tend to appear at the end of the abstract. Moreover, to tackle the second challenge, we employ the hypothesis that mismatch between any parts of the core contents of two articles may significantly reduce the similarity between the two articles, because such mismatch may indicate that the two articles do not focus on the same research issues. Based on the two hypotheses, CCSE separately estimates three kinds of similarity (*goal similarity, background similarity*, and *conclusion similarity*), and integrates them to produce the similarity between two articles.

More specifically, an article can be treated as a sequence of words, and Fig. 1 defines a linear way to estimate the relatedness of a term to the *goal* (R_{goal}), the *background* (R_{back}) and the *conclusion* (R_{conc}) of the article. Based on the relatedness definitions, similarity between two articles a_1 and a_2 is estimated by Eq. 3, which separately checks how the core content of a_1 appears in a_2 (i.e., *CoreMatch*(a_1, a_2)) and vice versa (i.e., *CoreMatch*(a_2, a_1)). CCSE thus conducts a *dual* match so that a_1 and a_2 are said to be quite similar to each other only if the core content of a_1 appears in a_2 *and* the core content of a_2 appears in a_1 as well. Any mismatch between the core contents will significantly reduce the similarity between a_1 and a_2.

$$Similarity_{CCSE}(a_1, a_2) = CoreMatch(a_1, a_2) \times CoreMatch(a_2, a_1) \qquad (3)$$

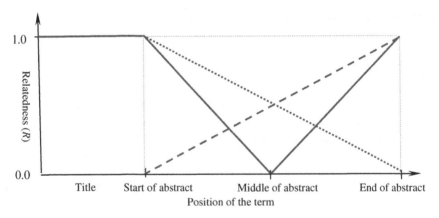

Fig. 1. Relatedness of a term to the *goal* (R_{goal}, the solid line), the *background* (R_{back}, the dotted line) and the *conclusion* (R_{conc}, the dashed line) of a biomedical article.

To estimate how the core content of a_1 appears in a_2, CCSE employs Eq. 4, which separately estimates how the three parts of the core content of a_1 (goal, background, and conclusion of a_1) appear in a_2. The degrees of match in the three parts are averaged to produce the degree of match of the core content.

$$CoreMatch(a_1, a_2) = \frac{Match_{goal}(a_1, a_2) + Match_{back}(a_1, a_2) + Match_{conc}(a_1, a_2)}{3} \qquad (4)$$

Equation 5 measures how the *goal* of a_1 appears in a_2. As the title of a_1 can provide the most reliable information about the goal of a_1, Eq. 5 checks how the terms in the title of a_1 appear in a_2. When a term t in the title of a_1 appears in a_2, it increases the degree of goal match between a_1 in a_2 (see the numerator of Eq. 5). The increment of the similarity is based on the degrees of relatedness of t to the goals of a_1 and a_2 (i.e., $R_{goal}(t, a_1)$ and $R_{goal}(t, a_2)$ defined in Fig. 1). If t appears at multiple positons in an article, its degree of relatedness is set to the maximum degree of relatedness at these positions.

$$Match_{goal}(a_1, a_2) = \frac{\displaystyle\sum_{t \in Title(a_1); t \in Title(a_2) \cup Abstract(a_2)} InterR(R_{goal}(t, a_1), R_{goal}(t, a_2)) \times Log_2 IDF(t)}{\displaystyle\sum_{t \in Title(a_1)} R_{goal}(t, a_1) \times Log_2 IDF(t)} \tag{5}$$

$$InterR(r_1, r_2) = \begin{cases} r_1, & \text{if } r_1 < r_2; \\ r_2, & \text{otherwise.} \end{cases} \tag{6}$$

Similarly, Eq. 7 measures how the *background* of a_1 appears in a_2 (i.e., $Match_{back}$), and Eq. 8 measures how the *conclusion* of a_1 appears in a_2 (i.e., $Match_{conc}$). Recall Fig. 1 for the degrees of relatedness of a term t to the background of a_1 and a_2 (i.e., $R_{back}(t, a_1)$ and $R_{back}(t, a_2)$) as well as the conclusion of a_1 and a_2 (i.e., $R_{conc}(t, a_1)$ and $R_{conc}(t, a_2)$).

$$Match_{back}(a_1, a_2) = \begin{cases} Match_{goal}(a_1, a_2), & \text{if } a_1 \text{ or } a_2 \text{ has no abstract} \\ \dfrac{\displaystyle\sum_{t \in Abstract(a_1); t \in Abstract(a_2)} InterR(R_{back}(t, a_1), R_{back}(t, a_2)) \times Log_2 IDF(t)}{\displaystyle\sum_{t \in Abstract(a_1)} R_{back}(t, a_1) \times Log_2 IDF(t)}, & \text{otherwise.} \end{cases} \tag{7}$$

$$Match_{conc}(a_1, a_2) = \begin{cases} Match_{goal}(a_1, a_2), & \text{if } a_1 \text{ or } a_2 \text{ has no abstract} \\ \dfrac{\displaystyle\sum_{t \in Abstract(a_1); t \in Abstract(a_2)} InterR(R_{conc}(t, a_1), R_{conc}(t, a_2)) \times Log_2 IDF(t)}{\displaystyle\sum_{t \in Abstract(a_1)} R_{conc}(t, a_1) \times Log_2 IDF(t)}, & \text{otherwise.} \end{cases} \tag{8}$$

Therefore, given the title and the abstract of a scholarly article, CCSE considers three parts of the core content of the article: goal, background, and conclusion of the article. CCSE estimates how each term is related to each of the three parts based on the positions of the term appearing in the title and the abstract of the article. A dual match is then conducted to estimate the similarity between two articles so that any mismatch between the core contents of the two articles will significantly reduce the similarity between them. CCSE is thus a pure text-based technique that works on titles and abstracts of scholarly articles, which are publicly available on the Internet.

4 Empirical Evaluation

4.1 Basic Experimental Settings

Given a scholarly article a, the experimental data needs to contain those articles whose core contents have been manually judged (by domain experts) to be *highly related* to the core content of a. To our knowledge, no benchmark data was built for this purpose, and hence following [12] we employ data from DisGeNET[3], which maintains a database of articles focusing on specific gene-disease associations. The database is maintained to facilitate the research of disease diagnosis and therapy.

More specifically, from DisGeNET we select those gene-disease pairs that had the largest number of articles annotated by Genetic Association Database (GAD[4]) or Comparative Toxicogenomics Database (CTD[5]) for human. Both GAD and CTD recruit domain experts to manually select articles to annotate each gene-disease pair [2, 16]. The articles used to annotate a gene-disease pair $<g, d>$ thus have the core contents focusing on the same research issue (i.e., the association between g and d). For each gene-disease pair $<g, d>$, we designate one article as the *target*, while the others as the highly related *candidates*. Given the target article, a better technique should rank high these highly related candidates, among other candidates that focus on *other* research issues (i.e., *not* dedicated to the association between g and d).

Therefore, for each gene-disease pair $<g, d>$, we also collect many candidate articles that are *not* dedicated to $<g, d>$. These candidate articles are "near-miss" articles for $<g, d>$, as they are collected by sending two queries to PubMed Central (PMC[6]): "g NOT d" and "d NOT g". The articles collected by this way mention g or d but not both, and hence they should not focus on the same research issue as the target article for $<g, d>$, although they may share a certain amount of contents with the target article. For each gene-disease pair, at most 200 near-miss candidate articles are collected. A gene-disease pair corresponds to a test in the experiment. We totally have 53 tests within which there are 53 target articles and 9,875 candidate articles. These articles totally have 435,786 out-link references.

4.2 Baselines and Evaluation Criteria

Based on the survey presented in Sect. 2, we design two experiments (Experiment I and Experiment II) in which state-of-the-art baselines are compared with CCSE. In Experiment I, there are two baselines: bibliographic coupling (BC) and BM25. BC is a link-based baseline that relies on out-link references (ref. Eq. 1). It can work on more articles than those techniques that work on in-link citations, as many scholarly articles have very few or even no in-link citations. Out-link references were also found to be more helpful than in-link citations in the clustering and classification of scholarly articles [4, 6]. BM25

---

[3] DisGeNET is available at http://www.disgenet.org/web/DisGeNET/menu/home.
[4] GAD is available at http://geneticassociationdb.nih.gov.
[5] CTD is available at http://ctdbase.org.
[6] PMC is available at http://www.ncbi.nlm.nih.gov/pmc.

(ref. Eq. 2) is a text-based technique that was found to be one of the best in finding related scholarly articles [3].

In Experiment II, the service of "Related Citations" provided by PubMed is the baseline. Given an article a, this service retrieves those articles that are related to a. It was found to be one of the best in finding related scholarly articles [3]. We aim at showing that CCSE can perform better than PubMed by focusing on the recognition of core contents of scholarly articles. For each target article a, PubMed retrieves a sequence S of related articles. We remove from S all articles that are *not* candidate articles. As noted in Sect. 4.1, the candidate articles include those that share similar core contents with a (judged by domain experts), as well as those that should *not* share similar core contents with a. By focusing on these candidate articles in S, performance of CCSE and PubMed can be objectively compared.

To measure the performnace of CCSE and the baselines, we employ two evaluation criteria: *Mean average precision* (MAP) and *avergage precision at top-X positions* (average P@X), which are routinely employed as evaluation criteria in previous studies (e.g., [11]). MAP measures how highly related articles are ranked above non-highly related articles, while average P@X measures how related articles are ranked at top-X positions. To verify whether the performance differences between CCSE and each of the baselines are *statistically significant*, we conduct significance tests by two-sided and paired t-tests with 95% confidence level.

4.3 Results

Figure 2 compares performance of CCSE, BM25 and BC. CCSE performs better than all the baselines, especially in MAP and average P@1. CCSE is thus more capable of ranking highly related articles at top-1. MAP of CCSE is significantly better than MAP of each baseline. The baseline BC, which employs out-link references to rank scholarly articles (ref. Eq. 1), achieves better MAP than BM25, which is a text-based technique. However CCSE performs 27% better than BC in MAP (0.5068 vs. 0.3980). CCSE performs significantly than BC even though CCSE only works on titles and abstracts of articles, which are more publicly obtainable than out-link references in the articles. Moreover, BM25, which is one of the best techniques in retrieving scholarly articles [3], achieves better average P@1 than BC. However CCSE performs 23% better than BM25 in average P@1 (0.5094 vs. 0.4151). These contributions are of practical significance to the identification of scholarly articles that focus on specific research issues.

Figure 3 compares performance of CCSE and PubMed. CCSE performs better than PubMed in all evaluation criteria, with statistically significant improvement in average P@3 (7% improvement, 0.5472 vs. 0.5094). The contribution is of practical significance as well, as PubMed is a popular search engine of biomedical articles. They are also of technical significance, as PubMed performs best in recommending related scholarly articles [3], and it has employed domain-specific thesauri and several typical text-based factors about terms and articles (as noted in Sect. 2). CCSE achieves better performance by core content recognition. It can be applied to various domains, as it achieves the better performance without relying on any domain-specific thesauri.

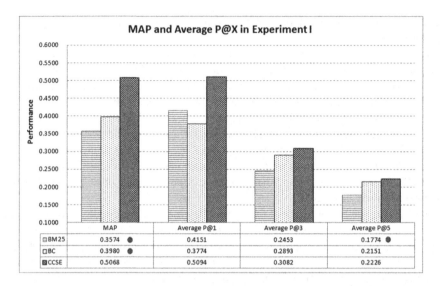

Fig. 2. MAP and average P@X in Experiment I: CCSE performs significantly better than both baselines in MAP (a dot on a system indicates that performance difference between the system and CCSE is statistically significant).

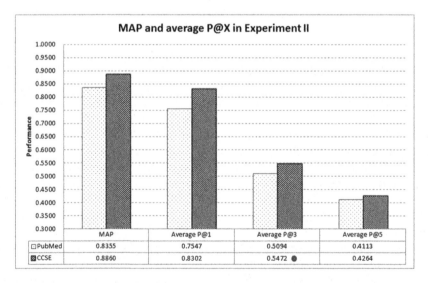

Fig. 3. MAP and average P@X in Experiment II: CCSE performs better than PubMed in all evaluation criteria (a dot on a system indicates that performance difference between PubMed and CCSE is statistically significant).

5   Conclusion and Future Extension

We have presented a novel technique CCSE to retrieve those biomedical articles that share similar core contents. Two articles that share similar core contents should have similar research goals, backgrounds, and conclusions. Given an article a, identification of those articles that share similar core contents with a is a routine job that biomedical researchers strive to do to survey and cross-validate the evidence already published in literature. As the terms related to the core content of an article may scatter in an article, it is challenging to recognize the core content and estimate inter-article similarity based on the core contents of the articles. To tackle the challenge, CCSE is developed based on two domain-independent hypotheses: (1) degrees of relatedness of a term t to the core content of an article a may depend on the *positions* where t appears in the title and the abstract of a; and (2) *mismatch* between any parts of the core contents of two articles a_1 and a_2 may significantly reduce the similarity between a_1 and a_2.

Performance of CCSE is investigated using real-world data in which articles that focus on highly related evidence about specific gene-disease associations have been manually selected and confirmed by domain experts. CCSE performs significantly better than several state-of-the-art baselines. Moreover, CCSE works on titles and abstracts of articles only, which are more publicly obtainable than bibliography that is employed by link-based baselines. The contribution is essential for the retrieval, clustering, mining, validation, and curation of the research findings published in literature.

The ideas of CCSE can be extended to extract *core entity terms* in a biomedical article. Core entity terms in an article a are those terms (in a) that are related to the core content of a (e.g., the target genes, diseases, and chemicals of a). By extracting the core entity terms, several interesting applications can be supported, including the retrieval of those articles that focus on a given entity, as well as the navigation on a network of interacting entities already published in literature, with specific scholarly articles annotated for readers to check.

Acknowledgment.   This research was supported by the Ministry of Science and Technology (grant ID: MOST 105-2221-E-320-004) and Tzu Chi University (grant IDs: TCRPP103020 and TCRPP104010), Taiwan.

References

1. Aljaber, B., Stokes, N., Bailey, J., Pei, J.: Document clustering of scientific texts using citation contexts. Inf. Retrieval **13**(2), 101–131 (2010)
2. Becker, K.G., Barnes, K.C., Bright, T.J., Wang, S.A.: The genetic association database. Nat. Genet. **36**(5), 431–432 (2004)
3. Boyack, K.W., Newman, D., Duhon, R.J., Klavans, R., Patek, M., Biberstine, J.R., et al.: Clustering more than two million biomedical publications: comparing the accuracies of nine text-based similarity approaches. PLoS ONE **6**(3), e18029 (2011)
4. Boyack, K.W., Klavans, R.: Co-citation analysis, bibliographic coupling, and direct citation: Which citation approach represents the research front most accurately? J. Am. Soc. Inform. Sci. Technol. **61**(12), 2389–2404 (2010)

5. Calado, P., Cristo, M., Moura, E., Ziviani, N., Ribeiro-Neto, B., Goncalves, M.A.: Combining link-based and content-based methods for web document classification. In: Proceedings of the 2003 ACM CIKM International Conference on Information and Knowledge Management, New Orleans, Louisiana, USA (2003)
6. Couto, T., Cristo, M., Gonçalves, M.A., Calado, P., Nivio Ziviani, N., Moura, E., Ribeiro-Neto, B.: A comparative study of citations and links in document classification. In: Proceedings of the 6th ACM/IEEE-CS Joint Conference on Digital Libraries, pp. 75–84 (2006)
7. Gipp, B., Beel, J.: Citation proximity analysis (CPA) – a new approach for identifying related work based on co-citation analysis. In: Proceedings of the 12th International Conference on Scientometrics and Informetrics, vol. 2, pp. 571–575 (2009)
8. Janssens, F., Glänzel, W., De Moor, B.: A hybrid mapping of information science. Scientometrics **75**(3), 607–631 (2008)
9. Kessler, M.M.: Bibliographic coupling between scientific papers. Am. Doc. **14**(1), 10–25 (1963)
10. Lin, J., Wilbur, W.J.: PubMed related articles: a probabilistic topic-based model for content similarity. BMC Bioinformatics **8**, 423 (2007)
11. Liu, R.-L.: Citation-based extraction of core contents from biomedical articles. In: Proceedings of the 29th International Conference on Industrial, Engineering & Other Applications of Applied Intelligent Systems (IEA/AIE 2016), pp. 217–228 (2016)
12. Liu, R.-L.: Passage-based bibliographic coupling: an inter-article similarity measure for biomedical articles. PLoS ONE **10**(10), e0139245 (2015)
13. PubMed: Computation of Related Citations. http://www.ncbi.nlm.nih.gov/books/NBK3827/#pubmedhelp.Computation_of_Similar_Articl. Accessed: Nov 2014
14. Robertson, S.E., Walker, S., Beaulieu, M.: Okapi at TREC-7: automatic ad hoc, filtering, VLC and interactive. In: proceedings of the 7th Text REtrieval Conference (TREC 7), Gaithersburg, USA, pp. 253–264 (1998)
15. Small, H.G.: Co-citation in the scientific literature: a new measure of relationship between two documents. J. Am. Soc. Inform. Sci. Technol. **24**(4), 265–269 (1973)
16. Wiegers, T.C., Davis, A.P., Cohen, K.B., Hirschman, L., Mattingly, C.J.: Text mining and manual curation of chemical-gene-disease networks for the Comparative Toxicogenomics Database (CTD). BMC Bioinf. **10**, 326 (2009)

An Efficient Hybrid Model for Vietnamese Sentiment Analysis

Thanh Hung Vo, Thien Tin Nguyen, Hoang Anh Pham[✉],
and Thanh Van Le[✉]

Ho Chi Minh City University of Technology, Ho Chi Minh, Vietnam
{vthung,thientin,anhpham,ltvan}@hcmut.edu.vn

Abstract. Sentiment analysis from the text is an exciting and challenging task which can be useful in many applications of exploiting people interests for improving the quality of services. Especially, text collected from social networks, websites or forums is usually represented by spoken language that is unstructured and difficult to handle. In this paper, we present a novel hybrid model that is based on Hierarchical Dirichlet Process (HDP) and adopts a combination of lexicon-based and Support Vector Machine (SVM) methods in the task of topic-based sentiment classification for Vietnamese text. The proposed model has been evaluated on five different topic-datasets, and the experimental results show the efficiency of our proposed model when the average accuracy is nearly 87%. Although this proposed model is initially designed for Vietnamese language, it is applicable and adaptable to other languages.

Keywords: Topic modeling · Lexicon-based · SVM · Sentiment classification · Unstructured text

1 Introduction

The World Wide Web has become the core medium for introducing new products and services. It is one huge big forum for customers or people who have interests in the newly launched products/services, to discuss and give their opinions. For businesses, tracking public opinions on their products/services are crucial. It helps them to adjust their products/services to better suit the customers' needs, and target the new products/services to the right customer groups. From a technical point of view, as the opinions exist in many forms (news, feeds, blogs, forums, etc.) and may be scattered in every corner of the World Wide Web, gathering public opinions on particular product/service is challenging. Additionally, understanding the semantic meanings of the opinions (positive, negative, or supportive) is another challenging issue because these opinions are commonly expressed in natural languages.

Sentiment analysis is widely used in various applications of text mining for extracting and summarizing of opinions in the tourism business, mining user opinions for customer-care services, cyber security, etc. Based on sentiment

© Springer International Publishing AG 2017
N.T. Nguyen et al. (Eds.): ACIIDS 2017, Part I, LNAI 10191, pp. 227–237, 2017.
DOI: 10.1007/978-3-319-54472-4_22

words defined in SentiWorldNet, Khan *et al.* [1] proposed a lexicon-based method for sentiment polarity detection. In another work, Khalid *et al.* [2] introduced BiSAL, a bilingual sentiment analysis lexical Youngjoong *et al.* [3] studied on term weighting schemes and applied machine learning method for sentiment analysis. In [4], Khan *et al.* proposed Twitter opinion mining framework using hybrid classification scheme which uses the output of lexicon-based classifier to produce the input for the machine learning algorithm.

In this paper, we focus on sentiment classification which determines the polarity (negative, positive) of a sentiment existed in an opinioned document. Firstly, by taking advantages of the probabilistic model of hierarchical Dirichlet allocation and discovering data clusters, we could explore automatically hidden topics in documents without needs of determining a priori the number of topics. Then we build a hybrid model that is a combination of lexicon-based and corpus-based models for each topic explored in the previous step to tackle the problem of cross-domain for improving the performance in terms of the accuracy.

The rest of this paper is organized in the following manner. Section 2 begins with a discussion of the related works. Section 3 presents the preliminary of Hierarchical Dirichlet Allocation HDP for topic modeling. In Sect. 4, we propose our hybrid sentiment classification by combining the lexicon-based model with the corpus-based model. We carry out some experiments with data come from some social sites to evaluate our proposed model in Sect. 5. The last section discusses the conclusions and ongoing studies.

2   Related Work

In the form of text mining, topic modeling has been studying as one of the most attractive fields. One of its popular applications is to uncover semantic structures of a text document. It can also statistically analyze and discover the topic distribution throughout a whole corpus. Given a collection of documents, e.g. newspapers or scientific papers, we can automatically get some insights about their content by using algorithms that come with topic modeling.

Many different approaches have emerged to deal with this problem. Most dated methods try to describe a collection of data in efficient ways while still preserving important information. For example, Salton and McGill introduced tf-idf formula which ideally aims at measuring how much a word might contribute to form topics of documents [5]. This approach is then significantly enhanced to form a method named Latent Semantic Indexing (LSI) [6]. In summary, LSI uses singular value decomposition to notably decrease output size from tf-idf.

Recent approaches take advantages of the probabilistic model. It models each topic as a distribution over a bag-of-words (i.e. collection of distinct words present in the corpus of documents). In other words, each word appearing in a document is a sample from its corresponding topic with a certain probability. Additionally, each document can describe a mixture of different topics. Based on these assumptions, David M. Blei et al. derived a well-known model called Latent Dirichlet Allocation (LDA) [7] that is a probabilistic model of a corpus.

LDA assumes a prior number of topics distributed as Dirichlet distribution through the entire document collection. Furthermore, multinomial distribution is used to illustrate the proportion of topics in each document.

Recently topic modeling techniques have been successfully used to classify the sentiment. In [8], Liang et al. proposed a new sentiment model called Auxiliary-Sentiment Latent Dirichlet Allocation (AS-LDA) for sentiment classification, in which they focus on sentiment element and auxiliary words and fed them as the word distribution for LDA model. Assumed that the sentiment polarity is dependent on the topic or domain, Li et al. [9] integrated the dependency of local context in sentiment analysis and proposed a joint sentiment and topic model Sentiment-LDA by adding a sentiment layer to the original LDA. Some other research groups also used topic modeling to identify the meaningful review aspects [10,11], or proposed a generative probabilistic topic model based on Latent Dirichlet Allocation (LDA) and Hidden Markov Model (HMM), which emphasizes on extracting multi-word topics from text data for automatically extracting topics or aspects in sentiment reviews [12].

Topic modeling is extensively used in text mining for exploring the hidden topics in sentiment analysis. However, almost all of the existing models do not consider to analyze the data structure, and these models are required to know the number of topics in advance. Therefore, it is inefficient to adopt LDA to new corpus and new topics. Hierarchical Dirichlet Process (HDP), an extension of LDA, has been emerged by using Dirichlet process instead of Dirichlet distribution [13] to overcome the drawback of LDA. Using HDP to determine the appropriate number of topics is the most attracted work in our proposed model.

3 Topic Modeling

The goal of our study is to infer the topic which a user might be interested in based on his comment messages in social networks and forums. Additionally, the study will be applied to a huge number of users, so it is more efficient to have a shared common set of topics. Each user's interests can be one or more topics that come from the common topic set. Moreover, the number of topics is unknown in advance and it can grow in the future. Therefore, topic modeling approach has to not only deal with an unknown number of clusters but also adapt to new topics generated in future. Due to its attracted features, HDP mixture model is taken as topic modeling method in our proposed model. The followings will briefly discuss HDP and its usage in our proposed model.

3.1 Hierarchical Dirichlet Process

Knowledge of user's interested subjects may be achieved by applying Hierarchical Dirichlet Process (HDP) model. Its basic instance is a two-level model in which the lower level represents the topic proportion of individual documents, and the upper level represents the overall topic distribution through the whole corpus. Furthermore, each document consists of a number of topics with a certain proportion that is considered as a sample from a distribution of general topic proportion. Therefore, these proportions of all documents aggregate to the overall topic distribution of the whole corpus.

Figure 1 graphically shows a HDP model representing a corpus that has J documents with lengths being $n_j, (j = 1...J)$ words. In this model, G_0 and G_j are the topic distributions of the upper level and lower level, respectively. Additionally, Dirichlet processes are used as prior distribution for both G_0 and G_j. The parameters of any Dirichlet process are the base probability distribution and the concentration number. As shown in Fig. 1, the parameters of G_0 are H and γ respectively meanwhile G_j will take G_0 and α_0 as its parameters. In each document j, word x_{ji} is considered to be drawn form topic θ_{ji} which is in turn the multinomial distribution with probabilities from G_j. The more details about HDP model can be found in [13].

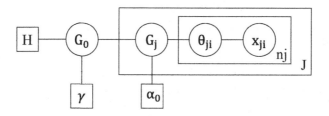

Fig. 1. A Hierarchical Dirichlet Process model

3.2   Topic Modeling Framework

This framework presents the general flow design that is adopted in our model including three main phases: (1) Pre-processing, (2) Training, and (3) Inferring. There are three steps in the phase of preprocessing. Firstly, the raw data will be extracted into meaningful words and phrases by using VnTagger tokenizer [14] that is a suitable tool for tokenizing Vietnamese or English language. Secondly, due to the nature of collected data from social networks, it might have meaningless words (e.g. typos) that will reduce the system performance. Therefore, these words will be added to a so-called Stop_words collection and removed out. Finally, a tf-idf calculator is used to eliminate words that have a very limited contribution to the topic distributions. They are either the most popular words which show up in most of the comments or extremely rare words which are considered as noise.

In the training phase, the preprocessed dataset will be used to train the HDP model for analyzing the topics distributions. The Gibbs sampling based on Markov Chain Monte Carlo algorithm is used to estimate the joint distribution in the HDP model. Additionally, Gibbs sampling method is wise in initiating the model's parameters so that it can well converge to a reasonable representation of training data. In detail, the base probability distribution (e.g. H) describes the aggregate distribution of all topics in the whole data. It is intuitive to see that each document can have a different proportion of topics despite the overall distribution. Besides, the concentration parameter (e.g. γ) indicates how closely each of that individual proportions is to the base probability distribution. Finally, the inferring phase takes advantages of the generated topics distributions to infer which topic a new message is most related to. In our proposed model, the well-known approach Jaccard has been applied with some extensions that are presented as follows.

Jaccard Inference. For each topic t_i in trained model, a set $\text{Top}(t_i)$ that includes a predefined number of words which have the highest probabilities are stored for inference. Given a document d which consists of N_d words, a collection of unique words and terms is measured. Then, the probability that document d relates to topic t is calculated by Eq. (1).

$$p_{J\_d}(t_k) = \sum_{i=0}^{N} \sigma_{Top(t_k)}(term_i) \tag{1}$$

$$\sigma_{Top(t_k)}(term_i) = \begin{cases} 1, & term_i \in Top(t_k) \\ 0, & term_i \notin Top(t_k) \end{cases} \tag{2}$$

The Jaccard formula only takes account of a number of words in $\text{Top}(t_k)$ that are included in document d without regarding how often they present in d. It also ignores the probabilities of these words in the topic distribution. These features should be considered as important factors for analyzing the document's topic, therefore, we introduce some adjustments to come up with wpJaccard formula.

wpJaccard Inference. Both repetition and probability of each term are taken into account in this method. The contribution of a term to the overall probability is calculated by its individual possibility in the topic as Eq. (3).

$$p_{wpJ\_d}(t_k) = \sum_{i=0}^{N} f_d(term_i) \times p_{t_k}(term_i) \times \sigma_{Top(t_k)}(term_i) \tag{3}$$

4 Topic-Based Modeling for Sentiment Analysis

4.1 Sentiment Analysis

The development of mobile devices and Web 2.0 helps people easily express their emotions and attitudes about what they had read, hear, or watch on blogs, forums, or social sites in the form of unstructured text and icons. The sentiment information extracted by analyzing these texts is called sentiment analysis or opinion mining. There are three main approaches to sentiment analysis such as manual approach, lexicon-based and corpus-based approaches. The manual approach is the most accurate one because it is directly based on people intuition and decision. However, it costs too much human labor and time consumption for handling a large volume text crawled from the Internet. Therefore lexicon-based and corpus-based approaches are widely used in sentiment analysis.

4.2   Lexicon-Based Method for Sentiment Analysis

Using a dictionary to extract sentiment words is one of the main approaches in opinion mining. To handle the Vietnamese language, we have built a Vietnamese dictionary containing about 26000 meaningful opinion-bearing words. Then each sentence-level document is analyzed in phrases and extract sentiment word to classify the sentiment polarity.

Data Pre-processing: In this step, texts and words must be standardized because people usually use a combination of standard spelling, apparently accidental mistakes, and interjections in their reviews. Then each sentence will be tokenized as an exact Vietnamese word that might contain a morpheme or a compound of two or many ones. Observing by carrying out experiments, we realized that phrase analysis is more accurate than sentence analysis. Therefore each sentence will be segmented into chunks for determining grammatical roles of noun phrases, adjective phrases, and verb phrases.

Sentiment Analysis by Phrase-Level Document: After preprocessing data, we use our dictionary built in the previous step to extract opinion terms and then analyze opinion for each comment that comprises of a set of phrases.

4.3   SVM-Based Method for Sentiment Analysis

Support vector machine (SVM) studied by Vapknik become an efficient and popular method for data mining since the 1990s. By combining with the kernel function, SVM provides effective models for classification problems, linear and nonlinear regression. Leveraging the accuracy and efficiency of SVM in classification problems, we will apply and build a SVM model to classify sentiments for each sentence-level document. A SVM model requires a training input data in which each data must be labeled. Our study will deal with the unstructured textual data, therefore to convert data to a structured one, we use bag-of-words model (BOW) in which a word can be an unigram, bigram or trigram due to the complexity of Vietnamese language structure.

Another problem with our data is the existence of noise words and emoticons as we handle data crawled from the social network, retail forums. These words may reduce the accuracy of our classification model. Therefore, a stop word list and TFxIDF have been used to remove the insignificant word of our training data. A word in our BOW model is reserved if its $TFxIDF$ respects to a meaningful threshold determined by statistical analysis.

After cleaning data, a SVM model with Radial basis function kernel has been building for sentiment classification. We define a problem of two classes (positive, negative) to determine the sentiment polarity of a sentence. The accuracy of our model is rather good and will be further discussed in the section of experiments.

4.4   A Combination of Lexicon-Based and SVM-Based Methods

Both of above-discussed methods already show their efficiency in sentiment analysis and these methods are adopted in our system of product/services evaluation application. The lexicon-based method is useful for classifying data with slang language, even with the rare word, because our sentiment dictionary is easy to rebuild and extend. However, the dictionary may be lacked knowledge because our Vietnamese sentiment words are too generous. Therefore, it will lead to the poor accuracy when it encounters an unknown word in our dictionary.

Our built-in SVM model allows more accuracy results as our training data is probably well-collected, good preprocessed and also we take advantages of SVM method characteristics in classification problem. This model is not strictly dependent on a set of sentiment words as the lexicon-based model be. Nevertheless, by observing and analyzing our experiments, if a sentiment word that is rarely appeared in our training data or not yet trained, this will drop down our model accuracy.

For minimizing trade-offs and breaking down technical barriers to getting maximum benefit and improved performance for each method, a hybrid model of both called LSVM has been proposed. We define the following equation to get the final score for a comment from the output of lexicon-based and SVM model.

$$score_{final} = (score_{Lex} - 5) * score_{Lex} + (score_{SVM} - 5) * score_{SVM} \qquad (4)$$

where:

$score_{final}$: the final score of a comment.

$score_{Lex}$: the score computed by lexicon-based approach.

$score_{ML}$: the score computed by SVM.

Our system use a 10-point grading scale, with 10 being the highest positive polarity and 0 being the lowest negative one. It is noted that $(score_i - 5)$ value represents the coefficient of sentiment polarity trend, if this coefficient is smaller than 0, it determine the negative direction and positive otherwise.

4.5   A Hybrid Model of Topic-Based Model and LSVM

When observing results obtained from experiments, we realized that our hybrid model are affected by domains. For example, the text "the service time is long" always returns negative polarity of sentiment without considering the context. The result must be positive if concerning a mobile or computer product and negative if a client complains by saying this text in a restaurant. Therefore, we propose a hybrid model to combine the topic modeling with our sentiment framework for tackling this problem.

In this task, we firstly build a topic model with HDP to explore automatically hidden topics of our text collection. As the latter was crawled from several Facebook fan-pages, website retailers, it is hard to define manually topics or number of topics existed in our collection. For this reason, HDP model has been chosen because it can determine automatically the number of hidden topics based on data distributions and characteristics of data clusters. After building HDP model, we retrieved a set of topics but it cannot directly be used because of noisy topics which are represented by a set of the unrelated words. We must tackle this problem to get clear and qualified topics for the task of building the sentiment model. The next step we infer data of our collection by using our topic model. The result of this inference leads to produce the training datasets by topics which are considered as the input of sentiment model for each topic.

For the task of sentiment classification, a new text (review) firstly will be inferred to get its topic by our HDP model. Then we apply the mixed sentiment model to compute the sentiment score for this text as we described in the previous section. Our proposed hybrid model demonstrated that it reduce the affection by domain to improve the performance of the task of sentiment classification. Moreover, this model also could deduce hidden topics from our data collection which is a big challenge in processing data from various social sources.

5   Experiments

In order to evaluate the performance of our proposed model, we have performed experiments on datasets crawled from website retailers and forums of travels and restaurants where clients and readers usually write comments to give their opinions. These messages are typically short and unstructured so they apparently just mention a very few parts of user's perspectives. Therefore, the concentration parameter α_0 in the HDP model should be small. In our experiments, it is 0.2.

The training phase inferred 16 distributions of words, some distributions referring to the same idea will be classified to one topic. Some well describing topics are shown in Table 1. They are manually named as Mobile Phone, Car, Computer, Camera, and Restaurant. For each topic, seven most frequent words (**Word** column) are listed together with their probabilities (**Prob.** column). The most suitable English translations of the corresponding Vietnamese words are provided in the italic form.

It is noticeable that these words are excellent presentations of the corresponding topic. For example, all seven words of the first topic describe the most

Table 1. Topic distributions

MOBILE PHONE		CAR		COMPUTER		CAMERA		RESTAURANT	
Word	Prob.	Word	Prob.	Word	Prob.	Word	Prob.	Word	Prob.
màn hình *screen*	0.0149	xe *car*	0.0786	máy *related to machine*	0.0582	chụp *photograph*	0.0644	quán *restaurant*	0.0240
hỗ trợ *support*	0.0127	xăng *gas*	0.0199	cấu hình *configuration*	0.0227	ảnh *photo*	0.0402	ngon *tasty*	0.0226
điện thoại *phone*	0.0122	chạy *run*	0.0180	đồ họa *graphic*	0.0180	zoom *zoom*	0.0198	món *dish, food*	0.0210
camera *camera*	0.0098	kiểu dáng *design*	0.0118	card *card*	0.0122	máy ảnh *camera*	0.0197	ăn *eat*	0.0170
sử dụng *use*	0.0090	phù hợp *suitable*	0.0115	game *game*	0.0118	nét *clear*	0.0183	bánh *cake*	0.0113
tính năng *specification*	0.0067	đẹp *nice*	0.0115	dòng *version*	0.0106	chế độ *mode*	0.0176	ko *no*	0.0081
thiết kế *design*	0.0062	tiết kiệm *save*	0.0098	chạy *run*	0.0097	hình ảnh *photo*	0.0130	uống *drink*	0.0079

common things that might be discussed when talking about mobile phone such as its design, display, specification, front and rear camera, application, etc. This result has partly proved the efficiency of the HDP model when it comes to automatically inferring latent topic.

Furthermore, the same word in different topics can have different probabilities. The word (chạy) (*run*), for example, is in top frequent words of topics (Car) and (Computer) with proportions of 0.0180 and 0.0090 respectively. This means that when the word (chạy) shows up in a comment, one might deduce that the user more likely talks about Car than Computer. This is also an excellent explanation of our chosen inference method - wpJaccard which takes into account both word's number of occurrences in document and probability on a topic.

For experiments of sentiment analysis, Fig. 2 compares the percentage of accuracy of five different methods on five topics detected from HDP model. It can be clearly seen that our proposed model (HDP_LSVM) give the best performance. SVM method gives a better result than lexicon-based one in almost cases, except for the restaurant. When rechecking with the restaurant dataset, we realized that SVM accuracy is drop down because it classifies wrongly sentences which do not contain sentiment. Besides, reviewers use more rare slang word and long text, that why's lexicon-based method is more suitable than SVM for tackling this problem.

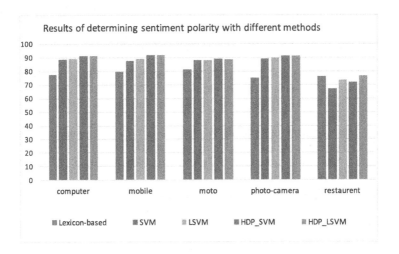

Fig. 2. Results of Determining Sentiment Polarity with Different Methods

6   Conclusion

This paper introduces a hybrid model for sentiment analysis from Vietnamese text using the HDP modeling with SVM and lexicon-based methods. The experimental results show that our proposed solution can determine the sentiment polarity although the text is represented by spoken language and rare slang words. The proposed model can also detect hidden topics in people interests represented by reader reviews and then help us to build sentiment models in restricting the dependency of the domain. The combination of SVM and lexicon-based method can take advantages of both for sentiment classification so that it can handle with the natural language, slang word of Vietnamese text collected from social networks. Currently, the proposed model has been designed for processing Vietnamese text, but it can be extended to be well adapted to other languages. The current weakness of the proposed model is that it has limits in processing long text which may exist phrases of both negative and positive sentiments.

Acknowledgments. We would like to thank TIS Inc. (www.tis.com) for supporting and funding this research.

References

1. Khan, F.H., Qamar, U., Bashir, S.: SentiMI: introducing point-wise mutual information with SentiWordNet to improve sentiment polarity detection. Appl. Soft Comput. J. **39**, 140–153 (2016)
2. Al-Rowaily, K., Abulaish, M., Haldar, N.A.-H., Al-Rubaian, M.: BiSAL - a bilingual sentiment analysis lexicon to analyze Dark Web forums for cyber security. Digital Invest. **14**, 53–62 (2015)
3. Ko, Y.: A study of term weighting schemes using class information for text classification. In: Proceedings of the 35th International ACM SIGIR Conference on Research and Development in Information Retrieval - SIGIR 2012 (2012)
4. Khan, F.H., Bashir, S., Qamar, U.: TOM: twitter opinion mining framework using hybrid classification scheme. Decis. Support Syst. **57**, 245–257 (2014)
5. Salton, G., McGill, M.: Introduction to Modern Information Retrieval. McGraw-Hill, New York (1983)
6. Scott, D., Dumais, S.T., Furnas, G.W., Landauer, T.K., Harshman, R.: Indexing by latent semantic analysis. J. Am. Soc. Inf. Sci. **41**, 391–407 (1990)
7. Blei, D.M., Ng, A.Y., Jordan, M.I.: Latent Dirichlet allocation. J. Mach. Learn. Res. **3**, 993–1022 (2003)
8. Liang, J., Liu, P., Tan, J., Ba, S.: Sentiment classification based on AS-LDA model. In: Proceeding of 2nd International Conference on Information Technology and Quantitative Management, vol. 31, pp. 511–516 (2014)
9. Li, F., Huang, M., Zhu, X.: Sentiment analysis with global topics and local dependency. In: Proceedings of the Twenty-Fourth AAAI Conference on Artificial Intelligence (2010)
10. Raja Mohana, S.P., Umamaheswari, K., Karthiga, R.: Sentiment classification based on latent Dirichlet allocation. Int. J. Comput. Appl. 14–16 (2015). International Conference on Innovations in Computing Techniques
11. Bin, L., Ott, M., Cardie, C., Tsou, B.: Multi-aspect sentiment analysis with topic models. In: Proceedings of the 2011 IEEE 11th International Conference on Data Mining Workshop, pp. 81–88 (2011)
12. Bagheri, A., Saraee, M.: Latent Dirichlet Markov allocation for sentiment analysis. In: Fifth European Conference on Intelligent Management Systems in Operations, Salford, UK, pp. 90–96 (2013)

13. Whye, T.Y., Jordan Michael, I., Beal Matthew, J., Blei, D.M.: Hierarchical Dirichlet processes. J. Am. Stat. Assoc. **101**, 1566–1581 (2006)
14. Hông Phuong, L., Thi Minh Huyên, N., Roussanaly, A., Vinh, H.T.: A hybrid approach to word segmentation of Vietnamese texts. In: Martín-Vide, C., Otto, F., Fernau, H. (eds.) LATA 2008. LNCS, vol. 5196, pp. 240–249. Springer, Heidelberg (2008). doi:10.1007/978-3-540-88282-4_23

Simple and Accurate Method for Parallel Web Pages Detection

Alibi Jangeldin[(⊠)] and Zhenisbek Assylbekov

School of Science and Technology, Nazarbayev University, Astana, Kazakhstan
{alibi.jangeldin,zhassylbekov}@nu.edu.kz

Abstract. This paper presents language independent method for measuring structural similarity between web pages from bilingual websites. First we extract a new feature from those which are used by the STRAND architecture and combine it with the existing one. Next we analyze properties of this feature and develop an iterative algorithm to infer the parameters of our model. Finally, we propose an unsupervised algorithm for detecting parallel pairs of web pages based on these features and parameters. Our approach appears to benefit the structural similarity measure: in the task of distinguishing parallel web pages from five different bilingual websites the proposed method is competitive with other unsupervised methods.

Keywords: Parallel corpora · Document alignment · Structural similarity · STRAND · Unsupervised learning

1 Introduction

World Wide Web is an important source of parallel corpora (bitexts), since many websites are available in two or more languages. Many approaches have been therefore proposed for trying to exploit the Web as a parallel corpus: STRAND [11], PTMiner [3], BITS [8], WPDE [15], Bitextor [5], ILSP-FC [9], etc. The task of extracting bitexts from the Web typically involves the following four major consecutive steps: (i) data collection (crawling), (ii) document alignment, (iii) sentence splitting and (iv) sentence alignment.

Very often websites are already aligned on document level, i.e. web pages in one language contain links to the translated versions of themselves. However, even in this case one often encounters misaligned pairs, e.g. when a translated web page is missing (dead link) or documents are not exact translations of each other (comparable texts). To address these issues one usually needs to perform additional filtering on document level in order to detect and discard non-parallel document pairs. For this task three main strategies can be found in the literature – they exploit: (i) similarities in URLs; (ii) structural similarity of HTML files; (iii) content-similarity of texts. In this paper we address the second strategy, i.e. measuring structural similarity between HTML documents. We *do not* involve content-similarity metrics here as our goal is to have a language-agnostic tool.

© Springer International Publishing AG 2017
N.T. Nguyen et al. (Eds.): ACIIDS 2017, Part I, LNAI 10191, pp. 238–247, 2017.
DOI: 10.1007/978-3-319-54472-4_23

In this paper we develop a simple yet accurate language-independent technique for measuring structural similarity between HTML pages, which uses the same amount of information as previous approaches to distinguish parallelism of web pages and can be applied in unsupervised manner.

2 Related Work

Measuring structural similarity between HTML files was first introduced in [10], where a linearized HTML structure of candidate pairs was used to confirm parallelism of texts. Later approaches combined structural similarity metrics with other measures. E.g. Shi et al. [13] additionally used a file length ratio and a sentence alignment score. Zhang et al. [15] used file length ratio and content translation to train k-nearest-neighbors classifier for parallel pairs verification. Esplà-Gomis and Forcada [5] used text-language comparison, file size ratio, total text length difference for preliminary filtering and then HTML tag structure and text block length were used for deeper filtering. In [12] the bitext detection module runs three major filters: link follower filter, URL pattern search, and a combination of an HTML structure filter and a content filter. In [9] structural filtering is based on length ratios and edit distances between linearized versions of candidate pairs. Liu et al. [7] proposed a link-based approach in conjuction with content-based similarity and page structural similarity to distinguish parallel web pages from bi-lingual web sites.

All of these works require labeled data for model training or thresholds estimation in a supervised way, but there is usually not enough labeled parallel corpora for every language pair to do it and manual labeling is expensive. Our work is mainly motivated by unsupervised approach of Assylbekov et al. [1], who developed a statistical model for measuring structural similarity between web pages; they were using raw counts for text lengths and misalignment of HTML tags. In this paper we show that it is more reasonable to consider normalized quantities; we perform a detailed analysis of their distributions, propose an iterative algorithm to infer the parameters of our model in unsupervised manner and introduce two tuning parameters for it. Due to its simplicity the proposed approach can be used both for structural filtering and as a competitive baseline.

3 Methodology

3.1 Features Extraction

Let us assume that candidate pairs are linearized as in STRAND and linearized sequences are aligned using a standard dynamic programming technique [6]. For example, consider two HTML-files, in English and in Kazakh[1], that begin as in Fig. 1. Then the aligned linearized sequences would be as in Fig. 2.

---

[1] Kazakh is a Turkic language belonging to the Kipchak branch, with approximately 11 million native speakers.

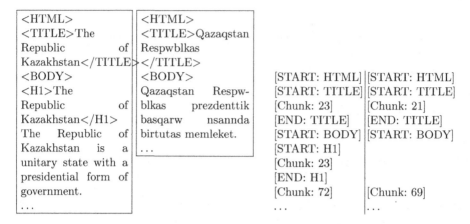

Fig. 1. Raw HTML files

Fig. 2. Aligned and linearized sequences

Let M_1 and M_2 denote the total numbers of alignment tokens in the first and the second documents; let L_1 and L_2 denote the total character lengths of the texts in the first and the second documents respectively. Let W be the total number of alignment tokens that are present in one file, but not the other (alignment cost). In our example, $M_1 = 9$, $M_2 = 6$, $L_1 = 118$, $L_2 = 90$ and $W = 3$.

It is critical to note that we are dealing with the task of unsupervised learning, since initially it is unknown which pairs of web pages are translated in a proper way, and thus meaningful patterns that are inherent to parallel documents should be extracted from the available data. We construct two new features that we believe to be helpful to discriminate between parallel and nonparallel pairs:

$$P_d = \frac{W}{M_1 + M_2} \quad \text{and} \quad L_d = \frac{L_1 - L_2}{L_1 + L_2}.$$

Furthermore, we discuss in more details why they are chosen by constructing illustrative examples for each feature. `akorda.kz`[2] and `pm.gc.ca`[3] websites will be used to visualize our work process. Our basic assumption throughout the paper is that *most of the web pages on a bilingual website are translated correctly.*

Let us consider two pairs of web pages with the following features: $M_1' = M_2' = 10$ and $W' = 2$, and $M_1'' = M_2'' = 100$ and $W'' = 2$. Comparing their parallelism only in terms of alignment cost gives equality ($W' = W'' = 2$), but with such difference in total number of tokens this conclusion may be misleading. To solve this issue, W is adjusted by the total number of the alignment tokens: $P_d = W/(M_1 + M_2)$. Now $P_d' = 0.1$ and $P_d'' = 0.01$ are more trustworthy to differentiate between the two cases. Since alignment cost is a measure of difference between a pair of web pages, the closer P_d of a pair to 0 the more

---

[2] Official site of the President of the Republic of Kazakhstan.

[3] Official site of the Prime-minister of Canada.

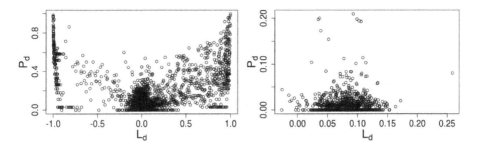

Fig. 3. L_d versus P_d for akorda.kz   **Fig. 4.** L_d versus P_d for pm.gc.ca

parallel it should be. This is supported by our basic assumption that most of the web pages are parallel and can be noticed in Figs. 3 and 4.

Furthermore, can we use $|L_d| = \frac{|L_1 - L_2|}{L_1 + L_2}$ as it was done in [5] to measure parallelism in terms of scaled difference in total text length? To answer this question suppose that there are two pairs of web page translations such that their character length features are: $L_1' = 400$ and $L_2' = 600$, $L_1'' = 600$ and $L_2'' = 400$. Thus, the equality $|L_d'| = |L_d''| = 0.2$ means that these two pairs should be equally parallel in terms of difference in number of characters. However, in this case we are losing some information on linguistic difference between languages, e.g. if texts in the first language are on average shorter than in the second language, i.e. $E(L_1) < E(L_2)$, then $L_d \to 0^-$ should be a better evidence towards parallelism than $L_d \to 0^+$. Thus, $|L_d|$ may be properly used only for language pairs with $E(L_d)$ closer to 0 as in Fig. 5 with Kazakh-English language pair. Thus, we may need more robust estimation of $E(L_d)$. The difference between language pairs was the most prominent when we analyzed `pm.gc.ca` website where total character length in French language was on average 17% longer than in English ($E(L_d) = 0.082$), illustrated in Fig. 6. Thus, it's important to use L_d instead of $|L_d|$ in order to differentiate between language pairs.

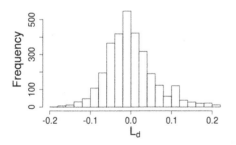
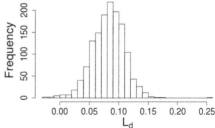

Fig. 5. Histogram of L_d for akorda with (3)

Fig. 6. Histogram of L_d for pm.gc.ca

As mentioned above, initially we do not know which pairs of web pages are parallel: if it was known we could estimate proper thresholds for P_d and L_d from such data. Hence we need another way of finding threshold values for P_d and L_d, that will allow us to discriminate between parallel and non-parallel pairs of web pages.

3.2   Thresholds Estimation

Let's formally derive our insight for L_d. For parallel pairs of web pages in a given language pair there is almost linear relationship between L_1 and L_2:

$$L_1 \approx c \cdot L_2,$$

where the constant c depends on the relationship between two languages. Under such assumption, L_d can be approximated as:

$$L_d = \frac{L_1 - L_2}{L_1 + L_2} \approx \frac{c - 1}{c + 1}. \tag{1}$$

By comparing P_d and L_d one may notice that L_d will have different distributions for different language pairs due to their linguistic differences and thus, will depend on c. Therefore we will need to adapt it to every language pair. On the other hand, P_d has the same interpretation for any language pair since number of alignment tokens is independent of language pairs. Thus, threshold proposed by the author of STRAND method [11] for P_d will be used in our work in accordance with our basic assumption. The rationale for this can be noticed from Figs. 3 and 4 in which majority of the points are below 0.2-level on P_d-axis.

One more thing to notice in L_d- P_d scatter plot is that parallel pairs of web pages are grouped around mean value of L_d which, according to (1), should be close to $(c - 1)/(c + 1)$. We will need this value for our further analysis and if we try to estimate it using the average of L_d across *all* pairs, it may be severely biased by the values of nonparallel pairs (see Fig. 3). Hence we need a more robust estimation of $E(L_d)$ for *parallel* pairs. For this purpose we extract a subset of pairs using the following rule

$$D = \{\text{document pairs for which } P_d = 0\}, \tag{2}$$

and we calculate sample mean of L_d on D, let us denote it by μ, since the condition $P_d = 0$ provides strong evidence towards parallelism of pairs supported by our basic assumption. Comparison of mean estimation methods for `akorda.kz` website shows that estimation of the mean for all pairs gives $E(L_d) = 0.033$ and from the Fig. 5 it seems that the bias comes from the right tail. Subset of pairs with $P_d = 0$ condition has $E(L_d) = -0.005$ that is much closer to the center of symmetry. For `pm.gc.ca` the new value is almost the same as previous mean with $E(L_d) = 0.083$ because most of the pairs are already parallel there before subsetting. Using the obtained estimate μ we will try to model the distribution of L_d around it.

For the illustrative purposes let us choose 0.2 thresholds for both P_d and $|L_d - \mu|$, i.e.

$$P_d < 0.2 \quad \text{and} \quad |L_d - \mu| < 0.2, \tag{3}$$

and look at the behavior of L_d on pairs which satisfy (3) – its distribution is given in Fig. 5.

One can see that it is symmetric around the mean and this property will be used to estimate threshold for L_d in unsupervised way. Essential part of our work is described in Algorithm 1. The main idea is to start with a small threshold for L_d which reliably separates parallel pairs from nonparallel and iteratively increase it while we are confident enough in parallelism of added pairs.

Algorithm 1. L_d threshold estimation

Require: $S \leftarrow$ subset of pairs with $P_d < 0.2$
Ensure: *threshold* – threshold for L_d
   $D \leftarrow \{s \in S : P_d = 0\}$
   $\mu \leftarrow E_D(L_d)$
   $threshold \leftarrow 0.01$
   $step \leftarrow 0.01$
   $N \leftarrow \#\{s \in S : |L_d - \mu| < threshold\}$
   $\Delta \leftarrow 1$
   while $\Delta \geq 0.01$ **do**
     $threshold \leftarrow threshold + step$
     $N_{new} \leftarrow \#\{s \in S : |L_d - \mu| < threshold\}$
     $\Delta \leftarrow N_{new}/N - 1$
     $N = N_{new}$
   end while

After threshold estimation, our suggestion is to predict the following subset of pairs as parallel:

$$P_d < 0.2 \qquad \text{and} \qquad |L_d - \mu| < threshold, \tag{4}$$

The loop described above should be finite because majority of the candidate pairs in all websites are parallel and thus have $P_d < 0.2$ and are located around the mean μ of L_d from which we start iteration. There are two main parameters that may be tuned in the algorithm: Δ threshold and *step* value. The trade-off between precision and recall may be addressed by tuning threshold for Δ. Precision may be maximized by increasing threshold value and thus decreasing threshold for L_d whereas recall can be maximized conversely. On the other hand, *step* value may be tuned to optimize the trade-off between computational efficiency and accuracy of the *threshold* value estimation for L_d. Thus, higher *step* value will lead to faster convergence, whereas lower *step* value will be beneficial for *threshold* estimation. Nevertheless, these two parameters are interdependent and tuning either of them will affect both of the trade-offs described above.

4   Experiments and Results

4.1   Data Sets and Evaluation Criteria

To evaluate the performance of our algorithm we used web pages from 5 different sites. To obtain candidate pairs from those websites we used GNU wget[4] tool. Since the obtained pairs are *candidate* pairs, there is no guarantee that all of them are parallel. General information about the websites, including URLs, short description, languages and total numbers of candidate pairs, is given in Table 1. From each of the websites we extracted representative samples (sample sizes were calculated using the Cochran's formula [4]) and manually checked them for parallelism. These samples are used as test sets with sizes given in Table 1.

Table 1. Information about websites

Websites:	Description	Languages	Number of pairs	Sample size
akorda.kz	President of Kazakhstan	kk-en	4135	352
egov.kz	Electronic government of Kazakhstan	kk-en	2400	312
mfa.kz	Ministry of Foreign Affairs of Kazakhstan	kk-en	180	180
presidencia.pt	President of Portugal	pt-en	960	275
pm.gc.ca	Prime-minister of Canada	fr-en	1397	302

We used precision, recall and F_1-score for performance evaluation:

$$prec = \frac{tp}{tp + fp}, \quad rec = \frac{tp}{tp + fn}, \quad F_1 = \frac{2 \cdot prec \cdot rec}{prec + rec},$$

where tp is the number of true positives (i.e. number of pairs in a sample correctly labeled as parallel) and fn is the number of false negatives (i.e. number of pairs in a sample incorrectly labeled as non-parallel).

4.2   Baseline and Other Approaches

In the following we describe a baseline and other approaches for parallel web pages detection which will be contrasted to our method.

Baseline. We use the STRAND's default thresholds from [11]:

$$P_d < 0.2 \quad \text{and} \quad p\text{-value} < 0.05,$$

where p-value corresponds to the significance of correlations between the lengths of aligned text chunks.

---

[4] http://www.gnu.org/software/wget.

Hierarchical Clustering (HC). We apply hierarchical clustering [14] with average linkage using the same features P_d and L_d as in Algorithm 1 and different ways of choosing the number of clusters:

HC1. setting two clusters in a belief that document pairs will naturally split into parallel and non-parallel,

HC2. using CH-index [2] to automatically calculate the appropriate number of clusters based on within-cluster and between-cluster variances,

In all of these approaches we assumed the largest cluster to contain parallel pairs and other clusters to contain nonparallel pairs according to our basic assumption.

Statistical Model (Algorithm 0). In this approach described in detail in [1] raw features were used as inputs and parameters were estimated in unsupervised way using EM algorithm. This work derives a joint distribution for W, M, N, L_1, and L_2 in a rigorous way throwing in independence assumptions along the derivation. Our approach is more empirical and much more simple.

4.3   Results

We applied all the methods mentioned above, i.e. baseline STRAND, HC1, HC2, Algorithm 0 and Algorithm 1, to the websites from Table 1, and resulting precision, recall and F_1-scores are provided in Table 2. As we see from Table 2 there is no single method which consistently outperforms all others. Our suggested Algorithm 1 did best on websites from the .kz domain which use Kazakh-English language pair, however HC2 was better on presidencia.pt (Portuguese-English) where Algorithm 1 showed good results as well. One can see that clustering showed good performance only on the websites with high quality of translation where signal-to-noise ratio is lower, but it still was not stable in those. The main reason why it outperformed proposed approach may be the fact that clustering algorithm is nonlinear, whereas the proposed algorithm has conservative linear boundaries used to maximize precision which may result in losing some of the parallel web pages in the boundaries.

Percentages of parallel pairs in test sets, estimated threshold values and number of iterations to converge in Table 3 will be used to analyze effectiveness of

Table 2. Results

Method	akorda.kz			egov.kz			mfa.gov.kz			presidencia			pm.gc.ca		
	$prec$	rec	F_1	$prec$	rec	F_1	$prec$	rec	F_1	$prec$	rec	F_1	$prec$	rec	F_1
Baseline	92.5	87.5	89.9	100	76.3	86.6	96.0	97.6	96.8	96.6	91.3	93.9	99.0	95.3	97.0
HC1	79.5	100	85.6	75.3	100	86.2	95.8	94.1	95.0	91.8	100	95.7	99.3	100	**99.7**
HC2	87.5	100	93.3	87.2	98.9	92.7	100	31.1	47.5	98.8	99.4	**99.1**	99.4	58.1	73.3
Algorithm 0	94.1	97.1	95.6	91.5	96.9	94.1	94.4	100	**97.1**	99.1	95.0	97.0	99.0	1.00	99.5
Algorithm1	94.5	98.6	**96.5**	100	90.5	**95.0**	95.5	98.8	**97.1**	97.4	97.7	97.5	99.3	100	**99.7**

the proposed method in more details. Process of the while loop convergence in Algorithm 1 for the five data sets is illustrated with corresponding Δ values in Fig. 7. One may notice that websites with .kz domain require more iterations to converge than others due to linguistic differences between language pairs, this explains why Δ values of these websites are fluctuating. Nevertheless, the values for all websites are steadily declining at the last three steps before convergence.

Percentages of parallel pairs in the test sets allow us to estimate quality of translations on the considered web sites. Empirical results show that the proposed algorithm is demonstrating better results on websites with higher quality of translation. The main limitations of our algorithm arise from the structural similarity measures. In most of the false positive pairs there are missing sentences in one or both sides and we believe that it would be easier to detect such page pairs with other approaches. On the contrary, some of false negatives have P_d and L_d values that are close to threshold values and therefore can be reached by increasing the thresholds.

Table 3. Properties of websites

Websites:	% of parallel pairs	*Threshold*	Iterations
akorda.kz	0.740	0.15	13
egov.kz	0.751	0.24	22
mfa.kz	0.944	0.16	14
pr-cia.pt	0.915	0.09	7
pm.gc.ca	0.990	0.08	6

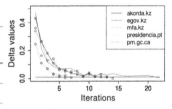

Fig. 7. Iterations versus Δ values

5   Conclusion and Future Work

In this paper we propose unsupervised method for parallel web pages detection. Feature engineering and analysis of properties of the new feature allowed us to efficiently estimate a threshold for it in unsupervised manner by considering inherent differences between language pairs and tuning parameters of the iterative algorithm. Empirical results show that the proposed approach is competitive with the previous unsupervised approaches and reproducible for further work.

Next step in this work may be to combine the proposed approach with content-similarity of texts as a deeper filtering method. Additionally, it would be good for robustness to make sure that pages in a candidate pair are in different languages, so that we are not measuring similarity of two identical pages. We are planning to use it for compiling large-scale Kazakh-Russian and Kazakh-English parallel corpora. Once it is done, modern approaches in statistical machine translation will be used to build competitive machine translation systems for Kazakh.

References

1. Assylbekov, Z., Nurkas, A., Mouga, I.R.: A statistical model for measuring structural similarity between webpages. In: Recent Advances in Natural Language Processing, pp. 24–31 (2015)
2. Caliński, T., Harabasz, J.: A dendrite method for cluster analysis. Commun. Stat. Theory Methods **3**(1), 1–27 (1974)
3. Chen, J., Nie, J.-Y.: Automatic construction of parallel English-Chinese corpus for cross-language information retrieval. In: Proceedings of the Sixth Conference on Applied Natural Language Processing, pp. 21–28. Association for Computational Linguistics (2000)
4. Cochran, W.G.: Sampling Techniques. Wiley, New York (2007)
5. Esplà-Gomis, M., Forcada, M.: Combining content-based and URL-based heuristics to harvest aligned bitexts from multilingual sites with bitextor. Prague Bull. Math. Linguist. **93**, 77–86 (2010)
6. Hunt, J.W., MacIlroy, M.: An Algorithm for Differential File Comparison. Bell Laboratories, Murray Hill (1976)
7. Liu, L., Hong, Y., Lu, J., Lang, J., Ji, H., Yao, J.: An iterative link-based method for parallel web page mining. In: Proceedings of EMNLP, pp. 1216–1224 (2014)
8. Ma, X., Liberman, M.: A method for bilingual text search over the web. In: Machine Translation Summit VII, pp. 538–542 (1999)
9. Papavassiliou, V., Prokopidis, P., Thurmair, G.: A modular open-source focused crawler for mining monolingual and bilingual corpora from the web. In: Proceedings of the Sixth Workshop on Building and Using Comparable Corpora, pp. 43–51 (2013)
10. Resnik, P.: Parallel strands: A preliminary investigation into mining the web for bilingual text. In: Farwell, D., Gerber, L., Hovy, E. (eds.) AMTA 1998. LNCS (LNAI), vol. 1529, pp. 72–82. Springer, Heidelberg (1998). doi:10.1007/3-540-49478-2_7
11. Resnik, P., Smith, N.A.: The web as a parallel corpus. Comput. Linguist. **29**(3), 349–380 (2003)
12. San Vicente, I., Manterola, I.: A fully automated tool for gathering parallel corpora from the web. In: LREC, pp. 1–6 (2012)
13. Shi, L., Niu, C., Zhou, M., Gao, J.: A DOM tree alignment model for mining parallel data from the web. In: Proceedings of the 21st International Conference on Computational Linguistics and the 44th annual meeting of the Association for Computational Linguistics, pp. 489–496. Association for Computational Linguistics (2006)
14. Ward Jr., J.H.: Hierarchical grouping to optimize an objective function. J. Am. Stat. Assoc. **58**(301), 236–244 (1963)
15. Zhang, Y., Wu, K., Gao, J., Vines, P.: Automatic acquisition of Chinese–English parallel corpus from the web. In: Lalmas, M., MacFarlane, A., Rüger, S., Tombros, A., Tsikrika, T., Yavlinsky, A. (eds.) ECIR 2006. LNCS, vol. 3936, pp. 420–431. Springer, Heidelberg (2006). doi:10.1007/11735106_37

Combining Latent Dirichlet Allocation and K-Means for Documents Clustering: Effect of Probabilistic Based Distance Measures

Quang Vu Bui[1,2], Karim Sayadi[2(✉)], Soufian Ben Amor[3,4], and Marc Bui[2]

[1] Hue University of Sciences, Hue, Vietnam
[2] CHArt Laboratory EA 4004, EPHE, PSL Research University, Paris, France
karim.sayadi@ephe.sorbonne.fr
[3] LI-PARAD Laboratory, University of Versailles-Saint- Quentin-en-Yvelines, Versailles, France
[4] Paris-Saclay University, Paris, France

Abstract. This paper evaluates through an empirical study eight different distance measures used on the LDA + K-means model. We performed
our analysis on two miscellaneous datasets that are commonly used.
Our experimental results indicate that the probabilistic-based distance
measures are better than the vector based distance measures including Euclidean when it comes to cluster a set of documents in the topic
space. Moreover, we investigate the implication of the number of topics
and show that K-means combined to the results of the Latent Dirichlet
Allocation model allows us to have better results than the LDA + Naive
and Vector Space Model.

Keywords: Latent Dirichlet Allocation · Topic modeling · Document
clustering · K-means · Similarity measure · Probabilistic-based distance ·
Clustering evaluation

1 Introduction

Clustering a set of documents is a standard problem addressed in data mining, machine learning, and statistical natural language processing. Document
clustering can automatically organize many documents into a small number of
meaningful clusters and find latent structure in unlabeled document collections.

K-means is one of the most used partitioned-based clustering algorithms. It
became popular among information retrieval tasks [12]. For clustering a set of
documents with K-means, each document is firstly quantified as a vector where
each component indicates a corresponding feature in the document. Then, a
distance is used to measure the difference between two documents. The collection of documents is represented by a sparse and high-dimensional matrix. The
use of this matrix raises an issue known as the "curse of dimensionality" [14].
Thus, using K-means require reducing the documents dimensionality and using
a "good" distance measure to get the most accurate clusters.

© Springer International Publishing AG 2017
N.T. Nguyen et al. (Eds.): ACIIDS 2017, Part I, LNAI 10191, pp. 248–257, 2017.
DOI: 10.1007/978-3-319-54472-4_24

In our work, we first reduce the dimensionality by decomposing the document matrix into latent components using the Latent Dirichlet Allocation (LDA) [2] method. Each document is represented by a probability distribution of topics and each topic is characterized by a probability distribution over a finite vocabulary of words. We use the probability distribution of topics as the input for K-means clustering. This approach called LDA + K-means was proposed by [3,17]. We note that [17] proposed LDA + K-means but only used Euclidean distance.

We then compare the efficiency of eight distance measures [5]. These measures are based on two approaches: (i) Vector based approach (VBM) with Euclidean distance, Sørensen distance, Tanimoto distance, Cosine distance and (ii) Probabilistic-based approach (PBM) with Bhattacharyya distance, Probabilistic Symmetric χ^2 divergence, Jensen-Shannon divergence, Taneja divergence.

In order to come up with a sound conclusion, we have performed an empirical evaluation of the eight distance measures according to a labeled clustering. We compared the clusters with the two evaluation criteria: Adjusted Rand Index (ARI) [9] and Adjusted Mutual Information (AMI) [16]. We used two common datasets in the NLP community: the 20NewsGroup dataset contains newsgroup posts and the WebKB contains texts extracted from web pages.

Our experiments can be compared to the work of [8,11,17]. The key differences are the following: In comparison with the VBM we conducted our experiments with a PBM, we show that in the case of LDA + K-means where the input is a probability distribution the use of PBM leads to better results. Then, our results show that the Euclidean distance may not be suitable for this kind of application. Finally, by evaluating the results of the VBM and PBM with ARI and AMI criteria we have investigated the implication of the number of topics in the clustering processing.

This paper is organized as follows. The next section describes the methodology in which we present K-means algorithms and document clustering, similarity measures in probabilistic spaces and evaluation indexes used in the experiments. We explain the experiment, discuss the results in Sect. 3 and also conclude our work in Sect. 4.

2   Methodology

2.1   Document Clustering

Vector Space Model. Most current document clustering methods choose to view text as a bag of words. In this method, each document is represented by word-frequency vector $d_{wf} = (wf_1, wf_2, \ldots, wf_n)$, where wf_i is the frequency of the ith word in the document. This gives the model its name, the vector space model (VSM) [15].

The two disadvances of VSM are the high dimensionality because of the high number of unique terms in text corpora and insufficient to capture all semantics. Latent Dirichlet Allocation [2] proposed a good solution to solve these issues.

Latent Dirichlet Allocation. Latent Dirichlet Allocation (LDA) [2] is a generative probabilistic model for topic discovery. In LDA, each document may be considered as a mixture of different topics and each topic is characterized by a probability distribution over a finite vocabulary of words. The generative model of LDA, described with the probabilistic graphical model in Fig. 1, proceeds as follows:

1. Choose distribution over topics θ_i from a Dirichlet distribution with parameter α for each document.
2. Choose distribution over words ϕ_k from a Dirichlet distribution with parameter β for each topic.
3. For each of the word positions i, j:
   3.1. Choose a topic $z_{i,j}$ from a Multinomial distribution with parameter θ_i
   3.2. Choose a word $w_{i,j}$ from a Multinomial distribution with parameter $\phi_{z_{i,j}}$

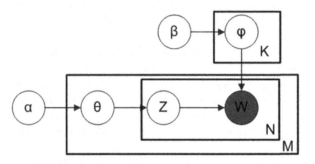

Fig. 1. Probabilistic graphical model of LDA

For posterior inference, we need to solve the following equation:

$$p(\theta, \phi, z|w, \alpha, \beta) = \frac{p(\theta, \phi, z, w|\alpha, \beta)}{p(w|\alpha, \beta)}$$

There are some inference algorithms available including variational inference used in the original paper [2] and Gibbs Sampling. Please refer to the work of [1] for more details.

K-Means Algorithm. K-means which proposed by Forgy [6] is one of the most popular clustering algorithms. It provides a simple and easy way to classify objects in k groups fixed a priori. The basic idea is to define k centroids and then assign objects to the nearest centroid. A loop has been generated. In each step, we need to re-calculate k new centroids and re-assign objects until no more changes are done. The algorithm works as follows:

1. Selecting k initial objects called centroids of the k clusters.
2. Assigning each object to the cluster that has the closest centroid.
3. Computing the new centroid of each cluster.
4. Repeat step 2 and 3 until the objects in any cluster do no longer change.

2.2 Combining LDA and K-Means

The output of LDA is two probability distributions: the document-topic distribu-tion θ and the word-topic distribution ϕ. To use as much as possible information from LDA result, we can combine Latent Dirichlet Allocation and K-means, denoted LDA + K-means, by using document-topic distributions θ extracted from LDA as the input for K-means clustering algorithms. For a matter of space, we invite the readers to find more details in the work of [3].

2.3 Similarity Measures

Since LDA represents documents as probability distributions, we need to con-sider the "good" way to choose a distance or similarity measure for comparing two probability distributions. Eight distances families as categorized by [5] were used in K-means + LDA. These families can be divided into two groups:

- Vector-Based Measurements (VBM): Euclidean distance, Sørensen distance, Tanimoto distance, Cosine distance
- Probabilistic-Based Measurements (PBM): Bhattacharyya distance, Prob-abilistic Symmetric χ^2 divergence, Jensen-Shannon divergence, Taneja divergence

Let $A = (a_1, a_2, \ldots, a_k)$ and $B = (b_1, b_2, \ldots, b_k)$ be two vectors with k dimen-sions. The eight distances between A and B are defined as:

Euclidean distance: $d_{Euc} = \sqrt{\sum_{i=1}^{k} |a_i - b_i|^2}$

Sørensen distance: $d_{Sor} = \frac{\sum_{i=1}^{k} |a_i - b_i|}{\sum_{i=1}^{k} (a_i + b_i)}$

Tanimoto distance: $d_{Tani} = \frac{\sum_{i=1}^{k} (max(a_i, b_i) - min(a_i, b_i))}{\sum_{i=1}^{k} max(a_i, b_i)}$

Cosine distance: $d_{Cos} = 1 - Sim_{Cos} = 1 - \frac{\sum_{i=1}^{k} a_i b_i}{\sqrt{\sum_{i=1}^{k} a_i^2} \sqrt{\sum_{i=1}^{k} b_i^2}}$

Jensen-Shannon Divergence. The Jensen-Shannon (JS) divergence, known as a total divergence to the average, is based on Kullback-Leibler (KL) diver-gence, which is related to Shannon's concept of uncertainty or "entropy" $H(A) = \sum_{i=1}^{k} a_i ln a_i$.

$$d_{JS} = \frac{1}{2} \sum_{i=1}^{k} a_i ln(\frac{2a_i}{a_i + b_i}) + \frac{1}{2} \sum_{i=1}^{k} b_i ln(\frac{2b_i}{a_i + b_i})$$

Bhattacharyya Distance. Bhattacharyya distance is a divergence-type measure between distributions, defined as,

$$d_{Bhat} = -ln \sum_{i=1}^{k} \sqrt{a_i b_i}$$

Probabilistic Symmetric χ^2 Divergence. Probabilistic Symmetric χ^2 divergence is a special case of χ^2 divergence. It is a combination of Pearson χ^2 divergence and Newman χ^2 divergence.

$$d_{PChi} = 2 \sum_{i=1}^{k} \frac{(a_i - b_i)^2}{a_i + b_i}$$

Taneja Divergence. Taneja divergence is a combination between KL divergence and Bhattacharyya distance, using KL-divergence with $a_i = \frac{a_i+b_i}{2}, b_i = \sqrt{a_i b_i}$

$$d_{TJ} = \sum_{i=1}^{k} (\frac{a_i + b_i}{2}) ln(\frac{a_i + b_i}{2\sqrt{a_i b_i}})$$

2.4 Evaluation Methods

For each dataset, we obtained a clustering result from the K-means algorithm. To measure the quality of the clustering results, we used two evaluation indexes: Adjusted Rand Index (ARI) [9] and Adjusted Mutual Information (AMI) [16], which are widely used to evaluate the performance of unsupervised learning algorithms.

Adjusted Rand Index: Adjusted Rand Index (ARI) [9], an adjusted form of Rand Index (RI), is defined as:

$$ARI = \frac{\sum_{ij} \binom{n_{ij}}{2} - [\sum_i \binom{n_{io}}{2} \sum_j \binom{n_{oj}}{2}]/\binom{n}{2}}{\frac{1}{2}[\sum_i \binom{n_{io}}{2} + \sum_j \binom{n_{oj}}{2}] - [\sum_i \binom{n_{io}}{2} \sum_j \binom{n_{oj}}{2}]/\binom{n}{2}} \tag{1}$$

where $n_{ij}, n_{io}, n_{oj}, n$ are values from the contingency Table 1.

Adjusted Mutual Information. The Adjusted Mutual Information (AMI) [16], an adjusted form of mutual information (MI), is defined:

$$AMI(P,Q) = \frac{MI(P,Q) - E\{MI(P,Q)\}}{\max\{H(P), H(Q)\} - E\{MI(P,Q)\}} \tag{2}$$

where

$$H(P) = -\sum_{i=1}^{k} \frac{n_{io}}{n} \log \frac{n_{io}}{n}; MI(P,Q) = \sum_{i=1}^{k} \sum_{j=1}^{l} \frac{n_{ij}}{n} \log \frac{n_{ij}/n}{n_{io}n_{oj}/n^2}.$$

Table 1. The Contingency Table, $n_{ij} = |P_i \cap Q_j|$

$P \setminus Q$	Q_1	Q_2	$\cdots$	Q_l	Sums
P_1	n_{11}	n_{12}	$\cdots$	n_{1l}	n_{1o}
P_2	n_{21}	n_{22}	$\cdots$	n_{2l}	n_{2o}
$\vdots$	$\vdots$	$\vdots$	$\ddots$	$\vdots$	$\vdots$
P_k	n_{k1}	n_{k2}	$\cdots$	n_{kl}	n_{ko}
Sums	n_{o1}	n_{o2}	$\cdots$	n_{ol}	$\sum_{ij} n_{ij} = n$

Both ARI and AMI have a boundary above by 1. Higher values of ARI or AMI indicate more agreement between the two partitions. Please refer to the work of [9], [16] for more details.

3 Experiments and Results

3.1 Datasets

The proposed methodology is evaluated on 2 miscellaneous datasets that are commonly used for the NLP community regarding the task of document clustering. Table 2 describes some statistics about the used datasets. The 20Newsgroup collect has 18821 documents distributed across 20 different news categories. Each document corresponds to one article with a header that contains the title, the subject, and quoted text. The WebKB dataset contains 8230 web pages from the computer science department of different universities (e.g. Texas, Wisconsin, Cornell, etc.).

Table 2. Statistics of the datasets. Where #Docs refers to the number of documents in the dataset, #Classes refers to the number of classes in the dataset and < Class, > Class, refers to the minimum number of documents and the maximum number of document in a class.

Dataset	#Docs	#Classes	< Class	> Class
News20	18821	20	628	999
WebKB	8230	4	504	1641

3.2 Setup

In our experiments, we compared eight distances used with LDA + K-means divided into the two categories: the Probabilistic-Based Measurements (PBM) and the Vector-Based Measurements (VBM). We run LDA with Gibbs sampling method using the `topicmodels` R package[1]. The prior parameters α and β are

---

[1] https://cran.r-project.org/web/packages/topicmodels/index.html.

respectively set to 0.1 and 0.01. These parameters were chosen according to the state-of-the-art standards [7]. The number of iterations of the Gibbs sampling is set to 5000. The input number of topics for the 20NewsGroups dataset is set to 30 and for the WebKB dataset is set to 8. This number of topics will be confirmed in our experiments by testing different values. For each of the eight distances, we run the K-means 20 times with a maximum number of iterations equal to 1000. We compute the ARI and AMI on the results of each K-means iteration and report the average values.

3.3   Results

Comparing Effectiveness of Eight Distance Measures for LDA + K-Means. The average values of the ARI and AMI are reported in Table 3. The average ARI and AMI values of the PBM group are better than the average values of the VBM group. We notice that the Euclidean distance has the worst results regarding the ARI and AMI criteria. In the PBM group, the best average values are obtained by the two distances Bhattacharyya and Taneja. Thus, we propose to work with Taneja or Bhattacharyya distance for LDA + K-means. For a better understanding of the results, we additionally provide a bar plot illustrated in Fig. 2.

Table 3. The average values of ARI, AMI for VSM, LDA Naive, LDA + K-means with eight different distance measures for two datasets

Distances	20NewsGroups		WebKB	
	ARI	AMI	ARI	AMI
Euclidean	0,402	0,608	0,436	0,432
Sorensen	0,592	0,698	0,531	0,479
Tanimoto	0,582	0,691	0,531	0,48
Cosine	0,552	0,678	0,519	0,468
Bhattacharyya	0,619	0,722	0,557	0,495
ChiSquared	0,602	0,708	0,545	0,487
JensenShannon	0,614	0,717	0,551	0,488
Taneja	0,642	0,739	0,559	0,489
VSM	0,128	0,372	0,268	0,335
LDA + Naive	0,434	0,590	0,171	0,197

The Role Played by the Number of Topics for LDA + K-Means. We chose the number of topics based on the Harmonic mean of Log-Likelihood (HLK) [4]. We notice in the Fig. 3(a), that the best number of topics are in the range of [30, 50] of a maximum value of HLK. We run the LDA + K-means with a different number of topics and four distances: two from the PBM group, two from the VBM group including the Euclidean distance. We plot the evaluation with AMI and ARI in the Fig. 3(b) and (c).

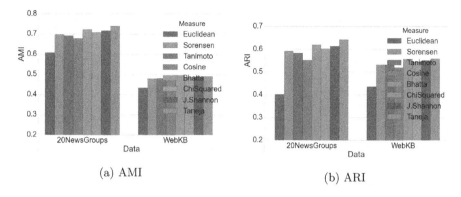

(a) AMI   (b) ARI

Fig. 2. The average values of ARI, AMI for LDA + K-means with eight different distance measures for two datasets

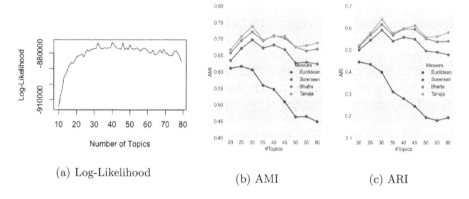

(a) Log-Likelihood   (b) AMI   (c) ARI

Fig. 3. The harmonic mean of the log-likelihood and ARI, AMI values with four distances for 20NG dataset with different # of topics.

As the number of topics increases, the LDA + K-means with Euclidean distance decreases in performance. The Euclidean distance is clearly not suitable for the LDA + K-means. The other three used distances (i.e. Sorensen, Bhattacharyya, and Taneja) kept a steady behavior with a slight advantage for the Taneja distance. This is due to the fact that these distance were defined for probability distribution and thus are more suitable for the kind of input provided by LDA. We notice that after 50 topics the performance of the three distances decreases.

Comparing LDA + K-Means, LDA + Naive, VSM. In order to study the role played by topic modeling, we compare three document clustering methods. The first is Vector space model (VSM) that uses a word-frequency vector $d_{wf} = (wf_1, wf_2, \ldots, wf_n)$, where wf_i is the frequency of the ith word in the document as input for K-means [13]. The second is proposed in [10], which

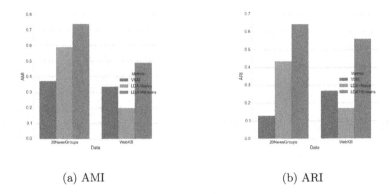

(a) AMI   (b) ARI

Fig. 4. ARI, AMI values for three methods: VSM, LDA + Naive, LDA + K-means with Taneja distance computed on 20NGNewsGroups and WebKB datasets

considers each topic as a cluster. In fact, document-topic distribution θ can be viewed as a mixture proportion vector over clusters and thus can be used for clustering as follows. Suppose that x is a cluster, a document is assigned to x if $x = argmax_j \theta_j$. Note that this approach is a simple solution, usually referred to as a naive solution to combine topic modeling and document clustering. This approach is denoted in our experiments as LDA + Naive. The third one is the LDA + Kmeans with the probabilistic-based distance measure (eg. Bhattacharyya, Taneja). The results are plotted in Fig. 4, we notice that the LDA + Kmeans used with Taneja distance obtains the best average results for both of the used datasets.

4   Conclusion

In this paper, we compared the effect of eight distance or similarity measures represented to eight distance measure families for clustering document using LDA + K-means. Experiments on two datasets with two evaluation criteria demonstrate the fact that the efficiency of Probabilistic-based measurement clustering is better than the Vector based measurement clustering including Euclidean distance. Comparing among LDA + K-means, LDA + Naive, Vector Space Model, the experiments also show that if we choose the suitable value of a number of topic for LDA and Probabilistic-based measurements for K-means, LDA + K-means can improve the effect of clustering results.

References

1. Blei, D.M.: Probabilistic topic models. Commun. ACM **55**(4), 77–84 (2012)
2. Blei, D.M., Ng, A.Y., Jordan, M.I.: Latent dirichlet allocation. J. Mach. Learn. Res. **3**, 993–1022 (2003)
3. Bui, Q.V., Sayadi, K., Bui, M.: A multi-criteria document clustering method based on topic modeling and pseudoclosure function. Informatica **40**(2), 169–180 (2016)

4. Buntine, W.: Estimating likelihoods for topic models. In: Zhou, Z.-H., Washio, T. (eds.) ACML 2009. LNCS (LNAI), vol. 5828, pp. 51–64. Springer, Heidelberg (2009). doi:10.1007/978-3-642-05224-8_6

5. Cha, S.-H.: Comprehensive survey on distance/similarity measures between probability density functions. City **1**(2), 1 (2007)

6. Gordon, A.: Classification. Chapman & Hall/CRC Monographs on Statistics & Applied Probability, 2nd edn. CRC Press, Boca Raton (1999)

7. Griffiths, T.L., Steyvers, M.: Finding scientific topics. Proc. Natl. Acad. Sci. U.S.A. **101**(Suppl 1), 5228–5235 (2004)

8. Huang, A.: Similarity measures for text document clustering. In: Proceedings of the Sixth New Zealand Computer Science Research Student Conference (NZCSRSC 2008), Christchurch, New Zealand, pp. 49–56 (2008)

9. Hubert, L., Arabie, P.: Comparing partitions. J. Classif. **2**(1), 193–218 (1985)

10. Lu, Y., Mei, Q., Zhai, C.: Investigating task performance of probabilistic topic models: an empirical study of PLSA and LDA. Inf. Retrieval **14**(2), 178–203 (2010)

11. Maher, K., Joshi, M.S.: Effectiveness of different similarity measures for text classification and clustering. Int. J. Comput. Sci. Inf. Technol. **7**(4), 1715–1720 (2016)

12. Manning, C.D., Raghavan, P.: An Introduction to Information Retrieval. Cambridge University Press, Cambridge (2009)

13. Modha, D.S., Spangler, W.S.: Feature weighting in k-means clustering. Mach. Learn. **52**(3), 217–237 (2003)

14. Pestov, V.: On the geometry of similarity search: dimensionality curse and concentration of measure. Inf. Process. Lett. **73**(1), 47–51 (2000)

15. Salton, G., Buckley, C.: Term-weighting approaches in automatic text retrieval. Inf. Process. Manage. **24**(5), 513–523 (1988)

16. Vinh, N.X., Epps, J., Bailey, J.: Information theoretic measures for clusterings comparison: Variants, properties, normalization and correction for chance. J. Mach. Learn. Res. **11**, 2837–2854 (2010)

17. Xie, P., Xing, E.P.: Integrating Document Clustering and Topic Modeling, September 2013. arXiv:1309.6874

A Hybrid Method for Named Entity Recognition on Tweet Streams

Van Cuong Tran[1], Dinh Tuyen Hoang[1], Ngoc Thanh Nguyen[2],
and Dosam Hwang[1(✉)]

[1] Department of Computer Engineering,
Yeungnam University, Gyeongsan, South Korea
vancuongqbuni@gmail.com, hoangdinhtuyen@gmail.com, dosamhwang@gmail.com
[2] Faculty of Computer Science and Management,
Wrocław University of Science and Technology, Wrocław, Poland
Ngoc-Thanh.Nguyen@pwr.edu.pl

Abstract. Information extraction from microblogs has recently attracted researchers in the fields of knowledge discovery and data mining owing to its short nature. Annotating data is one of the significant issues in applying machine learning approaches to these sources. Active learning (AL) and semi-supervised learning (SSL) are two distinct approaches to reduce annotation costs. The SSL approach exploits high-confidence samples and AL queries the most informative samples. Thus they can produce better results when jointly applied. This paper proposes a combination of AL and SSL to reduce the labeling effort for named entity recognition (NER) from tweet streams by using both machine-labeled and manually-labeled data. The AL query algorithms select the most informative samples to label those done by a human annotator. In addition, Conditional Random Field (CRF) is chosen as an underlying model to select high-confidence samples. The experiment results on a tweet dataset demonstrate that the proposed method achieves promising results in reducing the human labeling effort and that it can significantly improve the performance of NER systems.

Keywords: Named entity recognition · Active learning · Semi-supervised learning · Hybrid method · Tweet streams

1 Introduction

Social networking services (SNSs) like Twitter, Facebook, Google+ have attracted millions of users who publish and share the most up-to-date information, emergent social events, and their personal opinions, resulting in large volumes of data produced everyday. For example, Twitter has more than 313 million monthly active users and 500 million tweets are sent per a day[1]. The nature of data in SNSs is that they are usually short, incomplete, noisy, but up-to-date.

---

[1] https://about.twitter.com/company.

© Springer International Publishing AG 2017
N.T. Nguyen et al. (Eds.): ACIIDS 2017, Part I, LNAI 10191, pp. 258–268, 2017.
DOI: 10.1007/978-3-319-54472-4_25

Thus many applications in natural language processing (NLP) suffer from these kinds of data. Twitter is typical; unlike well-formatted text, processing tweets presents difficult challenges [1,5,13]. A big challenge for mining streaming data is the lack of labeled data due to rapid changes in distribution and the high cost of labeling [11]. The traditional NER methods require prior labeling of the training data, and NER processes can be done offline. This is not suitable for SNSs where the data are dynamically updated [3]. To deal with the brief and up-to-date information in tweets, different NER methods have been proposed in recent years. The majorities of them have primarily focused on recognizing named entities in a pool of tweets [4,10]. However, the big problem is how to effectively conduct the NER task in tweet streams.

Supervised learning methods require a large amount of labeled training data with high accuracy to construct a good statistical model. They achieve high performance if applied them to well-formatted texts. However, the achieved results are not satisfying when applying to short and noisy messages like tweets [9]. On the other hand, annotating data can be rather tedious and time-consuming. It takes much of human effort, and it is hard to annotate a large corpus covering a wide-domain, whereas unlabeled data are often easily obtained. Therefore, semi-supervised learning (SSL) has been suggested to decrease the annotation effort. This method can utilize unlabeled data for the training phase to mitigate the impact of insufficiently labeled data by supplying high-confidence machine-labeled data. CRF is often used as an underlying model to extract high-confidence data [8,9]. Active learning (AL) is an attractive method that can be used in conjunction with SSL [6,15,16]. It can address the shortage and incomprehensiveness of training data. Instead of relying on the samples being selected randomly from a large corpus, the AL method chooses significant samples to be labeled by an expert via optimal query algorithms [12]. This method is quite suitable for streaming data, such as those from SNSs where updating the training data is frequently required.

To solve the weakness of methods mentioned above, a hybrid method of the NER task is proposed to exploit both labeled data and unlabeled data to train high-performance classifiers. It is a combination of the AL and SSL methods. The proposed method aims to minimize annotation costs while maximizing the desired performance from the model. The AL query algorithms are applied to unlabeled data to select the most informative samples labeled by means of an expert. A classifier trained on the current training data is used to classify the remaining unlabeled data. High-confidence samples are selected. Both manually-labeled and machine-labeled samples are added to the training data. The classifier is then retrained on the updated training data and applied to newly arriving data. Experiments have been conducted on tweet data to assess the proposed method. The AL and SSL algorithms have also been implemented separately as a baseline for comparing results with the hybrid method. We will show that the proposed method achieves better results than the baseline.

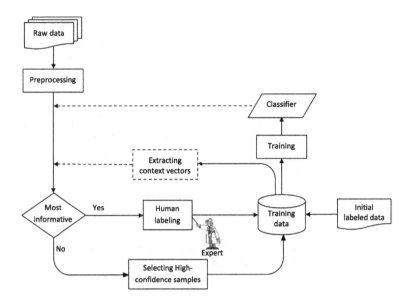

Fig. 1. The workflow of the system for training the classifiers

The organization of this paper is as follows. Section 2 presents the proposed method. The experiments and results are presented in Sect. 3. The conclusion and future work are drawn in the last section.

2 Stream-Based Hybrid Method for NER

This section presents a hybrid method for NER that is a combination of AL and SSL method. Unlabeled data is exploited to find the manually-labeled and machine-labeled samples for training the classifiers.

2.1 Problem Description

The NER system labels sequences of tokens in texts which are the names of entities, such as personal names, locations, organizations, etc. This study takes into account the classification of the named entities from tweet streams.

Definition 1 (tweet). *A tweet is a short message on the social media website Twitter. The maximum length of a tweet is 140 characters. A tweet tw is a sequence of tokens defined as*

$$tw = \langle S, \tau \rangle \tag{1}$$

where S is the sequence of tokens and τ is the time of the tweet appearance.

A classifier is trained on the training data used to classify the tokens. A named entity is characterized by the token sequence and the category. A named entity NE is represented as

$$NE = \langle S, \xi \rangle \tag{2}$$

where S is the token sequence of NE, and ξ is the label of NE. A label of a token is the category of the named entity in which the token is an element. The output results R of a classifier for a sequence of tokens is described as

$$R = \langle S, \chi, \psi \rangle \tag{3}$$

where χ is a set of labels of the tokens in S, and ψ is a set of real numbers denoting the probabilities of the labels. The probability of the label describes the ability to belong to the particular category of the token. The probability of labels belongs to the interval $[0,1]$, where 0 indicates that the classifier is not at all confident about its decision, and 1 indicates complete confidence.

Each element $r_i \in R$ is a triple, $r_i = (tk_i, \xi_i, \psi_i)$, where ξ_i is the label of token tk_i, and ψ_i is the probability of label ξ_i. The label has a low-confidence level if the probability of the label is less than predefined threshold α; otherwise, it has high-confidence. The goal of updating the model is to reduce the samples in which the labels are low confidence.

The main task of this work is how to update the model to improve the performance of a system. We focus on supplying the training data with new labeled data. Given a set of unlabeled tweets $U = \{tw_1, tw_2, ..., tw_n\}$ and a set of labeled tweets L as initial training data. L is updated by the machine-labeled and manually-labeled data gotten from U. In the following sections, we present the detail of the proposed methods.

2.2 Semi-supervised Learning for Selecting Confident Samples

In SSL, unlabeled data is exploited to find data that can be labeled by a machine. Machine-labeled data is added to the training data in order to improve the performance of the model. Initially, a small set of labeled data is provided to train an initial classifier. It is utilized to automatically classify the unlabeled data, and then the machine-labeled data is added directly to the training data to retrain the model. The performance of the classifier has improved due to the machine-labeled data; thus it reduces the human labeling effort. However, the performance of this approach can be affected adversely by the quality of the machine-labeled data if the training data are not perfect and large enough [7].

Definition 2 (high-confidence sample). *A high-confidence sample indicates the certainty of predicted label sequence of a classifier. A sample is called high confidence if the predicted label sequence of tokens has high confidence.*

In order to reduce the noise of the machine-labeled data because of wrong predictions, the high-confidence samples as defined in Definition 2 will be considered for addition to the training data. We only take into account the probability of the whole label sequence of a sample. The confidence level of a label sequence describes the reliability of a predicted label sequence. A label sequence has high confidence if the probability of the whole label sequence is greater than predefined threshold β.

A set of high-confidence samples L_h is determined as

$$L_h = \{tw_i : Prob(tw_i) \geq \beta, i = 1, .., m\} \tag{4}$$

where $Prob(tw_i)$ is the probability of the whole label sequence of a tweet tw_i.

2.3 Active Learning for Selecting Informative Samples

AL is a supervised learning approach in which the learner controls the selection of data necessary for the training phase. The samples are considered labeling by a human annotator relying on less certain about the labeling of the current model. The primary purpose of AL is to create a classifier that is as good as possible without supplying more labeled samples and without needing more human effort to annotate the data. The AL algorithms find the data necessary for improving the classifier after adding that data to the training data. The most informative samples are selected based on less reliable predictions of the learned model. An expert in the related domain is asked to annotate the informative samples.

The query strategy is the most important task in AL. The query algorithms can select suitable data needed to supplement the training data. In the following subsection, we present three query strategies that were used in our proposed AL method [14].

Query by Committee. (QBC) is a sampling method based on using a multi-classifier to select the most informative samples that need for training the model. The different classifiers are trained on the same training data by individual models. We used two models: CRF provided by Stanford[2] and maximum entropy provided by OpenNLP[3]. The classifiers are then used to classify unlabeled data. The disagreement between the classifiers with respect to the value and the category of a named entity is utilized to decide whether that sample is labeled by a human annotator.

Uncertainty-Based Sampling. Examines samples where the current model is less certain with respect to the probability of the labels. We take into account the individual labels other than the label sequence while looking for uncertainty samples.

Definition 3 (uncertainty sample). *An uncertainty sample indicates the uncertainty of predicted labels of a classifier. A sample is called uncertainty if it contains at least a label which has a low-confidence level.*

The uncertainty samples should be selected to update the model to increase the certainty of the classifier dealing with those samples. The classifier performance is directly proportional to the value of threshold α and the number of selected samples. If the threshold is set low, it takes less annotation effort. Otherwise, the threshold is set high, much more samples may be chosen, and the annotation effort also increases.

---

[2] http://nlp.stanford.edu/software/CRF-NER.shtml.
[3] https://opennlp.apache.org/documentation/1.5.3/manual/opennlp.html.

Diversity-Based Sampling. (DBS) is a sampling method based on context similarity of tweets. A vector model is used to measure the context similarity between an unlabeled tweet and all labeled tweets in the training data. A context vector represents each proper noun and named entity.

Definition 4 (context vector). *A context vector represents the context information of words surrounding a particular phrase, such as a named entity, or a proper noun. The size of the context vector is the size of the considered window that covers the phrase in the middle (i.e., the number of words considered on both sides of the phrase). The elements of the vector are POS of words. It can be represented as*

$$V = (..., p_{-3}, p_{-2}, p_{-1}, p_{+1}, p_{+2}, p_{+3}, ...) \tag{5}$$

where p_i is POS of the word at location i from the phrase. The negative sign and the positive sign, respectively, mean that the words are to the left-hand side and the right-hand side of the phrase.

An unlabeled tweet that contains the proper nouns is examined for context similarity with the current training data. Tweets where the similarity score is less than or equal to threshold θ are selected for labeling. The similarity level of two context feature vectors is measured by comparing their corresponding elements. The similarity score of two vectors is the number of their similar elements.

Algorithm 1. Active semi-supervised learning algorithm

Input: T - Set of time intervals; L - Set of initial labeled data
$\qquad\quad$ γ, β - The thresholds of most informative sample and high-confidence sample
Output: C - Classifier

1: Train an initial classifier C using L;
2: **repeat**
3: $\quad$ Get raw data according to time given $t \in T$ giving U;
4: $\quad$ Preprocess U;
5: $\quad$ Calculate the informativeness of samples $tw_i \in U, i = 1, .., n$;
6: $\quad$ Select the most informative samples from U,
$\qquad$ let $U_{mi} = \{tw_i : tw_i \in U \wedge MI(tw_i) \geq \gamma, i = 1, .., k\}$;
7: $\quad$ Ask an expert to label U_{mi} to obtain L_{mi};
8: $\quad$ $L = L + L_{mi}; U = U - U_{mi}$;
9: $\quad$ Classify the samples $tw_i \in U, i = 1, .., n - k$, using C giving R_u;
10: $\quad$ Select high-confidence samples from R_u,
$\qquad$ let $L_h = \{tw_i : tw_i \in R_u \wedge Prob(tw_i) \geq \beta, i = 1, .., m\}$;
11: $\quad$ $L = L + L_h$;
12: $\quad$ Retrain the classifier C using L;
13: **until** Out of T or other stop criteria met;
14: **return** C.

2.4   Combining Active Learning and Semi-supervised Learning

The main purpose of using SSL and AL is to reduce the annotation effort by learning from a small set of annotated data and achieving a classifier with high performance. In AL, the unlabeled data are the most informative used for updating the classifier. The most informative samples are selected based on the predictions of valuable data that are helpful in evolving the model. To exploit the remaining unlabeled data, SSL is a suitable approach. The unlabeled data are exploited to find credibility data that can be labeled by a machine. This approach only considers high-confidence samples. The samples are classified with high confidence by the current model chosen for training.

SSL and AL support each other, and the combination is an effective method for exploiting the unlabeled data. The workflow of the proposed hybrid method is depicted in Fig. 1 and a brief algorithm is shown in Algorithm 1. A small set of unlabeled data is selected randomly to be labeled initially. These training data are used to train an initial classifier. The training data will be dynamically updated to improve classifier performance. This work is applied to tweet streams where the tweets are produced continually. The time of a tweet appearance is a criterion for the querying order. To increase the training data, the samples that are the most informative are queried by the AL sampling strategies (as mentioned in Sect. 2.3). In the algorithm, the threshold of selecting most informative samples is denoted by γ. A human annotator labels the selected samples. The classifier is applied to the remaining unlabeled data to predict their labels. The high-confidence samples are selected automatically and added to the training data along with the manually-labeled samples. Training a new model is conducted when the sampling time is satisfied. The training processing is iterated until the stopping criterion is met.

3   Experimental Results

3.1   Dataset and Evaluation Measures

The proposed method was applied to tweet data experimented with our previous work [14]. The dataset includes 10,813 unlabeled tweets and 4,716 labeled tweets. The labeled tweets were used as initial training data. The test set (TS) consisted of 1,153 tweets annotated as the gold standard (GS) to assess the performance. The dataset was annotated with three named entity categories: Person, Location, and Organization.

The systems were evaluated based on three measure scores following Making Sense of Microposts (#MSM2013) [2]. Precision, Recall, and F-measure were calculated for each entity category, and the final results for overall entity categories are the average performance of defined categories. The named entity is represented in a tuple (entity value, entity category), and strict matching was performed between the named entities in TS and the answers in GS for correct detections of the value and the category.

3.2 Systems and Experiment Setting

We conducted three separate experiments on the SSL algorithm, the AL algorithm, and the hybrid algorithm, called SSL, AL, and Hybrid, respectively. Three query algorithms for the AL method were implemented: QBC, uncertainty-based sampling (UBS), and DBS, called AL-QBC, AL-UBS, AL-DBS. The hybrid algorithm also used three AL query algorithms for selecting the samples, called Hybrid-QBC, Hybrid-UBS, Hybrid-DBS.

In the experiments, CRF was used as the underlying model. In addition, the maximum entropy model was also used in the QBC algorithm. The parameters of the models were set to default values. The time interval for querying tweets in each training time was set to 20 (i.e., 20 days).

Table 1. The performance of the systems

System	Precision	Recall	F_1
SSL	65.1	**61.5**	63.2
AL-QBC	81.9	52.5	64.0
AL-UBS	80.5	54.2	64.8
AL-DBS	**82.0**	51.9	63.6
Hybrid-QBC	77.3	60.9	**68.1**
Hybrid-UBS	74.0	59.2	65.8
Hybrid-DBS	76.2	59.0	66.5

3.3 Evaluation Results and Discussion

Three methods were experimented with separately on the dataset described in Sect. 3.1. The parameters of the systems were tuned in the different experiments. For SSL, we set the threshold $\beta = 0.4$ (i.e., the confidence threshold of the whole label sequence). For the AL strategies, the parameters were set as follows: $\alpha \in [0.1, 0.4]$; the window size of the context vector $w = 6$; and the similarity threshold of the context vectors, $\theta = 3$. The classifier was trained based on the CRF model used for classifying TS. The performance of the systems is shown in Table 1. Although the SSL method does not need additional manually-labeled data, the F_1 score of SSL was not better than the others. The performance of the AL method was better than the SSL method by about 1%, but it needed additional manually-labeled data. The combination of AL and SSL significantly improved the performance. The F_1 score of Hybrid-QBC is the best, in comparison with the others. The F_1 score of Hybrid-UBS was not better than the others, though UBS is the best among the AL methods. The performance of all hybrid methods outperformed AL and SSL methods.

The number of additional annotated tweets in AL-QBC, AL-UBS, and AL-DBS were 20%, 19.7%, and 30.1%, respectively. Despite selecting 30.1% of the

Table 2. The performance of AL-DBS and hybrid-DBS with threshold $\beta = 2$

System	Precision	Recall	F_1
AL-DBS	74.7	47.7	58.2
Hybrid-DBS	70.0	62.2	65.9

unlabeled data, the achieved performance of AL-DBS was not as desired. The number of selected tweets was decided by the similarity threshold of context vectors. By setting $\theta = 2$, the performance of two systems is shown in Table 2. The F_1 score of Hybrid-DBS after changing the θ value was better than Hybrid-UBS by 0.1%; in particular, the number of additionally annotated tweets was only 1,225 tweets (i.e., 11.3% of the unlabeled data). The tweets in which context is the least similar to the others can improve the performance of the classifier.

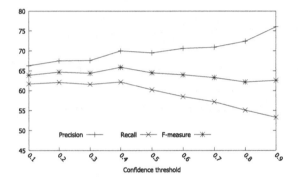

Fig. 2. The performance of Hybrid-DBS with respect to the confidence threshold β

The number of machine-labeled tweets was selected based on confidence threshold β. The impact of the value of β on system performance is shown in Fig. 2. The precision score is directly related to the value of β, whereas the recall score is inversely related to the value of β. The performance of the system was adversely affected by increasing β for selecting the machine-labeled data. The F_1 score of Hybrid-DBS was best at $\beta = 0.4$.

4   Conclusions and Future Work

This study conducted experiments based on the combination of AL and SSL for NER dealing with tweet streams. The proposed hybrid method can significantly improve the performance of the classifier. Unlabeled data were exploited to complement the training data. By using different query strategies (i.e., QBC, UBS, and DBS), the most informative samples were selected for labeling by an expert.

Also, high-confidence samples were chosen automatically, based on the results of applying the classifier to the remaining unlabeled data.

The experiments demonstrated that the proposed method achieved better performance than AL and SSL methods. The hybrid method has the potential to reduce annotation costs and exploit high-confidence samples from the unlabeled data. The performance of the Hybrid-QBC method was the best, compared to the others.

In the future, it is very important to conduct more insightful analyses of the effectiveness of the proposed method. Furthermore, an ontology will also be considered to extract the confidence of the sample selection.

Acknowledgment. This work was supported by the BK21+ program of the National Research Foundation (NRF) of Korea.

References

1. Baldwin, T., Cook, P., Lui, M., Mackinlay, A., Wang, L.: How noisy social media text, how diffrnt social media sources? In: Proceedings of IJCNLP, pp. 356–364 (2013)
2. Basave, A.E.C., Varga, A., Rowe, M., Stankovic, M., Dadzie, A.S.: Making sense of microposts (# MSM2013) concept extraction challenge (2013)
3. Delcea, C., Bradea, I.A.: Grey clustering in online social networks. Vietnam J. Comput. Sci., 1–9 (2016). doi:10.1007/s40595-016-0087-8
4. Derczynski, L., Maynard, D., Rizzo, G., van Erp, M., Gorrell, G., Troncy, R., Petrak, J., Bontcheva, K.: Analysis of named entity recognition and linking for tweets. Inf. Process. Manage. **51**(2), 32–49 (2015)
5. Eisenstein, J.: What to do about bad language on the internet. In: Proceedings of HLT-NAACL, pp. 359–369 (2013)
6. Hassanzadeh, H., Keyvanpour, M.: A two-phase hybrid of semi-supervised and active learning approach for sequence labeling. Intell. Data Anal. **17**(2), 251–270 (2013)
7. Korecki, J.N., Banfield, R.E., Hall, L.O., Bowyer, K.W., Kegelmeyer, W.P.: Semi-supervised learning on large complex simulations. In: Proceedings of ICPR 2008, pp. 1–4. IEEE (2008)
8. Liao, W., Veeramachaneni, S.: A simple semi-supervised algorithm for named entity recognition. In: Proceedings of the NAACL HLT 2009 Workshop on Semi-supervised Learning for Natural Language Processing, pp. 58–65. ACL (2009)
9. Liu, X., Zhang, S., Wei, F., Zhou, M.: Recognizing named entities in tweets. In: Proceedings of the 49th Annual Meeting of the Association for Computational Linguistics: Human Language Technologies, vol. 1, pp. 359–367. ACL (2011)
10. Liu, X., Zhou, M.: Two-stage ner for tweets with clustering. Inf. Process. Manage. **49**(1), 264–273 (2013)
11. Nguyen, N.T.: Using consensus methods for solving conflicts of data in distributed systems. In: Hlaváč, V., Jeffery, K.G., Wiedermann, J. (eds.) SOFSEM 2000. LNCS, vol. 1963, pp. 411–419. Springer, Heidelberg (2000). doi:10.1007/3-540-44411-4_30
12. Settles, B.: Active learning literature survey. University of Wisconsin, Madison, **52**(55–66), 11 (2010)

13. Tran, V.C., Hwang, D., Jung, J.J.: Twisner: Semi-supervised method for named entity recognition from text streams on twitter. J. Univ. Comput. Sci **22**(6), 782–801 (2016)
14. Tran, V.C., Nguyen, T.T., Hoang, D.T., Hwang, D., Nguyen, N.T.: Active learning-based approach for named entity recognition on short text streams. In: Zgrzywa, A., Choroś, K., Sieminski, A. (eds.) Multimedia and Network Information Systems. AISC, vol. 506, pp. 321–330. Springer, Cham (2017). doi:10.1007/978-3-319-43982-2_28
15. Yao, L., Sun, C., Wang, X., Wang, X.: Combining self learning and active learning for chinese named entity recognition. J. Softw. **5**(5), 530–537 (2010)
16. Zhang, Y., Wen, J., Wang, X., Jiang, Z.: Semi-supervised learning combining co-training with active learning. Expert Syst. Appl. **41**(5), 2372–2378 (2014)

A Method for User Profile Learning in Document Retrieval System Using Bayesian Network

Bernadetta Maleszka[(⊠)]

Faculty of Computer Science and Management,
Wrocław University of Science and Technology,
Wybrzeze Wyspianskiego 27, 50-370 Wrocław, Poland
Bernadetta.Maleszka@pwr.edu.pl

Abstract. User modeling methods are developed by many researches in area of document retrieval systems. The main reason is that the system can not present the same results for every user. Each user can have different information needs even if he uses the same terms to formulate his query. In this paper we present the solution for the problem. We propose a method for user profile building and updating using Bayesian network approaches which allows to discover dependencies between terms. Additionally, we use domain ontology of terms to simplify the calculations. Performed experiments have shown that the quality of presented methods is promising.

Keywords: Document retrieval · Knowledge integration · User profile adaptation · Ontology-based user profile · Bayesian network

1 Introduction

Due to the information overload and ambiguity of words meaning, it is difficult to find relevant information quickly. Usually, user submits a few words to the browser and expects that presented results are accurate. Unfortunately, many users have problems with formulating a query. It can be a problem with vocabulary, specific name or simply the user can not know how to express his information needs.

It can be easier to present to a user more accurate results when the system knows the user interests. Such information can be collected in a profile. One can find reach literature connected with user modeling: starting from simple bag of words, through vector-based model to more sophisticated profile construction based on ontological structures.

We consider a personalized document retrieval system where documents are described by a set of terms from domain ontology. In the ontology of Main Library and Scientific Information Centre in Wroclaw University of Science and Technology [17] we can differentiate two relations: "is-a" (generation-specification) and "see-also" relation. The user profile is built based on the set of

© Springer International Publishing AG 2017
N.T. Nguyen et al. (Eds.): ACIIDS 2017, Part I, LNAI 10191, pp. 269–277, 2017.
DOI: 10.1007/978-3-319-54472-4_26

documents that user has read. To learn user model we use methods of Bayesian network. Devitt et al. [4] claim that Bayesian networks are probabilistic structured representations of domains which have been applied to monitoring and manipulating cause and effects for modeled systems. In our approach it is very important to combine existing relations between terms in the domain ontology with the connection in Bayesian network.

The aim of this paper is to present a method for learning user profile using Bayesian network. A user reads some documents and marks which of them are relevant for him. We propose a method where based on these information the system develops a model of user profile using Bayesian network. The network consists of nodes – terms of user interests and edges which reflect dependencies between the terms. Each connections between two nodes is described by a probability table which can be interpreted as a degree of user interests of term A in condition that he is interested in term B.

The rest of the paper is organized as follows. In Sect. 2 we present a survey of approaches to learning user profile and capturing knowledge from data using Bayes networks. Section 3 contains information about the model of documents set and user profile. The developed method for building user profile is described in Sect. 4. Section 5 shows the results of performed evaluations. In the last Sect. 6 we gather the main conclusions and future works.

2   Related Works

This Section contains information about existing approaches of using Bayesian networks and ontology to model user profiles. We present also methods that use Bayesian networks for discovering connections between concepts in ontology. Such approaches were applied initially in biology to discover interactions between genes. The authors of [6] have proposed to use Bayesian networks for representing statistical dependencies. Between advantages of this method they mention ability to describe complex stochastic processes and clear methodology for learning from (noisy) observations.

A Bayesian network can be represented as a directed acyclic graph $D = (V, E)$, where V is a finite set of nodes and E is a finite set of directed edges (arrows) between the nodes [1]. Each node $v \in V$ reflects a random variable X_v and is connected with its parent node $pa(v)$. It is also attached a a local probability distribution, $p(x_v|x_{pa(v)})$. Let us use P for the set of local probability distributions for all variables in the network. A Bayesian network for a set of random variables X is then the pair (D, P).

When two nodes in the graph are not connected (there is no edge between them), it means that these nodes are independent. Then we can use the following formula to calculate the joint probability distribution:

$$p(x) = \prod_{v \in V} p(x_v|x_{pa(v)}).$$

The main advantages of using Bayesian networks are presented by Margaritis [9]:

1. graphical models enable to display relationships clearly and intuitively;
2. representing cause-effect relationships;
3. handle uncertainty though the established theory of probability;
4. can be used to represent indirect in addition to direct causation.

In the area of user modeling one can find reach literature. Middleton et al. [11] describe a typical user profiling approach for behaviour-based recommender systems. They investigate a binary model representing what users find interesting and uninteresting. To obtain the result model of user they propose some machine-learning techniques. A survey of the most frequently used methods is also presented by Sebastiani [14] and De Roure [3].

Devitt et al. [4] have highlighted that Bayesian networks are notoriously difficult to build accurately and efficiently which has somewhat limited their application to real world problems. To obtain better structure of the network it is worth to use ontology. The information saved in the ontology (such as facts and rules about a given domain) can be helpful during process of developing Bayesian network. The paper [2] takes a broad look at the literature on learning Bayesian networks – in particular their structure from data. A knowledge-engineering methodology for building and maintaining Bayesian networks is presented in [7]. The aim of this paper is to model the knowledge of an application domain into an ontology. Using these results the authors propose a method for obtaining structure of a Bayesian network.

An exemplary combination of Bayesian network and ontology in problem of document recommendation can be found in [15]. The authors have proposed a method for constructing a user profile using either a Bayesian or Markov network. Similarly as in many approaches, user needs to view a list of sample documents and marks relevant versus irrelevant documents. The probabilistic inference is then found and used to judge if the next document is relevant or not. The result is presented in the term of conditional probability defined by the network.

More sophisticated approach is presented by Zhang et al. [16] where Bayesian hierarchical model is considered. Due to the fact that recommendation systems can work for millions of users, the authors have proposed EM algorithm to parameters learning. The main problem with processing documents is the fact that they are represented by a very high dimensional space, in which each document is represented by a very sparse vector. The results of the paper have shown that proposed modified EM algorithm is less computationally expensive for thousands or millions of users.

Bayesian networks are also often combined with another machine learning paradigm, e.g. with case-based reasoning [13]. The network is used to model qualitative and quantitative relationships among the different elements the user is interested in. Next, the paradigm of case-based reasoning allows to solve a new problem by remembering a previous similar situation and by reusing information and knowledge of that situation. Cases also provide information used to detect patterns in a user's behavior and determine his routine.

In this paper we focus on learning user profile combining domain ontology with Bayesian network approach. Fenz et al. [5] present a method for the ontology-based generation of Bayesian networks. They use the following steps:

- using ontology concepts to create the nodes of the Bayesian network;
- using ontology relations to link the Bayesian network nodes;
- exploiting the ontological knowledge base to support the conditional probability table calculation for each node.

In our personalization system we will use idea of above presented approaches:

- concepts are represented in the nodes;
- relations are represented by directed links;
- instances are represented by documents that user was viewing.

Unfortunately, the paper [5] presents only a short example of using ontology and Bayesian network to model user profile. In this paper we present a method for learning user profile using Bayesian network. Additionally, we would like to consider a problem of comparing the structure of domain ontology with the structure of obtained Bayesian network.

3   Model of Personalized Document Retrieval System

We consider a personalized document retrieval system which consists of library – database of documents and a user that would like to view some documents from the library. Each document is described by the set of terms coming from domain ontology. The system learns the user interests based on his search history and tries to recommend him some documents that are relevant for his information needs.

In the following subsections we present formal description of the system and its elements.

3.1   Model of Document

A library is a set of documents:

$$D = \{d_i : i = 1, 2, \ldots, n_d\} \tag{1}$$

where n_d is a number of documents and each document d_i is described by the set of weighted terms:

$$d_i = \{(t_j^i, w_j^i) : t_j^i \in T \wedge w_j^i \in [0.5, 1), j = 1, 2, \ldots, n_d^i\} \tag{2}$$

where t_j^i is index term coming from assumed set of terms T (domain ontology), w_j^i is appropriate weight and n_d^i is a number of index terms that describe document d_i.

3.2 Model of User Profile

In our approach we use ontological structure of user profile. The definition of ontology and its components are taken from [12].

Ontology-based user profile is defined as a triple:

$$O = (C, R, I) \tag{3}$$

where C is a finite set of concepts, R is a finite set of relations between concepts $R = \{r_1, r_2, ..., r_n\}$, $n \in N$ and $r_i \subset C \times C$ for $i \in \{1, n\}$ and I is a finite set of instances.

In our model of ontology-based profile we consider index term and its synonyms as a concept, relations between concepts are relations between the terms (eg. relation "is-a-part" or generalization – specification relation) and instance is a set of documents that are described using particular term.

4 Method of Determining the User Profile

In our previous works [8,10] we have presented a method for determining user profile based on his activities. The assumption there was as follows. The user had m documents that were relevant for his information needs. Each document was an instance of domain ontology. The system gathered information about viewed documents and for each concept it calculated average weight from user history. If the weight was greater then assumed value, we have added this concept to user profile.

In our current approach we would like to consider more sophisticated methodology for determining user profile. Let us assume that the system still gathers information about user activities (viewed documents) and saves relevance for each document (if document was relevant or not). Based on the set of these documents, a Bayesian network is used to obtain the probability distribution for each concept from domain ontology. The detailed description of building Bayesian network from data can be found in many papers, e.g. [1,6]. We are interested mainly in concepts that contains terms from user queries.

In the next step, we can try to use domain ontology as a basis for structure of Bayesian network.

Additionally, it is worth to considered also another concepts which occur in Bayesian network. Occurrence of such concepts can mean that user is interested in this concept but he does not use some specific terminology. We can use these terms e.g. extend his next queries.

5 Experimental Evaluation

In this section we present an idea of experiments. The aim of the experiments is to check if built profile becomes more similar to user interests. It is not possible to build Bayesian network for user preferences (real information needs and

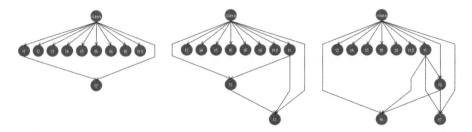

Fig. 1. The result Bayesian networks for increasing set of documents in user history. The figure presents results for the second, third and fourth iteration of experiments. In the first iteration, the algorithm has not discover any dependencies between terms. In the last iteration the result was the same as in fourth iteration.

interests). We can only try to build such a network based on growing set of viewed documents. We can divide user activities into some sessions (one session contains a set of k user queries). Let us denote D_l for the set of relevant and irrelevant documents that were viewed in first l sessions. N is a total number of experiment series. In the i-th series user had w_i sessions, $i \in \{1, 2, \ldots, N\}$.

We propose to perform the series of experiments according to the following plan:

1. Build Bayesian network for the first l sessions (using set D_l)
2. $i \leftarrow 1$
3. While $i \leq N$ do:
   (a) Calculate the quality of built model using w_i following sessions
   (b) $l = l + w_i$
   (c) Build Bayesian network using set of D_l
   (d) Compare networks built in last two iterations
   (e) $i \leftarrow i + 1$

To compare two Bayesian networks we check how many nodes (concepts) and how many links (relations) occur in both networks.

Below we present the results for an exemplary set of user activity in above described personalization document retrieval system.

We have assumed the following values of variables: $k = 10$, $l = 15$, $w = 3$. After 5 iterations of building Bayesian network, the results become stable. This can mean that user profile was not changing in a meaningful way: there were differences in probability distribution tables for different terms but the structure of network was not changed. In Fig. 1 we can compare the structure of network in subsequent iterations.

In Fig. 2 we present the comparison of correctly and incorrectly classified instances in subsequent steps of experiments. When the system has more information about the user interests, the built model is better (more new instances are classified into proper class).

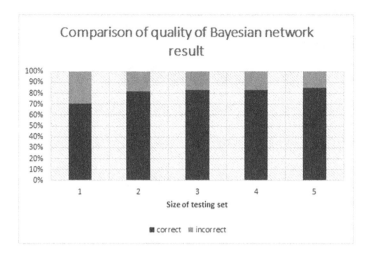

Fig. 2. Correctly and incorrectly classified instances in subsequent steps of experiments.

5.1 Discovering Ontology Structure Using Bayesian Network

When there is lack of information connected with the dependencies between concepts (there is no domain ontology that can be used), it is worth to build Bayesian network. It can be used then to improve search methods when the system takes into account information about dependencies between the terms.

We have performed exemplary experiments where for a given set of values for 10 terms, a Bayesian network was built. Each document were described by a terms from t_1 to t_{10}. The value of each term could be 0 or 1 (with the following interpretation: when value of $t_i, i \in \{1, \ldots, 10\}$ is equal to 1 it means that this document is described by the term t_i). As the decision attribute we have two classes (relevant – class A or irrelevant – class B). The result of Bayesian network was a graph presented in Fig. 3 and each node were described by the probability distribution table containing probabilities for dependent terms.

An exemplary probability distribution table for term t_3 is presented in Table 1. The obtained results can be interpreted as follows. Term t_3 is influenced by terms t_1 and t_4 and the class. For example: the probability for the situation, when term t_3 has value 1 in condition that class has value A, t_1 has value 1 and term t_4 has value 0, equals to 0.917.

Using a Bayesian network to discover dependencies between user interests can find an application for many databases of documents where the domain ontology is not delivered.

6   Summary and Future Works

In this paper we have proposed a method for building ontology-based user profile and a method for updating user profiles in personalized document

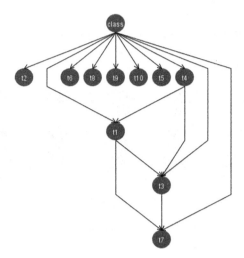

Fig. 3. Graphical representation of Bayesian network.

Table 1. Probability distribution table for t_3. Reference Bayesian network is presented in Fig. 3.

class	t_1	t_4	0	1
A	0	0	0.500	0.500
A	0	1	0.083	0.917
A	1	0	0.130	0.870
A	1	1	0.277	0.723
B	0	0	0.547	0.453
B	0	1	0.652	0.348
B	1	0	0.581	0.419
B	1	1	0.486	0.514

retrieval system. The methods use Bayesian network approach to learn the structure of user profile and to calculate the probability distribution for different terms used to describe documents. Additionally, we have shown that obtained directed graph in Bayesian network can be helpful where domain ontology can not be used.

In our further research we would like to improve the results for user modeling method and perform experimental evaluations using real data sets.

Acknowledgments. This research was partially supported by Polish Ministry of Science and Higher Education.

References

1. Bottcher, S.G., Dethlefsen, C.: Learning Bayesian networks with R. In: Proceedings of the 3rd International Workshop on Distributed Statistical Computing (DSC 2003) (2003)
2. Dalu, R., Shen, Q., Aitken, S.: Learning Bayesian networks: approaches and issues. Knowl. Eng. Rev. **26**(2), 99–157 (2011). doi:10.1017/S0269888910000251. Cambridge University Press
3. De Roure, D., Hall, W., Reich, S., Hill, G., Pikrakis, A., Stairmand, M.: MEM-OIR - an open framework for enhanced navigation of distributed information. Inf. Process. Manage. **37**, 53–74 (2001)
4. Devitt, A., Danev, B., Matusikova, K.: Ontology-driven automatic construction of bayesian networks for telecommunication network management (2006)
5. Fenz, F., Tjoa, M., Hudec, M.: Ontology-based generation of Bayesian networks. In: Proceedings of International Conference on Complex, Intelligent and Software Intensive Systems. IEEE (2009). doi:10.1109/CISIS.2009.33
6. Friedman, N., Linial, M., Nachman, I., Pe'er, D.: Using Bayesian networks to analyze expression data. J. Comput. Biol. **7**(3/4), 601–620 (2000)
7. Helsper, E.M., van der Gaag, L.C.: Building Bayesian networks through ontologies. In: Proceedings of ECAI European Conference on Artificial Intelligence (2002)
8. Maleszka, B.: A method for determining representative of ontology-based user profile in personalized document retrieval systems. In: Nguyen, N.T., Trawiński, B., Fujita, H., Hong, T.-P. (eds.) ACIIDS 2016. LNCS (LNAI), vol. 9621, pp. 202–211. Springer, Heidelberg (2016). doi:10.1007/978-3-662-49381-6_20
9. Margaritis, D.: Learning Bayesian network model structure from data. Ph.D. thesis. School of Computer Science. Carnegie Mellon University (2003)
10. Mianowska, B., Nguyen, N.T.: Tuning user profiles based on analyzing dynamic preference in document retrieval systems. Multimed. Tools Appl. **65**, 93–118 (2012). doi:10.1007/s11042-012-1145-6
11. Middleton, S.E., De Roure, D.C., Shadbolt, N.R.: Capturing knowledge of user preferences: ontologies in recommender systems. In: Proceedings of the 1st International Conference on Knowledge Capture, pp. 100–107 (2001)
12. Pietranik, M., Nguyen, N.T.: A multi-attribute based framework for ontology aligning. Neurocomputing **146**, 276–290 (2014)
13. Schiaffino, S.N., Amandi, A.: User profiling with case-based reasoning and Bayesian networks. In: Proceedings of International Joint Conference IBERAMIA-SBIA, pp. 12–21 (2000)
14. Sebastiani, F.: Machine learning in automated text categorization. ACM Comput. Surv. **34**, 1–47 (2002). Consiglio Nazionale delle Ricerche, Italy
15. Wong, S.K.M., Butz, C.J.: A Bayesian approach to user profiling in information retrieval. Technol. Lett. **4**(1), 50–56 (2000)
16. Zhang, Y., Koren, J.: Efficient Bayesian hierarchical user modeling for recommendation systems. In: proceedings of SIGIR 2007. ACM 978-1-59593-597-7070007 (2007)
17. Main Library and Scientific Information Centre in Wroclaw University of Science and Technology (2016). http://aleph.bg.pwr.wroc.pl/

Intelligent Database Systems

Answering Temporal Analytic Queries over Big Data Based on Precomputing Architecture

Nigel Franciscus$^{(\boxtimes)}$, Xuguang Ren, and Bela Stantic

Institute for Integrated and Intelligent Systems, Brisbane, QLD, Australia
{n.franciscus,x.ren,b.stantic}@griffith.edu.au

Abstract. Big data explosion brings revolutionary changes to many aspects of our lives. Huge volume of data, along with its complexity poses big challenges to data analytic applications. Techniques proposed in data warehousing and online analytical processing (OLAP), such as precomputed multidimensional cubes, dramatically improve the response time of analytic queries based on relational databases. There are some recent works extending similar concepts into NoSQL such as constructing cubes from NoSQL stores and converting existing cubes into NoSQL stores. However, only few works are studying the precomputing structure deliberately within NoSQL databases. In this paper, we present an architecture for answering temporal analytic queries over big data by precomputing the results of granulated chunks of collections which are decomposed from the original large collection. By using the precomputing structure, we are able to answer the *drill-down* and *roll-up* temporal queries over large amount of data within reasonable response time.

Keywords: NoSQL · Data warehouse · Precompute · Temporal

1 Introduction

With the development of data-driven applications in many aspects of our daily lives, it is worthy to praise again the significance of big data whose value has already been recognized by both industry and academia. A significant amount of data is being collected and analyzed to support various decision makings, and that amount is expected to grow by 64% per year according to a recent report [1]. One of the major tasks in mining big data is to answer the analytic queries efficiently. Analytic queries often involve sophisticated aggregations which demand significant computing powers. The huge volume of big data, along with its complexity poses big challenges in the processing of these queries. Aiming to tackle these challenges and to enhance the performance of analytic query processing, the concept of data warehouse [8] and OLAP [4] have been introduced since 1960 s and many further works around these concepts have been proposed later on.

Data warehouses integrate data from different data sources into large repositories. The data within data warehouses are normally structured as denormalise

© Springer International Publishing AG 2017
N.T. Nguyen et al. (Eds.): ACIIDS 2017, Part I, LNAI 10191, pp. 281–290, 2017.
DOI: 10.1007/978-3-319-54472-4_27

multidimensional cubes to reduce the cost of heavy table joins. A large part of modern OLAP systems are built on top of data warehouses which are stored with additional information. An essential and widely used technique in OLAP systems is *precomputation* where analytic results are precomputed and materialized in the data warehouses. When a user submits queries, the system simply retrieves the corresponding precomputed results, conduct proper merging tasks and then quickly return the final result to the user.

Previous techniques proposed in data warehouses and OLAP systems are mainly focusing on relational data structure and relational databases, however, it is evident that relational database is struggling in handling large amount of data [10]. The rise of NoSQL [11] databases has attracted the attention of database community due to its flexibility in providing schema-later architecture and its scalability for handling huge amount of big data. Various NoSQL databases have been chosen and applied in many domains, which leads to more and more data being collected into NoSQL databases. Consequently, it has become an urgent demand to process analytic queries based on NoSQL databases efficiently. Some recent works are extending the techniques of data warehouses and OLAP into NoSQL. The work of [9] present strategies for constructing cubes from NoSQL stores. In contrast, the work in [3] gives the rules in converting existing cubes into NoSQL stores. However, there are few works focusing on the precomputing structure deliberately for NoSQL databases.

Motivated by the above issue, in this paper we present an architecture for answering temporal analytic queries over big data. As the time is an essential dimension for most of data analytic platforms, we choose temporal queries aspect as our focus in this paper and as the starting point of our work. We may extend our work into other dimensions in future works. The basic idea of our architecture is to divide the original NoSQL data into separated and smaller chunks and then precompute the results for each chunk. The precomputed results are then materialized in the NoSQL database. We process the upcoming analytic queries based on the precomputed results.

Contribution. In this paper, we present a precomputing architecture for answering analytic queries for NoSQL data stores. Based on our architecture, we are able to answer the *drill-down* and *roll-up* temporal queries over large amount of data within fast response time. To be specific,

(1) We proposed the technique to index raw data into separated and smaller chunks based on temporal interval.
(2) We designed the storage structures for the precomputed results within MongoDB and Redis.
(3) We design three types of query models along with the strategies to answer each query type.
(4) We conducted extensive experiments to demonstrate the performance of answering three types of queries based on our architecture.

Organization. The rest of the paper is organised as follows: in Sect. 2, we give some related works; in Sect. 3, we present the details of our precomputing architecture; in Sect. 4, we provide the experiment results; and finally in Sect. 5 we conclude the paper and indicate some future work.

2   Related Work

In this section, we present some related works which are into classed into two categories.

(1) *NoSQL Database* According to a survey in [7], there are more than 100 No-SQL databases developed for various purposes. Specifically, No-SQL databases can be classified into four classes:

(a) *Key-Value* stores the data as key-value pairs where the value can be anything and is treated as opaque binary data, the key is transformed into an index using a hash function. Redis is one of the widely used key-value databases.

(b) *Column-Family* applies an column-oriented architecture which is contrast to the row-oriented architecture in RDBMS. Cassandra and HBase are two most used column-family databases.

(c) *Document-Database* treats the document as the minimum data unit and is designed deliberately for managing document-oriented information, such as JSON, XML documents. MongoDB is a typical document database which is designed to handle JSON documents.

(d) *Graph-Database* models the data as graphs and focuses more on the relationships between data units. There are over 30 graph database systems such as Neo4j, Titan, and Sparksee.

In our work, we used two NoSQL databases, MongoDB and Redis. The pros and cons are not our focus as MongoDB and Redis are using entirely different mechanisms. However, we present the query processing performance for those two databases based on our pre-computing structure.

(2) *Data Warehouse and OLAP* The concept of data warehouse and OLAP have been proposed very early aiming to answer analytic queries efficiently. The key structure in data warehouse is the cube which is normally stored as a denormalised multidimensional table in relational database [2][5]. A large part of modern OLAP systems are built on top of data warehouses and utilize the cubes when processing analytic queries [6]. The work presented in [12] focuses on the time-range queries on relational cubes. There are some recent works extending the techniques of data warehouses and OLAP into NoSQL. The work of [9] present strategies for constructing cubes from NoSQL stores. In contrast, the work in [3] gives the rules in converting existing cubes into NoSQL stores. However, only few works studying the precomputing structure deliberately within NoSQL databases.

Contrast to previous work, we focus on the processing of analytic queries for NoSQL databases where no data is stored in relational database. We propose an index structure based on which we can answer the *drill-down* and *roll-up* queries over large amount of data within fast response time. Similar to work in [12], we particularly focus on the temporal queries with the time-range as the query parameter.

3   Precomputing Architecture

In this section, we present our detailed design of the precomputing architecture. We first give an overview of the architecture. Then we study the components respectively, which are: *(1) Raw Data Indexing, (2) Precomputed Results Structure* and *(3) Query Answering.*

3.1   Overview

In this subsection, we give an overview of our precomputing architecture as shown in Fig. 1. It can be divided into several inter-related components: (i) *Raw Data Indexing*, where we collect the raw Twitter data and store them into NoSQL database(MongoDB) as time-indexed collections. (ii) *Precompute Results Structure*, where we execute analytic jobs (MapReduce based on Hadoop) and then store the precomputed results into NoSQL database (MongoDB and Redis). (iii) *Query Answering*, where we apply efficient strategies to answer queries by utilizing the precomputed results through merging. As a case study, we demonstrate our architecture by using specific data sources, database platforms, analytic jobs and processing techniques in this paper, as indicated within the above parentheses. However, it is worth noting that our architecture is quite flexible and can be easily extended to other use cases.

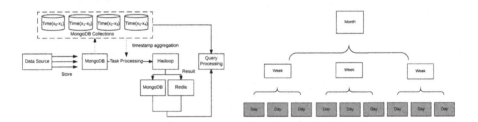

Fig. 1. Pre-computing architecture   **Fig. 2.** Time interval index structure

We present more details about each chosen specific ingredient.

Data Source. As shown in Fig. 1, we use Twitter as the data source in our case study. Twitter is an online social networking service that enables users to post short 140-character messages called "tweets". Twitter is widely used in monitoring society trends and user behaviors due to its large user pool [13]. The tweets are formatted into JSON and they include the textual content as well as the posted time.

Database Platform. The Twitter data is in JSON format which is naturally supported by the MongoDB. We choose MongoDB to store the raw data in our case study. MongoDB provides some features from relational databases like sorting, compound indexing and range/equal queries. Additionally, MongoDB has its own aggregation capability and in-house MapReduce operation.

Analytic Jobs. Computing word frequency is a widely used analytic job in many literatures. Its intuitive application is the word cloud which is intensively used to detect hot topics and trends in the society. Compared with word frequency, sometimes we are more interested in the frequency of word-pair(co-occurrence of two words) as it can help us to detect hidden patterns. Therefore, we choose the job of computing word-pair frequency in our case. That is given a set of tweets and any word-pair, we compute the number of tweets in which this word-pair co-occurred.

Processing Techniques. We choose Hadoop as the processor to execute the word-pair jobs. Hadoop is an open source implementation of MapReduce framework. Although MongoDB ships with in-house MapReduce, it has poor analytic libraries compared to Hadoop. We store the computed results into both Redis and MongoDB.

3.2 Raw Data Indexing

It is evident that tremendous amount of data is difficult to process without proper indexing. However indexing the high-cardinality attribute, such as timestamp, is not suitable due to excessive seek [12]. For example, there will be numerous index entries if we index every specific timestamp for the tweets, which will lead to a higher latency. To tackle this problem, in this subsection, we introduce the technique of time interval index inside the collection layer of MongoDB.

Specifically, we group tweets into a single collection where the time of those tweets are within the same interval. The length of the interval can be tuned based on the dataset; it can be an hour, a day or a month. We use the timestamp of this time interval as the name of the corresponding indexed collection. By utilizing the time interval index, we dramatically alleviate the cost of index seeking while still be able to support *drill-down* and *roll-up* temporal queries. Consider the example index structure in Fig. 2, we choose a day as the time interval. The tweets posted on the same day (grey box) will be grouped into the same collection. It is worth noting that a week is a super-interval of a day, however, we do not store a separated collection to group the tweets in the same week. As this will dramatically increase the storage size.

In order to support the *drill-down* and *roll-up* temporal queries, we precompute the analytic results for each indexed collection. For example, we precompute the results for the *day* collections in Fig. 2. For each time-range query, the system will answer the query using *bottom-up* merging approach. Specifically, given a time-range query whose range is more than one day, we first lock down to the tweets collections within this range and load their corresponding precomputed results. Then we merge these results together and get the final results for the query. By implementing this technique, we remove the necessity to pre-compute/store the result for super-interval collection such as weekly, monthly or yearly.

3.3   Precomputed Results Structure

As discussed in the above subsection, we precompute the analytic results for each indexed collection. In this subsection, we study the structure to store the precompute results which are the frequencies of word-pair in tweets. We present the structures for two NoSQL databases: *MongoDB* and *Redis*.

MongoDB Structure. In MongoDB, we use a separate *collection* to store the results of each indexed collection. Each result collection contains a list of frequency results for word-pairs. The format of each frequency result for any word-pair is in the following document format:

$$[\_id, word_1, word_2, frequency]$$

where *_id* is created automatically by MongoDB if not specified, $word_1$ and $word_2$ are the words in this word-pair and the *frequency* is the number of tweets in which this word-pair co-occurred. Consider the example in Fig. 3, the name of the result collection is *1475118067000* (29 Sep 2016). The *hello* and *world* co-occurred in 100 tweets which are posted on the day of 29 Sep 2016.

1475118067000			
_id:0001	word₁: "hello"	word₂: "world"	frequency: 100
_id:0002	word₁: "happy"	word₂: "world"	frequency: 70
_id:0003	word₁: "hello"	world₂: "great"	frequency: 60

Key	Value
1475118067000_hello_world	100
1475118067000_happy_world	70
1474315055000_hello_world	60

Fig. 3. MongoDB result structure   **Fig. 4.** Redis result structure

Redis Structure. Redis is an in-memory key-value database. We use a combination of timestamp and the word-pair as the key and the frequency as the value. The format is given as follows:

$$[time_x\_word_1\_word_2 : frequency]$$

Redis support searching based on *key pattern*, thus, we can quickly lock down to the corresponding *set* of records when given a specific timestamp and/or word-pair. As we can see in the example in Fig. 4, the *hello* and *world* co-occurred in 100 tweets which are posted on *1475118067000* (29 Sep 2016) while they co-occurred in 60 tweets which are posted on *1474315055000* (19 Sep 2016).

3.4   Query Answering

In above subsections, we presented the indexing strategy and the structures to store the precomputed results. Now we are ready to study the process of answering user queries. We classify the user queries into three types: (i) Single

Selectivity Query, (ii) Drill-down Query, (iii) Roll-up Query as we can see in the following models:

(i) Single Selectivity Query
QUERY data WHERE $time = T_x$ WITH $Gra(time, T_x) = \Phi$.

(ii) Drill-down Query
QUERY data WHERE $time > T_x$ AND $time < T_y$ AND T_y-$T_x < \Phi$ WITH $Gra(time, T_x, T_y) < \Phi$.

(iii) Roll-up Query
QUERY data WHERE $time > T_x$ AND $time < T_y$ WITH $Gra(time, T_x, T_y) = \Phi$.

In the above models, we use Φ to denote the interval when we index the raw data (as mentioned in Subsect. 3.2). The function $Gra(t)$ is to decide the granularity of the time parameter t, for example, $Gra(12am\ 15\ Sep\ 2016) = hour$ and $Gra(15\ Sep\ 2016) = day$. Intuitively, *Single Selectivity Query* aims to query the data falling into a single indexed data collection. *Drill-down Query* aims to query the data which are a subset of a single indexed collection. *Roll-up Query* aims to query the data involves multiple indexed collections. Consider the following example where each one query corresponds to one query type respectively.

(i) Word-pair frequency on 02/April/2016.

(ii) Word-pair frequency from 9:00pm of 08/April/2016 to 11:00pm of 08/April/2016.

(iii) Word-pair frequency from 18/April/2016 to 28/April/2016.

(1) The time is trivial to answer the single selectivity query, as we only need to navigate to the corresponding result collection of MongoDB (set of records of Redis) by the timestamp. *(2)* To answer the drill-down query, we need to navigate to the corresponding indexed data collection, fetch the tweets falling into the time range and then execute the word-pair job onto the filtered tweets. The time of this process depends on the complexity of the analytic job to be executed and can be very slow if size of the fetched tweets is large. *(3)* To answer the roll-up query, we need to merge multiple result collections in MongoDB (sets of records in Redis) falling into the time range. This process is similar to the table-join in the relational database. We present a basic algorithm to merge multiple result MongoDB collections here, as shown in Algorithm 1.

As we can see in the merging algorithm, the algorithm takes multiple results as input and output the word-pair result R. A hashmap H is used to temporarily save the frequency(value) of the *word_pair*(key) (Line 1). The algorithm iterates through each collection and visits each document inside the collection (Line 2 to 10). For each document, if there is no such *word_pair* in the hashmap, we add a new *word_pair* to the hashmap (Line 4 to 6). If there is already one, we just add up the frequency (Line 7 to 9). The above algorithm can be very fast if we tune the index interval properly. The merging algorithm for Redis is similar to Algorithm 1, we omit it here.

Algorithm 1. MERGERESULTS

    Input: Multiple precomputed results $T = \{T_1 \dots T_n\}$
    Output: final result R
1 HashMap $H \leftarrow \emptyset$
2 **for** *each collection* $T_k \in T$ **do**
3   **for** *each document* $w \in T_k$ **do**
4   **if** $w.word_1\_w.word_2$ *is not in* H **then**
5            $H(w.word_1\_w.word_2) \leftarrow w.frequency$
6   **end**
7   **else**
8            $H(w.word_1\_w.word_2) \leftarrow H(w.word_1\_w.word_2) + w.frequency$
9   **end**
10   **end**
11 **end**
12 Result $R \leftarrow JSON(H)$
13 **return** R

It is worth noting that a larger interval leads to a larger index collection. Many queries will fall into the drill-down type. When the number of tweets in one indexed collection is large, it will increase the time to answer a drill-down query. In contrast, a smaller interval will lead to many result collections(sets). Many queries will fall into the roll-up type. Excessive merge will increase the time to answer a roll-up query. Therefore, it is a trade-off between the performance of drill-down and roll-up when tuning the index interval.

4   Experiments

Our architecture has already demonstrated its effectiveness within a practical HumanSensor project[1]. In this section, we present our experiment results so as to study the response time of the query answering under different data settings.

4.1   Dataset and Environment

The twitter data in our experiment were downloaded though the public API provided by Twitter. We wrapped 5 datasets which contains 200×10^3, 400×10^3, 600×10^3, 800×10^3 and 1 million tweets respectively. For each dataset, we index the data according to a *day* interval. Bigger dataset will lead to more indexed collections and more documents within each collection. We synthetically generated three query sets for each query type, each of which contains 100 queries. The process of answering selectivity query and roll-up query utilized the precomputed results in MongoDB and Redis, thus we report the average response time of these two types for MongoDB and Redis respectively. As drill-down query

---

[1] HumanSensor project is a data mining application conducted by Griffith University and sponsored by Gold Coast City Council.

only involves indexed data collections which are stored in MongoDB, we simply report the average response time for it.

Our experiments were conducted on a cluster with 20 nodes, each node is equipped with quad core Intel(R) Core(TM) i5-2400 CPU @ 3.10GHz with 4GB RAM. We used Hadoop (2.6.0), MongoDB (3.2.9) and Redis (3.0.1).

4.2   Results and Analysis

The results of processing single selectivity query are given in Fig. 5. As we can see, the time cost of answering selectivity query almost keeps constant if precomputed results are stored in MongoDB. While it depicts a linear increment when utilizing the precomputed results stored in Redis. The reason for this phenomena is due to the storage structure of results in MongoDB and Redis. For MongoDB, we saved the results of an indexed data collection in a separated collection. Given a result collection name, the time to locate the corresponding collection is a hash-search which are trivial and almost constant. While the results of an indexed data collection for Redis are spread into the KEY. The internal pattern search of Redis takes a linear time in terms of the number of KEYS.

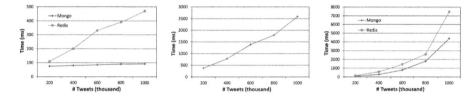

Fig. 5. Single selectivity   **Fig. 6.** Drill-down   **Fig. 7.** Roll-up

Figure 6 presents the results of answering drill-down query. As discussed in Sect. 3.4, the time cost by answering drill-down query depends on the size of the indexed data collection and the complexity of the analytic job. As we can see in Fig. 6, the time experience a linear increment in terms of the size of datasets. Note that, it takes linear time to execute *word_pair* job.

The results of processing roll-up queries were given in Fig. 7. Both the time cost for MongoDB and Redis demonstrates an sharper increment in terms of the size of the datasets. This is because of the merging process is similar to the table-join of the relational database whose time consumption may grow quickly when the data size gets bigger. However, through a proper tune of the index interval, we can achieve a reasonable response time in practice. The MongoDB shows a slightly better performance than Redis, this shares the same reason when answering selectivity query. It takes more time for Redis to assemble the precomputed results for a given indexed collection.

5   Conclusion

We presented a precomputing architecture based on NoSQL database to answer temporal analytic queries. Within the architecture, we propose the indexing techniques, results storage structures and the query processing strategies. Based on our architecture, we are able to answer the *drill-down* and *roll-up* temporal queries over large amount of data within fast response time. Through integration in practical project and the performance study in our experiments, we proved the effectiveness of our architecture and demonstrated its efficiency under different settings. Some future works include extending the precomputed result over spatial data and to enable the distribute join for the merging function. That way, we can do parallel join to reduce the time threshold per collection.

References

1. Knowledge Management. http://www.globalgraphics.com/technology/knowledge-management/
2. Chaudhuri, S., Dayal, U.: An overview of data warehousing and olap technology. ACM Sigmod Rec. **26**(1), 65–74 (1997)
3. Chevalier, M., El Malki, M., Kopliku, A., Teste, O., Tournier, R.: Implementing multidimensional data warehouses into NoSQL. In: International Conference on Enterprise Information Systems (ICEIS 2015), pp. 172–183 (2015)
4. Codd, E.F., Codd, S.B., Salley, C.T.: Providing OLAP (on-line analytical processing) to user-analysts: an it mandate. Codd Date **32**, 3–5 (1993)
5. Coronel, C., Morris, S.: Database Systems: Design, Implementation, & Management. Cengage Learning, Florence (2016)
6. Cuzzocrea, A., Bellatreche, L., Song, I.Y.: Data warehousing and OLAP over big data: current challenges and future research directions. In: Proceedings of the Sixteenth International Workshop on Data Warehousing and OLAP, pp. 67–70. ACM (2013)
7. Gudivada, V.N., Rao, D., Raghavan, V.V.: Renaissance in database management: navigating the landscape of candidate systems. IEEE Comput. **49**(4), 31–42 (2016). doi:10.1109/MC.2016.115
8. Inmon, W.H., Hackathorn, R.: Using the data warehouse. Wiley, New York (1994)
9. Scriney, M., Roantree, M.: Efficient cube construction for smart city data. In: Proceedings of the Workshops of the EDBT/ICDT 2016 Joint Conference (2016)
10. Stantic, B., Pokorný, J.: Opportunities in big data management and processing. In: Frontiers in Artificial Intelligence and Applications, vol. 270. IOS Press (2014)
11. Stonebraker, M.: SQL databases v. NoSQL databases. Commun. ACM **53**(4), 10–11 (2010)
12. Tao, Y., Papadias, D.: The mv3r-tree: a spatio-temporal access method for timestamp and interval queries. In: Proceedings of Very Large Data Bases Conference (VLDB), 11–14 September, Rome (2001)
13. Yu, Y., Wang, X.: World cup 2014 in the twitter world: a big data analysis of sentiments in us sports fans tweets. Comput. Hum. Behav. **48**, 392–400 (2015)

Functional Querying in Graph Databases

Jaroslav Pokorný[✉]

MFF UK, Malostranské nám. 25, 118 00 Praha, Czech Republic
pokorny@ksi.mff.cuni.cz

Abstract. The paper is focused on a functional querying in graph databases. An attention is devoted to functional modelling of graph databases both at a conceptual and data level. The notions of graph conceptual schema and graph database schema are considered. The notion of typed attribute is used as a basic structure both on the conceptual and database level. As a formal approach to declarative graph database querying a version of typed lambda calculus is used.

Keywords: Graph database · Querying graph database · Graph database modelling · Conceptual graph database model · Functional database model · Typed lambda calculus · Language of terms

1 Introduction

Graph databases are focused on efficient storing and querying highly connected data. They are a powerful tool for graph-like queries, e.g., computing the shortest path between two nodes in the graph. They reach an excellent performance for local reads by traversing the graph and can use various data models for graphs and their data extensions.

Graph databases are considered usually as NoSQL databases (e.g., [12]). One rather popular definition of a *graph database* (GDB), also called a *graph-oriented database*, says that it is a database that uses graph theory to store, map and query relationships. That is, the distinguished characteristics of the domain include:

- relationship-rich data,
- relationships are first-class citizens in graph databases.

Despite of the fact that there are various approaches to GDB implementation, native graph processing based on so called *index-free adjacency* is the most efficient means of processing data in a graph because connected nodes use physical "pointers" to neighbour nodes in the database.

A GDB can contain one (big) graph or a collections of graphs. The former includes, e.g., graphs of social networks such, Web graph, the latter is especially used in scientific domains such as bioinformatics and chemistry. Graph search occurs in other application scenarios, like recommender systems, complex software plagiarism detection, and traffic route planning. In line with similar concepts in other database technologies, we will talk about *Graph Data Management Systems* (GDBMS) and *Graph Database Systems* (GDBS).

© Springer International Publishing AG 2017
N.T. Nguyen et al. (Eds.): ACIIDS 2017, Part I, LNAI 10191, pp. 291–301, 2017.
DOI: 10.1007/978-3-319-54472-4_28

An important part of GDB technology is querying graphs. Always there is the intimate relationship between database modelling and querying. Most graph query languages use directly a structure of directed graphs or property graphs. Now, the most known declarative query language over property graphs is Cypher[1] of GDBMS Neo4j [10]. Cypher was the first pattern-matching query language to target the property graph data model. Cypher commands use partially SQL syntax and are targeted at ad hoc queries over the graph data.

Yet other approaches are possible, e.g., a functional approach. In the late 80s, there was the functional language DAPLEX [11]. The language only allowed nested applications of functions. The functional map was applied in context-oriented semantics of multivalued functions. A number of significant works using functional approach to data management are contained in the book [3]. In the current era of GDBMSs, we can mention the Gremlin[2] - a functional graph query language developed by Apache TinkerPop which allows to express complex graph traversals and mutation operations over property graphs. It is supported by many GDBMSs (e.g., Titan[3]).

Here, we will use a functional approach in which a database graph is represented by so called attributes, i.e. typed partial functions. We use for this approach the HIT Database Model, see, e.g., [6], as a functional alternative variant of E-R model. Then a typed lambda calculus can be used as a data manipulation language. This approach reflects the graph structure of a GDB and, on the other hand, provides powerful possibilities for dealing with properties, i.e. with the GDB content.

The rest of the paper is organized as follows. Section 2 introduces a graph data model based on (labelled) property graphs. Section 3 describes modelling GDBs, both on the conceptual and database level. The notions of graph conceptual schema and graph database schema are introduced including some integrity constraints (ICs). Section 4 shortly introduces a functional approach to GDB modelling and a version of typed lambda calculus appropriate for GDB querying. Details of these principles are explained in examples in Sect. 5. Section 6 gives the conclusion.

2   Graph Data Model

In general, database technologies are based on a database model. Here we will use a (*labelled*) *property graph data model* whose basic constructs include:

- entities (nodes),
- properties (attributes),
- labels (types),
- relationships (edges) having a direction, start node, and end node,
- identifiers.

---

[1] http://neo4j.com/developer/cypher-query-language/.
[2] http://tinkerpop.apache.org/.
[3] http://titan.thinkaurelius.com/.

Entities and relationships can have any number of properties, nodes and edges can be tagged with labels. Both nodes and edges are defined by a unique identifier (Id). Properties are expressed in the key:value style. In graph-theoretic notions we also talk about *labelled and directed attributed multigraphs* in this case. These graphs are used both for GDB and its database schema (if any). An example of a GDB is in Fig. 1.

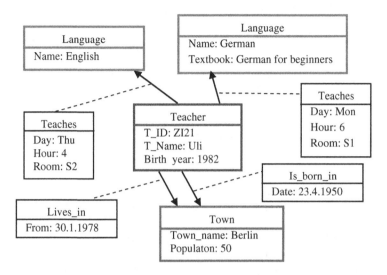

Fig. 1. Example of a GDB

3   Modelling and Querying Graph Databases

Current commercial GDBMSs need more improvements to meet traditional definitions of conceptual and database schema known, e.g., from the relational databases world. The graph database model is usually not presented explicitly, but it is hidden in constructs of a data definition language (DDL) which is at disposal in the given GDBMS. These languages also enable to specify some simple ICs. Conceptual modelling of graph databases is not used at all. An exception is the GRAD database model [5], which although schema-less, uses conceptual constructs occurring in E-R conceptual model and some powerful ICs. Both *graph conceptual schema* and *graph database schema* can provide effective communication medium between users of any GDB. They can also significantly help to GDB designers.

3.1   Graph Conceptual and Database Schemas

In [9] we proposed a binary E-R model as a variant for graph conceptual modelling considering strong entity types, weak entity types, relationship types, attributes, identification keys, partial identification keys, ISA-hierarchies, and min-max ICs. Figure 2 uses for min-max ICs well-known notation with dotted lines and crow's foots used for the start node and the end node of some edges. The perpendicular line denotes the identification and existence dependency of weak entity types. Subtyping (ISA-hierarchies) are simply expressed by arrow to the entity supertype.

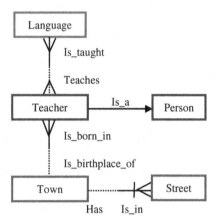

Fig. 2. Graph conceptual schema

A correct graph conceptual schema may be mapped into an equivalent (or nearly equivalent) graph database schema with the straightforward mapping algorithm [9] but with a weaker notion of a database schema, i.e. some inherent ICs from the conceptual level have to be neglected to satisfy usual notation of labelled and directed attributed multigraphs. Consequently, we can propose several different graph database schemas from a graph conceptual schema. For example, the edges `Teaches` and `Is_born_in` provide only a partial information w.r.t. the associated source conceptual schema. For example, the inverted arrow `Is_taught_by` could be used as well. Due to the loss of the inherent ICs on the conceptual level, we should put some explicit ICs into the GDB schema, e.g., that "A teacher can teach more languages" and "A teacher is born in exactly one town".

Figures 2 and 3 give examples of graph conceptual schema and graph database schema, respectively. Concerning properties, identification key of `Teacher` would be `#Person_ID`. On the database schema level, the identification key of `Street` would be `{Town_name, Street_name}`. Details of mapping of graph conceptual schemas to graph database schemas can be found in [9].

Due to the graph structure of data in GDB, associated explicit ICs can have also a graph form. Very simple IC is a *functional dependency* (FD) or *conditional functional dependency* (CFD) introduced in [9]. For example, "Each teacher is born in one town" and "Teachers born later than in 1994 teach at most one language" are examples of FD and CFD, respectively.

On the other hand, NoSQL databases often does not require the notion of database schema at all. Strict application of schemas is sometimes considered disadvantageous by those who develop applications for dynamic domains, e.g., domains working with user-generated content, where the data structures are changing very often [1]. Consequently, many GDBMSs are schema-less. OrientDB[4] even distinguishes three roles of graph database schema: schema-full, schema-less, and schema-hybrid.

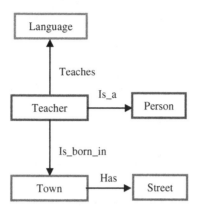

Fig. 3. Graph database schema

3.2 Graph Querying

In this section we focus on basics of graph querying (for more details see [8]). Its simplest type uses the index-free adjacency. In practice, the basic queries like k-hop queries are the most frequent. Looking for a node, looking for its neighbours (1-hop), scan edges in several hops, retrieval of property values, etc., belong in this category.

More complex queries are *subgraph* and *supergraph* queries. They belong to traditional queries based on exact matching. Other typical queries include *breadth-first/depth-first search*, *path* and *shortest path finding*, *least-cost path finding*, *finding cliques* or *dense subgraphs*, *finding strong connected components*, etc.

Very useful are *regular path queries* (RPQ). RPQs have the form:

$$RPQ(x, y) := (x, R, y)$$

where R is a regular expression over the vocabulary of edge labels. RPQs provide couples of nodes connected by a path conforming to R. With the closure of RPQs under conjunction and existential quantification we obtain *conjunctive RPQs*.

For example, the Cypher working with Neo4j databases lacks some fundamental graph querying functionalities, namely, RPQs and graph construction. In [2] an interesting newer approach is offered by the language G-Path. G-Path is an RPQ language working on graphs, which supports mostly all useful regular expression operators (grouping, alternation, Kleene operator, etc.). PGQL [13] is based also on the paradigm of graph pattern matching, closely follows syntactic structures of SQL, and provides RPQs with conditions on labels and properties.

---

[4] http://www.orientechnologies.com/

4   Database Modelling with Functions

A conceptual modelling can be based on the notion of attribute viewed as an empirical typed function that is described by an expression of a natural language [6]. A lot of papers are devoted to this approach studied mainly in 90ties (see, e.g., [7]).

4.1   Types

A hierarchy of types is constructed as follows. We assume the existence of some (*elementary*) *types* $S_1,..., S_k$ ($k \geq 1$). They constitute a *base* **B**. More complex types are constructed in the following way.

If $S, R_1,..., R_n$ ($n \geq 1$) are types, then

(i) $(S:R_1,...,R_n)$ is a (*functional*) *type*,
(ii) $(R_1,...,R_n)$ is a (*tuple*) *type*.

The set of *types* **T** over **B** is the least set containing types from **B** and those given by (i)-(ii). When S_i in **B** are interpreted as non-empty sets, then $(S:R_1,...,R_n)$ denotes the set of all (total or partial) functions from $R_1 \times ... \times R_n$ into S, $(R_1,...,R_n)$ denotes the Cartesian product $R_1 \times ... \times R_n$.

The elementary type Bool = {*TRUE, FALSE*} is also in **B**. The type Bool allows to type some objects as sets and relations. They are modelled as unary and *n*-ary characteristic functions, respectively. The notion of set is then redundant here.

The fact that X is an object of type $R \in$ **T** can be written as X/R, or "X is the R-object". For each typed object o the function type returns type(o) $\in$ **T** of o. Logical connectives, quantifiers and predicates are also typed functions: e.g., **and**/(Bool:Bool,Bool), R-identity $=_R$ is (Bool:R,R)-object, universal R-quantifier Π_R, and existential R-quantifiers Σ_R are (Bool:(Bool:R))-objects. R-singularizer I_R/(R:(Bool:R)) denotes the function whose value is the only member of an R-singleton and in all other cases the application of I_R is undefined. Arithmetic operations $+$, $-$, $*$, $/$ are examples of (Number:Number,Number)-objects. The approach also enables to type functions of functions, etc.

4.2   Attributes

Object structures usable in building a database can be described by some expressions of a natural language. For example, "the language thought by a teacher at a school" (abbr. LTS) is a (Language:Teacher, School)-object, i.e. a (partial) function f:Teacher $\times$ School$\rightarrow$Language, where Teacher, School, and Language are appropriate elementary types. Such functions are called *attributes* in [6].

More formally, attributes are functions of type ((S:T):W), where W is the logical space (possible worlds), T contains time moments, and S $\in$ **T**. M_w denotes the application of the attribute M to w/W, M_{wt} denotes the application of M_w to the time moment t. We can omit parameters w and t in type(M). In the case of LTS attribute we consider possible worlds, where teachers teach at most one language in each school. For GDBs we can elementary entity types conceive as sets of node IDs.

Attributes can be constructed according to their type in a more complicated way. For example, "the classes in a school" could be considered as an attribute *CS* of type ((Bool:(Bool:Student)):School), i.e. the classes contain sets of students and the *CS* returns a set of classes (of students) for a given school.

We can also consider other functions that need no possible world. For example, aggregate function like COUNT, + (adding) provides such function. These functions have the same behaviour in all possible worlds and time moments.

Consequently, we can distinguish between two categories of functions: empirical (e.g. attributes) and analytical. The former are conceived as partial functions from the logical space. Range of these functions are again functions. Analytical functions are of type R, where R does not depend on W and T. In the conceptual modelling, each base **B** consists of descriptive and entity types. Descriptive types (String, Number, etc.) serve for domains of properties.

The notion of attribute applied in GDBs could be restricted on attributes of types (R:S), (Bool(R):S), or Bool(R,S), where R and S are entity types. This strategy simply covers binary functional types, binary multivalued functional types, and binary relationships described as binary characteristic functions. The last option corresponds to *m:n* relationship types. For modelling directed graphs the first two types are sufficient, because *m:n* relationships types can be expressed by two "inverse" binary multivalued functional types. Here we will consider always one of them.

Now we add properties. Properties describing entity types $S_1,\ldots,S_m$ can be of types $(S_1, \ldots, S_m:R)$, where S_i are descriptive elementary types and R is an entity type. So we deal with functional properties. Similarly, we can express properties of edges. They are of types $(S_1, \ldots, S_m, R_1:R_2)$ or $(Bool(S_1, \ldots, S_m, R_1):R_2)$.

Then a functional database schema describing GDB in Fig. 1 can look as:

La/((Name, Textbook):Language)
Te/((T_ID, T_Name, Birth_year):Teacher)
Tw/((Town_name, Population):Town)
Teaches/(Bool:(Day, Hour, Room, Language):Teacher)
Is_Born_in/((Date, Town):Teacher)

We remark, however, that our functional GDBs with such schemas can contain isolated nodes with at least one property. IDs of edges are not necessary, because edges are not explicitly considered.

4.3 Manipulating Functions

A manipulating language for functions is traditionally a typed lambda calculus. Our version of the typed lambda calculus uses tuple types and directly supports manipulating objects typed by **T**. We will suppose a collection **Func** of constants, each having a fixed type, and denumerable many variables of each type. Then the *language of (lambda) terms* LT is defined as follows:

Let types R, S, $R_1, ..., R_n$ ($n \geq 1$) are elements of **T**.

(1) Every variable of type R is a term of type R.
(2) Every constant (a member of **Func**) of type R is a term of type R.
(3) If M is a term of type $(S:R_1, ..., R_n)$, and $N_1, ..., N_n$ are terms of types $R_1, ..., R_n$, respectively, then $M(N_1, ..., N_n)$ is a term of type S. /*application*/
(4) If $x_1, ..., x_n$ are different variables of the respective types $R_1, ..., R_n$ and M is a term of type S, then $\lambda x_1, ..., x_n(M)$ is a term of type $(S:R_1, ..., R_n)$ /*lambda abstraction*/
(5) If $N_1, ..., N_n$ are terms of types $R_1, ..., R_n$, respectively, then
     $(N_1, ..., N_n)$ is a term of type $(R_1, ..., R_n)$. /*tuple*/
(6) If M is a term of type $(R_1, ..., R_n)$, then
     $M[1], ..., M[n]$ are terms of respective types $R_1, ..., R_n$. /*components*/

Instead of the position notation we can use more readable dot notation for components. Consider the Prague object (entity, node). Then instead of *Tw(Prague)* [2], where *Prague*/Town, we can write *Tw(Prague).Population*.

Terms are interpreted in a standard way by means of an interpretation assigning to each function symbol from **Func** an object of the same type, and by a semantic mapping [ ] from LT into all functions given by **T**. **Func** influences the power of LT. It can contain usual arithmetic and aggregation functions, etc. Of a special importance is *R*-identity $=_{Bool(S)}$ usable for a comparison of two sets of *S*-objects. Consider Teacher-objects *John* and *Frank*. Then with

$$\lambda l_1^{Language} \exists d_1, h_1, r_1, \textit{Teaches}(John)\left(d_1, h_1, r_1, l_1^{Language}\right)$$

$$=_{Bool(Language)} \lambda l_2^{Language} \exists d_2, h_2, r_2, \textit{Teaches}(Frank)\left(d_2, h_2, r_2, l_2^{Language}\right)$$

we can test whether John and Frank teach the same set of languages. Similarly to domain relational calculus, we can simplify this expression by omitting some existential quantifiers and associated variables. Then the resulted expression can look as

$$\lambda l_1^{Language} \textit{Teaches}(John)\left(l_1^{Language}\right) =_{Bool(Language)} \lambda l_2^{Language} \textit{Teaches}(Frank)\left(l_2^{Language}\right)$$

A further simplification could be made by omitting lambda abstraction supposing that information about objects considered is in the Bool(Language)-identity. Then we can write

$$\textit{Teaches}(John) =_{Bool(Language)} \textit{Teaches}(Frank)$$

In other words, an application is evaluated as the application of an associated function to given arguments, a lambda abstraction "constructs" a new function. In the conventional approach a valuation δ is used. Supposing this mapping we can assign objects to every variable occurring in a term.

For example, *CS(Oxford_House)(AM_training)* is *TRUE*, if there is the class AM_training in the Oxford_house school (*AM_training* is a constant of type

(Bool:Student)). It is true while *AM_training* contains all students of the class AM_training.

In accordance to the semantics of the quantifiers and the singularizer, we can write simply $\forall x(M)$ instead of $\Pi(\lambda\ x(M))$. Similarly, $\exists x(M)$ replaces $\Sigma(\lambda\ x(M))$. Finally, we write $I(\lambda\ x(M))$ shortly as *onlyone x(M)* and read "the only x such as *M*". Certainly, *M*/ *Bool*.

Similarly, aggregation functions as COUNT/Number:(Bool:S), SUM/Number: (Bool:Number), etc. can be defined with usual meaning.

The deductive power of LT is given by *β-reduction*. The main rule is called *β-rule* in the system:

$$\lambda x_1, \dots, x_n(M)(N) = M[x_i \leftarrow N[i]]$$

where the right side of equality represents substitution of $N[i]$ for x_i in *M*.

From the database point of view, we have at disposal a powerful declarative language for formulating schema transformation, ICs, etc., even querying GDBs, as we shall see in Sect. 5.

5   Querying with Functions

The LT language can be used as a theoretical tool for building a functional database language. A query in such language is expressed by a term of LT, e.g.,

$$\lambda t, n(n = \text{COUNT}(\lambda l(\exists d, h, rTeaches(t)(d, h, r, l))))$$

with *d*, *h*, *r*, and *l* of types Teacher, Day, Hour, Room, Language, respectively.

Using the syntactic abbreviation similar to that one used in Sect. 4.3, the resulted lambda expression

$$\lambda t^{\text{Teacher}}, n^{\text{Number}}\left(n^{\text{Number}} = \text{COUNT}\left(\lambda l\left(Teaches\left(t^{\text{Teacher}}\right)\left(l^{\text{Language}}\right)\right)\right)\right)$$

denotes the query "Give a set of couples associating to each teacher the number of languages teaching by him/her". Obviously, the query could be also reformulated as λ *t* $(\lambda\ n\ (\dots))$, or more conventionally $\lambda\ t$ *onlyone n* $(\dots)$.

To consider only teachers born in Prague the lambda expression might look like

$$\lambda t^{\text{Teacher}}, n^{\text{Number}}\left(n^{\text{Number}} = \text{COUNT}\left(\lambda l\left(Teaches\left(t^{\text{Teacher}}\right)\left(l^{\text{Language}}\right)\right)\right) \textbf{ and }$$
$$Is - born\_in\left(t^{\text{Teacher}}\right).Town\_name = Prague\right)$$

Remark: In our conceptual framework we could conceive each query as an attribute, i.e., a lambda expression dependent on possible worlds and time moments. In practice, we omit λw ($\lambda\ t$ … in its head. Clearly, the resulted lambda expressions can express more complex graph structures than *k*-hop queries, i.e. contribute to constructions of new graphs from the original GDB.

In querying GDBs, RPQs are of user importance. We will consider expressions

$$\lambda x, y Reg(x, y)$$

where *Reg* is an expression simulating concatenation and closure. There are two styles how to construct *Reg*. First, consider the attribute *Friends_Of_Friend*((Bool:Person): Person), abbreviated as *FOF*. Then

$$FOF^*(p_1, p_k)$$

will denote the expression $FOF(p_1)(p_2)$ **and** ... **and** $FOF(p_{k-1},)(p_k)$, for a k ($k>1$), and

$$\lambda x, y FOF^*(x, y)$$

provides a set of couples (p_1, p_k), where there is a directed path from p_1 to p_k along edges *FOF*.

Now we will consider a single-valued function, e.g., *Manager_of*/Person (Person), abbreviated as *MO*. $MO^*(p)$ will denote the expression $MO(...(MO(p)...))$ with k applications of *MO*, for a k ($k>1$), and

$$\lambda x, y (MO^*(x) = y)$$

provides a set of couples (p_1, p_k), where there is a directed path from p_1 to p_k along edges *MO*.

We remind that the LT is not computationally complete. But it makes possible to increase its computational power by adding new built-in functions into **Func**.

6   Conclusions

The objective of this paper was to provide an alternative approach to GDB querying based on a functional approach. Comparing to other graph query languages the functional language designed here is based on the notion of graph database schema using the notion of attribute.

All the techniques associated to GDBMS and supported in a graph search engine should fulfil so called *FAE rule* [4]. The FAE rule says that the quality of search engines includes three key factors: *F*riendliness, *A*ccuracy and *E*fficiency, i.e. that a good search engine must provide the users with a friendly query interface and highly accurate answers in a fast way. Clearly, friendliness of our functional language is missing till now. This is the main challenge for future work.

Acknowledgments. This work was supported by the Charles University project P46.

References

1. Angels, R.: A comparison of current graph database models. In: IEEE 28th International Conference on Data Engineering Workshops, pp. 171–177 (2012)
2. Bai, Y., Wang, Ch., Ning, Y., Wu, H., Wang, H.: G-path: flexible path pattern query on large graphs. WWW (Companion Volume), pp. 333–336 (2013)
3. Gray, P.M.D., Kerschberg, L., King, P.J.H., Poulovassilis, A.: The Functional Approach to Data Management. Springer, Berlin (2004)
4. Ma, S., Li, J., Hu, Ch., Lin, X., Huai, J.: Big graph search: challenges and techniques. Front. Comput. Sci. **10**(3), 387–398 (2016)
5. Ghrab, A., Romero, O., Skhiri, S., Vaisman, A., Zimányi, E.: GRAD: on graph database modeling. cornel university library, arXiv:1602.00503 (2014)
6. Pokorný, J.: A function: unifying mechanism for entity-oriented database models. In: Batini, C. (ed.) Entity-Relationship Approach, pp. 165–181. Elsevier Science Publishers B.V, North-Holland (1989)
7. Pokorný, J.: Database semantics in heterogeneous environment. In: Jeffery, K.G., Král, J., Bartošek, M. (eds.) SOFSEM 1996. LNCS, vol. 1175, pp. 125–142. Springer, Heidelberg (1996). doi:10.1007/BFb0037401
8. Pokorný, J.: Graph databases: their power and limitations. In: Saeed, K., Homenda, W. (eds.) CISIM 2015. LNCS, vol. 9339, pp. 58–69. Springer, Heidelberg (2015). doi:10.1007/978-3-319-24369-6_5
9. Pokorný, J.: Conceptual and database modelling of graph databases. In: Desai, B. (Ed.) Proceedings of IDEAS 2016, pp. 370–377. ACM (2016)
10. Robinson, I., Webber, J., Eifrém, E.: Graph Databases. O'Reilly Media, Sebastopol (2013)
11. Shipman, D.W.: The functional data model and the data languages DAPLEX. ACM Trans. Database Syst. (TODS) **6**(1), 140–173 (1981). ACM
12. Tivari, S.: Professional NoSQL. Wiley/Wrox (2011)
13. van Rest, O., Hong, S., Kim, J., Meng, X., Chafi, H.: PGQL: a property graph query language. In: Proceedings of GRADES 2016: Redwood Shores, CA, USA, p. 7 (2016)

Online Transaction Processing (OLTP) Performance Improvement Using File-Systems Layer Transparent Compression

Suharjito[✉] and Adrianus B. Kurnadi

Magister in Computer Science, Binus Graduate Program,
Bina Nusantara University, Jakarta, Indonesia
suharjito@binus.edu, akurniadi@gmail.com

Abstract. In this research, we use three Swingbench OLTP benchmark scenarios to examine that three compression algorithm in ZFS file-systems namely LZ4, LZJB and ZLE can improve OLTP database performance. Beside the database performance, we also compare how much storage can be saved, impact to maximum response time, and the increase of CPU utilization from the three compression algorithms. The acquired data were then be analyzed using Analytic Hierarchy Process to find out the highest ranking compression in terms of benefits and benefits to cost ratio. The result indicates that LZJB achieved the highest performance improvement, LZ4 achieved the highest storage saving and ZLE achieved the smallest CPU utilization overhead. The safest algorithm that did not experience any reduction in database performance or increase of maximum response time in this research is LZJB.

Keywords: OLTP database · Performance · Filesystems layer · Compression

1 Introduction

The popularity of the internet and online commerce has resulted in the explosion of the number of electronic transactions. The increase of the size of transactions data results in a huge increase of storage requirements and its associated costs including space, electricity and cooling. Two techniques that are often used to reduce data size are compression and reduplication. Both techniques result in the increase of CPU utilization traded with saving of storage capacity that has to be physically provided. This CPU utilization overhead of compression and decompression is one of the deterrents that prevent the use of database compression [1]. The other effects of compression is that it will reduce the amount of I/O that has to be done by the application and this opens the possibility that compression besides saving the storage space might also improve performance.

To be able to increase application performance, the amount of time saving for I/O operations must be bigger than the additional time needed to do compression or decompression of data. There are a several studies on the effect of compression implemented on database layer or storage layer. The effect of file-systems compression on database performance, on the other hand, has not been much studied.

© Springer International Publishing AG 2017
N.T. Nguyen et al. (Eds.): ACIIDS 2017, Part I, LNAI 10191, pp. 302–311, 2017.
DOI: 10.1007/978-3-319-54472-4_29

2   Related Works

Several studies on the effect of compression to database performance have been done before. However, most of them focus on compression implementation at the database layer itself [1–3] or at the storage layer [4]. Only one study [5] was found that focus on compression implementation at the file-systems layer. Compression implementation at the database layer has a unique benefit. It can also benefits to database backup and database replication process, but it will only work for that specific database and sometimes translates to higher software license cost or complex modification of the database software. Compression implementation on the file-systems level, on the other hand, will work on any database and in the case that the file-systems are available on several operating systems, the applicability of the study will be higher.

With regard to the type of the database for related works, some use database that is not widely used commercially or they did not mention the database type. One of the earliest studies [1] discusses compression implementation for Scientific and Statistical database, compression benefits and disadvantages and compares several compression algorithms suitable for the type of data in Scientific and Statistical Database.

The second related work still deals with compression implementation at the database layer for a generic database [2] and discusses the query algorithm and compression characteristic that can enable query processing without decompression and its effect on I/O performance, transaction processing and query processing. Performance comparison in the second study is done only through theoretical calculation for hybrid hash join. The third related work [3] also still discusses compression implementation at database layer. It discusses the characteristic of a compression algorithm that can increase database performance. The compression must be fast and fine grained. The study uses TPC-D benchmark to shows that light weight compression can increase most query performance, in the extreme case up to 2 times and that performance only reduced for some update operations. The database used for this study was also not named.

The fourth related work [4] discusses how H-HIBASE compression implemented in the storage layer can increase performance for all kinds of query operations compared to DHIBASE and uncompressed Oracle 10 g database. The last study [5] is the only one that discusses the impact of transparent compression at the file-system layer (ZFS) to the performance of a data warehouse application. This last study employs the widely used Oracle database combined with ZFS file-systems compression and SwingBench to show up to 92% performance improvement of compressed database. The last study uses Sales History scenario in Swingbench which is an OLAP scenario, and the algorithm compared was LZJB, ZLE and GZIP. The study also compares the performance improvement with Oracle Advanced Compression Option (ACO) which is implemented at the database layer. The advantage of this study is that it can be directly applied since Oracle database is widely used and ZFS file-system is available from various operating systems such as Oracle Solaris, Linux and FreeBSD. Although the study has successfully shown the performance improvement for data warehouse workload, it was limited in the sense that it did not compare against Oracle Hybrid Columnar Compression which was claimed to have much higher compression ratio and performance improvement for OLAP workload.

Our study aims to extend the usability and applicability of the previous research by testing OLTP database workload. OLAP workload benchmark using Swingbench will only test decompression impact on performance, while an OLTP workload that has various read write ratios will test the combination of both compression and decompression impact to database performance. The compression algorithm compared is changed to LZ4, LZJB and ZLE. GZIP was not tested in our study because it was the compression algorithm with the highest CPU overhead in the last study. LZ4 is a new compression algorithm in ZFS that was newly incorporated in Oracle Solaris 11.3 operating systems used for the experiment.

Our study also aims to interpret three technical measurement results namely compression ratio, transaction per second and maximum response time into business benefit variables like storage savings, increased productivity, and SLA improvement. The business benefits will be ranked using Analytic Hierarchy Process and compared with the cost ranking for each scenario. With three read write ratios, it is also possible to see the correlation between read write ratio and the level of performance improvement. There are also risks that with certain read write ratios; compression might have a negative impact to performance in terms of transaction per second or the increase in maximum response time. This study will be able to find out which algorithm is most suitable for a given scenario and which algorithm is the safest to use in case we do not have knowledge of read write ratio or application characteristics.

3   Research Method

Experiment on the impact of file-system compression was done using the following environment:

- SPARC T4-4 server logical domain with the specification as follows:
  - 4 core SPARC T4 2.85 GHz
  - 32 GB RAM
  - 450 GB HDD for OS
  - 450 GB HDD for data
- Solaris 11.3 operating systems
- Oracle Database 12c
- Swingbench version 2.5.971

The server used actually had 4×8 core SPARC T4 CPU and 256 GB, but it was find out early that 4 core was already very powerful and it was easier to measure CPU utilization when the number of core was reduced. The memory allocated to the logical domain was also reduced to reduce cache effect at ZFS file-system. The logical domain was set with bare metal I/O so there was no I/O overhead.

After the operating system had been installed, we need to create user, group and project settings as the requirements of Oracle Database installation. We also need to create several ZFS file-systems with different compression settings. Before creating a ZFS file-system we need to create a ZFS pool with the following command:

```
# zpool create dpool cxtxdx
```

The command above created a zfs pool called dpool from the disk cxtxdx. After the pool was created, we created several file-systems with the following commands:

```
# zfs create dpool/baseline
# zfs create dpool/zle
# zfs create dpool/lzjb
# zfs create dpool/lz4
```

To set appropriate compression algorithm to the file-systems, we used the following commands:

```
# zfs set compression = zle dpool/zle
# zfs set compression = lzjb dpool/lzjb
# zfs set compression = lz4 dpool/lz4
```

The benchmark tool Swingbench was created by an Oracle UK employee named Dominic Giles. It was created to provide a realistic benchmark to test Oracle RAC [6]. Swingbench was chosen as the benchmark tool for this study because it is often used to benchmark Oracle database performance both for vendor sponsored white paper [7, 8] and also for academic research published in international conferences especially about database performance on virtualized environment [9–11]. Swingbench software has four built-in benchmark scenarios. This experiment will use three Swingbench OLTP benchmark scenarios to provide data namely: Order Entry with 60/40 read write ratio, Calling Circle with 70/30 read write ratio and Stress Test with 50/50 read write ratio.

3.1 Data Collection Method

Swingbench is not only a benchmark tool, it also comes with wizards to create the schema and populate data required for each benchmark scenario. The wizard can be run interactively with GUI to select such parameters as database network address, username, password, scale of data size, location of data file and other parameters. The wizard can also be run in lights out mode by providing command line parameter or referring to an xml configuration file.

We did some trial run with some parameter settings with each scenario to check if all data especially CPU utilization could easily be measured. Based on the trial run the following is the parameter settings for each scenario: (1) Order Entry scenario was set with 1 GB data size and 10 min runtime; (2) Calling Circle scenario was set with 10 GB raw data size and 5 min runtime; (3) Stress Test scenario was set with 10 GB raw data size and 5 min runtime. All scenario was set with 500 users

To ensure that the compression ratio and other measurable results can be compared between compression algorithms, initial data for the benchmark must only be generated once for each scenario. The compression ratio is only valid when we compare the exact same data. The wizard for each scenario should then be used to populate the data located in the dpool/baseline file-systems.

Order Entry benchmark and Calling Circle benchmark have its own setup wizard named oewizard and ccwizard, but Stress Test scenario uses the same schema and data population as Order Entry. Stress Test only differs from Order Entry in the type of transactions and its relative weight during the benchmark run. Before the benchmark was run, compression ratio was measured and the data file along with Oracle spfile and control file should be backed up. The compression ratio for a given ZFS file-system can be measured with this command:

```
# zfs get compressratio dpool/filesystemname
```

To run the benchmark Swingbench provides three options, we can use swingbench, minibench and charbench. The first is a full blown GUI, where we can set benchmark duration, number of users and so so on, minibench is a minimalist GUI for the same purpose and charbench is using command line options to provide parameters for the benchmark. All options can use xml configuration files to set the transactions that will be run and its weight during the benchmark. We will use charbench to run the benchmark because it can provide more detailed output including transaction dump and cpu utilization monitoring result to output files. The type of transaction and its relative weight for Order Entry benchmark that we used can be seen in Table 1.

Table 1 Order entry transactions and weight

Transaction name	Weight	Enabled
Customer registration	15	TRUE
Update customer details	10	TRUE
Browse products	40	TRUE
Process orders	5	TRUE
Browse orders	5	TRUE
Sales rep query	2	FALSE
Warehouse query	2	FALSE
Warehouse activity query	2	FALSE

Order Entry has some transactions type that by default is not enabled. It can be used if we want to increase the read ratio, but for this study we left the settings as default. Type of transactions and its weight settings for Calling Circle and Stress Test are shown in Tables 2 and 3. One notable difference of Calling Circle scenario is that during data population there is a setting to specify how much transactions should be prepared. We prepared 8000 transactions that took approximately 5 min to run.

Table 2. Calling circle transactions and weight

Transaction name	Weight	Enabled
New customer	25	TRUE
Update customer details	100	TRUE
Retrieve customer details	50	TRUE

Table 3. Stress test transactions and weight

Transaction name	Weight	Enabled
Insert transaction	15	TRUE
Simple select	40	TRUE
Update transaction	30	TRUE
Delete transaction	10	TRUE

After the benchmark was run for the baseline, we need to shut down the database and restore the backup to another file-system with compression. Oracle database should be set to Noarchivelog mode so that after the restore, the transactions from the first benchmark run will not be replayed.

3.2 Data Analysis Method

To analyze the result for each scenario we used Analytic Hierarchy Process that is used to make decisions when we have several criteria. Analytic Hierarchy Process (AHP) was developed by Saaty [12]. AHP decomposed a decision objective into several criteria that can be decomposed further to sub criteria. Each criterion will be assigned ranking which is calculated from how many times a criteria is more important than the other criteria. The alternatives that will be chosen for the decision will also be ranked for each criterion. For our work the decision goal or objective is to choose which algorithm has the most benefit for each benchmark scenario. There are four criteria in making the decision: (1) Storage saving is derived from compression ratio; (2) Productivity increase is derived from transaction per second improvement; (3) Improvement of SLA is derived from the improvement of maximum response time; (4) Compression cost is derived from CPU utilization overhead.

From the four criteria above, we separated the first three as benefit criteria and the last as the cost criterion. Early when AHP was first implemented, people tended to lump together positive and negative criteria together. However, it was recognized that positive and negative priorities in nature are not directly comparable. [13] By grouping positive criteria together, we will be able to rank the benefit of each scenario without the cost. Of course, we then calculate cost ranking separately and later can create benefit to cost ratio ranking for each scenario.

4   Results and Discussion

The data results from each benchmark scenario were based on different data, so they were analyzed separately by, first, using AHP method. AHP method decomposed decision making process into hierarchy of objective, criteria and alternatives. The criteria were cost saving, productivity and maximum response time. The objective was choosing the most optimum compression algorithm for order entry scenario.

First, we need to determine relative importance of the benefit criteria. The relative importance of the criteria is subjective. In this research, based on experience the writer's subjective judgment is that storage saving has equal importance to performance improvement and business SLA. The relative importance sum of all criteria is 1. After the ranking of the criteria has been decided, we need to calculate the relative importance of the alternatives for each criterion using pair wise comparison matrices. Because the matrices values are taken from the measurement of each alternatives value for related criteria, the resulting ranking will be consistent. AHP can combine both subjective and objective factors in the decision making process. After calculating the highest ranking for benefits and benefits to cost ratio for each scenario, then we can do cross scenario comparison to see if there is any correlation between the results and the change of read/ write ratio and other characteristics of each scenario.

4.1 The Result of Order Entry Scenario

Based on the calculation of relative benefit ranking of compression algorithm for performance improvement, storage saving and SLA improvement using pair-wise comparison matrices, Order Entry scenario LZJB has the highest benefit, followed by LZ4 and ZLE. For Order Entry scenario the compression with the highest cost ranking is LZ4, followed by LZJB and ZLE. The calculation of benefit to cost ratio and comparison chart for it is shown in Fig. 1.

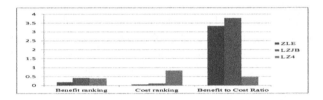

Fig. 1. Order entry benefit to cost ranking

The compression with highest benefit to cost ranking is LZJB followed by ZLE and LZ4. LZ4 comes last in the ranking mainly because of its high CPU utilization.

4.2 The Result of Calling Circle Scenario

The benchmark results between the baseline and the three compression algorithm for Calling Circle scenario show increased performance. We can see that Calling Circle transaction is heavier and more CPU intensive than Order Entry, shown from the lower average transaction per second achieved. Data in Calling Circle are more compressible, shown by the higher compression ratio achieved compared to compression ratio in Order Entry. All compression in Calling Circle scenario improves maximum response time. So we do not need to use relative response speed. We will be able to use speed improvement column for Response Time Improvement comparison matrix because there are no negative or zero numbers.

Based on the calculation, in Calling Circle scenario, LZ4 has the highest benefit, followed by LZJB and ZLE. For Calling Circle scenario the compression with the highest cost ranking is LZ4, followed by LZJB and ZLE. The calculation of benefit to cost ratio and comparison chart for it is shown in Fig. 2.

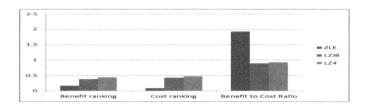

Fig. 2. Calling circle benefit to cost ranking

The compression with the highest benefit to cost ranking for Calling Circle is ZLE followed by LZ4 and LZJB. ZLE comes first in the ranking mainly because of its very low CPU overhead.

4.3   Result and Analysis from Stress Test Scenario

Stress Test is the benchmark with the lightest type of transactions shown by the highest average transaction per second achieved, but it is also the benchmark with the highest write ratio which results in the lowest performance improvement. LZ4 caused decrease in performance in Stress Test scenario. We had to use Relative Performance value to avoid the negative comparison later. As usual, LZ4 achieved the highest storage saving followed by LZJB and ZLE. In Stress Test scenario, ZLE achieved the highest perform-ance improvement but made the maximum response time worse. We had to use Relative Response Speed in the comparison matrix. LZ4 consistently showed the highest CPU utilization overhead followed by LZJB and ZLE. CPU overhead for ZLE is so small it measured zero. We used Relative CPU utilization to avoid comparison with zero.

Based on the calculation, in the Stress Test scenario, LZ4 has the highest benefit, followed by LZJB and ZLE. Cost ranking for Calling Circle can be calculated using comparison matrix with data from Table 16. For Calling Circle scenario, the compres-sion with the highest cost ranking is LZ4, followed by LZJB and ZLE. Comparison chart for cost ranking and benefit ranking can be seen as shown in Fig. 3.

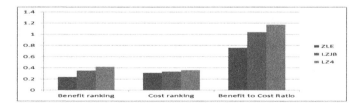

Fig. 3. Stress test benefit to cost ranking

The compression with the highest benefit to cost ranking for Stress Test scenario is LZ4 followed by LZJB and ZLE. This might be because when the transaction types are very light the cost differences will not change the benefit ranking.

4.4   Result and Analysis from Stress Test Scenario

We compared the results between scenario to see if there was any correlation between read/write ratio and the amount of performance improvement.

Comparison Chart of Performance Improvement for All Scenario in Fig. 4 shows trends of decreasing performance improvement for almost all compression algorithm when the write ratio increases. The exception is ZLE which improves between 60/40 to 50/50 read/write ratio.

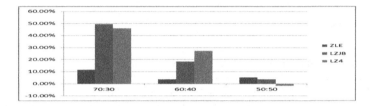

Fig. 4. Performance improvement all scenario

From the comparison chart in Fig. 5, we can see that LZJB is the only compression that did not cause any increase in maximum response time. Meanwhile, from Fig. 5 we can see that LZJB compression also never caused any performance decrease in this study. This observation makes LZJB the safest compression algorithm to choose from the three algorithm in case we have not enough knowledge about read write ratio or other characteristics of the application accessing the database.

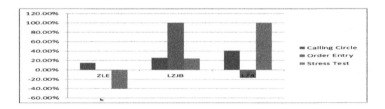

Fig. 5. Maximum response time improvement all scenario

5   Conclusion and Future Works

The results of the benchmark and its analysis have shown that compression at the file systems layer can improve OLTP database performance with the following things to note: (1) Performance improvement tends to be higher for OLTP applications with higher read ratio; (2) Among the algorithm studied here, LZJB is the safest to implement for OLTP and seems to strike the right balance between compression ratio and CPU overhead; (3) Applications with more complex query like Calling Circle scenario will yield more performance using light weight compression rather than higher compression ratio.

We hope that the results of this study can help to increase the adoption of file systems compression in general and ZFS file systems particularly to help to save storage cost and improve OLTP application performance.

This study is limited by time and equipment available and there are still a lot of topics that can be pursued further for future works related to performance improvement using file systems compression such as: (1) OLTP performance improvement using other file systems besides ZFS; (2) OLTP performance improvement using compression on storage based on ZFS filesytems; (3) OLTP performance improvement using file system compression on flash storage.

References

1. Bassiouni, M.A.: Data compression in scientific and statistical databases. IEEE Trans. Softw. Eng. **SE-11**(10), 1047–1058 (1985)
2. Graefe, G., Shapiro, L.D.: Data compression and database performance. In: ACM/IEEE-CS Symposium on Applied Computing, Kansas City (1991)
3. Westman, T., Kossmann, D., Helmer, S., Moerkotte, G.: The implemmentation and performance of compressed databases. ACM SIGMOD Rec. **29**(3), 55–67 (2000)
4. Habib, A., Hoque, A.S.M.L., Rahman, M.S.: High performance query operations on compressed database. Int. J. Database Theor. Appl. **5**(3), 1–14 (2012)
5. Wenas, A.: Meningkatkan Kinerja Data Warehouse Menggunakan Teknologi Filesystem Dengan Kompresi GZIP, LZJB dan ZLE, Jakarta (2015)
6. Giles, D.: Swingbench (2015). http://www.dominicgiles.com/Swingbench.pdf. Accessed 2015
7. IBM, Oracle Database 11g and 12c on IBM Power Systems built with IBM Power8 processor technology and IBM FlashSystem 840, IBM Oracle International Competency Center (2014)
8. VMware, Oracle Databases on vSphere Workload Characterization Study, VMware, Palo Alto (2010)
9. Almari, F.N., Zavarsky, P., Ruhl, R., Lindskog, D., Aljaedi, A.: Performance analysis of oracle database in virtual environments. In: 2012 26th International Conference on Advanced Information Networking and Applications Workshops (WAINA), Fukuoka (2012)
10. Tope, I.E., Zavarsky, P., Ruhl, R., Lindskog, D.: Performance evaluation of oracle VM server virtualization software 64 bit linux environment. In: 2011 Third International Workshop on Security Measurements and Metrics (Metrisec), Banff, AB (2011)
11. Ye, D., Pavuluri, A., Waldspurger, C., Tsang, B., Rychlik, B., Woo, S.: Prototyping a hybrid main memory using a virtual machine monitor. In: IEEE International Conference on Computer Design, ICCD 2008, Lake Tahoe, CA (2008)
12. Saaty, T.L.: Decision making with analytic hierarchy process. Int. J. Serv. Sci. **1**(1), 83–98 (2008)
13. Saaty, T.L., Ozdemir, M.: Negative priorities in the analytic hierarchy process. Math. Comput. Model. **37**(9), 1063–1075 (2003)

A Multi-database Access System with Instance Matching

Thanapol Phungtua-Eng[1] and Suphamit Chittayasothorn[2(✉)]

[1] Faculty of Business Administration and Information Technology,
Rajamangala University of Technology Tawan-Ok Chakrabongse Bhuvanarth Campus,
Bangkok 10400, Thailand
thanapol.phu@cpc.ac.th
[2] Faculty of Engineering, King Mongkut's Institute of Technology Ladkrabang,
Bangkok 10520, Thailand
suphamit.ch@kmitl.ac.th

Abstract. Organizations that use several separately-developed information systems face a common problem. The data which are used by different systems have no standard. Different databases that keep information of same entity instances use different representations. Attribute names are different. Attribute values are different. Even unique identifiers which are used to identify object instances are different. Yet the data need to be referred to and used by some mission-critical applications. This paper presents a multi-database instance matching system which is developed to bring data from separate sources that refer to different unique identifiers and attribute details. Entity resolution techniques are employed to match the database instances. After matched entity instances are identified, an ontology is used to keep the matched identifiers. Queries from the users then refer to the ontology and are rewritten to refer to the correct instances of the original database.

Keywords: Multi-databases · Schema matching · Entity resolution · Query rewriting

1 Introduction

This paper reports the development of a multi-database access system that refers to databases from separated information systems. The system matches database instances from different databases and provides their access to applications that need to refer to the information from the matched instances.

The databases are known to contain information that refer to object types with almost the same set of object instances. These database object types, which actually describe the same real-world object type, from the two databases employ different unique identifiers. Applications would like to access both databases and obtain information from them.

A straight-forward but tedious and hard to implement solution is to manually match all individual data from the two databases and migrate them to a central database. In practice, there are both managerial and technical constraints which prevent the physical and permanent integration to take place. Existing databases are expected to remain

© Springer International Publishing AG 2017
N.T. Nguyen et al. (Eds.): ACIIDS 2017, Part I, LNAI 10191, pp. 312–321, 2017.
DOI: 10.1007/978-3-319-54472-4_30

separated as they are. A mechanism must be provided to match their instances and refer to the matched information.

A major problem is therefore the entity resolution problem. How to identify instances from the two databases which use different identifiers that refer to the same entity instances in the real world? After the matching and identification steps, there must be an identifier mapping that refers to the correct instance of the other database and allow applications to access both databases.

Another problem is the system integration architecture. There are several application domains that require multi-database access in this manner such as medical information systems, logistics systems and warehouses management systems [1–4]. These information systems comprise separately developed subsystems, some of which are from different developing teams or purchased from different vendors. The users may want to access the separated databases which use different identifiers, attribute names and values to refer to identical instances. They would like to access the rows that represent the same entity instances.

According to the literatures, in order to allow applications to access other applications' databases, the databases can be transformed or mapped directly to the other databases at the database level [1, 5–8]. Alternatively, applications can be integrated at the applications/process level. In this latter case, existing applications are called from a new main application that operates on top of them. A good example is the Service Oriented Architecture [4, 9–11]. Also, applications can be integrated at the User Interface level where a central GUI refers to other existing application GUIs.

In this project, the first approach is chosen because a new application is the one that needs the data from the existing databases. Applications which are using the existing databases are not touched and remain in production. Since the databases use different identifiers to identify database instances, database rows which refer to the same instances have to be first matched. Due to the completeness requirement of the matching process, this is done semi automatically. Our system helps match most rows but there are still cases that human experts are required to make final decisions. An ontology which refers to entity instances from the two databases and identifies clearly the identical instances are built. Queries which are entered into the integrated database system will consult the ontology, and are rewritten to refer to correct entity instances of the original databases.

2   Reviews and Case Study

To match database instances, there are three major problems [12, 13]. The first one is the data model difference. Each data model has different logical data structures. Basically there are the tabular format of the relational database model, trees and virtual pointers of the hierarchical database models, graphs of the network model, encapsulated objects of the object data model, non-encapsulated tabular objects of the object relational data model, and so on. To integrate those databases with different data models is a real challenge and requires data model transformations. A middle data model may be required to be used as the middle layer between two source databases with different data models.

This problem is commonly referred to as the structural problem or model difference problem [1, 8, 14–17].

The use of the same attribute name or attribute value to refer to two or more different things from different databases is also a challenge in database integration. This is commonly referred to as the semantic problem [15, 18, 19]. Very common attribute names such as ID#, NAME, ADDRESS could appear as attributes of different entity types. They could be attributes of EMPLOYEE or CUSTOMER or SUPPLIER. Common attribute values such as 'Eng' could represent Engineer or English. It is therefore very challenging to identify two database objects from different databases which actually refer to the same thing.

The third problem is the syntax problem [2, 3, 14, 20, 21]. This is the case where two different attribute values actually have the same meaning. A classic example is the different values of the masculine gender 'M' and 'Male' which actually represents the same gender.

This project concentrates on the latter two problems. We limit the scope of our work to the relational database model and therefore refer to the tabular structure. The work is demonstrated by a high-level integration of two relational databases from two mission-critical information systems. There are sematic and syntactic problems that have to be encountered. Several entity resolution techniques are applied to match tuples which belong to entity instances. Since the applications are mission critical, a single incorrect match cannot be tolerated. All uncertain cases are then resolved by domain experts.

Figure 1 shows the table structures of the two tables which are to be matched and accessed. They both keep information about faculty members of a university department. The Table 1(a) is from the Human Resource Department (HRD) and the Table 1(b) is from the Registration Office (REG).

COLUMN_NAME	COLUMN_NAME	COLUMN_NAME
1 PERSON_ID	9 TYPE	
2 TITTLE	10 ADD_NUMBER	
3 NAME	11 STUDY	
4 SURNAME	12 START_DATE	1 ID
5 FAC	13 BIRTH_DATE	2 NAME_SURNAME
6 DEPART	14 STATUS	3 FAC
7 PRO_TYPE	15 EMAIL	4 DEPART
8 WORKING_TYPE	16 END_DATE	

a)   b)

Fig. 1. Relational database schemas of (a) the HRD and (b) the REG tables

The HR table keeps information of all employees in the university; including the lecturers. It keeps personal details of the university employees. This table uses the national identification number PERSON_ID as the primary key. The REG table shows some information about university personals who teach in the university courses. This table uses ID as its primary key. This ID is a system generated surrogate key. There are other tables in the registration system which keeps information on the courses that they are teaching in each semester. Those tables also use the surrogate key ID to uniquely identify the teachers. The two systems were separately developed and operated.

Table 1. Result of the matching process

		Actual	
		True	False
Classifier	True	980 row	11 row
	False	38 row	1512 row

3   Schema Matching and Entity Resolution

Schema matching is a process for finding out if two database schemas refer to the same object type and entity resolution identifies matched entity instances. The two processes help match tables and rows in relational databases. Entity types have to be matched first before rows are matched to find the same entity instances. Most similar attributes of the entity types or tables which are to be matched have to be identified first before the attribute values are compared. Since our goal is to find attributes that can be used to identify instances, attributes from the two schemas which are to be matched and are found not likely to be the same attribute, are ignored.

At first, the attribute names of the matching schemas seem to be relevant. However, some attribute names are very common and different entity types may have the same attribute names such as ID, NAME, ADDRESS, etc. Instead, attribute values are the ones that can be used to determine more accurately if two attributes from different schemas should actually be the same attribute; or refer to the same set of attribute values. After previous investigations [22], we found that the application of the Kulczynski Cody index [23] yields the most relevant and accurate result.

This method considers attribute values in order to determine if two sets of attribute values refer to the same attribute. The relationship between two attribute values sets are found by using the Kulczynski Cody index [23] calculation Eq. (1):

$$Kulczynski; Cody(x, y) = \left(\frac{|x \cap y|}{2|x|} \cdot \frac{|x \cap y|}{2|y|} \right) \tag{1}$$

Attributes from the HR and REG schemas are matched and evaluated using the Kulczynski Cody index. This index gives the indication if the x set and y set have many common members. The value closes to 1 indicates high common members. As an example, consider the attribute FAC whose values are the name of the faculty (school) that the teacher in the REG table belong to. We examine which attributes of the HR table refers to it. First, FAC is matched against NAME from the HR table. This NAME attribute contains the employee names and therefore should not be considered the same attribute as FAC.

The FAC attribute has 12 distinct values. The NAME attribute has 918 distinct values. The two sets of values are disjointed. The index calculation result is 0.

Each attribute is tested against all attributes from the other table. The highest index value obtained is 0.9166 when the FAC attribute of REG is paired with the attribute FAC of HR. There are cases that some attributes from a schema does not have a matched attribute in the other schema. Those unmatched attributes are ignored. There are also cases that two attributes which actually represent the same thing are not found to be matched by this Kulczynski Cody index technique due to the lack of enough duplicate values. These attributes are therefore not used in the instance matching stage. At the end of this stage, attributes which are to be used in the instance matching stage are identified.

After the attributes to be matched are identified, the next step is to determine attribute values from different databases which represent the same property. The concept of functional dependency is used to determine such values.

In the case that $a_1, a_2, \ldots, a_m$ are attribute values whose Key $\{a_1, a_2, \ldots, a_n\} \rightarrow b (1 \leq n \leq m)$ is in the first relation, and $\{a_1, a_2, \ldots, a_n\} \rightarrow c (1 \leq n \leq m)$ is in the second relation. It can therefore be concluded that $\{a_1, a_2, \ldots, a_n\} \rightarrow \{b, c\} (1 \leq n \leq m)$.

Based on similar attributes dependency, identifiers can be determined if they identify the same entity instances and are considered to identify the same entity instance. Let a, b are values of attributes A and B respectively, if $a \rightarrow \{c, d, e\}$ and $b \rightarrow \{c, d, e\}$ then a and b are likely to identify the same thing.

Figure 2 shows the matching of two tuples, one from REG and the other one from HR. The REG tuple has the surrogate key 70004 as its primary key. The HR tuple uses the government identification number 5100444515860. Since some major attribute values are the same, the primary key values are assumed to identify the same entity instance. In this case, they refer to the same employee.

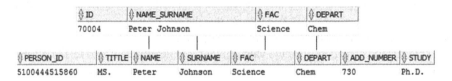

Fig. 2. Two tuples from different databases that represent the same instance.

4   More Instance Matching Problems

Most database instances from the two databases can be automatically matched using the above mentioned techniques described in the previous section. However, there are some database instance matching problems that are found to require human experts' assistance. They can be categorized into four cases, namely the multiple identifier problem, the multiple identifier problem, the missing value problem, and the multiple languages problem. These problems require human experts' assistance.

(a) The mistyping problem is a human error problem in the data entry of an object instance. An object instance incorrectly has multiple unique identifiers due to the error in data entry operations. In our case, some lecturers have multiple database rows with different primary key values. This is due to the fact that some attribute values are mistyped such as the name of the lecturer as shown in Fig. 3. The name 'Tanaka Shinji' is mistyped to 'Tanaka Chinji' and caused the application to generate another ID for the same person. Noted that it is possible that the two rows belong to different persons with very similar names.

ID	NAME_SURNAME	Faculty	Department
90460	Tanaka Shinji	Engineer	Computer
90399	Tanaka Chinji	Engineer	Computer

Fig. 3. Two rows with different identifiers of the same person (name mistyped)

(b) The multiple identifier problem is the problem where two identical rows have different primary key values. This problem is similar to the previous one. The difference is that the two rows are more likely to belong to the same person than the ones in the previous problem. This is often the case where the data are entered into the system repeatedly by different staff members. Noted also that it is still possible, even though rare, that the two rows belong to different persons with the same name. In Fig. 4, two rows of John Smith are in the same table.

ID	NAME_SURNAME	Faculty	Department
90300	John Smith	Science	Computer Science
90297	John Smith	Science	Computer Science

Fig. 4. Two identical rows are found to have different ID.

(c) The missing value problem is the problem where some important values are missing thus leads to the confusion of identities. Figure 5 shows two rows of a lecturer whose name is Tony but the surnames of both rows are missing. Again, this could either be the same or different persons.

ID	NAME_SURNAME	Faculty	Department
90187	Dr. Tony	Science	Physics
72702	Dr. Tony	Science	Physics

Fig. 5. Two identical rows are found to have different ID and the Surnames are missing.

(d) The multiple languages problem is the problem where information of the same entity instance is entered into the database more than once using different languages. This is a typical data preparation error and because the two rows of the same person are entered using different languages. Figure 6 shows two rows of this case.

ID	NAME_SURNAME	Faculty	Department
90122	Peter Jackson	Engineer	Computer
10508	ปีเตอร์แจ็คสัน	Engineer	Computer

Fig. 6. Two rows of the same object instance in different languages

5   Implementation

After the schema matching and entity resolution steps, a central ontology is created to record and relate different identifiers. Database integration using ontologies have been previously carried out by our team [22]. In that work an ontology is created for each source database. Their ontologies are then combined to be an integrated global ontology. The global ontology contains information from all source databases, thus becomes a large one. Query processing is done directly on the ontology using a version of the Ontology Web Language (OWL).

The previous version has a large ontology which contains all facts and axioms. The ontology is in a text file format. Queries on the ontology yield slow response time due to the lack of an efficient DBMS and query optimization for the ontology.

In this project, a different approach is introduced. Only matched unique identifiers are stored in the ontology. The system architecture of the multi database system is shown in Fig. 7.

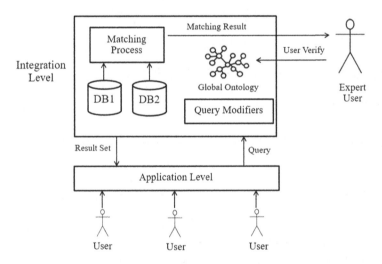

Fig. 7. The system architecture of the multi database access system.

Initially, database tables from the source databases are registered to the system. Database schemas are then checked for possible matching using schema matching techniques. Two mission-critical database tables from separate information systems are matched. The HR table that contains 1031 employee rows and the REG tables which contains 2541 teaching rows are matched. In the HR table there are employees which are not lecturers and therefore do not appear in the REG table. Also, in the REG tables, there are lecturers who are outsiders and are not employees. The most significant difference between the two tables is the use of different unique identifiers. The matching result is shown below. Accuracy is 98.07%. Precision is 98.89%. Recall is 96.26%.

After the matching, an ontology is created. Queries from applications are entered into the system, checked with the ontology for possible rewriting. The final queries are sent to the appropriate databases with correct naming and identification. The final result is returned to the applications. Figure 8 shows the query rewriting process.

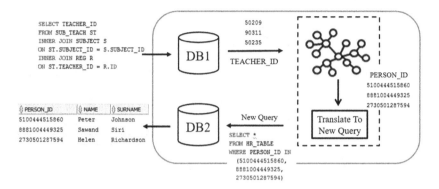

Fig. 8. The query rewriting process

6   Conclusions

Multi-database access is an organization-level problem which becomes more and more serious when the databases are required to support business intelligence and management decision making. In many organizations, information systems are developed independently. Different naming conventions, data types, and values in such different databases make it hard to consider if the data items from different databases refer to the same real-world objects.

This paper presents an approach to the database integration problem. Ontology is used as a central identifier matching structure after data items and relationships are identified and resolved. Since the multi-database access process must yield perfect or close to perfect result, any mismatches or errors are not acceptable and user involvements are required. Hence, a semi-automatic approach is adopted.

The multi-database access problem becomes harder in the case that databases employ different data models. Polyglot environments may be a solution for operational systems but may turn to be problems for decision support systems.

Acknowledgement. The authors would like to express our sincere thanks to the anonymous reviewers whose constructive comments gave us an opportunity to clarify the points that we did not previously explain. Many thanks also to Dr. Pitak Thumwarin, Vice President for Human

Resource Evaluation of King Mongkut's Institute of Technology Ladkrabang Thailand, for providing the actual working datasets from the university databases.

References

1. Mate, S., Köpcke, F., Toddenroth, D., Martin, M., Prokosch, HU., Bürkle, T., Ganslandt, T.: Ontology-based data integration between clinical and research systems. In: PLOS One 2015, pp. 1–20. PLOS One, San Francisco (2015)
2. Jung, Y., Yoon, Y.: Data integration for clinical decision support. In: 2016 8th International Conference on Ubiquitous and Future Networks (ICUFN), pp. 164–166 (2016)
3. Anjum, A., Bloodsworth, P., Branson, A., Hauer, T.: The requirements for ontologies in medical data integration: a case study. In: IEEE Database Engineering and Applications Symposium (IDEAS 2007), pp. 308–314 (2007)
4. Anjum, A., Bloodsworth, P., Branson, A., Hauer, T.: The requirements for ontologies in medical data integration: a case study. In: 11th Database Engineering and Applications Symposium 2007 (IDEAS 2007), pp. 308–314 (2007)
5. Liu, Z., Du, X., Ishii, N.: Integrating database in internet. In: 1998 Second International Conference on Knowledge-Based Intelligent Electronic System, pp. 381–385 (1998)
6. Karasneh, Y., Ibrahim, H., Othman, M., Yaakob, R.: A model for matching and integrating heterogeneous relational biomedical databases schemas. In: Proceedings of the 2009 International Database Engineering & Applications Symposium (IDEAS 2009), pp. 242–250 (2009)
7. Kavitha, C., Sadasivam, G.S., Shenoy, S.N.: Ontology based semantic integration of heterogeneous databases. Eur. J. Sci. Res. **64**, 115–122 (2011)
8. Kiran, V.K., Vijayakumar, R.: Ontology based data integration of NoSQL datastores. In: 9th International Conference on Industrial and Information Systems (ICIIS), pp. 1–6 (2014)
9. Kantere, V.: Approximate queries on big heterogeneous data. In: Proceedings of the 2015 IEEE International Congress on Big Data, pp. 712–715 (2015)
10. Niang, C., Marinica, C., Leboucher, É., Bouiller, L., Capderou, C., Bouchou, B.: Ontology-based data integration system for conservation-restoration data (OBDIS-CR). In: Proceedings of the 20th International Database Engineering & Applications Symposium (IDEAS 2016), pp. 218–223 (2016)
11. Garg, B., Kaur, K.: Integration of heterogeneous databases. In: 2015 International Conference on Advances in Computer Engineering and Applications (ICACEA), pp. 1034–1038 (2015)
12. Alexiev, V., Breu, M., Bruijin, J., Fensel, D., Lara, R., Lausen, H.: Information Integration with Ontologies Experiences from an Industrail Showcase. Wiley, New Jersey (2005)
13. Kopcke, H., Thor, A., Rahm, E.: Evaluation of entity resolution approaches on real-world match problems. In: Conference on Very Large Databases (VLDB) Proceedings of the VLDB Endowment, pp. 484–493 (2010)
14. Lee, M.L., Ling, T.W.: A methodology for structural conflicts resolution in the integration of entity-relationship schema. Knowl. Inf. Syst. J. **5**, 225–247 (2003)
15. Giunchiglia, F., Shvaiko, P., Yatskevich, M.: Semantic schema matching. In: Meersman, R., Tari, Z. (eds.) OTM 2005. LNCS, vol. 3760, pp. 347–365. Springer, Heidelberg (2005). doi: 10.1007/11575771_23
16. Kılıç, Y.O., Aydin, M.N.: Automatic XML schema matching. Eur. Mediterr. Conf. Inf. Syst. **2009**, 1–7 (2009)
17. Raunich, S., Rahm, E.: Towards a benchmark for ontology merging. In: Interoperability and Networking (EI2 N'2012), pp. 1–10 (2012)

18. Lawrence, R., Barker, K.: Integrating relational database schemas using a standardized dictionary. In: Proceedings of the 2001 ACM symposium on Applied computing (SAC 2001), pp. 225–230 (2001)
19. Giunchiglia, F., Yatskevich, M., Giunchiglia, E.: Efficient semantic matching. In: Gómez-Pérez, A., Euzenat, J. (eds.) ESWC 2005. LNCS, vol. 3532, pp. 272–289. Springer, Heidelberg (2005). doi:10.1007/11431053_19
20. Karasneh, Y., Ibrahim, H., Othman, M., Yaakob, R.: Integrating schemas of heterogeneous relational databases through schema matching. In: Proceedings of the 11th International Conference on Information Integration and Web-based Applications & Services (WAS 2009), pp. 209–216 (2009)
21. He, B., Chang, K.C.: Object matching for information integration: a profiler-based approach. ACM Trans. Database Syst. **31**, 1–45 (2006)
22. Phungtua-Eng, T., Chittayasothorn, S.: Semi-automatic relational databases integration using ontology. In: The 17th World Multi-conference on Systemics, Cybernetics and Informatics, pp. 203–208 (2013)
23. Chao, A., Chazdon, R.L., Colwell, R.K., Shen, T.: Abundance-based similarity indices and their estimation when there are unseen species in samples. Biometrics **26**, 361–371 (2006)

Intelligent Information Systems

Enhancing Product Innovation Through Smart Innovation Engineering System

Mohammad Maqbool Waris[1]([✉]), Cesar Sanin[1], and Edward Szczerbicki[2]

[1] The University of Newcastle, Callaghan, NSW, Australia
MohammadMaqbool.Waris@uon.edu.au, cesar.sanin@newcastle.edu.au
[2] Gdansk University of Technology, Gdansk, Poland
edward.szczerbicki@newcastle.edu.au

Abstract. This paper illustrates the idea of Smart Innovation Engineering (SIE) System that helps in carrying the process of product innovation. The SIE system collects the experiential knowledge from the formal decisional events. This experiential knowledge is collected from the set of similar products having some common functions and features. Due to the fact that SIE system collects, captures and reuses the experiential knowledge of all the similar products apart from the knowledge about new technological advancements, it behaves like a group of experts in its domain. Through this system, the innovation process of manufactured products can be greatly enhanced. Moreover, entrepreneurs and manufacturing organizations will be able to take proper, enhanced decisions and most importantly at appropriate time. The expertise of SIE System is ever increasing as every decision taken is stored in the form of set of experience that can be used in future for similar queries.

Keywords: Smart Innovation Engineering · Product innovation · Product design · Set of experience · Decisional DNA

1 Introduction

A new product is designed and manufactured considering the required features/functions, technology, resources available, manufacturing processes and other such factors at that time. The properly designed and manufactured product leaves an impact in the market initially. But with time, due to technological advancements, development of new materials having enhanced properties/lower costs, improved/cost-effective manufacturing processes and similar other factors, the entrepreneurs and manufacturing organizations have to introduce new features in the product leading to innovation. Product innovation process has to be repeated continuously after a particular time in order to survive in this competitive market. Product innovation is very difficult and complex process as it requires not only the new technological advancements but also the complete knowledge of all the past and present similar products. This knowledge is possessed by a group of experts/innovators. For the survival and prosperity of the manufacturing unit, entrepreneurs need to find out new ideas that can be implemented in the products leading to innovation. Both knowledge and experience are essential attributes of an innovator

© Springer International Publishing AG 2017
N.T. Nguyen et al. (Eds.): ACIIDS 2017, Part I, LNAI 10191, pp. 325–334, 2017.
DOI: 10.1007/978-3-319-54472-4_31

that are necessary to find the best possible solution for the required changes leading to achieve innovation. These changes are based on the innovation objectives reapplied to the established, existing product. Due to the enormous amount of ever evolving and increasing knowledge and rapid changes in the dynamic environment of product design and manufacturing, the product innovation process is difficult to practice. Innovators not only need to take proper decisions, they have to do this quickly and systematically so that the changes in the product may be implemented at the required time. We try to address this problem by proposing a system that uses a collective, team-like knowledge developed by past innovation related experiences. Through this systematic approach, product innovation process can be performed semi-automatically. We call this system as Smart Innovation Engineering (SIE) System [1–3].

In this approach, SIE System comprising Set of Experience Knowledge Structure (SOEKS or SOE in short) and Decisional DNA (DDNA) [4] captures and stores experiential knowledge from the formal decisional events related to product innovation. This experiential knowledge is stored in the form of sets of experience and recalled whenever a similar query is presented during the innovation problem solving process. Such system quickly provides a list of proposed optimal solutions, based on priorities set by user, to a particular innovation objective. The Smart Innovation Engineering System is a prominent tool to support the innovation processes in a quick and efficient way. The SIE system stores the experiential knowledge or sets of experience from the past decisions related to product innovation which enhances product innovation process significantly. Manufacturing organizations and entrepreneurs can take improved decisions systematically and at an appropriate time by implementing the SIE system in the process of product innovation. The system gains more and more expertize with time in a particular domain as it stores data, relevant information and knowledge related to formal decision events. The SIE System is based on SOE Decisional DNA, which were first presented by Sanin and Szczerbicki [1, 5, 6]. SOE and DDNA have been applied successfully in various fields of application like virtual engineering processes, virtual engineering factory, industrial maintenance, semantic enhancement of virtual engineering applications, virtual organization, diagnosis of Alzheimer's disease by decision support medical system, banking activities involving periodic decision making and storing information, digital control system of the geothermal and renewable energy, e-decisional community, and smart interactive TV to name a few. More details can be find in Shafiq et al. [7].

Different authors have defined innovation in various ways. In the context of manufactured products it can be defined as the process of making required changes to the already established product by introducing something new that adds value to users and also provide expertise knowledge that can be stored in the organization [8]. It was reported by Frishammar [9] that a strong emphasis on user information leads to incremental rather than radical product innovation. In incremental innovation, forecasting of user need are easier as it is repeated over a short span of time. Level of risk is also lower and have a less complex product development process. As compared to radical innovation, incremental innovation requires relatively low investment thus providing financial advantage. Revenue and profits are also additional positive factors as they show up faster [10].

On the other hand, radical innovation involves lower chances of success resulting from highly unpredictable outcome in general [11]. This is due to the fact that it depends on new assumptions about the product satisfying the users as well as assumption about the ability of manufacturing organization to satisfy those values. Commercial success is not the deciding factor always. In some certain cases, radical innovation of products has enhanced the communicative value of the manufacturing organization leading to the acceptance of its brand name [12]. This leads to commercial success in future rather that in the current innovated product.

2 Introduction to Smart Innovation Engineering (SIE) System

Every innovation process is based on some objectives that necessitate product innovation. Well defined and clearly stated innovation objectives triggers the product innovation process. Various techniques are used for determining these initial innovation objectives. Some of these techniques are lead user analysis [13], beta testing [14], consumer idealized design [15], use of online user toolkits [16] to name a few. New technological achievements in materials or processes are also considered apart from experts' advice before stating the initial innovation objectives. These objectives are directly or indirectly linked to functions or features of the product that are ultimately technically attributed to one or more components of the product.

Bryant et al. [17] presented a methodology that clearly defines the systematic placements of components on the basis of functionality in the form of hierarchical ontology. More than 100 distinct generically listed component terms are provided in the work by Kurtoglu et al. [18]. Based on this functionality-based hierarchy, one can find and select the required product fulfilling that particular function. Figure 1 shows this extended functional hierarchy that is used for selecting the Screw Jack model 2 which is selected as a product for case study illustrating and explaining this approach. Screw Jack model 2 is further represented as a Virtual Engineering Object (VEO). The concept of VEO was presented by Shafiq et al. [19, 20]. VEO represents both the virtual and real world exemplification of manufactured products or components.

The decomposition of the product under consideration is structured as hierarchical nested parts are shown in Fig. 1. This product (in this case the Screw Jack Model 2) if represented as a VEO and is further divided into number of sub-VEOs as subsystems representing/performing specific VEO features/functions that are represented as sub-VEOs level 1 as shown in Fig. 1. Similarly, further decomposition of sub-VEOs level 1 into lower level subsystems are represented as sub-VEOs level 2, that perform collectively the function at subsystem level 1 and so on. This decomposition of VEOs continues until we reach the basic component level. The same approach can be applied for selecting a product on the basis of functionality and then decomposing it on the basis of hierarchy.

2.1   SIE-DDNA Architecture

SIE-DDNA architecture is consists of eight modules that stores all the relevant information about the product under consideration and is also linked with architecture of similar products. It is an efficient structure for representing the SOE based experiential knowledge and also has the capability to capture, add, improve, store, share as well as reuse this knowledge in decision making and process of product innovation whenever required. This is done in a similar way as performed by group of experts or innovators. SIE-DDNA architecture contains all the relevant knowledge and experience about the product and its features. The eight basic modules of a SIE-DDNA structure are: Characteristics, Requirements, Functionality, Connections, Process, Systems, Usability, and Cost [2, 3]. Out of these, the information contained in the five modules (Characteristics, Requirements, Functionality, Connections, and Process) is linked to and can be extracted from the VEO-DDNA [19, 20]. As these five modules contain the knowledge about the manufacturing scenario of the product. These are required only for the purpose of selecting relevant manufacturing process and better material for required quality. For purpose of innovation process, the Systems and Usability modules are most important and crucial modules of the SIE system as they contain very important knowledge about the product and links with other similar products. These modules are described below.

Systems represent the knowledge about the relationships between various sub-VEOs at the same level or cross levels like hierarchy and dependability of various components on each other. This provides complete information about the logical relationships among all the components. The information about the components (VEOs) that were used in the past for performing a particular function is also stored in it. The potential alternative VEOs present in other similar products performing the same function/sub-function as well as the possible alternative VEOs that have the potential of replacing the current one are also linked with this module. Systems module is also updated continuously with the alternative VEOs used in advanced and similar new products apart from new inventions, new technological practices, and new advanced materials. From the above discussion, it clear that this module is very crucial for the working of SIE system for supporting innovation process as it represents all the structural knowledge of the product along with the link to similar products and/or similar sub-VEOs/components used in other products. This provides the systematic approach in selecting a most suitable sub-VEOs/components in place of the current one and thus fulfilling the desired objectives leading to efficient product innovation.

Usability represents the knowledge about the uses of a particular component of this product/VEO in other similar products performing the same function. This is very useful for calculating its performance of a component in other products for calculating its specific/overall performance. Other information like which products have stopped using the given component, its recent applications in other products, and the effect of inclusion of this component on the performance, popularity, sales or price of the other products.

Cost represents the knowledge about the cost of the materials, manufacturing, assembling, maintenance and total cost of each component of the product.

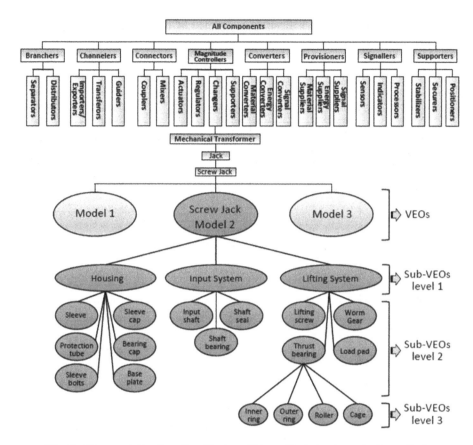

Fig. 1. Representation Functionality-based hierarchical structure of a Screw Jack.

2.2   Implementation of SIE System

Proper implementation of SIE System requires its integration with the Decisional DNA. Main components SOE are Variables, Functions, Constraints and Rules. Moreover, we discussed in Sect. 2.1 that a structure of SIE System include modules like Characteristics, Requirements, Functionality, Connections, Process, Systems, Usability and Cost. For each module of the SIE system, Sets of Experience are created individually for providing more scalable setting, this is similar to the one used for representing a wide range of manufactured products. Sets of experience are generated for each individual modules having specific weightages for the variables of the product. Combination of all the individual Sets of experience are combined under the SIE system that represents complete knowledge and experience necessary for supporting innovation process of manufactured products.

The purpose of the present test case study is to apply our proposed architecture of SIE System to gather, store and reuse information related to the innovation process of manufactured products. Table 1 illustrates a detailed SOE of SIE System structure for screw jack model 2 (see Fig. 1).

Table 1. SOE as a combination of Variables, Functions, Constraints and Rules.

Variables	Functions	Constraints	Rules
sie_Efficiency	sie_Efficiency = sie_WorkDone /	sie_Efficiency ≥ 0	IF sie_Efficiency > 0.5 THEN
sie_ForceApplied	sie_WorkInput	sie_Efficiency ≤ 0.5	sie_Safety = 9
sie_WorkInput			
sie_WorkDone	sie_WorkInput =		
sie_Friction	2 * π *sie_Size_Radius		IF sie_Friction_Bearing > 0.4 THEN
sie_Friction_Screw-Nut	*sie_ForceApplied		sie_Instructions = "Change Lubricant"
sie_Friction_Shaft-Gear			
sie_Friction_Bearing			
sie_Maintenance			IF sie_Maintenance_Bearing = 4 AND
sie_Maintenance_Bearing			sie_Maintenance_Shaft = 2 AND
sie_Maintenance_Shaft			sie_Maintenance_Gear = 3 THEN
sie_Maintenance_Gear			sie_Maintenance = 3
sie_Lubrication			
sie_Lubrication_Type		sie_Lubrication_MaxPr ≤ 60	
sie_Lubrication_MaxPr			
sie_Portability			
sie_Size	sie_Size_Area = 2 * π *sie_Size_Radius		
sie_Size_Area			
sie_Size_Radius			
sie_Portability			
sie_Corrosion			IF sie_Corrosion_Shaft = 4 AND
sie_Corrosion_Shaft			sie_Corrosion_Nut = 5 AND
sie_Corrosion_Nut			sie_Corrosion_Gear = 4 THEN
sie_Corrosion_Gear		sie_Cost_Gear ≤ 0.07 * sie_Cost	sie_Corrosion = 4
sie_Cost			
sie_Cost_Bearing		sie_Cost_Maintenance_Bearing ≤ 0.3 *	
sie_Cost_Gear		sie_Cost_Bearing	
sie_Cost_Maintenance			
sie_Cost_Maintenance_Bearing		sie_Cost_Maintenance ≤ 0.2 * sie_Cost	
sie_Safety			
sie_Instructions			

As the decisional DNA is constructed in JAVA [21] and has been applied successfully in various other fields of application, the structure of the SIE system is implemented using JAVA as a programing language. For illustration purpose the code for sie_Maintenance (see code below) is shown below that is stored as a SOE variable:

```
<set_of_variables>
<!-- Variables included in the model -->
  <variable>
    <var_name>sie_Maintenance</var_name>
    <var_type>NUMERICAL</var_type>
    <var_cvalue>3</var_cvalue>
    <var_evalue>3</var_evalue>
    <unit></unit>
    <internal>true</internal>
    <weight>0.8</weight>
    <l_range>4.0</l_range>
    <u_range>7.0</u_range>
    <priority>0.8</priority>
  </variable>
```

3   Design of a Test Case Study

The algorithm for the working of SIE system is shown in Fig. 2.

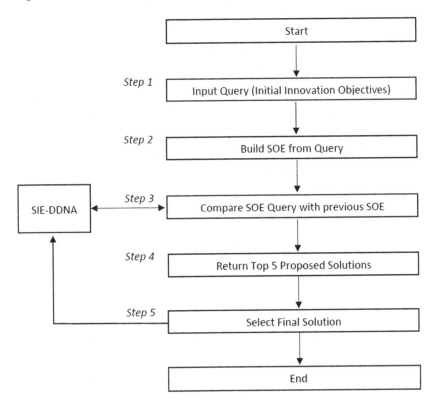

Fig. 2. Algorithm of SIE system.

The description of the working of the SIE system is described below by five main steps (Fig. 2):

Step 1. Input Query (based on Initial innovation Objectives): Innovation process is started by inputting the query into the SIE-DDNA system based on initial innovation objectives. Let us consider for the present case study, the initial innovation objectives to be Low maintenance, More Stability, Ease of Operation and, Portability.

Step 2. Build SOE from Query: The above query based on innovation objectives is converted to a SOE in which the variables, functions, constraints and rules are uniquely combined.

Step 3. Compare SOE Query with previous SOE: this SOE query is compared with the similar previous SOE. An outline of the information inside the SOE is shown in Fig. 3 containing the list of initial innovative objectives, final innovation objectives (leading to innovative changes), changes in the product as per the priorities set by the user and, value of SOE performance factor representing the effective impact of that change on the product innovation.

The input query is then compared with the previous SOE that are ranked according to the common Initial Innovation Objectives and SOE Performance Factor (SPF).

Step 4. List of Proposed Solutions (for example top ten): According to information in *Step 3*, the SIE system finds the Final Innovation Objectives. As shown in Table 2, the final objective 'ease of operation' is decomposed further into finer objectives.

For example 'ease of operation' can be divided into finer objectives viz: easy handling, smooth movement, less friction, fatigue of users and, ambient conditions. It is also shown in Table 2 how the finer objective (less friction) depends on various components of the product.

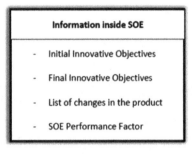

Fig. 3. Typical information outline inside the SOE.

Table 2. Decomposition of Innovation Objectives' and its connection with components of product.

Initial innovation objectives	Final innovation objectives	Technical breakdown of objectives	Connection with parts/ components
More Stability	More Stability		
Low maintenance			
Portability			
Ease of operation	Ease of operation	Smooth movement	
		Easy handling	
		Fatigue of user	
		Ambient conditions	
		Less friction	Friction between screw and nut
			Type of lubrication
			Friction between input shaft and worm gear
			Friction in thrust bearing

Different weightages are assigned to these parts/components that represents friction percentage depending on particular component. In the same way all the Objectives are linked to the respective parts/components with respective weightages.

Alternative components are selected from the *System* module. This selection is based on the value of the performance factor calculated by collecting the information about the particular component (in all the similar products) from the *Usability* module for that particular function/feature. Once the similar analysis is done for all the objectives, list of proposed solutions are presented by the SIE system.

Step 5. Final Solution: Final solution is selected from the list of proposed solutions by the user based on priorities to complete the process of product innovation. This final solution is then stored in the SIE-DDNA that can be used in future.

In this way, the product innovation process of the manufactured products can be enhanced by using the semi-automatic Smart Innovation Engineering System. This will help in increasing the life of the manufactured products.

4 Conclusion

For the survival and prosperity of manufactured products, it has to pass through repeated innovation processes after certain time. The concept of Smart Innovation Engineering is presented that enhances the process of product innovation systematically and semi-automatically. This concept of smart innovation engineering system based on knowledge representation platform stores the experiential knowledge from the formal decisional events and reuses it in future when a similar query is presented. Through this system, the innovation process can be performed at a proper time. The proposed decision support system is dynamic in nature as it updates itself every time a new decision is taken. The concept is illustrated with the example of a Screw Jack as a case study that helps to understand the architecture and the working of the proposed system.

The SIE system will benefit the entrepreneurs and manufacturing organizations at it facilitates product innovation process, thus reducing the extent of dependability on experts. Product innovation process can be performed quickly due to fast computational capabilities of the system. In this way, Smart Innovation Engineering system is one step forward in the direction of automation and Industrie 4.0. The next step will be the refinement of the algorithm in more detail and its translation into JAVA platform.

References

1. Waris, M.M., Sanin, C., Szczerbicki, E.: Smart innovation management in product life cycle. In: Borzemski, L., Grzech, A., Świątek, J., Wilimowska, Z. (eds.) ISAT 2015 – Part I. AISC, vol. 429, pp. 183–192. Springer, Cham (2016). doi:10.1007/978-3-319-28555-9_16
2. Waris, M.M., Sanin, C., Szczerbicki, E.: Framework for product innovation using SOEKS and decisional DNA. In: Nguyen, N.T., Trawiński, B., Fujita, H., Hong, T.-P. (eds.) ACIIDS 2016. LNCS (LNAI), vol. 9621, pp. 480–489. Springer, Heidelberg (2016). doi: 10.1007/978-3-662-49381-6_46

3. Waris, M.M., Sanin, C., Szczerbicki, E.: Toward smart innovation engineering: decisional DNA-based conceptual approach. Cybern. Syst. **47**(1–2), 149–159 (2016)
4. Sanin, C., Szczerbicki, E.: Towards decisional DNA: developing a holistic set of experience knowledge structure. Found. Control Manage. Sci. **9**, 109–122 (2008)
5. Sanin, C., Szczerbicki, E.: Set of experience: a knowledge structure for formal decision events. Found. Control Manage. Sci. **3**, 95–113 (2005)
6. Sanin, C., Szczerbicki, E.: A complete example of set of experience knowledge structure in XML. In: Szuwarzynski, A. (ed.) Knowledge Management: Selected Issues 2005, pp. 99–112. Gdansk University Press, Gdansk (2005)
7. Shafiq, S.I., Sanin, C., Szczerbicki, E.: Set of experience knowledge structure (SOEKS) and decisional DNA (DDNA): past, present and future. Cybern. Syst. **45**(2), 200–215 (2014)
8. O'Sullivan, D., Dooley, L.: Applying Innovation. Sage Publications, London (2008)
9. Frishammar, J.: Managing information in new product development: a literature review. Int. J. Innov. Technol. Manage. **2**(3), 259–275 (2005)
10. Smith, P.G., Reinertsen, D.G.: Developing Products in Half the Time, 2nd edn. Wiley, New York (1998)
11. Capello, R.: Knowledge, innovation, and regional performance: toward smart innovation policies introductory remark to the special issue. Growth Change **44**(2), 185–194 (2013)
12. Rampino, L.: The innovation pyramid: a categorization of the innovation phenomenon in the product-design field. Int. J. Des. **5**(1), 3–16 (2011)
13. Herstatt, C., von Hippel, E.: From experience: developing new product concepts via the lead user method: a case study in a "low-tech" field. J. Prod. Innov. Manage. **9**(3), 213–221 (1992)
14. Dolan, R.J., Matthews, J.M.: Maximizing the utility of customer product testing: beta test design and management. J. Prod. Innov. Manage. **10**(4), 318–330 (1993)
15. Ciccantelli, S., Magidson, J.: From experience: consumer idealized design: involving consumers in the product development process. J. Prod. Innov. Manage. **10**(4), 341–347 (1993)
16. Piller, F.T., Walcher, D.: Toolkits for idea competitions: a novel method to integrate users in new product development. R&D Manage. **36**(3), 307–318 (2006)
17. Bryant, C.R., Stone, R.B., Greer, J.L., McAdams, D.A., Kurtoglu, T., Campbell, M.I.: A function-based component ontology for systems design. In: International Conference on Engineering Design, Paris, France, 28–31 August 2007
18. Kurtoglu, T., Campbell, M.I., Bryant, C.R., Stone, R.B., McAdams, D.A.: Deriving a component basis for computational functional synthesis. In: Proceedings of the International Conference on Engineering Design, Melbourne, Australia, 15–18 August 2005
19. Shafiq, S.I., Sanin, C., Szczerbicki, E., Toro, C.: Virtual engineering objects/virtual engineering processes: a specialized form of cyber physical systems and industry 4.0. Procedia Comput. Sci. **60**, 1146–1155 (2015)
20. Shafiq, S.I., Sanin, C., Toro, C., Szczerbicki, E.: Virtual engineering (VEO): towards experience-based design and manufacturing for industry 4.0. Cybern. Syst. **46**, 35–50 (2015)
21. Sanin, C., Szczerbicki, E.: Towards the construction of decisional DNA: a set of experience knowledge structure Java class within an ontology system. Cybern. Syst. **38**, 859–878 (2007)

Analyses of Aspects of Knowledge Diffusion Based on the Example of the Green Supply Chain

Anna Maryniak[1] and Łukasz Strąk[2(✉)]

[1] Poznań University of Economics and Business, Al. Niepodległości 109,
61-875 Poznań, Poland
anna.maryniak@ue.poznan.pl
[2] University of Silesia, Będzińska 39, 41-205 Sosnowiec, Poland
lukasz.strak@us.edu.pl

Abstract. The diffusion of knowledge in the area of green supply chain (GSC) is connected with the outcomes obtained as a result. This article concentrates on the issues of knowledge diffusion and is based on own research of manufacturing companies on the example of the GSC. As a result of the research, it was established that exchanging knowledge between a supplier and a manufacturer has a higher measurement than between a manufacturer and a receiver, while, generally speaking, a low level of knowledge exchange is observed within the area of GSC. Companies need a wider access to sources of knowledge in the field of GSC and need the development of the culture of measurement of the executed undertakings. The knowledge diffusion in the area of GSC is connected with outcomes obtained as a result.

Keywords: Green supply chain · Sustainable supply chain · Information sharing · Knowledge diffusion

1 Introduction

The theories of diffusion concentrate around the assessment of transfer of some resource - knowledge, ideas, skills, innovations and technology in companies. It is an interdisciplinary concept that constitutes an object of interest of IT specialists, sociologists and economists, from a great many points of view. The process of diffusion can be compared to the phenomenon of an epidemic, which is often used while creating the theoretical model of this phenomenon [9]. The elementary terminology comes from E.M. Rogers and E.F. Shoemaker [19]. The theory includes such issues as: communication channel, responsible for the transfer of resources (e.g. knowledge), adaptation, responsible for implementation (acquisition), source, responsible for the creation of resources, and a network of connection between the examined objects. Diffusion is a subject of theoretical research connected with mathematic modeling of this phenomenon [7,11] and practical research as well [21,23]. In the first case scientists are attempting to reconstruct

© Springer International Publishing AG 2017
N.T. Nguyen et al. (Eds.): ACIIDS 2017, Part I, LNAI 10191, pp. 335–344, 2017.
DOI: 10.1007/978-3-319-54472-4_32

the real flow of information in a mathematical model (using various topologies of connections between objects). In the second case we are witnessing an examination of diffusion in real objects, together with an identification of barriers that limit it, as well as an examination of the intensity of resource flow. This article concerns the second approach, with a particular emphasis on the knowledge diffusion in a green supply chain. The selection of exemplifications of the research subject is motivated by both the current knowledge in the field of knowledge management, as well as the trends connected with the development of supply chains.

The wide array of primary sources, concerning knowledge management in a supply chain, justifies an assumption, that there is shortage of research within the area of these issues, especially within the scope of the influence of this process on the performance of horizontally integrated companies [13]. Since the supply chains evolve from classic chains, not only toward lean, resilient and agile supply chains, but also more and more often toward green supply chains, [4,18,26] attempts were undertaken to face this still not exhausted area of research, which is still in its primary phase.

Currently it is more and more frequently acknowledged that green supply chain management is one of the elementary pillars of achieving and maintaining competitive advantage. On the basis of a great many research results, it is argued that the results connected with its implementation are favorable for both the internal and external stakeholders [1,16,25,27]. On the other hand, what draws attention is a low level of awareness in companies amid the links jointly creating the supply chains, within the scope of green management and a low level of pro-environmental practices [2,12]. Therefore, an assumption was adopted, that the current situation is connected with a low level of knowledge diffusion within the scope of GSCM. In connection with the above, what seems just is to examine the level of knowledge diffusion within the area of these issues and indicate any troublesome elements connected with it.

It was assumed in the study that the green chain is a chain in which traditional principles connected with a supply chain management are used, however, particular attention is given to the environmental aspects and recycling [24]. In the study it was assumed that the green supply chain management consists in: designing products and managing their flow up and down the supply chain by entities that participate in it, taking particularly into consideration the needs of environment protection [14]. The range of the possible activities within this area is unusually wide. For instance it can concern environmentally conscious design (ECD), conducting life-cycle assessment (LCA), following environmental directives (like Restriction of Hazardous Substances - RoHS, Waste of Electrical and Electronic Equipment - WEEE), introducing Leadership in Energy and Environmental Design (LEED) certificates, using materials suitable for reusing, recycling and reassembling (3R – reuse, recycle, reassemble), undertaking joint activities between a supplier and a manufacturer, a manufacturer and a receiver, in order to increase the environmental efficiency. Despite having many programs, business and external initiatives (like Coalition for Environmentally Responsible

Economies, Global Reporting Initiative), as well as permanent promotion of green practices on international platforms (like South Asian Association for Regional Cooperation, World Health Organization), there is still a shortage of knowledge and exemplars, as well as research concerning environmental aspects at the level of links and the supply chain, in particular, as it was already mentioned, there is a lack of studies that would connect the concept of knowledge diffusion and green supply chains.

2   Literature Review

In order to define the current stage of research, concerning knowledge diffusion in the green supply chain, bibliometric studies were carried out. In this study the following retrieval query was applied: TITLE-ABS-KEY ("sustainab* supply chain*" OR "environmental* supply chain*" OR "green supply chain*") AND ("information sharing" OR "knowledge"). As a result of the query, it was established that there are few studies concerning the subject matter of our contemplations. In total 143 records were selected. Taking into consideration years of publishing it can be established that from year to year the interest in this subject is increasing[1]. On the basis of bibliometric studies, it can be established that the discussed subject matter includes a great many areas. These are, among others: Business, Management and Accounting Engineering, Decision Sciences, Environmental Science, Computer Science. The most frequently quoted authors within the field of the discussed subject matter are: [6,8,10,20,22]. The studies are of a theoretical and conceptual, as well as empirical character. They are varied and it is difficult to reduce them to a common denominator.

Hall and Matos [8] argue, based on the example of the biofuel business branch, that larger knowledge diffusion within the field of pro-environmental activities, especially aimed at the initial supply chain links represented by farmers, is essential. In turn, Kim et al. [10] established, on the basis of research results, that better results, connected with the implementation of green activities, are obtained owing to trust and knowledge sharing in a supply chain. Meanwhile, Solér et al. [22] believe that environmental information is perceived and used differently by purchasers in the supply chain depending on where (in the supply chain) they are situated in relation to other chain actors. Sarkis et al. [20] conducted a survey of literature, which addresses the issues of pro-environmental diffusion practices, and outlined future research directions. Baresel et al. [3] established that while good management of green knowledge can lead to added value, the definition and implementation of relevant measure items and targets seems indispensable for the improvement of green knowledge quality and seems to be an important driver for implementation of green practices. In turn, Cheng et al. [6] examined how trust interacts with factors affecting interorganizational knowledge sharing in green supply chains, where cooperation and competition coexist. Nani et al. [17] present guidance for practical managers in evaluating and measuring green

---

[1] Owing to the volume of this article we resigned from a detailed presentation of data.

practices by developing knowledge based on the balanced scorecard and evolutionary game theory. The findings implemented by Cheng [5] provide useful insights into how green supply chain actors should reinforce their relational benefits and guanxi activities that would improve their value-based relationships, in order to enhance the environmental knowledge sharing for the green supply chain as a whole. Referring to precise conclusions, together with a register of the remaining items generated as part of the bibliometric tests, can constitute an interesting area of separate contemplation.

On the basis of a review of primary sources, it was established that there is not much research and relevant contemplation dedicated to the issue of knowledge diffusion in an environmentally sustainable supply chain, and in particular there is a lack of research based on the data obtained from Polish entities. For this reason, as well as the previously mentioned arguments, research was designed and the following questions were asked:

1. Is the knowledge exchange at both ends of the chain even?
2. Is the GSCM activity measured and what is the access to the sources of knowledge regarding GSCM?
3. Is there a connection between the level of knowledge transfer and the results obtained through the implementation of pro-environmental activities?

3   Research

3.1   Research Methodology

The research results presented in the study are a component part of a broader research project, which concerns GSCM activity, motivators, barriers and implementation results. The participants of the research were medium and large manufacturing companies located in Poland, which were selected on the basis of the Polish Central Statistical Office database. The research was conducted on the basis of an original questionnaire. Since a comparative analysis with other research results cannot be performed. A five-point Likert scale or a single choice answers were used in the questionnaire. 1087 companies participated in the statistical draw. This research was conducted among 73 companies. Author's previous research experience gave an assumption to acknowledge that the quality of data and the number of obtained questionnaires is much higher when the questionnaires are distributed personally, rather than by electronic mail. Therefore, in order to have access to the companies that were qualified for the research, all selected companies were located in the region of Wielkopolska. Therefore, the results are representative of only this region. In the future conducting research on a broader group of respondents seems purposeful. Hence, other ways of obtaining data, which are of an intermediary character (for instance CATI - Computer Assisted Telephone Interview, or CAWI - Computer Assisted Web Interviews), were not applied. The questionnaire was, therefore, the main method of examination and the primary source of empirical materials. Other source materials, including statistical instruments, do not include the data essential for conducting the assumed analyses. In the future it is worth to apply focus tests in order to deepen the current research and grasp behavioral aspects[2]. Simultaneously, the research constitutes a deepening of the already conducted research in Poland, which concerned the knowledge diffusion in a sustainable supply chain [15][3].

3.2 Information Availability

The aim of the first group of questions was to evaluate to what extent companies exchange information about the pro-environmental activities in a supply chain[4]. As a result of the conducted research, it was established that companies retain majority of information inside the organization. What is characteristic, a significant range of information is retained at the executive level. Only one fourth of the questioned companies transfer information to suppliers and one fifth transfer information to receivers, where receivers are, first and foremost, production and trading companies (Fig. 1). At the same time the companies declared that half of the entities on the supply side and one-forth on the receiver's side have access to the information in question (Fig. 2). Taking into consideration the flow up and down the supply chain we can state that only a dozen or so percent companies have a full access to information from both ends of the supply chain. What is characteristic for 41% of the examined companies is a one direction flow of information on the side of purchasing (to or from the supplier), and for 23% of companies a single-sided flow of information on the side of sale (to or

Fig. 1. Access to announcements concerning GSCM in the examined companies for the participants of the organization, supply chain and surrounding

---

[2] Statistical inference will be applied with a view to analyzing the data in a more in-depth manner. Especially data mining and machine learning methods.

[3] This research was conducted among 103 Polish manufacturing companies. The research was conducted on the basis of a questionnaire.

[4] Transferring announcement/ reports connected with GSCM concerns non-formalized information (placed on the company websites in various documents), as well as formalized information (preparing reports based on the Global Reporting Initiative guidelines, ISO 26000, International Integrated Reporting Council reports and all other type reports about the influence on environment, both voluntary and required by the law as well).

Fig. 2. Examined companies' access to announcements concerning GSCM of the participants of the supply chain and surrounding

from receivers) (Fig. 3). Particularly noticeable is a low level of knowledge flow between a manufacturer and a receiver, since in as many as 66% of companies even a one direction transfer of information was not observed.

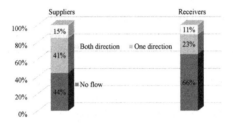

Fig. 3. The exchange of information, concerning GSCM, between a supplier and a manufacturer (entrance), as well as a manufacturer and a receiver "exit"

3.3 System Activity Measures and Sources of Knowledge About GSCM

During the research the respondents were asked to provide information whether the undertaken pro-environmental activities and the results connected with them are measured. It turned out that only 26% of respondents undertake measurement (Fig. 4). Therefore, we can say that a majority of companies undertake ad hoc activities, without a systemic approach and without knowledge about any influence on costs and intangible results. So probably the activities are not conducted in a professional and strategic manner.

As part of the research it was diagnosed whether any branch reports undertaking GSCM aspects are available. Among the companies which suggested an availability of branch reports, concerning environment protection in a supply chain, only 5% stated that the reports are readily available, since many companies prepare reports. The remaining 25% indicated toward an availability only in a limited scope (Fig. 5). Among the entities which suggested that there are no reports, one in ten believes that they will appear in the future. Suggestion "it's difficult to say" and "we do not have a need to report" are the worst possible answers. The fact that as many as 45% of respondents could not make an

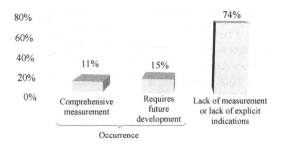

Fig. 4. Occurrence of a system of activity development and result measurement within the scope of environment protection in the supply chain

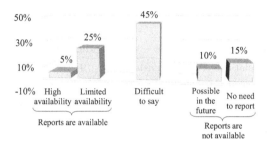

Fig. 5. Availability of branch reports concerning pro-environmental activities in a supply chain

unambiguous comment on the posed question, proves a lack of awareness and knowledge concerning pro-environmental balance in a supply chain.

3.4 Level of Knowledge Diffusion and Results Obtained from GSCM Implementation

During another analysis the companies were divided into three goups that reflected the level of diffusion within the area of GSCM:[5]

1. The highest level of diffusion - companies in this group report GSCM activity and simultaneously acquire external knowledge concerning this subject - 30.1% of all examined companies,
2. Medium level of diffusion - companies in this group report or acquire external knowledge (they do not belong to group 1) - 42.5% of all examined companies,
3. Low or zero level of diffusion - companies in this group neither report nor acquire external knowledge - 27.4% of all examined companies.

---

[5] Level of diffusion in this study is measured by the condition of reporting and acquiring information from the outside. However, it is important to highlight that in the future matrices allowing for different measures can be successfully applied.

Table 1 presents groups that reflect the level of knowledge diffusion advancement with the obtained results as a result of GSCM implementation[6]. The most frequently observed effect for the highest diffusion level group is a "reduction of hazardous substances usage". The achieved result means that every company in this group noticed this effect. To a lesser extent the observed effects were: improvement of image, reduction of CO2 emission and reduction of soil contamination (91%). The companies in the remaining two groups do not feel the effects of GSCM implementation to such an extent as the group of companies classified as the highest diffusion level group.

Table 1. Selected GSCM implementation results and the level of diffusion within the area of pro-environmental activities

Result	GSCM level of diffusion		
	Highest	Medium	Low or zero
Improvement of relations with end customers (companies, institutions, consumers)	91%	58%	20%
Increase in market share	45%	35%	5%
Decrease in the usage of hazardous materials	100%	84%	60%
Higher innovativeness of business processes	86%	52%	15%
Limitation of materials usage	82%	81%	45%
Introduction of environment protection policy	82%	81%	35%
Reduction of CO2 emission to environment	91%	71%	60%
Reduction of soil contamination	91%	65%	45%
Reduction of manufacturing costs	23%	19%	15%
Improved effectiveness indicators regarding logistics processes	32%	32%	5%

4   Discussion

On the basis of the conducted research we can establish that a much bigger difference in the level of transferred information, on both ends of the supply chain, is noticeable in case of information obtained from the outside. The fact that as much as half of the companies have access to the suppliers' knowledge probably stems from the arrangement of tender effectiveness in the supply chain. The research demonstrates that the entities transfer information outside to a larger extent to receivers than to suppliers. However, the percentage difference

---

[6] Every answer in the questionnaire could adopt the following value: 1- undoubtedly yes, 2 – yes, 3 difficult to say, 4 no, 5 absolutely no. In case when the companies answered 1 or 2, it was assumed that the effect of implementing GSCM included in the question was achieved. The results presented in Table 1 were rescaled to match the count of the group.

is significantly lower. Therefore, we can establish that in the supply chain we notice a higher balance in the area of knowledge flow in case of outgoing information. Thus it can be concluded that companies are not precisely informed what requirements the market sets. Therefore, in the future the role of receivers in creating green flow should be increased.

Insignificant percentage of companies characterized by a two-way flow of knowledge allows an assumption that the possibility of working out mutual solutions regarding GSCM, with such a level of information flow, is little. Therefore, there is an urgent need to increase the level of information flow, which would constitute a sine qua non condition for the development of pro-environmental cooperation in the supply chain. It seems important since it was established that knowledge diffusion in the area of GSCM is related to the obtained results.

Better discipline concerning reporting and measurement of the pro-environmental undertakings in progress would facilitate cooperation in a more systemic way.

5   Conclusions

The group with the greatest diffusion of knowledge seen the most positive impact of the GSCM like "decrease in the usage of hazardous materials" or "improvement of relations with end customers". As a result of the undertaken work, it was established, among others, that the level of flow of the information mentioned is low. The GSCM knowledge in companies is chiefly preserved at the executive level. Therefore, in the future it is advisable to develop research toward this direction, diagnose the reason for this situation and indicate corrective directions.

References

1. Azevedo, S.G., Carvalho, H., Machado, V.C.: The influence of green practices on supply chain performance: a case study approach. Transp. Res. Part E: Logist. Transp. Rev. **47**(6), 850–871 (2011). n/a
2. Balaji, M., Velmurugan, V., Prasath, M.: Barriers in green supply chain management: an indian foundry perspective. Int. J. Res. Eng. Technol. **3**(7), 423–429 (2014)
3. Baresel-Bofinger, A.C.R., Ketikidis, P.H., Lenny Koh, S.C., Cullen, J.: Role of 'green knowledge' in the environmental transformation of the supply chain: the case of greek manufacturing. Int. J. Knowl. Based Dev. **2**(1), 107–128 (2011)
4. Carvalho, H., Duarte, S., Machado, V.C.: Lean, agile, resilient and green: divergencies and synergies. Int. J. Lean Six Sigma **2**(2), 151–179 (2011)
5. Cheng, J.-H.: Inter-organizational relationships and knowledge sharing in green supply chains-moderating by relational benefits and guanxi. Transp. Res. Part E: Logist. Transp. Rev. **47**(6), 837–849 (2011)
6. Cheng, J.-H., Yeh, C.-H., Chia-Wen, T.: Trust and knowledge sharing in green supply chains. Supply Chain Manage. Int. J. **13**(4), 283–295 (2008)
7. Cointet, J.P., Roth, C.: How realistic should knowledge diffusion models be? J. Artif. Soc. Soc. Simul. **10**(3), 5 (2007)

8. Hall, J., Matos, S.: Incorporating impoverished communities in sustainable supply chains. Int. J. Phys. Distrib. Logist. Manage. **40**(1/2), 124–147 (2010)
9. Imrišková, E., Mravcová, Z.: Diffusion processes in the knowledge economy in terms of theoretical approaches. In: CERS 3rd, Central European Conference in Regional Science, Košice, Slovak Republic (2009)
10. Kim, J.H., Youn, S., Roh, J.J.: Green supply chain management orientation and firm performance: evidence from South Korea. Int. J. Serv. Oper. Manage. **8**(3), 283–304 (2011)
11. Klarl, T.: Knowledge diffusion and knowledge transfer revisited: two sides of the medal. J. Evolu. Econ. **24**(4), 737–760 (2014)
12. Lee, S.-Y.: Drivers for the participation of small and medium-sized suppliers in green supply chain initiatives. Supply Chain Manage. Int. J. **13**(3), 185–198 (2008)
13. Marra, M., Ho, W., Edwards, J.S.: Supply chain knowledge management: a literature review. Expert Syst. Appl. **39**(5), 6103–6110 (2012)
14. Maryniak, A.: Zielony łańcuch dostaw. Wydawnictwo Uniwersytetu Ekonomicznego w Poznaniu, Poznań (2017, to appear)
15. Maryniak, A., Stefańska, M.: Diffusion of knowledge about sustainable supply chain. Int. J. Arts Sci. **8**(4), 173–184 (2015)
16. Muma, B.O., Nyaoga, R.B., Matwere, R.B., Nyambega, E.: Green supply chain management and environmental performance among tea processing firms in kericho county-kenya. Int. J. Econ. Finan. Manage. Sci. **2**(5), 270–276 (2014)
17. Naini, S.G.J., Aliahmadi, A.R., Jafari-Eskandari, M.: Designing a mixed performance measurement system for environmental supply chain management using evolutionary game theory and balanced scorecard: a case study of an auto industry supply chain. Resour. Conserv. Recycl. **55**(6), 593–603 (2011)
18. Nelson, D., Marsillac, E., Rao, S.: Antecedents and evolution of the green supply chain. J. Oper. Supply Chain Manage. **1**(1), 29–43 (2012)
19. Rogers, E.M., Shoemaker, F.F.: Communication of innovations; a cross-cultural approach (1971)
20. Sarkis, J., Zhu, Q., Lai, K.: An organizational theoretic review of green supply chain management literature. Int. J. Prod. Econ. **130**(1), 1–15 (2011)
21. Scarbrough, H., Swan, J.: Explaining the diffusion of knowledge management: the role of fashion. Brit. J. Manage. **12**(1), 3–12 (2001)
22. Solér, C., Bergström, K., Shanahan, H.: Green supply chains and the missing link between environmental information and practice. Bus. Strat. Environ. **19**(1), 14–25 (2010)
23. Sorenson, O., Fleming, L.: Science and the diffusion of knowledge. Res. Policy **33**(10), 1615–1634 (2004)
24. Tundys, B.: Zielony łańcuch dostaw w gospodarce o okrężnym obiegu-założenia, relacje, implikacje. In: Research Papers of the Wroclaw University of Economics/Prace Naukowe Uniwersytetu Ekonomicznego we Wroclawiu (383) (2015)
25. Vachon, S., Klassen, R.D.: Environmental management and manufacturing performance: the role of collaboration in the supply chain. Int. J. Prod. Econ. **111**(2), 299–315 (2008)
26. Wei, B.S., Solvang, D., Deng, Z.: A closed-loop supply chain model for managing overall optimization of eco-efficiency. In: POMS 18th, Annual Conference Dallas, Texas, USA (2007)
27. Zhu, Q., Sarkis, J.: The moderating effects of institutional pressures on emergent green supply chain practices and performance. Int. J. Prod. Res. **45**(18–19), 4333–4355 (2007)

Comparative Evaluation of Bluetooth and Wi-Fi Direct for Tablet-Oriented Educational Applications

Keiichi Endo[✉], Ayame Onoyama, Dai Okano, Yoshinobu Higami,
and Shinya Kobayashi

Graduate School of Science and Engineering,
Ehime University, Matsuyama 790-8577, Japan
endo@cs.ehime-u.ac.jp

Abstract. This study conducted a survey to implement educational applications that can share information even in environments where access points cannot be used. In particular, we investigated whether Bluetooth (widely used for many years) or Wi-Fi Direct (developed recently) is more suitable when creating educational applications using an ad hoc network. To survey the influence of hand movements on delay time while operating tablets, we created a paint application that shares a drawing screen across two tablets and conducted an experiment. In addition, to survey the influence of human presence on delay time, we conducted an experiment in which we changed the number of students seated between the two tablets in the classroom. From the results of these experiments, we conclude that Bluetooth is less influenced by hand movements and human presence than Wi-Fi Direct.

Keywords: Ad hoc network · Education · Tablet · RTT

1   Introduction

Recently, lessons that use tablets have become popular in elementary and middle school education [1–4]. When it is necessary to exchange information among tablets in group learning, Wi-Fi access points are usually used for the communication. However, since access point installation and management entails considerable personnel and monetary costs, it is difficult to establish a wireless communication environment that enables Wi-Fi communication via access points in every classroom of each school.

Therefore, we are considering the use of ad hoc networks. Ad hoc networks are composed of devices with wireless communication capabilities, such as laptops, tablets, and smartphones [5]. Each device communicates directly with other devices. With ad hoc network use, it is possible to share information even when infrastructures such as access points are unavailable; this is why ad hoc networks have been attracting attention in recent years as a means of communication during large-scale disasters [6–9].

© Springer International Publishing AG 2017
N.T. Nguyen et al. (Eds.): ACIIDS 2017, Part I, LNAI 10191, pp. 345–354, 2017.
DOI: 10.1007/978-3-319-54472-4_33

Bluetooth is the typical communication standard for performing direct wireless communication between devices such as tablets. Recently, Wi-Fi Direct was developed and the number of devices that support Wi-Fi Direct has been increasing. In this study, we assumed scenarios of education utilizing ad hoc networks in classrooms and performed a comparative evaluation of the two communication standards mentioned above. In particular, we surveyed the influence of hand movements on delay times while using a paint application sharing a drawing screen between two tablets. In this survey, we not only used Bluetooth and Wi-Fi Direct but also communicated via Wi-Fi access points and conducted experiments by sending data to other tablets through a server. Furthermore, we investigated the influence of human presence on delay times by conducting a wireless communication experiment where we changed the number of students seated between two tablets in the classroom.

The rest of this paper is composed of the following sections. First, in Sect. 2, we describe the communication standards used in the survey. In Sect. 3, we describe the survey on the influence of hand movements on delay times while operating tablets. In Sect. 4, we describe the survey on the influence of human presence on delay times. Finally, in Sect. 5, we summarize the study and describe future challenges and implications.

2   Communication Standards

In this section, we explain the communication standards used in the survey – Wi-Fi, Wi-Fi Direct, and Bluetooth.

2.1   Wi-Fi

Wi-Fi is the name of interconnection among devices using the IEEE 802.11 standard recognized by the Wi-Fi Alliance [10]. In Wi-Fi connection, communication can be encrypted using a security protocol such as Wi-Fi Protected Access 2 (WPA2), and security is ensured. At the end of IEEE 802.11, one or more alphabets are added to distinguish the frequency and communication speed. The main ones are IEEE 802.11a, 11g, and 11ac.

2.2   Wi-Fi Direct

Wi-Fi Direct is the name of the standard that allows direct wireless connection among Wi-Fi devices [11]. The Wi-Fi Alliance established this standard in 2010, and it performs certification of Wi-Fi Direct-enabled devices. Wi-Fi Direct-enabled devices can connect with Wi-Fi-enabled devices. The connection is secured through WPA2. The maximum communication speed is 250 Mbps, and connections are possible even at distances of 90 m.

2.3 Bluetooth

Bluetooth is a short-range wireless communication technology that uses IEEE 802.15.1 standard certified by Bluetooth Special Interest Group (Bluetooth SIG) [12]. The 2.4 GHz band is divided into 79 frequency channels, and frequency hopping is performed to change the frequency used randomly. The maximum possible distance for communication is up to 10 m in Class 2 and greater than 10 m in Class 1. Encryption is performed with 128-bit Advanced Encryption Standard (AES). Compared to Wi-Fi, Bluetooth's power consumption is low [13]. It is possible to connect up to seven devices to one device directly.

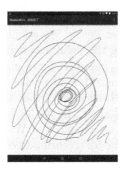

Fig. 1. Examples of wave and spiral drawings.

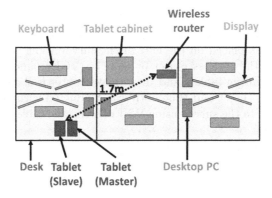

Fig. 2. Environment for survey on influence of hand movements.

3 Survey on the Influence of Hand Movements

In this section, we describe the survey on the influence of hand movements on delay times while operating tablets. After explaining the metrics used in the survey, survey methods, and survey environment, we describe and discuss the results.

3.1   Survey Overview

In the survey, we used two Nexus 9, Android-powered tablets. We created an application with a drawing screen shared between two tablets to be used during group work at school. Round Trip Time (RTT) was surveyed based on the return response when one tablet received a packet.

In this section, RTT refers to the time between when a packet (maximum payload of 28 bytes) containing touch coordinate data is sent in a frequency of 60 times per second from one tablet (Tablet A) to another tablet (Tablet B) and when an immediate return response sent from Tablet B is received by Tablet A. The RTT is measured in the Application Layer of the Open Systems Interconnection (OSI) model. Therefore, the RTT includes retransmission delay and processing time in the lower layers. We adopted this measurement method because we aim at investigating the delay recognized by the students using the application. The following two types of RTT are surveyed: "all-RTT" is measured when Tablet B returns a response for all packets received, and "last-RTT" is measured when a response is returned only when the touching finger is released.

Three types of lines are drawn in the all-RTT survey: Hold (the hand does not move), Wave, and Spiral. A screenshot of Wave and Spiral lines are shown in Fig. 1. One line was drawn for 5 s (coordinate data were sent 300 times), and each type of line was drawn five times. In the last-RTT survey, short lines of approximately 3 cm were drawn 100 times.

When connecting between tablets via Bluetooth or Wi-Fi Direct, the tablet sending the connection request is referred to as Master (M), and the tablet receiving the request is Slave (S). The survey was conducted in Building 5, Room 9-7 of the Faculty of Engineering of Ehime University. Two tablets were placed adjacent to each other. The Master tablet was placed on the right, while the Slave tablet was placed on the left. In the survey using Wi-Fi, the tablets were connected to the wireless router by 802.11a (5 GHz band). The positional relationship between the tablets and the wireless router is shown in Fig. 2. The survey using Wi-Fi Direct and Bluetooth was conducted without connection to the wireless router. In the survey using Wi-Fi Direct, the tablets were connected via a 2.4 GHz band.

There are some types of Bluetooth, such as High Speed (HS) and Low Energy (LE). In this study, we used Basic Rate (BR), which is the earliest type of Bluetooth. The Nexus 9 tablets used in the survey were Bluetooth Class 1 enabled (output 4.05 mW).

3.2   Results and Discussion

In this section, we present and discuss the results of the surveys conducted in real environments with the RTT cumulative probability distribution.

The all-RTT cumulative probability distributions for Wi-Fi, Wi-Fi Direct, and Bluetooth are shown in Figs. 3, 4, and 5, respectively.

The difference in RTT among drawing operations is particularly noticeable in Wi-Fi (Fig. 3). When the hand does not move (Hold), there is a 90% probability

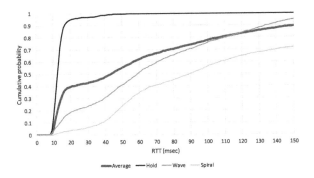

Fig. 3. All-RTT in Wi-Fi (comparison based on drawing operations).

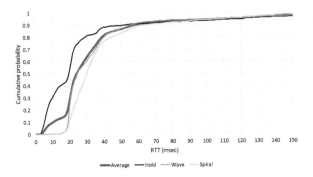

Fig. 4. All-RTT in Wi-Fi Direct (comparison based on drawing operations).

that the RTT is shorter than 20 msec. However, when the hand moves (Wave and Spiral), the RTT often exceeds 150 msec. Thus, it can be said that with Wi-Fi, hand movements influence communication greatly.

With Wi-Fi Direct (Fig. 4), the difference in RTT among drawing operations is smaller compared with Wi-Fi. One reason for the less influence of hand movements in the case of Wi-Fi Direct is thought to be that communication is performed between adjacent tablets unlike in the case of Wi-Fi where the tablets communicate via a wireless router. Another reason could be that, since communication is performed using a 2.4 GHz band in Wi-Fi Direct, it is less susceptible to obstruction influences compared with Wi-Fi communication using a 5 GHz band.

With Bluetooth (Fig. 5), the difference in RTT among drawing operations is even smaller than with Wi-Fi Direct. Bluetooth shows the smallest influence of hand movements among the three standards used in this study's surveys. Bluetooth uses Frequency Hopping Spread Spectrum (FHSS), and since communication is performed while switching frequencies between 2402 to 2480 MHz, it is unlikely for multipath fading to occur. Although Orthogonal Frequency-Division Multiplexing (OFDM) used in Wi-Fi and Wi-Fi Direct is also known to be resistant to multipath fading, it is considered to be more susceptible to multipath fading compared to FHSS used in Bluetooth because the channel

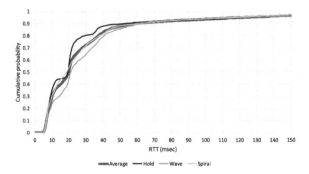

Fig. 5. All-RTT in Bluetooth (comparison based on drawing operations).

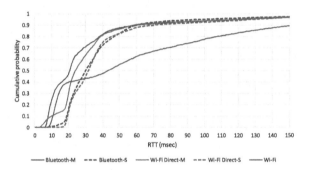

Fig. 6. All-RTT for all drawing operations (comparison based on communication standards).

widths are narrower. In addition, it is possible that the above-described RTT differences occurred due to differences such as header and trailer sizes associated with the data, error correction method, and retransmission protocol in Wi-Fi Direct and Bluetooth.

Figure 6 demonstrates the all-RTT cumulative probability distribution for sending from Master via Bluetooth (Bluetooth-M), sending from Slave via Bluetooth (Bluetooth-S), sending from Master via Wi-Fi Direct (Wi-Fi Direct-M), sending from Slave via Wi-Fi Direct (Wi-Fi Direct-S), and using Wi-Fi (Wi-Fi). With Bluetooth and Wi-Fi Direct, there is an 80% probability that the RTT is shorter than 50 msec. With Wi-Fi, due to the previously described influence of hand movements, the probability that the RTT is shorter than 50 msec is no higher than approximately 55%.

With Bluetooth, RTT when sending from Master (Bluetooth-M) is approximately 10 msec shorter than when sending from Slave (Bluetooth-S). This is considered to be due to the fact that data can be sent at arbitrary timing when sending from Master, while Slave must wait for polling by Master to send data.

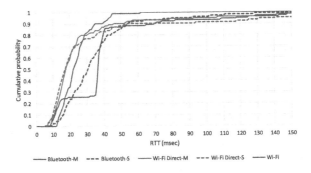

Fig. 7. Last-RTT for all drawing operations (comparison based on communication standards).

There are more instances of smaller RTT with Bluetooth than with Wi-Fi Direct when sending from Master, while the distribution is mostly the same when sending from Slave.

Figure 7 shows the last-RTT cumulative probability distribution. With all the communication standards, there is an over 80% probability that the RTT is shorter than 50 msec.

With Bluetooth, a significant difference between sending from Master (Bluetooth-M) and sending from Slave (Bluetooth-S) is observed in last-RTT cumulative probability distribution. This is believed to be because for Bluetooth, if there is no data to be sent from the Slave, the Slave shifts to a low-power consumption mode called Sniff mode, and it cannot receive data from Master immediately.

4   Survey on the Influence of Human Presence

In this section, we describe the survey on the influence of human presence on delay times.

4.1   Survey Overview

This survey was conducted in Building 5, Room E591 of the Faculty of Engineering of Ehime University. As indicated in Figs. 8 and 9, we conducted an experiment changing the number of students seated between two tablets (Nexus 9) from 9, 5, 3, to 0. Figure 10 shows the scene of the experiment with 9 students.

In the experiment, the two tablets were wirelessly connected using either Bluetooth or Wi-Fi Direct. Then, packets (with 1 byte payloads) to measure RTT were sent from Slave to Master at 200 msec intervals. The experiment was conducted by sending the packet 300 times for each condition while changing the conditions (communication standard or number of students). RTT was measured in the Application Layer of the OSI model 900 times in total for each condition, and we calculated the averages and standard deviations. Rather than measuring

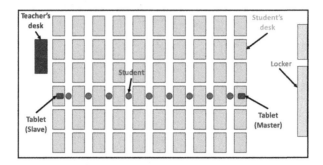

Fig. 8. Environment for survey on influence of human presence (9 students).

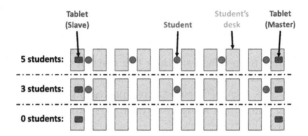

Fig. 9. Environment for survey on influence of human presence (5, 3, 0 students).

the 900 times at a time, the conditions were changed every 300 times to suppress the influence of interference due to radio waves unrelated to the experiment.

4.2 Results and Discussion

The result of the experiment is shown in Fig. 11. In this figure, "Ave." represents RTT average, and "SD" represents RTT standard deviation.

For communication via Wi-Fi Direct, it is observed that the human body as an obstacle influences communication because the RTT average and RTT standard deviation grew bigger by increasing the number of students seated between the two tablets. In contrast, with communication via Bluetooth, we find hardly any influence of the human presence.

Furthermore, regardless of the number of students, when using Bluetooth, RTT average and standard deviation were smaller than when using Wi-Fi Direct.

Based on the above results, we conclude that Bluetooth has shorter delay times and is less susceptible to influence from human presence when sending small packets. Therefore, when creating educational applications for students in a classroom that can share some data in real time, Bluetooth is more suitable than Wi-Fi Direct.

Fig. 10. Scene of survey on influence of human presence (9 students).

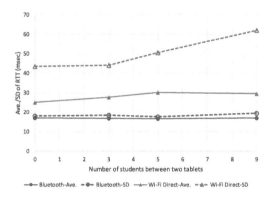

Fig. 11. Result of survey on influence of human presence.

5   Conclusion

In this study, we conducted a comparative evaluation of Bluetooth and Wi-Fi Direct, assuming an education in a classroom using ad hoc networks. We conducted surveys on the influence of hand movements while operating the tablets and the influence of human presence on delay times. From the results, we found that Bluetooth is less susceptible to the above-mentioned influences compared to Wi-Fi Direct. Bluetooth can be said to be more suitable for educational applications used on tablets for group work in a classroom where a large number of students are seated.

Future challenges include investigating not only delay characteristics but also throughput characteristics. Additionally, it would be useful to create an educational application that can run on an ad hoc network configured with dozens of tablets and verify its practicality.

354 K. Endo et al.

Acknowledgments. This work was supported by JSPS KAKENHI Grant Number JP15K16105. The authors are grateful to the members of Laboratory of Computational Science in Ehime University for the cooperation on the survey on the influence of human presence.

References

1. Olson, T., Olson, J., Olson, M., Thomas, A.: Exploring 1:1 tablet technology settings: a case study of the first year of implementation in middle school mathematics classrooms. In: Proceedings of the 26th Annual Conference of the Society for Information Technology and Teacher Education (SITE 2015) (2015)
2. Paek, S., Fulton, L.A.: Elementary students using a tablet-based note-taking application in the science classroom. J. Digit. Learn. Teacher Educ. **32**(4), 140–149 (2016)
3. Aşkar, P., Altun, A., Şimşek, N., Özdemir, S.: How teachers and students depict interactive whiteboards and tablet PCs in a 9th grade classroom? In: Zhang, J., Yang, J., Chang, M., Chang, T. (eds.) ICT in Education in Global Context. LNET, pp. 19–35. Springer, Singapore (2016). doi:10.1007/978-981-10-0373-8_2
4. Tront, J.G., Prey, J.C.: Tablet PCs and slate devices can improve active learning classroom experiences. In: Proceedings of the 18th International Conference on Interactive Collaborative Learning (ICL 2015) (2015)
5. Taneja, S., Kush, A.: A survey of routing protocols in mobile ad hoc networks. Int. J. Innov. Manage. Technol. **1**(3), 279–285 (2010)
6. Reina, D.G., Askalani, M., Toral, S.L., Barrero, F., Asimakopoulou, E., Bessis, N.: A survey on multihop ad hoc networks for disaster response scenarios. Int. J. Distrib. Sens. Netw. **11**(10) (2015). Article ID: 647037
7. Garg, V., Kataoka, K., Talluri, S.S.R.: Performance evaluation of wireless ad-hoc network for post-disaster recovery using linux live USB nodes. In: Proceedings of the 11th IEEE International Conference on Wireless and Mobile Computing, Networking and Communications (WiMob 2015) (2015)
8. Saha, S., Nandi, S., Paul, P.S., Shah, V.K., Roy, A., Das, S.K.: Designing delay constrained hybrid ad hoc network infrastructure for post-disaster Communication. Ad Hoc Netw. **25**, 406–429 (2015). Part B
9. Morreale, P., Goncalves, A., Silva, C.: Mobile ad hoc network communication for disaster recovery. Int. J. Space Based Situated Comput. **5**(3), 178–186 (2015)
10. Wi-Fi Alliance: Wi-Fi Alliance. https://www.wi-fi.org/
11. Wi-Fi Alliance: Wi-Fi Direct. https://www.wi-fi.org/discover-wi-fi/wi-fi-direct
12. Bluetooth SIG: Bluetooth Technology Website. https://www.bluetooth.com/
13. Xu, N., Zhang, F., Luo, Y., Jia, W., Xuan, D., Teng, J.: Stealthy video capturer: a new video-based spyware in 3G smartphones. In: Proceedings of the Second ACM Conference on Wireless Network Security (WiSec 2009), pp. 69–78 (2009)

Comparing TPC-W and RUBiS via PCA

Markus Lumpe[✉] and Quoc Bao Vo

Faculty of Science, Engineering and Technology,
Swinburne University of Technology, Hawthorn, VIC, Australia
{mlumpe,bvo}@swin.edu.au

Abstract. We aim to understand the fundamental design correspondences between TPC-W and RUBiS, two benchmark applications modeled after the well-known E-commerce solutions Amazon and eBay, respectively. Furthermore, we investigate how these benchmarks reflect the design principles of real-world applications by comparing them against *Qualitas Corpus*, offering an effective domain context of curated Java software systems. To perform this study, we employ Principal Component Analysis (PCA) to distill the important information (*i.e.*, the principal components) from a set of observations of possibly correlated variables (*i.e.*, software metrics). The results of our analysis reveal that TPC-W and RUBiS are comprised of surprisingly dissimilar features that clearly show that TPC-W and RUBiS do not share too many design commonalities. Moreover, we demonstrate that PCA is a powerful tool to uncover key software quality attributes.

1 Introduction

Benchmark applications have been used extensively within IT industry to measure and characterize performance of an IT system under test (e.g., a computer system, a database system, a server or a cloud computing system). In particular, application-level benchmarks emulate an end user's application software to measure the performance of a system as typically perceived by a user of the application. TPC-W and RUBiS are two well-known application-level benchmarks that have been used widely to measure the performance of computer systems, whether they are on-premise physical hardware or a cloud system [2]. However, for a benchmark to truly reflect the performance of real-world applications, it has to share similar (if not the same) design principles as being used to construct real-world software systems. Moreover, as both TPC-W and RUBiS are modeled after E-commerce solutions having a common purpose, namely transactional web benchmark, and with many common features we theorize that TPC-W and RUBiS are naturally of *"similar design"*.

To test the above two hypotheses, namely, that (i) TPC-W and RUBiS follow similar design principles as real-world software systems, and (ii) TPC-W and RUBiS share similar designs, we carry out exploratory data analysis (EDA) into these benchmarks both in isolation and in comparison with *Qualitas Corpus* [16]. EDA is an effective tool to explore whether the software systems under analysis

© Springer International Publishing AG 2017
N.T. Nguyen et al. (Eds.): ACIIDS 2017, Part I, LNAI 10191, pp. 355–366, 2017.
DOI: 10.1007/978-3-319-54472-4_34

are of similar design and posses features that fit the domain attributes. The focus of EDA is to study the *resemblances* and *differences* between observations assessed from a multidimensional point of view, and to provide a descriptive summary and visualization of the underlying data set [6]. One kind of EDA is *Principal Component Analysis* (PCA) [6], which is a multivariate technique to analyze a data table where the observations are described by inter-correlated quantitative dependent variables. The aim of PCA is to draw conclusions from the linear relationships between those variables by reducing the data to the principal dimensions of variability [6]. The result of PCA can be visualized by aligning the viewpoints of the data with the *principal components* (usually only the first and second dimension) so that resemblances and differences between observations can be easily recognized [6].

In this study, we apply PCA as a primary means to conduct a comparative analysis of the benchmarks TPC-W and RUBiS. In terms of software measurement, we explore both the scale and the distribution of features separately via PCA in order to determine whether TPC-W and RUBiS are closely related, and also to investigate how the designs of these two benchmarks relate to the systems collated in *Qualitas Corpus*. The latter, in particular, aims at considering TPC-W and RUBiS in the wider context of representative Java applications and their design.

The results of our analysis reveal that TPC-W and RUBiS share little common design features, and the design features of TPC-W and RUBiS also signify more atypical design choices compared to the systems in *Qualitas Corpus*. PCA projects the data tables of TPC-W and RUBiS largely onto different quadrants on the principal plane (*i.e.*, the plane obtained from the first and second principal components). We have to conclude that TPC-W and RUBiS are of fundamentally different designs. With respect to the systems in *Qualitas Corpus*, the design features of RUBiS are the least similar to those in *Qualitas Corpus*. There is little, if any, overlap with the data representing the design choices for the systems collated in *Qualitas Corpus*. Consequently, exchanging TPC-W with RUBiS and *vice versa* may result is a significant disruption of the underlying software ecosystem [10], and maintenance efforts might not yield desired outcomes.

The rest of this paper is organized as follows: in Sect. 2 we briefly discuss the concepts underpinning software measurement and PCA, and proceed with a presentation of our experimental setup. We present the results of performing Principal Component Analysis on our data set in Sect. 3. We conclude this presentation with a summary of our main observations in Sect. 4.

2   Experimental Setup

Measurement [16] provides an important engineering component to improve the quality of software development and to monitor and assess the resulting product [1,3]. We use software metrics [5] as a practical means for software measurement. Metrics allow for assigning numbers (or symbols) to software attributes that describe them according to pre-defined assessment criteria [13]. The specific aims vary and, consequently, lead to different kinds of investigations. Typically, we seek a mix of *direct* and *indirect* measures [5] that relate, for example, to *size* [11], *complexity* [9], or *quality* [3]. The ultimate aim of software measurement is to design a *model of reality* related to actual software development practices and products. This model should provide an effective and unequivocal tool [7] to formulate and test theories that would allow one to reason about software systems and the process being used to build them.

The essential idea of Principal Component Analysis is to describe a data set with I individuals and K possibly correlated variables using a smaller number $S \leq K$ of uncorrelated and orthogonal variables [6]. This *dimensionality reduction* aims at making the important information contained in the data set explicit. The new variables are called the *principal components* and capture the largest variance in the data set in decreasing order. In summary, the goals of PCA are to (i) retain as much information as possible, (ii) compress the data set by keeping only the important information, (iii) simplify the description and visualization of the data set, and (iv) analyze the structure of individuals and variables [6].

2.1   Data Selection

We use the binary versions of RUBiS 1.4.1,[1] TPC-W 1.0,[2] and *Qualitas Corpus* [16] (edition 20101126) as reference data set for this study. In order to distill the core features of RUBiS 1.4.1 and TPC-W 1.0, we applied the approach suggested by Tempero *et al.* [16] and eliminated any infrastructure and third-party code. Processing the original system distributions, we obtained 84 classes for RUBiS 1.4.1 and 86 classes for TPC-W 1.0.

Qualitas Corpus (edition 20101126) provides a curated collection of 106 non-trivial open-source Java software systems. The actual system sizes vary over a wide spectrum. For example, Jasml 1.4.2 is the smallest system with just 49 classes, whereas NetBeans 6.9.1 is the largest comprising 32,475 classes. The median system size in *Qualitas Corpus* is 901 classes. The corpus has been designed to allow for repeatable studies and to assemble a representative sample of how Java software is being constructed in practice [16]. Any statement about the nature of Java software and its development is expected to be generalizable from observations made based on *Qualitas Corpus*.

2.2   Software Metrics Selection

To perform our analysis and to capture commonly shared features among systems, we draw on three previously tested metrics sets: the Chidamber and Kemerer (CK) metrics suite [3] for measuring object-oriented software systems,

---

[1] http://rubis.ow2.org.

[2] http://www.tpc.org/tpcw/.

Table 1. Analyzed class-level metrics.

Name	Description	Name	Description
MCC	Cyclomatic complexity [9]	COUT	Number of class couplings
LIC	Load instruction count	SOUT	Number of intra-class couplings
SIC	Store instruction count	FOUT	Number of inter-class couplings
TCC	Type construction count	NOSA	Number of static attributes
NAR	Number of field (including static) reads	NUSA	Number of public static attributes
NAW	Number of field (including static) writes	NISA	Number of private static attributes
WMC	Weighted methods (including static) per class	NOA	Number of attributes
DIT	Depth of inheritance	NOIA	Number of private attributes
NOC	Number of children (direct)	NOSM	Number of static methods
RFC	Response for a class	NUSM	Number of public static methods
LCOM	Lack of Cohesion of Methods	NOM	Number of methods
NDC	Number of all descendant classes (transitive)	NOUM	Number of public methods
IND	Number of class dependencies	NOIM	Number of private methods
CIND	Cumulative number of class dependencies		

the size and complexity measures studied in the context of software evolution [17], and the field metrics used to understand how fields and properties are used in Java [15].

The Chidamber and Kemerer (CK) metrics suite [3] has become a *de facto* standard for measuring object-oriented software systems. As such, it has been extensively studied before (*e.g.*, [4,14]) in order to formally assert its validity. These studies revealed evidence for the presence of *collinearity* between some measures [14] and a *confounding with size* [4]. Collinearity of measures provides a means for a *dimensionality reduction*, a key feature of Principal Component Analysis [6], whereas size-related effects give rise to *outlier detection* in PCA [6].

The measures studied by Vasa *et al.* [17] seek to reveal those classes in a given system that are the most active. Active classes undergo frequent modification, typically via enhancement or correction. The evolution metrics comprise size and complexity measures that quantify, for example, the degree of popularity, responsibility, and delegation of a class. As in case of the CK metrics suite, some evolution measures exhibit, possibly strong, collinearity. However, Vasa *et al.* [17] observed that a mere occurrence of such a linear relationship might not justify sacrificing one measure or the other *per se*. For example, the metrics for load and store instructions are strongly correlated, yet their distributions (aggregated in terms of their Gini coefficients) vary independently [17].

Tempero [15] suggested a set of field metrics to study how fields are used in Java applications. Fields metrics allow for a systematic analysis of data encapsulation and possible breaches thereof. For this reason, field metrics quantify not only the amount but also the type of field definitions and uses. For example, Tempero discovered that most systems in *Qualitas Corpus* contain non-private fields, but only 12% of the exposed classes [15] are subject to non-private field access. The existence of possible occurrences of correlations between field metrics was, however, not explored.

By combining these three metrics sets, we obtained 27 distinct measures in total (cf. Table 1). The sets partially overlap and, hence, we had to eliminate duplicates in order to obtain a viable aggregate. For example, the CK and the evolution metrics suite [17] both contain a measure for the coupling between object classes. We kept only the evolution metrics variant, as this suite also contains an associated measure to quantify couplings to library classes. The aggregate metrics set gives rise to a correlation matrix with 351 coefficients between pairs of metrics to perform PCA. To collect theses measures, we used jCT [8], an extensible data mining infrastructure for Java code, that is readily equipped with the necessary metrics modules in question.

2.3 Tooling

We use the **FactoMinerR** package, a dedicated collection of abstractions to perform exploratory data analysis. It includes a versatile **PCA** function that produces two graphs: a variables factor map and an individuals factor map. The variables factor map visualizes the projections of the variables to the principal plane. The length and direction of the projection vector for each variable represents the strength and direction of the correlation of a variable to a principal component. Consider, for example, Fig. 1 (left) that illustrates a concrete variable mapping (metrics aggregated via *mean*). The metrics NOM projects to the first principal component. Vector length and direction signify a strong positive correlation (*i.e.*, 0.8696). The metrics IND and FOUT are projected to the second principal component, but in opposite directions. IND is positively

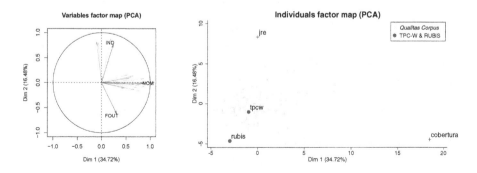

Fig. 1. Metrics and system mappings.

correlated (*i.e.*, 0.7645), whereas FOUT is negatively correlated (*i.e.*, −0.6254). Hence, when considering the projection onto the first plane (individuals factor map), systems with high values for NOM are projected to the right along the first dimension. On the other hand, systems with high values for IND and lows values for FOUT are projected at the top quadrants along the second dimension and *vice versa*, as shown in Fig. 1 (right). PCA maps TPC-W and RUBiS to the lower left quadrant. This means, compared with the systems collated in *Qualitas Corpus*, TPC-W and RUBiS have a lower the average number of methods (NOM) per class, lack strong cohesion (low IND), and depend heavily on external services (high FOUT). Moreover, the design features of RUBiS are more extreme (dissimilar) than those of TPC-W.

3   Software System Comparison

A comparative analysis of software systems aims at identifying similarities between software systems based on a set of pre-selected and attainable system attributes (or software metrics). PCA yields a topology of software systems that allows us to assess their closeness with respect to the chosen variable perspective. We use Standardized Principal Component Analysis, that is, all variables are centered and reduced (*i.e.*, $(x_{ik} - \overline{x}_k)/s_k$, where $\overline{x}_k$ is the sample mean and s_k is the standard deviation of the sample of variable k). In addition, all variables are weighted equally. In the following, we first perform PCA for a data table being comprised of the class-level metrics distilled from TPC-W and RUBiS, one row per class. In a second step, we summarize the class-level metrics data via either *arithmetic mean* or the *Gini coefficient* and run PCA on the resulting data table. Here, a row represents a system.

3.1   TPC-W vs. RUBiS

If the design characteristics of TPC-W and RUBiS are similar, then their respective data clouds should coincide. Conceptually, PCA *"rotates"* a given data table (*i.e.*, a set of multivariate data points) so that the facet with the highest variability is put into focus [6]. The highest variability is associated with the first principal plane. If TPC-W and RUBiS are of different design, then this should manifest most prominently on the first plane. The other dimensions capture variance also, but its degree is strictly smaller compared with that represented by the first and second dimension. In the comparison of TPC-W and RUBiS, we shall limit ourselves to four dimensions.

The PCA variable mappings of the class-level metrics for TPC-W and RUBiS are shown in Fig. 2. The first principal plane (cf. Fig. 2(a)) separates mostly WMC/RFC/TCC (positively correlated to first dimension) from NOM/COUT/FOUT (positively correlated to second dimension) and NOSM/NUSM (negatively correlated to second dimension). Hence, the first dimension captures the size and complexity of methods in TPC-W and RUBiS, whereas the second dimension encapsulates interface size and class coupling.

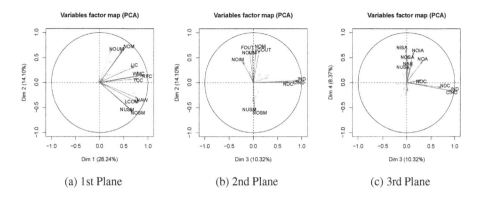

(a) 1st Plane   (b) 2nd Plane   (c) 3rd Plane

Fig. 2. Metrics mappings for TPC-W and RUBiS.

The other dimensions represent class dependency (cf. Fig. 2(b) Dim 3) and state size (*i.e.*, number of attributes, cf. Fig. 2(c) Dim 4).

The variable mappings, as shown in Fig. 2, denote how PCA rotates the data table and how the attributes are projected to the principal dimensions of variability. We need to inspect the individuals maps to determine whether TPC-W and RUBiS are of the same design. The results are shown in Fig. 3 (only the 10 highest contributors are explicitly labeled). The design features of TPC-W and RUBiS are quite dissimilar. The projections of the individuals (*i.e.*, system classes) diverge significantly, in particular, in the second and fourth principal dimension. We find that RUBiS entails a greater number of classes with large interfaces and class coupling. In addition, RUBiS contains a large number of factory classes that define many attributes. In general, the state size in RUBiS is more extreme compared with TPC-W. Overall, the rotated data tables of TPC-W and RUBiS do not coincide in any principal plane (the centers of gravity are disjoined). As a result, we conclude that the design features of TPC-W and RUBiS follow fundamentally different approaches.

3.2   Mean-Based Analysis

TPC-W and RUBiS are dissimilar, but which system is placed closer to the average design characteristics of general Java systems? To answer this question, we first explore the scale-facet of software metrics for TPC-W, RUBiS, and the systems collated in *Qualitas Corpus*, and aggregate the system metrics data via the *arithmetic mean*. The outcomes are summarized in Fig. 4. When using a two-dimensional view only, most metrics positively correlate with the first component, the principal axis of variability. Seven metrics tie to the second component: CIND, NDC, DIT, NOC, SOUT, IND, FOUT, of which only FOUT exhibits a negative correlation (cf. Fig. 4, variables factor mapping).

The projection of NOM, RFC, and LCOM to the same principal component agrees with an earlier observation by Succi *et al.* [14], which recommended that prediction models should never combine NOM, RFC, and LCOM. Similarity,

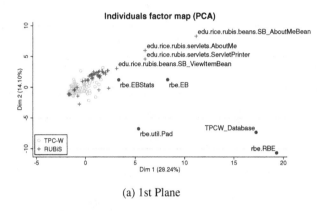

(a) 1st Plane

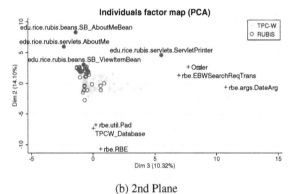

(b) 2nd Plane

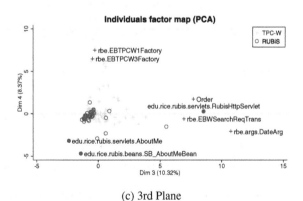

(c) 3rd Plane

Fig. 3. Class mappings for TPC-W and RUBiS.

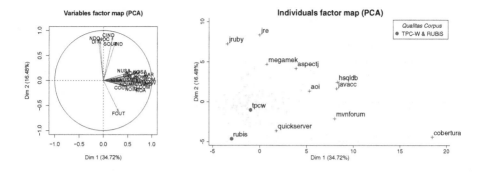

Fig. 4. Mean-based mappings to the principal plane.

the projection of NOM to the first and SOUT/FOUT to the second component also supports the second recommendation by Succi *et al.* that prediction models should test CBO (SOUT in our study) and NOM for collinearity. The projection to different principal components reveals that NOM and CBO are not linearly correlated (under arithmetic mean aggregation). In general, the first principal component separates systems with high average values for interface size and complexity from those with low values, whereas the second principal component isolates those with great popularity (*i.e.*, large inter-class coupling) and low library dependencies and *vice versa*.

Most importantly, the epicenter of the mappings of the benchmark applications TPC-W and RUBiS is outside that of *Qualitas Corpus*. TPC-W and RUBiS are not only quite different from each other, but also quite different from the systems in *Qualitas Corpus*. However, whereas RUBiS is noticeably more different (it even qualifies as an outlier with a contribution of 4.50% to the construction of the second component), TPC-W's mapping suggests that its basic design approach matches the average for systems collated in *Qualitas Corpus*, when we consider the scale of features.

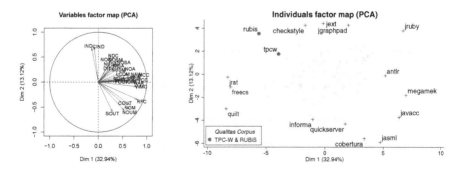

Fig. 5. Gini-based mappings to the principal plane.

3.3 Gini-Based Analysis

The Gini coefficient belongs to a class of *socio-economic inequality indices* [12] that have been invented in econometrics research to provide a viable alternative to central tendency statistics. For a discrete population with non-negative values x_i, $1 \leq i \leq n$, the Gini coefficient is one-half of the *relative mean difference* of every pair (x_i, x_j), $1 \leq i, j \leq n$, in the population [12]:

$$G = (1/2n^2\mu) \sum_{i=1}^{n} \sum_{j=1}^{n} |x_i - x_j|$$

All our selected metrics are applicable to be aggregated with the Gini coefficient. The result provides us with a distribution-facet of the software metrics data. Using standardized PCA again, we obtain a principal plane as shown in Fig. 5. The mappings are quite different compared to those produced by mean-based PCA (cf. Fig. 4). The Gini coefficient is a scale and size agnostic summary statistic. It also sensitive to small variations in a data set that are otherwise *mean-invariant*. Hence, when aggregating metrics data with the Gini coefficient, we quantify how different the metrics distribution is compared to an even distribution (*i.e.*, how the load is being shared among the classes in a system). This naturally leads to a different viewpoint of the variables (*i.e.*, software metrics). In particular, metrics capturing complexity, responsibility, and interface become the prime source for variability and are, consequently, projected to the first principal component.

Returning to our question, whether TPC-W's or RUBiS's design characteristics are more similar to the features of the systems collated in *Qualitas Corpus*, we find the focal point of the mapping for TPC-W and RUBiS is noticeably outside the center of gravity for *Qualitas Corpus*. When considering concentrations of features, TPC-W and RUBiS occupy the fringes in the principal plane. We conclude that TPC-W and RUBiS posses quite unique (atypical) design feature distribution profiles compared with those for the systems collated in *Qualitas Corpus*. Again, RUBiS's mapping suggest it is significantly more different in its design.

4 Conclusion

Our primary objective in this study has been to assess the similarity of TPC-W and RUBiS. To provide a context for this analysis, we tested the connection between these systems with respect to *Qualitas Corpus* and 27 metrics for actually observed shared features. The results indicate that TPC-W and RUBiS share little common attributes. There respective projections to the principal plane signals that their design attributes make them more dissimilar than similar. Moreover, when compared with the systems contained in *Qualitas Corpus*, RUBiS, in particular, is mapped more or less to an "extreme" location, indicating that its design attributes are quite unique.

Principal Component Analysis is an effective tool to compare software systems and to develop an appreciation of the design characteristics governing software systems. The aim of this study was to explore the system features of TPC-W and RUBiS and relate them to the features of the systems collated in *Qualitas Corpus*. TPC-W and RUBiS are two E-commerce benchmarking systems. Targeting the same domain and performing similar tasks, one would expect those system to share common system attributes. An analysis via mean-based PCA and Gini-based PCA revealed that TPC-W and RUBiS are quite dissimilar with respect to their design characteristics. Moreover, we discovered that both system also share little similarities with comparable systems in *Qualitas Corpus*.

In general, our study demonstrates that different viewpoints chosen in PCA can yield complementary insights to enhance our understanding of software systems. The classical approach to use mean to aggregate software metrics data has merits and allows us to make inferences based on scale. Amending a study with a distribution-based perspective adds feature relocation sensitivity to the analysis that is more powerful than a purely scale-based inspection alone.

We based the motivation for this study on the hypotheses that TPC-W and RUBiS follow real-world software system design principles as they aim at emulating an end user's application software, and share similar design features as both have a common purpose: transactional web benchmark. This is not the case, at least not for the features considered in this study. Replacing TPC-W with RUBiS and *vice versa* might not preserve the experiences and characteristics of the underlying software ecosystem.

References

1. Basili, V.R.: The role of experimentation in software engineering: past, current, and future. In: Boehm, B., Rombach, D.H., Zelkowitz, M.V. (eds.) Foundations of Empirical Software Engineering, pp. 1–13. Springer, Heidelberg (2005)
2. Chhetri, M.B., Chichin, S., Vo, Q.B., Kowalczyk, R.: Smart CloudMonitor-providing visibility into performance of black-box clouds. In: Proceedings of 2014 IEEE 7th International Conference on Cloud Computing (CLOUD 2014), pp. 777–784. IEEE, July 2014
3. Chidamber, S.R., Kemerer, C.F.: A metrics suite for object oriented design. IEEE Trans. Softw. Eng. **20**(6), 476–493 (1994)
4. El Emam, K., Benlarbi, S., Goel, N., Rai, S.N.: The confounding effect of class size on the validity of object-oriented metrics. IEEE Trans. Softw. Eng. **27**(7), 630–650 (2001)
5. Fenton, N.E., Bieman, J.: Software Metrics: A Rigorous and Practical Approach, 3rd edn. CRC Press, Boca Raton (2015)
6. Husson, F., Lê, S., Pagès, J.: Exploratory Multivariate Analysis by Example Using R. CRC Press, Boca Raton (2011)
7. Kitchenham, B., Pfleeger, S.L., Fenton, N.E.: Towards a framework for software measurement validation. IEEE Trans. Softw. Eng. **21**(12), 929–944 (1995)
8. Lumpe, M., Mahmud, S., Goloshchapova, O.: jCT: java code tomograph. In: Proceedings of 26th IEEE/ACM International Conference on Automated Software Engineering (ASE 2011), Lawrence, Kansas, USA, pp. 616–619, November 2011

9. McCabe, T.J.: A complexity measure. In: ICSE Proceedings of 2nd International Conference on Software Engineering, USA, pp. 407–419 (1976)
10. Messerschmitt, D.G., Szyperski, C.: Software Ecosystem: Understanding an Indispensable Technology and Industry. The MIT Press, Cambridge (2003)
11. Mordal, K., Anquetil, N., Laval, J., Serebrenik, A., Vasilescu, B., Ducasse, S.: Software quality metrics aggregation in industry. Softw. Evol. Proc. **25**, 1117–1135 (2012)
12. Sen, A.K.: On Economic Inequality. Oxford University Press, Oxford (1973)
13. Stevens, S.: On the theory of scales of measurement. Sci. New Ser. **103**(2684), 677–680 (1946)
14. Succi, G., Pedrycz, W., Djokic, S., Zuliani, P., Russo, B.: An empirical exploration of the distributions of the chidamber and kemerer object-oriented metrics suite. Empirical Softw. Eng. **10**(1), 81–103 (2005)
15. Tempero, E.: How fields are used in java: an empirical study. In: Proceedings of 20th Australian Software Engineering Conference (ASWEC 2009), Gold Coast, Queensland, IEEE Computer Society, pp. 91–100 (2009)
16. Tempero, E., Anslow, C., Dietrich, J., Han, T., Li, J., Lumpe, M., Melton, H., Noble, J.: The qualitas corpus: a curated collection of java code for empirical studies. In: Proceedings of 17th Asia Pacific Software Engineering Conference (APSEC 2010), Sydney, Australia, pp. 336–345, December 2010
17. Vasa, R., Lumpe, M., Branch, P., Nierstrasz, O.: Comparative analysis of evolving software systems using the Gini coefficient. In: Proceedings of 25th IEEE International Conference on Software Maintenance (ICSM 2009), Edmonton, Alberta, pp. 179–188. IEEE Computer Society, September 2009

Analysis and Solution Model of Distributed Computing in Scientific Calculations

Josef Horalek and Vladimír Soběslav[✉]

Faculty of Informatics and Management, University of Hradec Kralove,
Hradec Kralove, Czech Republic
{josef.horalek,vladimir.sobeslav}@uhk.cz

Abstract. Processing huge amounts of data is currently of concern in various fields of science and commercial data processing, such as pharmaceutical drug development, astronomical probe data processing, security analysis of large amounts of communication data, etc. Generally, centrally administered methods are used, but their employment and operation are very expensive. The aim of this paper is to present a model of high-capacity data processing that is based on the technology of Apache Hadoop with emphasis on use of volunteer host devices with the service distribution via the Internet.

Keywords: Hadoop · Cluster · HDFS · Apache Hadoop · Distributed computing

1 Introduction

The most frequent approach to processing high-capacity data is the use of one cluster containing hundreds or thousands of nodes that process all the user requests. If the nodes are used to their limit, additional nodes can be added to the cluster. Figure 1 depicts traditional access of one user to the cluster [3]. Any computing performed in the cluster is conducted by the manager responsible for their administration, running, and terminating. Once the program is terminated, the output data is stored to Hadoop distributed file system (HDFS) and the client is informed about the computation being completed. However, given cluster is not accessed by a single user, but by tens, hundreds, thousands, or millions of them at the same time, depending on the given cluster use. Therefore, with higher numbers of users, time required to process all the tasks elongates. Moreover, in this case it is necessary for the client to provide not only data to be processed, but also a program written in the supported language. Many users using the performance of distributed computing centres are scientific workers that care mostly about the correct final result and primarily are not concerned about the optimisation of the code used or additional scripts. All these imperfections lead to prolonging the time needed to execute the tasks, but it can also be the cause of incorrect calculations, or it can lead to a crash of the whole cluster.

The aim of this article is to present an innovative approach to high-capacity data processing, including the application draft. The purpose of the application is to improve the work of the users using automation and simplifying the communication between the

© Springer International Publishing AG 2017
N.T. Nguyen et al. (Eds.): ACIIDS 2017, Part I, LNAI 10191, pp. 367–376, 2017.
DOI: 10.1007/978-3-319-54472-4_35

used and Hadoop Clusters. Therefore, the application allows for higher computing performance and redistributing the data processing between multiple clusters. Rules for the user interface are also defined, as well as basic functionality and methods required to process the data [1, 2].

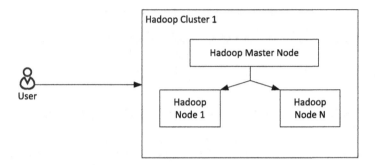

Fig. 1. Standard user access to a Hadoop Cluster

2   Principal Solution Draft

The suggested approach of processing high-capacity data [4, 5] lies in interconnecting distributed computing technologies with volunteer computing through Apache Hadoop. In volunteer computing, users are engaged as computing nodes that provide processing power necessary for complicated calculations. The node is defined as any device capable of performing required calculations, along with hardware allowing for being integrated into the cluster and with being compatible with Apache Hadoop, i.e. device with any supported OS (Linux, Windows, BSD, Mac OS/X, or OpenSolaris) [6]. All these nodes log in a larger unit that represents one Hadoop cluster. Logging into the cluster is performed via client application installed on the device. Every Hadoop cluster then handles its nodes and makes use of their computing power. Additional nodes can be added to the cluster dynamically, without the need of restarting or shutting it down, as referenced in [6]. This ability allows for adding nodes into the cluster when the devices are not being used, and can, therefore, process the data only in time intervals when the user does not use them actively. All the Hadoop clusters are then logged in the designed application. The application serves as a link between the users and the Hadoop clusters, as a means of adding nodes into the clusters, and as a way of managing the Hadoop clusters logged in [7]. The placement of the application, Hadoop clusters, and nodes, is depicted in Fig. 2.

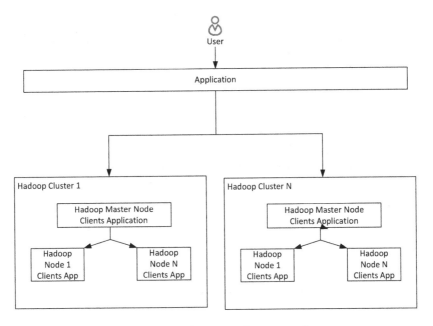

Fig. 2. Extended data processing access scheme

3   Application and Basic Functions

From the scheme of our approach to high-capacity data processing above it is apparent that two main applications are being used. The main server application is designated for the users that have registered into it, and after that it allows them to download the respective client application, which is designated for the nodes that are to be connected to the Hadoop cluster maintained by the server application. Figure 3 depicts the basic scheme of communication between the server application, the client application, and the Hadoop clusters (Hadoop Master Node or Hadoop Data Node). The server application is built on two foundations: Web interface and App Core. As can be seen in Fig. 3, the user communicated only with the server application via web interface, which provides them with all the requisite information. The user's main concern is whether the processed tasks have been computed or if an error has occurred during the calculation, which tasks have been executed, how long has their computation taken etc. All the user's communication with the Hadoop cluster is managed by the App Core, which substitutes tasks that formerly had to be performed by the user, such as data upload, task execution, or gathering information about the running tasks.

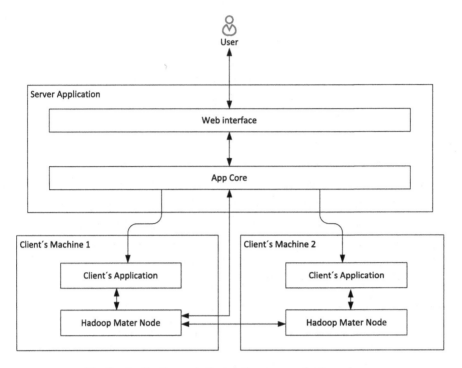

Fig. 3. Application and client station communication scheme

App Core also secures directing the tasks to the clusters for the optimal performance usage of the particular nodes. Another advantage of the users' access to the clusters using the application is that the users do not overload the Hadoop cluster. As Hadoop Master Node is designated for common hardware, it cannot be guaranteed that all of the queries will be successfully processed should many of them occur at the same time. The server application also communicates with the client device via the client application or directly with the Hadoop cluster (Hadoop Master Node), which provides the requisite information about the tasks being processed. Should a failure of the Hadoop Master Node occur and should there be a Hadoop Secondary Master Node available, the server application would start to communicate with the latter as if it were the Master Node. App Core communicates with Hadoop and the client application via XML/JSON object exchange, which is procured via the Internet network using HTTP/HTTPS, which is commonly used for the Internet communication.

3.1   Client Application

The purpose of the client device application is to maintain the connection of the given device to the cluster. After being installed, the client checks whether Hadoop is installed on the device and then prompts the user to set up configuration settings using a graphical interface. If the user does not have Hadoop installed, the application automatically prompts the user to install it. If the user chooses to install Hadoop, the application automatically prompts the user whether to download all the requisite files from the server and guides the user through the installation. After validating or during the installation of Hadoop, Hadoop cluster configuration settings pop up. The user can choose to select the preferred Hadoop cluster to which their device will connect, to connect automatically, or to act as a NameNode, i.e. become a Hadoop cluster. If the user chooses the option of the preferred Hadoop cluster, the device will always connect to that cluster. During the first use, the user is prompted with a list of available Hadoop clusters they can connect to. Once the user confirms their selection, the client application takes care of the connection. When repeatedly logging in the chosen cluster, it is not necessary to contact the server application again as the connection of the given DataNode to the cluster is secured by the client application. Should an error occur during connecting – e.g. maximum number of nodes in the cluster reached, NameNode offline etc. – log is sent to the server and the user is informed about the problem and prompted by the application with the list of available Hadoop clusters. If the user chooses to connect automatically, whenever they log in, the client application contacts the server application, which selects a suitable Hadoop cluster, then passes the information to the client and secures the connection of the computing node in a way similar to connecting to a manually chosen cluster. The advantage of the automatic cluster selection is primarily an ability to balance the cluster workload. Another factor is the distance of the connected node from the NameNode, which communicates with the newly connected node in fixed intervals, and files to be processed are interchanged between them. The last option is to become the NameNode to which other DataNodes will connect, which creates another Hadoop cluster. The user can specify the maximum usage of memory, CPU, and data storage. This information is stored into Hadoop configuration of the given node.

3.2   Server Application

The purpose of the server counterpart of the application is to balance workload, manage Hadoop clusters, and handle addition of new clusters, DataNodes (computing nodes), to the existing clusters. The App Core is the core of the server application and its purpose is to handle users' requests and users' task termination notifications, manage Hadoop clusters, balance workload, and handle errors that occur throughout the whole system. Using web interface, it is possible to log into the application with one of the two basic account types (User type or Admin type). User type allows the user to manage the programs installed and to display the statistics of the running tasks. Before being able to use this account, the user has to register using the application web interface (if this option is allowed by the application admin) or has an account created by the admin. After the account is created, the user receives an email with the required information. After logging in, the user can access a portal, where they can view an **Overview** with the statistics of the running tasks and the finished tasks, including their successful or failed completion. Another option can be found in **My programs**, where the user can obtain a list of MapReduce programs that have been uploaded by the user.

Application server part methods. The App Core on the server side offers several essential methods for the implementation of our model. The main method is **Task**

routing, i.e. choosing the cluster to execute the given task. **Secure communication with Hadoop clusters** support is required as well as it, upon choosing which cluster to use to execute the task, secures the connection and uploads the required data to the cluster. **Uploading and file processing** method serves for uploading source codes and data to be processed through the web interface. Should any error occur while executing the calculations, the information is logged on the given cluster, which is secured by **Logging and error handling** method, which also serves for gathering the data about the error from the cluster and informing the user with the description of the error that had occurred. **Create cluster** method serves to add required entries to the database while creating a cluster for the cluster to be usable for calculations. **Delete cluster** method serves to terminate all the running tasks on the given cluster. If there are any pending tasks, they are moved to the active clusters. If there is a running task, its state is set to error, and after set time, the task is started in another active cluster. The method also secures removal (permanently deleting information) or deactivation (temporary removing) of the cluster. Another significant method is **Add computing node**, which performs the requisite configuration in the database and sets the required configuration to the Hadoop cluster main node. Selection of the Hadoop cluster can be fixed or automatic.

Databases. In the application, databases (DB) are used to store data. For the model application, it is possible to use either of the two lead types, i.e. SQL and NoSQL. In terms of databases, the principal requirements for the application are good scalability, data loading/saving speed, high-capacity data handling, and their diversity. With similar qualities, an application is designed in NoSQL database as those have more independent structure than relation databases, which offers better scalability and performance capabilities of the database [10, 11]. The application model is going to store the required data in the database and as it will manipulate with large amounts of data, document storage type has been chosen. The choice of NoSQL is supported by the fact that the communication between the application and the Hadoop clusters is performed via exchange of XML or JSON objects, which can be stored in the database without any extensive modifications [3, 12]. As for technical solution, MongoDB open source solution has been chosen. For the purposes of the model, a basic collection that will be used to store the data in the application has been defined. In particular, Users collection stores basic information about users that log into the system. Clusters collection serves to store requisite information about the clusters (cluster address, total number of nodes, number of active nodes) from Resource Manager REST API. ClusterScheduler collection contains information about tasks running on a cluster, which is used by the forwarding algorithm. This information is being refreshed periodically and during typical events (finishing a calculation on the cluster). Program collection contains information about programs that have been uploaded and contains information who uploaded it, description of the program, name of the program, whether the program is private or public, how many times it has been run, and average time required to process 1 MB by the program per cluster.

4   Forwarding Algorithms and Workload Balancing

The main aim of our application is chiefly work with higher number of clusters and it is, therefore, to ensure proper forwarding of the tasks to free Hadoop clusters, where the calculations will be performed. Forwarding of the tasks to the clusters is determined by the following rules:

- Information about the number of available nodes within Hadoop clusters and their available memory is gathered from Resource Manager API, if no available node exists or all available memory of the cluster is used, the given cluster is removed from executing further calculations.
- Basic calculation for a suitable Hadoop cluster, where the task is to be run, is time required to process the amount of data uploaded by the user and this information is gathered by repeatedly running the given tasks on the cluster:

$$V \cdot \left(\frac{1}{v_{avg_n}} + t_{avg_n} \right)$$

V – amount of data being processed in MB, **n** – chosen Hadoop cluster, v_{avg_n} – total average speed (MB/s) of data transfer between the server application and the Hadoop cluster (n) in distance of **z** is calculated by:

$$v_{avg_n} = \frac{v_1 + v_2 + \cdots + v_{z-1} + v_z}{z}$$

z – number of connection between the application and the Hadoop cluster, v_l – average speed of data transfer between **z-1** and **z** nodes, speed is calculated when the node is logging into the application in MB/s. t_{avg_n} – average time required to process 1 MB on the chosen cluster for the chosen task:

$$t_{avg_n} = \frac{\dfrac{t_1 + 2 + \cdots + t_{j-1} + t_j}{j}}{V_1 + V_2 + \cdots + V_{j-1} + V_j}$$

If the program has not yet been run on the given cluster, t_{avg_n} is set to average value of performance of other clusters, and if the program has not yet been run on any Hadoop cluster, t_{avg_n} is set to a value that had been recorded during functionality testing after user's uploading of the program. This configuration is used to prevent eliminating or discriminating clusters where the given program had not been run yet as the average time would be 0, which would significantly influence the cluster's position in the list. t_j – time required to execute the task on the cluster. This time is acquired as the difference of ending from beginning of the execution of the task from API Hadoop Cluster for the tasks that have been successfully executed (state set to "SUCCEEDED"). j - number of tasks run in the past.

The outcome of the formula above is a list of potential clusters where the calculation can be performed. The list is ordered from the lowest value to the highest. From this list, the first cluster in order is picked and then the required operations for data transfer and task execution are performed on the given cluster. To ensure the shortest distances between the clusters and the application's server, we will use one of the common algorithms for calculating the shortest path, such as Dijkstra's or Floyd–Warshall algorithm. The required information about the cluster and its nodes is obtained using Resource Manager Api, which provides the requisite functions. The gathered information is saved to the database and subsequently used by the application. One method of adding data nodes to the clusters is automatic addition. For this method, two criteria are evaluated. One is the number of nodes in the cluster, where the value is compared to an average number of nodes. If the cluster contains below-average number of nodes, it is put into a list of candidates for addition of a node.

5   Data Error Rate and Validation

Another significant merit of the application is error handling and data validation. Data validation can be divided into two groups: data to be processed by the chosen program and user entries. Data validation/cleansing during processing by the given program is reserved solely to the programmer of that program. For such situations, the application contains a method that displays the necessary information about an occurring validation error to the user and logs information about the error to a text file reserved for this purpose [13]. Entry validation, which the user fills in a web browser, is handled in two ways: client validations and server validations.

Client validation is for mandatory fields, correct value types in the forms, i.e. validation of the form field for number, correct date format, email, etc. Secondary validation is performed at the server, where more complicated validations are used. Should an error occur during the validation process, the user is informed via information in the web browser. We distinguish between data transfer errors and errors that occur during processing within Hadoop clusters. Data transfer errors can occur during the data transfer from the user to the application servers, as well as during the file transfer from the application to the clusters where they are to be processed. As already existing protocols, such as http/https or ftp/ftps, are used for data storing, basic error codes returned by these protocols are used. The application responds to the errors by notification and logging relevant information into a special error log. Handling errors of the second type is reserved to Hadoop, which handles them based upon its model. Hadoop then differentiates errors [2] into several categories. One category is **Task errors**, whose errors are most frequently thrown during program runtime. Next category is **Application Master Failure**, where similarly to the tasks run by MapReduce procedures, the main application is provided with multiple attempts for successfully executing the program. **Resource Manager Failure** error is essentially an error within the resource manager, which is a grave error as without resource manager no programs are run nor task containers assigned. The last category is **HDFS Failure**. This error can be handled by using replication, but it is necessary to keep in mind that NameNode is still the weakest component as the requisite configurations are stored locally and if this node fails, it results in the inability to perform any tasks or communicate with the cluster or the user.

6   Conclusion

The realization of the application itself takes requirements and logics of the model itself into account. For that very reason, an implementation using Spring framework, which is designed for development of Java-based enterprise application, has been used for the app core. For the web interface implementation, Angular 2 framework or React library can be used. The client application is required to be multi-platform and able to use GUI.

We have already mentioned several times that Apache Hadoop is very popular and that it is used in many large institutions for work with high-capacity data. There are several new approaches and differences of our model from the standard one. First, we group multiple Hadoop clusters in order to reach higher performance. Second, our model is built upon using computing devices that are already present in given institutions and, therefore, it is not necessary to spend large amounts of money on cluster establishment. The model is also focused on facilitating the users' work for the application to be an asset for them, both in data processing effectiveness and in terms of presenting results. If people do not use some of the Hadoop modifications, presentation of the result data will cost them more time to process the data further.

To conclude, it should be stated that the presented model allows for expanding the model which is currently discussed. The expansion is focused on the model of assembling MapReduce programs by the user directly from the web interface. That could be achieved chiefly by creating templates after which a MapReduce program code would be generated and templates for interpretation of outcome data. Therefore, the user does not have to understand the language of the code in which the given program is implemented, and only specifies the method of the program behaviour. This functionality would greatly speed up the process of program implementation and facilitate the users' work. Additionally, we would gain an opportunity to create templates regardless of the interpreted language that is supported by Hadoop.

Acknowledgement.   This article has been produced in cooperation with Petr Volf, a graduate at FIM of UHK, to whom we hereby give our thanks for extraordinary assistance with proposing the solution using Hadoop in effective distributed computing. The support of Czech Science Foundation GACR 15-11724S DEPIES is gratefully acknowledged.

References

1. Lu, Q., Li, S., Zhang, W., Zhang, L.: A genetic algorithm-based job scheduling model for big data analytics. Eurasip J. Wireless Commun. Networking (1), 1–9, art. no. 152 (2016)
2. Hashem, I.A.T., Anuar, N.B., Gani, A., Yaqoob, I., Xia, F., Khan, S.U.: MapReduce: review and open challenges. Scientometrics **109**(1), 389–422 (2016)
3. Kobayashi, K., Kaito, K.: Big data-based deterioration prediction models and infrastructure management: towards assetmetrics. Struct. Infrastruct. Eng. **13**(1), 84–93 (2017)

4. Govindarajan, K., Somasundaram, T.S., Boulanger, D., Kumar, V.S., Kinshuk: A framework for scheduling and managing big data applications in a distributed infrastructure, ICoAC 2015 – 7th International Conference on Advanced Computing, art. no. 7562784 (2016)

5. Alekseev, A.A., Osipova, V.V., Ivanov, M.A., Klimentov, A., Grigorieva, N.V., Nalamwar, H.S.: Efficient data management tools for the heterogeneous big data warehouse. Phys. Part. Nucl. Lett. **13**(5), 689–692 (2016)

6. Hadoop wiki - faq. http://wiki.apache.org/hadoop/faq

7. Kranjc, J., Orač, R., Podpečan, V., Lavrač, N., Robnik-Šikonja, M.: ClowdFlows: online workflows for distributed big data mining. Future Gener. Comput. Syst. **68**, 38–58 (2017)

8. Sobeslav, V., Maresova, P., Krejcar, O., Franca, T.C.C., Kuca, K.: Use of cloud computing in biomedicine. J. Biomol. Struct. Dyn. **34**(12), 1–10 (2016). Article in Press

9. Sobeslav, V., Komarek, A.: Opensource automation in cloud computing. Lect. Notes Electr. Eng. **355**, 805–812 (2015)

10. Bao, X., Xiao, N., Lu, Y., Chen, Z.: A configuration management study to fast massive writing for distributed NoSQL systém. IEICE Transactions on Information and Systems, E99D (9), pp. 2269–2282 (2016)

11. Li, C.-S., Franke, H., Parris, C., Abali, B., Kesavan, M., Chang, V.: Composable architecture for rack scale big data computing. Future Gener. Comput. Syst. **67**, 180–193 (2017)

12. Tran, M.C., Nakamura, Y.: Classification of HTTP automated software communication behaviour using NoSQL database. In: International Conference on Electronics, Information, and Communications, ICEIC 2016, art. no. 7562957 (2016)

13. Holik, F., Horalek, J., Neradova, S., Zitta, S., Novak, M.: Methods of deploying security standards in a business environment. In: Proceedings of 25th International Conference Radioelektronika, RADIOELEKTRONIKA 2015, art. no. 7128984, pp. 411–414 (2015)

Decision Support Control Systems

Failures in Discrete Event Systems and Dealing with Them by Means of Petri Nets

František Čapkovič[✉]

Institute of Informatics, Slovak Academy of Sciences, Bratislava, Slovakia
Frantisek.Capkovic@savba.sk

Abstract. An approach based on Petri nets pointing to the manner how to deal with failures in discrete-event systems is presented. It uses the reachability tree and/or graph of the Petri net-based model of the real system as well as the synthesis of a supervisor to remove the possible deadlock(s).

Keywords: Detection · Discrete event systems · Failure · Modelling · Petri nets · Recovery

1 Introduction

A failure can be defined [2,3,9] as a deviation of a system from its intended (normal) behavior. The process of detecting a potential failure in the system behavior followed by isolating the cause or the source of the failure, is called as the system diagnosis. In the discrete-event systems (DES) diagnosis, faults may correspond to any discrete event. Unfortunately, the predisposition of systems to fail increases with their complexity. The research effort has been spent in the development of diagnostic systems. It is necessary to distinguish (according to the manner in which faults are reset after they occur) [2,13] between permanent and intermittent faults. In case of the permanent fault the recovery event occurs only due to repairing the controllable and observable fault. In case of the inter- mittent fault the recovery event occurs either spontaneously or due to repairing, and such event has a tendency to be uncontrollable and unobservable. The fault diagnosability [25] is interested in whether the system is diagnosable or not - i.e. in the fact whether the system can detect the occurrence of the fault in a finite number of steps or not.

Error recovery is [11,18,23] the set of actions that must be performed in order to return the system to its normal state. At least one sequence of actions should exist in order to bring the system into its normal operation. When there exist more sequences, the best one is chosen with respect to a prescribed criterion. Usually it is the sequence of actions which minimally disorganizes the system.

For systems without failures using Petri nets (PN) for modelling, analysing and control synthesis is very useful. However, practically any system is not

F. Čapkovič—Partially supported by the grants VEGA 2/0039/13 and 2/0029/17.

N.T. Nguyen et al. (Eds.): ACIIDS 2017, Part I, LNAI 10191, pp. 379–391, 2017.
DOI: 10.1007/978-3-319-54472-4_36

failure-free. Failures can emerge in any device or software. It is gratifying that PN can be used [3, 6, 9, 10, 14–16, 18, 20–22, 24] also for systems where failures occur. There failures can be categorized into hardware failures and software ones. To minimize the hardware failures of devices, it is necessary to timely execute their maintenance, test and/or check as well as timely replace their components. To decrease occurrence of software failures, fault-tolerant software techniques are necessary. Error recovery is possible only for the so called soft failures [5]. Hard (catastrophic) failures in systems are classified as functional and/or structural failures.

Strategies and forms for detection and recovery of system soft failures are based on the so called *error treatment* and *failure treatment*. The error treatment contains error detection, damage assessment and error recovery. The failure treatment includes localization, identification, system repair and continued service. However, the hard failures are overcome in most systems by means of redundancy [1].

Here, in this paper, failures in DES and their recovery will be examined by means of utilizing Petri nets (PN). DES are systems discrete by nature. They persist in a steady state until the occurrence of a discrete event which will cause their transition into another state. Typical representatives of DES are discrete manufacturing systems, transport systems, communication systems, etc. PN are frequently used for DES modelling, analysing and control synthesizing.

PN [7, 17, 19] are (as to their structure) bipartite directed graphs - i.e. graphs with two kinds of nodes (places and transitions) and two kinds of edges (arcs directed from places to transitions and arcs directed contrary) - $\langle P, T, F, G \rangle$, where P, $|P| = n$, is a finite set of places and T, $|T| = m$, is a finite set of transitions; $F \subseteq P \times T$, $G \subseteq T \times P$ are subsets of the directed arcs. The set $B = F \cup G$ contains all directed arcs. As to dynamics, the PN can be formally defined as $\langle X, U, \delta, \mathbf{x}_0 \rangle$, where X is a set of PN states, U is a set of discrete events; $\delta : X \times U \rightarrow X$ symbolizes the fact that the new state depends on existing state and an occurred discrete event; $\mathbf{x}_0 \in X$ is the initial state. The state equation (DES model) $\mathbf{x}_{k+1} = \mathbf{x}_k + \mathbf{B}.\mathbf{u}_k$, $k = 0, 1, \ldots, N$, where $\mathbf{B} = \mathbf{G}^T - \mathbf{F}$ and $\mathbf{F}.\mathbf{u}_k \leq \mathbf{x}_k$, expresses PN dynamics. Here, $\mathbf{x}_k = (\sigma_{p_1}^k, \ldots, \sigma_{p_n}^k)^T$ with entries $\sigma_{p_i}^k \in \{0, 1, \ldots, \infty\}$, representing the states of particular places, is the PN state vector in the k-th step of the dynamics development; $\mathbf{u}_k = (\gamma_{t_1}^k, \ldots, \gamma_{t_m}^k)^T$ with entries $\gamma_{p_i}^k \in \{0, 1\}$, representing the states of particular transitions (either enable - when 1, or disable - when 0) is the control vector; $\mathbf{F}$, $\mathbf{G}^T$ are incidence matrices of arcs corresponding to sets F, G.

The PN reachability tree (RT) express all states reachable from $\mathbf{x}_0$ as well as how (by means of firing which transitions) they can be reached. Thus, the nodes of the RT are labeled with the actual PN marking (state vectors) and the arcs are labeled with the transitions between the states. The RT root is represented by the initial state $\mathbf{x}_0$ and the RT leafs are expressed by the states reachable from $\mathbf{x}_0$. All these vectors create columns of $\mathbf{X}_{reach}$. Connecting the leafs with the same name the reachability graph (RG) arises. The PN T-invariants and P-invariants [8, 12, 17] are important too, respectively, at diagnosability [15] and supervision [4] (and subsequently for deadlocks elimination). While T-invariants restore an initial state, P-invariants ensure the token preservation. A T-invariant $\mathbf{v}$ is a solution of the equation $\mathbf{Bv} = \mathbf{0}$. A P-invariant $\mathbf{y}$ is a solution of the equation $\mathbf{B}^T\mathbf{y} = \mathbf{0}$. For any state $\mathbf{x}$ reachable from $\mathbf{x}_0$ the relation $\mathbf{y}^T.\mathbf{x} = \mathbf{y}^T.\mathbf{x}_0$ is valid. This fact was utilized at the supervisor synthesis [4] based on P-invariants.

To illustrate the PN-based approach to the detection and recovery of failures in DES modelled by PN let us introduce the following case study.

2 Case Study on Simple Railroad Crossing

Consider the simple railroad crossing where the railroad crossing gate prevents a direct contact of vehicles on the road with trains. The PN model of such system consists of three cooperating sub-models expressing in Fig. 1 (left) the behaviour of the train, crossing gate and control system. Here, the sense of the places in the failure-free case is the following: (i) the train has the states: p_1 = approaching to the crossing, p_2 = being before the crossing, p_3 = being within the crossing, p_4 = being after the crossing; (ii) the barrier of the crossing gate has the states: p_{11} = it is up, p_{12} = it is down. The transitions t_6 and t_7 model, respectively, the events of raising and lowering the barrier; (iii) the control system has the states: p_5, p_6, p_7, p_8, p_9, p_{10}; (iv) the place p_{13} represents the interlock giving the warning signal for the train that the barrier is still up. The reachable states $\mathbf{x}_i, i = 0,\ldots,7$ (RT/RG nodes N_{i+1}), of the failure-free system are expressed as the rows of the following matrix

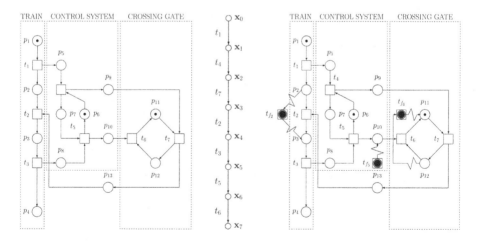

Fig. 1. The PN model of the failure-free case together with its RT (left) and the PN model with three potential failures (right)

$$
\mathbf{X}_{reach}^{T} = \begin{pmatrix}
1\,0\,0\,0\,0\,1\,0\,0\,0\,0\,1\,0\,0 \\
0\,1\,0\,0\,1\,1\,0\,0\,0\,0\,1\,0\,0 \\
0\,1\,0\,0\,0\,0\,1\,0\,1\,0\,1\,0\,0 \\
0\,1\,0\,0\,0\,0\,1\,0\,0\,0\,0\,1\,1 \\
0\,0\,1\,0\,0\,0\,1\,0\,0\,0\,0\,1\,0 \\
0\,0\,0\,1\,0\,0\,1\,1\,0\,0\,0\,1\,0 \\
0\,0\,0\,1\,0\,1\,0\,0\,0\,1\,0\,1\,0 \\
0\,0\,0\,1\,0\,1\,0\,0\,0\,0\,1\,0\,0
\end{pmatrix} \tag{1}
$$

The RT is displayed just by the failure-free PN model in Fig. 1. It is simple, without any branching.

However, there can occur three potential failures, one in each subsystem. They are expressed by means of the failure transitions t_{f_2}, t_{f_5}, t_{f_6} given in Fig. 1 (right). The transition t_{f_2} takes a token from p_2 and puts a token into p_3 out of the correct sequence, t_{f_6} does the same for p_{12} and p_{11}, and t_{f_5} involves an erroneous generation of a token in p_{10} which directly influences the position of the barrier. Thus, t_{f_2} represents a human failure (when the engine-driver omits or ignores the warning signal), t_{f_6} expresses the failure of the crossing gate (when a premature gate raising occurs or the gate is mechanically damaged), and t_{f_5} represents a control system failure (when an illegitimate signal occurs).

It is practically impossible to recover the human failure of the engine-driver. Likewise, the technical problem in the crossing gate caused by a wrong function of the barrier raising/lowering can be hardly recovered. However, the erroneous function of the control system can be detected and recovered. Consequently, let us consider in Fig. 1 (right) only the failure represented by t_{f_5} and neglect the failures represented by the transitions t_{f_2} and t_{f_6}. Then the coverability tree and graph are given in Fig. 2. The reachable states of this model (nodes of the RT/RG) are given as the columns of the following matrix where

$$
\mathbf{X}_{reach} = \begin{pmatrix}
1\,0\,1\,0\,0\,0\,0\,0\,0\,0\,0\,0\,0\,0\,0\,0\,0\,0\,0\,0\,0\,0 \\
0\,1\,0\,1\,1\,1\,1\,0\,1\,1\,0\,0\,0\,1\,0\,0\,0\,0\,0\,0\,0\,0 \\
0\,0\,0\,0\,0\,0\,0\,1\,0\,0\,1\,1\,0\,0\,0\,0\,1\,0\,0\,0\,0\,0 \\
0\,0\,0\,0\,0\,0\,0\,0\,0\,1\,0\,0\,0\,1\,1\,1\,0\,1\,1\,1\,1 \\
0\,1\,0\,0\,1\,0\,0\,0\,0\,0\,0\,0\,0\,0\,0\,0\,0\,0\,0\,0\,0\,0 \\
1\,1\,1\,0\,1\,0\,0\,0\,0\,0\,0\,0\,1\,0\,0\,0\,1\,1\,0\,1 \\
0\,0\,0\,1\,0\,1\,1\,1\,1\,1\,1\,1\,1\,0\,1\,1\,1\,0\,0\,1\,0 \\
0\,0\,0\,0\,0\,0\,0\,0\,1\,0\,0\,0\,0\,1\,1\,0\,0\,0\,1\,0 \\
0\,0\,0\,1\,0\,0\,1\,0\,0\,0\,0\,0\,0\,0\,0\,0\,0\,0\,0\,0\,0 \\
1\,1\,\omega\,1\,\omega\,1\,\omega\,1\,0\,\omega\,1\,0\,\omega\,\omega\,2\,0\,\omega\,\omega\,1\,\omega\,\omega\,\omega \\
1\,1\,1\,1\,1\,0\,1\,0\,1\,0\,0\,1\,0\,1\,0\,1\,0\,1\,1\,0\,1\,1 \\
0\,0\,0\,0\,0\,1\,0\,1\,0\,1\,1\,0\,1\,0\,1\,0\,1\,0\,0\,1\,0\,0 \\
0\,0\,0\,0\,0\,1\,0\,0\,1\,1\,0\,0\,0\,1\,0\,0\,0\,0\,0\,0\,0
\end{pmatrix} \tag{2}
$$

It can be seen that at the infinity number of t_{f_5} occurrences, one half of the 22 states have the self-loops (see Fig. 2 right) which are expressed by the symbol ω.

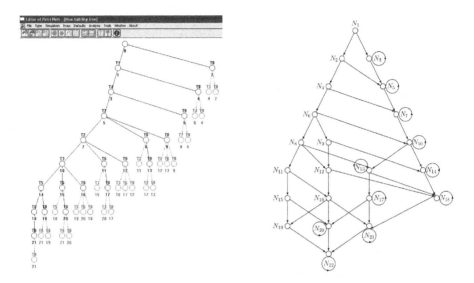

Fig. 2. The coverability tree (left) and coverability graph (right) of the PN model with t_{f_5} at the infinite number of possible occurrences of the failure

In order to generate only the finite number of the failure t_{f_5} occurrences, the place p_{14} was added to the previous 13 places - see Fig. 3. In general, the failure can occur more times. The more times the failure occurs, the more complicated will be the structure and dimensionality of RT (RG). Therefore, here we will suppose its occurrence only once as it is displayed in Fig. 3 (left) in order to demonstrate how to deal with the failure. In case of more failures such process will be more complicated. The model parameters are

$$
\mathbf{F} = \begin{pmatrix}
1 & 0 & 0 & 0 & 0 & 0 & 0 & 0 & 0 & 0 \\
0 & 1 & 0 & 0 & 0 & 0 & 0 & 0 & 0 & 0 \\
0 & 0 & 1 & 0 & 0 & 0 & 0 & 0 & 0 & 0 \\
0 & 0 & 0 & 0 & 0 & 0 & 0 & 0 & 0 & 0 \\
0 & 0 & 0 & 1 & 0 & 0 & 0 & 0 & 0 & 0 \\
0 & 0 & 0 & 1 & 0 & 0 & 0 & 0 & 0 & 0 \\
0 & 0 & 0 & 0 & 1 & 0 & 0 & 0 & 0 & 1 \\
0 & 0 & 0 & 0 & 1 & 0 & 0 & 0 & 0 & 0 \\
0 & 0 & 0 & 0 & 0 & 0 & 1 & 0 & 0 & 0 \\
0 & 0 & 0 & 0 & 0 & 1 & 0 & 0 & 1 & 0 \\
0 & 0 & 0 & 0 & 0 & 1 & 0 & 1 & 1 & 1 \\
0 & 0 & 0 & 0 & 0 & 1 & 0 & 0 & 0 & 0 \\
0 & 1 & 0 & 0 & 0 & 0 & 0 & 0 & 0 & 0 \\
0 & 0 & 0 & 0 & 0 & 0 & 0 & 1 & 0 & 0
\end{pmatrix} ;\ \mathbf{G}^T = \begin{pmatrix}
0 & 0 & 0 & 0 & 0 & 0 & 0 & 0 & 0 & 0 \\
1 & 0 & 0 & 0 & 0 & 0 & 0 & 0 & 0 & 0 \\
0 & 1 & 0 & 0 & 0 & 0 & 0 & 0 & 0 & 0 \\
0 & 0 & 1 & 0 & 0 & 0 & 0 & 0 & 0 & 0 \\
1 & 0 & 0 & 0 & 0 & 0 & 0 & 0 & 0 & 0 \\
0 & 0 & 0 & 0 & 1 & 0 & 0 & 0 & 0 & 0 \\
0 & 0 & 0 & 1 & 0 & 0 & 0 & 0 & 0 & 1 \\
0 & 0 & 1 & 0 & 0 & 0 & 0 & 0 & 0 & 0 \\
0 & 0 & 0 & 1 & 0 & 0 & 0 & 0 & 0 & 0 \\
0 & 0 & 0 & 0 & 1 & 0 & 0 & 1 & 0 & 0 \\
0 & 0 & 0 & 0 & 0 & 1 & 0 & 0 & 1 & 0 \\
0 & 0 & 0 & 0 & 0 & 0 & 1 & 0 & 0 & 1 \\
0 & 0 & 0 & 0 & 0 & 0 & 1 & 0 & 0 & 0 \\
0 & 0 & 0 & 0 & 0 & 0 & 0 & 0 & 0 & 0
\end{pmatrix} ;\ \mathbf{x}_0 = \begin{pmatrix}
1 \\ 0 \\ 0 \\ 0 \\ 0 \\ 1 \\ 0 \\ 0 \\ 0 \\ 0 \\ 1 \\ 0 \\ 0 \\ 1
\end{pmatrix} . \quad (3)
$$

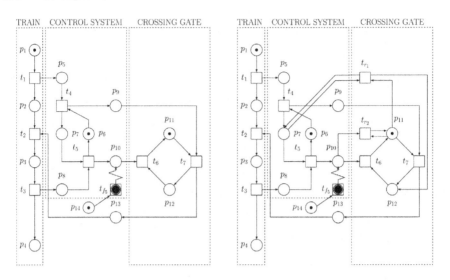

Fig. 3. The PN model of the system with the failure represented by t_{f_5} (left) and the PN model of the final recovered system (right)

Then, RT and RG of the failed system are given in Fig. 4. It can be seen that the number of states as well as the RT/RG structure are completely different in comparison with RT of the failure-free system in Fig. 1. Namely, the branching occurs here. The states (nodes of the RT/RG) are the columns of the matrix $\mathbf{X}_{reach}$.

$$\mathbf{X}_{reach} = \begin{pmatrix} 1\,0\,1\,0\,0\,0\,0\,0\,0\,0\,0\,0\,0\,0\,0\,0\,0\,0\,0 \\ 0\,1\,0\,1\,1\,1\,1\,0\,1\,0\,0\,1\,0\,0\,0\,0\,0\,0\,0 \\ 0\,0\,0\,0\,0\,0\,0\,1\,0\,0\,1\,0\,0\,0\,1\,0\,0\,0\,0 \\ 0\,0\,0\,0\,0\,0\,0\,0\,0\,1\,0\,0\,1\,1\,0\,1\,1\,1\,1 \\ 0\,1\,0\,0\,1\,0\,0\,0\,0\,0\,0\,0\,0\,0\,0\,0\,0\,0\,0 \\ 1\,1\,1\,0\,1\,0\,0\,0\,0\,0\,0\,1\,0\,0\,1\,1\,0\,1 \\ 0\,0\,0\,1\,0\,1\,1\,1\,1\,1\,1\,0\,1\,1\,0\,0\,1\,0 \\ 0\,0\,0\,0\,0\,0\,0\,0\,1\,0\,0\,0\,1\,0\,0\,0\,1\,0 \\ 0\,0\,0\,1\,0\,0\,1\,0\,0\,0\,0\,0\,0\,0\,0\,0\,0\,0 \\ 0\,0\,1\,0\,1\,0\,1\,0\,1\,0\,1\,0\,1\,1\,0\,0\,2\,0\,1 \\ 1\,1\,1\,1\,1\,0\,1\,0\,0\,0\,0\,1\,0\,0\,1\,1\,0\,1\,1 \\ 0\,0\,0\,0\,0\,1\,0\,1\,1\,1\,1\,0\,1\,1\,0\,0\,1\,0\,0 \\ 0\,0\,0\,0\,0\,1\,0\,0\,1\,0\,0\,1\,0\,0\,0\,0\,0\,0 \\ 1\,1\,0\,1\,0\,1\,0\,1\,0\,1\,0\,0\,1\,0\,0\,1\,0\,0\,0 \end{pmatrix} \qquad (4)$$

The RT has 19 nodes. However, with the accruing number of occurrences of the failure, the RT/RG dimensionality and complexity escalate. When $\sigma_{p_{14}} = 2$ RT has 30 nodes, when $\sigma_{p_{14}} = 5$ RT has 63 nodes, when $\sigma_{p_{14}} = 10$ RT has 118 nodes, etc. Although the procedure of RT computation is the same, computational time correspondingly increases.

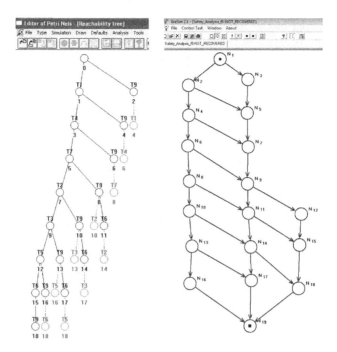

Fig. 4. The RT of the system with the finite number (namely only once in this case) of possible occurrences of the failure represented by t_{f_5} (left) and the corresponfing RG (right)

To detect and recover the failure(s) we have to distinguish whether the barrier is down or up. When the train is approaching, in the standard situation (without any failure) the barrier is down. However, in the non-standard situation (when the failure t_{f_5} occurs) the barrier is going up. This is very dangerous situation, critical as to safety. To detect the failure it is necessary to have redundant information. It must be contained in the control system itself. Because p_6 and p_7 in the control system correspond to p_{11} and p_{12} in the real crossing gate, the failure is detected by checking if p_7 and p_{11} are active simultaneously. If yes, there exists a contradiction between the real (i.e. fault) situation and standard one. After detecting the failure a kind of recovery can be applied. It depends on which case is accepted as the true state. When it is supposed that the barrier is up and drops down the recovery is realized by means of t_{r_1}. More detailed analysis is possible by means of the RT and/or RG in Fig. 5 using information about the nodes given in the matrix given by the relation (5). When the barrier is up and none train is approaching, the situation is considerably simpler. Namely, by virtue of t_{r_2} the fail signal p_{10} from the control system and the activity of p_{11} guarantee that the fail signal can be practically ignored.

The PN model of the recovered system is given in Fig. 3 (right). It has 30 states being the columns of the following matrix

$$
\mathbf{X}_{reach} =
\begin{pmatrix}
1\,0\,1\,0\,0\,1\,0\,0\,0\,0\,0\,0\,0\,0\,0\,0\,0\,0\,0\,0\,0\,0\,0\,0\,0\,0\,0\,0\,0\,0 \\
0\,1\,0\,1\,1\,0\,1\,1\,1\,1\,0\,1\,1\,1\,0\,0\,1\,1\,1\,0\,0\,0\,0\,0\,0\,0\,0\,0\,0\,0 \\
0\,0\,0\,0\,0\,0\,0\,0\,0\,0\,1\,0\,0\,0\,0\,1\,0\,0\,0\,0\,0\,1\,1\,0\,0\,0\,0\,0\,0\,0 \\
0\,0\,0\,0\,0\,0\,0\,0\,0\,0\,0\,0\,0\,0\,1\,0\,0\,0\,0\,1\,1\,0\,0\,1\,1\,1\,1\,1\,1\,1 \\
0\,1\,0\,0\,1\,0\,0\,0\,0\,1\,0\,0\,0\,0\,0\,0\,0\,0\,0\,0\,0\,0\,0\,0\,0\,0\,0\,0\,0\,0 \\
1\,1\,1\,0\,1\,1\,0\,0\,0\,1\,0\,0\,0\,0\,0\,0\,0\,1\,0\,0\,0\,1\,1\,0\,0\,1\,1\,1 \\
0\,0\,0\,1\,0\,0\,1\,1\,1\,0\,1\,1\,1\,1\,1\,1\,1\,1\,0\,1\,1\,1\,0\,0\,1\,1\,0\,0\,0 \\
0\,0\,0\,0\,0\,0\,0\,0\,0\,0\,0\,0\,0\,1\,0\,0\,0\,0\,0\,1\,0\,0\,0\,0\,1\,1\,0\,0\,0 \\
0\,0\,0\,1\,0\,0\,0\,1\,1\,0\,0\,0\,1\,1\,0\,0\,0\,1\,0\,0\,0\,0\,0\,0\,0\,0\,0\,0\,0 \\
0\,0\,1\,0\,1\,0\,0\,1\,0\,0\,0\,1\,1\,0\,0\,1\,0\,0\,0\,1\,1\,0\,0\,0\,2\,0\,0\,1\,1\,0 \\
1\,1\,1\,1\,1\,1\,0\,1\,0\,1\,0\,0\,0\,1\,0\,0\,1\,0\,0\,0\,0\,1\,0\,1\,0\,1\,0\,1\,0\,1 \\
0\,0\,0\,0\,0\,0\,1\,0\,1\,0\,1\,1\,1\,0\,1\,1\,0\,1\,1\,1\,1\,0\,1\,0\,1\,0\,1\,0\,1\,0 \\
0\,0\,0\,0\,0\,0\,1\,0\,0\,0\,0\,1\,0\,0\,0\,0\,1\,1\,0\,0\,0\,0\,0\,0\,0\,0\,0\,0\,0 \\
1\,1\,0\,1\,0\,0\,1\,0\,1\,0\,1\,0\,0\,0\,1\,0\,0\,0\,0\,1\,0\,0\,0\,1\,0\,0\,0\,0\,0\,0
\end{pmatrix}
\tag{5}
$$

The RT and RG of the recovered system are given in Fig. 5. But the deadlock N_{19} $(\mathbf{x}_{18})$ occurs there.

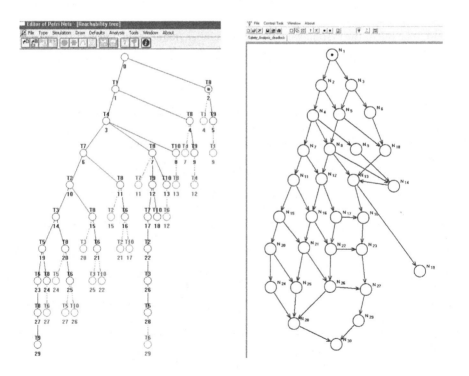

Fig. 5. The RT of the recovered system (left) and the corresponding RG (right)

2.1 Supervisor Synthesis for Deadlock(s) Elimination

A general approach to the supervisor synthesis based on P-invariants of PN was presented in [4]. Suitable linear combinations of entries of the state vector $\mathbf{x}$ (i.e. $\mathbf{L}.\mathbf{x}$) are restricted by means of entries of the constant vector $\mathbf{b}$ ($\mathbf{L}.\mathbf{x} \leq \mathbf{b}$). Then, in a nutshell, the supervisor synthesis is as follows

$$\mathbf{L}.\mathbf{x} \leq \mathbf{b} \text{ i.e. } \mathbf{L}.\mathbf{x} + \mathbf{I}_s.\mathbf{x}_s = \mathbf{b}; \ \mathbf{Y}^T.(\mathbf{B}^T \ \mathbf{B}_s^T)^T = \mathbf{0}$$
$$\text{i.e. when } \mathbf{Y}^T \triangleq (\mathbf{L} \ \mathbf{I}_s) \text{ then } \mathbf{L}.\mathbf{B} + \mathbf{I}_s.\mathbf{B}_s = \mathbf{0}$$
$$\mathbf{B}_s = -\mathbf{L}.\mathbf{B} = \mathbf{G}_s^T - \mathbf{F}_s; \ \mathbf{x}_0^s = \mathbf{b} - \mathbf{L}.\mathbf{x}_0 \tag{6}$$

where $\mathbf{L}$ is a $(n_s \times n)$ matrix of integers, $\mathbf{b}$ is a $(n_s \times 1)$ vector of integers, $\mathbf{I}_s$ is the identity matrix, $\mathbf{Y}$ is the matrix of invariants, $\mathbf{B}_s$ is the supervisor structure and $\mathbf{x}_0^s$ is its initial state, $\mathbf{F}_s$, $\mathbf{G}_s$ are the incidence matrices of the supervisor. Consequently, ${}^s\mathbf{B} = (\mathbf{B}^T \ \mathbf{B}_s^T)^T$ is the structure of the supervised system (i.e. original system plus supervisor) and ${}^s\mathbf{x}_0 = (\mathbf{x}_0^T \ (\mathbf{x}_0^s)^T)^T$ is its initial state.

Let us deal with the deadlock state $\mathbf{x}_{18}$ by means of synthesizing a suitable supervisor. Because the deadlock state is N_{19} i.e. $\mathbf{x}_{18} = (0, 1, 0, 0, 0, 0, 1, 0, 1, 0, 0, 1, 0, 0)^T$, we have to avoid its activation.

Consider $\mathbf{L} = (0, 1, 0, 0, 0, 0, 1, 0, 1, 0, 0, 1, 0, 0)$ and $\mathbf{b} = 3$, i.e. at most three of p_2, p_7, p_9, p_{12} can be active together. Then, the supervisor structure is given as $\mathbf{B}_s = (-1, 1, 0, -2, 1, 1, 0, 0, 0, -1)$. After the break-up of $\mathbf{B}_s$ the incidence matrices of arcs are acquired. $\mathbf{F}_s = (1, 0, 0, 2, 0, 0, 0, 0, 0, 1)$ and $\mathbf{G}_s^T = (0, 1, 0, 0, 1, 1, 0, 0, 0, 0)$. The initial state of the supervisor is ${}^s\mathbf{x}_0 = 3 - 0 = 3$. The supervisor is incorporated into the PN model of the recovered system given in Fig. 3 (right). Consequently, the form of the PN model is changed into the form given in Fig. 6. Its RT and RG are in Fig. 7. The reachable states of the deadlock-free recovered system are given as the columns of the matrix

$$\mathbf{X}_{reach}^s = \begin{pmatrix}
1\ 0\ 1\ 0\ 0\ 1\ 0\ 0\ 0\ 0\ 0\ 0\ 0\ 0\ 0\ 0\ 0\ 0\ 0\ 0\ 0\ 0\ 0\ 0\ 0\ 0\ 0\ 0 \\
0\ 1\ 0\ 1\ 1\ 0\ 1\ 1\ 1\ 0\ 1\ 1\ 0\ 0\ 1\ 1\ 0\ 0\ 0\ 0\ 0\ 0\ 0\ 0\ 0\ 0\ 0\ 0 \\
0\ 0\ 0\ 0\ 0\ 0\ 0\ 0\ 0\ 1\ 0\ 0\ 0\ 1\ 0\ 0\ 0\ 0\ 1\ 1\ 0\ 0\ 0\ 0\ 0\ 0\ 0\ 0 \\
0\ 0\ 0\ 0\ 0\ 0\ 0\ 0\ 0\ 0\ 0\ 0\ 1\ 0\ 0\ 0\ 1\ 1\ 0\ 0\ 1\ 1\ 1\ 1\ 1\ 1\ 1\ 1 \\
0\ 1\ 0\ 0\ 1\ 0\ 0\ 0\ 1\ 0\ 0\ 0\ 0\ 0\ 0\ 0\ 0\ 0\ 0\ 0\ 0\ 0\ 0\ 0\ 0\ 0\ 0\ 0 \\
1\ 1\ 1\ 0\ 1\ 1\ 0\ 0\ 1\ 0\ 0\ 0\ 0\ 0\ 0\ 1\ 0\ 0\ 0\ 1\ 1\ 0\ 0\ 1\ 1\ 1 \\
0\ 0\ 0\ 1\ 0\ 0\ 1\ 1\ 0\ 1\ 1\ 1\ 1\ 1\ 1\ 1\ 0\ 1\ 1\ 1\ 0\ 0\ 1\ 1\ 0\ 0\ 0 \\
0\ 0\ 0\ 0\ 0\ 0\ 0\ 0\ 0\ 0\ 0\ 0\ 1\ 0\ 0\ 0\ 0\ 1\ 0\ 0\ 0\ 0\ 1\ 1\ 0\ 0\ 0 \\
0\ 0\ 0\ 1\ 0\ 0\ 0\ 1\ 0\ 0\ 0\ 1\ 0\ 0\ 0\ 0\ 0\ 0\ 0\ 0\ 0\ 0\ 0\ 0\ 0\ 0 \\
0\ 0\ 1\ 0\ 1\ 0\ 0\ 1\ 0\ 0\ 1\ 0\ 0\ 1\ 0\ 0\ 1\ 1\ 0\ 0\ 0\ 2\ 0\ 0\ 1\ 1\ 0 \\
1\ 1\ 1\ 1\ 1\ 1\ 0\ 1\ 1\ 0\ 0\ 1\ 0\ 0\ 1\ 0\ 0\ 0\ 1\ 0\ 1\ 0\ 1\ 0\ 1\ 0\ 1 \\
0\ 0\ 0\ 0\ 0\ 0\ 1\ 0\ 0\ 1\ 1\ 0\ 1\ 1\ 0\ 1\ 1\ 1\ 0\ 1\ 0\ 1\ 0\ 1\ 0\ 1\ 0 \\
0\ 0\ 0\ 0\ 0\ 0\ 1\ 0\ 0\ 0\ 1\ 0\ 0\ 0\ 1\ 1\ 0\ 0\ 0\ 0\ 0\ 0\ 0\ 0\ 0\ 0\ 0 \\
1\ 1\ 0\ 1\ 0\ 0\ 1\ 0\ 0\ 1\ 0\ 0\ 1\ 0\ 0\ 0\ 1\ 0\ 0\ 0\ 1\ 0\ 0\ 0\ 0\ 0\ 0 \\
3\ 2\ 3\ 0\ 2\ 3\ 0\ 0\ 2\ 1\ 0\ 0\ 1\ 1\ 1\ 0\ 2\ 1\ 2\ 1\ 3\ 2\ 2\ 1\ 3\ 2\ 3
\end{pmatrix} \tag{7}$$

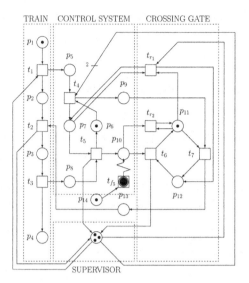

Fig. 6. The PN model of the system with the recovered failure and the deadlock removed by means of the supervisor

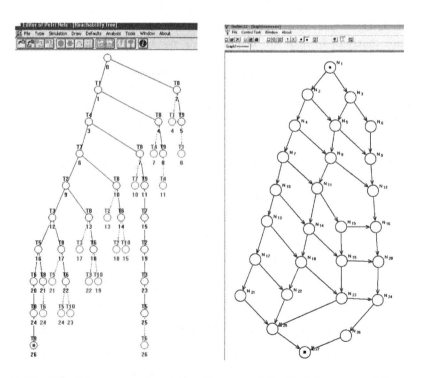

Fig. 7. The RT of the recovered system with removed deadlock by means of the supervisor (left) and the corresponding RG (right)

3   Conclusion

The PN-based approach to dealing with failures in DES was presented. It is based on utilizing RT/RG of the PN-based model of DES. Moreover, the elimination of deadlock(s) by means of supervision (synthesizing of the suitable supervisor) based on P-invariants of PN, introduced in [4], was utilized.

The presented approach consists of the following steps: (i) creating the PN model of the investigated kind of DES; (ii) finding its behaviour in the standard (failure-free) situation; (iii) analyzing the model with respect to possible failures (in general, each system has its specificity and it is practically impossible to find a unified approach for all systems); (iv) selecting the failures which can be successfully recovered (because there are different kinds of failures and some of them cannot be recovered - e.g. human failures of the engine-driver or a mechanical problem in the crossing gate); (v) finding the structure of the recovered PN model; (vi) testing its behaviour with respect to deadlocks; (vii) removing deadlocks and finding the deadlock-free PN model.

PN were used in all of the steps. They make possible to create the uniform model of a system and compute its RT/RG. However, in different systems different states can fail. Hence, the model recovering process is individual. As to the computational complexity of the approach, it corresponds especially to that of computing RT, that depends on the structure of the PN model in question.

To illustrate the soundness of the procedure, the case study on the simple railroad crossing was introduced. Finally, the deadlock-free recovery model was found. It is necessary to emphasize that there are also the failures in DES which cannot be recovered by means of the procedure. They depend on human failures, bad properties and mistakes and/or on bad technical state of devices. They must be precluded either by means of the better preparation of human operators and/or by means of better executing maintenance of devices, their routine testing and/or checking, early replacing their components, etc.

In future a possibility of generalization of the recovery process by means of PN will be investigated.

References

1. Bernardi, S., Flammini, F., Marrone, S., Merseguer, J., Papa, C., Vittorini, V.: Model-driven availability evaluation of railway control systems. In: Flammini, F., Bologna, S., Vittorini, V. (eds.) SAFECOMP 2011. LNCS, vol. 6894, pp. 15–28. Springer, Heidelberg (2011). doi:10.1007/978-3-642-24270-0_2
2. Cabasino, M.P., Giua, A., Pocci, M., Seatzu, C.: Discrete event diagnosis using labeled Petri nets. An application to manufacturing systems. Control Eng. Pract. **19**(9), 989–1001 (2011)
3. Cabasino, M.P., Giua, A., Lafortune, S., Seatzu, C.: New approach for diagnosability analysis of Petri nets using verifier nets. IEEE Trans. Autom. Control **57**(12), 3104–3117 (2012)

4. Čapkovič, F.: Petri net-based synthesis of agent cooperation by means of modularity and supervision principles. In: Dimirovski, G.M. (ed.) Complex Systems: Relationships Between Control, Communications and Computing. Studies in Systems, Decision and Control, pp. 429–450. Springer International Publishing, Heidelberg (2016)

5. Chang, S.J., DiCesare, F., Goldbogen, G.: Failure propagation trees for diagnosis in manufacturing systems. IEEE Trans. SMC **21**(4), 767–776 (1991)

6. Chung, S., Wu, C., Jeng, M.: Failure diagnosis: A case study on modeling and analysis by Petri nets. In: Proceedings of the IEEE International Conference on Systems, Man & Cybernetics, Washington, D.C., USA, 5-8 October 2003, pp. 2727–2732 (2003)

7. Desel, J., Reisig, W.: Place/transition Petri nets. In: Reisig, W., Rozenberg, G. (eds.) ACPN 1996. LNCS, vol. 1491, pp. 122–173. Springer, Heidelberg (1998). doi:10.1007/3-540-65306-6_15

8. Desel, J., Esparza, J.: Free Choice Petri Nets. Cambridge Tracts in Theoretical Computer Science, vol. 40. Cambridge University Press, Cambridge (1995)

9. Fanni, A., Giua, A., Sanna, N.: Control and error recovery of Petri net models with event observers. In: Proceedings of the Second International Workshop on Manufacturing and Petri Nets, Toulouse, France, pp. 53–68 (1997)

10. Giua, A.: State estimation and fault detection using Petri nets. In: Kristensen, L.M., Petrucci, L. (eds.) PETRI NETS 2011. LNCS, vol. 6709, pp. 38–48. Springer, Heidelberg (2011). doi:10.1007/978-3-642-21834-7_3

11. Guo, Z., et al.: Failure recovery: when the cure is worse than the disease. In: Proceedings of the 14th Workshop on Hot Topics in Operating Systems, Santa Ana Pueblo, New Mexico, USA, May 13–15 2013, USENIX, Berkeley, USA, 6 pages (2013). URL:https://www.usenix.org/conference/hotos13/failure-recovery-when-cure-worse-disease

12. Haar, S.: Types of Asynchronous Diagnosability and the Reveals-Relation in Occurrence Nets. Research Report RR-6902, INRIA, France (2009)

13. Huang, Z., Chandra, V., Jiang, S., Kumar, R.: Modeling discrete event systems with faults using a rules based modeling formalism. Math. Comput. Model. Dyn. Syst. **9**(3), 233–254 (2003)

14. Leveson, N.G., Stolzy, J.L.: Safety analysis using Petri nets. IEEE Trans. Softw. Eng. **SE–13**(3), 386–397 (1987)

15. Li, B., Khlif-Bouassida, M., Toguyéni, A.: On-the-fly diagnosability analysis of labeled Petri nets using T-invariants, pp. 064–070 . IFAC-PapersOnLine 48-7, Science Direct, Elsevier (2015)

16. Liu, B.: An efficient approach for diagnosability and diagnosis of DES based on labeled Petri nets - untimed and timed contexts. Ph.D. Thesis, Laboratoire d' Automatique, Génie Informatique et Signal, École Centrale de Lille, Lille, France (2014)

17. Murata, T.: Petri nets: Properties, analysis and applications. Proc. IEEE **77**, 541–580 (1989)

18. Odrey, N.G.: Error recovery in production systems: a Petri net based intelligent system approach. In: Kordic, V. (ed.) Petri Net, Theory and Applications, pp. 302–336. I-Tech Education and Publishing, Vienna (2008)

19. Peterson, J.L.: Petri Net Theory and the Modeling of Systems. Prentice-Hall Inc., Englewood Cliffs, New Jersey (1981)

20. Ramaswamy, S., Valavanis, K.P.: Modeling, analysis and simulation of failures in a materials handling system with extended Petri nets. IEEE Trans. Syst. Man Cybernet. **24**(9), 1358–1373 (1994)

21. Ramírez-Treviño, A., Ruiz-Beltrán, A.E., Arámburo-Lizárraga, J., López-Mellado, E.: Structural diagnosability of DES and design of reduced Petri net diagnosers. IEEE Trans. Syst. Man Cybernet. Part A Syst. Hum. **42**(2), 416–429 (2012)
22. Ramírez-Treviño, A., Ruiz-Beltrán, A.E., Rivera-Rangel, I., López-Mellado, E.: Online fault diagnosis of discrete event systems. A Petri net-based approach. IEEE Trans. Autom. Sci. Eng. **4**(1), 31–39 (2007)
23. Urban, S.D., et al.: The assurance point model for consistency and recovery in service composition. In: Innovations, Standards and Practices of Web Services: Emerging Research Topics, Chapt. 12, pp. 250–287, IGI Global, USA (2012)
24. Wen, Y., Jeng, M.: Diagnosability analysis based on T-invariants of Petri nets. In: Proceedings of 2005 IEEE International Conference on Networking, Sensing and Control, March 2005, pp. 371–376 (2005)
25. Zaytoon, J., Lafortune, S.: Overview of fault diagnosis methods for discrete event systems. Annu. Rev. Control **37**, 308–320 (2013)

Defining Deviation Sub-spaces for the A*W Robust Planning Algorithm

Igor Wojnicki and Sebastian Ernst[(✉)]

Department of Applied Computer Science, AGH University of Science
and Technology, Al. Mickiewicza 30, 30-059 Kraków, Poland
{wojnicki,ernst}@agh.edu.pl

Abstract. The paper presents further results from the development of
the A*W hybrid planning algorithm aimed at determining robust plans
for multiple entities co-existing in a common environment under uncertain conditions. The main focus is on strategies to determine deviation
sub-spaces, i.e. the areas for which multi-variant plans are generated, as
that selection determines the balance between computational efficiency
and robustness. A general strategy is presented, followed by examples
used to discuss the influence of the parameters on the behaviour of the
algorithm. Guidelines for sub-space identification are provided, and further directions for research are outlined.

Keywords: Robust planning · Heuristic search · Uncertainty

1 Introduction

Automated planning is a very important area of Artificial Intelligence, as it
can help to solve many practical problems, from robot or vehicle navigation,
to arrangement of tasks in highly complex industrial applications. In general, a
planning problem involves a set of possible states and known transitions between
them; its solution is usually in the form of a sequence of actions which lead from
the start state to the goal state. Many problem domains can be expressed as
graphs, usually with vertices representing possible states, and edges indicating
possible actions or decisions, along with their cost.

As such, these problems can be solved with general graph-search algorithms,
such as BFS, DFS (breadth/depth-first search) [5] or Dijkstra's algorithm [2]. For
more complex or larger problems, heuristic search is often used [1]. Algorithms
such as A* [3] use a *heuristic function* which estimates the distance from a given
state to the goal state, thus directing the search process towards the finish. As it
may be difficult to define a good heuristic function for some problems, domain-
independent planning methods are also being developed [4,6].

Moreover, in real world, there are uncertainties. We may not know everything
about the world (i.e. have an incomplete or outdated "map"), we may not be
able to fully control plan execution or accurately observe its progress. Thus, it
is often advisable to have not one, but many alternative plans which can be

© Springer International Publishing AG 2017
N.T. Nguyen et al. (Eds.): ACIIDS 2017, Part I, LNAI 10191, pp. 392–399, 2017.
DOI: 10.1007/978-3-319-54472-4_37

used in case a deviation from the initial plan (for whatever reason) occurs. This, however, greatly increases the complexity of the necessary computations.

In addition, if planning is performed for multiple entities (e.g. a number of robots moving in a common environment), the complexity grows exponentially. This is because running the same algorithm once for every entity individually (which would yield linear growth) may lead to collisions. Instead, one needs to generate one plan encompassing all entities, which means constructing one global state consisting of the individual states of each agent.

2   The A*W Algorithm

The A*W (A*/Wavefront) hybrid robust planning algorithm [7,8] addresses the aforementioned problem, i.e. computation of plans for multiple entities existing in a common environment. To avoid collisions, the states of all individual agents (consisting, for instance, of two dimensions) are collected into a single, multidimensional global state (for n agents in a two-dimensional domain, that will result in a $2n$-dimensional structure). Two agents cannot occupy a single state at a given time (in a single algorithm step). Therefore, each state where all components for more than one entity have equal values is forbidden and excluded at the domain definition step.

If, at the same time, the problem features limited controllability, predictability or observability, it is most advisable to use an algorithm which generates plans leading from any possible state to the goal. That way, plan execution will not be interrupted if deviations occur.

The Wavefront algorithm is often used in robotics to guarantee such robustness. It consists in performing a breadth-first search [5] from the goal and labelling all states with the estimated cost of reaching the goal. Plan execution can then follow a simple gradient-descent strategy, starting at any available state.

Unfortunately, the complexity of the algorithm renders it hardly usable in case of multi-entity (i.e. multidimensional) planning. Therefore, the A*W algorithm joins the efficiency of A* with the robustness of Wavefront by using the following strategy:

1. A global plan for all entities ($\mathcal{P}_\mathcal{V}$, consisting of a sequence of global states[1]) is generated using heuristic A* search.
2. Areas with potential for collisions (so-called *deviation sub-spaces* are identified.
3. Wavefront is used to generate robust plans in these subspaces to allow for quick recovery if a deviation leading to a collision occurs.

Deviation sub-spaces are identified using a possibility-of-collision detection function $cd : N \rightarrow S$, where $N = (v_{i+1}, \ldots, v_{i+m})$ is a neighbourhood of states (a subsequence of the plan), and S, the deviation sub-space is a subgraph of the

---

[1] A single glboal state consists of the states/positions of all entities.

planning domain. The goal state for the Wavefront algorithm is v_{i+m}, being the last state in the neighborhood of states. More than one deviation sub-space can exist for a given plan.

In essence, the cd function is a heuristic, which keeps the balance between computational performance (A* part) and execution robustness (Wavefront part). As such, it should reflect the features of agents which affect the probability of deviations and resulting collisions. In case of robots, that can be their tendency to slip, their motion accuracy, weight and size.

The goal of this paper is to analyse the possible strategies of identifying deviation sub-spaces, which is described in Sect. 3. Experimental results for various problem sizes and dimensionalities are provided in Sect. 4.

3   Identifying Deviation Sub-spaces

Let us assume that:

- $G = (E, V, c)$ is a state space represented as a graph, where V is the set of vertices, with each vertex representing a single possible state, E is the set of edges with each edge representing a possible transition between two states, and c is a cost function, assigning a cost to each possible transition,
- $\mathcal{P}_V = (v_1, \ldots, v_n)$ is a plan generated by the A-star algorithm over a state space given as $G = (E, V, c)$,
- N is a neighbourhood of states such as $N = (v_{i+1}, \ldots, v_{i+m})$ such that $N \subset \mathcal{P}_V$ and
- $S = (V_S, E_S, c_S, g_S)$, being a deviation sub-space, is a state sub-space for the Wavefront algorithm.

To identify a deviation sub-space, neighbourhoods N have to be identified and a $cd : N \to S$ function needs to be provided.

The proposed strategy is as follows:

1. Identify a neighbourhood being a sequence $N \subset \mathcal{P}_V$ for which a distance between any two agents are less than or equal to a given d_c; such a neighbourhood is referred to as Q (there are two separate symbols N and Q since in general case number of dimensions covered by vertices belonging to Q can be less than number of dimensions covered by vertices belonging to N).
2. Make deviation sub-space S by growing Q by a distance g; for each location in Q add all locations within a distance of g.
3. Go to 1.

It is assumed that g is calculated as $g = \left\lceil \frac{d_c}{2} \right\rceil$. The performance of this strategy is presented in the following section.

4   Examples

This section provides examples of the strategy described in Sect. 3 in different cases.

4.1 Four-Dimensional Case

Let us consider a plan which involves two agents: A and B which is given in Fig. 1a. A starts at $(1,3)$ while B at $(2,1)$, having their goals at $(6,6)$ and $(5,6)$ respectively. Coordinates for each of the agents are given in Table 1 Each of them operates within a two-dimensional space, 7×7 grid world. The planning problem regards a four-dimensional state space. Only one agent is allowed to occupy the same coordinates at the same time, thus at the same planing step. Distances among pairs of agents are given in Table 1).

Assuming and assuming $d_c = 1$ there are the following sequences of steps for which Q exist: $T_1 = \{2,3\}$, $T_2 = \{8,9\}$. There are the following Qs: $Q_1 = \{(2,2,2,3),(2,3,3,3)\}$, $Q_2 = \{(4,6,5,6),(5,6,6,6)\}$. The plan, Qs and Ss are given in Fig. 1b. Grayed areas represent Qs, their growth is indicated as hatched.

Growing Q_1 and Q_2 by $g = 1$ results in $|S_1| = 10*9 = 90$, $|S_2| = 11*10 = 110$ (it is two agents, agent A can occupy one of 10 states while agent B can occupy $10 - 1$ since one of the states is already taken by agent A). It gives in total:

$$O_S = |SS| = \sum_i |S_i| = 200$$

Number of operations to generate the initial plan with A-star algorithm is $O_A = 156$. Total number of operations to calculate:

$$O_H = O_A + O_S = 156 + 200 = 356$$

It is significantly less than a complete wavefront coverage of the proposed space, which is:

$$S_{7 \times 7, 4d} = 49 * 48 = 2,352$$

Table 1. Deviation sub-space identification, four-dimensional; distances between agents.

Step	AX	AY	BX	BY	Distance A-B
0	2	1	1	3	3
1	2	2	2	3	1
2	2	3	3	3	1
3	2	4	4	3	3
4	2	5	5	3	5
5	2	6	5	4	5
6	3	6	5	5	3
7	4	6	5	6	1
8	5	6	6	6	1

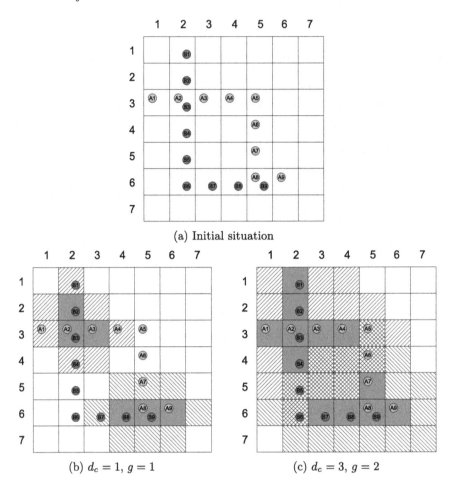

(a) Initial situation

(b) $d_c = 1, g = 1$ (c) $d_c = 3, g = 2$

Fig. 1. Deviation sub-space identification, four-dimensional space projected on two-dimensions.

Let us change d_c to increase robustness assuming the same plan. Assuming and assuming $d_c = 3$ there are the following sequences of steps for which Q exist: $T_1 = \{1, 2, 3, 4\}$, $T_2 = \{7, 8, 9\}$. There are the following Qs: $Q_1 = \{(2, 1, 1, 3), (2, 2, 2, 3), (2, 3, 3, 3), (2, 4, 4, 3)\}$, $Q_2 = \{(3, 6, 5, 5), (4, 6, 5, 6), (5, 6, 6, 6)\}$. Growing Q_1 and Q_2 by $g = 2$ results in $|S_1| = 25 * 24 = 600$, $|S_2| = 24 * 23 = 552$. The plan, Qs and Ss are given in Fig. 1c. Grayed areas represent Qs, their growth is indicated as hatched. It gives in total:

$$O_S = |SS| = \sum_i |S_i| = 1,152$$

Number of operations to generate the initial plan with A-star algorithm is $O_A = 156$. Total number of operations to calculate:

$$O_H = O_A + O_S = 156 + 1,152 = 1,308$$

Comparing with $S_{7\times7,4d}$ indicates that number of calculations reduction is not that significant in this case. It confirms that proper cd function and its parameters is crucial.

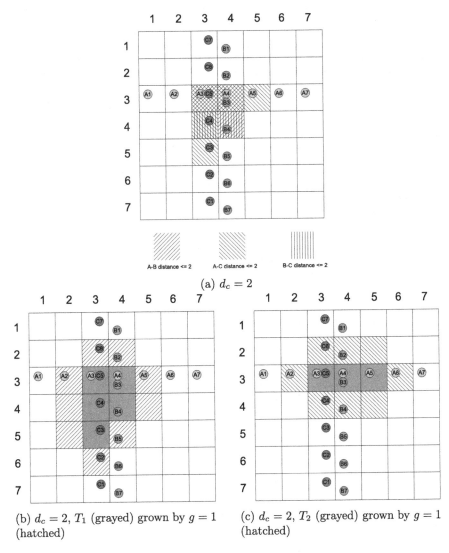

(a) $d_c = 2$

(b) $d_c = 2$, T_1 (grayed) grown by $g = 1$ (hatched)

(c) $d_c = 2$, T_2 (grayed) grown by $g = 1$ (hatched)

Fig. 2. Deviation sub-space identification, six-dimensional space projected on two-dimensions.

4.2 Six-Dimensional Case

Let us consider a more complex example. There are three agents. Each operates within a two-dimensional space, 7×7 grid world. The planning problem regards a six-dimensional state space. Only one agent is allowed to occupy the same coordinates at the same time, thus at the same planing step. Their planned coordinates are given in Table 2. The plan projected on a two-dimensional space is given in Fig. 2a. Calculating distances among agents (see Table 2) and assuming $d_c = 2$ there are the following sequences of steps for which Q exist: $T_1 = \{3, 4\}$ (all agents involved), $T_2 = \{5\}$ (agents A and C involved). Thus, there are corresponding Qs: $Q_1 = \{(3, 3, 4, 3, 3, 5), (4, 3, 4, 4, 3, 4)\}$, $Q_2 = \{(5, 3, 3, 3)\}$. There is a dimension reduction for Q_2 since only two agents are involved, thus first two coordinates correspond to agent A, while latter correspond to agent C.

Growing all Qs' locations (having $g = 1$ each agent can move according to von Neumann neighborhood or not move at all, which gives 5 states for each agent) gives an estimate of state space size, see Fig. 2b and c. $|S_1| = 14 * 13 * 12 = 2, 184$, while $|S_2| = 11 * 10 = 110$. In this case:

$$O_S = |SS| = \sum_i |S_i| = 2,294$$

The number of operations to generate the initial plan with A-star algorithm is $O_A = 603$. Total number of operations to calculate:

$$O_H = O_A + O_S = 603 + 2,294 = 2,897$$

It is significantly less than the complete wavefront coverage of the given 7×7 space for three agents, which is: $|S_{7 \times 7, 6d}| = 49 * 48 * 47 = 110, 544$.

Table 2. Deviation sub-space identification, six-dimensional; distances between agents.

Step	AX	AY	BX	BY	CX	CY	Distance A-B	Distance A-C	Distance B-C
1	1	3	4	1	3	7	5	6	7
2	2	3	4	2	3	6	3	4	5
3	3	3	4	3	3	5	1	2	3
4	4	3	4	4	3	4	1	2	1
5	5	3	4	5	3	3	3	2	3
6	6	3	4	6	3	2	5	4	5
7	7	3	4	7	3	1	7	6	7

5 Summary

This paper presents further work related to development of the A*W hybrid robust planning algorithm, aimed at generating robust (multi-variant) plans for multiple entities in a common environment. The focus of the described research is placed on determination of the so-called *deviation sub-spaces* – areas, for which robust plans are generated. The selection of these areas affects computational complexity, but at the same time determines the robustness in case deviations occur.

A short summary of the A*W algorithm is followed by a proposal of a general strategy for identification of deviation sub-spaces. That strategy is discussed and analysed using examples of various sizes and dimensionalities.

Future work involves development of more sophisticated heuristic (*cd*) functions and experimental analysis using computationally-intensive, large-scale cases.

An experimental implementation of the A*W algorithm has already been developed, but has yet to be released to the public. Experimental results will be published in follow-up papers, and the implementation will be released after it has been positively validated.

References

1. Bonet, B., Geffner, H.: Planning as heuristic search. Artif. Intell. **129**(1–2), 5–33 (2001). http://linkinghub.elsevier.com/retrieve/pii/S0004370201001084
2. Cormen, T.H., Leiserson, C.E., Rivest, R.L., Stein, C.: Introduction to Algorithms, 3rd edn. The MIT Press, Cambridge (2009)
3. Dechte, R., Pearl, J.: Generalized best-first search strategies and the optimality of A*. J. ACM **32**(3), 505–536 (1985). http://portal.acm.org/citation.cfm?id=3830
4. Haslum, P., Bonet, B., Geffner, H.: New admissible heuristics for domain-independent planning. In: Proceedings of the National Conference on Artificial Intelligence, vol. 20, p. 1163. AAAI Press, MIT Press, Menlo Park, CA, Cambridge, MA, London 1999 (2005). http://scholar.google.com/scholar?hl=en&btnG=Search&q=intitle:New+Admissible+Heuristics+for+Domain-Independent+Planning#0
5. Russell, S.J., Norvig, P.: Artificial Intelligence: A Modern Approach, 3rd edn. Pearson Education, Upper Saddle River (2010)
6. Srivastava, B., Kambhampati, S., Nguyen, T.A., Do, M.B., Gerevini, A., Serina, I.: Domain independent approaches for finding diverse plans. In: IJCAI 2007, Proceedings of the 20th International Joint Conference on Artifical Intelligence (2007)
7. Wojnicki, I., Ernst, S., Turek, W.: A robust heuristic for the multidimensional a-star/wavefront hybrid planning algorithm. In: Rutkowski, L., Korytkowski, M., Scherer, R., Tadeusiewicz, R., Zadeh, L.A., Zurada, J.M. (eds.) ICAISC 2015. LNCS (LNAI), vol. 9120, pp. 282–291. Springer, Cham (2015). doi:10.1007/978-3-319-19369-4_26
8. Wojnicki, I., Ernst, S., Turek, W.: A robust planning algorithm for groups of entities in discrete spaces. Entropy **17**(8), 5422 (2015). http://www.mdpi.com/1099-4300/17/8/5422

Creative Expert System: Comparison of Proof Searching Strategies

Bartlomiej Sniezynski[1]([⊠]), Grzegorz Legien[1], Dorota Wilk-Kołodziejczyk[1,2], Stanislawa Kluska-Nawarecka[2], Edward Nawarecki[1], and Krzysztof Jaśkowiec[2]

[1] AGH University of Science and Technology,
Al. Mickiewicza 30, 30-059 Krakow, Poland
`bartlomiej.sniezynski@agh.edu.pl`

[2] Foundry Research Institute in Krakow, Zakopianska Street 73, Krakow, Poland

Abstract. This paper presents comparison of time cost of three proof searching strategies in a creative expert system. Initially, model of the creative expert system and inference algorithm are proposed. The algorithm searches for a proof up to a given maximal depth, using one of the following strategies: finding all possible proofs, finding the first proof by depth-first and finding the first proof by breadth-first. Calculation time is measured in inference scenarios from a casting domain. Creativity of the expert system is achieved thanks to integration of inference and machine learning. The learning algorithm can be automatically executed during inference process, because its execution is formalized as a complex inference rule. Such a rule can be fired during inference process. During execution, training data is prepared from facts already stored in the knowledge base and new implications are learned from it. These implications can be used in the inference process. Therefore, it is possible to infer decisions in cases not covered by the knowledge base explicitly.

Keywords: Creative reasoning · Expert system · Machine learning · Knowledge representation and processing

1 Introduction

Traditionally, systems are based on inference engines, which apply reasoning on fixed Knowledge Bases (KB). Popular knowledge representation techniques are rule-based systems [8] or Bayesian networks [12]. Machine learning algorithms can be used to provide knowledge stored in the KB, but they are not integrated with the inference process. In this paper we consider a Creative Expert System, which integrates reasoning and machine learning. The learning algorithm can be automatically executed during inference process, because its execution is formalized as a complex inference rule. Such a rule can be fired during inference process. During execution, training data is prepared from facts already stored in the knowledge base and new implications are learned from it. These implications can be used in the inference process. Therefore, it is possible to infer decisions in cases not covered by the knowledge base explicitly.

© Springer International Publishing AG 2017
N.T. Nguyen et al. (Eds.): ACIIDS 2017, Part I, LNAI 10191, pp. 400–409, 2017.
DOI: 10.1007/978-3-319-54472-4_38

This paper is a continuation of [15,16], where inference algorithm searches for all proof trees up to a given maximal depth. Several strategies may be applied to direct the proof search and save time. In this paper we experimentally compare time complexity of three strategies: finding all possible proofs, finding the first proof by depth-first and finding the first proof by breadth-first. Reasoning algorithm is executed on a knowledge base prepared for a material choice for casting.

In the paper related research is discussed and a model for Creative Reasoning is proposed. Then, inference algorithm, the software developed and knowledge base are described. Next, experimental results are presented.

2   Machine Learning in Expert Systems

There are several examples of machine learning and expert system integrations. One of the examples is presented in [3]. This system applies Neural logic networks, what corresponds to three-valued logic. The system collects experience, which is used for adaptive learning of new rules.

Another adaptive expert system is presented in [18]. Its goal is aircraft maintenance support. Its user reports symptoms and system recommends the most accurate action. In this case historical data are used for learning too. They consist of repairs cases. Weights describing associations between symptoms and actions are updated to achieve appropriate certainty of suggested diagnosis. In case of unsuccessful prediction weights are decreased, in case of successful they are increased.

In [19] supervised learning with back propagation neural network helps the expert system to choose questions, which should be asked in the current case. Training data consists of examples collected during user's interaction with the system. Experimental results show that after training performance is much better than unaided expert system. However, knowledge bases tested were very small.

CoMES system presented in [2] applies agent-based approach for knowledge integration. It joins several popular AI and Software Engineering techniques. As a result, several algorithms can access the knowledge base. It can be also updated by machine learning algorithms.

In works cited above integration of expert systems and machine learning algorithms is ad hoc. Execution of learning is not fully integrated with a formal inference system.

Integration presented in this paper is deeper. Its source is Inferential Theory of Learning (ITL) developed by Ryszard Michalski and presented in [9]. ITL defines learning as a sequence of knowledge transmutation applications. INTERLACE system [1] developed by Michalski et al. is ITL implementation. However, implemented knowledge transmutations are related to inference only, not machine learning.

For knowledge representation in our implementation we have chosen Logic of Plausible Reasoning (LPR). It is formalisation of patterns extracted from answering common life question scenarios collected by Collins [4]. These patterns are LPR inference rules. LPR was formulated by Collins and Michalski [5]. Such an origin has serious consequences: LPR contains many inference patterns and several parameters are applied to deal with knowledge uncertainty. Reasoning engine we developed in our system origins from AUTOLOGIC system [11].

3   Creative Inference Model and Inference Algorithm

To apply *Creative Inference Model* we propose, the knowledge representation and reasoning patterns must be represented as a labeled deductive system (LDS) [7]. As a result, we assume that knowledge is represented by formulas and inference process is a sequence of *Knowledge transmutation* applications.

Knowledge transmutation is defined as the following triple:

$$kt = (p, c, a), \tag{1}$$

where p is a (possibly empty) premise or precondition, c is a consequence (pattern of formula(s) that can be inferred after transmutation application) and a is an action (empty for simple transmutations) that should be executed if premises are true according to the knowledge base.

Because of its characteristics, knowledge transmutations can be partitioned into three groups:

- simple – representing pure logic proof rules (like Modus Ponens);
- complex – representing algorithmic knowledge transformations (e.g. machine learning algorithms such as rule induction or clustering);
- search – representing search procedures (in web or database).

Simple knowledge transmutation have empty action. For example, for Modus Ponens $p = a_1, a_2, ..., a_n \rightarrow b$, $c = b$, and $a = \emptyset$. For C4.5 rule induction algorithm [13] p is a condition that checks if there is enough information about objects of some type T; e.g. patient, c represents implications with conclusion assigning *Class* attribute to objects of type T; a is execution of C4.5 algorithm to learn a classifier predicting attribute *Class* for T. If $T = patient$ and $Class = disease$ then rules predicting *disease* attribute for patients based on other patient's attributes will be generated.

We assume that cost (represented by $cost_{kt}$) is assigned for every knowledge transmutation. The cost represents computational complexity of the transmutation. It can also take into account other factors like fees for resources used. Usually, the highest cost is assigned to complex transmutations, moderate to search and low to simple ones.

It is also possible to model uncertainty using *label algebra*:

$$\mathcal{A} = (A, \{f_{kt}\}), \tag{2}$$

where set A contains labels estimating uncertainty of formulas. Therefore, if f is a formula and $l \in A$ is a label, we can define a *Labeled formula* as the following pair $f : l$. A knowledge base is a finite set of labeled formulas. To calculate label of some inference conclusion, functions f_{kt} are used. If $kt = (p, c, a)$ and p is a conjunction of premises α_i (of length n) then the plausible label of the conclusion c is equal to $f_{kt}(l_1, ..., l_n)$, where l_i is a label of α_i.

To find proofs of hypothesis formulas, we may apply *Creative Inference Algorithm* (see Algorithm 1), which is based on AUTOLOGIC system developed by Morgan [11]. This algorithm is a modification of LPR proof algorithm [14]. The main difference is that proof rules are replaced by more general knowledge transmutations. For the purpose of this research, we have added *STRATEGY* parameter, which controls how the proof tree is searched.

Input: φ – hypothesis formula, KB – finite set of labeled formulas, d – max depth, $STRATEGY$ – strategy of the proof tree search

Output: Proofs of $\varphi : l$ from KB

$T :=$ tree with one node (root) $s = [\varphi]$;

$OPEN := [s]$;

$PROOFS := \emptyset$;

while $OPEN \neq \emptyset$ **do**

    $n :=$ the first element from $OPEN$;

    Remove n from $OPEN$;

    if n *is empty* **then**

        Add to $PROOFS$ the path from s to n;

        if $STRATEGY$ *is not exhaustive* **then**

            Return $PROOFS$;

        end

    end

    if *the first formula of n represents action act* **then**

        Execute act; **if** *act was successful* **then**

            add results of act to KB;

            $E :=$ node generated from n by removing act;

        end

    else

        $K :=$ knowledge transmutations, which consequence can be unified with the first formula of n;

        $E :=$ nodes generated by replacing the first formula of n by premises and actions of transmutations from K;

        if *the first formula from n can be unified with element of KB* **then**

            Add to E node obtained from n by removing the first formula;

        end

    end

    Remove from E nodes generating loops;

    Remove from E nodes with proofs longer than d;

    Append E to T connecting nodes to n;

    Add nodes from E to $OPEN$;

    Order $OPEN$ using $STRATEGY$;

end

Return $PROOFS$;

Algorithm 1. Creative Inference Algorithm

The creative inference algorithm gets the following input data: KB – a knowledge base (a set of labeled formulas), φ – a hypothesis (formula), which should be derived from KB, $d \in \mathbb{R}$ – maximal allowed proof depth, and $STRATEGY$ mentioned above. Goal of the algorithm is to find an inference chain of $\varphi : l$ from KB if the hypothesis can be derived from KB, else empty proof is returned.

During inference process, the Creative Inference Algorithm creates a proof tree T. Its nodes N are labeled by formula sequences. Its edges may be labeled by knowledge transmutations, which consequence can be unified with the first formula of a parent node. They may be also labeled by the term $kb(l)$ if the first formula of a parent node can be unified with $\psi : l \in KB$. The root of the proof tree is designated by s and labeled by $[\varphi]$.

If $STRATEGY$ is exhaustive algorithm finds all possible proof up to the given maximal depth. In other case it stops when a node labeled by the empty set of formulas is generated.

$STRATEGY$ is also used to direct the search. Order corresponding to $STRATEGY$ is applied the $OPEN$ list at the end of the main loop. $STRATEGY$ can use various criteria for node ordering. It may estimate confidence of proof which could be obtained on current path or their cost. In this research, we have implemented three strategies. The first one (used in previous versions of the system) is exhaustive – all proofs for reasoning tree of restricted maximal depth are found and returned. In iterative variant nodes are ordered by proof certainty and algorithm returns the first proof found as best. User may continue reasoning and the tree can be extended, bringing remaining proofs sorted by confidence. The second order criteria takes into account proof depth. As a result, two such strategies are implemented: Iterative Depth-First and Iterative Breadth-First.

4   Experiments

We have performed experiments to compare proof searching strategies. It was obvious that exhaustive search will be the slowest. However, we wanted to check which iterative strategy is faster. The comparison was made on a decision support domain. The goal of the decision support system developed to test Creative Inference Model is to recommend a casting procedure and material that fits the user's requirements, reducing the casting production costs.

4.1   LIIS System

Experiments were executed using LPR Intelligent Information System (LIIS), developed for knowledge-based applications. The main component of LIIS is its inference module. Input data represents hypothesis in a form of statement possibly with variables that should be determined. If the hypothesis can be proved from KB, an appropriate proof is presented to the user. If hypothesis may be inferred in several ways, all proofs can be listed with information concerning their credibility.

Maximum proof tree depth, proof searching strategy and knowledge transmutations that are to be used in the inference process can be chosen by the user.

There are two complex knowledge transmutations implemented in the system: AQ [10] and C4.5 [13] induction algorithms. All attributes can be used as a category during learning. However, the user may select smaller *category set* to save the computation time. The learning algorithm is executed only if category attribute is element of the category set. Before action execution (in p) system checks if enough examples may be generated from the knowledge base. Training data generated is spitted into testing and training examples. The last one allows to calculate learned rules' strength.

LIIS is a browser-based application created with GWT (Google Web Toolkit), solution supporting the development of web applications. It is executed on Apache Tomcat web server. Its three main modules are described below.

Back-end of the system implements the main functionalities, such as expert system module, reasoning engine and machine learning module. Two main services related to reasoning and expert systems are provided. Administrative tasks are performed by App and User Services.

All formulas, expert system scenarios, user data and knowledge bases' metadata are stored in a relational database. Object-relational mapping provided by Hibernate and Data Access Objects are used.

Client part of application is responsible for visualization and contains JavaScript forms built from Java classes. It cooperates with back-end using RPC. All necessary data is sent via HTTP as plain, serializable Java classes (Data Transfer Objects) shared by both sides.

Expert system functionality provided by LIIS is used in experiments presented below. Decision support scenarios are based on answers (facts) provided by a user and formulas stored in the knowledge base.

LIIS provides also user GUI for knowledge base edition. Every formula can be developed in a window, where object names appearing as arguments are suggested using names (constant symbols) already entered to the knowledge base. Filtering tool may be applied to find formulas containing entered character sequences in arguments.

4.2   Knowledge Base

The domain chosen is complex enough to present advantages of the model developed. It contains object hierarchies and some parameters with intuitive nature parameters which are difficult to measure. The goal of the system is to support the choice of metal products manufacturing technology, including casting. Over 700 formulas are stored in the knowledge base.

As a base knowledge representation and reasoning we have chosen LPR. However, other techniques may be also used. Details of the LPR and knowledge base can be found in [15,16].

Important part of the knowledge are hierarchies. They were constructed after consultations with domain experts. They are used for example to define casting materials and its subtypes.

In statements properties of materials like minimum elongation or tensile strength are represented. Some statements may be missing if corresponding properties are not known.

Important group of formulas have implication form. Four of them are used to recommend a material for production. The more premises the rule has (and more material properties are checked), the more certain the answer is. The rest of implications are used to predict the production costs taking into account the batch size and the product weight.

4.3   Setup

All tests were performed on a computer with 4 GB of RAM and 4×3.33 GHz processor. Times were measured on the server side of application. As mentioned, three proof searching strategies were tested:

- **Exhaustive** - exhaustive implementation of reasoning algorithm as a control sample.
- **Iterative Depth-First** - iterative algorithm which returns one proof. Firstly it process paths which could give proof with highest confidence, estimating it according to applied proof rules. As a secondary comparison factor the strategy takes depth of path on which node is located, setting higher priorities to longer paths.
- **Iterative Breadth-First** algorithm, identical as Iterative Depth-First, but promoting shorter paths first.

Four usage scenarios were used for the comparison. They are characterized by the questions answered by a user. If some question dos not appear in the description, it means that the answer was skipped by the user.

- Scenario I - Trivial
  Question: *What would you like to produce?* Answer: *rake*
- Scenario II - All questions
  Question: *What would you like to produce?* Answer: *rake*
  Question: *What is the production volume?* Answer: *Large*
  Question: *What is the application weight?* Answer: *Medium*
  Question: *What is the maximum cost required?* Answer: *15*
  Question: *Give minimal tensile strength:* Answer: *1000*
  Question: *What is the application hardness?* Answer: *high*
- Scenario III - Half of questions
  Question: *What would you like to produce?* Answer: *rake*
  Question: *What is the production volume?* Answer: *Large*
  Question: *What is the application weight?* Answer: *Medium*
  Question: *What is the maximum cost required?* Answer: *15*

– Scenario IV - Empty results
  Question: *What would you like to produce?* Answer: *rack*
  Question: *What is the application weight?* Answer: *light*
  Question: *What is the application hardness?* Answer: *low*

4.4 Results

Each test was performed ten times for each algorithm. Average times in milliseconds are presented in Table 1.

Table 1. Average times for algorithms in scenarios I - IV.

Scenario	Exhaustive algorithm	Iterative depth-first	Iterative breadth-first
I	52	42	41
II	3738	57	431
III	2521	1546	1660
IV	31	33	35

To check if differences are statistically significant, Student's t-test between algorithms for each Scenario were calculated. Results of the tests are presented in Table 2.

Table 2. Results of student's t-test (p-value).

Scenario	Exhaustive algorithm vs. Iterative depth-first	Exhaustive algorithm vs. Iterative breadth-first	Iterative depth-first vs. Iterative breadth-first
I	0,0385	0,0148	0,637
II	$1,629 * 10^{-12}$	$3,729 * 10^{-12}$	$1,351 * 10^{-16}$
III	$5,734 * 10^{-8}$	$9,332 * 10^{-9}$	0,0038
IV	0,5770	0,2817	0,5409

As we can see, Iterative algorithms are significantly faster than exhaustive implementation in 3 scenarios. Exhaustive version of algorithm achieved lower time only in the first scenario, with empty list of proofs found. Iterative Depth-First reached better time than Iterative Breadth-First in two cases: it was significantly faster in scenario II (more then 7 times) and subtly in scenario III (less then 7%). Both variants should be deeply evaluated with further tests, considering different sizes of knowledge base and maximal allowed tree depth parameter.

Tests results were similar for scenarios IV and I. Times differs the most for scenario II, which suggest that optimization could bring performance profits during complex inferences.

5   Conclusions

In this work we have presented Creative Expert System model, inference algorithm and LIIS system that is implementation of the proposed ideas. The solution we propose allows to build creative expert systems. The main advantage of this solution is that the system autonomously tries to create intrinsically new knowledge to continue the reasoning process instead of getting stuck in a case in which there is no appropriate knowledge in the knowledge base.

To test the system we have developed knowledge base providing technology choice support. We showed in experiments that system works well and applies machine learning when needed. We have also compared three proof strategies and confirmed that iterative proof search strategies are better than exhaustive one, implemented in the first version of the system. Results show that the best is Iterative depth-first strategy and it will be chosen in the system by default.

In the further works we would like to develop procedures for checking knowledge base consistency. We would also like to add other complex transmutations (e.g. clustering). We are also working on applications in different domains, like detection of money laundering [6] and telemetry systems [17].

Acknowledgments. The research reported in the paper was supported by the grant of The National Centre for Research and Development (LIDER/028/593/L-4/12/NCBR/2013) and by the Polish Ministry of Science and Higher Education under AGH University of Science and Technology Grant 11.11.230.124.

References

1. Alkharouf, N.W., Michalski, R.S.: Multistrategy task-adaptive learning using dynamically interlaced hierarchies. In: Michalski, R.S., Wnek, J. (eds.) Proceedings of the Third International Workshop on Multistrategy Learning (1996)
2. Althoff, K., Bach, K., Deutch, J., Hanft, A., Manz, J., Muller, T., Newo, R., Reichle, M., Schaaf, M., Weis, K.: Collaborative multi-expert-systems realizing knowledge-lines with case factories and distributed learning systems. In: Proceedings of the 3rd Workshop on Knowledge Engineering and Software Engineering (2007)
3. Low, B.T., Lui, H.C., Tan, A.H., Teh, H.H.: Connectionist expert system with adaptive learning capability. IEEE Trans. Knowl. Data Eng. **3**(2), 200–207 (1991)
4. Collins, A.: Human plausible reasoning. Technical Report 3810, Bolt Beranek and Newman Inc. (1978)
5. Collins, A., Michalski, R.S.: The logic of plausible reasoning: A core theory. Cogn. Sci. **13**, 1–49 (1989)
6. Drezewski, R., Sepielak, J., Filipkowski, W.: The application of social network analysis algorithms in a system supporting money laundering detection. Inf. Sci. **295**, 18–32 (2015)
7. Gabbay, D.M.: LDS - Labeled Deductive Systems. Oxford University Press (1991)
8. Ligeza, A.: Logical Foundations for Rule-Based Systems. Springer, Heidelberg (2006)

9. Michalski, R.S.: Inferential theory of learning: Developing foundations for multistrategy learning. In: Michalski, R.S. (ed.) Machine Learning: A Multistrategy Approach, Vol. IV. Morgan Kaufmann Publishers (1994)

10. Michalski, R.S., Larson, J.: Aqval/1 (aq7) user's guide and program description. Technical Report 731, Department of Computer Science, University of Illinois, Urbana, June 1975

11. Morgan, C.G.: Autologic. Logique et Anal. **28**(110–111), 257–282 (1985)

12. Neapolitan, R.E.: Probabilistic Reasoning in Expert Systems: Theory and Algorithms. CreateSpace Independent Publishing Platform, USA (2012)

13. Quinlan, J.: C4.5: Programs for Machine Learning. Morgan Kaufmann, San Francisco (1993)

14. Sniezynski, B.: Proof searching algorithm for the logic of plausible reasoning. In: Kłopotek, M.A., Wierzchoń, S.T., Trojanowski, K. (eds.) Intelligent Information Processing and Web Mining. AINSC, pp. 393–398. Springer, Heidelberg (2003). doi:10.1007/978-3-540-36562-4_41

15. Sniezynski, B., Legien, G., Wilk-Kołodziejczyk, D., Kluska-Nawarecka, S., Nawarecki, E., Jaśkowiec, K.: Creative expert system: Result of inference and machine learning integration. In: Hartmann, S., Ma, H. (eds.) DEXA 2016. LNCS, vol. 9827, pp. 257–271. Springer, Cham (2016). doi:10.1007/978-3-319-44403-1_16

16. Sniezynski, B., Wilk-Kołodziejczyk, D., Legien, G., Kluska-Nawarecka, S., Nawarecki, E.: Reasoning algorithm for a creative decision support system integrating inference and machine learning. Submitted to International Journal of Applied Mathematics and Computer Science (2016)

17. Szydlo, T., Nawrocki, P., Brzoza-Woch, R., Zielinski, K.: Power aware MOM for telemetry-oriented applications using gprs-enabled embedded devices - levee monitoring use case. In: Proceedings of the 2014 Federated Conference on Computer Science and Information Systems. vol. 2, pp. 1059–1064. IEEE, September 2014

18. Tran, L.P., Hancock, J.P.: An adaptive-learning expert system for maintenance diagnostics. In: Proceedings of the IEEE 1989 National Aerospace and Electronics Conference, NAECON 1989, vol. 3, pp. 1034–1039, May 1989

19. Wiriyacoonkasem, S., Esterline, A.C.: Adaptive learning expert systems. In: Proceedings of the IEEE Southeastcon 2000, pp. 445–448. IEEE (2000)

Spatial Planning as a Hexomino Puzzle

Marcin Cwiek[1,2] and Jakub Nalepa[1,2,3(✉)]

[1] Deadline24, Gliwice, Poland
{mcwiek,jnalepa}@deadline24.pl
[2] Future Processing, Gliwice, Poland
{mcwiek,jnalepa}@future-processing.com
[3] Silesian University of Technology, Gliwice, Poland
jakub.nalepa@polsl.pl

Abstract. Exact cover problem is a well-known NP-complete decision problem to determine if the exact cover really exists. In this paper, we show how to solve a modified version of the famous Hexomino puzzle (being a noteworthy example of an exact cover problem) using a Dancing-links based algorithm. In this modified problem, a limited number of gaps in the rectangular box may be left uncovered (this is a common scenario in a variety of spatial planning problems). Additionally, we present the benchmark generator which allows for elaborating very demanding yet solvable problem instances. These instances were used during the qualifying round of Deadline24—an international 24-h programming marathon. Finally, we confront our baseline solutions with those submitted by the contestants, and elaborated using our two solvers.

Keywords: Exact cover · Dancing links · Hexomino puzzle · Benchmark generation

1 Introduction

In the standard Hexomino puzzle, a *polyomino* is known as a pattern created by the connection of a given number of equally sized squares along common edges. Then, a *hexomino* is a polyomino composed of six squares [1]. To solve this puzzle, it is necessary to find the positions of the available hexominoes which cover all cells exactly once in the rectangular box (no cells can be left uncovered, and no cells can be covered more than once). Hexominoes—being the building blocks—may be only rotated and translated (no other shape modifications are allowed). This Hexomino puzzle, along with the Pentomino puzzle (in which the building blocks contain five squares) are fairly good examples of the exact cover problems which can be applied to model various spatial planning scenarios. These problems are known for their NP-completeness [2].

There are two other real-life and well-established examples of the exact cover problems. In the *Sudoku*[1] puzzle (being a constraint satisfaction problem [3]),

---

[1] From Japanese—a *singular number*.

© Springer International Publishing AG 2017
N.T. Nguyen et al. (Eds.): ACIIDS 2017, Part I, LNAI 10191, pp. 410–420, 2017.
DOI: 10.1007/978-3-319-54472-4_39

the grid of 9×9 cells (divided into 3×3 blocks) is to be filled in with distinct numbers. Three rules are imposed on the feasible solution to the Sudoku puzzle (each puzzle has only one solution): (i) each row contains all numbers without repetitions, (ii) each column contains all numbers without repetitions, and (iii) each block contains all numbers without repetitions. Generating valid sudoku instances is a demanding optimization task—a good sudoku instance should be solvable without any guessing, in a logical sequence of steps [4]. A number of algorithms [5], including metaheuristics (also allowing for backtracking during the search) have been proposed to tackle the sudoku instances of various difficulties [3,6]. Another well-known example of the generalized exact cover problem is the N queens problem, in which N non-attacking queens should be feasibly placed on an $N \times N$ chessboard. This problem (and some of its variants) has been tackled by several metaheuristic approaches [7–9]. The exact cover formulations have been successfully applied in several other real-life problems, ranging from designing automation systems [10], to implementing web services [11].

Generating benchmark instances for the challenging and computationally-intensive optimization problems [12–15] is a very demanding task [16]. Such benchmark-generating routines should allow for not only retrieving difficult tests, but also for elaborating the baseline solutions for such instances. These baseline solutions can be later used to verify how "close" to the optimum the solutions obtained using other algorithms are. In this paper, we show how to generate the instances which are quite challenging (yet solvable) for a modified version of the Hexomino puzzle. Also, we present its real-life application (in the spatial planning scenario), along with the quality function used to assess the elaborated solutions. This problem was exploited as a task during the qualifying round of the Deadline24 marathon[2]. Finally, we present confront the baseline solutions (found while generating the benchmark instances) with those retrieved using our two solvers ones, and those submitted by the marathon participants.

This paper is structured as follows. Section 2 gives a modified version of the well-established Hexomino puzzle (as presented to the Deadline24 contestants). In Sect. 3, we discuss our benchmark generator which allows for retrieving difficult (but solvable) problem instances. The experiments along with the comparisons with the solutions elaborated by the Deadline24 marathon participants are discussed in Sect. 4. Section 5 summarizes the paper and serves as an outlook to the future work.

2   Modified Hexomino Puzzle

The Deadline24 programming marathon is famous worldwide thanks to the challenging tasks and their interesting plots. Since the very beginning, the optimization problems have been hidden behind lines in various adventures of brave beetlejumpers—creatures living in the Universum. The first task is therefore always to uncover the "real" problem to be solved. In this section, we show the

---

[2] http://www.deadline24.pl.

task (*Developer*) presented to the contestants of the 2015 edition of Deadline24 qualifying round. Additionally, we highlight the differences between the standard Hexomino puzzle and its modified version, which became the underlying problem in *Developer*—these differences are given in parentheses and they are **boldfaced**. For the sake of completeness, we provide the full task description (along with the entire beetlejumper story).

2.1 Developer Problem (DP)

Introduction. One of the most recently seized planets occupied by beetlejumpers has very friendly living conditions for these creatures. Construction of modern residential estates for war veterans on its surface may significantly improve morale of the team conquering the farthest corners of Universum.

Your team must success as a construction developer. At your disposal you will have plots of land (each being a grid $N \times M$ of fields) to construct houses with gardens. Each plot of land may include inaccessible fields which cannot be used (such **inaccessible fields do not exist** in the standard Hexomino formulation) as well as trees which may influence the final value of the investment. The plot of land must be developed in the best possible way while observing the applicable provisions of law.

Construction of Households. Local construction provisions clearly regulate the possible shape and arrangement of households (these constitute the **additional constraints** to the Hexomino puzzle):

- each house must have its own garden,
- each house adjoins its garden (such neighborhood means that at least one side of the grid is joint),
- the area of each house and each garden is the same and it equals 6 fields (one household is 12 fields large: 6 fields for a house and 6 fields for a garden),
- surfaces of the houses and gardens are separate (they cannot overlap),
- surfaces of the houses and gardens cannot include inaccessible fields,
- each house and each garden must be of a feasible shape—ten possible shapes are presented below (only translations and rotations are admissible):

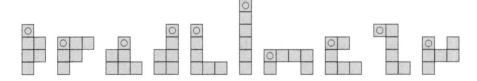

Real Estate Evaluation. Differences between particular ways of land development are extremely important for future residents. Beetlejumpers wish to enjoy the natural richness of the planet. At the same time, they prize their privacy, thus the value of each constructed household is calculated as follows:

- for the construction of a house together with the garden: +3,
- for each cut down tree for the purposes of the construction of a house: −2,
- for each tree left in the garden: +2,
- for each joint wall of the house with another house: −1.

The total investment value S of a given plot of land ($N \times M$ fields) is the sum of all household investment values constructed on its surface (the **quality function** used for assessing the elaborated solutions). The task is to plan the construction of the houses on the plot in accordance with the provisions so that the total investment value S is as high as possible.

Example. In Fig. 1, we present (a) an example *Developer* problem instance, along with (b) its possible solution. Here, the 'X' character visualizes a tree, and '#' is an inaccessible fields. It is easy to note that in this example solution, one potentially accessible field (without a tree) has not been utilized (**this is in contrary to a standard Hexomino puzzle**, where all cells must be covered). Taking into account the scoring scheme discussed above, the total investment value (S) can be calculated as follows:

- for constructed houses: $4 \cdot 3 = 12$
- for cut down trees: $4 \cdot (-2) = -8$
- for trees in gardens: $11 \cdot 2 = 22$
- for joint walls: $-1 - 1 - 3 - 1 = -6$

$S = 12 - 8 + 22 - 6 = 20$

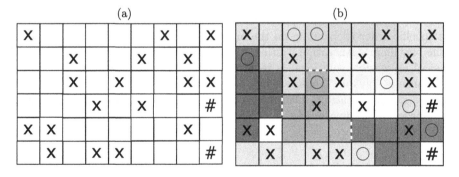

Fig. 1. An example (a) *Developer* problem instance, along with (b) a possible solution.

3   Generating Benchmark Tests

In this section, we present the algorithm to generate and validate the benchmark instances for the DP. However, the proposed technique may be easily adapted to handle other exact-cover based problems (with other modifications reflecting various real-life circumstances), especially those tailored to the spatial planning scenarios. In Sect. 3.1, we present our modifications to the Dancing links algorithm for tackling this kind of puzzles, and then we present our reversed-order approach of generating benchmark tests (Sect. 3.2), along with our two solvers (Sect. 3.3).

3.1 Modified Dancing Links

Dancing links (DLX) is a well-established technique of implementing the famous Algorithm X efficiently [17]. It is a recursive and backtracking algorithm for finding all solutions to the exact cover problem. Although the problem introduced in the previous section is a specific variant of the exact cover problem, the original implementation of DLX cannot be used directly to tackle it (and to generate the benchmark instances for this problem). Hence, we propose the Modified Dancing Links (MDL) algorithm, encompassing the following modifications:

- **Termination condition.** In DLX, it is assumed that a feasible solution is found only when all fields are covered (therefore, the exact cover problem is solved). Since we slightly mitigate this necessity (see Sect. 2 for details), the termination condition can be modified as well (Algorithm 1, line 2). In MDL, the optimization proceeds until the maximum (or a smaller) number of fields that may remain uncovered is reached (note that the number of uncovered cells may significantly change after the insertion of just a single building block composed of 6 squares). This allows the algorithm to find the sets of households which do not necessarily occupy every single cell in the rectangular grid (hence, not all of the cells are finally covered).
- **Search starting point.** In the original Algorithm X, all possible solutions are traversed during the optimization. Then, these solutions are returned in a specific order (depending on the order of the analyzed building blocks). In the benchmark generation routine, only one random solution (from the entire set of all feasible solutions) is required, to show that the generated problem instance is solvable. Thus, we randomly change the search starting point to find a random (feasible) placement of households, which is independent from the order of building blocks.

3.2 Reversed Order Approach

The main goal of creating any new benchmark dataset is an ability to assess the quality of the solutions retrieved using the emerging techniques. Therefore, this benchmark dataset should contain challenging instances which are known to be solvable (at least one feasible solution exists).

The benchmark generation routine creates the instances along with their baseline solutions (Algorithm 2). We propose a reversed order approach (a similar technique that we exploited in our work on generating instances for rich routing problems [16]). First—after creating the initial empty rectangular grid of $w \times h$ cells (line 1)—we randomly exclude a subset of cells from further processing

Algorithm 1. Modified Dancing Links (MDL) algorithm.

```
 1: procedure search(p)
 2: if (termination condition is reached) then
 3:     Store solution and return it
 4: end if
 5: c ← choose column; cover(c); t ← c
 6: if randomize then
 7:     for i = 1 to rand() do                           ▷ Move search starting point
 8:         t ← t.down
 9:     end for
10: end if
11: for all r ← t.down, t.down.down, ..., do
12:     while r ≠ t do
13:         O_p ← r                                      ▷ Add r to the partial solution
14:         for all m ← r.right, r.right.right, ..., do
15:             while m ≠ r do
16:                 cover(m.column)
17:             end while
18:         end for
19:         search(p + 1)
20:         for all m ← r.left, r.left.left, ..., do
21:             while m ≠ r do
22:                 uncover(m.column)
23:             end while
24:         end for
25:     end while
26: end for
27: uncover(c)
```

(line 2). These cells are unavailable and cannot be exploited in a solution. Then, we determine the feasible positions of the households in this grid (line 3) using MDL—these households constitute the baseline solution. We randomly place trees in each garden of each household (line 4) (this operation may be interpreted as imposing new constraints on a feasible solution). These trees boost the value of the quality function of this baseline solution (see Sect. 2.1). Finally, the benchmark instance is stored, along with the baseline solution (lines 5–6).

Algorithm 2. Generating the DP instance (of size $w \times h$) with a baseline solution.

```
1: Prepare empty area of a selected size (w × h)
2: Make some random fields unavailable
3: Use MDL to split the available area into households
4: Place randomly trees in each garden of each household
5: Save benchmark data (area with trees and unavailable field)
6: Save baseline solution (households determined from MDL)
```

It is possible to generate benchmark instances without the baseline solutions (in this case, the line 3 in Algorithm 2 is not executed). In this operation mode, trees are randomly scattered around the entire grid (excluding the unavailable cells). Clearly, verifying if such instance is solvable (and how difficult it is) requires further processing (MDL can be run to retrieve a feasible solution).

3.3 Solving the Generated Benchmark Instances

The simplest way of solving the generated problem instances (note that for most of the tests, we elaborated the baseline solutions, thus we already showed that they are solvable) is to use a single building block (out of 10 available ones presented in Sect. 2.1), and to elaborate a feasible solution using a brute-force approach. Here, we exploited the building block composed of 6 squares connected in row. In this solution, all houses and gardens are aligned vertically or horizontally. This trivial solver is able to successfully determine feasible solutions for all generated problem instances (however, these solutions are quite low-quality).

Searching for an optimal solution for the benchmark tests requires analyzing all possible placements of households. For all solutions, the value of the quality function should be found. Clearly, browsing through these solutions is a computationally-demanding task, and cannot be easily applied in practice due to the NP-completeness of the exact cover problem. In this work, we utilized the divide-and-conquer paradigm, and split the problem into sub-problems. The entire area which should be covered with households is divided into sub-regions (containing at most 48 fields), and the full exhaustive search is executed for each sub-region using the MDL algorithm. Then, the results for each sub-region are merged into the final solution. This optimization technique obviously does not involve checking all possibilities (e.g., the households positioned at the borders of neighboring sub-regions), but still it may provide a very good estimate of how the solution can be boosted compared to those retrieved using a naive solver.

4   Experimental Results

In this section, we analyze the results obtained for benchmark tests (distinguished by their unique names, r1–r10) generated using our algorithm (note that for each set of parameters we create a number of distinct test instances—see the number of tests column in Table 1)[3]. We compare the baseline solutions (generated along with the benchmark instances), with the solutions retrieved using our two solvers, and those submitted by the contestants during the qualifying round (which lasted 4 clock hours) of the Deadline24 marathon. The settings used for elaborating the benchmark tests are summarized in Table 1. The size of the board ($N \times M$) was drawn randomly from the [min, max] range (see Table 1). The benchmark generation routine, along with the solvers discussed in Sect. 3.3 were implemented in C++, and run on a computer equipped with an Intel Core i7-3740QM 2.7 GHz (16 GB RAM) processor.

---

[3] For the details of these tests see: http://sun.aei.polsl.pl/~jnalepa/Developer/.

Table 1. The settings used to generate the benchmark tests.

Id	Mode	Number of tests	min	max	τ (in sec.)
r1	Random	4	5	30	8
r2	DL	5	25	50	101
r3	Random	8	10	70	350
r4	DL	10	10	90	399
r5	DL	6	70	100	971
r6	DL	5	70	100	626
r7	DL	3	40	200	999
r8	Random	2	110	200	69
r9	DL	2	10	200	410
r10	DL	4	110	200	495

Generating benchmark tests along with the baseline solutions and executing the basic solver were extremely fast, and took less than 1 min. in total for all problem instances. It is worth mentioning, that for three sets of parameters we executed the variant of the benchmark routine (the Rand mode) which randomly scatters the trees across the map (see Sect. 3.2)—for these tests, the baseline solution is not known (it is known for the DL mode). The execution time of our divide-and-conquer solver (denoted as τ, in seconds—summed up for all instances retrieved using the corresponding generator parameters) are reported in Table 1. These times give a rough estimate on the difficulty of generated instances—the r7 and r5 problems appeared the most challenging ones, whereas r1 is relatively easy to solve, and can be tackled even using a brute force approach. Hence, this naive technique would allow for scoring some points in the *Developer* task during the qualifying round of the marathon.

In Table 2, we present the best results retrieved for all benchmark tests (the values of the quality function are summed up for each group of tests created using the same settings) using the investigated techniques. The solutions include those obtained along with the benchmark instances (for seven tests), those found using our naive and divide-and-conquer algorithms, and the solutions submitted by the Deadline24 contestants. The total scores for the naive solutions are smallest, however they show that the benchmark tests are solvable using even quite simple approaches (the solutions are of very low quality). Although the divide-and-conquer approach gives notably better results compared with the naive solver, these solutions are still much worse than those created in the benchmark generation routine.

In all cases, the results submitted by the Deadline24 contestants appeared to be the best known results for our benchmark instances (the algorithms exploited by the contestants, and the total execution times of their implementations are not known—only the final solutions were assessed, and their feasibility was verified). The quality-function values of our baseline solutions were a very rough estimates of the values obtained for the submitted solutions—all scores are around $20-40\%$ higher than the values of the quality functions obtained for the baseline solutions.

Table 2. The best results obtained for all benchmark tests: (a) the results obtained using our naive solver, (b) those elaborated using our divide-and-conquer algorithm, (c) the baseline solutions—generated along with instances, and (d) the solutions submitted by the Deadline24 contestants.

Id	(a)	(b)	(c)	(d)
r1	194	232	—	**780**
r2	989	1519	3151	**4499**
r3	2063	2157	—	**10184**
r4	2754	4688	10544	**14567**
r5	5108	8512	20419	**26889**
r6	5974	8874	19222	**24587**
r7	5664	8885	21751	**28884**
r8	5580	9796	—	**25884**
r9	2787	4479	10971	**15002**
r10	11553	20998	37535	**45177**

5 Conclusions and Outlook

In this paper, we introduced a new variant of the famous Hexomino puzzle. This version of the problem resembles a real-life problem of determining the plan for buildings, and incorporates several real-life constraints. We proposed a benchmark generation routine for this problem which allows for creating difficult yet solvable problem instances, along with the baseline solutions. These tests were used as the input data in the qualifying round of the 24-h programming marathon, and were to be solved by the participants. Also, we presented a divide-and-conquer algorithm for solving the new Hexomino variant. Finally, we compared the results elaborated using the proposed solvers with the baseline solutions, and those submitted by the contestants—the baseline solutions appeared much worse than the submitted ones (however, they gave a rough estimate on the difficulties of these instances). The algorithms used by the contestants (along with their execution times) are not known.

The number of real-life applications of exact cover formulations started increasing recently. We believe that numerous common scenarios (especially from the field of the spatial planning [18] and robot navigation [19]) can be modelled using different variants of the polyomino-like problems. Hence, applying new constraints in our benchmark generation routine should allow for handling new circumstances relatively easily. Also, we plan to create a massive benchmark dataset (along with the baseline solutions) for the Pentomino and Hexomino problems. It could be utilized to assess the efficacy of emerging algorithms for solving these challenging puzzles.

References

1. Gravier, S., Moncel, J., Payan, C.: A generalization of the pentomino exclusion problem: Dislocation of graphs. Discrete Math. **307**(35), 435–444 (2007)
2. Wachira, K., Mwangi, E., Jeon, G.: A pentomino-based path inspired demosaicking technique for the bayer color filter array. In: AFRICON 2015, pp. 1–5, September 2015
3. Ansótegui, C., Béjar, R., Fernández, C., Gomes, C., Mateu, C.: Generating highly balanced sudoku problems as hard problems. J. Heuristics **17**(5), 589–614 (2011)
4. Lewis, R.: Metaheuristics can solve sudoku puzzles. J. Heuristics **13**(4), 387–401 (2007)
5. Crawford, B., Castro, C., Monfroy, E.: Solving sudoku with constraint programming. In: Shi, Y., Wang, S., Peng, Y., Li, J., Zeng, Y. (eds.) MCDM 2009. CCIS, vol. 35, pp. 345–348. Springer, Heidelberg (2009). doi:10.1007/978-3-642-02298-2_52
6. Geem, Z.W.: Harmony search algorithm for solving sudoku. In: Apolloni, B., Howlett, R.J., Jain, L. (eds.) KES 2007. LNCS (LNAI), vol. 4692, pp. 371–378. Springer, Heidelberg (2007). doi:10.1007/978-3-540-74819-9_46
7. Sosic, R., Gu, J.: Efficient local search with conflict minimization: a case study of the n-queens problem. IEEE Trans. Knowl. Data Eng. **6**(5), 661–668 (1994)
8. Hu, X., Eberhart, R.C., Shi, Y.: Swarm intelligence for permutation optimization: a case study of n-queens problem. In: 2003 Proceedings of the IEEE Swarm Intelligence Symposium, SIS 2003, pp. 243–246, April 2003
9. Mohabbati-Kalejahi, N., Akbaripour, H., Masehian, E.: Basic and hybrid imperialist competitive algorithms for solving the non-attacking and non-dominating n-queens problems. In: Madani, K., Correia, A.D., Rosa, A., Filipe, J. (eds.) Computational Intelligence. SCI, vol. 577, pp. 79–96. Springer, Cham (2015). doi:10.1007/978-3-319-11271-8_6
10. Lehmann, M., Mai, T.L., Wollschlaeger, B., Kabitzsch, K.: Design approach for component-based automation systems using exact cover. In: Proceedings of the IEEE Emerging Technology and Factory Automation (ETFA), pp. 1–8, September 2014
11. Ba, C.: An exact cover-based approach for service composition. In: 2016 IEEE International Conference on Web Services (ICWS), pp. 631–636, June 2016
12. Nalepa, J., Blocho, M.: Co-operation in the parallel memetic algorithm. Int. J. Parallel Prog. **43**(5), 812–839 (2015)
13. Cekała, T., Telec, Z., Trawiński, B.: Truck loading schedule optimization using genetic algorithm for yard management. In: Nguyen, N.T., Trawiński, B., Kosala, R. (eds.) ACIIDS 2015. LNCS (LNAI), vol. 9011, pp. 536–548. Springer, Cham (2015). doi:10.1007/978-3-319-15702-3_52
14. Nalepa, J., Blocho, M.: Adaptive memetic algorithm for minimizing distance in the vehicle routing problem with time windows. Soft Comput. **20**(6), 2309–2327 (2016)
15. Nalepa, J., Blocho, M.: Enhanced guided ejection search for the pickup and delivery problem with time windows. In: Nguyen, N.T., Trawiński, B., Fujita, H., Hong, T.-P. (eds.) ACIIDS 2016. LNCS (LNAI), vol. 9621, pp. 388–398. Springer, Heidelberg (2016). doi:10.1007/978-3-662-49381-6_37
16. Cwiek, M., Nalepa, J., Dublanski, M.: How to generate benchmarks for rich routing problems? In: Nguyen, N.T., Trawiński, B., Fujita, H., Hong, T.-P. (eds.) ACIIDS 2016. LNCS (LNAI), vol. 9621, pp. 399–409. Springer, Heidelberg (2016). doi:10.1007/978-3-662-49381-6_38

17. Knuth, D.E.: Dancing links (2000). http://arxiv.org/abs/cs.DS/0011047
18. Stewart, T.J., Janssen, R.: A multiobjective GIS-based land use planning algorithm. Comput. Environ. Urban Syst. **46**, 25–34 (2014)
19. Lau, B., Sprunk, C., Burgard, W.: Efficient grid-based spatial representations for robot navigation in dynamic environments. Robot. Auton. Syst. **61**(10), 1116–1130 (2013)

Ramp Loss Support Vector Data Description

Vo Xuanthanh[1(✉)], Tran Bach[1], Hoai An Le Thi[1], and Tao Pham Dinh[2]

[1] Laboratory of Theoretical and Applied Computer Science EA 3097,
University of Lorraine, Ile du Saulcy, 57045 Metz, France
{xuan-thanh.vo,bach.tran,hoai-an.le-thi}@univ-lorraine.fr
[2] Laboratory of Mathematics, INSA-Rouen, University of Normandie,
Avenue de l'Université, 76801 Saint-Etienne-du-Rouvray, Cedex, France
pham@insa-rouen.fr

Abstract. Data description is an important problem that has many applications. Despite the great success, the popular support vector data description (SVDD) has problem with generalization and scalability when training data contains a significant amount of outliers. We propose in this paper the so-called ramp loss SVDD then prove its scalability and robustness. For solving the proposed problem, we develop an efficient algorithm based on DC (Difference of Convex functions) programming and DCA (DC Algorithm). Preliminary experiments on both synthetic and real data show the efficiency of our approach.

Keywords: Support vector data description · Ramp loss · DC Programming · DCA

1 Introduction

Data description (or one-class classification) refers to finding a compact characterization of a training dataset and see whether newly arrived data resemble to the training data or not. This technique is very useful when most of training data are belong to one class while the other classes are severely undersampled, or when we want to detect uncharacteristic objects (outliers) from the dataset.

The classical and most successful methods for data description are the one-class support vector machine (OC-SVM) [9] and the support vector data description (SVDD) [10]. The OC-SVM uses a hyperplane to separate the data set from the origin with maximal margin in the feature space. Thus, it requires to map data into the feature space usually via kernel trick. The SVDD, on the other hand, seeks a hypersphere to enclose the regions of high density. Due to the nonlinearity nature, the SVDD can work directly in the input space.

Despite the great success, the conventional SVDD that uses an ℓ_1-norm cost function (hinge loss) may fail to capture the representative region of the target data in the presence of outliers and noisy data. There have been attempts to address the limitations of the SVDD. Most of these works focus on reducing the influence of outliers on the descriptive model. By taking into account the distribution of the training dataset, Lee et al. [4] proposed using density-induced

© Springer International Publishing AG 2017
N.T. Nguyen et al. (Eds.): ACIIDS 2017, Part I, LNAI 10191, pp. 421–431, 2017.
DOI: 10.1007/978-3-319-54472-4_40

distance measurements. The authors in [5] associated each data point with a confidence score based on its distance to the dataset's center and then solved a weighted version of the conventional SVDD. These versions of the SVDD are still convex problems and can be optimally solved. Azami et al. [1] considered the so-called L_0-SVDD where an ℓ_0-norm cost function is used in the place of the ℓ_1-norm cost function. The resulting problem is nonconvex but can be solved efficiently by means of DC (Difference of Convex functions) programming and DCA (DC Algorithm) [3,7,8]. Experiments in [1] show promising performance of the L_0-SVDD.

In the context of support vector machines (SVMs), it has been widely known that the use of bounded losses such as ramp loss produces models more robust to outliers and more scalable than the use of hinge loss [2,6]. Motivated by this success, we propose another version of the SVDD that employs the ramp loss for measuring errors. We will show that the proposed model is resistant to outliers during learning process and has less support vectors than existing methods. Finally, we develop a DCA based algorithm for solving the proposed problem.

The paper is organized as follows. In the next section, we review the conventional SVDD problem and propose the ramp loss SVDD formulation. In Sect. 3, we give an overview of DC programming and DCA, then develop a DCA-based algorithm for solving the proposed R-SVDD problem. Preliminary numerical experiments are presented in Sects. 4 and 5 draws some conclusions.

2   Problem Formulation

Given a data set of n objects $T = \{\mathbf{x}_1, \ldots, \mathbf{x}_n\} \subset \mathbb{R}^d$. The problem is to find a minimal hypershpere centered at $\mathbf{c}$ with radius $R \geq 0$ that captures a majority of the training data (representing the distribution of normal data). The most popular method for this problem is the support vector data description (SVDD) which has the following mathematical formulation (with $z = R^2$) [10]

$$\min_{z \geq 0, \mathbf{c} \in \mathbb{R}^d} \left\{ z + \lambda \sum_{i=1}^{n} \xi_i : \|\mathbf{c} - \mathbf{x}_i\|_2^2 - z \leq \xi_i, \xi_i \geq 0 \ \forall i = 1, \ldots, n \right\}, \quad (1)$$

where $\| \cdot \|_2$ is the Euclidean norm and $\lambda > 0$ is a parameter controlling the trade-off between the volume of the hypersphere and the errors. More generally, we consider the following problem

$$\min_{z \geq 0, \mathbf{c} \in \mathbb{R}^d} z + \lambda \sum_{i=1}^{n} L(\|\mathbf{c} - \mathbf{x}_i\|_2^2 - z), \quad (2)$$

where L is an increasing loss function. We define the hinge loss with a hinge point $\eta \in \mathbb{R}$ as $H_\eta(u) = \max\{u - \eta, 0\}$. It is clear that if $L = H_0$ we recover the SVDD formulation.

Problem (1) is often solved via its Lagrange dual problem [10], which is convex quadratic,

$$\min_{\alpha \in \mathbb{R}^n} \left\{ \alpha^T K \alpha - \langle q, \alpha \rangle : \sum_{i=1}^{n} \alpha_i = 1; \ 0 \leq \alpha_i \leq \lambda \ \forall i = 1, \ldots, n \right\}, \qquad (3)$$

where $K \in \mathbb{R}^{n \times n}$ defined by $K_{ij} = \langle \mathbf{x}_i, \mathbf{x}_j \rangle$ $(\forall i, j = 1, \ldots, n)$ and q is the diagonal of K. Here, notation $\langle \cdot, \cdot \rangle$ is the canonical scalar product of two vectors. Optimal solutions to problems (1) and (3) have the following relationships

$$\mathbf{c}^* = \sum_{i=1}^{n} \alpha_i^* \mathbf{x}_i, \quad \alpha_i^* \begin{cases} = 0 & \text{if } \|\mathbf{c}^* - \mathbf{x}_i\|_2^2 < z^* \\ = \lambda & \text{if } \|\mathbf{c}^* - \mathbf{x}_i\|_2^2 > z^* \\ \in [0, \lambda] & \text{if } \|\mathbf{c}^* - \mathbf{x}_i\|_2^2 = z^* \end{cases}, \qquad (4)$$

$$z^* = \min \left\{ \|\mathbf{c}^* - \mathbf{x}_i\|_2^2 : \alpha_i^* > 0 \right\}. \qquad (5)$$

(4) shows that the optimal center is a convex combination of training objects, and only objects $\mathbf{x}_i$ with $\alpha_i^* > 0$ contribute to defining the description. These objects are called *support vectors*. Since all miscaptured objects are support vectors and present in the expression of $\mathbf{c}^*$ with a maximal weight λ, the description becomes unscalable and even degenerated if the training dataset contains a huge amount of outliers. The same phenomenon is also encountered in the support vector machines (SVMs). To address this issue in SVMs, numerous researches have suggested using bounded losses to prevent the bad impact of outliers [2,6].

In this paper, we propose a robust version of the SVDD by using the ramp loss defined as

$$L_\eta(u) = \min \{H_0(u), \eta\} = H_0(u) - H_\eta(u), \qquad (6)$$

where $\eta > 0$ is a parameter. Then the ramp loss SVDD problem takes the form

$$\min_{z \geq 0, \mathbf{c} \in \mathbb{R}^d} f(\mathbf{c}, z) := z + \lambda \sum_{i=1}^{n} \min \left\{ \max \left\{ \|\mathbf{c} - \mathbf{x}_i\|_2^2 - z, 0 \right\}, \eta \right\}. \qquad (7)$$

In (7), objects $\mathbf{x}_i$ whose distances to the boundary of the descriptive hypershpere $\|\mathbf{c} - \mathbf{x}_i\|_2^2 - z$ are larger than the threshold η will have a constant error η. This is different from the SVDD formulation (1) where the classification error of an object is linearly monotonic with respect to its squared distance to the hypershpere's center. Therefore, with the formulation (7), the impact of objects far way from the boundary could be eliminated.

Equation (7) is a nonconvex, nonsmooth optimization problem. We will develop below an algorithm based on DC programming and DCA for solving this problem.

3 Solving the Ramp Loss Support Vector Data Description

3.1 Overview of DC Programming and DCA

DC programming and DCA are well-known as powerful tools for nonconvex programming and global optimization. In the standard form, a DC program is given by

$$\min \left\{ f(\mathbf{x}) := g(\mathbf{x}) - h(\mathbf{x}) : \mathbf{x} \in \mathbb{R}^d \right\}, \tag{P}$$

where g, h are lower semicontinuous proper convex functions on $\mathbb{R}^d$. Such a function f is called a DC function, $g - h$ is a DC decomposition of f, while g and h are DC components. A DC program with convex constraint $\mathbf{x} \in \Omega$ can be equivalently expressed as an unconstrained DC program by adding the indicator function χ_Ω ($\chi_\Omega(\mathbf{x}) = 0$ if $\mathbf{x} \in \Omega$ and $+\infty$ otherwise) to the first DC component g. When one of DC components is polyhedral convex, (P) is called a polyhedral DC program.

Recall that, for a convex function θ and $\mathbf{x}_0$ such that $\theta(\mathbf{x}^0) < +\infty$, the subdifferential of θ at $\mathbf{x}_0$, denoted as $\partial\theta(\mathbf{x}_0)$, is the closed convex set defined by

$$\partial\theta(\mathbf{x}_0) = \{\mathbf{y} \in \mathbb{R}^d : f(\mathbf{y}) \geq f(\mathbf{x}_0) + \langle \mathbf{y}, \mathbf{x} - \mathbf{x}_0 \rangle \ \forall \mathbf{x} \in \mathbb{R}^d\}.$$

An element $\mathbf{y} \in \partial\theta(\mathbf{x}_0)$ is called a subgradient of θ at $\mathbf{x}_0$. The concept of subdifferential generalizes the derivative notion for convex functions in the sense that θ is differentiable at $\mathbf{x}_0$ iff $\partial\theta(\mathbf{x}_0)$ is singleton, and in such case $\partial\theta(\mathbf{x}_0) = \{\nabla\theta(\mathbf{x}_0)\}$.

DCA is based on the local optimality condition for (P), namely $\partial g(\mathbf{x}^*) \cap \partial h(\mathbf{x}^*) \neq \emptyset$. This is also called a generalized Karus-Kuhn-Tucker (KKT) condition for (P), and such a point $\mathbf{x}^*$ is called a *critical* point of (P).

Starting from an initial point $\mathbf{x}^0$, the DCA consists in constructing two sequences $\{\mathbf{x}^t\}$ and $\{\mathbf{y}^t\}$ such that, for any $t = 0, 1, 2, \ldots$

$$\mathbf{y}^t \in \partial h(\mathbf{x}^t) \quad \text{and} \quad \mathbf{x}^{t+1} \in \arg\min_{\mathbf{x}\in\mathbb{R}^d}\{g(\mathbf{x}) - \langle \mathbf{y}^t, \mathbf{x}\rangle\}.$$

The solid mathematical foundation for DCA was developed in [7,8]. Important convergence properties worth being mentioned are

(i) DCA is a descent method without line search: the sequence $\{f(\mathbf{x}^t)\}$ is decreasing. If $f(\mathbf{x}^{t+1}) = f(\mathbf{x}^t)$, then $\mathbf{x}^t$ and $\mathbf{x}^{t+1}$ are critical points of (P). In such a case, DCA terminates after a finite number of iterations.
(ii) Any limit point of the sequence $\{\mathbf{x}^t\}$ is a critical point of (P).
(iii) If (P) is a polyhedral DC program, there is a finite t such that $f(\mathbf{x}^{t+1}) = f(\mathbf{x}^t)$, i.e., DCA has a finite convergence.

3.2 DCA for Solving the Ramp Loss SVDD

It is obvious that the hinge losses H_η are increasing convex functions and $L_\eta = H_0 - H_\eta$. Thus, the objective function of (7) is a DC function with DC decomposition $f(\mathbf{c}, z) = \bar{g}(\mathbf{c}, R) - \bar{h}(\mathbf{c}, R)$ where

$$\bar{g}(\mathbf{c}, z) := z + \lambda \sum_{i=1}^{n} H_0(\|\mathbf{c} - \mathbf{x}_i\|_2^2 - z), \quad \bar{h}(\mathbf{c}, z) := \lambda \sum_{i=1}^{n} H_\eta(\|\mathbf{c} - \mathbf{x}_i\|_2^2 - z).$$

Then DCA is applicable. However, we will present below another DC reformulation of problem (7) for which the resulting DCA is more elegant.

By introducing the slack variables $\xi = (\xi_1, \ldots, \xi_n)$, we can reformulate problem (7) as follows

$$\min_{\mathbf{c}, z, \xi} \left\{ z + \lambda \sum_{i=1}^{n} \min \{\xi_i, \eta\} : z \geq 0, \|\mathbf{c} - \mathbf{x}_i\|_2^2 - z \leq \xi_i, 0 \leq \xi_i \; \forall i \right\}. \tag{8}$$

The following result is straightforward.

Proposition 1. *Problems (7) and (8) are equivalent in the following sense:*

(i) *If $(\mathbf{c}^*, z^*)$ is a solution to problem (7), by letting $\xi_i^* = \max\{\|\mathbf{c}^* - \mathbf{x}_i\|_2^2 - z^*, 0\}$ ($i = 1, \ldots, n$), then $(\mathbf{c}^*, z^*, \xi^*)$ is a solution to (8).*

(ii) *Inversely, if $(\mathbf{c}^*, z^*, \xi^*)$ is a solution to problem (8), then $(\mathbf{c}^*, z^*)$ is a solution to problem (7).*

Notice that the objective function of (8) is concave, so (8) can be reformulated as a DC program

$$\min \left\{ \chi_\Omega(\mathbf{c}, z, \xi) - h(\mathbf{c}, z, \xi) : (\mathbf{c}, z, \xi) \in \mathbb{R}^d \times \mathbb{R} \times \mathbb{R}^n \right\}, \tag{9}$$

where

$$\Omega = \left\{ (\mathbf{c}, z, \xi) : z \geq 0; \|\mathbf{c} - \mathbf{x}_i\|_2^2 - z \leq \xi_i, 0 \leq \xi_i \; \forall i = 1, \ldots, n \right\}$$

$$h(\mathbf{c}, z, \xi) = -\left(z + \lambda \sum_{i=1}^{n} \min \{\xi_i, \eta\} \right) = -z + \lambda \sum_{i=1}^{n} \max \{-\xi_i, -\eta\}.$$

Applying DCA to the DC program (9), at each iteration t, we compute $(\bar{\mathbf{c}}^t, \bar{z}^t, \bar{\xi}^t) \in \partial h(\mathbf{c}^t, z^t, \xi^t)$ explicitly given by

$$\bar{\mathbf{c}}^t = 0, \quad \bar{z}^t = -1, \quad \bar{\xi}_i^t = \begin{cases} -\lambda & \text{if } \xi_i^t < \eta \\ 0 & \text{otherwise} \end{cases} \quad \forall i = 1, \ldots, n. \tag{10}$$

Then we compute $(\mathbf{c}^{t+1}, z^{t+1}, \xi^{t+1})$ as a solution to the convex problem

$$\min \left\{ \chi_\Omega(\mathbf{c}, z, \xi) - \langle \bar{\mathbf{c}}^t, \mathbf{c} \rangle - \bar{z}^t z - \sum_{i=1}^{n} \bar{\xi}_i^t \xi_i : (\mathbf{c}, z, \xi) \in \mathbb{R}^d \times \mathbb{R} \times \mathbb{R}^n \right\},$$

or precisely,

$$\min \left\{ z + \lambda \sum_{i \in I^t} \xi_i : z \geq 0; \ \|\mathbf{c} - \mathbf{x}_i\|_2^2 - z \leq \xi_i, \ 0 \leq \xi_i \ \forall i = 1, \ldots, n \right\}, \quad (11)$$

where $I^t = \{1 \leq i \leq n : \bar{\xi}_i^t = -\lambda\} = \{1 \leq i \leq n : \xi_i^t < \eta\}$.

Similarly to Proposition 1, we can prove that problem (11) is equivalent to

$$\min_{z \geq 0, \mathbf{c} \in \mathbb{R}^d} \ z + \lambda \sum_{i \in I^t} \max\{\|\mathbf{c} - \mathbf{x}_i\|_2^2 - z, 0\}. \quad (12)$$

We observe that problem (12) is an instance of the SVDD problem (1) where the training set is $\mathcal{T}^t = \{\mathbf{x}_i : i \in I^t\}$. Thus, for computing $(\mathbf{c}^{t+1}, z^{t+1}, \xi^{t+1})$, we can solve the following quadratic problem and get its optimal solution α^t

$$\min_{\alpha \in \mathbb{R}^n} \ \alpha^T K \alpha - \langle q, \alpha \rangle \quad (13)$$

$$\text{s.t.} \ \sum_{i=1}^{n} \alpha_i = 1; \ 0 \leq \alpha_i \leq \lambda \ \forall i \in I^t; \ \alpha_i = 0 \ \forall i \notin I^t,$$

then

$$\mathbf{c}^{t+1} = \sum_{i=1}^{n} \alpha_i^t \mathbf{x}_i, \quad z^{t+1} = \min\{\|\mathbf{c}^{t+1} - \mathbf{x}_i\|_2^2 : \alpha_i^t > 0\} \quad (14a)$$

$$\xi_i^{t+1} = \max\{\|\mathbf{c}^{t+1} - \mathbf{x}_i\|_2^2 - z^{t+1}, 0\} \quad \forall i = 1, \ldots, n. \quad (14b)$$

Since $\xi = (\xi_1, \ldots, \xi_n)$ are dummy variables, we can ignore them from calculation. Finally, the DCA for solving the ramp loss SVDD problem can be described as follows.

Algorithm R-SVDD: DCA for solving the ramp loss SVDD problem
Initialization: $(\mathbf{c}^0, z^0) \in \mathbb{R}^d \times \mathbb{R}_+, \ t = 0$.
repeat
    1. Compute $I^t = \{1 \leq i \leq n : \|\mathbf{c}^t - \mathbf{x}_i\|_2^2 - z^t < \eta\}$.
    2. Solve the quadratic problem (13) and get its solution α^t.
    3. Compute $(\mathbf{c}^{t+1}, z^{t+1})$ via (14a).
    4. $t = t + 1$.
until: $f(\mathbf{c}^t, z^t) = f(\mathbf{c}^{t-1}, z^{t-1})$.

Note that the second DC component h of DC program (9) is polyhedral convex. Thus, the following result is a direct consequence of property (iii) stated in Sect. 3.1.

Theorem 1. *Algorithm R-SVDD is terminated after a finite number of iterations.*

Remark 1. The reformulation (8) of the R-SVDD is closely related with the L_0-SVDD [1], where the ramp loss $\min\{\xi_i, \eta\}$ is replaced by a logarithmic loss $\log(\gamma + \xi_i)$ ($\gamma > 0$ is a parameter). DCA-based algorithm developed in [1] for solving the L_0-SVDD is also similar to the R-SVDD. However, the quadratic subproblems in [1] are weighted versions of the SVDD and have n variables, meanwhile our quadratic subproblems have $|I^t|$ variables. Moreover, the use of the logarithmic loss can reduce, but cannot eliminate, the impact of outliers since the weight α_i is computed as the inverse of the error, i.e., $\alpha_i = \frac{1}{\xi_i + \gamma}$. It means that our proposed R-SVDD is more scalable than the L_0-SVDD since the former produces less support vectors and smaller quadratic subproblems. This observation can be seen in the experiments below. Having finite convergence is another advantage of our method over the L_0-SVDD.

Remark 2. Similar to the SVDD, kernel trick can be applied to the R-SVDD by replacing the scalar product $\langle \mathbf{x}_i, \mathbf{x}_j \rangle$ with a kernel function $K(\mathbf{x}_i, \mathbf{x}_j) = \langle \Phi(\mathbf{x}_i), \Phi(\mathbf{x}_j) \rangle$, where Φ is an implicit feature mapping. The representation of $\mathbf{c}$ now becomes $\mathbf{c} = \sum_{i=1}^{n} \alpha_i \Phi(\mathbf{x}_i)$. Then the distance from an object $\mathbf{x}_j$ to the center $\mathbf{c}$ is explicitly given by $\|\mathbf{c} - \Phi(\mathbf{x}_j)\|_2^2 = \alpha^T K \alpha - 2(K\alpha)_j + K_{jj}$.

4   Experiments

In this section, we conduct some experiments on both synthetic and real datasets to validate our proposed method. We will compare our R-SVDD with the SVDD [10] and the L_0-SVDD [1]. All algorithms were implemented on Matlab, and all quadratic convex problems were solved by Cplex 12.6.

For using algorithm R-SVDD, we need to specify initial solution $(\mathbf{c}^0, z^0) \in \mathbb{R}^d \times \mathbb{R}_+$. In our experiments, we simply take

$$\mathbf{c}^0 = \frac{1}{n} \sum_{i=1}^{n} \mathbf{x}_i \quad \text{and} \quad z^0 = \frac{1}{n} \sum_{i=1}^{n} \|\mathbf{c}^0 - \mathbf{x}_i\|_2^2.$$

The parameter η is fixed to $0.1 \times z^0$. Thus, only the parameter λ need to be tuned. For all algorithms, we take $\lambda = \frac{1}{\nu n}$ with $\nu \in (0,1)$ as in [10] and tune the value of ν. For the L_0-SVDD, the parameter γ is set to be 1 as in [1]. A data point will be considered as support vector if its weight α_i exceeds 10^{-3}.

Experiment on synthetic datasets. This experiment aims at investigating the robustness of three compared methods in making a description of a spherically distributed dataset with the presence of noise. We first generated $n = 100$ target data points (to be enclosed) in two-dimensional space. Each dimension is randomly drawn following the Gaussian distribution with mean $\mu = 1$ and standard deviation $\sigma = \frac{1}{2}$. Then we added $x\% \times n$ outliers to the target data. Outliers are uniformly drawn in the region $[0, 5] \times [0, 5]$. For each algorithm, we considered three values of ν and examined changes in the number of support vectors and the compactness of the description. To study the influence of the amount of outliers, we considered two cases $x = 5$ and $x = 20$. Numerical results are reported in

Figs. 1 and 2. In these figures, the black pluses represent data points, the red dot is the mean of the target data which is $(1,1)$, the red circled points are support vectors, the blue solid line and the blue dot are the decision boundary and its center respectively.

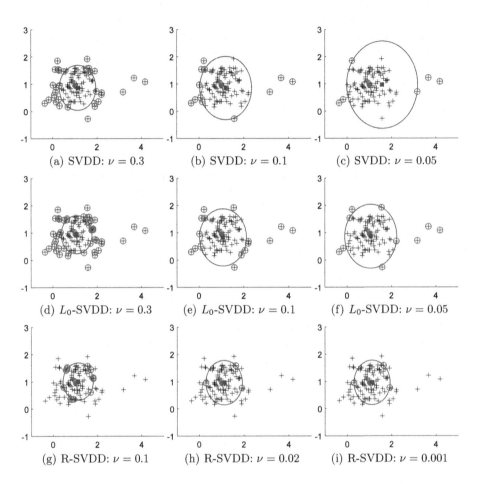

Fig. 1. Synthetic dataset with 5% outliers. (Color figure online)

From numerical experiments, we see that the R-SVDD outperforms the two others. In all cases, despite the wide range of ν, the R-SVDD is very robust to outliers, captures well the target region, and produces very few support vectors. The SVDD can only capture the target region when there is less outliers and ν is small, so λ is large (Fig. 1 (a) and (b)). However, its description has many support vectors. When the percentage of outliers is high (Fig. 2), the SVDD fails to captures the target region for all considered values of ν. The L_0-SVDD is better than the SVDD at deriving good description, but it is still influenced by outliers and sensitive to the selection of the trade-off parameter (Fig. 2 (f)). Moreover, it produces as many support vectors as the SVDD does.

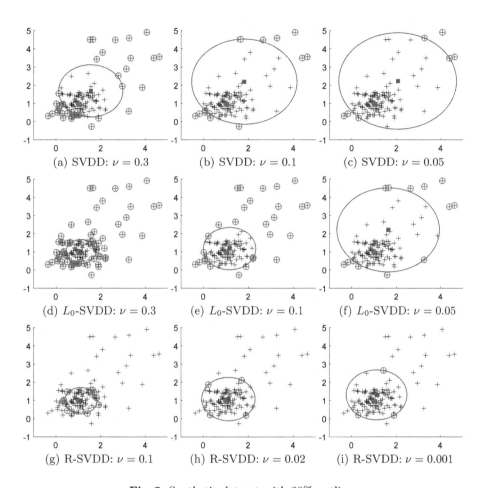

Fig. 2. Synthetic dataset with 20% outliers.

Experiment on real datasets. We used three real datasets for binary classification from UCI repository[1]. Their information is given in Table 1.

The experiment is set up as follows. For each dataset, we choose one class as positive data (to be enclosed), the other class is regarded as negative data. We randomly select 60% of positive data, which is supposed to have m objects, then we randomly select $10\% \times m$ objects from negative data. These selected data form the training set. The rest of the dataset is treated as testing set. For each value of ν, we train each algorithm on the training set and evaluate the BER

---

[1] http://archive.ics.uci.edu/ml/.

(balanced error rate) on the testing set. The process is repeated 10 times and the average BER of each algorithm corresponding to a value of ν is recorded. Performance of each algorithm is the performance corresponding to the value of ν that has the best average BER. Experimental results for real datasets are reported in Table 1. Note that each dataset has two classes and their roles are not symmetric, as it is probable that a class is spherically separable from the other but the reverse is not. In our experiment, we try both cases of labeling data and report the more reasonable case.

Observing from Table 1, it is clear that, on all datasets, the R-SVDD has the best performance and the L_0-SVDD appears as the second best.

Table 1. Average of BER. The bold-face numbers indicate the best results

Dataset	n	d	SVDD	L_0-SVDD	R-SVDD
ionosphere	351	34	24.77%	19.22%	**18.40%**
pima	768	8	32.31%	31.01%	**27.66%**
wdbc	569	30	42.14%	19.89%	**16.66%**

5   Conclusion

We have proposed in this paper a robust version of the SVDD by using the ramp loss for measuring errors. The resulting problem is nonconvex, nonsmooth and can be efficiently solved by a DCA-based algorithm. The proposed method has been theoretically and experimentally proven to be more robust and scalable than the SVDD and the L_0-SVDD. In future works, we will focus on searching for better initial solution and exploiting kernel trick.

References

1. Azami, M.E., Lartizien, C., Canu, S.: Robust outlier detection with L_0-SVDD. In: ESANN 2014 Proceedings, European Symposium on Artificial Neural Networks, Computational Intelligence and Machine Learning, pp. 389–394 (2014)
2. Collobert, R., Sinz, F., Weston, J., Bottou, L.: Trading convexity for scalability. In: Proceedings of the 23rd International Conference on Machine Learning, pp. 201–208 (2006)
3. Le Thi, H.A., Phan Dinh, T.: The DC (difference of convex functions) programming and DCA revisited with DC models of real world nonconvex optimization problems. Ann. Oper. Res. **133**(1–4), 23–48 (2005)
4. Lee, K., Kim, D.W., Lee, K.H., Lee, D.: Density-induced support vector data description. IEEE Trans. Neural Netw. **18**(1), 284–289 (2007)
5. Liu, B., Xiao, Y., Cao, L., Hao, Z., Deng, F.: SVDD-based outlier detection on uncertain data. Knowl. Inf. Syst. **34**(3), 597–618 (2013)
6. Ong, C.S., Le Thi, H.A.: Learning sparse classifiers with difference of convex functions algorithms. Optim. Meth. Softw. **28**(4), 830–854 (2013)

7. Phan Dinh, T., Le Thi, H.A.: Convex analysis approach to DC programming: theory, algorithms and applications. Acta Mathematica Vietnamica **22**(1), 289–355 (1997)

8. Phan Dinh, T., Le Thi, H.A.: A DC optimization algorithm for solving the trust-region subproblem. SIAM J. Optim. **8**(2), 476–505 (1998)

9. Schölkopf, B., Platt, J.C., Shawe-Taylor, J., Smola, A.J., Williamson, R.C.: Estimating the support of a high-dimensional distribution. Neural Comput. **13**, 1443–1471 (2001)

10. Tax, D.M., Duin, R.P.: Support vector data description. Mach. Learn. **54**(1), 45–66 (2004)

The Temporal Supplier Evaluation Model Based on Multicriteria Decision Analysis Methods

Jarosław Wątróbski[1], Wojciech Sałabun[1(✉)], and Grzegorz Ladorucki[2]

[1] Faculty of Computer Science and Informtaion Technology,
West Pomeranian University of Technology in Szczecin,
Żołnierska 49, 71-210 Szczecin, Poland
{jwatrobski,wsalabun}@wi.zut.edu.pl
[2] Faculty of Economics and Management, University of Szczecin,
Mickiewicza 64, 71-101 Szczecin, Poland
gladorucki@wneiz.pl

Abstract. An important element of the provided chain managing process is the rating of suppliers. The paper presents a new framework to identify a temporal supplier evaluation model by using Multi-Criteria Decision Analysis (MCDA) methods. Proposed approach extends classical MCDA paradigm with aspects of temporal evaluation and dedicated aggregation strategies. Afterwards, the proposed framework is used in the identification process for as an illustrative example. Finally, the accuracy of the obtained results is compared and discussed.

Keywords: MCDA · TOPSIS · Supplier selection

1 Introduction

The selection of market suppliers play an important role in effective management of an organization [17]. These activities included in decisions science made on strategic, tactic and operative levels of a company are of particular importance both in the realization of a basic activity of the company and providing services (often outsourcing ones) in auxiliary departments of the company (e.g. HR or accountancy) [14]. Correct selection of suppliers has a direct influence on raising effectiveness of the company (e.g. a decrease of costs) [15], and then, in consequence, on a subject's market position [7]. Effective selection of market collaborators determines efficient introduction to many various modern methods and techniques of management, for example Lean Management, Total Quality Management, and at the same time being the fundament of concentrating market subject on core competitions. Also, important is the influence of suppliers' selection on competitiveness improvement or effectiveness of Supply Chain Management (SCM) [7]. Correct selection of providers has significantly taken part in fast reactions to innovations dynamic and requirements set by the market [2].

In many works, negative results of an incomplete process or a lack of an evaluation process of providers are being analyzed. Even in early works, Meade (1998)

© Springer International Publishing AG 2017
N.T. Nguyen et al. (Eds.): ACIIDS 2017, Part I, LNAI 10191, pp. 432–442, 2017.
DOI: 10.1007/978-3-319-54472-4_41

observed that inappropriate selection of suppliers might have a bad influence on several processes in economic organization [19]. Other authors also confirm this statement, at the same time extending negative consequences on all integrated supply chain, so as the increase in the risk level and dangers of running current business activity [3]. Nowadays, a very important role in the realization of a company's basic business activity has effective SCM management. Selection and rating collaborators are the most significant aspects in correct functioning of the whole SCM [32] and are the base of its stability [24]. Zou points out that these are necessary actions for all stakeholders of the integrated SCM [36], at the same time being the tools of its effectiveness economic increase, or even strongly influencing its creation [18]. Direct consequences of correct evaluation of providers also include accurate amount of services or product obtainment, their planned realization and supply [35], possible flexibility in contractor action [30] or also minimization of risk in the supply process [18].

The development of the Internet and information technologies has taken part in a significant increase in the effectiveness of processing information. In consequence, current monitoring and supplier evaluation as well as the ability to judge repeatedly have been made feasible. The development of intelligent IT management systems made the enlargement of the rating possible to temporal level, opening new research areas with regard to the construction of time based models of provider evaluation.

The main purpose of this paper is to build a framework of dynamic evaluation and selection of suppliers. The model identification is based on MCDA methods. It guarantees that in the decision making process all necessary criteria will be included, which often can be have a opposite character. Main contributions are: construction of the author's procedure for the dynamic evaluation of suppliers, selection of an appropriate MCDA method, define criteria for the suppliers evaluation (based on the literature studies), and a practical verification of proposed procedure in a real enterprise environment (extracts data from the ERP system and expert surveys, and perform calculations).

Rest of the paper is organized as follows. Literature review is presented in Sect. 2. In Sect. 3, we introduced the proposed framework. In Sect. 4, an illustrative example is given to demonstrate the practicality and effectiveness of the proposed approach. Finally, we conclude the paper and give some remarks in Sect. 5.

2 State of Art

Selection of suppliers is an important element, which is of great significance for all involved members. Subject literature analysis gives a variety of theoretical solutions and practical studies, based on which authors make ratings of providers in a number of economic subjects. A vast majority of researches has been made within production companies [13]. Currently, authors pay more and more attention to rating and selection of providers in green supply chains as well as to green SCM which takes into consideration environmental issues [1]. For instance, Kusi-Sarpong judged providers in a green supply chain in the mining industry [16].

Hsu carried out research into rating on providers, who use recycled materials [10]. The authors point out the importance of judging providers as a strategy [18], which has an important influence on improving the efficiency of a supply chain [6] and on its improvement [29]. The researches into rating and selection of suppliers, apart from the above mentioned, include: the pharmaceutical [23], textile [28], railway [2], electric energy [20,34], detergent production [24] industries.

When analyzing the research methodology, it may be noted that the evaluation process is carried out with the use of a number of analytical methods. A thorough discussion of different approaches can be found in [31]. Among them, a very significant role is played by multi-criteria decision support methods (MCDA). By using them, the authors take successful attempts to build models of supplier evaluation in SCM. It is worth noting that the works differ in the nature of the presented models. For example, the number of criteria and the form of a hierarchy model used in the evaluation are often very different. The diversity of individual decision-making problems, differences in the form of input, inaccuracies measuring the same data or, above all, different objectives support, e.g., selection of the best supplier, a ranking of suppliers or their grouping, make even the same specific discipline issue can be examined using different methods MCDA. The literature review indicates the wide applicability of MCDA methods, in particular the Analytic Hierarchy Process (AHP), the Analytic Network Process (ANP), and TOPSIS (The Technique for Order of Preference by Similarity to Ideal Solution), their hybrid and fuzzy expansions in the problem of selection or evaluation of suppliers [11,33].

We can indicate numerous works using the AHP method in the problem of evaluation of suppliers. For example, Chan and Kumar [3] based their assessment process on the development of the fuzzy AHP method. Similarly, Tam, Tahriri, Lady, and Sivrikaya also used the AHP method in their works [20]. A deterministic version of the ANP method was used in the researches of Malmir and Jharkharia [18]. Palanisamy and Dargie used the fuzzy version of the ANP method [21].

Another popular method of MCDA widely used in the evaluation of suppliers is the TOPSIS method. Plenty of works also devote plenty of time to a combination of several methods MCDA (including TOPSIS) in one model of decision-making choice of supplier. For example, Shyur proposed a hybrid approach by using the ANP and TOPSIS methods [27]. Shemshadi based the process of selecting suppliers on the TOPSIS (fuzzy version) and ANP methods [26]. Analogous developments of the AHP and TOPSIS methods were used by Tyagi [29].

To carry out the evaluation and selection of suppliers, other methods of the MCDA family are also successfully used. For example, one should mention the VIKOR method [10], the rough sets theory [36], a fuzzy version of the VIKOR [26], Fuzzy Axiomatic Design [12], DEA [17], ELECTRE I [8], DEMATEL [31] and Vikor [1].

3   Framework for Temporal Supplier Evaluation

In the current studies on the evaluation and selection of suppliers, the dominant trend is a multi-criteria approach, which is based on the search for compromise or simply "good" solutions [25]. The literature review indicates that the MCDA methods are suitable to assess suppliers [17]. It is worth noting that the use of MCDA methods requires building decision-making model for each decision task, according to the individual preferences of the decision maker, and the result has limited range. When analysing the models available in the literature, one can notice that the authors assumed the immutability of the components of the decision support process, e.g., an unchanging set of suppliers, the criteria for their evaluation as well as partial and global assessments.

In the paper we propose a framework for dynamic process of evaluation and selection of suppliers. Methodically, achieving this objective requires the expanding classical MCDA assumptions contained for example in the work of Guitoni [9] and Roy [25]. The classic MCDA procedure assumes constancy of both the set of alternatives considered and the criteria for their evaluation. The result of the construction of the model is a one-off recommendation to the decision maker. Thus, the aggregation of the preferences takes place once. It is also easy to note that the transfer of direct changes in the domain of decision support to the same model is difficult, e.g., a variable number of suppliers, the increase in the number of returns of products, the decline in the quality of precast, delayed deliveries at specified time periods. It should be noted that the validity of the same ratings increases with the approach of the timeline to the last evaluation period. Of course, those criteria, and in particular, their value can be aggregated using, for example, fuzzy sets theory. However, this affects significantly to the accuracy of the process of aggregation assessments and could also lead to over simplifying of the model and the decline in the quality of decision support.

Modeled aspects of the dynamics of the evaluation process include the variability of partial assessments of decision variants over time and the analysis of the impact of that variability on the final outcome of the decision-support process. However, the research procedure is based on the classic assumptions made by Guitoni [9] and Roy [25]. It was decided that modelling preferences and aggregation of the data source would be completed the use of the AHP method in the form of a classical and fuzzy approach. The complex nature of the hierarchical set of criteria for evaluating suppliers is well reflected in the AHP method.

As noted above, the procedure (Fig. 1) is consistent with the classical model of Roy. However, the concept of "time-based aggregation strategies" requires clarification. This concept is based on the construction of the function taking into account an additional attribute called "Forgetting" or possible "Recall". In practical terms at this stage, the previously obtained results of partial evaluations of suppliers of t periods should be aggregated with the function of forgetting [4]:

$$V(a^i) = \sum_{k=1}^{n} CC_{ik} \cdot p(t_k) \tag{1}$$

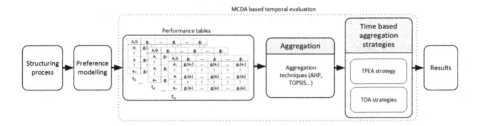

Fig. 1. The proposed framework for temporal supplier evaluation.

where $p(t_k)$ means significance for k period in time t, CC_{ik} is utility of the $i-th$ decisional variant (here it is assessment of the $i-th$ supplier) in k period, which is determined by any process of aggregation, eg., AHP, TOPSIS, etc. Finally, $V(a^i)$ means the general utility for the $i-th$ decisional variant (supplier) on the basis of n periods.

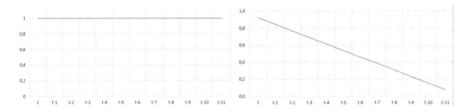

Fig. 2. Two examples of forgetting function (TPEA (left) and TDA1 (right) - Forgetting during 12 months).

In the illustrated procedure the concepts of TDA and TPEA also require clarification. TDA (Time Depreciated Aggregation) means that the final aggregation of the assessment results of the with the function of forgetting is carried out taking into account the declining influence of individual ratings with increasing distance of the historical data of the current period. An example of the function of forgetting is shown in Fig. 2. In TPEA (Time Period Equal Aggregation) the impact of individual ratings on the final outcome of the assessment is balanced.

4   Practical Verification of Proposal Approach

The verification of presented approach was accomplished with data from actually functioning company, where a group of experts has selected 18 criteria for evaluating suppliers. The criteria are divided into six types: logistics, product quality, service efficiency, profile of supplier, economic, and risk. Criteria such as time to confirm delivery, delivery time, MOQ, delivery on-time, complaints, turnover, cost of transport, terms of payment and currency were downloaded

Table 1. The hierarchy of criteria and exemplary partial assessments

Main criteria		Subcriteria		Unit	Supplier A_i			
No	name	No	name		A_1	A_2	...	A_{30}
C_1	Logistic	C_{11}	Delivery on time	%	85.2	89.0	...	80.9
		C_{12}	AVG MOQ	PCS	1227.3	1690.8	...	1618.5
		C_{13}	AVG lead time	days	18.3	25.3	...	20.6
		C_{14}	Order confirmation ($< 48H$)	%	88.6	77.5	...	96.4
		C_{15}	Professionalism of sales person	ling	MG	G	...	G
		C_{16}	Quality of relationship	ling	MG	G	...	G
C_2	Product quality	C_{21}	Complains	%	0.2	0.0	...	0.0
		C_{22}	Environmental control	ling	F	VG	...	VG
		C_{23}	Remedy for quality problems	ling	MG	MG	...	VG
C_3	Service efficienty	C_{31}	Response	ling	G	G	...	MG
		C_{32}	Response to change	ling	MG	F	...	G
		C_{33}	I& DT	ling	F	G	...	MG
C_4	Supplier profile	C_{41}	Yearly amount	1000 pln	4634.1	3067.5	...	234.0
		C_{42}	The level of capital	ling	VG	VG	...	VG
		C_{43}	Company size	ling	VG	MG	...	MG
C_4	Economic	C_{51}	Avg transp cost	%	2.2	0.0	...	0.1
		C_{52}	Delivery time	days	60	70	...	35
C_6	Risk	C_{61}	Currency	%	EUR	EUR	...	EUR

Table 2. The input data for the forgetting function of supplier evaluation model using the AHP method

tk	t	t-1	t-2	t-3	t-4	t-5	t-6	t-7	t-8	t-9	t-10	t-11
TPEA pk	1.000	1.000	1.000	1.000	1.000	1.000	1.000	1.000	1.000	1.000	1.000	1.000
TDA1 pk	0.857	0.714	0.571	0.429	0.286	0.143	0.000	0.000	0.000	0.000	0.000	0.000
TDA2 pk	0.923	0.846	0.769	0.692	0.615	0.538	0.462	0.385	0.308	0.231	0.154	0.077

automatically from the ERP system using a SQL query. The remaining nine were collected by surveys, which were filled by experts working closely with the customer. Table 1 presents the complete hierarchy of criteria and exemplary values of the partial assessments for one period and selected vendors.

Data contained in the table were normalized and values of linguistic variables were mapped according to the Likert scale. Finally, general assessments were calculated for a set of alternatives using the AHP method for each period. When analysing the obtained results, it can be seen high volatility rankings resulting in subsequent periods. Figure 3 presents the dynamics of the vendors position, which occupying the first three positions in the period of 12 months.

Table 3. The final rankings for the considered suppliers and their positions

A_i	Assessment results			Position in the rankings		
	TPEA	TDA1	TDA2	TPEA	TDA1	TDA
A_1	0.5695	0.1469	0.2888	1	1	1
A_2	0.4608	0.1143	0.2308	5	6	5
A_3	0.3866	0.1008	0.1958	16	12	14
A_4	0.4456	0.1008	0.2185	6	13	8
A_5	0.3293	0.0817	0.1649	20	22	22
A_6	0.3726	0.0899	0.1832	17	19	19
A_7	0.3232	0.0913	0.1731	21	18	20
A_8	0.2505	0.0655	0.1265	28	27	28
A_9	0.2387	0.0601	0.1195	29	29	30
A_{10}	0.3936	0.0929	0.1904	14	16	17
A_{11}	0.5374	0.1385	0.2691	2	2	3
A_{12}	0.3578	0.1077	0.1921	19	10	16
A_{13}	0.3942	0.0964	0.1974	13	14	13
A_{14}	0.3198	0.0738	0.1673	22	25	21
A_{15}	0.1966	0.0875	0.1235	30	20	29
A_{16}	0.4145	0.0925	0.1953	10	17	15
A_{17}	0.2582	0.0639	0.1299	27	28	27
A_{18}	0.2987	0.0785	0.1553	25	23	23
A_{19}	0.4298	0.0870	0.1989	8	21	12
A_{20}	0.2984	0.0557	0.1343	26	30	26
A_{21}	0.4302	0.1095	0.2173	7	8	9
A_{22}	0.4013	0.1035	0.2054	12	11	11
A_{23}	0.3015	0.0762	0.1497	24	24	24
A_{24}	0.4081	0.1219	0.2248	11	4	6
A_{25}	0.5170	0.1123	0.2452	4	7	4
A_{26}	0.4178	0.1166	0.2237	9	5	7
A_{27}	0.3041	0.0675	0.1461	23	26	25
A_{28}	0.3922	0.1084	0.2082	15	9	10
A_{29}	0.3710	0.0937	0.1863	18	15	18
A_{30}	0.5277	0.1371	0.2702	3	3	2

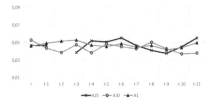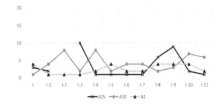

Fig. 3. The assessment values (left side) and place in the ranking (right side) for the top three suppliers in the analyzed period by months

In the next step, the final aggregation was done one more time by using the AHP method, which included the additional two attribute of forgetting (see Eq. 1). Calculation carried out for 2 different parameters power of the forgetting function (TDA1 - linear function for 12 months period and TDA2 - linear function for 6 months period). Aggregation was also done for a constant value of the forgetting parameter in the whole evaluation period. A set of input parameters of these functions are presented in Tables 2 and 3 shows the final rankings for the considered suppliers and their positions in the rankings. This table presents results for 3 forms of used functions of forgetting.

By analyzing the obtained results, it should be noted that regardless of the used form of the forgetting functions the impact on the final ranking is relatively low for the first four positions of the ranking. However, further positions are changed (for example, A4, A12 and further). Big changes in the rankings, where we used TDA1 and TDA2 strategies, are the result of the rapid response of these models to a relatively very recent historical data. It is clearly visible, when compared results of these models with the results obtained by using the classical TEPA model. Similarly to the best suppliers, only slight changes occur in all three rankings for suppliers recognized for the worst (position 26 and further).

5   Conclusions

The presented framework extends the classical MCDA model to temporal variants assessment and time-based aggregation. The multiple evaluation and time based aggregation allow to more accurate reflection of the source data, and thus the construction of a model. This model will be quickly react to changes of the attributes values with respect to the classical forms of data aggregation. The obtained results confirm these conclusions and show high dynamics of changes in the domain that is the supplier evaluation problem. This justifies the need for frequent and thorough evaluation of the considered suppliers.

Research has revealed broader possibilities of analysis and interpretation of the aggregated data. However during the research areas of possible improvement have been identified. Especially interesting, it seems to be identifying the value of the function $p(t_k)$ on the basis of other parameters and available in a real business environment data, e.g., the value of the contract with the supplier in a given period of time and use it as an attribute in the final aggregation.

References

1. Akman, G.: Evaluating suppliers to include green supplier development programs via fuzzy c-means and VIKOR methods. Comput. Ind. Eng. **86**, 69–82 (2015). doi:10.1016/j.cie.2014.10.013
2. Bruno, G., Esposito, E., Genovese, A., Passaro, R.: AHP-based approaches for supplier evaluation: problems and perspectives. J. Purchasing Supply Manag. **18**, 159–172 (2012). doi:10.1016/j.pursup.2012.05.001
3. Chan, F.T.S., Kumar, N.: Global supplier development considering risk factors using fuzzy extended AHP-based approach. Omega **35**, 417–431 (2007). doi:10.1016/j.omega.2005.08.004
4. Chen, Y., Li, K.W., He, S.: Dynamic multiple criteria decision analysis with application in emergency management assessment. In: 2010 IEEE International Conference on Systems, Man and Cybernetics, pp. 3513–3517. IEEE (2010). doi:10.1109/ICSMC.2010.5642410
5. Chen, Y.-H., Chao, R.-J.: Supplier selection using consistent fuzzy preference relations. Expert Syst. Appl. **39**, 3233–3240 (2012). doi:10.1016/j.eswa.2011.09.010
6. Chow, P.S., Choi, T.M., Cheng, T.C.E.: Impacts of minimum order quantity on a quick response supply chain. IEEE Trans. Syst. Man Cybern. Part A Syst. Hum. **42**, 868–879 (2012). doi:10.1109/TSMCA.2012.2183351
7. Dargi, A., Anjomshoae, A., Galankashi, M.R., Memari, A., Tap, M.B.M.: Supplier selection: a Fuzzy-ANP approach. Procedia Comput. Sci. **31**, 691–700 (2014). doi:10.1016/j.procs.2014.05.317
8. de Boer, L., van der Weng, L., Telgen, J.: Outranking methods in support of supplier selection. Eur. J. Purchasing Supply Manag. **4**, 109–118 (1998). doi:10.1016/S0969-7012(97)00034-8
9. Guitouni, A., Martel, J.M.: Tentative guidelines to help choosing an appropriate MCDA method. Eur. J. Oper. Res. **109**, 501–521 (1998)
10. Hsu, C.H., Wang, F.K., Tzeng, G.H.: The best vendor selection for conducting the recycled material based on a hybrid MCDM model combining DANP with VIKOR. Resour. Conserv. Recycl. **66**, 95–111 (2012). doi:10.1016/j.resconrec.2012.02.009
11. Jankowski, J., Kolomvatsos, K., Kazienko, P., Wątróbski, J.: Fuzzy modeling of user behaviors and virtual goods purchases in social networking platforms. J. Univ. Comput. Sci. **22**, 416–437 (2016)
12. Kannan, D., Govindan, K., Rajendran, S.: Fuzzy Axiomatic Design approach based green supplier selection: a case study from Singapore. J. Clean. Prod. **96**, 194–208 (2015). doi:10.1016/j.jclepro.2013.12.076
13. Kaur, P.: Vendor selection in intuitionistic fuzzy sets environment: a comparative study by MADM methods. Int. J. Appl. Eng. Res. **10**, 38146–38153 (2015)
14. Kawa, A., Koczkodaj, W.W.: Supplier evaluation process by pairwise comparisons. Math. Prob. Eng. **2015**, 1–9 (2015). doi:10.1155/2015/976742
15. Khodaverdi, R., Olfat, L.: A fuzzy MCDM approach for supplier selection and evaluation: a case study in an automobile manufacturing company. In: 2011 IEEE International Conference on Industrial Engineering and Engineering Management, pp 25–27. IEEE (2011)
16. Kusi-Sarpong, S., Bai, C., Sarkis, J., Wang, X.: Green supply chain practices evaluation in the mining industry using a joint rough sets and fuzzy TOPSIS methodology. Resour. Policy **46**, 86–100 (2015). doi:10.1016/j.resourpol.2014.10.011
17. Ng, W.L.: An efficient and simple model for multiple criteria supplier selection problem. Eur. J. Oper. Res. **186**, 1059–1067 (2008). doi:10.1016/j.ejor.2007.01.018

18. Malmir, R., Hamzehi, E., Farsijani, H.: A multi stage decision making model to evaluate suppliers by using MOLP and ANP in a strategic approach. Int. J. Appl. Innov. Eng. Manag **2**, 563–577 (2013)
19. Meade, L., Sarkis, J.: Strategic analysis of logistics and supply chain management systems using the analytical network process. Transp. Res. Part E: Logistics Transp. Rev. **34**, 201–215 (1998). doi:10.1016/S1366-5545(98)00012-X
20. Noorollahi, E., Fadai, D., Akbarpour Shirazi, M., Ghodsipour, S.H.: Land suitability analysis for solar farms exploitation using GIS and fuzzy analytic hierarchy process (FAHP)–a case study of Iran. Energies **9**, 643 (2016). doi:10.3390/en9080643
21. Palanisamy, P., Abdul Zubar, H.: Hybrid MCDM approach for vendor ranking. J. Manufact. Technol. Manag. **24**, 905–928 (2013). doi:10.1108/JMTM-02-2012-0015
22. Pani, A.K., Kar, A.K.: A study to compare relative importance of criteria for supplier evaluation in e-procurement. In: 44th Hawaii International Conference on Systems Science, pp. 4–7. IEEE (2011). doi:10.1109/HICSS.2011.35
23. Pourghahreman, N., Qhatari, A.: Supplier selection in an agent based pharmaceutical supply chain: an application of TOPSIS and PROMETHEE π. Uncertain Supply Chain Manag. **3**, 231–240 (2015). doi:10.5267/j.uscm.2015.4.001
24. Roshandel, J., Miri-Nargesi, S.S., Hatami-Shirkouhi, L.: Evaluating and selecting the supplier in detergent production industry using hierarchical fuzzy TOPSIS. Appl. Math. Model. **37**, 10170–10181 (2013). doi:10.1016/j.apm.2013.05.043
25. Roy, B., Vanderpooten, D.: The European school of MCDA: emergence, basic features and current works. J. Multi-Criteria Decis. Anal. **5**, 22–38 (1996). doi:10.1002/(SICI)1099-1360(199603)5:1⟨22::AID-MCDA93⟩3.0.CO;2-F
26. Shemshadi, A., Shirazi, H., Toreihi, M., Tarokh, M.J.: A fuzzy VIKOR method for supplier selection based on entropy measure for objective weighting. Expert Syst. Appl. **38**, 12160–12167 (2011). doi:10.1016/j.eswa.2011.03.027
27. Shyur, H.J., Shih, H.S.: A hybrid MCDM model for strategic vendor selection. Math. Comput. Model. **44**, 749–761 (2006). doi:10.1016/j.mcm.2005.04.018
28. Sivrikaya, B.T., Kaya, A., Dursun, E., Çebi, F.: Fuzzy AHP-goal programming approach for a supplier selection problem. Res. Logistics Prod. **5**, 271–285 (2015)
29. Tyagi, M., Kumar, P., Kumar, D.: A hybrid approach using AHP-TOPSIS for analyzing e-SCM performance. Procedia Eng. **97**, 2195–2203 (2014). doi:10.1016/j.proeng.2014.12.463
30. Vinodh, S., Ramiya, R.A., Gautham, S.G.: Application of fuzzy analytic network process for supplier selection in a manufacturing organisation. Expert Syst. Appl. **38**, 272–280 (2011). doi:10.1016/j.eswa.2010.06.057
31. Wang, C.H.: Using quality function deployment to conduct vendor assessment and supplier recommendation for business-intelligence systems. Comput. Ind. Eng. **84**, 24–31 (2015). doi:10.1016/j.cie.2014.10.005
32. Wątróbski, J., Sałabun, W.: Green supplier selection framework based on multicriteria decision-analysis approach. In: Setchi, R., Howlett, R.J., Liu, Y., Theobald, P. (eds.) Sustainable Design and Manufacturing 2016. Smart Innovation, Systems and Technologies, vol. 52, pp. 361–371. Springer International Publishing, Cham (2016). doi:10.1007/978-3-319-32098-4_31
33. Wątróbski, J., Sałabun, W.: The characteristic objects method: a new intelligent decision support tool for sustainable manufacturing. In: Setchi, R., Howlett, R.J., Liu, Y., Theobald, P. (eds.) Sustainable Design and Manufacturing 2016. Smart Innovation, Systems and Technologies, vol. 52, pp. 349–359. Springer International Publishing, Cham (2016). doi:10.1007/978-3-319-32098-4_30

34. Wątróbski, J., Ziemba, P., Jankowski, J., Zioło, M.: Green energy for a green city–a multi-perspective model approach. Sustainability **8**, 702 (2016). doi:10.3390/su8080702

35. Yu, C., Wong, T.N.: An agent-based negotiation model for supplier selection of multiple products with synergy effect. Expert Syst. Appl. **42**, 223–237 (2015). doi:10.1016/j.eswa.2014.07.057

36. Zou, Z., Tseng, T.L.B., Sohn, H., Song, G., Gutierrez, R.: A rough set based approach to distributor selection in supply chain management. Expert Syst. Appl. **38**, 106–115 (2011). doi:10.1016/j.eswa.2010.06.021

Machine Learning and Data Mining

A Novel Entropy-Based Approach to Feature Selection

Chia-Hao Tu and Chunshien Li[✉]

Laboratory of Intelligent Systems and Applications, Department of Information Management,
National Central University, Taoyuan City, Taiwan
jamesli@mgt.ncu.edu.tw

Abstract. The amount of features in datasets has increased significantly in the age of big data. Processing such datasets requires an enormous amount of computing power, which exceeds the capability of traditional machines. Based on mutual information and selection gain, the novel feature selection approach is proposed. With Mackey-Glass, S&P 500, and TAIEX time series datasets, we investigated how good the proposed approach could perform feature selection for a compact subset of feature variables optimal or near optimal, through comparing the results by the proposed approach to those by the brute force method. With these results, we determine the proposed approach can establish a subset solution optimal or near optimal to the problem of feature selection with very fast calculation.

Keywords: Feature selection · Probability density estimation · Information entropy · Time series dataset

1 Introduction

In the field of machine learning, feature selection is an issue that has always received much attention, especially in the past few years, due to the amount of features in datasets increased significantly. A good feature selection method not only can select relatively important feature subsets but also can reduce the amount of calculation required for the model while maintaining forecast/estimation accuracy. Feature selection refers to using certain designated methods or rules to select an important subset of feature variables from a dataset, of which the corresponding data are to be used as training and testing data for establishing machine learning models, in the hope of producing a model with good capabilities in estimation or prediction. With rapid advances in computer and database technologies, datasets with thousands of features are now ubiquitous in pattern recognition, data mining, and machine learning [1, 5, 10, 12]. Processing such datasets requires an enormous amount of processing power, which exceeds the capability of traditional machines. The use of feature selection eliminates irrelevant, redundant, and noisy data and allows machine learning models to operate normally.

In this study, we propose a novel feature selection algorithm based on mutual information and selection gain, which uses Claude Shannon's information theory as its foundation. With selection gain, the proposed method can select features from a large amount of data to form a compact feature subset to the target variable. The proposed method is intuitive, easy to manage and explain, and is expected to be fast with good performance.

© Springer International Publishing AG 2017
N.T. Nguyen et al. (Eds.): ACIIDS 2017, Part I, LNAI 10191, pp. 445–454, 2017.
DOI: 10.1007/978-3-319-54472-4_42

Hence, we designed experiments to test the algorithm for feature selection on time series datasets, and compared the results to those by the brute force method to investigate how good the proposed algorithm could quickly achieve near optimal solution.

The rest of this paper is organized as follows. Section 2 introduces previous studies in literature on feature selection and probability distribution. Section 3 describes a novel approach to feature selection based on mutual information and selection gain. The experimental results are presented in Sect. 4. Finally, Sect. 5 concludes the paper.

2 Literature Review

Using a correlation between feature selection and machine learning procedures, John et al. [11] first categorized feature selection methods into two types: "filter" or "wrapper". Later researchers also adopted these classifications. In addition to filter and wrapper, Guyon and Elisseeff [9] proposed a new type known as "embedded". The following is a brief overview of the filter, wrapper, and embedded types:

- For filter methods, feature selection is completely an independent procedure from machine learning for computational models. The feature subsets generated through feature selection are assessed using certain metrics.
- In wrapper methods, various subsets of feature variables generated by the feature selection are inputted into machine learning models for assessment, and the best subset is kept. Loughrey [13] mentioned that wrappers are much more computationally expensive and have a risk of overfitting to the model considered.
- Embedded methods take the feature selection procedure and integrate it with the machine learning procedure. This type is only used in a few specific models.

Within filter methods, certain metrics need to be defined as evaluation functions and used to assess the feature subsets selected to decide whether to keep or eliminate them. In different applications, the selection metrics will also be different. For example, in the field of text classification applications, Forman [7] listed 12 metrics that had been used in previous studies and compared them. These metrics include: Chi-Squared, Information Gain (IG), Odds Ratio, Bi-Normal Separation (BNS), etc. Dash and Liu [4] categorized evaluation functions into four types according to metrics evaluation content: Distance Measures, Information Measures, Dependence Measures and Consistency Measures.

Shannon proposed the concept of entropy and mutual information in his 1948 paper on information theory. *Entropy* is used to measure the total amount of disordered information of a single random variable, while *mutual information* is used to measure mutual dependence between two random variables. These types of intuitive concepts have been widely used in applications across different fields. Naghibi et al. [15], Peng et al. [17], and Torkkola [20] all had used this method in their feature selection studies, obtaining good results.

A random variable is a variable whose value is subject to variations due to chance. It can take on a set of possible different values, each with an associated probability, in contrast to other mathematical variables. A random variable can be either discrete or continuous, depending on its legal values. A probability distribution, usually defined by a mathematical function, describes all possible events within a given range and their likelihoods, with the constraint that the integration (or summation) of likelihoods of all possible events must be equal to one. If a random variable is continuous, the distribution function is called a probability density function. Otherwise, it is called a probability mass function (or simply probability function). In the real world, probability distribution can be estimated using observations.

The methods of density estimation can be classified into parametric and non-parametric estimation methods. In parametric estimation methods, the probability density function is assumed to be given, and the parameter values are unknown. Hence, the parameter values are estimated. When the assumptions are correct, parametric methods will produce more accurate and precise estimates than non-parametric methods. However, if the assumptions are wrong, a larger error will result. Parzen [16] and Rosenblatt [18] proposed the kernel density estimation method, also known as the Parzen window estimation, a non-parametric estimation method that is commonly used. In the non-parametric estimation methods, only the assumption that is *similar inputs have similar outputs* was made, where selecting the probability density function is not required. This can avoid false assumption when used in probability density function, required by parametric estimation methods.

Although selecting the kernel function is required in the kernel density estimation, the final curve shape is not closely correlated with the selected kernel function [6], which is due to the fact that the neighboring wave crests generated by the kernel density estimation may be synthesized. Considering the ease of use of a function of wave synthesis, generally the Gauss function is used as the kernel function. The kernel density estimation method is very widely used. In the studies of feature selection, Alibeigi et al. [2], Azmandian et al. [3], Geng and Hu [8], Supriyanto et al. [19] and Zhang and Wang [21] had developed some feature selection methods based on the concept of kernel density estimation.

3   Proposed Approach

3.1   Entropy-Based Information

Assume that there are n discrete events as possible results for a random variable X, denoted as $\{x_i, i = 1, 2, \ldots, n\}$. The corresponding probability of each event is denoted as $\{p(x_i), i = 1, 2, \ldots, n\}$. In 1948, Shannon defined the concept of *entropy* to measure the amount of uncertainty. If additional information to X is added, the corresponding entropy will be decreased, because uncertainty of X is subsided. Thus, the change of entropy of X can be viewed as additional information to X, and vice versus. For a discrete random variable X, the entropy is denoted by $H(X)$, defined below.

$$H(X) = -\sum_{x_i \in U_X} p(x_i) \log p(x_i) \tag{1}$$

If there are two discrete random variables X and Y, the joint entropy can be defined below for the total amount of uncertainty between the two random variables, denoted as $H(X, Y)$.

$$H(X, Y) = - \sum_{x_i \in U_X} \sum_{y_j \in U_Y} p(x_i, y_j) \log p(x_i, y_j) \tag{2}$$

The joint entropy must be not greater than the sum of the individual entropies of the two variables, as shown below.

$$H(X, Y) \leq H(X) + H(Y) \tag{3}$$

If X is already known, then conditional entropy can be used to calculate the amount of uncertainty for Y, denoted as $H(Y|X)$.

$$H(Y|X) = - \sum_{x_i \in U_X} \sum_{y_j \in U_Y} p(x_i, y_j) \log p(y_j|x_i) \tag{4}$$

The relationship for joint entropy, and conditional entropy can be written below.

$$\begin{aligned} H(Y|X) &= H(X, Y) - H(X) \\ \text{or } H(X, Y) &= H(X) + H(Y|X) \end{aligned} \tag{5}$$

From Eqs. (3) and (5), the following result can be established.

$$H(Y) \geq H(Y|X) \tag{6}$$

From formula (6), if X is known, the amount of uncertainty in Y can be reduced. The *mutual information* is denoted as $I(X, Y)$, defined as follows.

$$I(X, Y) = H(Y) - H(Y|X) \tag{7}$$

Equations (1) to (7) are used for discrete cases. If continuous random variables considered, entropy equations for $H(Y)$, and $H(Y|X)$ are given, respectively, as follows.

$$H(Y) = - \int_{y \in U_Y} p(y) \log p(y) dy \tag{8}$$

$$H(Y|X) = - \int_{x \in U_X} \int_{y \in U_Y} p(y|x)p(x) \log p(y|x) dx\, dy \tag{9}$$

The purpose of entropy-based feature selection is to find a feature subset, denoted as FS, with maximum dependency and minimum redundancy. Previous studies just applied the mutual information value between X and Y, denoted as Eq. (7), for this purpose [17]. In this study, we take a new approach, using negative and positive values of X and Y, to break mutual information into four types: $I(X_+, Y_+), I(X_+, Y_-), I(X_-, Y_+)$ and $I(X_-, Y_-)$. Then the values of the four types of mutual information are averaged to become an element of influence information matrix (IIM), as shown in the algorithm given in Sect. 3.2.

3.2 Proposed Method

For a dataset, the data can be re-organized into a matrix, each column of which is viewed as a variable. These variables are then expressed in the paired form of $(\vec{X}, Y)$, where $\vec{X}$ is an n-dimensional vector of feature variables, $\vec{X} = [X_1, X_2, \ldots, X_n]$, and Y is the corresponding target variable. The objective of feature selection is to select a compact subset of feature variables from $\vec{X}$ so that the target variable can have as effective information supplied by the compact subset as possible. All or part of the variables of the subset can be used as input (feature) variables, when utilized afterwards in machine learning applications. The subset is denoted by FS $= \{f_k, k = 1, 2, \ldots, m\}$, where m, $m \leq n$, refers to the amount of feature variables selected into FS and f_k is the feature variable at the kth selection. Note that the parameters m and n are usually pre-given, depending on applications.

The procedure of feature selection is given as follows.

Step 1. Calculate the influence information matrix (IIM) for all feature and target variables, using the dataset corresponding to these variables. Each component of IIM is calculated using mutual information given in Eq. (7). Note that IIM can be an asymmetry matrix because the influence from one variable to another can be different and vice versus.

Step 2. Select the feature variable, which is with the best influence information to the target variable in IIM, to the empty FS initially.

Step 3. Calculate selection gains for feature variables $\{X_i\}$ that are not in FS, respectively. And, select the feature variable that is with the best selection gain to FS. The function of selection gain is composed of two parts. The 1st part is the influence information from X_i to the target variable. The 2nd one is the average influence information between X_i and each variable in FS. The selection gain function for the feature variable X_i is given below.

$$\text{gain}(X_i) = \text{IIM}(X_i, Y) - \frac{1}{2|\text{FS}|} \sum_{k=1}^{|\text{FS}|} \left(\text{IIM}(X_i, f_k) + \text{IIM}(f_k, X_i) \right) \tag{10}$$

where $\text{IIM}(X_i, Y)$ represents the influence information from X_i to the target variable Y; f_k is the feature variable at the kth selection to FS; $|\text{FS}|$ indicates the size of FS so far. Repeat Step 3 until $|\text{FS}| = m$ or $\text{gain}(X_i) < 0$.

4 Experiments

For experiments, we collected three datasets of time series: Mackey-Glass, S&P 500, and TAIEX. S&P 500 and TAIEX daily exchange data were collected from Google Finance, and in order to simplify the experiment, only the closing indices were used, and all feature and target variables were generated according to close stock index.

In 1977, Mackey and Glass used first-order differential-delay equations to describe physiological control systems with the Mackey-Glass differential equation [14].

$$\frac{dx(t)}{dt} = \beta \frac{x(t-\tau)}{1 + \{x(t-\tau)\}^n} - \gamma x(t) \tag{11}$$

where $\{\beta, \gamma, \tau, n\}$ are positive real-valued parameters, and $x(t-\tau)$ represents the value of the variable x at time $(t-\tau)$. At different τ the equation will exhibit different physical behaviors. If τ is small, a periodic phenomenon appears. If τ is no less than 17, a chaotic phenomenon appears. Chaotic Mackey-Glass time series has been extensively applied to testing criteria for the degree of accuracy of various models using neural network, fuzzy logic, and others. The parameter settings for the Mackey-Glass equation in (11) were given as $\beta = 0.2$, $\gamma = 0.1$, $\tau = 17$, and $n = 10$. The sampling time was 1 s. For x(t <= tau), values were set to be random in [0,1]. For the Mackey-Glass dataset in the study, 1000 data points for $t = 1001$ to 2000 were used.

Standard & Poor's 500 (S&P 500) is an American stock market index based on the market capitalizations of 500 major companies having common stock listed on the NYSE or NASDAQ. The constituents of the S&P 500 index are selected by a committee and are reviewed over time. Because of the strict selection process for the S&P 500 index, many consider it one of the best representations of the U.S. stock market as well as a bellwether for the U.S. economy. For the S&P 500 dataset in the study, 283 close stock index data from the 18th June 2015 to the 1st August 2016 were collected.

Taiwan Stock Exchange Capitalization Weighted Stock Index (TAIEX) uses the Passche Formula and uses the market capitalization of listed stocks for weighting to calculate the stock price index. This is similar to the US S&P 500, and it is regarded as an index that reflects the market value of the overall stock exchanges. For the TAIEX dataset in the study, 274 close stock index data from the 17th June 2015 to the 1st August 2016 were collected.

For the proposed approach, all the original datasets must be pre-processed and converted to datasets with 30 features to comply with the experimental setup. For each dataset, we calculated its difference series in time order, denoted as $\{\Delta y(k), k = 1, 2, \ldots, n_D\}$, where $\Delta y(k) = y(k+1) - y(k)$; $y(k)$ the datum at the kth time order; n_D the size of the dataset transformed. The dataset was then reorganized as $\{(\vec{x}(i), d(i)), i = 1, 2, \ldots\}$, where $\vec{x}(i) = [\Delta y(i + 30 - j), j = 30, 29, \ldots, 1]^T$ and $d(i) = \Delta y(i + 30)$. After pre-processing, the Mackey-Glass dataset was re-organized to be with 969 data pairs; the S&P 500 dataset with 252 data pairs; the TAIEX dataset with 243 data pairs.

The proposed approach was compared to the brute force method for performance comparison in terms of computing time and contents of compact subsets for feature selection. The brute force method could exhaustively form all possible subsets of features, for each of which it then computed to see its information gain contribution to the target variable. In this way, this compared method surely could find out the optimal subset of feature variables. However, it is computationally intensive as its name shows. By comparing the results by both the proposed approach and the brute force method, we investigated how good the proposed approach could establish the optimal or sub-optimal subset of feature variables. Note that to consider the computing time consumption by the brute force method we used a maximum of eight feature variables only in the three experiments. The device used for the experiments was a desktop with Intel Core i7-6500U processor, 16 GB of DDR3 memory and 256 GB SATA SSD.

For the Mackey-Glass, S&P 500, and TAIEX data preprocessed, each was used to test both the proposed algorithm (denoted as EBAFS for short henceforth) and the brute force method. The results are shown below. The feature subsets by both methods are shown in Tables 1 to 3 for Mackey-Glass, S&P 500, and TAIEX, respectively. The curve of feature selection for S&P 500 by the proposed approach is shown in Fig. 1.

Table 1. Feature selection by the EBAFS and the brute force method (Mackey-Glass)

EBAFS (proposed)		Brute force method	
Feature ID	Selection gain	Feature ID	Selection gain
1	0.8276	1	0.8276
23	0.0998	23	0.0998
20	0.0691	20	0.0691
14	0.0698	14	0.0698
16	0.0488	16	0.0488
12	0.0546	12	0.0546
18	0.0390	18	0.0390
13	0.0333	13	0.0333

Table 2. Feature selection by the EBAFS and the brute force method (S&P 500)

EBAFS (proposed)		Brute force method	
Feature ID	Selection gain	Feature ID	Selection gain
9	0.4932	9	0.4932
3	0.0328	3	0.0328
20	0.0196	7	0.0167
16	0.0206	22	0.0226
22	0.0175	16	0.0175
18	0.0156	1	0.0181
1	0.0134	18	0.0139
7	0.0132	13	0.0124

Table 3. Feature selection by the EBAFS and the brute force method (TAIEX)

EBAFS (proposed)		Brute force method	
Feature ID	Selection gain	Feature ID	Selection gain
22	0.5875	22	0.5875
30	0.0674	30	0.0674
28	0.0530	28	0.0530
5	0.0438	5	0.0438
3	0.0381	3	0.0381
21	0.0326	21	0.0326
1	0.0320	1	0.0320
20	0.0286	20	0.0286

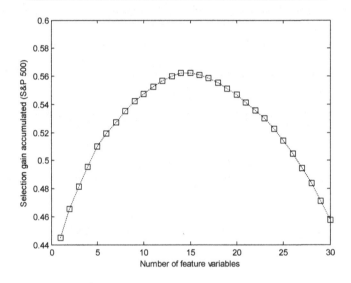

Fig. 1. Curve of feature selection (S&P 500)

As shown in Tables 1 and 3, for Mackey-Glass and TAIEX datasets, the subsets of feature variables by the proposed approach are completely consistent with the results found by the brute force method. As shown in Table 2, for S&P 500 dataset, the result by the EBAFS (the proposed approach) shows petite nonconformity from that by the brute force method (viewed as optimal solution). By comparing to the brute force method, we can see that the results of the three experiments by the proposed EBAFS are to be optimal or near optimal for feature selection.

The computing time spent by the proposed and compared methods is shown in Table 4, showing that the proposed method is much faster. Through the experimental results, the proposed approach can indeed achieve excellent performance in terms of the contents of compact feature subsets optimal or near optimal and faster computing time spent when compared to the brute force method.

Table 4. Computing time spent by the proposed method and the compared brute force method

Method	Mackey-Glass	S&P 500	TAIEX
Brute force method	3,543 s	3,234 s	3,197 s
EBAFS (proposed)	0.018 s	0.021 s	0.016 s
Times faster by EBAFS	196,833 times	154,000 times	199,813 times

5 Conclusion and Future Work

Feature selection has become more and more important for the problems faced in big data. We have proposed a novel entropy-based approach in the paper to feature selection, by which some influentially important feature variables to target variable can be effectively selected to form a compact subset. Such a subset can be utilized in the future for machine learning of intelligent computing models, such as fuzzy systems, neural networks, and others. This proposed approach is with the merits in terms of much less computing time spent when compared to the brute force method and capability of finding optimal or near optimal solution to feature selection.

For verifying the research idea of the proposed approach, we designed and performed experiments on chaotic Mackey-Glass, S&P 500, and TAIEX time series datasets. The results by the proposed method were compared to those by the brute force method for performance comparison, giving promising consequences, as shown in Tables 1 to 4. We have determined that the proposed approach can establish the subset solution optimal or near optimal in feature selection with very fast calculation. Furthermore, the method is intuitive and easy to understand.

Research topics to extend for the study in the future include finding a more efficient way to reduce further computational resources when establishing an influence information matrix for feature selection and applying on real-world problems using the proposed approach.

Acknowledgments. This study was supported by the research project with funding no. MOST 104-2221-E-008-116, Ministry of Science & Technology, Taiwan.

References

1. Aksakalli, V., Malekipirbazari, M.: Feature selection via binary simultaneous perturbation stochastic approximation. Pattern Recogn. Lett. **75**, 41–47 (2016)
2. Alibeigi, M., Hashemi, S., Hamzeh, A.: Unsupervised feature selection using feature density functions. Int. J. Electr. Electron. Eng. **3**(7), 394–399 (2009)
3. Azmandian, F., Dy, J.G., Aslam, J.A., Kaeli, D.R.: Local kernel density ratio-based feature selection for outlier detection. In ACML, pp. 49–64, November 2012
4. Dash, M., Liu, H.: Feature selection for classification. Intell. Data Anal. **1**(3), 131–156 (1997)
5. Dash, M., Liu, H.: Consistency-based search in feature selection. Artif. Intell. **151**(1), 155–176 (2003)

6. De Smith, M.J.: STATSREF: Statistical Analysis Handbook - a web-based statistics resource. The Winchelsea Press, Winchelsea (2015)

7. Forman, G.: An extensive empirical study of feature selection metrics for text classification. J. Mach. Learn. Res. **3**, 1289–1305 (2003)

8. Geng, X., Hu, G.: Unsupervised feature selection by kernel density estimation in wavelet-based spike sorting. Biomed. Sig. Process. Control **7**(2), 112–117 (2012)

9. Guyon, I., Elisseeff, A.: An introduction to variable and feature selection. J. Mach. Learn. Res. **3**, 1157–1182 (2003)

10. Jain, A., Zongker, D.: Feature selection: evaluation, application, and small sample performance. IEEE Trans. Pattern Anal. Mach. Intell. **19**(2), 153–158 (1997)

11. John, G.H., Kohavi, R., Pfleger, K.: Irrelevant features and the subset selection problem. In Proceedings of the Eleventh International Conference on Machine Learning, pp. 121–129 (1994)

12. Kohavi, R., John, G.H.: Wrappers for feature subset selection. Artif. Intell. **97**(1), 273–324 (1997)

13. Loughrey, J., Cunningham, P.: Overfitting in wrapper-based feature subset selection: the harder you try the worse it gets. In: Bramer, M., Coenen, F., Allen, T. (eds.) Research and Development in Intelligent Systems XXI, pp. 33–43. Springer, London (2005)

14. Mackey, M., Glass, L.: Oscillation and chaos in physiological control systems. Science **197**(4300), 287–289 (1977)

15. Naghibi, T., Hoffmann, S., Pfister, B.: A semidefinite programming based search strategy for feature selection with mutual information measure. IEEE Trans. Pattern Anal. Mach. Intell. **37**(8), 1529–1541 (2015)

16. Parzen, E.: On estimation of a probability density function and mode. Ann. Math. Stat. **33**(3), 1065–1076 (1962)

17. Peng, H., Long, F., Ding, C.: Feature selection based on mutual information criteria of max-dependency, max-relevance, and min-redundancy. IEEE Trans. Pattern Anal. Mach. Intell. **27**(8), 1226–1238 (2005)

18. Rosenblatt, M.: Remarks on some nonparametric estimates of a density function. Ann. Math. Stat. **27**(3), 832–837 (1956)

19. Supriyanto, C., Yusof, N., Nurhadiono, B.: Two-level feature selection for Naive Bayes with kernel density estimation in question classification based on Bloom's cognitive levels. In: 2013 International Conference on Information Technology and Electrical Engineering (ICITEE), pp. 237–241, October 2013

20. Torkkola, K.: Feature extraction by non-parametric mutual information maximization. J. Mach. Learn. Res. **3**, 1415–1438 (2003)

21. Zhang, J., Wang, S.: A novel single-feature and synergetic-features selection method by using ISE-based KDE and random permutation. Chin. J. Electron. **25**(1), 114–120 (2016)

An Artificial Player for a Turn-Based Strategy Game

Filip Maly, Pavel Kriz[✉], and Adam Mrazek

Department of Informatics and Quantitative Methods, Faculty of Informatics
and Management, University of Hradec Kralove, Hradec Kralove, Czech Republic
{Filip.Maly,Pavel.Kriz}@uhk.cz

Abstract. This paper describes the design of an artificial intelligent
opponent in the Empire Wars turn-based strategy computer game. Several approaches to make the opponent in the game, that has complex
rules and a huge state space, are tested. In the first phase, common
methods such as heuristics, influence maps, and decision trees are used.
While they have many advantages (speed, simplicity and the ability
to find a solution in a reasonable time), they provide rather average
results. In the second phase, the player is enhanced by an evolutionary
algorithm. The algorithm adjusts several parameters of the player that
were originally determined empirically. In the third phase, a learning
process based on recorded moves from previous games played is used.
The results show that incorporating evolutionary algorithms can significantly improve the efficiency of the artificial player without necessarily
increasing the processing time.

Keywords: Artificial intelligence · Computer games · Influence maps ·
Evolutionary algorithms · Heuristics

1 Introduction

Artificial intelligence (AI) is one of the most emerging fields. Each year there are
new advances in this field, but we still fail to create an artificial intelligent entity
that would correspond to human intelligence. Artificial players in (computer)
games are one of the use-cases for the artificial intelligence. We have seen some
successful solutions in the field of the game AI [1]. For example, an intelligent
computer program that is able to defeat the chess world champion has been
created [2]. But such intelligent opponents have not been successful in many
complex games with complicated rules and aspects of a random chance. Creating
an artificial intelligent opponent also faces time constraints, especially in games
taking place in real time.

In this paper we focus on the development of AI opponents in a turn-based
strategy game called *Empire Wars*, which has complex rules and extensive state
space. The aim is to design an artificial player that will be able to make reasonable moves in the game in real time and will be a suitable opponent to a human

© Springer International Publishing AG 2017
N.T. Nguyen et al. (Eds.): ACIIDS 2017, Part I, LNAI 10191, pp. 455–465, 2017.
DOI: 10.1007/978-3-319-54472-4_43

player. We will use different approaches that will be compared in terms of their effectiveness and suitability.

The rest of this paper is organized as follows. Section 2 describes related work. We formulate the problem in Sect. 3. Section 4 describes the proposed solution. We present the results of the testing in Sect. 5. Section 6 concludes the paper.

2   Related Work

There are many different approaches to development of the opponent AI. **Heuristic algorithms** [3,4] are ones of the first historical approaches used. A greedy algorithm [5–7] is a typical example of heuristic algorithms. The simplicity of implementation and low computational complexity were the reasons for their use in the past. However, heuristic algorithms are used today as well, as they are on par with many more sophisticated algorithms and sometimes more effective compared to them. Especially in computer games running in real-time, CPU time can be a very important factor [8–10].

Cheating AI [11–13] is another approach that eliminates solving complex tasks. A cheating artificial player can use information about the game situation, which is not available to him/her according to the rules. The cheating AI player is an effective and popular way to easily create a sufficiently strong opponent without a complicated processing huge amounts of data. On the other hand, the cheating artificial opponent is unfair to the human player. Additionally, it cannot be used in games where both players have a complete overview of the situation (gameboard) and possible moves (eg. in Chess).

State space search [14,15] is another approach. This method generates possible moves in the game and searches the resulting game situation. Subsequently it evaluates which move leads to the best result. But this approach is time-consuming. For many complex tasks it is impossible to search such a large state space.

Decision trees [16,17] are another popular and frequently used method. They represent the formalization of the decision-making process in the *If-Then* form [18]. In one step of the algorithm, the condition assigned to the current node (root in the beginning) of the decision tree is tested. Depending on the result, the algorithm proceeds to one or the other branch from the node. These steps are repeated until a leaf node is reached. Then the action in the leaf node is taken.

Influence map [19,20] is an auxiliary tool for players created using the above described approaches. It provides a simplified representation of complex large amounts of data that would otherwise be difficult to process. Essentially, it is a model of an environment that serves as a basis for decision-making or evaluation of the benefits of some area in the environment.

Evolutionary algorithms [21,22] or **genetic algorithms** in particular represent the newer trend in the artificial intelligence. Instead of looking for a specific solution for a particular game, we create artificial players who are able to take random moves or use only a basic AI such as heuristics. They then evolve as they learn the best strategy either by playing against each other or against human opponents [23, 24]. This approach is applicable to virtually any situation, which is its main advantage. Like a human player, an artificial player gradually learns game strategy no matter how complex the game is.

3   Problem Formulation

The aim of this work is to compare different approaches to the game AI and to try to improve or combine them to achieve more efficient and better solutions. Particular solutions will be applied to a new turn-based strategy game for up to four players called *Empire Wars* aimed at building a civilization.

The game contains typical elements that can be found in many other games, such as warfare, economic management, mining, construction of buildings or inventing new technologies. The game also includes a limited use of a random chance. Thus, the modeled world will not be completely deterministic.

3.1   Empire Wars Game Rules

The Empire Wars game is played on the gameboard composed of 11 by 11 hexagonal cells, see Fig. 1. Each cell is assigned one resource in a certain amount. There are five types of resources: wheat, oil, stone, iron and gold. Gold can be exchanged for any other resource.

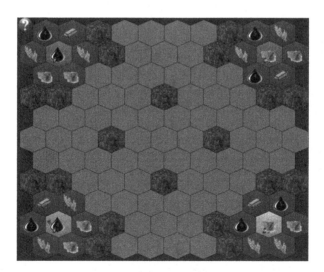

Fig. 1. Empire Wars gameboard

A player can build one building in each cell. There are 5 different types of buildings; barracks, laboratories, factories, houses, cathedrals. Barracks increase defensive military strength in the cell, a laboratory can invent a new technology, a factory consumes iron and produces weapons for troops and houses increase the number of people. People are needed for mining and producing resources and for operation of buildings. A cathedral produce the *Victory Points*. The player who first reaches 100 Victory Points wins the game. A player can use troops to attack a neighboring cell. Military units (troops) consume oil. During the attack, the number of attacking and defending troops are taken into account. A random chance also slightly influences the outcome of the attack.

Players take turns when playing the Empire Wars. In each turn a player can take up to 3 actions. There are several types of actions; an attack, moving troops, buying troops, mining resources, inventing new technologies and building a building. At the end of the turn, the player may also destroy his/her military unit(s) (especially when there is not enough oil for them) and deploy people into buildings and on the cells to mine and produce resources.

The "fog of war" is also a natural part of the game mechanics. Players cannot see details in all cells, but only in cells that belong to them or to an opponent or in cells adjacent to these already occupied cells. A part of the gameboard is thus completely covered at the beginning of the game.

4   Artificial Players

A design of the AI opponent will be conducted in three phases. In the first phase, we will design a player called *Alpha*, which will utilize decision trees, influence maps and heuristic algorithms.

4.1   Alpha Player

The Alpha player considers all feasible actions in a given turn and then assigns individual priorities to them. These priorities are recalculated in every turn.

Movement of the troops is one of the actions. The following factors affect the importance of the movement: distance of a cell that has to be defended (as it contains a valuable building or a weak military unit), or whether there is a reachable cell from which we can further attack. These and other factors are combined into a decision tree.

In the absence of foreign adjacent cells (and therefore the impossibility of an attack) an auxiliary influence map is created in order to improve decision-making. The map calculates the importance of every cell from the global point of view. When a player occupies some of the important cells, he/she will be able to defend a broad area of the gameboard. An influence map is shown in Fig. 2. The red color indicates the strategically important cells, brown color indicates inaccessible cells and white color denotes unimportant accessible cells. The artificial player prefers strategically important cells.

Furthermore, based on the quantity and necessity of resources and buildings, which are located in the target cell, the AI player calculates the value of the cell. It chooses a target cell with the highest value with regard to the chances of a

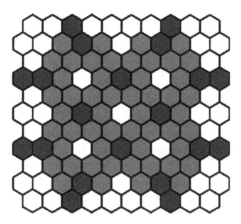

Fig. 2. Influence map (Color figure online)

successful attack. The player is actually using the greedy algorithm. It chooses a local optimum regardless the values of new cells that would become accessible in case of a successful attack.

Building a building is another possible action. The artificial player always calculates priorities for all five possible types of buildings. Then it begins to build a building with the highest priority. But only if it has sufficient resources for it. Target cell for a new building is chosen based on influence maps, decision trees and heuristics. Strategically important cells and the possibility of enemy troops attack are taken into account. A similar principle applies when **inventing new technologies**. A new technology with the highest current priority is chosen to be invented.

Buying troops is also controlled by priorities based on decision trees. Reasons for the purchase of military units are either imminent shortage of troops of the player, or a need to increase the defense of important cells.

At the end of the turn, the artificial player deploys people into buildings and on the cells to mine and produce resources. Decision-making is also based on a combination of decision trees and some heuristics. The necessity of production of the resource (and also the quantity produced by the cell) and the importance of the building are taken into account. The heuristics focus on increasing the number of people and having as much cathedrals as possible in order to achieve a victory.

4.2 Beta Player

We used coefficients (to calculate priorities) when designing the Alpha player. These coefficients had been determined empirically rather than by an exact calculation; therefore they may cause suboptimal decisions. The Beta player is based on the Alpha player, while an evolutionary algorithm is used to adjust some parameters influencing the decision-making process.

A chromosome of the Beta player consists of 31 variables affecting different parts of the original decision-making process implemented in the Alpha player. For example, coefficients which affect the resulting priorities of the various types of events and the usefulness of each type of resources are adjusted.

At the beginning we generate a population of 24 Beta players having random parameter values in their chromosomes. The description of one cycle of the evolutionary algorithm follows. Players play against each other. At the end we select players who have won at least one of the matches. Other players are discarded. We create six new players by crossover-based breeding of the selected players. In addition, we create clones of players who won at least two matches. These clones are then slightly modified by mutation. Six more players are randomly generated. This completes the first cycle of the evolutionary algorithm.

These cycles are then repeated using the newly established generation of players. Total run time of the algorithm is limited in order to get results in reasonable time. At the end we evaluate successful players from which we pick values of the parameters.

4.3   Gamma Player

The Gamma player is based on the original Alpha player. Some parts of the decision-making process will be modified by learning based on recorded moves from previous games played.

In each game played, a log of all moves was created no matter what kind of players are participating in it; artificial, human or a combination. The Gamma player continuously analyzes these logs and adjusts its behavior based on successful players' moves. We assume that if a strategy has led to victory in the past, it will be similar to the optimal strategy.

Based on the logs, we generate influence maps describing the positions of the buildings and people. These maps are used as a decision support tool when choosing where to build a building and which cell to conquer. This advanced method of decision-making replaces some auxiliary heuristic algorithms used by the Alpha player.

Furthermore, we analyze the number of resources produced and the number of buildings owned by the winning player in every turn. We also analyze actions it took in every turn. Then we slightly adjust priorities of individual actions of the Gamma player, so it achieves similar numbers of resources and buildings and takes similar actions in general.

5   Testing and Results

We have implemented the Empire Wars game and the artificial players in the Java language using the JMonkeyEngine gaming framework. The moves in the games played are recorded into the Derby SQL database for the purpose of the Gamma player. All tests have been performed on a PC equipped with Intel Core 2 Duo E6400 @ 2.13 GHz CPU and 3.5 GB of RAM.

First, we tested the Alpha players based on decision trees, influence maps, greedy algorithms and heuristics. The goal was to create an AI player in a short time that would be able to make decisions in few seconds. In a sample of 10 games each played by four Alpha players, an average move took 5.92 s. Moves at the beginning of the match took significantly less time than moves at the end because of the more complex arrangement of the game entities at the end.

During testing, the Alpha player did not make any significant mistakes. It was able to achieve the required number of 100 Victory Points in approximately 29 to 33 turns. But in complex game situations, it made some mistakes. It would probably lose to an experienced human player.

The Beta players were created using the evolutionary algorithm. Regarding the time constraints, the algorithm was limited to five generations. The most successful individual (let's call it X) was created in the fourth generation and won all four matches, which it participated in. Another player (let's call it Y) is also worth mentioning. It was generated randomly in the first cycle and was always promoted to the next generation in each subsequent cycle. Note that the X player is a descendant of the Y player. Players created by breeding with X or Y had better average results than most other players in the last cycles. The resulting values of the parameters of both X and Y players are shown in Table 1 and compared with the original Alpha player. The Gamma player starts every game with the same values as the Alpha player does, but adjusts them at the beginning of each move (thus omitted from Table 1).

In a sample of 10 games each played by four Beta-X players, an average move took 5.96 s. As expected, due to the similar approach to decision-making process, the speed was virtually identical to the speed of the Alpha player. The Gamma player needed the longest time for one move. On average, it took 7.49 s to make a move, because of the huge amount of the data fetched from the database and processed. By design, the Gamma player's success rate depends on the learning data set. In our tests it learned from 100 games played. Most game logs belonged to matches played by four Alpha or Beta players. Some matches in the learning data set were played by a human player against three artificial players Alpha or Beta.

We have made 30 test games in order to compare all three players – Alpha, Beta and Gamma. Table 2 shows the various combinations of opponents and the results of 10 matches played in any combination. The results show that the Beta and Gamma players are significantly better than the original Alpha player. The Beta player defeats the Gamma player quite clearly. This can be explained by the learning set of games for the Gamma player, which comes from relatively weak players.

Table 1. Parameters of X and Y Beta players compared to the Alpha player

Parameter	Alpha	Beta-Y	Beta-X
Attack coefficient	1	0.84	0.82
Troops movement coefficient	1	0.92	0.92
Building coefficient	1	1.07	1.07
Troops buying coefficient	1	1.08	1.04
Invention coefficient	1	0.93	0.93
Resource acquisition coefficient	1	0.96	0.96
Need for wheat coefficient	0.05	0.05	0.13
Need for oil coefficient	0.05	0.11	0.13
Need for iron coefficient	0.1	0.22	0.13
Need for stone coefficient	0.005	0.03	0.061
Barracks building coefficient	1	0.94	0.91
Laboratory building coefficient	1	1.03	1.2
Factory building coefficient	1	1.03	0.79
Preference for obtaining 4 units of wheat	1	0.88	0.88
Preference for obtaining 2 units of oil	1	1.13	0.83
Preference for obtaining 2 units of iron	1	1	1
Preference for obtaining 4 units of stone	1	1.15	1.15
Preference for occupying cells in the center of the gameboard	0.05	0.21	−0.1
Coefficient for a maximum value of an opponent's cell adjacent to the cell we are attacking	0.05	0.18	0.09
Preference for attacking opponent's cells	0.05	0.2	0.25
Preference for attacking most successful opponent	0.05	−0.16	−0.16
Preference for attacking strategically important cells	0	0.1	−0.19
Priority of the Genetically modified wheat invention	0	0.1	0.1
Priority of the Deep mining invention	0	0.9	0.7
Priority of the Metallurgy invention	0	0	1
Priority of the Dynamite invention	0	−0.8	0.6
Priority of the Fighter plane invention	0	0.2	0.2
Priority of the Fortification invention	0	0.7	0.7
Priority of the Battle tactics invention	0	0.8	0.5
Maximum number of troops; useless troops above this limit will be self-destroyed	20	22	20
Threshold difference between the number of our defensive troops and opponent's threatening troops, when we tend to buy more troops	7	4	4

Table 2. Matches among players and their results

Players in matches		# of matches won by		
		Alpha	Beta-X	Gamma
Alpha	Beta-X	1	9	
Alpha	Gamma	2		8
Beta-X	Gamma		7	3

6 Conclusion

In this paper, we have described the design, implementation and testing of artificial intelligent players for the new Empire Wars game. Three players called Alpha, Beta and Gamma have been designed. The Alpha player is based on decision trees, influence maps and heuristics. The Beta player is based on the Alpha player while its parameters have been adjusted using the evolutionary algorithm. The Gamma player is also based on the Alpha players and is designed to learn and to improve its decision-making process based on logs of previous games stored in the database.

We had created the Alpha player relatively quickly, but it reached rather average results. We have shown that incorporating evolutionary algorithms or machine learning can significantly improve the efficiency of the artificial player without necessarily increasing the processing time.

The artificial players do not make fundamental mistakes and make decisions in a matter of seconds, in contrast to the approach that searches the entire state space. The Beta and Gamma players achieve significantly better results against the Alpha player through a combination of the advanced AI approaches. The Beta Player gives the best results. It has won 9 out of 10 matches against the Alpha player without exceeding the Alpha's processing time. In the future, we plan to further improve the Gamma player by increasing the number of games played by good players recorded in the database.

Acknowledgements. The authors of this paper would like to thank Tereza Krizova for proofreading. This work and the contribution were also supported by project of Students Grant Agency — FIM, University of Hradec Kralove, Czech Republic (under ID: UHK-FIM-SP-2017).

References

1. Billings, D., Davidson, A., Schaeffer, J., Szafron, D.: The challenge of poker. Artif. Intell. **134**(1), 201–240 (2002)
2. Schaeffer, J., Burch, N., Björnsson, Y., Kishimoto, A., Müller, M., Lake, R., Lu, P., Sutphen, S.: Checkers is solved. Science **317**(5844), 1518–1522 (2007)

3. Liu, R., Xie, X., Augusto, V., Rodriguez, C.: Heuristic algorithms for a vehicle routing problem with simultaneous delivery and pickup and time windows in home health care. Eur. J. Oper. Res. **230**(3), 475–486 (2013)

4. Kanal, L., Kumar, V.: Search in Artificial Intelligence. Springer Science & Business Media, Heidelberg (2012)

5. Liu, E., Temlyakov, V.N.: The orthogonal super greedy algorithm and applications in compressed sensing. IEEE Trans. Inf. Theory **58**(4), 2040–2047 (2012)

6. Pan, Q.K., Ruiz, R.: An effective iterated greedy algorithm for the mixed no-idle permutation flowshop scheduling problem. Omega **44**, 41–50 (2014)

7. Hingston, P.: A turing test for computer game bots. IEEE Trans. Comput. Intell. AI Games **1**(3), 169–186 (2009)

8. Yannakakis, G.N.: Game AI revisited. In: Proceedings of the 9th Conference on Computing Frontiers, pp. 285–292. ACM (2012)

9. Rogers, K.D., Skabar, A.A.: A micromanagement task allocation system for real-time strategy games. IEEE Trans. Comput. Intell. AI Games **6**(1), 67–77 (2014)

10. Graham, R., McCabe, H., Sheridan, S.: Pathfinding in computer games. ITB J. **4**(2), 6 (2015)

11. Cowley, B.U., Charles, D.: Adaptive artificial intelligence in games: issues, requirements, and a solution through behavlets-based general player modelling. arXiv preprint arXiv:1607.05028 (2016)

12. Lara-Cabrera, R., Cotta, C., Fernández-Leiva, A.J.: A review of computational intelligence in RTS games. In: 2013 IEEE Symposium on Foundations of Computational Intelligence (FOCI), pp. 114–121. IEEE (2013)

13. Pistono, F., Yampolskiy, R.V.: Unethical research: how to create a malevolent artificial intelligence. arXiv preprint arXiv:1605.02817 (2016)

14. Narang, A., Srivastava, A., Jain, R., Shyamasundar, R.K.: Dynamic distributed scheduling algorithm for state space search. In: Kaklamanis, C., Papatheodorou, T., Spirakis, P.G. (eds.) Euro-Par 2012. LNCS, vol. 7484, pp. 141–154. Springer, Heidelberg (2012). doi:10.1007/978-3-642-32820-6_16

15. Kojima, H., Nagashima, Y., Tsuchiya, T.: Model checking techniques for state space reduction in manet protocol verification. In: 2016 IEEE International Parallel and Distributed Processing Symposium Workshops, pp. 509–516. IEEE (2016)

16. Phillips, C.R.: Employing an efficient and scalable implementation of the Cost Sensitive Alternating Decision Tree algorithm to efficiently link person records. Ph.D. thesis, Texas State University (2015)

17. Oliver, J.J., Hand, D.J.: On pruning and averaging decision trees. In: Machine Learning: Proceedings of the Twelfth International Conference, pp. 430–437 (2016)

18. Xu, Z.: Uncertain Multi-attribute Decision Making: Methods and Applications. Springer, Heidelberg (2015)

19. Park, H., Kim, K.J.: Mcts with influence map for general video game playing. In: 2015 IEEE Conference on Computational Intelligence and Games (CIG), pp. 534–535. IEEE (2015)

20. Mark, D.: Modular tactical influence maps. In: Game AI Pro 2: Collected Wisdom of Game AI Professionals, p. 343 (2015)

21. Fiondella, L., Rahman, A., Lownes, N., Basavaraj, V.V.: Defense of high-speed rail with an evolutionary algorithm guided by game theory. IEEE Trans. Reliab. **65**(2), 674–686 (2016)

22. Mariano, P., Correia, L.: Population dynamics of centipede game using an energy based evolutionary algorithm. In: Advances in Artificial Life, ECAL 2013, pp. 1116–1123 (2013)

23. Ura, A., Miwa, M., Tsuruoka, Y., Chikayama, T.: Comparison training of Shogi evaluation functions with self-generated training positions and moves. In: Herik, H.J., Iida, H., Plaat, A. (eds.) CG 2013. LNCS, vol. 8427, pp. 208–220. Springer, Cham (2014). doi:10.1007/978-3-319-09165-5_18

24. Liapis, A., Martínez, H.P., Togelius, J., Yannakakis, G.N.: Adaptive game level creation through rank-based interactive evolution. In: 2013 IEEE Conference on Computational Intelligence in Games (CIG), pp. 1–8. IEEE (2013)

Recognizing the Pattern of Binary Hermitian Matrices by a Quantum Circuit

Joanna Wiśniewska[1](✉) and Marek Sawerwain[2]

[1] Faculty of Cybernetics, Institute of Information Systems,
Military University of Technology, Kaliskiego 2, 00-908 Warsaw, Poland
jwisniewska@wat.edu.pl
[2] Institute of Control and Computation Engineering, University of Zielona Góra,
Licealna 9, 65-417 Zielona Góra, Poland
M.Sawerwain@issi.uz.zgora.pl

Abstract. The chapter contains a description of quantum circuit for pattern recognition. The task is to distinguish Hermitian and Non-Hermitian matrices. The quantum circuit is constructed to accumulate the elements of a learning set and with use of Hamming distance and some Hamiltonian designed for so-called quantum summing operation it is able to distinguish if a tested element fits to the pattern from the learning set. The efficiency of the this solution is shown in a computational experiment.

Keywords: Quantum circuits · Pattern recognition · Hamming distance

1 Introduction

The conception of solving problems with use of k-nearest neighbor algorithm was created in fifties of the previous century [1] and it is still very popular and develops constantly. As the researchers working on some aspects of quantum computing we were inspired by artificial neural networks and classical k-nearest neighbor algorithms to apply the idea of pattern recognition in quantum circuits [2,4,11,12].

We refer to [7–9] where it was shown that it is possible to build a quantum circuit which works as a classifier. The basic idea was prepared to be utilized in the field of image processing [6,10]. It was constructed with visible inspiration flowing out of methods used in artificial neural networks: the data describing the patterns is enclosed in learning and testing sets, the circuit obtains the learning set and, with use of Hamming distance, is able to recognize if a tested element fits to the pattern emerging from the learning set, but in contrast to neural networks the circuit itself does not change inside (no adaptive weights).

In this chapter a quantum circuit was used to classify the Hermitian and non-Hermitian matrices. In Sect. 2 we present some basic definitions helpful in understanding the main idea of this work. Section 3 is dedicated to the detailed description of the quantum circuit for pattern recognition. Section 4 contains the results of computational experiment which was carried out. A summary and conclusions are presented in Sect. 5.

© Springer International Publishing AG 2017
N.T. Nguyen et al. (Eds.): ACIIDS 2017, Part I, LNAI 10191, pp. 466–475, 2017.
DOI: 10.1007/978-3-319-54472-4_44

2   Definitions

To explain the analyzed matter we need to introduce a few definitions corresponding to quantum information processing [3]. First notion concerns a quantum bit – so-called qubit which is a normalized vector in a two-dimensional Hilbert space $\mathcal{H}_2$.

Two orthogonal qubits constitute a computational base. Of course, we can imagine the infinite number of such paired qubits. One of the most known bases is so-called standard base. This base is created by qubits:

$$|0\rangle = \begin{bmatrix} 1 \\ 0 \end{bmatrix}, \quad |1\rangle = \begin{bmatrix} 0 \\ 1 \end{bmatrix}. \tag{1}$$

where the form $|\cdot\rangle$ is a Dirac notation. The difference between classical bit and qubit is that qubit may be a "mixture" of orthogonal vectors. This phenomenon is called superposition. Hence, the state of quantum bit $|\psi\rangle$ we can present as:

$$|\psi\rangle = \alpha_0|0\rangle + \alpha_1|1\rangle, \text{ where } |\alpha_0|^2 + |\alpha_1|^2 = 1. \tag{2}$$

The coefficients α_0 and α_1 are complex numbers and they are termed amplitudes. A character of quantum state is non-deterministic. The probability that the state $|\psi\rangle$ equals $|0\rangle$ is $|\alpha_0|^2$ and, adequately, the probability that the state $|\psi\rangle$ is $|1\rangle$ is expressed by value $|\alpha_1|^2$. Of course, it is also possible that one of the amplitudes equals to 0 and the other to 1 (in this case state $|\psi\rangle$ is one of basic states).

If we need more than one quantum bit to perform any calculations, we can use a quantum register which is a system of qubits, joined by a tensor product (denoted by symbol $\otimes$) in a mathematical sense. For example, the state $|\phi\rangle$ of 3-qubit register containing qubits $|0\rangle, |1\rangle$ and $|1\rangle$ is

$$|\phi\rangle = |0\rangle \otimes |1\rangle \otimes |1\rangle = \begin{bmatrix} 1 \\ 0 \end{bmatrix} \otimes \begin{bmatrix} 0 \\ 1 \end{bmatrix} \otimes \begin{bmatrix} 0 \\ 1 \end{bmatrix}. \tag{3}$$

Usually, we omit the symbols $\otimes$, so the above state may be also denoted as $|\phi\rangle = |011\rangle$.

In case of any n-qubit state $|\varphi\rangle$ its form can be expressed as a superposition of basic states:

$$|\varphi\rangle = \alpha_0|00\ldots000\rangle + \alpha_1|00\ldots001\rangle + \alpha_2|00\ldots010\rangle + \ldots + \alpha_{(2^n-1)}|11\ldots111\rangle \tag{4}$$

where the normalization condition must be fulfilled:

$$\sum_{i=0}^{2^n-1} |\alpha_i|^2 = 1, \quad \alpha_i \in \mathbb{C}. \tag{5}$$

To perform the calculations on quantum states we use quantum gates. Quantum gates have to preserve a quantum state – that is the performed operation preserves the state's normalization condition – so they have to be unitary operators to ensure this feature.

The gate which is very useful in many algorithms is a Hadamard gate H. Let $|x\rangle$ be a n-qubit state as in example Eq. (3), but with labeled qubits:

$$|x\rangle = |x_0\rangle \otimes |x_1\rangle \otimes \cdots \otimes |x_{n-2}\rangle \otimes |x_{n-1}\rangle. \tag{6}$$

The impact of Hadamard gate on state $|x\rangle$ is:

$$H|x\rangle = H(|x_0\rangle \otimes |x_1\rangle \otimes \cdots \otimes |x_{n-2}\rangle \otimes |x_{n-1}\rangle) =$$

$$= H|x_0\rangle \otimes H|x_1\rangle \otimes \cdots \otimes H|x_{n-2}\rangle \otimes H|x_{n-1}\rangle = \tag{7}$$

$$= \bigotimes_{i=0}^{n-1} H|x_i\rangle = \frac{1}{\sqrt{2^n}} \left[\bigotimes_{i=0}^{n-1} \left(|0\rangle + (-1)^{x_i}|1\rangle \right) \right].$$

As we can see the Hadamard gate makes all absolute values of state's amplitudes even (equal to $\frac{1}{\sqrt{2^n}}$), so it causes the phenomenon of superposition in a quantum register.

It is important to remember that quantum gates, in contrast to classical ones, always have the same number of inputs and outputs. It is so because they were designed as reversible operators the matrix form of quantum gate is unitary and Hermitian.

Another basic gate is an exclusive-or (XOR) gate, called also a controlled negation $(CNOT)$ gate. This gate has two entries and, naturally, two outputs. The operation XOR is realized on the second qubit (the first qubit remains not changed):

$$XOR|00\rangle = |00\rangle, \quad XOR|01\rangle = |01\rangle,$$
$$XOR|10\rangle = |11\rangle, \quad XOR|11\rangle = |10\rangle. \tag{8}$$

The matrix form of this gate is:

$$XOR = \begin{bmatrix} 1 & 0 & 0 & 0 \\ 0 & 1 & 0 & 0 \\ 0 & 0 & 0 & 1 \\ 0 & 0 & 1 & 0 \end{bmatrix}. \tag{9}$$

3   Quantum Pattern Detection

In this work we would like to use a quantum circuit to check if some matrices are Hermitian or not. The whole approach is based on the k-nearest neighbor algorithm for pattern classification [7]. The matrices we focus on are binary and unitary, sized 4×4 as at Fig. 1. In this case all analyzed matrices are column (or row, interchangeably) permutations of identity matrix $I_{4\times 4}$. Let us remind that a matrix U is Hermitian when it is equal to its conjugate transpose:

$$U = U^\dagger. \tag{10}$$

Because the analyzed set of matrices does not contain any matrices with complex elements, it would be some simplification to say that a Hermitian matrix is just equal to its transpose what can be observed at Fig. 1.

Hermitian matrices

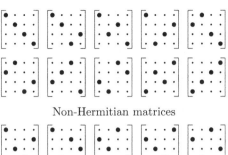

Non-Hermitian matrices

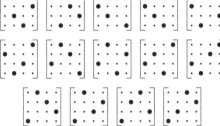

Fig. 1. The set of all binary and unitary matrices sized 4×4, divided into Hermitian matrices and non-Hermitian matrices. The bigger dots stand for 1 and the smaller dots for 0

The quantum circuit to perform a Hermitian matrices' recognition is shown at Fig. 2. Although in this work we only test a circuit for matrices sized 4×4, the presented circuit is universal in terms of matrix dimensions. Before we will describe the operations performed by the mentioned circuit, let us explain the form of data in learning and testing sets.

To describe the matrices we use binary series. If a matrix is Hermitian then a value of every element above the diagonal, having the coordinates (a, b), is equal to the value of element below the diagonal with coordinates (b, a), so we take into consideration only elements on and below the diagonal. For example, one of Hermitian matrices at Fig. 1 is

$$\begin{bmatrix} 0 & 0 & 1 & 0 \\ 0 & 1 & 0 & 0 \\ 1 & 0 & 0 & 0 \\ 0 & 0 & 0 & 1 \end{bmatrix}, \qquad (11)$$

so we reduce this matrix to:

$$\begin{matrix} 0 \\ 0\ 1 \\ 1\ 0\ 0 \\ 0\ 0\ 0\ 1 \end{matrix} \qquad (12)$$

and we transform it to a series in a column-wise way:

$$0010100001. \qquad (13)$$

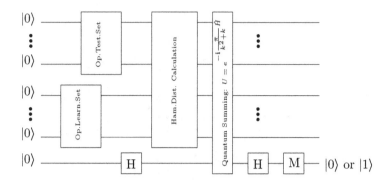

Fig. 2. The quantum circuit classifying Hermitian and non-Hermitian matrices

If the matrix is not Hermitian the transformation is slightly more different. To ensure that the series describing matrices have the same number of elements, the value of each element in position (a, b) above the diagonal is added to the value of element with coordinates (b, a), e.g.

$$\begin{bmatrix} 1\,0\,0\,0 \\ 0\,0\,1\,0 \\ 0\,0\,0\,1 \\ 0\,1\,0\,0 \end{bmatrix} \rightarrow \begin{matrix} 1 \\ 0\ 0 \\ 0\ 1\ 0 \\ 0\ 1\ 1\ 0 \end{matrix} \rightarrow 1000011010. \tag{14}$$

In carried out computational experiment matrices 4×4 were taken into account, so the generated series have 10 elements. In general, the length of series is $\frac{k^2+k}{2}$ if the matrices are sized $k \times k$. Consequently, the number of quantum circuit's (Fig. 2) inputs/outputs is:

$$\frac{k^2 + k}{2} \times 2 + 1 = k^2 + k + 1. \tag{15}$$

The first $\frac{k^2+k}{2}$ inputs serve to enter the succeeding elements of testing set. The next $\frac{k^2+k}{2}$ inputs serve to constitute a learning set. There is also one ancilla input on which the Hadamard operation is performed.

Firstly, we would like to describe a block labeled Op.Learn.Set. It is in circuit to accumulate the elements of learning set. In analyzed example it has to contain 10 series describing Hermitian matrices. These series have 10 elements with 2 to 4 elements equal to one and the rest elements equal to zero. We can roughly call these series permutative – the number of ones is similar, but they occupy other positions in series. It means that the block Op.Learn.Set is a subsystem built of gates XOR and H, because the XOR gates ensure the permutations and H gates allow to accumulate 10 different series in some qubits of quantum register.

The block Op.Test.Set is also constructed of XOR and H gates, but is not the same subsystem as Op.Learn.Set. This block contains only one element of testing set, representing Hermitian or non-Hermitian matrix, in each computational experiment. It means that, in case of matrices 4×4, we have to perform 24 experiments, always with different block Op.Test.Set.

Summarizing this part of the circuit, we have the initial quantum state entering the system:

$$|\psi_0\rangle = |0\rangle^{\otimes k^2 + k + 1}. \tag{16}$$

Then, the first $\frac{k^2+k}{2}$ qubits will be affected by the block Op.Test.Set and successive $\frac{k^2+k}{2}$ qubits by the block Op.Learn.Set to produce the superposition of elements from a learning set:

$$\frac{1}{\sqrt{L}} \sum_{p=1}^{L} |l_1^p, \ldots, l_{\frac{k^2+k}{2}}^p\rangle, \tag{17}$$

where L denotes the number of elements in learning set and l represents the successive qubits.

The first $\frac{k^2+k}{2}$ qubits describe an element from a testing set. It means that after the operations caused by blocks Op.Learn.Set, Op.Test.Set and Hadamard gate on ancilla qubit the system's state is:

$$|\psi_1\rangle = |t_1, \ldots, t_{\frac{k^2+k}{2}}\rangle \otimes \frac{1}{\sqrt{L}} \left(\sum_{p=1}^{L} |l_1^p, \ldots, l_{\frac{k^2+k}{2}}^p\rangle \right) \otimes \frac{1}{\sqrt{2}}(|0\rangle + |1\rangle) \tag{18}$$

where t stands for the qubits of element from the testing set.

The next block serves to calculate the Hamming distance between one test series and elements of learning set. The Hamming distance is a measure expressing the difference between two series. It tells how many elements, occupying the same positions, differ from each other. The block Ham.Dist.Calculation uses the XOR operations on every couple of qubits (t_i, l_i^p) and saves their results as d in the part of register which previously contained elements of a learning set:

$$XOR(t_i, l_i^p) = (t_i, d_i^p), \quad i = 1, \ldots, \frac{k^2 + k}{2}. \tag{19}$$

After this operation the system's state is:

$$|\psi_2\rangle = |t_1, \ldots, t_{\frac{k^2+k}{2}}\rangle \otimes \frac{1}{\sqrt{L}} \left(\sum_{p=1}^{L} |d_1^p, \ldots, d_{\frac{k^2+k}{2}}^p\rangle \right) \otimes \frac{1}{\sqrt{2}}(|0\rangle + |1\rangle). \tag{20}$$

If the differences are already computed then they must be summed to obtain the Hamming distance. The block Quantum Summing stands for the operation U:

$$U = e^{-i\frac{\pi}{k^2+k}\hat{H}}, \quad \hat{H} = I_{\frac{k^2+k}{2} \times \frac{k^2+k}{2}} \otimes \begin{bmatrix} 1 & 0 \\ 0 & 0 \end{bmatrix}^{\otimes \frac{k^2+k}{2}} \otimes \begin{bmatrix} 1 & 0 \\ 0 & -1 \end{bmatrix}, \tag{21}$$

where $\mathbf{i}$ represents the imaginary unit. This results with the state $|\psi_3\rangle$:

$$|\psi_3\rangle = \frac{1}{\sqrt{2L}} \sum_{p=1}^{L} \left(e^{\mathbf{i}\frac{\pi}{k^2+k}d(t,l^p)} |t_1, \ldots, t_{\frac{k^2+k}{2}}\rangle \otimes |d_1^p, \ldots, d_{\frac{k^2+k}{2}}^p\rangle \otimes |0\rangle \right.$$
$$\left. + e^{-\mathbf{i}\frac{\pi}{k^2+k}d(t,l^p)} |t_1, \ldots, t_{\frac{k^2+k}{2}}\rangle \otimes |d_1^p, \ldots, d_{\frac{k^2+k}{2}}^p\rangle \otimes |1\rangle \right). \qquad (22)$$

The last step is designed to reverse the Hadamard operation on the last qubit (of course with use of reversible H gate). That will allow to measure the last qubit in a standard base and obtain $|0\rangle$ or $|1\rangle$ with quite high probability. The whole final state of the system is:

$$|\psi_4\rangle = \frac{1}{\sqrt{L}} \sum_{p=1}^{L} \left(\cos\left(\frac{\pi}{k^2+k}d(t,l^p) \right) |t_1, \ldots, t_{\frac{k^2+k}{2}}\rangle \otimes |d_1^p, \ldots, d_{\frac{k^2+k}{2}}^p\rangle \otimes |0\rangle \right.$$
$$\left. + \sin\left(\frac{\pi}{k^2+k}d(t,l^p) \right) |t_1, \ldots, t_{\frac{k^2+k}{2}}\rangle \otimes |d_1^p, \ldots, d_{\frac{k^2+k}{2}}^p\rangle \otimes |1\rangle \right). \qquad (23)$$

Now the measurement needs to be done only on the last ancilla qubit. If the result is $|0\rangle$ that means the tested matrix is close to the pattern described in learning set (is Hermitian) and if the result is $|1\rangle$ the matrix should not be Hermitian.

4   Results of the Experiment

We conducted a computational experiment to examine if a quantum system is able to recognize the pattern of Hermitian and non-Hermitian matrices. The calculations were based on Eq. (23). The script written in Python programming language delivered the results which are presented in Table 1.

In Sect. 3 it is stated that the learning set should contain only the patterns describing Hermitian matrices and the testing set contains both Hermitian and non-Hermitian matrices. If the Hermitian matrices were the elements of testing set then the lowest probability of obtaining $|0\rangle$ on the last qubit of the circuit (Fig. 2) was about 0.6242, so that means the presented circuit is able to recognize the pattern of Hermitian matrix. In case of non-Hermitian matrices in the testing set, the situation is different. The values of probabilities are closer to 0.5 in both $P_{|0\rangle}$ and $P_{|1\rangle}$ columns of the table. In some cases the probability of recognizing a non-Hermitian matrix as Hermitian one is even higher than the probability of the correct result. One may notice that these values of probabilities may be used to classify the matrix as Hermitian or non-Hermitian – in the analyzed case we know the frequency featuring the appearance of both patterns of matrices. In addition, the sums of Hamming distances' values, calculated between all series representing Hermitian matrices and tested matrices, are greater if the tested matrix was non-Hermitian what can be used to introduce a discrimination line separating the patterns (see: Fig. 3).

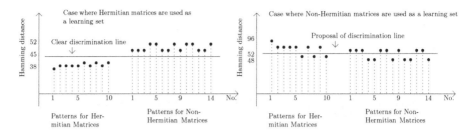

Fig. 3. The discrimination lines between analyzed patterns. If the learning set contains Hermitian matrices, then the discrimination line may be pointed clearly. In case when the learning set contains non-Hermitian matrices, the summed Hamming distances do not allow to precisely separate the patterns

We were also curious if we would obtain a similar results when the learning contains only non-Hermitian matrices, so we carried out the second experiment (the results of the first experiment are shown in the left part of Table 1 and the results of the second experiment in the right part). As it may be noticed (Fig. 3) in the second case the circuit classifies the patterns slightly worse.

Of course, it should be emphasized that the presented approach with use of quantum version of k-nearest neighbor algorithm has a probabilistic character. That implies the fact that the calculations have to be repeated (e.g. by duplicating the instances of circuit from Fig. 2) to obtain the most accurate approximation of distribution shown in Table 1. The probabilities of measuring $|0\rangle$ or $|1\rangle$ on the last qubit are different, so it allows to indicate the distance of analyzed case from the learning set.

5   Summary

The computational experiment showed that if the learning set contains the patterns of Hermitian matrices, the quantum circuit should be able to correctly classify the matrices of the same type. The non-Hermitian matrices are not classified so well, but the nature of quantum algorithms' majority is probabilistic, so even working with a physical system the experiments have to be carried out many times and the obtained results must be averaged. That shows the importance of probability values and we managed to calculate the frequency with which the circuit is able to recognize the non-Hermitian matrix (about ≈ 0.5).

In previous works [7] the authors carried out the experiments on learning and testing sets with different number of elements and they noticed that the better results were obtained for bigger sets. In this chapter we were working on a quite small set and we obtained satisfactory results.

It is also important to understand that the main factor which causes the recognition in this case is the number of ones and zeros in series describing the matrices. We had conducted a few experiments with other representation of matrices than described in this chapter and results of these experiments were far more chaotic.

Table 1. The computational experiment's results. The denotation D_H stands for the sum of Hamming distances between a tested pattern and series from the learning set (LS). The column $P_{|0\rangle}$ contains the probabilities of state's $|0\rangle$ occurrence as the result of measurement on the last qubit of the system and the column $P_{|1\rangle}$ contains, respectively, the probabilities of obtaining $|1\rangle$ on the last qubit

Hermitian matrices in a LS			Tested pattern	Non-Hermitian matrices in a LS			
No	D_H	$P_{|0\rangle}$	$P_{|1\rangle}$		No	D_H	$P_{|0\rangle}$	$P_{|1\rangle}$

No	D_H	$P_{|0\rangle}$	$P_{|1\rangle}$	Tested pattern	No	D_H	$P_{|0\rangle}$	$P_{|1\rangle}$
1	36	0.6799	0.3200	1010010001	1	96	0.2383	0.7616
2	38	0.6551	0.3448	1010000010	2	74	0.4580	0.5419
3	38	0.6551	0.3448	1000100001	3	74	0.4580	0.5419
4	38	0.6551	0.3448	1000010100	4	74	0.4580	0.5419
5	38	0.6551	0.3448	0100010001	5	74	0.4580	0.5419
6	40	0.6242	0.3757	0100000010	6	52	0.6817	0.3182
7	38	0.6551	0.3448	0011000001	7	74	0.4580	0.5419
8	40	0.6242	0.3757	0001000100	8	52	0.6817	0.3182
9	38	0.6551	0.3448	0010011000	9	74	0.4580	0.5419
10	40	0.6242	0.3757	0000101000	10	52	0.6817	0.3182

Hermitian matrices in a LS			Tested pattern	Non-Hermitian matrices in a LS			

No	D_H	$P_{|0\rangle}$	$P_{|1\rangle}$	Tested pattern	No	D_H	$P_{|0\rangle}$	$P_{|1\rangle}$
1	48	0.5309	0.4690	1000100110	1	60	0.5714	0.4285
2	48	0.5309	0.4690	1000100110	2	60	0.5714	0.4285
3	48	0.5309	0.4690	0101100001	3	60	0.5714	0.4285
4	52	0.4662	0.5337	0100101010	4	48	0.7038	0.2961
5	52	0.4662	0.5337	0101000110	5	48	0.7038	0.2961
6	48	0.5309	0.4690	0100011100	6	60	0.5714	0.4285
7	48	0.5309	0.4690	0101100001	7	60	0.5714	0.4285
8	52	0.4662	0.5337	0101000110	8	48	0.7038	0.2961
9	48	0.5309	0.4690	0011001010	9	60	0.5714	0.4285
10	52	0.4662	0.5337	0001101100	10	48	0.7038	0.2961
11	52	0.4662	0.5337	0100101010	11	48	0.7038	0.2961
12	48	0.5309	0.4690	0100011100	12	60	0.5714	0.4285
13	48	0.5309	0.4690	0011001010	13	60	0.5714	0.4285
14	52	0.4662	0.5337	0001101100	14	48	0.7038	0.2961

The further work on the idea of using quantum circuit as a classifier should develop in direction of deeper analysis of Hamiltonian $\hat{H}$ used in (23) – especially to formulate the matrix U given there. The other issue is to simulate the behaviour of quantum circuit and check if the averaged results will overlap with calculated values of probability from Table 1. Finally, using the SVM approach [5] and checking the quality of the classification would be also very interesting.

References

1. Fix, E., Hodges, J.L.: Discriminatory analysis, nonparametric discrimination: Consistency properties. Technical report 4, USAF School of Aviation Medicine, Randolph Field, Texas (1951)
2. Mateus, P., Omar, Y.: Quantum pattern matching. arXiv preprint: arXiv:quant-ph/0508237v1 (2005)
3. Nielsen, M.A., Chuang, I.L.: Quantum Computation and Quantum Information. Cambridge University Press, New York (2000)
4. Pinkse, P.W.H., Goorden, S.A., Horstmann, M., Škorić, B., Mosk, A.P.: Quantum pattern recognition. In: 2013 Conference on and International Quantum Electronics Conference on Lasers and Electro-Optics Europe (CLEO EUROPE/IQEC), Munich, p. 1 (2013)
5. Rebentrost, P., Mohseni, M., Lloyd, S.: Quantum support vector machine for big data classification. Phys. Rev. Lett. **113**, 130503 (2014)
6. Schaller, G., Schtzhold, R.: Quantum algorithm for optical-template recognition with noise filtering. Phys. Rev. A **74**, 012303 (2006)
7. Schuld, M., Sinayskiy, I., Petruccione, F.: Quantum computing for pattern classification. In: Pham, D.-N., Park, S.-B. (eds.) PRICAI 2014. LNCS (LNAI), vol. 8862, pp. 208–220. Springer, Cham (2014). doi:10.1007/978-3-319-13560-1_17
8. Trugenberger, C.A.: Quantum pattern recognition. Quantum Inf. Process. **1**(6), 471–493 (2002). Springer
9. Trugenberger, C.A.: Phase transitions in quantum pattern recognition. Phys. Rev. Lett. **89**, 277903 (2002)
10. Ruan, Y., Chen, H., Tan, J., Li, X.: Quantum computation for large-scale image classication. Quantum Inf. Process. **15**(10), 4049–4069 (2016). Springer
11. Wiebe, N., Kapoor, A., Svore, K.M.: Quantum algorithms for nearest-neighbor methods for supervised and unsupervised learning. Quantum Inf. Comput. **15**(3–4), 316–356 (2015)
12. Yoo, S., Bang, J., Lee, C., Lee, J.: A quantum speedup in machine learning: finding an N-bit Boolean function for a classification. New J. Phys. **16**(10), 103014 (2014)

Fuzzy Maximal Frequent Itemset Mining Over Quantitative Databases

Haifeng Li[✉], Yue Wang, Ning Zhang, and Yuejin Zhang

School of Information, Central University of Finance and Economics, Beijing, China
{mydlhf,wangyue,zhangning,zhangyuejin}@cufe.edu.cn

Abstract. Fuzzy frequent itemset mining is an important problem in quantitative data mining. In this paper, we define the problem of fuzzy maximal frequent itemset mining, which, to the best of our knowledge, has never been addressed before. A simple tree-based data structure called *FuzzyTree* is constructed, in which the fuzzy itemsets are sorted dynamically based the supports. Then, we propose an algorithm named *FMFIMiner* to build the *FuzzyTree*. In *FMFIMiner*, we can ignore processing the other children nodes once the supports between the parent node and one child node are equal; moreover, we conduct pruning the certain support computing by checking whether an itemset is in the final results. Theoretical analysis and experimental studies over 4 datasets demonstrate that our proposed algorithm can efficiently decrease the runtime and memory cost, and significantly outperform the baseline algorithm *MaxFFI-Miner*.

Keywords: Fuzzy frequent itemset · Data mining · Fuzzy maximal frequent itemset · Quantitative database

1 Introduction

The development of the internet brings us new applications, in which the fuzzy set [15] can be used for better decision-making. Many fuzzy mining algorithms, such as classification [13], clustering [14], web mining [8], neural networks [1], have been introduced. Frequent itemset mining is a very important method in data mining, which together with the fuzzy techniques, have been evaluated more powerful to obtain the accurate mining results [7]. Then, several methods were proposed to improve the performance. [4] concentrated on the development of a general model to discover association rules, which can be used in relational databases that contain quantitative data. [5] addressed the problem of the increasing use of very large quantitative-value databases, and proposes a new fuzzy mining algorithm based on the *AprioriTid* approach to find fuzzy association rules from

This research is supported by the National Natural Science Foundation of China (61100112,61309030), Beijing Higher Education Young Elite Teacher Project (YETP0987), Granted by Discipline Construction Foundation of Central University of Finance and Economics 2016XX05).

ⓒ Springer International Publishing AG 2017
N.T. Nguyen et al. (Eds.): ACIIDS 2017, Part I, LNAI 10191, pp. 476–486, 2017.
DOI: 10.1007/978-3-319-54472-4_45

given quantitative transactions; each item only used the linguistic term with the maximum cardinality in later mining processes, and thus made the number of fuzzy regions to be processed the same as that of the original items. [6] introduced a *FP-tree* based data structure *FUFP-tree*, which made the tree updating process easier; an incremental algorithm was also proposed for reducing the execution time in reconstructing the tree when new transactions were inserted, and thus can achieve a trade-off between runtime and space complexity. [10] designed a novel tree structure called *CFFP-tree* to store the related information in the fuzzy mining process, where each node maintained the membership value of the contained item and the membership values of its super-itemsets in the path; moreover, the *CFFP-tree* was built by the proposed algorithm CFFP-growth. The author then proposed a simple tree structure called the *UBFFP-tree* [9]; the designed two-phase fuzzy mining approach can easily derive the upper-bound fuzzy counts of itemsets using the tree; thus, it can prune the unpromising itemsets in the first phase, and then finds the actual fuzzy frequent itemsets in the second phase. [11] developed a fuzzy frequent itemset algorithm *FFI-Miner* to mine the complete set of fuzzy frequent itemsets without candidate generation. *FFI-Miner* used a novel fuzzy-list structure to keep the essential information for later mining process; plus, it employed an efficient pruning strategy to reduce the search space, and thus reduce the runtime cost.

Motivation: The discovered fuzzy frequent itemsets are massive when the threshold is small. Traditional frequent itemset mining methods use the itemset representation, such as the maximal itemsets [2], the closed itemsets [12] and the non-derivable itemsets [3] to reduce the itemset count, which can not only improve the performance, but also enable the user to understand the mining results easier. The maximal itemset is the most compacted representation of the itemset, and thus can be better used in real applications. In this paper, we focus on the problem of how to discover the fuzzy maximal frequent itemsets with an effective and efficient method over quantitative databases. The main contributions are as follows.

- We define the concept of fuzzy maximal itemset, which is the most effective representation to compress the fuzzy itemsets. To our best knowledge, the fuzzy maximal itemset mining problem has never been addressed before.
- We design a compacted prefix-tree named *FuzzyTree* to maintain the mining results in memory. In this tree, we "wisely" maintain the idlist for each itemset to compute the support efficiently with a much reduced memory usage.
- We present an algorithm *FMFIMiner* (**F**uzzy **M**aximal **F**requent **I**temset **Miner**) to discover the fuzzy maximal frequent itemsets. In *FMFIMiner*, we generate the idlist based on the parent information; further, the parent support pruning and the superset pruning are used, which, when together used with *FuzzyTree*, can improve the performance significantly.
- We evaluate FMFIMiner on 4 datasets. The experimental results show that our algorithm can much reduce the count of the fuzzy frequent itemsets, and it can significantly outperform the state-of-the-art algorithm *FFI-Miner*.

The rest of this paper is organized as follows: In Sect. 2 we present the preliminaries of the fuzzy itemset mining and define the fuzzy maximal frequent itemset, and state the problem addressed in this paper. Section 3 presents the data structures, and illustrates our algorithm in detail. Section 4 evaluates the performance with theoretical analysis and experimental results. Finally, Sect. 5 concludes this paper.

2   Preliminaries and Problem Statement

2.1   Preliminaries

Let $\Gamma = \{i_1, i_2, \cdots, i_m\}$ be a set of m distinct items in a quantitative database $QD = \{QT_1, QT_2, \cdots, QT_n\}$, where each $QT_i \subseteq \Gamma$ is called a quantitative transaction, which contains items with the quantities. An example is shown in Table 1.

The membership functions μ are used to fuzzy up the database, which is shown in Fig. 1.

(1) The quantitative value of each item will be converted to a fuzzy value; thus, each item Z can be represented by a special notation $\sum \frac{\mu_i}{z_i}$, where μ_i denotes the membership grade, and z_i denotes the fuzzy region of Z. As an example, item A:5 in transaction 1 can be denoted as $\frac{0.2}{A.L} + \frac{0.8}{A.M}$, in which 0.2 is the membership grade of item A when computing with membership function $\mu.Low$.

Table 1. A quantitative database

TID	QTransaction
1	A:5, C:10, D:2, E:9
2	A:8, B:2, C:3
3	A:5, B:3, C:9
4	A:5, B:3, C:10, E:3
5	A:7, C:9, D:3
6	A:5, B:2, C:10, D:3
7	A:5, B:2, C:5
8	A:3, C:10, D:2, E:2

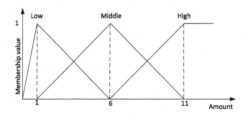

Fig. 1. The membership functions of linguistic 3-terms

(2) The membership grades will be summed up in the quantitative database according to different fuzzy areas, and the area with maximal summed membership grade will be choose as the final one. Again, taking item A as an example, the summary of $A.L$ is 1.6, $A.M$ is 5.8, and $A.H$ is 0.6. Then, we regard $A.M$ as the final fuzzy area of item A, with summed membership grade 5.8 and occurrence number 8. Other items are $B.L$, $C.H$, $D.L$ and $E.L$.

(3) The items in quantitative database will be updated by the fuzzy areas. For an instance, item A:5 in transaction 1 will be replaced by $\frac{0.8}{A.M}$. The updated fuzzy database FD is shown in Table 2.

Table 2. Fuzzy database

TID	FTransaction			
1	$\frac{0.8}{A.M}$,	$\frac{0.8}{C.H}$,	$\frac{0.8}{D.L}$	
2	$\frac{0.6}{A.M}$,	$\frac{0.8}{B.L}$		
3	$\frac{0.8}{A.M}$,	$\frac{0.6}{B.L}$,	$\frac{0.6}{C.H}$	
4	$\frac{0.8}{A.M}$,	$\frac{0.6}{B.L}$,	$\frac{0.8}{C.H}$,	$\frac{0.6}{E.L}$
5	$\frac{0.8}{A.M}$,	$\frac{0.6}{C.H}$,	$\frac{0.6}{D.L}$	
6	$\frac{0.8}{A.M}$,	$\frac{0.8}{B.L}$,	$\frac{0.8}{C.H}$,	$\frac{0.6}{D.L}$
7	$\frac{0.8}{A.M}$,	$\frac{0.8}{B.L}$		
8	$\frac{0.4}{A.M}$,	$\frac{0.8}{C.H}$,	$\frac{0.8}{D.L}$,	$\frac{0.8}{E.L}$

An itemset $X = \{x_1, x_2, \cdots, x_k\}$ is a k-items set. If X is covered by a fuzzy transaction FT, then we regard the minimum membership grade of x_i as the membership grade of X. The summed grade in fuzzy database FD is called the support of X, denoted $\Lambda(X)$. Also, we use $ct(X)$ to denote the count of occurrence of X. Given a minimum support λ, X is called frequent if $\Lambda(X) \geq \lambda*|FD|$. For an instance, suppose the minimum support $\lambda = 0.2$, itemset $\{C.H, D.L\}$ has the membership grade $\min(0.8,0.6)=0.6$ in fuzzy transaction 6, and its support in FD is $0.8+0.6+0.6+0.8 = 2.8$; thus, it is frequent because $2.8 > 0.2*8 = 1.6$.

2.2 Problem Statement

Fuzzy Maximal Frequent Itemset. A fuzzy frequent itemset X is a fuzzy maximal frequent itemset if no other frequent itemsets cover it.

Example 1. In the fuzzy database FD shown in Fig. 2, if the minimum support $\lambda = 0.2$, then itemset $\{B.L, A.M\}$ is covered by fuzzy frequent itemsets $\{B.L, C.H, A.M\}$ with support 2; thus, it is not a fuzzy maximal frequent itemset. $\{B.L, C.H, A.M\}$, on the other hand, is a fuzzy maximal frequent itemset since it is not covered by any frequent itemset. As can be seen, there are 11 fuzzy frequent itemsets, in which only 2 are fuzzy maximal frequent itemsets.

Given the minimum support λ and the membership functions μ, the problem in this paper is to discover the fuzzy maximal frequent itemsets over quantitative databases.

3   Fuzzy Maximal Frequent Itemset Mining method

Vertical data style has been testified can reduce the mining cost [11], that is, the data is organized like $\{item : (id, \nu)\}$, where ν denotes the membership grade. In this paper, we will follow this data representation. Table 3 shows the vertical fuzzy database of Table 2. In our method, we use an independent array to store the database. Because the vertical fuzzy database can be generated initially without any input threshold, the computing cost can be ignored.

Table 3. Vertical fuzzy database

Item	IdList
A.M	(1, 0.8)(2, 0.6)(3, 0.8)(4, 0.8)(5, 0.8)(6, 0.8)(7, 0.8)(8, 0.4)
B.L	(2, 0.8)(3, 0.6)(4, 0.6)(6, 0.8)(7, 0.8)
C.H	(1, 0.8)(3, 0.6)(4, 0.8)(5, 0.6)(6, 0.8)(8, 0.8)
D.L	(1, 0.8)(5, 0.6)(6, 0.6)(8, 0.8)
E.L	(4, 0.6)(8, 0.8)

3.1   Data Structure

FuzzyTree. We build an in-memory prefix-tree named *FuzzyTree* to store the fuzzy itemsets, which are denoted by the tree nodes. A node n_X is a 4-tuple $< X, ct, sup, cr >$. X denotes the fuzzy itemset. ct and sup are the count and the support of X in the database. cr are the children nodes. Each node has a pointer to its parents except the root node. As a result, the child node represents an itemset that covers the parent node. The nodes are sorted by the support-ascending order of the itemsets. By this manner, the search space can be reduced. To speedup the computing, we need to maintain the idlist for the node, that is, the ids of transactions that cover X. The idlist of the child node can be achieved by directly interacting it with the idlist of the append item. For an example, the itemset $\{D.L, A.M, C.H\}$ has the idlist by interacting the idlist of $\{D.L, A.M\}$ and $\{C.H\}$, that is, $\{1, 5, 6, 8\} \cap \{1, 3, 4, 5, 6, 8\} = \{1, 5, 6, 8\}$, corresponding to the membership grades $\{0.8, 0.6, 0, 6, 0.4\}$, and the support is the summary, 2.4. To speedup the interacting, we use a hash map to store the id list. On the other hand, if we record the idlist for each node, the memory usage will be huge. We notice once the support is computed and the idlists of the children are generated, the idlist of the node itself is not yet useful, which provides us an "wisely" implementation, that is, we only store the idlist in memory temporarily before the idlists of its children nodes are generated. This method has been verified can significantly save the memory usage in our experiments. Using the database

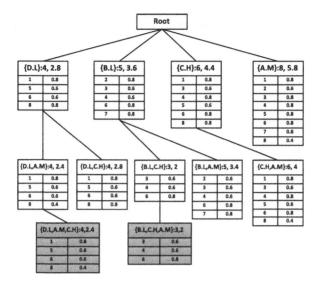

Fig. 2. FuzzyTree for $\lambda = 0.2$

from Table 3, we show the *FuzzyTree* in Fig. 2 when the minimum support is set to 0.2. For an instance, the itemset $D.L$ has count 4, support 2.8, and has the idlist with 4 ids.

Fuzzy Maximal Frequent Itemset Collection. The final mining results do not have to maintain the fuzzy value of each item, we employed a collection named *FMFIC* to store the fuzzy maximal frequent itemsets accordingly, which helps us conduct the superset pruning. We employ a two-level hash table to implement the *FMFIC*. In the hash table, we use the size of the itemset as a key, which, however, may create frequent hash collisions. We know if $X \in Y$, then the sum of items in X is smaller than that of Y, which supplies us a new key, that is, the sum of items. As a result, for an itemset $X = \{x_1, \cdots, x_n\}$, we use the $(n, \sum_{i=1}^{n}(x_i))$ as the keys; plus, the value is a list that stores the itemsets with the same keys. To add X in the *FMFIC*, we compute the size and the sum and insert it into the corresponding list, and the time complexity is $O(1)$. Thus, the large size of X, the earlier to determine whether X is maximal.

Example 2. For the data in Fig. 2, we use $\{1, 2, 3, 4\}$ to denote the fuzzy items $\{D.L, B.L, C.H, A.M\}$. The first fuzzy maximal frequent itemset $\{B.L, C.H, A.M\}$ is inserted into *FMFIC* with key (3,8). When new fuzzy itemset $\{D.L, A.M, C.H\}$ with key (3,9) is generated, it will be inserted into the *FMFIC* without being compared to $\{B.L, C.H, A.M\}$ since $3 = 3, 9 > 8$.

Algorithm 1. FMFIMiner Algorithm

Require: *root*: root of *FuzzyTree*; *VFD*: Vertical Fuzzy Database; Γ: Distinct items
    list; λ: minimum support; *FMFIC*: Fuzzy Maximal Frequent Itemset Collection;
 1: $ALLITEMS=\{\}$
 2: **for** each item X in Γ **do**
 3:    get the idlist of X(idlist(X));
 4:    compute the support($\Lambda(X)$) of X;
 5:    **if** $\Lambda(X) \geq \lambda$ **then**
 6:        add X into $ALLITEMS$;
 7: sort $ALLITEMS$ w.r.t. the support of each item;
 8: **for** each item X in $ALLITEMS$ **do**
 9:    generate new child node n_X of root;
10:    CALL Explore(n_X, idlist(X), *VFD*, λ, *FMFIC*);

3.2 Pruning Strategies

Theorem 1. *For a fuzzy itemset X and a fuzzy item y in the fuzzy database FD, if $\Lambda(X) = \Lambda(X \cup y) \geq \lambda$, then frequent fuzzy itemset Z covering X but not y has the frequent superset $Z \cup y$.*

Proof. Using $t(X)$ to denote the ids of transactions that cover X, we can get $t(X) = t(X \cup y)$ from $\Lambda(X) = \Lambda(X \cup y)$, that is, $\mu_i(X) \leq \mu_i(Y)$ for each i in $t(X)$. On the other hand, Z covering X means $t(Z) \subseteq t(X) = t(X \cup y)$; thus, $\min(\mu_i(Z\backslash X), \mu_i(X)) = \min(\mu_i(Z\backslash X), \mu_i(X \cup y))$ for each i in $t(Z)$, that is, $\Lambda(Z) = \Lambda(Z \cup y) \geq \lambda$.

    Theorem 1 supplies us a method to prune the support computation of itemsets that cover X but not y.

Theorem 2. *In the FuzzyTree, if a node n_X is a leaf and X is frequent but not in the FMFIC, then X is a fuzzy maximal frequent itemset.*

Proof. Since X is a leaf, then it will not covered by any itemsets from its children nodes and right sibling nodes; also, X is not covered by the generated fuzzy maximal frequent itemsets, then X is a fuzzy maximal itemset. Furthermore, X is frequent; thus, it is a fuzzy maximal frequent itemset.

    Theorem 2 provides the way to quickly find the fuzzy maximal frequent itemset, that is, given the sorted distict items $\Gamma = \{i_1, i_2, \cdots, i_m\}$, assuming itemset $X = x_1 x_2 \cdots x_n$ has the sub-itemset $x_t \cdots x_n$ which can exactly right-match the sub-items of Γ sequentially, that is, $x_n = i_m, \cdots, x_t = i_{m-n+t}$, then we can ignore the computation of the right sibling nodes of n_{x_t}.

3.3 Dynamically Reordering

Traditional method to build the tree is sorted the distinct items ascendent according to the support or the lexicographical order of the items. In this paper, we are inspired by [2] and dynamically sorted the children nodes w.r.t. the support ascendent. That is, we will compute the supports of the children, based on

---

Algorithm 2. Explore Algorithm

Require: n_X: node of *FuzzyTree*, denotes itemset X; *idlist(X)*: the idlist of X; *VFD*:
   Vertical Fuzzy Database; λ: minimum support; *FMFIC*: Fuzzy Maximal Frequent
   Itemset Collection;
1: $Z = X \cup Y_1 \cup \cdots \cup Y_n (Y_i$ is X's right sibling items);
2: **if** $Z \in FMFIC$ **then**
3:    return;
4: $ALLITEMS=\{\}$
5: **for** each frequent node n_X's right sibling node n_Y **do**
6:    $idlist(X \cup Y) = idlist(X) \cap idlist(Y)$;
7:    compute the support $\Lambda(X \cup Y)$;
8:    **if** $\Lambda(X \cup Y)==\Lambda(X)$ **then**
9:       clear the $ALLITEMS$;
10:       add Y into $ALLITEMS$;
11:       break;
12:    **else**
13:       **if** $\Lambda(X \cup Y) \geq \lambda$ **then**
14:          add Y into $ALLITEMS$;
15: sort $ALLITEMS$ according to the support of each item;
16: **for** each item Y in $ALLITEMS$ **do**
17:    generate new child node n_Y of n_X;
18:    CALL Explore(n_Y, idlist(Y), VFD, λ, FMFIC);
19: **if** n_X is a leaf node and X not in $FMFIC$ **then**
20:    add X into $FMFIC$;

---

which to sort the frequent children nodes with ascendent order. This method
can keep the search space as small as possible, which, together with the pruning
method in Theorem 1, can achieve a better performance. We can see this manner
is actually a hybrid traverse, but it can help us further prune more search space.

3.4   FMFIMiner Algorithm

We propose the *FMFIMiner* algorithm to bottom-up construct the *FuzzyTree*
with a hybrid traverse manner, which is because the fuzzy frequent itemset also
follows the a priori property. In our algorithm, the *FuzzyTree* is constructed in
five steps. First and the foremost, the quantitative database will be converted
to the vertical fuzzy style according to the given membership functions, and we
maintain the vertical fuzzy database in memory. Then, we conduct Step 2–3 in
Algorithm 1, and Step 4–5 in Algorithm 2.

4   Experimental Results

In this section, we evaluate the performance of our algorithm *FMFIMiner*. We
employed the state-of-the-art algorithm *FFI-Miner* [11] to discover the fuzzy
frequent itemsets, and we call this method the *MaxFFI-Miner*, which was used
as the evaluation method.

4.1 Running Environment and Datasets

We implemented the algorithms with Python 2.7 running on Microsoft Windows 7 and performed on a PC with a 3.60 GHZ Intel Core i7-4790M processor and 12 GB main memory. We used 2 synthetic datasets and 2 real-life datasets as the evaluation datasets. The detailed data characteristics are shown in Table 4.

Table 4. Fuzzy dataset characteristics

Dataset	Transaction count	Average size	Min size	Max size	Items count	Transaction correlation
T25I15D100K	100 000	26	4	67	1000	38
T40I10D100K	100 000	39	4	77	1000	25
KOSARAK	990 002	8	1	2498	41 270	5159
ACCIDENTS	340 183	33	18	51	468	14

4.2 Effect of Minimum Support

Figure 3 presented the runtime cost over different datasets. As can be seen, the runtime of both algorithms reduced with the increment of the minimum support. Our algorithm *FIMFIMiner* achieved better computing performance over the baseline algorithm *MaxFFI-Miner*. We can clearly see that when the minimum support is large, the runtime is similar. Nevertheless, when the minimum support turned smaller, the runtime of *MaxFFI-Miner* increased greatly; which was more significantly over the dense datasets. This is due to the fact that our algorithm pruned certain nodes, which will reduce the computing of the support, as well the search space. Note that in Fig. 3(c), even though the *KOSARAK* dataset was sparse, our algorithm outperformed *MaxFFI-Miner* with hundreds-fold speedups when the minimum support was low. We found this is because *KOSARAK* dataset has massive distinct items, which will result in a huge useless computing without our pruning methods. This suggested us that the items count is an effective factor that may have impact on the runtime cost.

Figure 4 shown the memory usage of the two algorithms. Clearly to see, when the minimum support became small, our algorithm had an almost unchanged memory cost; the *MaxFFI-Miner* algorithm, however, with a exponentially increased memory usage. Specifically, over the dense dataset like the *ACCIDENTS*, the memory cost can reach to several order of magnitude larger than our algorithm. This is for the reason that *MaxFFI-Miner* had to maintain the idlist for all the nodes; on the contrary, we remove the idlist of the parent node once the children idlists are generated.

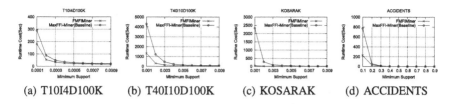

Fig. 3. Runtime cost vs. minimum support

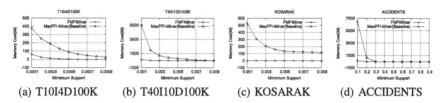

Fig. 4. Memory cost vs. minimum support

5   Conclusions

In this paper, we made a research on how to discover the fuzzy maximal frequent itemsets over quantitative databases, which, to our best knowledge, has never been studied before. We designed an in-memory data structure named *FuzzyTree* to store the dynamically sorted itemsets. An efficient *FMFIMiner* algorithm was presented to build and maintain the *FuzzyTree*. In *FMFIMiner*, we pruned the search space by comparing the support of the parent node and the child node, and employed the superset pruning method to reduce the computing cost. Furthermore, the local idlist was used to save the memory usage. The extensive experimental results shown that our algorithm outperformed the baseline algorithm significantly.

References

1. Buckley, J.J., Hayashi, Y.: Fuzzy neural networks: a survey. Fuzzy Sets Syst. **66**, 1–13 (1994)
2. Burdick, D., Calimlim, M., Gehrke, J.: MAFIA: a maximal frequent itemset algorithm for transactional databases (2001)
3. Calders, T., Goethals, B.: Mining all non-derivable frequent itemsets. In: Elomaa, T., Mannila, H., Toivonen, H. (eds.) PKDD 2002. LNCS, vol. 2431, pp. 74–86. Springer, Heidelberg (2002). doi:10.1007/3-540-45681-3_7
4. Delgado, M., Marin, N., Sanchez, D., Vila, M.A.: Fuzzy association rules: general model and applications. IEEE Trans. Fuzzy Syst. **11**(2), 214–225 (2003)
5. Hong, T., Kuo, C., Wang, S.: A fuzzy aprioriTid mining algorithm with reduced computational time. Appl. Soft Comput. **5**(1), 1–10 (2004)
6. Hong, T., Lin, C., Yulung, W.: Incrementally fast updated frequent pattern trees. Expert Syst. Appl. **34**(4), 2424–2435 (2008)

7. Kuok, C.M., Fu, A., Wong, M.H.: Mining fuzzy association rules in databases. Sigmod Rec. **27**(1), 41–46 (1998)

8. Lin, C.W., Hong, T.P.: A survey of fuzzy web mining. WIREs Data Min. Knowl. Discov. **3**, 190–199 (2013)

9. Lin, C., Hong, T.: Mining fuzzy frequent itemsets based on UBFFP trees. J. Intell. Fuzzy Syst. **27**(1), 535–548 (2014)

10. Lin, C., Hong, T., Wenhsiang, L.: An efficient tree-based fuzzy data mining approach. Int. J. Fuzzy Syst. **12**(2), 150–157 (2010)

11. Lin, J.C.W., Li, T., Fournierviger, P., Hong, T.: A fast algorithm for mining fuzzy frequent itemsets. J. Intell. Fuzzy Syst. **29**(6), 2373–2379 (2015)

12. Pei, J., Han, J., Mao, R.: CLOSET: an efficient algorithm for mining frequent closed itemsets (2000)

13. Wang, T., Li, Z., Yan, Y., Chen, H.: A survey of fuzzy decision tree classifier methodology. In: International Conference of Fuzzy Information and Engineering (2007)

14. Yang, M.S.: A survey of fuzzy clustering. Math. Comput. Model. **18**(11), 1–16 (1993)

15. Zadeh, L.A.: Fuzzy sets and systems. Int. J. Gen. Syst. **312**, 1–134 (2007)

A Method for Early Pruning a Branch of Candidates in the Process of Mining Sequential Patterns

Bac Le[1], Minh-Thai Tran[2(✉)], and Duy Tran[1]

[1] Department of Computer Science, University of Science, VNU-HCM,
Ho Chi Minh City, Vietnam
lhbac@fit.hcmus.edu.vn, tran.duy.huflit@gmail.com
[2] Faculty of Information Technology, University of Foreign Languages - Information Technology, Ho Chi Minh City, Vietnam
minhthai@huflit.edu.vn

Abstract. Mining patterns in sequence databases are one of the fields that many researchers study due to their high applicability in many areas such as trademarks, medical, education, or prediction. However, the main challenge of this is runtime and memory usage. Many approaches are proposed to improve the efficiency of mining algorithms. However, another issue needs to be studied further. In this paper, we propose a method of effectively solving the problem of sequential pattern mining, based on the event information represented in the vertical data format of the sequence dataset. With a proposed state table, a map table of co-occurrence of events and an early pruning search space technique, the paper proposes the efficiency mining algorithm called STATE-SPADE. The experimental results have shown the advantages of the proposed method in terms of execution time and memory usage.

Keywords: Sequential pattern · State table · Vertical data format · Early pruning

1 Introduction

Currently, all information related to the human being is expressed and stored through computer applications in the form of data. Data mining helps discover useful information hidden in large volumes of stored data. There are many different types of data, such as transaction data, sequence data, stream data, time-series data, image data, etc. Sequence data mining is one of the research areas of interest due to its high practical application. The application of data mining can be included in text mining [1], Web behavior mining [2], analysis of customer purchase behaviors [3], and natural disaster prevention [4]. Sequence data has two main characteristics: (i) The size or length of the sequence is not fixed. In the same database, the size of each sequence can vary and have even greater disparity; (ii) The position of the events in the sequence is important. This shows the relationship of the order or position of the events in the sequence.

The concept of sequence data mining was first introduced by Agrawal and Srikant in 1995 [5]. The first issue was mentioned in the field of business analysis, and the main goal was finding frequent patterns in the sequence of customer purchase in a supermarket

© Springer International Publishing AG 2017
N.T. Nguyen et al. (Eds.): ACIIDS 2017, Part I, LNAI 10191, pp. 487–497, 2017.
DOI: 10.1007/978-3-319-54472-4_46

where the bills are sorted by time. Then, the concept was widely used in many areas of health, education, military, or economics.

Many algorithms proposed different approaches in data organization and techniques to increase the mining efficiency as GSP [6], SPADE [7], PrefixSpan [8], SPAM [9], CloFS-DBV [11]. SPADE, SPAM or CloFS-DBV had a successful deal with the problem of multi-scanning the database by using the vertical data format. However, with a large database and a small support, a large number of candidates are generated in the process of mining. Among these candidates, some candidates do not exist in the database or does not satisfy the minimum support threshold. Therefore, these algorithms must take a lot of time to generate and detect the unnecessary candidates. Early pruning of such candidates will help save time and memory during the execution. Although there are effective methods to reduce the number of unnecessarily generated candidates, such as CMAP [10], a significant redundancy in candidates still remains. Therefore, how to detect and prevent the redundant candidates is a major challenge of studies in sequence data mining.

The main contribution of this paper is to suggest some additional steps and enabling improving execution time and memory usage on the datasets for SPADE and CM-SPADE [10]. This includes (i) improving the extraction step SPADE algorithm for matching technical proposals, while saving memory with additional data structures represents the current state of the sequential patterns; (ii) adding a test immediately following the CM-SPADE step that allows increasing the processing speed of the algorithm by combining with the state table.

The rest of this paper is structured as follows. Section 2 presents the problem definitions. Section 3 summarizes some related works. Section 4 shows the proposed algorithm. Section 5 presents the experimental results. Finally, the conclusions and recommendations for future research are given in Sect. 6.

2   Problem Definition

In this section, we define some basic concepts of sequence database mining and the issue of sequential pattern mining.

Let $I = \{i_1, i_2, i_3, \ldots, i_n\}$ is a set of distinct items (or events), where i_j is an item and $1 \leq j \leq n$. A set of unordered items is called an itemset. Each itemset is represented in parentheses. The parentheses are left out to simplify the notation for itemsets with only a single item. For example, (A, B, C) represents an itemset with three items A, B and C. A sequence $S = \langle e_1, e_2, e_3, \ldots, e_m \rangle$ is an ordered list of itemsets, where e_j is an itemset, $1 \leq j \leq m$. The size of a sequence is the number m of itemsets in the sequence. A sequence with length k, called a k-sequence, is the number of items in the sequence.

Definition 1 (sequence database). A sequence database SDB is a list of sequences and is denoted as SDB = $\{S_1, S_2, S_3, \cdots, S_{|SDB|}\}$, where |SDB| is the number of sequences in SDB and S_i $(1 \leq i \leq |SDB|)$ is the i-th sequence in SDB. For example, the database SDB in Table 1 includes four sequences, i.e., |SDB| = 4. Each sequence has a sequence identify (SID) is 1, 2, 3, 4 in the SID column. Sequence $\langle D, A, B \rangle$ is a sequence containing 3 itemsets, each itemset has one item, including the item D, in the first itemset, which occurred first, then the item A in the second itemset and the last one are item B in the third itemset.

Table 1. An example of a sequence database SDB

SID	Sequence
1	$\langle$A, B, C, A, D, B$\rangle$
2	$\langle$A, B, D, A, B$\rangle$
3	$\langle$D, B$\rangle$
4	$\langle$D, A, B$\rangle$

Definition 2 (subsequence and supersequence). Let $S_a = \langle a_1, a_2, ..., a_m \rangle$ and $S_b = \langle b_1, b_2, ..., b_n \rangle$ be two sequences. The sequence S_a is a subsequence of S_b if there exist m integers i_1 to i_m and $1 \leq i_1 < i_2 < ... < i_m \leq n$ such that $a_1 = b_{i1}, a_2 = b_{i2}, ..., a_m = b_{im}$. In this case, S_b is also called a supersequence of S_a, denoted as $S_a \subseteq S_b$.

Definition 3 (prefix). A sequence $S_a = \langle a_1, a_2, ..., a_m \rangle$ is called a prefix of a sequence $S_b = \langle b_1, b_2, ..., b_n \rangle$ if and only if $\forall m < n$ and $a_1 = b_1, a_2 = b_2, ..., a_m = b_m$.

Definition 4 (extending the length of a sequence). A sequence $S = \langle s_1, s_2, ..., s_n \rangle$ and an event i_k. Sequence S' is called the extending length of the sequence S with the event i_k if and only if $S' = \langle s_1, s_2, ..., s_n, s_{n+1} \rangle$ and $s_{n+1} = i_k$.

Definition 5 (sequential pattern). The support count of a sequence S in a sequence database SDB is calculated as the number of subsequences S in SDB, and is denoted as sup(S). Given a minimum support threshold, minSup, a sequence S is called a sequential pattern in SDB if sup(S) $\geq$ minSup.

In this paper, we would like to discover the full set of sequential patterns in a sequence database SDB with a given minSup value.

3   Related Work

AprioriAll [5] or GSP [6] was first proposed to solve the problem of sequential pattern mining. These algorithms are considered as the foundation of sequence data mining. However, these algorithms had to scan the database multiple times to find sequential patterns. Therefore, Zaki proposed the SPADE algorithm to fix the problem of accessing the database many times. In addition, he also pointed out ways to find sequential patterns based on the simple operation. The benefits of the SPADE algorithm include its ability to combine data characteristics to divide a problem into sub-problems, which can solve the problem effectively by filtering and joining patterns. The main advantage of SPADE algorithm as follows:

First, the data is presented in a vertical data format. The information of a sequence is stored in the list of a form {SID.pos(S)}, where, SID is the identifier of the sequence S, and pos(S) is the location of S appearing in each SID. For example, the data SDB in Table 1 can be presented in the vertical data format as Table 2. The support of each sequence is directly calculated based on the number of elements in the list: sup(A) = 3, sup(B) = 4, sup(C) = 1, and sup(D) = 4.

Table 2. SDB in vertical data format

A			B			C			D	
SID	Position		SID	Position		SID	Position		SID	Position
1	1, 4		1	2, 6		1	3		1	5
2	1, 4		2	2, 5		2			2	3
3			3	2		3			3	1
4	2		4	3		4			4	1

To extend the sequence, SPADE performs the intersection position and lists of two sequences. Table 3 shows an example of the extending item A to ⟨A, A⟩ .

Table 3. An example of extending the length of item A

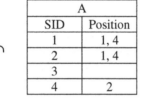

A			A			⟨A, A⟩	
SID	Position		SID	Position		SID	Position
1	1, 4		1	1, 4		1	4
2	1, 4		2	1, 4		2	4
3			3				
4	2		4	2			

Second, the lattice theory can be used to subdivide the search space. This algorithm only scans the database one time; therefore, it can minimize data access costs. Finally, it can be used in breadth-first search or depth-first search strategy.

However, the main drawback of SPADE depends on the tactics of generating candidates and checking the support of each candidate. With a large database, the number of candidates will be increased quickly. Among those, there are many candidates that do not appear in the database or their supports are very low. This will lead to storage costs and manipulation.

To overcome the weaknesses of SPADE, Philippe Fournier-Viger and et al. proposed a data structure called CMAP [10] and CM-SPADE [10] algorithm in 2014. This is the combination of SPADE and CMAP. The data stored in CMAP is a mapping of each item k on the set of items that these items are able to extend with k to form a new frequent sequential pattern. For example, Table 4 describes CMAP of the database SDB in Table 1 with a minSup value of 2. The items A, B, D belong to an extended set of item A to get new sequential patterns ⟨A, A⟩ , ⟨A, B⟩ and ⟨A, D⟩ . These patterns are potentially satisfying to the minSup.

In CM-SPADE, the ENUMERATE method of SPADE was changed to decide extending sequences by checking items in CMAP. The advantage of CM-SPADE is to utilize the advantages of the SPADE algorithm. The additional checking step, based on CMAP, allows for quick removal some of unpromising candidates and avoids the expense of manipulating redundant sequences. Thus, this will increase processing speed

Table 4. A CMAP with minSup = 2

Item	Set of extended items
A	A, B, D
B	A, B, D
D	A, B

and reduce memory usage while running the algorithm. However, the weakness of CM-SPADE is still using the strategies of SPADE. As mentioned above, it will generate more redundant candidates, leading to significant memory consumption and calculating costs. Although with the addition of CMAP, it greatly enables elimination of candidates; however, the number of redundant generated candidates still remains huge. In the next section, the paper will outline some remarks and suggestions to improve these algorithms.

4   The Proposed Algorithm

In this section, the paper proposes an enhanced method for overcoming the drawbacks of SPADE and CM-SPADE by preventing the generation of redundant candidates during the mining process. The improvements include proposing an additional state table for storing the necessary information on items and an enhanced method for early pruning of the search space.

4.1   A State Table Structure

The structure of a state table is similar to the positions list of SPADE. The only difference is that the state table only stores a single position for each SID that appears in the database. In the initialization phase, the state table stores the position of the first appearance of the item in each SID (Table 5).

Table 5. Comparison of data structures in SPADE and a state table

A			A	
SID	Position		SID	Position
1	1, 4		1	1
2	1, 4	→	2	1
3			3	
4	2		4	2
SPADE			A state table	

Based on the information of the state table and the position list of SPADE, the following remarks are relevant:

Remark 1. A candidate can be checked as frequent or not, based on the information of the state table and the position list, without performing the ENUMERATE method of SPADE. For example, suppose the minSup = 2, the state table of item ⟨A⟩ is {1.1, 2.1, 4.2}. We get the largest position corresponding to each SID of item ⟨B⟩ in position list, i.e., {1.6, 2.5, 4.3}. The three positions of ⟨B⟩ are larger than that of ⟨A⟩ ; therefore, sequence ⟨A, B⟩ will be frequent.

Remark 2. The cost for checking a frequent pattern will be lower than the intersection of the position list of candidates. For each intersection operation of two items x and y, at each SID, we must consider all of the positions of y to find the positions larger than the smallest position in that of x. Meanwhile, checking frequent operations based on the state table just took out the largest position in the list of positions in the same SID. In addition, the position list is already in ascending order; therefore, the largest position is the final position of the list. From Remarks 1 and 2, the redundant patterns will be detected based on the state table without using the extending sequence operator. In SPADE, the sequential patterns are mined by ENUMERATE procedure. Therefore, this operation requires that the entire patterns be stored in the equivalence class. Thereby, this leads to more memory consumption in the extraction process.

Thus, this paper proposes a different perspective of mining sequential patterns without storing the information of all patterns in the same equivalence class. It is a method based on a state table combined with the position list of items in vertical data format. The results obtained in the first step of the proposed algorithm is the database in vertical format and a set of state tables. Then, it builds CMAP. The result is a collection of items, and a list of items can be extended, respectively.

Remark 3. All sequential patterns starting with item x are patterns extended from the items in the set of expanding items of x. For example, considering a CMAP in Table 4, item A has a list of extending items A, B, and D. With minSup = 2, the set of 2-patterns starting with item A consists of ⟨A, A⟩ , ⟨A, B⟩ , and ⟨A, D⟩ . Based on CM-SPADE, the set of items that can expand pattern ⟨A, A⟩ is A, B, and D. Similarity, the set of items that can be extended for pattern ⟨A, B⟩ is also A, B, and D. The set of extending items that can be extended for pattern ⟨A, D⟩ is A, and B. Thus, the items that can be extended for patterns ⟨A, A⟩ , ⟨A, B⟩ , or ⟨A, D⟩ are the subsets of the extended items of pattern A. Therefore, based on the information of the state tables and the position lists, we can prune the extending events of patterns ⟨A, A⟩ , ⟨A, B⟩ , and ⟨A, D⟩ so that expanding these patterns with the remaining items will obtain the frequent sequential patterns.

4.2 STATE-SPADE Algorithm

The proposed algorithm, called STATE-SPADE, allows mining of a full set of sequential patterns, based on the state table with the input parameters as a sequence database SDB and a minimum support threshold. First, it converts the database to vertical database format V(SDB) in the set of sequential patterns of length 1 called F_1.

Algorithm: STATE-SPADE (SDB, minSup)
Input: A sequence database SDB and a minimum threshold support minSup
Output: Sequential 1-patterns
1. Scan SDB to create V(SDB) and store in F_1
2. Init CMAP
3. **FOR EACH** pattern p **IN** F_1
4.   Output p
5.   Init a state table of p
6.   **ENUMERATE** (p, StateTable(p), CMAP(p))

Then, the algorithm initializes the CMAP structure for F_1. For each pattern p in F_1, it outputs the patterns p. Then, it proceeds to initialize the state table of p. This operation involves copying information into p and retaining the smallest position in each sequence identifier. After that, the ENUMERATE procedure is called to enumerate all of the initial patterns with pattern p. First, the ENUMERATE scans all items in the list of events in the third parameter. The items in ListItem(x) can be successfully expanded with pattern x. For each item p in ListItem(x), it outputs pattern xp. The next step is the UPDATE-STATE procedure to update the state table of xp.

Algorithm: ENUMERATE (x, StateTable(x), ListItem(x))
Input: A pattern x, a state table StateTable(x), a list of extendable items ListItem(x)
Output: Sequential k-patterns
1. **FOR EACH** pattern p **IN** ListItem(x)
2.   Output xp
3.   **UPDATE-STATE**(StateTable(x), pos(p)) => StateTable(xp)
4.   **UPDATE-ITEMSET** (p, StateTable(xp), ListItem(x)) => ListItem(xp)
5.   **ENUMERATE** (xp, StateTable (xp), ListItem(xp))

UPDATE-STATE will consider each SID in both the state table of p and the position list of q. Initialize the state table for pattern pq so that the position of q is the smallest position that is greater than the position in the table p. UPDATE-ITEMSET is used to remove redundant items in a list of extensions of corresponding patterns by using 2-step checking. Step 1 is to check based on CMAP [10]. Step 2 is to check on the state table for the item that has passed Step 1. In Step 2, the procedure will take out the largest position in the list of event positions and compared to the corresponding position in the state table with the same SID. If the amount found is less than the minSup value, it removes this item from ArrayItem.

Algorithm: UPDATE-STATE(StateTable(p), pos(q))
Input: A state table of p and the position list of q
Output: A state table of pq
1. **FOR EACH** SID **IN** StateTable(p) which is in pos(p)
2.   pos_p=pos(StateTable(p), SID)
3.   **FOR EACH** pos_q **IN** pos(pos(q), SID)
4.   **IF** pos_q > pos_p **THEN** Set (StateTable(pq), SID, pos_q)
5. **RETURN** StateTable(pq)

Algorithm: UPDATE-ITEMSET (p, StateTable(xp), ArrayItem)
Input: A pattern p, a state table xp and an extended items list of p ArrayItem
Output: An update of ArrayItem
1. **FOR EACH** item **IN** ArrayItem
2.   **IF** item **NOT-PASS** CMAP-step **THEN**
3.   **REMOVE** item **FROM** ArrItem
4.   **CONTINUE**
5.   TMP=0;
6.   **FOR EACH** SID **IN** StateTable(xp) which is in info(item)
7.   pos_xp = pos(StateTable(xp), SID)
8.   **IF** pos_xp < **MAX**(pos(info(item), SID)) **THEN** TMP=TMP+1
9.   **IF** TMP<minSup **THEN**   **REMOVE** item **FROM** ArrItem
10. **RETURN** ArrayItem

---

5   Experiment Results

This section presents the experimental results of the proposed algorithm (STATE-SPADE). SPADE [7], the state-of-art method, and CM-SPADE [10] and is used to compare the performance. All algorithms were implemented on a personal computer with an Intel Core I3 3.2-GHz CPU and 4 GB of RAM running Fedora 24. The algorithms are evaluated to compare the execution time measured by second (s) and the memory usage measured by megabytes (MB). Three real-life datasets were used for comparisons that include Leviathan, BMSWebView1 and Kosarak10 k. Table 6 describes the characteristics of them.

Table 6. The characteristic of datasets

Dataset	Sequence count	Distinct item	Avg. seq. length	Type of data
Leviathan	5834	9025	33.81	Book
BMSWebView1	59601	497	2.51	Web click stream
Kosarak10 k	10000	10094	8.14	Web click stream

Experiments were conducted by running each algorithm on each set of data with decreasing minSup value until there was a clear difference between algorithms or the runtime was too long, so that it could not be done, or until it resulted in overflow memory. With each time on the experimental data, the results obtained included runtime, the obtained sequential patterns, and the amount of memory usage.

5.1 Comparisons of Runtime

The runtimes of three algorithms SPADE, CM-SPADE and STATE-SPADE are shown in Fig. 1. This figure shows that the runtime of the proposed algorithm is smaller than both SPADE and CM-SPADE with the smaller minSup values. STATE-SPADE removed the amount of redundant candidates. The redundant candidates increase quickly when the minSup values are reduced. Therefore, the proposed algorithm was more efficient. With higher minSup values, the less number of redundant candidates. Therefore, the time gap between the proposed algorithm and SPADE, CM-SPADE is not clear. However, with lower minSup values, the more number of redundant candidates generated, and the proposed algorithm performance is better.

5.2 Comparisons of Memory Usage

Results of comparing the memory consumption of three algorithms are shown in Fig. 2. Figure 2a and b show that the proposed algorithm uses less memory than SPADE and CM-SPADE in the two datasets BMSWebView1 and Kosarak10 k. This causes efficient memory usage because these two datasets are sparse, so when mining, the number of redundant candidates are raised more. By applying the cutting branch of candidate techniques, the manipulation of generating patterns of the proposed algorithm is significantly reduced. Therefore, memory usage of the proposed algorithm is much less than SPADE and CM-SPADE.

The experimental results with dataset Leviathan in Fig. 2c, the amount of memory used by the proposed algorithm is not significantly reduced versus SPADE or CM-SPADE. The main reason for this is the distribution of dataset Leviathan, which is dense and has longer sequences than BMSWebView1 and Kosarak10 k. Thus, for each sequential pattern generated, the proposed algorithm will have reduced efficiency, due to it using two checking stages before joining those candidates.

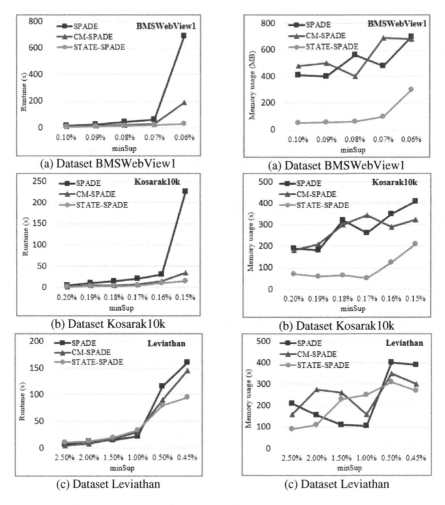

Fig. 1. Comparisons of runtime. (a) Dataset BMSWebView1 (b) Dataset Kosarak10 k (c) Dataset Leviathan

Fig. 2. Comparisons of memory usage. (a) Dataset BMSWebView1 (b) Dataset Kosarak10 k (c) Dataset Leviathan

6   Conclusion and Future Work

This paper presents the problem and suggests improvements for the SPADE and CM-SPADE algorithms that allow for mining sequential patterns more effectively. The main contribution of the paper includes proposals to use the state table to represent the state of patterns and the techniques to eliminate candidates early. The experimental results show that the proposed algorithm has better results than SPADE and CM-SPADE on three experimental datasets with a lower support threshold.

The method mentioned in this paper is limited to the itemset and has a single item. Therefore, it can be developed to mine sequential patterns in case the itemset has more than one item. The proposed algorithm can be developed to solve the problem of mining closed sequential patterns. It can combine with ClaSP [12] to increase the efficiency of mining of such patterns. Moreover, parallel processing techniques are developed quickly; therefore, we can apply parallel techniques to increase the mining process.

Acknowledgments. This work was funded by Vietnam's National Foundation for Science and Technology Development (NAFOSTED) under grant number 102.05-2015.07.

References

1. Lent, B., Agrawal, R., Srikant, R.: Discovering trends in text databases. KDD **97**, 227–230 (1997)
2. Vijaylakshmi, S., Mohan, V. Suresh Raja, S.: Mining of users access behavior for frequent sequential pattern from web logs. Int. J. Database Manag. Syst. (IJDM) **2**(3), 31–45 (2010)
3. Watada, J., Yamashiro, K.: A data mining approach to consumer behavior. IEEE Innovative Comput. Inf. Control **2**, 652–655 (2006)
4. Cervone, G., Kafatos, M., Napoletani, D., Singh, R.P.: An early warning system for coastal earthquakes. Adv. Space Res. **37**(4), 636–642 (2006)
5. Agrawal, R., Srikant, R.: Mining sequential patterns. In: Proceedings of IEEE International Conference on Data Engineering, pp. 3–14 (1995)
6. Srikant, R., Agrawal, R.: Mining sequential patterns: generalizations and performance improvements. In: Extending Database Technology, pp. 3–17 (1996)
7. Zaki, M.: SPADE: an efficient algorithm for mining frequent sequences. Mach. Learn. **42**(1–2), 31–60 (2001)
8. Pei, J., Han, J., Mortazavi-Asl, B., Pinto, H., Chen, Q. Dayal, U., Hsu, M.-C.: PrefixSpan: mining sequential patterns efficiently by prefix-projected pattern growth. In: Data Engineering, pp. 215–224 (2001)
9. Ayres, J., Flannick, J., Gehrke, J., Yiu, T.: Sequential pattern mining using a bitmap representation. In: ACM SIGKDD International Conference on Knowledge Discovery and Data Mining, pp. 429–435 (2002)
10. Tran, M., Le, B., Vo, B.: Combination of dynamic bit vectors and transaction information for mining frequent closed sequences efficiently. Eng. Appl. Artif. Intell. **38**, 183–189 (2015)
11. Fournier-Viger, P., Gomariz, A., Campos, M., Thomas, R.: Fast vertical mining of sequential patterns using co-occurrence information. In: Tseng, V.S., Ho, T.B., Zhou, Z.-H., Chen, A.L.P., Kao, H.-Y. (eds.) PAKDD 2014. LNCS (LNAI), vol. 8443, pp. 40–52. Springer, Heidelberg (2014). doi:10.1007/978-3-319-06608-0_4
12. Gomariz, A., Campos, M., Marin, R., Goethals, B.: ClaSP: an efficient algorithm for mining frequent closed sequences. In: Pei, J., Tseng, V.S., Cao, L., Motoda, H., Xu, G. (eds.) PAKDD 2013. LNCS (LNAI), vol. 7818, pp. 50–61. Springer, Heidelberg (2013). doi: 10.1007/978-3-642-37453-1_5

Analyzing Performance of High Frequency Currency Rates Prediction Model Using Linear Kernel SVR on Historical Data

Chanakya Serjam$^{(\boxtimes)}$ and Akito Sakurai

Graduate School of Science and Technology,
Keio University, Yokohama 223-8522, Japan
c.serjam.z3@keio.jp, sakurai@ae.keio.ac.jp

Abstract. We analyze the performance of various models constructed using linear kernel SVR and trained on historical bid data for high frequency currency trading. The bid tick data is converted into equally spaced (1 min) data. Different values for the number of training samples, number of features, and the length of the timeframes are used when conducting the experiments. These models are used to conduct simulated currency trading in the following year. We record the profits, hit ratios and number of trades executed from using these models. Our results indicate it is possible to obtain a profit as well as good hit ratio from a linear model trained only on historical data under certain pre-defined conditions. On examining the parameters for the linear models generated, we observe that a large number of models have all co-efficient values as negative while giving profit and good hit ratio, suggesting a simple yet effective trading strategy.

Keywords: Support vector regression (SVR) · Machine learning · Currency prediction · High frequency limit order book

1 Introduction

Rapid advances in technology, both hardware and software, have revolutionized the way markets function globally. Due to the widespread acceptance and usage of the latest electronic systems in worldwide markets, the processing time for placing and fulfilling orders has gone down exponentially as compared to older traditional markets. This has led to a rise in usage of high-frequency trading systems where human intervention is kept to a bare minimum, and all the transactions are handled by computer algorithms to keep overhead such as time and cost as low as possible. High-frequency trading systems have been playing an increasingly vital role in trading. One major form of trading is currency trading or foreign exchange (forex for short). Currency trading is a very huge and lucrative market, but is also prone to volatility. As per a report from the Bank of International Settlements, the results from a recent survey [1] show that trading in foreign exchange markets averaged \$5.1 trillion per day in just a single month

© Springer International Publishing AG 2017
N.T. Nguyen et al. (Eds.): ACIIDS 2017, Part I, LNAI 10191, pp. 498–507, 2017.
DOI: 10.1007/978-3-319-54472-4_47

(April) of 2016. Although this is down from an average of $5.3 trillion per day in April of 2013, it is still a very voluminous market.

Traders investing in the currency markets are particularly interested in predicting the direction of movement of the price for the currency pair which they are looking to trade. If the price of the currency is about to go up, the trader will want to take the buy position so he/she can sell the currency later at a higher price to turn a profit. Similarly, if the price of the currency is about to go down, the trader will want to take the sell position. Later, the trader can buy the currency again for a lower price and turn a profit. Finally, the trader may assume a neutral position, i.e., neither buy nor sell. Therefore, a prediction task can have three outputs based on the forecast: buy, sell, or do nothing. The advent of high-frequency trading necessitates the development and analysis of new trading strategies that can effectively capture the short-term behavior of the market.

In this paper, we conduct currency prediction experiments for Euro/US Dollar and Japanese Yen/US Dollar currency pairs using support vector machine for regression (SVR) [6,7] and examine the results to better understand the structure of currency trading in the forex market. Based on the forecast of the models, we perform simulated trading and record the profits or losses. We also examine the coefficient and intercept values and correlate them to the profit/loss and hit ratio metrics. The simulated trading is performed under some assumptions and defined pre-existing conditions that may not be representative of the real world but of an ideal scenario. Some interesting results are presented.

This paper is divided into the following major sections. Section 2 describes the method of research and examines the background and purpose of the research. Section 3 describes the experimental setup and discusses the process in detail. Section 4 presents the results of the experiments and is used for analysis and discussion of the results. Finally, Sect. 5 presents a conclusion to the research and this paper.

2   Method of Research

As mentioned earlier, the currency rates are volatile and prone to fluctuations. But they have also been shown to be deterministically chaotic [2]. While this may be due to a number of factors, it is generally believed that historical data captures this behavior most concretely & effectively. Concurrently, historical data usually becomes the primary input for any prediction model regardless of the technique used or the assumptions made. The primary aim of our research was to try and establish whether a linear model trained only on historical data can have good predictive performance. The choice of model to be linear was due to the fact that it would be easier to analyze a linear model as the parameters would relate to real and observable data values.

A variety of techniques have been used for prediction tasks depending on the mathematical foundation or the value of specific model parameters. There has been considerable research [2–4] done on applying Artificial Neural

Networks (ANNs) to forex forecasting. Deng & Sakurai et al. [10,11] applied complex hybrid prediction techniques including Multiple Kernel Learning (MKL) & Genetic Algorithms (GA) to currency prediction and achieved good results. Another technique utilized for currency rates and financial time-series prediction is Support Vector Machines [5–7], and it has also been applied successfully for high-frequency trading [8,9]. A study by Kamruzzaman and Sarker [5] with Australian forex data showed that SVM based models achieved better performance in forecasting of exchange rate as compared to NN based models. In our goal of analyzing the financial models which take historical data as input and produce relatively good performance, we planned to focus on the characteristics and structure of the model being generated. Hence, we decided on SVR with linear kernels to be the choice of technique for generating models.

In high-frequency trading, the limit order book is updated every time there is a change in the bid or ask price or in case of other events such as a transaction being executed. This data is called tick data. The limit order book, in a typical case, consists of the timestamp (year/month/date and hours/mins/sec), the best (highest) bid price, the bid volume, the best (lowest) ask price and the ask volume. Although, the volume data is relevant in prediction tasks, we only worked with the price data and eliminated the volume data. We also only dealt with only 1 price (bid price) rather than both the prices as there is not much qualitative difference between both prices. We also subjected the tick data to some pre-processing which included converting the tick data to equally spaced (1 min) data. Since the tick data is recorded every time there is a change in the order book, the data is unequally spaced and hence unsuitable for time-series analysis. We wanted to check whether some patterns might emerge which can be learned by a training models when the data is equally spaced. Converting the tick data to uniformly spaced data makes it easier to analyze as a time-series.

In our experiments, we wanted to analyze whether there is a correlation between performance metrics such as profits or hit ratio and initial parameters of the model such as size of the training set, the number of features to be used for prediction, and the length of timeframe (1 min, 2 min, 3 min, etc.). Therefore, we trained models for many different values of these parameters. The models were trained on one year, and then used for validation on the data from the next year. This is to establish the predictive value of the models, since validating the models on the same year they were learned would not have yielded any information about the predictive performance of the models on new unseen data. Then we examined the coefficients and intercept of the models generated to look for some basic learning rule or pattern in the models. Finally, we performed simulated trading based on the predictions of the model to see if a profit is obtained.

3   Experimental Setup

The currency rates data used in our experiments were acquired from ICAP. The experiments were performed on 2 different sets of currency pairs, the Euro/US Dollar (Euro/USD) dataset and the US Dollar/Japanese Yen (USD/JPY) dataset. As previously mentioned, the original datasets contain the best bid and ask prices as well as the volumes. The datasets are pre-processed to remove the volume data as well as the ask price data. Then the tick data is converted to equally spaced (1 min) data which is the last tick data in the minute. So we have datasets that contain the date and the last price at each minute. The datasets used were from 2001 to 2015 and separated by year. Since the model is trained on data from 1 year (or more specifically, it is trained on the training set of the specified size extracted from 1 year) and then used for prediction on the next year, the data of results for prediction analysis is from 2002 to 2015. For example, the models that were trained in the year 2001 were used for prediction in the year 2002, the models trained in 2002 were used for prediction in 2003, and so on.

3.1 Parameters for Training the Models

1. *Number of features*: The values used for the number of features were 1, 2, 3, 4, 5, and 6. Features used in our model are the difference of price between successive periods of time going back n periods from the current time (t). For instance, if the number of features is 1, it means the model predicts the next output based on just one previous difference of price. Consequently, that model will have two parameters (since we are using linear kernel SVR), the coefficient and the intercept, and we extract those parameters to do a qualitative analysis of the model. If the number of features is n, the model predicts the next output based on n previous time frames and therefore the model will have $n + 1$ parameters.
2. *Length of timeframes*: The lengths of timeframes (in minutes) used were 1, 2, 3, 4, 5, 7, 10, 20, 30, and 40. These values were used to see if there is any correlation between the length of the timeframes and the performance metrics such profits or hit ratio obtained. Although this could be extended to larger timeframes, we believe it might not be fully reflective of the structure of high-frequency trading, where trading is very fast and timeframes are inherently small. We also considered that in timeframes greater than 1 min, there may be multiple starting points from which the training set can begin. Therefore, we generate models for all the possible starting points (in minutes) within a timeframe and also average them.
3. *Size of training set*: The values used for the number of training samples are 2000, 3000, 4000, 5000, and 6000.

Models are generated for all possible combinations of these initial parameters.

3.2 Performance Metrics

1. *Hit ratio*: The hit ratio, also known as directional symmetry, is a measure of how many times the model predicted the change correctly. In other words, if the model predicts upward movement and the actual data used for validation confirms it, then it counts as a hit.

2. *Profits*: Profits are obtained as a result of simulated trading based on the predictions of our models. If the price at the closing of a timeframe t is $price(t)$, and the prediction at the closing of the timeframe t is $pred(t)$, then profit is given as

$$Profit = \sum [price(t+1) - price(t)] \times pred(t) \qquad (1)$$

For the Euro/USD currency pair, the profits were in US Dollars, and for the USD/JPY currency pair, the profits were in Japanese Yen.

For simulated trading, we put certain conditions in place. We assume that only 1 unit of the currency pair is being traded. This was done under the assumption that a small transaction of 1 unit will not change or alter the market prices condition substantially and thus the following dataset will not be disrupted. No fee is charged for transactions. In the real world there is a small fee charged for every transaction, but we have chosen to ignore that to focus solely on the time series properties of currency trading.

In the simulated trading, a trade is counted when we have a change in the predicted direction of movement of the currency. Since we are only trading 1 unit, if for instance the prediction of direction is downward movement more than one times in a row, we do not execute or count those trades. The next section discusses the results of the experiments.

4   Results and Analysis

The results of the experiments consisted of the profits per year, the hit ratios, and the no. of trades executed over the period of a year using those models, as well as the intercept and coefficients of the models. Since the models were grouped based on the number of features (1 to 6) used for the models, we calculated the average profits and hit ratios with respect to the length of timeframes and the size of training set (for each value of no. of features). This gave us 4 plots for each currency pair and gave insight into the performance of the models for different input parameters.

Figures 1 and 2 show the performance metrics (avg. profits per year and avg. hit ratio) as a function of the length of timeframe for all different values of number of features (1 to 6) for the Euro/USD pair and the USD/JPY pair respectively. It is interesting to note that as the length of timeframe increases, the avg. hit ratio increases too irrespective of the no. of features, meaning an increase in the accuracy of trend prediction. However, at the same time, the profits from simulated trading go down as the length of timeframe increases. This is an interesting result, because normally profit would be expected to rise when hit ratio rises and vice-versa. One reason for this might be that as the length of the timeframe increases, the no. of trades executed in our simulated trading decreases drastically. Thus, even if the hit ratio is higher, the number of trades executed might simply not be enough to generate profits comparable to lower timeframes, which have lower hit ratio but a large number of executed

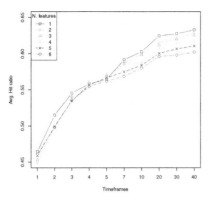
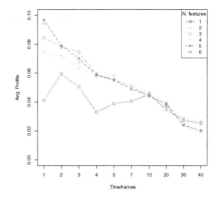

(a) Avg. hit ratio vs length of timeframe (Euro/USD)

(b) Avg. profit per year vs length of time-frame (Euro/USD)

Fig. 1. Plot for performance metrics vs. length of timeframe for Euro/USD pair

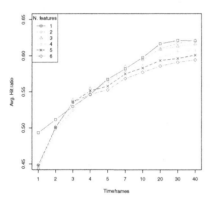
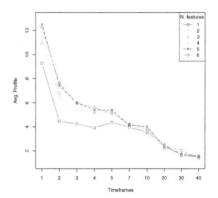

(a) Avg. hit ratio vs length of timeframe (USD/JPY)

(b) Avg. profit per year vs length of time-frame (USD/JPY)

Fig. 2. Plot for performance metrics vs. length of timeframe for USD/JPY pair

trades, and thus more average profit per year. Also, we can see that lower number of features results in higher hit ratio but lower profits on average.

Figures 3 and 4 show the performance metrics as a function of the size of the training set for all different values of number of features for the Euro/USD pair and the USD/JPY pair respectively. The plots show that there is an increase in both the hit ratio and the profits as the size of the training set increases. This might be because smaller training sets lead to over-fitting, whereas larger training sets can fine tune the parameters a bit better. Similar to the case with performance metrics as a function of timeframes, lower number of features results in higher hit ratio but lower profits on average in Fig. 4 (USD/JPY). However, in Fig. 3 (Euro/USD), in case of average profits a similar trend is not noticed.

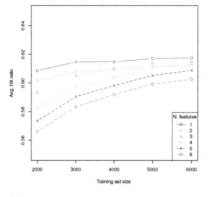
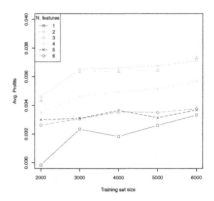

(a) Avg. hit ratio vs training set size (Euro/USD)

(b) Avg. profit per year vs training set size (Euro/USD)

Fig. 3. Plot for performance metrics vs. length of timeframe for Euro/USD pair

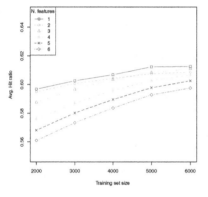
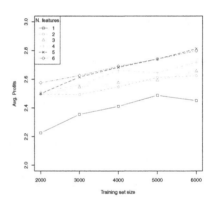

(a) Avg. hit ratio vs training set size (USD/JPY)

(b) Avg. profit per year vs training set size (USD/JPY)

Fig. 4. Plot for performance metrics vs. length of timeframe for USD/JPY pair

Instead, the middle values for the number of features perform much better and have larger profits.

While taking a cursory glance at our results, we noticed that a large number of models generated had similarities in the correlation between the values of the intercept and the coefficients. A large number of models had negative intercept as well as coefficients (although this decreased as the no. of features, and thus the no. of coefficients, increased). Moreover, a majority of models had extremely small absolute values for the intercept, especially when compared to the values of the coefficients. This indicated that the prediction of the models would not be influenced by the intercept (which was almost negligibly small) in a majority of the cases. This, coupled with the observation that the value of coefficients was

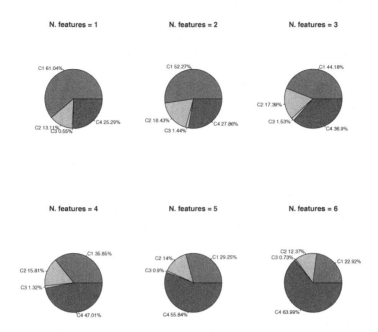

Fig. 5. Pie charts representing the number of models for cases C1 to C4 for different values of no. of features (Euro/USD pair)

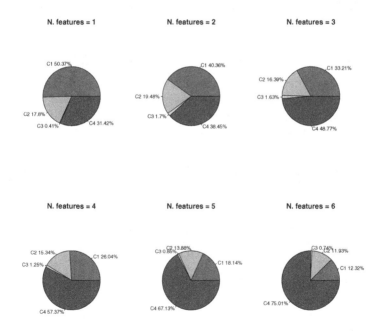

Fig. 6. Pie charts representing the number of models for cases C1 to C4 for different values of no. of features (USD/JPY pair)

very large and negative while still giving good profit and hit ratio, suggested a simple but effective trading rule. We checked for the number of models that satisfied the condition of very small intercept, negative coefficients, and positive profit and hit ratio. The results for both the Euro/USD pair and USD/JPY pair are shown above in Figs. 5 and 6 respectively. The cases C1, C2, C3, and C4 are described as follows:

1. *Case 1* (C1): Absolute value of intercept < 0.1, all coefficients < 0, profits > 0, and hit ratio $>= 60\%$
2. *Case 2* (C2): Absolute value of intercept < 0.1, all coefficients < 0, profits > 0, and hit ratio $>= 50\%$ $\&$ $<= 60\%$
3. *Case 3* (C3): Absolute value of intercept < 0.1, all coefficients < 0, profits > 0, and hit ratio $<= 50\%$
4. *Case 4* (C4): Rest of the models (where not all coefficients are negative, or absolute value of b > 0.1, or profits < 0)

The pie charts confirmed our initial observation that a large number of models had negative coefficients and negligible intercept values while giving profit and good hit ratio. The number of models satisfying the condition of all coefficients < 0 decreases when the number of features increases, as there are more corresponding coefficients in the linear model. As can be seen in Figs. 5 and 6, this effect is more pronounced in the Euro/USD pair as compared to the USD/JPY pair.

Thus, for models trained using linear SVR with low number of features (where features are defined as the difference of price between successive periods of time), we can give a simple rule which states: the next prediction will be the opposite of the most recent (previous) movement direction. Concretely, if the previous trend is down, the model will predict up for the next change, and if the previous trend is up, the model will predict down for the next change. Using this simple method as a training rule, we get profit and good hit ratio in our simulated trading. This property is called return reversal. Our experiments confirmed that the property holds in two different currency markets and linear kernel SVR was able to find it out. Moreover, since the results of the trained models were validated on the data from the year 2002 to the year 2015 for both currency pairs, we have shown that this property has been applicable for a very long period of time (at least 14 years), and as recently as last year (2015) when high frequency trading is more prevalent than ever.

5 Conclusion and Future Work

In this paper we conducted experiments to examine the performance of currency prediction models trained using linear kernel SVR on historical bid price data for high frequency currency trading. We created models using various values for input parameters such as the length of training set, number of features, and length of timeframe for prediction. We also validated the results by performing simulated trading and recording the profits and hit ratio on next years data and got good results. On examining the models, we found a simple rule that gave good results for models with low number of features, which is to predict opposite of the previous direction.

For future work, we plan to study the reasons for the correlation observed between no. of features and the performance metrics. We would also like to examine the return reversal property in greater detail and ascertain the causes behind it. We also plan to study the effect of fees charged for a successful trade on the metrics of profits and trades executed.

References

1. Bank of International Settlements: Triennial Central Bank Survey of foreign exchange and OTC derivatives markets in 2016. http://www.bis.org/publ/rpfx16.htm
2. Yao, J., Tan, C.L.: A case study on using neural networks to perform technical forecasting of forex. Neurocomputing **34**, 79–98 (2000)
3. Zimmerman, H., Neuneier, R., Grothmann, R.: Multi-agent modeling of multiple FX-markets by neural networks. IEEE Trans. Neural Netw. **12**(4), 735–743 (2001)
4. Zhang, G., Hu, M.Y.: Neural network forecasting of the British Pound/US Dollar exchange rate. OMEGA Int. J. Manage. Sci. **26**, 495–506 (1998)
5. Kamruzzaman, J., Sarker, R.: Application of support vector machine to forex monitoring. In: 3rd International Conference on Hybrid Intelligent Systems HIS03, Melbourne (2003)
6. Cortes, C., Vapnik, V.: Support vector networks. Mach. Learn. **20**(3), 273–297 (1995)
7. Smola, A., Vapnik, V., et al.: Support vector regression machines. In: Advances in Neural Information Processing Systems (NIPS), vol. 9, pp. 155–161 (1996)
8. Fletcher, T., Shawe Taylor, J.: Multiple kernel learning with Fisher kernels for high frequency currency prediction. Comput. Econ. **42**(2), 217–240 (2013)
9. Kercheval, A., Zhang, Y.: Modeling high-frequency limit order book dynamics with support vector machines. Quant. Finan. **15**(8), 1315–1329 (2015)
10. Deng, S., Sakurai, A., et al.: Hybrid method of multiple kernel learning and genetic algorithm for forecasting short-term foreign exchange rates. Comput. Econ. **45**(1), 49–89 (2015)
11. Deng, S., Sakurai, A.: Integrated model of multiple kernel learning and differential evolution for EUR/USD trading. Sci. World J. **2014**, article ID 914641, 12 p (2014)

Unsupervised Language Model Adaptation by Data Selection for Speech Recognition

Yerbolat Khassanov[✉], Tze Yuang Chong, Benjamin Bigot,
and Eng Siong Chng

Rolls-Royce@NTU Corporate Lab,
Nanyang Technological University, Singapore, Singapore
{yerbolat002,tychong,bbigot,aseschng}@ntu.edu.sg

Abstract. In this paper, we present a language model (LM) adaptation framework based on data selection to improve the recognition accuracy of automatic speech recognition systems. Previous approaches of LM adaptation usually require additional data to adapt the existing background LM. In this work, we propose a novel two-pass decoding approach that uses no additional data, but instead, selects relevant data from the existing background corpus that is used to train the background LM. The motivation is that the background corpus consists of data from the different domains and as such, the LM trained from it is generic and not discriminative. To make the LM more discriminative, we will select sentences from the background corpus that are similar in some linguistic characteristics to the utterances recognized in the first-pass and use them to train a new LM which is employed during the second-pass decoding. In this work, we examine the use of n-gram and bag-of-words features as linguistic characteristics of selection criteria. Evaluated on the 11 talks in the test-set of TED-LIUM corpus, the proposed adaptation framework produced a LM that reduced the word error rate by up to 10% relatively and the perplexity by up to 47% relatively. When the LM was adapted for each talk individually, further word error rate reduction was achieved.

Keywords: Language model adaptation · Unsupervised adaptation · Data selection · Speech recognition

1 Introduction

The language model (LM) is an important component of an automatic speech recognition (ASR) system. A LM is used to constrain the search to the most probable transcripts of an utterance. Typically, a basic LM is generic and trained on a large corpus consisting of text from various domains. Hence, when using such LM for a specific domain, the performance is sub-optimal [1].

Many approaches have been proposed to adapt the LM using additional data relevant to the target domain. In supervised adaptation approach, the adaptation data is acquired based on prior knowledge given by a human expert, e.g. a summary, a list of topic names or keywords. For example, in the lecture transcription

© Springer International Publishing AG 2017
N.T. Nguyen et al. (Eds.): ACIIDS 2017, Part I, LNAI 10191, pp. 508–517, 2017.
DOI: 10.1007/978-3-319-54472-4_48

systems, suitable adaptation data is extracted from the presentation slides [2–4]. In unsupervised adaptation approach, since prior knowledge is unavailable, the ASR transcripts are typically used as queries to retrieve new adaptation data by using information retrieval (IR) techniques [5–8].

This work belongs to the class of unsupervised adaptation approach in which the knowledge about the target domain is derived from the ASR transcripts. Here, we assume that the utterances presented to the ASR system are homogeneous, i.e. in the same domain, hence the transcripts can be gathered along the recognition session to provide knowledge about the target domain. However, when dealing with utterances in multiple domains, utterances are required to be identified into homogeneous segments before being used to select sentences [9].

Our work is closely related to the cache-based LM [10,11] and the topic-mixture models [12–14], as these two approaches similarly adapt the LM based on the ASR transcripts and require no external adaptation data. For example, the cache-based LMs update the probabilities of the words which have already been recognized based on the hypothesis that a word used in the recent past is more likely to be used soon. The topic-mixture models, on the other hand, use ASR transcripts to adjust the interpolation weights of the subtopic LMs. However, our method is different to the above two methods which only adjust the parameters of the LM, specifically, we will rebuild a new LM from the sentences selected from the background corpus. Inspired by [15], we believe that adapting LM by manipulating the data in the background corpus would provide better impact to the ASR performance, as compared to adapting the parameters in the LMs directly.

Specifically, we propose a LM adaptation framework where the adaptation data is selected from the background corpus that was used to build the generic background LM. We assume that the background corpus is large and comprises of text in various domains and will include the ones that are close to the target domain. We will examine the n-gram and BOW features to select adaptation data. Our approach of using only the background corpus as source data is different to previous works [5–8] where data is usually acquired from external sources. The experimental results show that the proposed approach is sufficient to improve the discriminative ability of the LM significantly.

This paper is organized as follows. Next section describes the proposed LM adaptation framework. Section 3 presents the experiment setups and obtained results. Section 4 concludes this paper and discusses future works.

2   LM Adaptation Framework

This section discusses the proposed LM adaptation framework which is realized on a two-pass recognition system, followed by the data selection technique which selects adaptation data from the background corpus based on two different linguistic features.

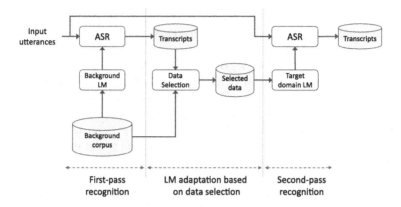

Fig. 1. LM adaptation framework.

2.1 Overview

The proposed LM adaptation framework is depicted in Fig. 1. The recognition process is carried out in three stages: (1) first-pass recognition, (2) LM adaptation, and (3) second-pass recognition, each stage is briefly described below.

During the first-pass recognition, utterances are recognized based on the generic background LM. As the background LM is trained on a background corpus which consists of text in various domains, the model is typically generic and is not optimal to recognizing specific domain utterances.

For LM adaptation, as prior knowledge is unavailable, the knowledge about the target domain is derived from the transcripts produced by the first-pass recognition and used for data selection. The data is selected from the background corpus and used to train the target domain LM.

During the second-pass recognition, the target domain LM replaces the background LM in the ASR system to re-evaluate the input utterances.

2.2 Data Selection

For data selection, each sentence in the background corpus is evaluated to check if it is similar to the decoded utterances generated in the first-pass. Specifically, two type of linguistic features to measure similarity are considered: the bag-of-words (BOW) and the n-gram. For each feature type, the selection criterion is discussed as follows.

BOW Feature. Data selection by using BOW features intends to acquire data with the similar word distribution as the ASR transcripts. A BOW is computed as a vector of vocabulary size where each element corresponds to the count, i.e. term frequency (TF), of the respective word. Each TF is further scaled by its respective inverse document frequency (IDF) to highlight the impact of the word to the meaning of the context.

Let's denote the BOW vector of the ASR transcripts as $\overrightarrow{t}$, while the BOW vector of an arbitrary sentence s in the background corpus as $\overrightarrow{s}$. How relevant sentence s to the target domain is determined by the cosine similarity score between the vectors $\overrightarrow{t}$ and $\overrightarrow{s}$, computed as follows:

$$score_{BOW}(s) = \cos(\overrightarrow{t}, \overrightarrow{s}) = \frac{\overrightarrow{t} \cdot \overrightarrow{s}}{\| \overrightarrow{t} \| \| \overrightarrow{s} \|} \tag{1}$$

Sentences with similar word distribution as the ASR transcripts would yield higher values of $score_{BOW}(\cdot)$ and sentences with similarity scores above a threshold value will be selected as the target domain data.

N-gram Feature. Data selection by using the n-gram features intends to select sentences with similar n-gram distribution as the ASR transcripts. We follow the approach proposed by Moore and Lewis [16], where a generic model is used to complement the data selection. To be more specific, besides computing the similarity between a candidate sentence to the ASR transcripts, the dissimilarity between the sentence with other sentences in the background corpus is also computed. Hence, how relevant an arbitrary sentence s to the target domain is determined as the cross-entropy difference between the background corpus and ASR transcripts, shown as follows:

$$score_{ngram}(s) = H_{BG}(s) - H_{TR}(s) \tag{2}$$

where $H_{BG}(s)$ and $H_{TR}(s)$ are the per-word cross-entropy values of sentence s according to the n-gram distributions in the background corpus and in the ASR transcripts, respectively.

The idea of incorporating the n-gram distribution in the background corpus for data selection is to select sentences, not only similar to the ASR transcripts, but also distinctive from the background corpus. This approach has been shown to select better quality data [16].

3 Experiments and Discussions

This section discusses the evaluation of the proposed LM adaptation framework, including the experimental data, the ASR system configurations, and the obtained results.

3.1 Data

The ASR system was developed by using the TED-LIUM corpus [17]. The corpus consists of monologue talks given on some specific topics, e.g. green technology. The demographics of the corpus are shown in Table 1. The train-set in the corpus was used to train the acoustic model (AM) of the ASR system, while the test-set was used for evaluation.

Table 1. TED-LIUM corpus characteristics.

Characteristics	Train	Test
Total duration	118 h 4 min	2 h 49 min
Number of talks	774	11
Number of words	1.6M	27.5k

The background corpus was adopted from '1 Billion Word Language Model Benchmark' project [18]. It consists of about 30 million sentences with a total of 700 million words after normalization. The corpus consists of text collected from different sources, including newswire text, news commentary text, and parliament speech. This corpus will be used to train the background LM and data selection.

3.2   The ASR System

The ASR system was built around the Kaldi speech recognition toolkit [19]. We followed the TED-LIUM's recipe for the training and the testing procedures. The AM was built with the train-set of the TED-LIUM corpus as the context-dependent deep neural network hidden Markov model (DNN-HMM). The mel-frequency cepstral coefficient (MFCC) features were transformed by using linear discriminant analysis (LDA), maximum likelihood linear transformation (MLLT), and then feature space maximum likelihood linear transformation (fMLLR). The DNN of the AM was pre-trained with restricted Boltzmann machine (RBM), and then trained with cross entropy training, and sequence discriminative method by using the state-level minimum Bayes risk (sMBR) criterion.

The background LM of the ASR system was trained as a trigram model with the '1 Billion Word Language Model Benchmark' corpus. The LM was smoothed by using the modified Kneser-Ney (KN) method available from the SRILM toolkit [20]. The lexicon contains 180k words provided by Kaldi's TED-LIUM recipe.

3.3   Results

This experiment was conducted to evaluate if the data selection within the background corpus can subsequently improve the ASR system performance. We considered the utterances in 11 talks of the test-set to be the same domain (e.g. TED talk domain), i.e. we first accumulated all the generated transcripts of 11 talks from the first-pass and used them jointly for data selection. Data was selected by using different feature types, i.e. BOW (scaled by TF-IDF weighting scheme), 1-gram, 2-gram and 3-gram. We examined the effects of different amounts of data selected, i.e. from 0.1%, 1%, until 100% of the background corpus. The amount of data selected was controlled by adjusting the threshold values in the

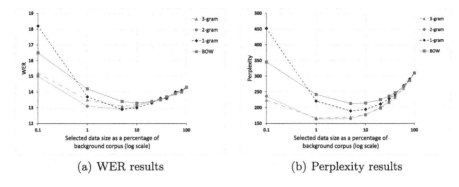

| | (a) WER results | (b) Perplexity results |

Fig. 2. WER and perplexity results obtained by the proposed LM adaptation framework. Data was selected based on different feature types. Note that 100% corresponds to the usage of entire background corpus which is identical to the result obtained from the first-pass recognition.

selection criteria, i.e. Eqs. 1 and 2. The word error rates (WER) in the transcripts after the second-pass recognition are shown in Fig. 2(a). Also, we evaluated the perplexities of the target domain LMs on the reference transcripts, as shown in Fig. 2(b), to see the 'fitness' of the models to the target domain. For all the cases considered, the target domain LMs were trained as trigram models with modified KN smoothing.

As shown in Fig. 2, for all feature types considered, $1\% - 1.4\%$ absolute WER reductions which are equivalent to $7\% - 10\%$ relative WER reductions, were obtained by selecting only $1\% - 10\%$ of the data in the background corpus to train the target domain LMs. Produced target domain LMs reduced the perplexity by up to $31\% - 47\%$ relatively as compared to the perplexity of the background LM. The WER after the first-pass recognition was 14.3%. The results indicate that the target domain data contained by the background corpus can be selected and used for adaptation to reduce the WER.

We noticed that when too little data was selected, e.g. 0.1%, the target domain LM deteriorated the ASR system performance, as can be observed from the left tails of the plots in Fig. 2(a) and (b) that indicate higher WERs and perplexities. In such cases, although the selected data matches the target domain well, i.e. higher similarity score according to Eqs. 1 and 2, the amount of data is insufficient for training the parameters in the LMs.

On the other hand, when too much data was selected, out-of-domain data will be selected as the adaptation data. In such cases, the WERs of the ASR system and the perplexities of the target-domain LMs became slightly higher, as shown by the right tails of the plots in Fig. 2(a) and (b).

In the test-set of TED-LIUM corpus, the talks are given by different speakers on different topics, e.g. technology, science, global issues, and etc. Such arrangement allows the data to be further categorized into different topics which is more specific domain. In next experiment, we evaluate the proposed adaptation framework in adapting LMs to the more specific domains.

LM Adaptation to more Specific Domain. In this experiment, the data selection technique was evaluated to see if the data in a more specific domain, i.e. topic of the talk, can be selected from the background corpus for LM adaptation. Here, the experiment discussed in previous section was repeated, except that for each talk the LM was adapted separately.

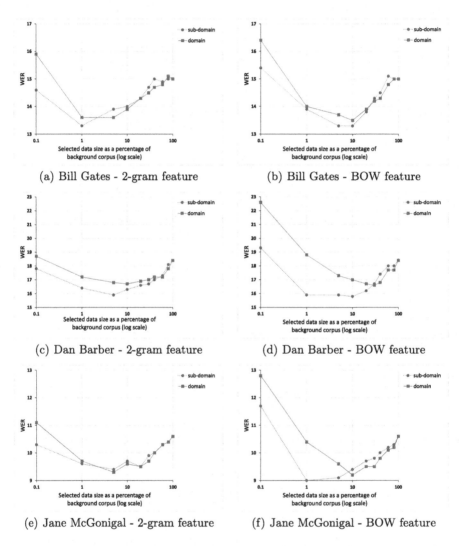

(a) Bill Gates - 2-gram feature   (b) Bill Gates - BOW feature

(c) Dan Barber - 2-gram feature   (d) Dan Barber - BOW feature

(e) Jane McGonigal - 2-gram feature   (f) Jane McGonigal - BOW feature

Fig. 3. The comparison of WER results obtained in the first experiment (referred to as 'domain') to the WER results of the second experiment (referred to as 'sub-domain') for three selected talks: (a)-(b) Bill Gates, (c)-(d) Dan Barber, (e)-(f) Jane McGonigal. Data was selected based on the BOW and the 2-gram features. 100% corresponds to the usage of the entire background corpus.

We picked three talks in different topics from the test-set. Selected talks and their corresponding topics are: (1) *Bill Gates - Innovating to zero*, (2) *Dan Barber - How I fell in love with a fish*, and (3) *Jane McGonigal - Gaming can make a better world*.

For this experiment, data was selected based on the BOW and the 2-gram features. The WER results obtained for three talks in this experiment (referred to as 'sub-domain') along with the WER results obtained in previous experiment (referred to as 'domain') are depicted in Fig. 3.

For the 2-gram features, achieved WER reductions are similar or slightly better than in the previous experiment, as shown in Fig. 3(a)(c)(e). On the other hand, the data selection by using the BOW feature showed much better WER performance as compared to previous experiment, as shown in Fig. 3(b)(d)(f). For example, the BOW feature further reduced the WER of the talk given by *Dan Barber* by 15% relatively for the case when 1% of the background corpus is selected.

We also noticed that, in this experiment, the WER reductions achieved by BOW features are slightly better than the ones achieved by 2-gram features. This observation is different in previous experiment where the BOW feature gave the least WER reduction as compared to other n-gram features, as shown in Fig. 2(a). This is explainable as it was the topic domain being considered in this case, and the BOW is a more effective feature to capture the topic information as compared to the n-gram features.

Such improvements suggest that the proposed LM adaptation framework was benefited by the data selected in a more specific domain.

4   Conclusions

This paper has presented a two-pass LM adaptation framework where the recognized utterances from the first-pass are used to select relevant sentences from the background corpus. Selected sentences are used to train target domain LM which is employed by the decoder during the second-pass recognition. Without using any additional data, the WER has been reduced by up to 10% relatively and perplexity has been reduced by up to 47% relatively, as evaluated on the 11 talks of the test-set from the TED-LIUM corpus. When the LM was adapted individually for each talk, which is more specific domain, further WER reductions have been achieved, e.g. 15% relative WER reduction for the talk given by *Dan Barber - How I fell in love with a fish* (for the case when 1% is selected with BOW feature). In addition, such improvements were achieved by selecting only 1% − 10% of the data from the background corpus.

For future work, implementing a technique to identify topic boundaries in ASR output to extract utterances into homogeneous segments is needed. As suggested by the results in Fig. 3, where data selection based on more specific domain has shown further improvement to the ASR system performance. Given the advancement in topic modeling [21] and distributed representation of text [22], domain can be defined at a finer level, such as sub-topic and latent topic. Data selection in such domains might further improve a LM adaptation approach.

Also, it is important to study the impact of mis-recognized words in the ASR transcripts to the performance of proposed adaptation framework, especially when the WER is high, such as in the conversational ASR system. As the knowledge about domain is derived from the ASR transcripts, high WER in the transcripts of the first-pass recognition might cause target domain data unable to be selected from the background corpus.

Acknowledgments. This work was conducted within the Rolls-Royce@NTU Corporate Lab with support from the National Research Foundation (NRF) Singapore under the Corp Lab@University Scheme.

References

1. Rosenfeld, R.: Two decades of statistical language modeling: where do we go from here? Proc. IEEE **88**, 1270–1278 (2000)
2. Munteanu, C., Penn, G., Baecker, R.: Web-based language modelling for automatic lecture transcription. In: INTERSPEECH, pp. 2353–2356 (2007)
3. Martnez-Villaronga, A., Miguel, A., Andrs-Ferrer, J., Juan, A.: Language model adaptation for video lectures transcription. In: 2013 IEEE International Conference on Acoustics, Speech and Signal Processing, pp. 8450–8454. IEEE (2013)
4. Akita, Y., Tong, Y., Kawahara, T.: Language model adaptation for academic lectures using character recognition result of presentation slides. In: 2015 IEEE International Conference on Acoustics, Speech and Signal Processing, pp. 5431–5435. IEEE (2015)
5. Suzuki, M., Kajiura, Y., Ito, A., Makino, S.: Unsupervised language model adaptation based on automatic text collection from WWW. In: 9th International Conference on Spoken Language Processing (2006)
6. Lecorv, G., Gravier, G., Sbillot, P.: An unsupervised web-based topic language model adaptation method. In: 2008 IEEE International Conference on Acoustics, Speech and Signal Processing, pp. 5081–5084. IEEE (2008)
7. Lecorv, G., Dines, J., Hain, T., Motlicek, P.: Supervised and unsupervised web-based language model domain adaptation. In: IDIAP (2012)
8. Schlippe, T., Gren, L., Vu, N.T., Schultz, T.: Unsupervised language model adaptation for automatic speech recognition of broadcast news using web 2.0. In: INTERSPEECH, pp. 2698–2702 (2013)
9. Watanabea, S., Iwataa, T., Horia, T., Sakob, A., Arikib, Y.: Topic tracking language model for speech recognition. Comput. Speech Lang. **25**(2), 440–461 (2011)
10. Kuhn, R., De Mori, R.: A cache-based natural language model for speech recognition. IEEE Trans. Pattern Anal. Mach. Intell. **12**(6), 570–583 (1990)
11. Jelinek, F., Merialdo, B., Roukos, S., Strauss, M.: A dynamic language model for speech recognition. In: HLT, pp. 293–295 (1991)
12. Kneser, R., Steinbiss, V.: On the dynamic adaptation of stochastic language models. In: 1993 IEEE International Conference on Acoustics, Speech and Signal Processing, pp. 586–589. IEEE (1993)
13. Chen, L., Gauvain, J.L., Lamel, L., Adda, G., Adda-Decker, M.: Using information retrieval methods for language model adaptation. In: INTERSPEECH, pp. 255–258 (2001)

14. Echeverry-Correa, J.D., Ferreiros-Lpez, J., Coucheiro-Limeres, A., Crdoba, R., Montero, J.M.: Topic identification techniques applied to dynamic language model adaptation for automatic speech recognition. Expert Syst. Appl. **42**(1), 101–112 (2015)
15. Chen, L., Huang, T.: An improved MAP method for language model adaptation. In: Sixth European Conference on Speech Communication and Technology (1999)
16. Moore, R.C., Lewis, W.: Intelligent selection of language model training data. In: Proceedings of the ACL 2010 Conference Short Papers, pp. 2200–224. ACL (2010)
17. Rousseau, A., Delglise, P., Esteve, Y.: TED-LIUM: an automatic speech recognition dedicated corpus. In: LREC, pp. 125–129 (2012)
18. Chelba, C., Mikolov, T., Schuster, M., Ge, Q., Brants, T., Koehn, P., Robinson, T.: One billion word benchmark for measuring progress in statistical language modeling. arXiv preprint arXiv:1312.3005 (2013)
19. Povey, D., Ghoshal, A., Boulianne, G., Burget, L., Glembek, O., Goel, N., Hannemann, M., Motlicek, P., Qian, Y., Schwarz, P., Silovsky, J.: The Kaldi speech recognition toolkit. In: IEEE Signal Processing Society (2011)
20. Stolcke, A.: SRILM - an extensible language modeling toolkit. In: INTERSPEECH, pp. 901–904 (2002)
21. Blei, D.M.: Probabilistic topic models. Commun. ACM **55**(4), 77–84 (2012)
22. Le, Q.V., Mikolov, T.: Distributed representations of sentences and documents. In: ICML, vol. 14, pp. 1188–1196 (2014)

Metaheuristic Optimization on Conventional Freeman Chain Code Extraction Algorithm for Handwritten Character Recognition

Muhammad A. Mohamad$^{(\boxtimes)}$, Habibollah Haron, and Haswadi Hasan

Faculty of Computing, Universiti Teknologi Malaysia, UTM, 81310 Johor Bahru, Johor, Malaysia
marif.utm@gmail.com, {habib,haswadi}@utm.my

Abstract. In Handwritten Character Recognition (HCR), interest in feature extraction has been on the increase with the abundance of algorithms derived to increase the accuracy of classification. In this paper, a metaheuristic approach for feature extraction technique in HCR based on Harmony Search Algorithm (HSA) was proposed. Freeman Chain Code (FCC) was used as data representation. However, the FCC representation is dependent on the route length and branch of the character node. To solve this problem, the metaheuristic approach via HSA was proposed to find the shortest route length and minimum computational time for HCR. At the end, comparison of the result with other metaheuristic approaches namely, Differential Equation (DE), Particle Swarm Optimization (PSO), Genetic Algorithm (GA) and Ant Colony Optimization (ACO) was performed.

Keywords: Harmony search algorithm · Freeman chain code · Handwritten character recognition

1 Introduction

Handwritten Character Recognition (HCR) is the ability of a computer to receive and interpret intelligible handwritten input then to analyze many automated process system [1]. The major problem in HCR system is the variation of the handwriting styles, which can be completely different for different writers [2]. The objective of HCR is to implement user friendly computer assisted character representation that will allow successful extraction of characters from handwritten documents and to digitalize and translate the handwritten text into machine readable text. However, after many years of intensive investigation and research, the main goal of developing character recognition system still remains unachieved [3].

Generally, HCR can be divided into three stages namely preprocessing, feature extraction and classification. Preprocessing stage is to produce a clean character image that can be used directly and efficiently by the feature extraction stage. Feature extraction stage is to remove redundancy from data. Classification stage is to recognize characters or words. Nowadays different methods are in widespread use for character recognition.

Recognition accuracy of the image depends on the sensitivity of selected features. Hence, number of feature extraction methods can be found in the literature [4]. Feature

© Springer International Publishing AG 2017
N.T. Nguyen et al. (Eds.): ACIIDS 2017, Part I, LNAI 10191, pp. 518–527, 2017.
DOI: 10.1007/978-3-319-54472-4_49

extraction can be defined as extracting the most representative information from the raw data, which minimizes the within class pattern variability while enhancing the between class pattern variability. For this purpose, a set of features are extracted for each class that helps distinguish it from other classes, while remaining invariant to characteristic differences within the class [5]. A good survey on feature extraction methods for character recognition can be found in [6].

This paper only concentrates on feature extraction stage of a HCR. Feature extraction stage in HCR is a very important field of the image processing and object recognition system because it is used by classifier to classify the data. The basic task of feature extraction and selection is to find out a group of the most effective features for classification that is by compressing from high dimensional feature space to low-dimensional feature space and to design classifier effectively [7]. Fundamental component of feature extraction is called features. As in many practical problems, it is often not easy to find those with most effective features [8]. This makes features extraction and selection as one of the most difficult and challenging tasks in pattern recognition system, data mining, and other fields.

There are many algorithms for feature extraction and selection. Recently, interest in feature extraction and selection has been on the increase with the abundance of algorithms derived. This algorithm can be classified into two namely heuristic and meta-heuristic approaches. Many heuristic algorithms have been proposed in the literature for finding near–optimal solutions [9, 10]. GA is one of metaheuristic approach and has been widely used to solve feature extraction and selection problems [11–13].

In this paper, feature extraction and selection technique based on metaheuristic algorithm is explored using a Harmony Search Algorithm (HSA). HSA were applied successfully in many areas such as computer science, electrical engineering, civil Engineering, mechanical engineering and biomedical application. Summarization of application of HSA can be found in [4].

In this paper, Freeman chain code (FCC) is selected as the representation of a character image. FCC is one of the techniques representations based on the boundary extraction which useful for image processing, shape analysis and pattern recognition. Chain code representation gives a boundary of character image where those codes represent the direction of where is the location of the next pixel. Unfortunately, the study about FCC construction using one continuous route and minimizing the length of chain code to FCC from a thinned binary image (TBI) has not been widely explored. To solve this problem, metaheuristic methods are used to extract the FCC that is correctly representing the characters. Therefore, this paper proposed the HS-FCC extraction algorithm in handwritten character recognition.

This paper is organized as follows. Section 2 presents the related work. Section 3 describes the methodology. Section 4 describes the HS-FCC extraction algorithms proposed. Section 5 presents the parameter value setting. Section 5 describes the result and discussion and followed by a conclusion in Sect. 6.

2 Related Work

Chain code is one of the representations technique based on the boundary extraction which useful for image processing, shape analysis and pattern recognition. The first approach of chain code was introduced by Freeman in 1961 that is known as Freeman Chain Code (FCC) [14]. There are many kinds of chain code algorithms, which have been developed through extension of FCC and enhancement of chain code. Previous work in the literature about chain code representation can be found in [15–17]. There are two type of directions of chain code, namely 4-neighborhood and 8-neighborhood. This paper utilizes 8-neighbourhood in extraction of characters. The challenge of the chain-coding process would be very much on the way of the image would be traversing and the starting point of the traversing method [18]. A start point of a character will produce a different chain code direction even though is the same image. Randomly, the start point in a character is selected and then the best solution is searched.

Metaheuristic method is used for minimizing the length of the chain code. The main problem in representation characters using FCC is the length of the FCC that is depends on the starting point, the branching node and the revisited walk. To solve this problems, metaheuristic is used to generate the FCC which has the ability to produce FCC correctly in representing the characters.

In [19] the metaheuristic approach has been used in FCC extraction for HCR. The main problems in FCC representation technique as previously stated have been a motivation in using the metaheuristic approach to solve the problem. Thus, FCC extraction technique via metaheuristic approach i.e. Differential Evaluation (DE), Particle Swarm Optimization (PSO), Genetic Algorithm (GA), and Ant Colony Optimization (ACO) has been proposed. The proposed algorithms were used to extract the FCC from handwritten character image that is Thinned Binary Image (TBI). In the proposed algorithm, three solution representations were used in representing a character to minimize the FCC length namely character transformation into graph, graph is a solution representation, and metaheuristic approach. The role of the proposed algorithms is to minimize the objective function of the solution representation. The objective function is to express the quality of FCC solution and is defined as the number of nodes which the FCC must visit from the starting node until all of the nodes are visited. Then, using their particular characteristics, every approach tries to find a collection of good FCC solutions which minimize the FCC length. Route length and computation time were selected in this experiment because they are depending on starting point and automatically affected on the route length and how many time it needs to solve the chain code. This method enables them to extract and recognize such difficult character in relatively shorter computational time and route length. The proposed algorithm is evaluated based on route length and computation time. The result shows that in term of route length, PSO obtained the lowest compared to DE, ACO and GA. Meanwhile, for the computational time, DE obtained the fastest computation time compared to others.

Harmony search algorithm (HSA) is one of the recent metaheuristic that inspired from the musician performance that search for the better state of harmony [20]. To date, HSA has been applied to many engineering optimization problems including structural engineering [21–23], structural materials [24–26], hydraulics [27, 28], cost optimization and construction management [29, 30], and structural vibration control [31]. Application of HSA in HCR that was implemented in [32] for recognition-based segmentation of online Arabic text was applied in the recognition stages instead of feature extraction phase using dominant point detection to extract the features. In contrast, in this paper HSA application was proposed to be applied in the feature extraction phase since as far as literature concern, HSA have not been implemented in feature extraction.

Therefore, in this paper the HS-FCC extraction algorithm is proposed in HCR that is harmony search (HS) algorithm is used for optimizing (minimize) the length of chain code. The proposed HS-FCC extraction algorithm is the similar study of application of metaheuristic (DE, PSO, GA and ACO) that was conducted by [19]. At the end of this study, the result in FCC extraction algorithms for HCR between the proposed meta-heuristic algorithm (DE, PSO, GA and ACO) in [19] and the proposed HSA is compared.

3   Methodology

There are three stages in the proposed methodology. First, by using input character that is digitized, thinning is performed as pre-processing stage in the HCR. The output of TBI is a skeleton is used in feature extraction stage. This work applied thinning algorithm that is proposed by Engkamat [33] in extracting the FCC. Second, the representation of pattern is needed in the feature extraction stage. In this case, FCC is selected and is used to represent the character. Finally, the desired output of chain code is obtained. The proposed HS-FCC extraction methodology of HCR is shown in Fig. 1.

Fig. 1. The Proposed HS-FCC Extraction Algorithm

The experiments on the HSA algorithms are performed based on the chain code representation derived from established previous works of Centre of Excellent for Document Analysis and Recognition (CEDAR) dataset. The scope area is isolated handwritten on upper-case characters (A–Z). The pixel input of original CEDAR and its output TBI is 50×50 pixels.

4   Proposed HS-FCC Extraction Algorithm

The HS algorithm is used to extract the FCC from handwritten character image that is Thinned Binary Image (TBI) as shown in Fig. 2. A generation of FCC from a binary image can be modelled as a route of a graph problem. Initially, the binary image is transformed into a digraph which consists of vertices and edges. The vertices of the graph is taken from node which has only one neighbor and the node which have neighbors more than two. In meantime, the edges of the graph are come from nodes which have two neighbors connecting the vertices from before. The lengths are obtained from the total number of nodes between two vertices. The complete graph can be seen in Fig. 3.

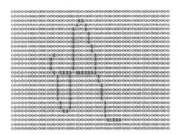 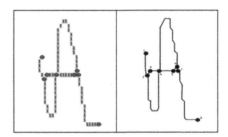

Fig. 2. Binary Image of "A" Handwritten Character

Fig. 3. Character Transformation of Fig. 2 into Graph

The proposed HS-FCC extraction algorithm uses a sequence of edges to represent the FCC solutions. The edge is used as the solution representation. An edge is derived and ended from the same node. Two different edges can be derived from the same node or can also ended in another same node too. Thus, one edge can visited twice and as a result the solution representation can have a complete tour since a revisit to the previous visited nodes is often needed. The objective function is defined as the number of nodes which the FCC must visit from the starting node until all of the nodes are visited (revisit is counted too). The HS-FCC extraction algorithm process is now summarized. The implementation of HS algorithm to generate the chain code as depict in Table 1.

Table 1. The Pseudocode Of HS-FCC Extraction Algorithm

The Pseudocode of HS-FCC Extraction Algorithm
Input data and settings parameter values
Clean image
Enumerate junction/end as nodes and interconnecting routes
Generate random node sequences.
Start HSA search with objective function( )=path length
Initialize the harmony memory (HM)
Repeat
Improvise a new harmony from the HM
Update the HM
Until stopping criterion is archive

The algorithm work as follows:

1. Clean and break an image into paths
2. Enumerate junctions/end as nodes and interconnecting routes. At this point, generate chain code for corresponding route.
3. Find vertex by obtain list of junctions by testing neighbor population (i = 2).

4. Connecting between the junctions by testing each connected edges. Record the edge and junction at the other end.
5. Generate random node sequences for HSA memory initialization.
6. Start HSA with objective function() = path length. Assume the goal is to find the shortest route possible. Path length is the sequence of routes to reach all nodes.

5 Result and Discussion

This section describes the experimental results of the experiment. HSA is used to generate the continuous FCC which acts as the image features. The proposed HS-FCC extraction algorithm consists of 10 replications where each replication is 100 FCC solutions for every TBI. The result is analyzed by comparing result of the proposed metaheuristic HSA for extraction of FCC extraction with other metaheuristic algorithm [19] as shown in Table 2 and Fig. 4. The comparisons are based on route length and computation time.

Table 2. Comparison of Proposed HSA with Method by [19]

Algorithm	Route Length			Average Computation Time(s)
	Best	Average	Worst	
DE	2334.74	2.359.42	2.386.32	865.70
PSO	2247.79	2370.89	2391.77	2012.11
GA	2318.04	2334.63	2349.73	1144.58
ACO	2343.42	2354.92	2380.19	1126.33
HSA	1880.28	1915.88	1934.13	1.10

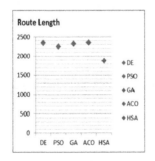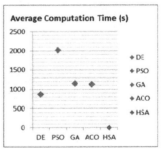

Fig. 4. Graph Comparison based on Route Length and Computation Time

Based on the result, the proposed HSA is better than other metaheuristic algorithm. HSA obtained the lowest in term of route length with 1880.28 compared to others. In addition, in term of computation time HSA is the fastest. To solve the whole thinned binary images, the proposed HSA only needs 1.10 s.

The efficiency of the HSA seem obvious by obtained the better performance compared to other metaheuristic technique. This is because of the ways of the HSA in handling the intensification and diversification. Diversification makes sure the search in the parameter space can explore as many locations and regions as possible in an efficient and effective manner. It also ensures that the evolving system will not be trapped in biased local optima. On the other hand, the intensification intends to exploit the history and experience of the search process. It aims to ensure to speed up the convergence when necessary by reducing the randomness and limiting diversification. The optimal balance of diversification and intensification is required optimization process.

In the HSA, diversification is essentially controlled by the pitch adjustment and randomization. There are two subcomponents for diversification, which might be an important factor for the high efficiency of the HSA method. The first subcomponent of composing or generating new solutions, via randomization would be at least at the same level of efficiency as other algorithms by randomization. However, an additional subcomponent for HSA diversification is the pitch adjustment characterized. Pitch adjusting is carried out by adjusting the pitch in the given bandwidth by a small random amount relative to the existing pitch or solution from the harmony memory. Essentially, pitch adjusting is a refinement process of local solutions. Both memory considering and pitch adjusting ensure that the good local solutions are retained while the randomization and harmony memory considering will explore the global search space effectively. The randomization explores the search space more efficiently and effectively; while the pitch adjustment ensures that the newly generated solutions are good enough, or not too far away from existing good solutions.

The intensification is mainly represented in the HSA by the harmony memory accepting rate. A high harmony acceptance rate means the good solutions from the history/memory are more likely to be selected or inherited. This is equivalent to a certain degree of elitism. Obviously, if the acceptance rate is too low, the solutions will converge more slowly. As mentioned earlier, this intensification is enhanced by the controlled pitch adjustment. Such interactions between various components could be another important factor for the success of the HSA over other algorithms.

6   Conclusion

In this paper, metaheuristic approach via harmony search algorithm for feature extraction in handwritten character recognition namely HS-FCC extraction in handwritten character recognition is proposed. The proposed HS-FCC extraction algorithm is used to produce one continuous route and to minimize the length of FCC. This work is similar study of application of metaheuristic which are DE, PSO, GA and ACO was conducted by [19]. The proposed HSA algorithm implemented for FCC extraction use a similar solution representation and objective function calculated method same as algorithm proposed by [19]. The objective function is to express the quality of an FCC solution and is defined as the number of nodes which the FCC must visit from the starting node until all of the nodes are visited. Using the solution representation, by assume that one edge can be visited twice to assure that the solution representation can make a complete tour since a revisit to the previously visited node is often needed. Then using the particular characteristics of HSA, this approach tries to find a collection of good FCC solutions which minimize the FCC length. Route length and computation time were selected in this experiment because they are depending on starting point and automatically affected on the route length and how many time it needs to solve the chain code. This method enables them to extract and recognize such difficult character in relatively shorter computational time and route length. The result of the proposed HSA was compared to the result of proposed method by [19] in term of FCC extraction in HCR based on route length and computation time.

The results show that, in term of route length, the proposed HSA obtained the lowest compared to the DE, PSO, GA and ACO proposed by [19]. Meanwhile in term of computational time, the proposed HSA also obtained the lowest means the HSA compute the fastest computation time in extracting the features compared to the DE, PSO, GA and ACO. The better performance of the HSA is controlling the optimal balance of diversification and intensification. In addition, the implementation of HSA is also easier. There is some evidence to suggest that HSA is less sensitive to the chosen parameters, which means that we do not have to fine-tune these parameters to get quality solutions. Furthermore, the HSA is a population-based metaheuristic, this means that multiple harmonics groups can be used in parallel. Proper parallelism usually leads to better implantation with higher efficiency. The good combination of parallelism with elitism as well as a fine balance of intensification and diversification is the key to the success of the HS algorithm, and in fact, to the success of any metaheuristic algorithms.

The resulting FCC will become the input to the classification stage. Every feature in chain code is fed to the classifier for recognition. The efficiency of FCC in the representation can be seen by the number of image characters that can be recognized. For future works, the HSA can be explored for enhancement by hybrid with other metaheuristic algorithm for instance, biogeography based optimization (BBO) and Particle Swarm Optimization (PSO). The advantages of characteristic of HSA make it very versatile to combine with other metaheuristic algorithms to produce hybrid metaheuristics and to apply in various applications.

Acknowledgment. The authors honorably appreciate to Ministry of High Education Malaysia (MOHE), Universiti Teknologi Malaysia (UTM), Research Management Centre (RMC) with grant vote number 4F860, Zamalah UTM and Soft Computing Research Group (SCRG) for their support.

References

1. Nasien, D., Haron, H., Yuhaniz, S.S.: Metaheuristics methods (GA & ACO) for minimizing the length of freeman chain code from handwritten isolated characters. World Acad. Sci. Eng. Technol. **62**, 230–235 (2010)
2. Patel, M., Thakkar, S.P.: Handwritten character recognition in english: a survey. Int. J. Adv. Res. Comput. Commun. Eng. **4**(2), 345–350 (2015)
3. Pithadia, N.J., Nimavat, V.D.: A review on feature extraction techniques for optical character recognition. Int. J. Innovative Res. Comput. Commun. Eng. **3**(2), 1263–1268 (2015)
4. Mohamad, M.A., Dewi, N., Hassan, H., Haron, H.: A review on feature extraction and feature selection for handwritten character recognition. Int. J. Adv. Comput. Sci. Appl. **6**(2), 204–212 (2015)

5. Oh, I.S., Lee, J.S., Suen, C.Y.: Analysis of class separation and combination of class-dependent features for handwriting recognition. IEEE Trans. Pattern Anal. Mach. Intell. **21**(10), 1089–1094 (1999)

6. Trier, D., Jain, A.K., Taxt, T.: Feature extraction method for character recognition - a survey. Pattern Recogn. **29**(4), 641–662 (1996)

7. Shi-Fei, D., Wei-Kuan, J., Chun-Yang, S., Zhong-Zhi, S.: Research of pattern feature extraction and selection. In: Proceedings of Machine Learning and Cybernetics. Kunming, China (2008)

8. Zhaoqi, B., Xuegong, Z.: Pattern recognition. In: 2nd Edition, Tsinghua University Press (2000)

9. Kwak, N., Choi, C.-H.: Input feature selection for classification problems. IEEE Trans. Neural Netw. **13**(1), 143–159 (2002)

10. Somol, P., Pudil, P., Kittler, J.: Fast branch and bound algorithms for optimal feature selection. IEEE Trans. Pattern Anal. Mach. Intell. **26**(7), 900–912 (2004)

11. Kudo, M., Sklansky, J.: Comparison of algorithms that select features for pattern recognition. Pattern Recogn. **33**(1), 25–41 (2000)

12. Oh, I.-S., Lee, J.-S., Moon, B.-R.: Hybrid genetic algorithms for feature selection. IEEE Trans. Pattern Anal. Mach. Intell. **26**(11), 1424–1437 (2004)

13. Cordella, L., De Stefano, C., Fontanella, F., Marrocco, C., Scotto di Freca, A.: Combining single class features for improving performance of a two stage classifier. In: 20th International Conference on Pattern Recognition (ICPR 2010), pp. 4352–4355 (2010)

14. Freeman, H.: Techniques for the digital computer analysis of chain-encoded arbitrary plane curves. In: Proceedings of Electron (1961)

15. Liu, Y.K., Zalik, B.: An efficient chain code with huffman coding. Pattern Recogn. **38**(4), 553–557 (2005)

16. Sánchez-Cruz, H., Bribiesca, E., Rodríguez-Dagnino, R.M.: Efficiency of chain code to represent binary objects. Pattern Recogn. **40**(6), 1660–1674 (2007)

17. Wulandhari, L.A., Haron, H.: The evolution and trend of chain code scheme. Graphics Vis. Image Process. **8**(3), 17–23 (2008)

18. Suliman, A., Shakil, A., Sulaiman, M.N., Othman, M., Wirza, R.: Hybrid of HMM and fuzzy logic for handwritten character recognition. In: Proceedings of Information Technology. Kuala Lumpur, Malaysia (2008)

19. Nasien, D., Haron, H., Yuhaniz, S.S., Najwa, A., Hassan, H.: Performance of metaheuristics technique in FreeMan chain code for handwritten characters recognition. Adv. Sci. Letters **4**(2), 270–274 (2011)

20. Erol, O.K., Eksin, I.: A new optimization method: big bang big crunch. Adv. Eng. Softw. **37**, 106–111 (2006)

21. Geem, Z.W., Kim, J.H., Loganathan, G.V.: A new heuristic optimization algorithm: harmony search. Simulation **76**, 60–68 (2001)

22. Lee, K.S., Geem, Z.W.: A new structural optimization method based on the harmony search algorithm. Comput. Struct. **82**, 781–798 (2004)

23. Saka, M.P.: Optimum geometry design of geodesic domes using harmony search algorithm. Adv. Struct. Eng. **10**(6), 595–606 (2007)

24. Degertekin, S.O.: Harmony search algorithm for optimum design of steel frame structures: a comparative study with other optimization methods. Struct. Eng. Mech. **29**(4), 391–410 (2008)

25. Saka, M.P.: Optimum design of steel sway frames to BS5950 using harmony search algorithm. J. Constr. Steel Res. **65**, 36–43 (2009)

26. Hasancebi, O.: Performance evaluation of metaheuristic search techniques in the optimum design of real size pin jointed structures. Comput. Struct. **87**, 284–302 (2009)
27. Degertekin, S.O., Hayalioglu, M.S., Gorgun, H.: Optimum design of geometrically non-linear steel frames with semi-rigid connections using a harmony search algorithm. Steel Compos. Struct. **9**(6), 535–555 (2009)
28. Erdal, F., Saka, M.P.: Harmony search based algorithm for the optimum design of grillage systems to LRFD-AISC. Struct. Multidiscip. O. **38**, 25–41 (2009)
29. Hasancebi, O., Carbas, S., Dogan, E., Erdal, F., Saka, M.P.: Comparison of non-deterministic search techniques in the optimum design of real size steel frames. Comput. Struct. **88**(17–18), 1033–1048 (2010)
30. Togan, V., Daloglu, A.T., Karadeniz, H.: Optimization of trusses under uncertainties with harmony search. Struct. Eng. Mech. **37**(5), 543–560 (2011)
31. Erdal, F., Dogan, E., Saka, M.P.: Optimum design of cellular beams using harmony search and particle swarm optimizers. J. Constr. Steel Res. **67**(2), 237–247 (2011)
32. Potrus, M.Y., Ngah, U.K.: A harmony search algorithm for recognition-based segmentation of online arabic text. In: Proceedings of International Conference on Engineering and Information Technology, pp. 205–210, 17–18 September 2012, Toronto, Canada (2012)
33. Engkamat, A.A.: Enhancement of parallel thinning algorithm for handwritten characters using neural network. M.Sc. Thesis, Universiti Teknologi Malaysia, November 2005

A Novel Learning Vector Quantization Inference Classifier

Chakkraphop Maisen[1,2], Sansanee Auephanwiriyakul[1,3(✉)], and Nipon Theera-Umpon[3,4]

[1] Department of Computer Engineering, Faculty of Engineering, Chiang Mai University, Chiang Mai, Thailand
chakkraphop_maisen@cmu.ac.th
[2] Graduate School, Chiang Mai University, Chiang Mai, Thailand
[3] Biomedical Engineering Center, Chiang Mai University, Chiang Mai, Thailand
{sansanee,nipon}@ieee.org
[4] Department of Electrical Engineering, Faculty of Engineering, Chiang Mai University, Chiang Mai, Thailand

Abstract. One of the popular tools in pattern recognition is a neuro-fuzzy system. Most of the neuro-fuzzy systems are based on a multi-layer perceptrons. In this paper, we incorporate learning vector quantization in a neuro-fuzzy system. The prototype update equation is based on the learning vector quantization while the gradient descent technique is used in the weight update equation. Since weights contain informative information, they are exploited to select a good feature set. There are 8 data sets used in the experiment, i.e., Iris Plants, Wisconsin Breast Cancer (WBC), Pima Indians Diabetes, Wine, Ionosphere, Colon Tumor, Diffuse Large B-Cell Lymphoma (DLBCL), and Glioma Tumor (GLI_85). The results show that our algorithm provides good classification rates on all data sets. It is able to select a good feature set with a small number of features. We compare our results indirectly with the existing algorithms as well. The comparison result shows that our algorithm performs better than those existing ones.

Keywords: Neuro-fuzzy classifier · Learning vector quantization · Prototype-based rule · Rule extraction · Feature selection

1 Introduction

Neural networks and fuzzy systems are two outstanding paradigms in the field of computational intelligence. The combination of both, called a neuro-fuzzy system [1], yields appreciated benefits. One of the major advantages is adaptive thinking from data, and revealed meaning of the knowledge. This advantage is appropriate for any application that needs the systems to give the reason of its analyzed results [2]. There are many techniques to combine neuro-fuzzy concepts. The neural-fuzzy system has been categorized into 3 groups [3, 4], i.e., cooperative models, concurrent models, and fused (or hybrid) models. There are several works on neuro-fuzzy systems. However, most of these research works are neuro-fuzzy systems based on a multi-layer feed

© Springer International Publishing AG 2017
N.T. Nguyen et al. (Eds.): ACIIDS 2017, Part I, LNAI 10191, pp. 528–544, 2017.
DOI: 10.1007/978-3-319-54472-4_50

forward networks [5–12]. Although, there are many types of neural networks, e.g., self organizing map (SOM), learning vector quantization (LVQ), etc., there are a few works involving with neuro-fuzzy systems based on the other types of neural networks. For example, the neuro-fuzzy system based on SOM was introduced in [13], or that based on SOM and LVQ was proposed in [14].

Hence, in this paper, we are focusing on a neuro-fuzzy system based on the learning vector quantization (LVQ). The LVQ was introduced by Kohonen [15]. Since then, there are many research works either incorporating the fuzzy set theory in the LVQ [16–21] or improving the performance of the LVQ [22–28]. There are a few research works on neuro-fuzzy systems based on LVQ [20, 21, 29, 30]. Although, these neuro-fuzzy systems show the ability of good classifiers, they cannot perform feature selection which is one of the important issues in pattern recognition [31, 32]. Therefore, in this paper, we develop a neuro-fuzzy system based on LVQ, called a learning vector quantization inference classifier, with the capability of feature selection. The details will be described in the next section.

2 Structure of Learning Vector Quantization Inference Classifier (LVQIC)

Our proposed model is based on the LVQ and gradient descent method in such a way that input prototypes and output weights are updated simultaneously. The model can cooperatively extract rules and select features after the training process is done. The structure of the proposed learning vector quantization inference classifier (LVQIC) is shown in Fig. 1.

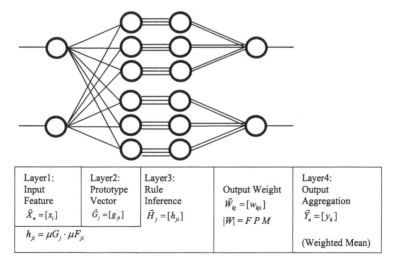

Layer1: Input Feature $\vec{X}_n = [x_i]$	Layer2: Prototype Vector $\vec{G}_j = [g_{ji}]$	Layer3: Rule Inference $\vec{H}_j = [h_{ji}]$	Output Weight $\vec{W}_{kj} = [w_{kji}]$ $|W| = F P M$	Layer4: Output Aggregation $\vec{Y}_n = [y_k]$
$h_{ji} = \mu G_j \cdot \mu F_{ji}$				(Weighted Mean)

Fig. 1. The structure of LVQIC

Suppose there are N input samples, each with F features. In addition, there are M classes, P prototypes per class, and Q prototypes in total. The details of our model are as follows:

Layer 1: Input features. Each feature is normalized to 0 to 1.
Layer 2: The closest prototype and the closest feature are determined. The closest prototype is calculated as

$$\text{for each feature } i, \text{ compute } D_j = \left\| \vec{X}_n - \vec{G}_j \right\|, \tag{1}$$

and

$$j_{win} = \arg\min_j \{D_j\}. \tag{2}$$

Then we need to find a winning prototype of each feature by,

$$\text{for each feature } i, \text{ compute } D_{ji} = \left| x_i - g_{ji} \right|, \ j = 1, \ldots, Q, \tag{3}$$

and

$$j_{win}^{(i)} = \arg\min_j \{D_{ji}\}. \tag{4}$$

Layer 3: There are 2 types of prototype closeness and feature closeness, i.e.,

1. Hard method (winner-takes-all)

$$\mu G(x) = \begin{cases} 1 & if \ x = D_{min} \\ 0 & else \end{cases}, \text{ where } x \in \{D_j\}, \ D_{min} \leq D_j \ \forall j. \tag{5}$$

$$\mu F_i(x) = \begin{cases} 1 & if \ x = D_{min}^{(i)} \\ 0 & else \end{cases}, \text{ where } x \in \{D_{ji}\}, \ D_{min}^{(i)} \leq D_{ji} \ \forall j. \tag{6}$$

2. Soft method (winner-takes-most)

$$\mu G(x) = \frac{D_{max} - x}{D_{max} - D_{min}}, \tag{7}$$

where $x \in \{D_j\}$, $D_{max} \geq D_j$ and $D_{min} \leq D_j, \forall j$,
    and

$$\mu F_i(x) = \frac{D_{max}^{(i)} - x}{D_{max}^{(i)} - D_{min}^{(i)}}, \tag{8}$$

where $x \in \{D_{ji}\}$, $D_{max}^{(i)} \geq D_{ji}$ and $D_{min}^{(i)} \leq D_{ji}, \forall j$.

Both closeness values are accounted as the output of this layer as follow:

$$h_{ji} = \mu G_j \times \mu F_{ji}, \tag{9}$$

where $\mu G_j = \mu G(D_j)$, $\mu F_{ji} = \mu F_i(D_{ji})$. Then the output vector of this layer will be $\vec{H}_j = [h_{ji}]_{i=1}^{F}$.

Layer 4: The output of this layer is calculated as

$$y_k = \sum_{j=1}^{P} \sum_{i=1}^{F} w_{kji} h_{ji}, \tag{10}$$

where $h_{ji} \in [0, 1]$, $y_k \in [0, 1]$, $w_{kji} \in [0, 1]$, $\sum_{j=1}^{P} \sum_{i=1}^{F} w_{kji} = 1$. The output node that gives the maximum output value will be recognized as the class label of the input vector as follow:

$$class = l \quad where \quad y_l \geq y_k \quad \forall k. \tag{11}$$

3 LVQIC Learning Algorithm and Feature Selection

In the learning process [15], we update the winning prototype

$$g_{ji}(t+1) = \begin{cases} g_{ji}(t) + \eta_1 \left[x_i - g_{ji}(t) \right] & \text{if } class(\vec{X}) = class(\vec{G}_{j_{win}}) \\ g_{ji}(t) - \eta_1 \left[x_i - g_{ji}(t) \right] & \text{else} \end{cases}. \tag{12}$$

We also update the winning prototype of each feature i by

$$g_{j_{win}^{(i)} k}(t+1) = \begin{cases} g_{j_{win}^{(i)} k}(t) + \eta_1 \left[x_k - g_{j_{win}^{(i)} k}(t) \right] & \text{if } class(\vec{X}) = class(\vec{G}_{j_{win}^{(i)}}) \\ g_{j_{win}^{(i)} k}(t) - \eta_1 \left[x_k - g_{j_{win}^{(i)} k}(t) \right] & \text{else} \end{cases}. \tag{13}$$

This method lets the winner prototype move by vector adaptation while any winning feature of the other prototypes is changed freely from their other dimensions.

Since the summation of all weights has to be 1, we modify the weight calculation to

$$w_{kji} = \frac{w_{kji}'^2}{\sum_{j=1}^{P} \sum_{i=1}^{F} w_{kji}'^2}. \tag{14}$$

The weight update equation is based on the error, i.e.,

$$E = \frac{1}{2} \sum_{k=1}^{M} (z_k - y_k)^2, \tag{15}$$

where z_k is the desired output, y_k is the calculated output from output node k, and M is the number of output nodes. Therefore, the weight update equation will be

$$w'_{kji}(t+1) = w'_{kji}(t) - \eta_2 \frac{\partial E}{\partial w'_{kji}(t)}. \tag{16}$$

Since,

$$\frac{\partial E}{\partial y_k} = -(z_k - y_k), \tag{17}$$

$$\frac{\partial y_k}{\partial w_{kji}} = h_{ji}, \tag{18}$$

and

$$\frac{\partial w_{kji}}{\partial w'_{kji}} = \frac{2w'_{kji}\left(\left(\sum_{n=1}^{P}\sum_{m=1}^{F} w'^2_{kji}\right) - w'^2_{kji}\right)}{\left(\sum_{n=1}^{P}\sum_{m=1}^{F} w'^2_{knm}\right)^2}, \tag{19}$$

the weights connecting to the winning prototype (j_{win}) are updated by

$$w'_{kji}(t+1) = w'_{kji}(t) + \eta_2(z_k - y_k)h_{ji}\frac{2w'_{kji}\left(\left(\sum_{n=1}^{P}\sum_{m=1}^{F} w'^2_{knm}\right) - w'^2_{kji}\right)}{\left(\sum_{n=1}^{P}\sum_{m=1}^{F} w'^2_{knm}\right)^2} \quad \text{for } i = 1, \ldots, F. \tag{20}$$

Then for each feature, we update the weights connecting to that feature of the winning prototype ($j_{win}^{(i)}$). However, if the winning prototype (j_{win}) is the same as the winning prototype of each feature ($j_{win}^{(i)}$), we only update the weights connecting to that feature for the first time.

After the algorithm is stable, we will then select a good feature set. Since, the weights give us the information regarding informative features. We rearrange the weight values in descendent order. Then we will be able to select feature according to the weight values. The algorithm for feature selection is shown as follows:

- Find $FR_i = \sum_{k=1}^{M} \sum_{j=1}^{Q} w_{kji}$ for all i.
- Initialize $topc = 0$, $slf = 0$, $count = 0$, $FS = ()$ ##no. selected features.
- Repeat
  - Pick i^{th} feature from the remaining features which FR_i is maximum.
  - Keep i^{th} feature in FS and set $count = count + 1$.
  - Set c to the classification rate of the training data set with features in FS
  - If $c > topc$ then $topc = c$, $slf = count$.
- Until no remaining feature.
- Set $FS' = \{k \mid k$ is a feature in the first selected slf features in $FS\}$.
- Set $FS = FS'$. ##FS contains only the first selected slf features.
- Initialize $topc = 0$, $slw = 0$, $count = 0$, $WS = ()$ ## no. selected weights.
  ##Consider only weights w_{kji} connecting to the first selected slf
  ##features in FS.
- Repeat
  - Pick w_{kji} connecting to the remaining features in the FS.
  - Keep weight w_{kji} in WS and set $count = count + 1$.
  - Set c to the classification rate of the training data set with weights in WS
  - If $c > topc$ then $topc = c$, $slw = count$.
- Until no remaining output weight.
- Set $WS' = \{l \mid l$ is a weight in the first selected slw weights in $WS\}$.
- Set $WS = WS'$ ##WS contains only the first selected slw weights.
- Return slf, slw, FS, WS.

The computational complexity of the algorithm is $O(F \times M \times P)$. In order to extract the rule from the system, the prototypes act as the rules set while the closeness value is taken as an inference process. There are 2 rule categories, i.e., crisp rule and fuzzy rule. However, in both categories, we only consider weights w_{kji} that are in WS. For input feature i (x_i) and prototype g_{ji} connecting to w_{kji}, the generated rule can be

Crisp rules: IF $(x_i$ is CLOSE TO $g_{ji})$ THEN class k
Fuzzy rules: IF $w_{kji} \times ($ x_i is CLOSE TO $g_{ji})$ THEN class k

The test vector is assigned to the consequence class of the rule that give maximum output of the antecedence. In the experiment, each feature is normalized using its own minimum and maximum values to eliminate the feature domination effect.

4 Experiment Results

We utilized the data sets from UCI [33], i.e., Iris Plants, Wisconsin Breast Cancer (WBC) [34], Pima Indians Diabetes, Wine, Ionosphere. We also utilized the Colon Tumor and Diffuse Large B-Cell Lymphoma (DLBCL) data sets from Kent Ridge Bio-Medical Dataset [35] and Glioma Tumor (GLI_85) from Feature Selection Datasets at Arizona State University [36]. Table 1 shows the characteristics of each of the 8 data sets. The k-fold cross validation (CV) is used in the experiments of LVQIC. The number of prototypes per class is varied from 1 to 4.

Table 1. Data sets used in this research

Name	Data types	Missing	Classes	Features	Samples
Iris	Real		3	4	150
WBC [34]	Integer	Yes	2	9	699
Pima	Integer, Real	Yes	2	8	768
Wine	Integer, Real		3	13	178
Ionosphere	Integer, Real		2	34	351
Colon	Real		2	2,000	62
DLBCL	Real		2	4,026	47
GLI_85	Real		2	22,283	85

The best validation classification results are shown in Table 2. We can see that the best validation result of the Iris with soft prototype and feature closeness and 2 selected features with 9 rules is 100%, whereas that of the GLI_85 with same closeness scheme and 15 selected features and 23 rules is at 94.12%. However, the WBC best validation result with hard prototype and soft feature closeness and 8 selected features with 14 rules is 100%. The same hard prototype and soft feature closeness scheme also provides the best validation classification results for the Pima with 5 selected features and 5 rules at 81.82%, for the Wine with 9 selected features and 13 rules at 100%, for the Ionosphere with 2 selected features and 2 rules at 97.06%, for the Colon with 17 selected features and 37 rules at 91.67%, and the DLBCL with 16 selected features and rules at 100%.

We can see that the number of selected features is reduced dramatically for the data sets with huge numbers of features, i.e., Colon (from 2,000 to 17 features), DLBCL (from 4,026 to 16 features), and GLI_85 (from 22,283 to 15 features) and the system still provides a good correct classification. The selected features from the best validation test are shown in Table 3. The Example of final rule set of each data set from the model that gives the best validation results are shown in Table 4.

One might want to know how good the performance of the LVQIC is when it is compared with the existing algorithms? We compare our LVQIC indirectly with the existing algorithms for each data set as shown in Tables 5, 6, 7, 8, 9, 10, 11 and 12.

Table 2. Best validation clasification result

Data set	Prototype closeness	Feature closeness	Prototypes	Features	Rule conditions	Classification rate (%)
Iris (10-fold CV)	**soft**	**soft**	**5**	**2**	**9**	**100**
	soft	hard	3	1	3	100
	hard	soft	5	3	7	100
	hard	hard	3	1	3	100
WBC (10-fold CV)	**hard**	**soft**	**3**	**8**	**14**	**100**
	soft	soft	2	7	14	98.57
	hard	hard	3	9	21	98.57
	soft	hard	2	6	7	100
Pima (10-fold CV)	**hard**	**soft**	**2**	**5**	**5**	**81.82**
	soft	soft	4	6	10	80.52
	hard	hard	3	8	23	79.22
	soft	hard	3	7	17	76.62
Wine (10-fold CV)	**hard**	**soft**	**5**	**9**	**13**	**100**
	soft	soft	8	11	35	100
	hard	hard	6	13	73	100
	soft	hard	3	7	13	100
Ionosphere (10-fold CV)	**hard**	**soft**	**2**	**2**	**2**	**97.06**
	hard	hard	2	10	11	97.06
	soft	soft	2	10	13	91.18
	soft	hard	2	2	3	91.18
Colon (5-fold CV)	**hard**	**soft**	**8**	**17**	**37**	**91.67**
	hard	hard	8	27	41	91.67
	soft	soft	8	11	77	83.33
	soft	hard	4	9	11	84.62
DLBCL (5-fold CV)	**hard**	**soft**	**2**	**16**	**16**	**100**
	soft	soft	4	69	103	100
	hard	hard	2	10	10	100
	soft	hard	4	144	415	100
GLI_85 (5-fold CV)	**soft**	**soft**	**4**	**15**	**23**	**94.12**
	hard	hard	5	938	955	93.75
	hard	soft	4	117	122	88.24
	soft	hard	2	11	11	88.24

In the tables, we not only show the best validation results from our algorithm but we also show the average validation results from our algorithm against the other algorithms. We can see that our LVQIC give a better result on all features experiments and selected features experiments on most of the data sets. There is only 1 full feature experiment on the GLI_85 that our algorithm gives a lower classification result than the existing algorithms.

Table 3. Selected features

Data set	Number of selected features	Selected features
Iris	2	{3, 2}
WBC	8	{1, 6, 2, 4, 5, 3, 7, 8}
Pima	5	{1, 0, 2, 6, 7}
Wine	9	{0, 3, 5, 12, 9, 11, 10, 6, 7}
Ionosphere	2	{1, 14}
Colon	17	{1495, 1836, 578, 380, 1947, 583, 1281, 248, 1882, 484, 1609, 1146, 1347, 1261, 1566, 1517, 406}
DLBCL	16	{884, 2447, 3731, 141, 2582, 515, 1356, 2554, 1784, 1949, 2036, 317, 3192, 959, 338, 473}
GLI_85	15	{3041, 7553, 15804, 11744, 4825, 3482, 6265, 13472, 8888, 9641, 9683, 17318, 1406, 3423, 15028}

Table 4. Example of the best validation result rule set for the Iris, WBC, Pima, Wine, Ionosphere, Colon, DLBCL, and GLI_85 data sets

Data set	Fuzzy rules	Crisp rules

Iris	IF 0.2666 (x2 is CLOSE TO 0.0860) THEN class 0
	IF 0.3840 (x3 is CLOSE TO 0.5330) THEN class 1
	IF 0.3817 (x3 is CLOSE TO 0.8075) THEN class 2
	(Class: 0 for iris-setosa, 1 for iris-versicolor, 2 for iris-virginica)
		IF (x2 is CLOSE TO 0.0860) THEN class 0
		IF (x3 is CLOSE TO 0.5330) THEN class 1
		IF (x3 is CLOSE TO 0.8075) THEN class 2
		(Class: 0 for iris-setosa, 1 for iris-versicolor, 2 for iris-virginica)

WBC	IF 0.1395 (x7 is CLOSE TO 0.0438) THEN class 0
	IF 0.1833 (x6 is CLOSE TO 0.6290) THEN class 1
	(Class: 0 for benign, 1 for malignant)
		IF (x7 is CLOSE TO 0.0438) THEN class 0
		IF (x6 is CLOSE TO 0.6290) THEN class 1
		(Class: 0 for benign, 1 for malignant)

Pima	IF 0.2024 (x7 is CLOSE TO 0.1310) THEN class 0
	IF 0.1359 (x0 is CLOSE TO 0.1476) THEN class 1
	(Class: 0 for negative, 1 for positive)
		IF (x7 is CLOSE TO 0.1310) THEN class 0
		IF (x0 is CLOSE TO 0.1476) THEN class 1
		(Class: 0 for negative, 1 for positive)

Wine	IF 0.0956 (x11 is CLOSE TO 0.7192) THEN class 0
	IF 0.0995 (x0 is CLOSE TO 0.3379) THEN class 1
	IF 0.1009 (x0 is CLOSE TO 0.5631) THEN class 2
	(Class: 0 for type1, 1 for type2, 2 for type3)
		IF (x11 is CLOSE TO 0.7192) THEN class 0
		IF (x0 is CLOSE TO 0.3379) THEN class 1
		IF (x0 is CLOSE TO 0.5631) THEN class 2
		(Class: 0 for type1, 1 for type2, 2 for type3)

(continued)

Table 4. (*continued*)

Data set	Fuzzy rules	Crisp rules

Ionosphere	IF 0.0542 (x14 is CLOSE TO 0.8869) THEN class 0
	 IF 0.1331 (x1 is CLOSE TO 0.0000) THEN class 1
	 (Class: 0 for good, 1 for bad)
		IF (x14 is CLOSE TO 0.8869) THEN class 0
		 IF (x1 is CLOSE TO 0.0000) THEN class 1
		 (Class: 0 for good, 1 for bad)

Colon	IF 0.0004 (x406 is CLOSE TO 0.7262)) THEN class 0
	 IF 0.0004 (x1495 is CLOSE TO 0.1870) THEN class 1
	 (Class: 0 for negative, 1 for positive)
		IF (x406 is CLOSE TO 0.7262) THEN class 0
		 IF (x1495 is CLOSE TO 0.1870) THEN class 1
		 (Class: 0 for negative, 1 for positive)

DLBCL	IF 0.000370 (x338 is CLOSE TO 0.5979) THEN class 0
	 IF 0.000368 (x2582 is CLOSE TO 0.4352) THEN class 1
	 (Class: 0 for germinal, 1 for activated)
		IF (x338 is CLOSE TO 0.5979) THEN class 0
		 IF (x2582 is CLOSE TO 0.4352) THEN class 1
		 (Class: 0 for germinal, 1 for activated)

GLI_85	IF 0.000065 (x1406 is CLOSE TO 0.6493) THEN class 0
	 IF 0.000063 (x7553 is CLOSE TO 0.1143) THEN class 1
	 (Class: 0 for grade III, 1 for grade IV)
		IF (x1406 is CLOSE TO 0.6493) THEN class 0
		 IF (x7553 is CLOSE TO 0.1143) THEN class 1
		 (Class: 0 for grade III, 1 for grade IV)

Table 5. The best validation result rule set for the IRIS data set

Method	CV	Classification rate (%)				Number of	
		All features	Selected features	Crisp rule	Fuzzy rule	Features	Rules
NEFCLASS [8]	50/50	97.3	96	–	–	**2**	**3**
SANFIS [9]	50/50	97.47	–	–	–	4	**3**
Novel NFS [11]	10 folds	95.33	93.33	86	–	**2**	**3**
MLP2LN [37]	100/100	–	–	98	–	**2**	**3**
DWLVQ1 [25]	10 folds	94.33	–	–	–	–	–
DWLVQ2 [25]	10 folds	95.53	–	–	–	–	–
SGLVQ [14]	67/33	90.2	–	98	–	4	8
Neuro-fuzzy scheme [10]	100/100	–	97.33	–	–	**2**	5
LVQIC (soft/soft) best	10 folds	**100**	**100**	**100**	**100**	**2**	9
LVQIC (soft/soft) average	10 folds	96	95.33	96	95.33		

Table 6. The best validation result rule set for the WBC data set

Method	CV	Classification rate (%)				Number of	
		All features	Selected features	Crisp rule	Fuzzy rule	Features	Rules
Multisurface [38]	67/33	95.9	–	–	–	–	–
NEFCLASS [8]	50/50	92.7	96.5	–	–	9	4
SANFIS [9]	50/50	96.28	–	–	–	9	2
Novel NFS [11]	10 folds	94.79	89.06	95.31	–	4	2
WLVQ [26]	10 folds	96	–	–	–	–	–
correlation criteria - SVM [31]	?	97	97	–	–	9	–
MI - SVM [31]	?	96	96	–	–	7	–
SFFS - SVM [31]	?	96	97	–	–	7	–
SFFS - RBF [31]	?	97	97	–	–	**2**	–
CHCGA - SVM [31]	?	–	97.36	–	–	5	–
CHCGA - RBF [31]	?	–	96.77	–	–	5	–
LVQIC (hard/soft) best	10 folds	**100**	**100**	**100**	**100**	8	14
LVQIC (hard/soft) average	10 folds	95.85	96.42	96.42	96.42		

Table 7. The best validation result rule set for the Pima data set

Method	CV	Classification rate (%)				Number of	
		All features	Selected features	Crisp rule	Fuzzy rule	Features	Rules
Novel NFS [11]	10 folds	71.61	69.79	72.01	–	5	2
WLVQ [26]	10 folds	76	–	–	–	–	–
correlation criteria - SVM [31]	?	79	80	–	–	5	–
MI - SVM [31]	?	79	79	–	–	8	–
SFFS - SVM [31]	?	79	80	–	–	5	–
SFFS - RBF [31]	?	68	78	–	–	**3**	–
CHCGA - SVM [31]	?	–	80.47	–	–	7	–
CHCGA - RBF [31]	?	–	76.82	–	–	7	–
LVQIC (hard/soft) best	10 folds	**81.82**	**81.82**	**81.82**	**81.82**	5	5
LVQIC (hard/soft) average	10 folds	72.79	76.04	76.04	76.04		

Table 8. The best validation result rule set for the Wine data set

Method	CV	Classification rate (%)				Number of	
		All features	Selected features	Crisp rule	Fuzzy rule	Features	Rules
SANFIS [9]	100/100	99.44	–	–	–	13	**3**
DWLVQ1 [25]	10 folds	96.84	–	–	–	–	–
DWLVQ2 [25]	10 folds	93.89	–	–	–	–	–
LVQIC (hard/soft) best	10 folds	**100**	**100**	**100**	**100**	**9**	13
LVQIC (hard/soft) average	10 folds	96.1	97.7	97.7	97.7		

Table 9. The best validation result rule set for the Ionosphere data set

Method	CV	Classification rate (%)				Number of	
		All features	Selected features	Crisp rule	Fuzzy rule	Features	Rules
DWLVQ1 [25]	10 folds	87.17	–	–	–	–	–
DWLVQ2 [25]	10 folds	88.22	–	–	–	–	–
WLVQ [26]	10 folds	87	–	–	–	–	–
correlation criteria - SVM [31]	?	90	92	–	–	15	–
MI - SVM [31]	?	90	90	–	–	15	–
SFFS - SVM [31]	?	90	95	–	–	9	–
SFFS - RBF [31]	?	84	94	–	–	10	–
CHCGA - SVM [31]	?	–	94.29	–	–	16	–
CHCGA - RBF [31]	?	–	94.29	–	–	16	–
LVQIC (hard/soft) best	10 folds	**97.06**	**97.06**	**97.06**	**97.06**	**2**	**2**
LVQIC (hard/soft) average	10 folds	87.77	86.66	86.66	86.66		

Table 10. The best validation result rule set for the Colon data set

Method	CV	Classification rate (%)				Number of	
		All features	Selected features	Crisp rule	Fuzzy rule	Features	Rules
CFS - naive Bayes [32]	5 folds	55	85	–	–	24	–
FCBF - naive Bayes [32]	5 folds	55	80	–	–	**14**	–
INT - naive Bayes [32]	5 folds	55	77	–	–	**14**	–
IG - naive Bayes [32]	5 folds	55	77	–	–	50	–
ReliefF - naive Bayes [32]	5 folds	55	84	–	–	50	–

(*continued*)

Table 10. (*continued*)

Method	CV	Classification rate (%)				Number of	
		All features	Selected features	Crisp rule	Fuzzy rule	Features	Rules
SVM-RFE - naive Bayes [32]	5 folds	55	76	–	–	50	–
mRMR - naive Bayes [32]	5 folds	55	80	–	–	50	–
CFS - SVM [32]	5 folds	77	81	–	–	24	–
FCBF - SVM [32]	5 folds	77	84	–	–	**14**	–
INT - SVM [32]	5 folds	77	81	–	–	**14**	–
IG - SVM [32]	5 folds	77	85	–	–	50	–
ReliefF - SVM [32]	5 folds	77	85	–	–	50	–
SVM-RFE - SVM [32]	5 folds	77	71	–	–	50	–
mRMR - SVM [32]	5 folds	77	84	–	–	50	–
LVQIC (hard/soft) best	5 folds	**91.67**	**91.67**	**91.67**	**91.67**	17	**37**
LVQIC (hard/soft) average	5 folds	77.69	77.69	77.69	77.69		

Table 11. The best validation result rule set for the DLBCL data set

Method	CV	Classification rate (%)				Number of	
		All features	Selected features	Crisp rule	Fuzzy rule	Features	Rules
CFS - naive Bayes [32]	5 folds	92	90	–	–	61	–
FCBF - naive Bayes [32]	5 folds	92	90	–	–	35	–
INT - naive Bayes [32]	5 folds	92	90	–	–	45	–
IG - naive Bayes [32]	5 folds	92	92	–	–	50	–
ReliefF - naive Bayes [32]	5 folds	92	92	–	–	50	–
SVM-RFE - naive Bayes [32]	5 folds	92	92	–	–	50	–
mRMR - naive Bayes [32]	5 folds	92	94	–	–	50	–
CFS - SVM [32]	5 folds	96	88	–	–	61	–
FCBF - SVM [32]	5 folds	96	81	–	–	35	–
INT - SVM [32]	5 folds	96	88	–	–	45	–
IG - SVM [32]	5 folds	96	94	–	–	50	–
ReliefF - SVM [32]	5 folds	96	93	–	–	50	–
SVM-RFE - SVM [32]	5 folds	96	88	–	–	50	–

(*continued*)

Table 11. (*continued*)

Method	CV	Classification rate (%)				Number of	
		All features	Selected features	Crisp rule	Fuzzy rule	Features	Rules
mRMR - SVM [32]	5 folds	96	96	–	–	50	–
LVQIC (hard/soft) best	5 folds	**100**	**100**	**100**	**100**	**16**	**16**
LVQIC (hard/soft) average	5 folds	95.78	89.56	89.56	89.56		

Table 12. The best validation result rule set for the GLI_85 data set

Method	CV	Classification rate (%)				Number of	
		All features	Selected features	Crisp rule	Fuzzy rule	Features	Rules
CFS - naive Bayes [32]	5 folds	84	82	–	–	141	–
FCBF - naive Bayes [32]	5 folds	84	85	–	–	116	–
INT - naive Bayes [32]	5 folds	84	82	–	–	117	–
IG - naive Bayes [32]	5 folds	84	85	–	–	50	–
ReliefF - naive Bayes [32]	5 folds	84	89	–	–	50	–
SVM-RFE - naive Bayes [32]	5 folds	84	88	–	–	50	–
mRMR - naive Bayes [32]	5 folds	84	80	–	–	50	–
CFS - SVM [32]	5 folds	**92**	88	–	–	141	–
FCBF - SVM [32]	5 folds	**92**	87	–	–	116	–
INT - SVM [32]	5 folds	**92**	88	–	–	117	–
IG - SVM [32]	5 folds	**92**	86	–	–	50	–
ReliefF - SVM [32]	5 folds	**92**	89	–	–	50	–
SVM-RFE - SVM [32]	5 folds	**92**	89	–	–	50	–
mRMR - SVM [32]	5 folds	**92**	89	–	–	50	–
LVQIC (soft/soft) best	5 folds	70.59	**94.12**	**88.24**	**88.24**	**15**	**23**
LVQIC (soft/soft) average	5 folds	71.59	86.9	78.65	78.65		

5   Conclusion

A neuro-fuzzy system has been a popular tool in many applications. Most of algorithms used in a neuro-fuzzy system are based on the gradient descent or back-propagation on the multi-layer neurons scheme. In this paper, we incorporate learning vector quantization in a neuro-fuzzy system. The update equation is based on the learning vector quantization algorithm and the gradient descent technique. We then select a good feature set based on the final weights. To show the performance of the algorithm, we implement our algorithm on 8 standard data sets, i.e., Iris Plants, Wisconsin Breast Cancer (WBC), Pima Indians Diabetes, Wine, Ionosphere, Colon Tumor, Diffuse Large B-Cell Lymphoma (DLBCL) and Glioma Tumor (GLI_85). We also indirectly compare our results with the results from several existing algorithms. We found out that our algorithm performs better than those existing algorithms. It is able to reduce the numbers of features dramatically and still produces the good classification results.

Acknowledgment. The breast cancer database was obtained from the University of Wisconsin Hospitals, Madison by Dr. William H. Wolberg.

References

1. Jang, J.-S.R., Sun, C.-T., Mizutani, E.: Neuro-Fuzzy and Soft Computing: A Computational Approach to Learning and Machine Intelligence. Prentice-Hall, Englewood Cliffs (1997)
2. Mitra, S., Hayashi, Y.: Neuro-fuzzy rule generation: survey in soft computing framework. IEEE Trans. Neural Netw. 11(3), 748–768 (2000)
3. Abraham, A.: Neuro fuzzy systems: state-of-the-art modeling techniques. In: International Work-Conference on Artificial and Natural Neural Networks, pp. 269–276 (2001)
4. Vieira, J., Dias, F.M., Mota, A.: Neuro-fuzzy systems: a survey. WSEAS Trans. Syst. 3(2), 414–419 (2004)
5. Keller, J., Hayashi, Y., Chen, Z.: Additive hybrid networks for fuzzy logic inference. Fuzzy Sets Syst. 66(3), 307–313 (1994)
6. Keller, J.M., Hayashi, Y., Chen, Z.: Interpretation of nodes in networks for fuzzy logic. In: IEEE International Conference on Fuzzy Systems, vol. 2, pp. 1203–1207 (1993)
7. Sun, C.-T., Jang, J.-S.: A neuro-fuzzy classifier and its applications. IEEE International Conference Fuzzy Systems, vol. 1, pp. 94–98 (1993)
8. Nauck, D., Nauck, U., Kruse, R.: Generating classification rules with the neuro-fuzzy system NEFCLASS. In: IEEE Conference on N.A. Fuzzy Information Processing Society, pp. 466–470 (1996)
9. Wang, J.-S., Lee, C.S.G.: Self-adaptive neuro-fuzzy inference systems for classification applications. In: IEEE International Conference on Fuzzy Systems, vol. 10, no. 6, pp. 790–802 (2002)
10. Chakraborty, D., Pal, N.R.: A neuro-fuzzy scheme for simultaneous feature selection and fuzzy rule-based classification. IEEE Trans. Neural Netw. 15(1), 110–123 (2004)
11. Eiamkanitchat, N., Theera-Umpon, N., Auephanwiriyakul, S.: A novel neuro-fuzzy method for linguistic feature selection and rule-based classification. In: IEEE International Conference on Computer and Automation Engineering, vol. 2, pp. 247–252 (2010)
12. Eiamkanitchat, N., Theera-Umpon, N., Auephanwiriyakul, S.: On feature selection and rule extraction for high dimensional data: a case of diffuse large B-cell lymphomas microarrays classification. Math. Probl. Eng. 2015, 1–12 (2015)
13. Mitra, S., Pal, S.K.: Fuzzy self-organization, inferencing, and rule generation. IEEE Trans. Syst. Man Cybern. Part A Syst. Hum. 26(5), 608–620 (1996)
14. Mikami, D., Hagiwara, M.: Self-growing learning vector quantization with additional learning and rule extraction abilities. In: IEEE International Conference on Systems, Man, and Cybernetics, vol. 4, pp. 2895–2900 (2000)

15. Kohonen, T.: Improved versions of learning vector quantization. In: IEEE International Joint Conference Neural Networks, vol. 1, pp. 545–550 (1990)
16. Bezdek, J.C., Pal, N.R.: Two soft relatives of learning vector quantization. Pergamon Neural Netw. **8**(5), 729–743 (1995)
17. Karayiannis, N.B., Pai, P.-I.: Fuzzy algorithms for learning vector quantization. IEEE Trans. Neural Netw. **7**(5), 1196–1211 (2000)
18. Wu, K.-L., Yang, M.-S.: A fuzzy-soft learning vector quantization. Elsevier Neurocomputing **55**(3–4), 681–697 (2003)
19. Ghalehnoie, M., Akbarzadeh-T, M.R., Naghibi-S, M.B.: A novel batch training algorithm for learning vector quantization networks using soft-labeled training data and prototypes. In: IEEE Iranian Conference on Fuzzy Systems, pp. 1–6 (2013)
20. Chung, F.-L., Lee, T.: Fuzzy learning vector quantization. In: Proceedings IEEE International Joint Conference Neural Networks, vol. 3, pp. 2739–2743 (1993)
21. Chung, F.-L., Lee, T.: A fuzzy learning model for membership function estimation and pattern classification. In: Proceedings of the IEEE International Conference on World Congress on Computational Intelligence, vol. 1, pp. 426–431 (1994)
22. Hammer, B., Villmann, T.: Generalized relevance learning vector quantization. Pergamon Neural Netw. **15**(8–9), 1059–1068 (2002)
23. Vakil-Baghmisheh, M.-T., Pavešić, N.: Premature clustering phenomenon and new training algorithmsfor LVQ. Pergamon Pattern Recogn. **36**(8), 1901–1912 (2003)
24. Shui-sheng, Z., Wei-weia, W., Li-huab, Z.: A new technique for generalized learning vector quantization algorithm. Elsevier Image Vis. Comput. **24**(7), 649–655 (2006)
25. Lv, C., An, X., Liu, Z., Zhao, Q.: Dual weight learning vector quantization. In: IEEE International Conference on Signal Processing, pp. 1722–1725 (2008)
26. Blachnik, M., Duch, W.: Improving accuracy of LVQ algorithm by instance weighting. In: Diamantaras, K., Duch, W., Iliadis, L.S. (eds.) ICANN 2010. LNCS, vol. 6354, pp. 257–266. Springer, Heidelberg (2010). doi:10.1007/978-3-642-15825-4_31
27. Villmann, T., Haase, S., Kaden, M.: Kernelized vector quantization in gradient-descent learning. Elsevier Neurocomputing **147**(1), 83–95 (2015)
28. Cruz-Vega, I., Escalante, H.J.: Improved learning rule for LVQ based on granular computing. In: Carrasco-Ochoa, J.A., Martínez-Trinidad, J.F., Sossa-Azuela, J.H., Olvera López, J.A., Famili, F. (eds.) MCPR 2015. LNCS, vol. 9116, pp. 54–63. Springer, Heidelberg (2015). doi:10.1007/978-3-319-19264-2_6
29. Kusumoputro, B., Budiarto, H., Jatmiko, W.: Fuzzy-neuro LVQ and its comparison with fuzzy algorithm LVQ in artificial odor discrimination system. ISA Trans. **41**, 395–407 (2002)
30. Lin, W.-S., Tsai, C.-H.: Self-organizing fuzzy control of multi-variable systems using learning vector quantization network. Elsevier Fuzzy Sets Syst. **124**(2), 197–212 (2001)
31. Chandrashekar, G., Sahin, F.: A survey on feature selection methods. Elsevier Comput. Electr. Eng. **40**(1), 16–28 (2014)
32. Bolón-Canedo, V., Sánchez-Maroño, N., Alonso-Betanzos, A., Benítez, J.M., Herrera, F.: A review of microarray datasets and applied feature selection methods. Elsevier Inf. Sci. **282**, 111–135 (2014)
33. UC Irvine Machine Learning Repository. http://archive.ics.uci.edu/ml/index.html
34. Wolberg, W.H., Mangasarian, O.L.: Multisurface method of pattern separation for medical diagnosis applied to breast cytology. In: Proceedings of the National Academy of Sciences, vol. 87, pp. 9193–9196 (1990)
35. Kent Ridge Bio-Medical Dataset. http://datam.i2r.a-star.edu.sg/datasets/krbd
36. Feature Selection Datasets at Arizona State University. http://featureselection.asu.edu/datasets.php

37. Duch, W., Adamczak, R., Grabczewski, K.: Extraction of crisp logical rules using constructive constrained back propagation networks. In: IEEE International Conference on Neural Network, vol. 4, pp. 2384–2389 (1997)
38. Klir, G.J., Yuan, B.: Fuzzy sets and fuzzy logic: theory and application. Prentice-Hall, New York (1997)

Mining Periodic High Utility Sequential Patterns

Tai Dinh[1], Van-Nam Huynh[2], and Bac Le[3(⊠)]

[1] Faculty of Information Technology,
HCM City Industry and Trade College, Ho Chi Minh City, Vietnam
duytai@fit-hitu.edu.vn
[2] Japan Advanced Institute of Science and Technology, Ishikawa, Japan
huynh@jaist.ac.jp
[3] Department of Computer Science, University of Science,
VNU-HCMC, Ho Chi Minh City, Vietnam
lhbac@fit.hcmus.edu.vn

Abstract. The aim of mining High Utility Sequential Patterns (HUSPs) is to discover sequential patterns having a high utility (e.g. high profit) based on a user-specified minimum utility threshold. The existing algorithms for mining HUSPs are capable of discovering the complete set of all HUSPs. However, they usually generate a large number of patterns which may be redundant in some cases. The periodic appearance of HUSPs can be regarded as an important criterion to consider the purchase behaviour of customers and measure the interestingness of HUSPs which is very common in real-life applications. In this paper, we focus on periodic high utility sequential patterns (PHUSPs) that are periodically bought by customers and generate a high profit. We proposed an algorithm named PHUSPM (Periodic High Utility Sequential Patterns Miner) to efficiently discover all PHUSPs. The experimental evaluation was performed on six large-scale datasets to evaluate the performance of PHUSPM in terms of execution time, memory usages and scalability. The experimental results show that the PHUSPM algorithm is very efficient by reducing the search space and discarding a considerable amount of non-PHUSPs.

Keywords: Data mining · Periodic high utility sequential patterns · Periodic patterns · High utility sequential patterns

1 Introduction

Sequential Patterns Mining (SPM) was first introduced by Agrawal (Agrawal et al. 1995). It has emerged as an important topic in data mining and knowledge discovery. The selection of interesting patterns is usually based on some measures such as frequency/support-based mining or utility-based mining. Frequent sequential pattern mining (FSPM) is widely seen as fundamental task of discovering informative knowledge in sequential databases. For simplicity, sequential patterns mining seeks to discover frequent subsequences as patterns in a sequence database (Srikant and Agrawal 1996), SPADE (Zaki 2001), PrefixSpan

© Springer International Publishing AG 2017
N.T. Nguyen et al. (Eds.): ACIIDS 2017, Part I, LNAI 10191, pp. 545–555, 2017.
DOI: 10.1007/978-3-319-54472-4_51

(Pei et al. 2001), SPAM (Ayres et al. 2002), FreeSpan (Han et al. 2000), PAID (Yang et al. 2006), LAPIN (Yang et al. 2007), CM-SPADE & CM-SPAM (Fournier-Viger et al. 2014)). FSPM is useful for promotions, makerting, retail store, telecommunications, web access pattern analysis, weather prediction. However, in practice, most frequent sequential patterns may not be informative for business decision-making, since they do not show the business value and impact. In some cases, such as fraud detection, some truly interesting sequences may be filtered because of their low frequency [1]. High utility sequential patterns mining (HUSPM) was introduced to tackle these problems (UL & US (Ahmed et al. 2010), Uspan (Yin et al. 2012), UP-Span (Wu et al. 2013), PHUS (Lan et al. 2014)). HUSPM considers the case where items can appear more than once in each sequence and where each item has a quality/weight (e.g. unit profit). Therefore, it can be used to discover sequential patterns having a high utility (e.g. high profit) based on a user-specified minimum utility threshold. HUSPM is more difficult and challenging than FSPM because the utility of a sequential pattern is neither monotonic or anti-monotonic, i.e. a high utility sequential pattern (HUSP) may have a superset or subset with lower, equal or higher utility. Thus the techniques to prune the search space developed in FSPM based on the anti-monotonicity of the support cannot be directly applied to HUSPM. HUSPM is able to discover the complete set of high utility sequential patterns with a pre-defined minimum utility threshold. However, if the minimum utility is too small, a huge number of patterns are generated and most of them might be found insignificant for applications or user requirements. Hence, several techniques such as top-k pattern mining, closed pattern mining, maximal pattern mining and correlated pattern mining are useful in discovering HUSPs of special interest. In retail store, among all high utility sold products, the managers may be interested in the regularly sold products compared to the rest. Detecting these purchase patterns is useful to better understand the behavior of customers and thus adapt marketing strategies [2]. For stock market, the set of high stocks indices that rise periodically may be of special interest to companies and individuals. The above examples show that the occurrence periodicity plays an important role in discovering the interestingness of HUSPs. It means we have to find the way to determine whether a high utility sequential pattern occurs periodically, irregularly, or mostly in specific time interval in a sequence database. In this paper, we integrated the concept of periodicity to HUSPM. The goal is to efficently discover all sequential patterns that are bought together periodically and generate a high profit. The contributions of this paper are threefold. We first combine the concept of periodic patterns with the concept of HUSPs to define a new type of patterns named periodic high utility sequential patterns (PHUSPs). Second, we proposed an efficient algorithm named PHUSPM (Periodic High Utility Sequential Patterns Miner) to discover all PHUSPs in a sequential database. Third, an extensive experimental evaluation is conducted to evaluate the algorithm in term of execution time, memory usage and scalability. The rest of this paper is organized as follows. Sections 2, 3, 4 and 5 respectively presents the background, the proposed algorithm, the experimental evaluation and the conclusion.

2   The Background

2.1   Related Work

High Utility Sequential Patterns Mining Algorithm. High-utility sequential patterns mining is an important task in data mining. Many algorithms have been proposed recently. In 2010, Ahmed et al. proposed UL and US algorithms for mining high utility sequential patterns where different significance values are assigned to each item. PHUS (Lan et al. 2014) use a maximum utility measure and a sequence-utility upper-bound (SUUB) model to mine HUSPs. Recently, Yin et al. proposed an efficient algorithm for mining HUSPs called USpan [1]. USpan relies on a lexicographic q-sequence tree (LQS-Tree), two concatenation mechanisms (called I-Concatenation and S-Concatenation) and two pruning strategies (Width Pruning and Depth Pruning) for mining HUSPs efficiently. The input of USpan is a database SDB and a given minimum utility threshold ξ. USpan use its depth pruning strategy to determine if a node is a leaf or not. If it is a leaf, the algorithm resumes the search from the parent node. Otherwise, USpan collects all promising items and puts them into two separate lists. USpan utilizes its width pruning to determine which unpromising items from these respective lists should be ignored for the rest of the mining process. USpan next uses the I-Concatenation and S-Concatenation mechanisms to calculate the utilities of sequences and output each pattern that is a high utility sequential pattern. USpan then calls itself recursively to explore other parts of the lexicographic q-sequence tree. In this paper, we adapt the concept of periodic patterns mining to USpan to propose the PHUSPM algorithm to achieve the purpose of mining periodic high utility sequential patterns.

Periodic Patterns Mining Algorithm. In the field of frequent-based patterns mining, some algorithms have been proposed for mining periodic frequent patterns (PFPs) in transaction databases [3–5], MTKPP (Amphawan et al. 2009) and in sequence database such as Max-Subpattern Hit-Set (Han et al. 1999), MAPB and MAPD (Wu et al. 2014), AP-PrefixspanM (Yu et al. 2015). In the field of utility-based patterns mining, in 2016, Fournier-Viger et al. proposed an algorithm named PHM [2] for mining periodic high utility itemset on transaction database. In this work, the authors defined a new type of pattern named periodic high utility itemset by combined the concept of periodic itemsets with the concept of high utility itemsets. They also introduced two novel measures called *minimum periodicity* and *averge periodicity* to assess the periodicity of patterns more precisely. High utility sequential pattern mining is much more diffirent and challenging than high utility itemsets mining (HUIM) and frequent sequential patterns mining (FSPM) since the downward closure property does not hold in utility-based sequence mining [1]. Therefore, we cannot apply the existing algorithms for HUIM and FSPM to mine HUSPs. More specifically, the algorithms for mining periodic high utilty itemsets, mining periodic frequent itemsets, mining periodic frequent sequential patterns can not be used for mining periodic high utility sequential patterns. In this paper, we proposed a new type of pattern named periodic high utility sequential patterns. The next section will present the framework for periodic high utility sequential pattern mining.

2.2 Problem Statement

The problem of HUSPM is defined as follows [1], UL & US (Ahmed et al. 2010), PHUS (Lan et al. 2014). Let $I = \{i_1, i_2, ..., i_n\}$ be a set of items. Let $SDB = \{s_1, s_2, ..., s_t\}$ be a quantitative sequence database or q-sequence database, such that SDB is a set of tuples of the form $\langle sid, s \rangle$ where s is a q-sequence and sid is its unique identifier (s_k means the sid of q-sequence s is k). A quantitative sequence or a q-sequence s is an ordered list of q-itemsets $s = \langle X_1 X_2 ... X_m \rangle$ where $X_k (1 \leq k \leq m)$ is a q-itemset. A quantitative itemset or q-itemset $X = [(i_1, q_1)(i_2, q_2)...(i_n, q_n)]$ is a set of one or more q-items where (i_k, q_k) is a q-item for $(1 \leq k \leq n)$. In the following, if a q-itemset contains only one q-item then brackets are omitted for the sake of brevity. Furthermore, without loss of generality, assume that q-items in a q-itemset are sorted according to a total order $\succ$ (e.g. $\succ$ is the alphabetical order). A quantitative item or q-item is a pair of the form (i, q) where $i \in I$ and q is a positive number representing the purchase quantity of i (also called internal utility), the quantity of a q-item i in a q-sequence s is denoted as $q(i, s)$. For the sake of readability, internal utility values are shown as integer values beside each item in itemsets/transactions of each sequence. Each item $i_k \in I$ $(1 \leq k \leq n)$ is associated with a weight (also called external utility) denoted as $p(i_k)$ representing the unit profit or importance of i_k. Tables 1 and 2 [1] are used for running our example. Table 1 shows the items and their respective weights or profit (quality) appearing in an online retail store. Specifically, the price (external utility) of each item a, b, c, d, e and f is respectively 2, 5, 4, 3, 1 and 1. Table 2 collects five shopping q-sequences with quantities. Each q-itemset/transaction in the q-sequence consists of one to multiple q-items, and each q-item is associated with a quantity showing how many of this item were purchased. For example, the q-sequence s_4 shows 3 q-itemsets/transactions $[(b, 2) (e, 2)], [(a, 7) (d, 3)], [(a, 4) (b, 1) (e, 2)]$ in which the quantity purchased (internal utility) of q-item a in the second q-itemset and the third q-itemset are respectively 7 and 4. The following are some definitions for utility-based sequential patterns mining [1] and periodic patterns mining [2,3].

The utility of a q-item (i, q) in a q-sequence s is denoted and defined as $u(i, q) = p(i) \times q(i)$. The utility of a q-itemset X in a q-sequence s is denoted and defined as $u(X) = \sum_{k=1}^{n} u(i_k, q_k)$. The utility of a q-sequence s is denoted and defined as $u(s) = \sum_{k=1}^{n} u(X_k)$.

Definition 1 (Q-Itemset Containing). *Given two q-itemsets $X_a = [(i_{a_1}, q_{a_1})$ $(i_{a_2}, q_{a_2})... (i_{a_n}, q_{a_n})]$ and $X_b = [(i_{b_1}, q_{b_1})$ $(i_{b_2}, q_{b_2})... (i_{b_m}, q_{b_m})]$. The q-itemset X_b is said to contain X_a (denoted as $X_a \subseteq X_b$) iff for any integer $k \in [1, n]$, there exists an integer $j \in [1, m]$ such that $i_{a_k} = i_{b_j}$ and $q_{a_k} = q_{b_j}$.*

Table 1. External utility values

Item	Quality
a	2
b	5
c	4
d	3
e	1
f	1

Table 2. A sequence database

SID	Q-sequence
1	$\langle (e,5)[(c,2)(f,1)](b,2)\rangle$
2	$\langle [(a,2)(e,6)][(a,1)(b,1)(c,2)][(a,2)(d,3)(e,3)]\rangle$
3	$\langle (c,1)[(a,6)(d,3)(e,2)]\rangle$
4	$\langle [(b,2)(e,2)][(a,7)(d,3)][(a,4)(b,1)(e,2)]\rangle$
5	$\langle [(b,2)(e,3)][(a,6)(e,3)][(a,2)(b,1)]\rangle$

Definition 2 (Q-Sequence Containing). *Given a q-sequence $s = \langle X_1, X_2,..., X_n\rangle$ and a q-sequence $s' = \langle X_1', X_2',...,X_{n'}'\rangle$. The q-sequence s is said to contain s' (denoted as $s' \subseteq s$) iff there exist integers $1 \leq j_1 < j_2 < ... < j_{n'} \leq n$ such that $X_k' \subseteq X_{j_k}$ for any integer $k \in [1, n']$. Moreover, iff $s' \subseteq s$, s' is said to be a q-subsequence of s and s is a q-supersequence of s'.*

Definition 3 (Matching). *Given a q-sequence $s = \langle (s_1, q_1)(s_2, q_2)...(s_n, q_n)\rangle$ and a sequence $t = \langle t_1 t_2 ... t_m\rangle$, s is said to match t iff $n = m$ and $s_k = t_k$ for $1 \leq k \leq n$, denoted as $t \sim s$.*

Definition 4 (Sequence Utility). *The sequence utility of a sequence $t = \langle t_1, t_2, ..., t_m\rangle$ in a q-sequence $s = \langle X_1, X_2, ..., X_n\rangle$ is denoted and defined as $v(t, s) = \bigcup_{s' \sim t \land s' \subseteq s} u(s')$. The utility of t in a q-sequence database SDB is denoted and defined as $v(t)$, which is also a utility set: $v(t) = \bigcup_{s \in SDB} v(t, s)$.*

Definition 5 (Sequence Weighted Utilization- SWU). *SWU of a sequence t in SDB is denoted and defined as $SWU(t) = \bigcup_{s' \sim t \land s' \subseteq s \land s \subseteq SDB} u(s)$.*

Definition 6 (High Utility Sequential Pattern). *In utility-based sequence mining, a sequence may match in many different ways with a given q-sequence as mentioned in Definition 3, the utility of a sequence can also be defined in many different ways, based on various utility functions [1], UL and US (Ahmed et al. 2010), PHUS (Lan et al. 2014). In this paper, the maximum utility function is used to calculate the utility of sequential patterns, as proposed in [1]. The maximum utility of a sequence t is denoted and defined as $u_{\max}(t) = \sum \max\{u(s')|s' \sim t \land s' \subseteq s \land s \in S\}$. A sequence t is said to be a high utility sequential pattern if $u_{\max}(t) \geq \xi$, where ξ is a given user-specified minimum utility threshold.*

The problem of *mining high utility sequential patterns* is to discover all high-utility sequential patterns (*HUSPs*). For instance, if *minutil* = 55, the complete set of HUSPs in the quantitative sequence database shown in Table 2 is

$\langle (be)(a)(ab)\rangle$: 73, $\langle (b)(a)(ab)\rangle$: 68, $\langle (be)(a)(a)\rangle$: 63, $\langle (be)(a)(b)\rangle$: 61, $\langle (a)(ab)\rangle$: 59, $\langle (b)(a)(a)\rangle$: 58, $\langle (b)(ae)\rangle$: 57, $\langle (ad)\rangle$: 57, $\langle (b)(a)(b)\rangle$: 56, $\langle (b)(a)\rangle$: 55, $\langle (e)(a)(a)\rangle$: 55 where each HUSP is annotated with its utility.

The conceptual framework of periodic pattern has been proposed for periodic frequent patterns mining in transaction databases [3–5] or sequence database Max-Subpattern Hit-Set (Han et al. 1999), MAPB and MAPD (Wu et al. 2014), AP-PrefixspanM (Yu et al. 2015), and periodic high utility itemsets [2]. Here, we base on the definitions in the previous works [2,3] to build the framework for periodic high utility sequential patterns mining.

Definition 7 (Periods of Sequence). *Given a q-sequence database* $SDB = \{s_1, s_2, ..., s_n\}$ *and a sequence t. The set of q-sequences containing t is denoted as* $X(t) = \{s_{x_1}, s_{x_2}, ..., s_{x_k}\}$ *where* $1 \leq x_1 < x_2 < ... < x_k \leq n$. *Two q-sequences* s_p *such that* $t \sim s' \wedge s' \subseteq s_p \wedge s_p \in X(t)$ *and* s_q *such that* $t \sim s' \wedge s' \subseteq s_q \wedge s_q \in X(t)$ *are said to be consecutive in relation to t if there does not exist a q-sequence* $s_r \in X(t)$ *such that* $p < r < q$. *The periodic of two consecutive q-sequence* s_p *and* s_q *is denoted and defined as* $pe(s_p, s_q) = (q-p)$ *which means the number of q-sequences between* s_p *and* s_q. *The periods of t is a list of periods denoted and defined as* $ps(t) = \{x_1 - x_0, x_2 - x_1, x_3 - x_2, ..., x_k - x_{k-1}, x_{k+1} - x_k\}$, *where* x_0 *and* x_{k+1} *are constants defined as* $x_0 = 0$ *and* $x_{k+1} = n$. *So* $ps(t) = \bigcup_{1 \leq z \leq k+1} (x_z - x_{z-1})$. *The maximum periodicity, minimum periodicity and average periodicity of a sequence t is denoted and defined respectively as* $maxper(t) = max(ps(t))$, $minper(t) = min(ps(t))$, $avgper(t) = \sum x \in ps(t)/|ps(t)|$.

For instance, given the sequence $\langle (ae)\rangle$, the list of q-sequences containing $\langle (ae)\rangle$ is $X(\langle (ae)\rangle) = \{s_2, s_3, s_4, s_5\}$. The periods of $\langle (ae)\rangle$ are $ps(\langle (ae)\rangle) = \{2, 1, 1, 1, 0\}$. So the $maxper(\langle (ae)\rangle) = 2$, the $minper(\langle (ae)\rangle) = 0$ and the $avgper(\langle (ae)\rangle) = 1$.

Property 1 (Relationship Between Average Periodicity and Support). Given a sequence t appearing in q-sequence database SDB. An alternative and equivalent way to calculate the *average periodicity* of t is $avgper(t) = |SDB|/(|X(t)| + 1)$.

Proof. Let $X(t) = \{s_{x_1}, s_{x_2}, ..., s_{x_k}\}$ be the set of q-sequences containing t such that $1 \leq x_1 < x_2 < ... < x_k$. From the definition of average periodicity of t in Definition 7, $avgper(t) = \sum x \in ps(t)/|ps(t)|$. We need to show $\sum x \in ps(t) = |SDB|$ and $|ps(t)| = |X(t)| + 1$. First, $\sum x \in ps(t) = (x_1 - x_0) + (x_2 - x_1) + (x_3 - x_2) + (x_k - x_{k-1}) + (x_{k+1} - x_k) = (x_{k+1} - x_0) = |SDB|$ (1). Second, by the definition of $ps(t)$ in Definition 7, $ps(t) = \bigcup_{1 \leq z \leq k+1} (x_z - x_{z-1})$, it means $ps(t)$ contains $k + 1$ elements. Since t appears in k sequences, $sup(t) = |X(t)| = k$, then $|ps(t)| = |X(t)| + 1$ (2). Since (1) and (2) hold, the property holds.

Definition 8 (Periodic High Utility Sequential Patterns). *Given positive user-specified numbers: minimum utility threshold (minutil)* ξ, *minAvg, maxAvg, minPer, maxPer. A sequence t is a periodic high utility sequential patterns iff* $minAvg \leq avgper(t) \leq maxAvg$, $minper(t) \geq minPer$, $maxper(t) \leq maxPer$, *and t satisfies the Definition 6.*

For example, if $\xi = 55$, $minPer = 1$, $maxPer = 3$, $minAvg = 1$, $maxAvg = 2$, the complete set of five PHUSPs which have the form $(t, u(t), X(t), minper(t), maxper(t), avgper(t))$ is $\{(\langle(b)(a)\rangle, 55, 3, 1, 2, 1.25), (\langle(b)(ae)\rangle, 57, 3, 1, 2, 1.25), (\langle(e)(a)(a)\rangle, 55, 3, 1, 2, 1.25), (\langle(ad)\rangle, 57, 3, 1, 2, 1.25), (\langle(a)(ab)\rangle, 59, 3, 1, 2, 1.25)\}$. In the next section, we will present PHUSPM algorithm - an efficient algorithm for mining periodic high utility sequential patterns.

3   PHUSPM Algorithm

The pseudo code of *PHUSPM* algorithm is shown in Algorithm 1. This algorithm inspired by the USpan algorithm for mining high utility sequential patterns [1]. *PHUSPM* scan the q-sequence database only one time to calculate the number of sequences in database SDB and the Sequence Weighted Utilization of each item

---

Algorithm 1. The PHUSPM Algorithm

---

input : SDB: a q-sequence database, ξ: the minimum utility threshold,
        $minPer, maxPer, minAvg, maxAvg$
output: the set of periodic high utility sequential patterns

1 *Scan SDB only one time to calculate $|SDB|$ & SWU for each item $i_k \in I$*
2 **if** *p is a leaf node* **then**
3 | return
4 **end**
5 *Scan the projected database SDB once to:*
    a. put I-Concatenation items into *i-list*, or
    b. put S-Concatenation items into *s-list*

6 *Remove unpromising items in i-list and s-list*
7 **foreach** *item $i \in$ i-list* **do**
8 | $(t', v(t')) \leftarrow$ I-Concatenate(p, i)
9 | *Add the s_{id} of the sequence s into $U(t)$ such that $s' \sim t' \wedge s' \subseteq s \wedge s \subset SDB$*
10 | **if** $(isPeriodic(U(t'), minPer, maxPer, minAvg, maxAvg) = true$ &
     $u_{\max}(t') \geq \xi)$ **then**
11 | | output t'
12 | **end**
13 | $PHUSPM(t', v(t'))$
14 **end**
15 **foreach** *item $i \in$ s-list* **do**
16 | $(t', v(t')) \leftarrow$ S-Concatenate(p, i)
17 | *Add the s_{id} of the sequence s into $U(t)$ such that $s' \sim t' \wedge s' \subseteq s \wedge s \subset SDB$*
18 | **if** $(isPeriodic(U(t'), minPer, maxPer, minAvg, maxAvg) = true$ &
     $u_{\max}(t') \geq \xi)$ **then**
19 | | output t'
20 | **end**
21 | $PHUSPM(t', v(t'))$
22 **end**
23 **return**;

Algorithm 2. isPeriodic Procedure

---

    input : $P(t)$ be a set of positions of t in SDB, $minPer$, $maxPer$, $minAvg$, $maxAvg$

    output: return $true$ if t is a periodic sequential patterns, inversely return $false$

1 *Calculate $minper(t) = min(ps(t))$ based on $U(t)$*
2 *Calculate $maxper(t) = max(ps(t))$ based on $U(t)$*
3 *Calculate $avgper(t) = |SDB|/(|X(t)| + 1) = |SDB|/(|P(t)| + 1)$*
4 **if** $minper(t) \geq minPer \wedge maxper(t) \leq maxPer \wedge minAvg \leq minavg(t) \leq maxAvg$ **then**
5 | **return** $true$;
6 **end**
7 **else**
8 | **return** $false$;
9 **end**

---

in I for Width prunning strategy. As mentioned in the Related Work, USpan algorithm use two prunning strategies: Depth Prunning and Width Prunnning. The detailed explanation of USpan algorithm was presented in [1]. Lines 2 to 4 are the depth prunning phase to indicate a pattern (also a node in LQS tree) whether is an unpromising pattern (leaf node) or not. Lines 5 to 6 is the width prunnning phase to collect the item into two seperate lists named i-list and s-list. In this phase, the unpromissing item will be rejected from two lists. Lines 7 to 13 is the I-Concatenation mechanism. In this phase, the algorithm will examine all items in i-list to expand the patterns. To achieve the purpose of assessing the periodicity of a pattern, we use a set named U to store the sequence identifier numbers (s_{id}) of all sequence containing sequence t' (line 9). Then *PHUSPM* will call *isPeriodic* Procedure (line 10) to check the pattern whether is periodic or not. The pseudo code of *isPeriodic* Procedure is shown in Algorithm 2. The input of this procedure is the set of sequence identifier numbers (s_{id}) of pattern t, the user-specified minPer, maxPer, minAvg, maxAvg. The procedure first calculates the minimum periodicity, maximum periodicity. This phase is quite easy, it just scans the set $U(t)$ to create the $ps(t)$. Then, *isPeriodic* chooses a minimum value and a maximum value from the $ps(t)$. The procedure calculates the average periodicity of t based on the Property 1. Finally, if t passes the condition of a periodic pattern (shown in line 4), the procedure *isPeriodic* will become true. Inversely, it is getting false. If the returned value of *isPeriodic* procedure is true, *PHUSPM* will calculate the maximum utility of pattern t'. If this value satisfies the minimum utility threshold, the pattern t' is a periodic high utility sequential pattern and then outputd it (lines 10, 11). *PHUSPM* recursively invokes itself to go deeper in the LQS-Tree to find other periodic high utility sequential patterns (line 13). Lines 15 to 22 is the S-Concatenation mechanism, the explanation for these codes is similar to I-Concatenation mechanism. By two concatenation mechanisms, the *PHUSPM* algorithm can find the complete set of periodic high utility sequential patterns. In next section, we will present the experimental results of *PHUSPM* algorithm on the large scale datasets.

4   Experimental Result

Experiments were performed to assess the performance of *PHUSPM* on a computer having a fifth generation 64 bit Core i3 CPU @ 2.00 GHz PC processor running Windows 10, and 4 GB of free RAM. The algorithm was implemented in C# with Visual Studio 2013. The performance of *PHUSPM* algorithm was compared with *USpan* algorithm [1]. The datasets were obtained from the SPMF library website [6]. The characteristics of these datasets are shown in Table 3.

Table 3. Datasets characteristics

Dataset	Size	#sequence	# item	Avg. seq length
Sign	375 KB	800	310	51.99
Kosarak10k	0.98 MB	10,000	10,094	8.14
Bible	8.56 MB	36,369	13,905	21.64
BMSWebView1	2.80 MB	59,601	497	2.51
BMSWebView2	5.46 MB	77,512	3,340	4.62
Kosarak990k	57.2 MB	990,000	41,270	8.14

In the experiments, the values for the periodicity thresholds have been found empirically for each dataset. The notation S-A-B-C-D in each dataset represents the name of dataset is S, $minPer = A$, $maxPer = B$, $minAvg = C$, $maxAvg = D$. For example, KOSARAK10K-1-25-1-20 means we set the $minPer$, $maxPer$, $minAvg$ and $maxAvg$ for KOSARAK10K are respectively 1, 25, 1, 20.

We first have to evaluate the execution time of the *PHUSPM* algorithm. In this study, we compare the running time of *PHUSPM* with the running time of *USpan* algorithm [1] to evaluate the efficiency of *PHUSPM* when we apply the concept of periodic patterns on high utility sequential patterns mining. The results are shown in Fig. 1a. In each sub figures, the x-axis presents the *minutil* values whereas the y-axis presents the running time in *second*. It can be observed that mining PHUSPs by *PHUSPM* algorithm can be much faster than mining HUSPs by *USpan* algorithm. The main reason is that *PHUSPM* can reject many unsatisfactory candidates and hence prune the search space. For example, on *KOSARAK10K* dataset and *minutil* is set from $100,000$ to $10,000$ with $minPer = 1$, $maxPer = 25$, $minAvg = 1$, $maxAvg = 20$ for *PHUSPM*, *PHUSPM* is respectively 1.2, 1.4, 1.3, 1.8, 2.4, 2.2, 3.4, 2.5, 38 and 68.6 times faster than *USpan*. In general, when *minutil* decreases, the gap between the runtime of *PHUSPM* and *USpan* increases. For example, for $minutil = 20,000$ on *SIGN*, *PHUSPM* is almost 16.3 times faster than *USpan*. The same explanation are applied for other sub figures.

A second observation is that the memory consumption of *PHUSPM* can be same or less than the memory consumption of *USpan*, especially when using at low *minutil* values. The results are shown in Fig. 1b. In each sub figures, the

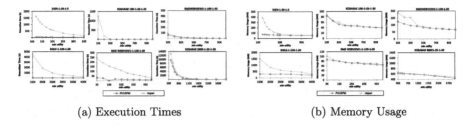

(a) Execution Times   (b) Memory Usage

Fig. 1. The experimental results

x-axis presents the *minutil* values whereas the y-axis presents the memory usage in *megabyte*. Generally, *PHUSPM* can use less memory than *USpan*, depending on datasets and *minutil* values. For example, on *KOSARAK10K* dataset and *minutil* is set from 100,000 to 10,000 with *minPer* = 1, *maxPer* = 25, *minAvg* = 1, *maxAvg* = 20 for *PHUSPM*. *PHUSPM* respectively consume 40 MB, 40 MB, 42 MB, 43 MB, 46 MB, 46 MB, 48 MB, 52 MB, 65 MB and 85 MB. Meanwhile, USpan consume respectively 40 MB, 41 MB, 43 MB, 43 MB, 49 MB, 49 MB, 50 MB, 60 MB, 67 MB and 86 MB. The same explanation is applied for other sub figures. We also test the scalibility of *PHUSPM* and change the values of *minPer*, *maxPer*, *minAvg* and *maxAvg* on each dataset. However, detailed results are not shown as a figure due to space limitation. In general, when the dataset size increases, overall mining time, and memory requirement increase. It can be observed from the scalability test that *PHUSPM* can mine the PHUSPs on the large datasets and distinct items with considerable amount of runtime and memory. These overall results show that the proposed *PHUSPM* algorithm is an efficient algorithm for mining periodic high utility sequential patterns.

5   Conclusion

In this paper, We applied the periodic patterns framework in some previous researches [2,3] on USpan [1] to make a new type of pattern named Periodic High Utility Sequential Patterns (PHUSPs). We also proposed an algorithm named PHUSPM (*Periodic High Utility Sequential Patterns Miner*) to efficiently discover all PHUSPs in a sequence database. Therefore, we achieve the goal of mining PHUSPs. An extensive experimental study shows that *PHUSPM* can mine PHUSPs faster than mining HUSPs by USpan. This algorithm is also efficient in the term of memory usage and scalability. For future work, we will consider introducing the parallel algorithms for mining periodic high utility patterns which outperform *PHUSPM*.

Acknowledgement. This research is funded by Vietnam National Foundation for Science and Technology Development (NAFOSTED) under grant number 102.05-2015.07.

References

1. Jin, J., Jan, Z., Cao, L.: USpan: an efficient algorithm for mining high utility sequential patterns. In: KDD 2012 Proceedings of the 18th ACM SIGKDD International Conference on Knowledge Discovery and Data Mining, pp. 660–668 (2012)
2. Fournier-Viger, P., Lin, J.C.W., Duong, Q.H., Dam, T.L.: PHM: mining periodic high-utility itemsets. ICDM **2016**, 64–79 (2016)
3. Tanbeer, S.K., Ahmed, C.F., Jeong, B.S., Lee, Y.K.: Discovering periodic frequent patterns in transactional databases. In: Proceeding 13th Pacific-Asia Conference on Knowledge Discovery and Data Mining, pp. 242–253 (2009)
4. Surana, A., Kiran, R.U., Reddy, P.K.: An efficient approach to mine periodic frequent patterns in transactional databases. In: Proceedings of the 2011 Quality Issues, Measures of Interestingness and Evaluation of Data Mining Models Workshop, pp. 254–266 (2012)
5. Uday, U.R., Kitsuregawa, M., Reddy, P.K.: Efficient discovery of periodic-frequent patterns in very large databases. J. Syst. Softw. **112**, 110–121 (2015)
6. Fournier-Viger, P., Gomariz, A., Gueniche, T., Soltani, A., Wu, C., Tseng, V.S.: SPMF: a java open-source pattern mining library. J. Mach. Learn. Res. (JMLR) **15**, 3389–3393 (2014). http://www.philippe-fournier-viger.com/spmf/

Mining Class Association Rules with Synthesis Constraints

Loan T.T. Nguyen[1,2(✉)], Bay Vo[3], Hung Son Nguyen[2],
and Sinh Hoa Nguyen[4]

[1] Faculty of Information Technology, Nguyen Tat Thanh University,
Ho Chi Minh City, Vietnam
nthithuyloan@gmail.com, nttloan@ntt.edu.vn
[2] Faculty of Mathematics, Informatics and Mechanics,
University of Warsaw, Warsaw, Poland
son@mimuw.edu.pl
[3] Faculty of Information Technology, Ho Chi Minh City University
of Technology, Ho Chi Minh City, Vietnam
bayvodinh@gmail.com
[4] Polish-Japanese Academy of Information Technology, Warsaw, Poland
hoa@mimuw.edu.pl

Abstract. Constraint-based methods for mining class association rules (CARs) have been developed in recent years. Currently, there are two kinds of constraints including itemset constraints and class constraints. In this paper, we solve the problem of combination of class constraints and itemset constraints are called synthesis constraints. It is done by applying class constraints and removing rules that do not satisfy itemset constraints after that. This process will consume more time when the number of rules is large. Therefore, we propose a method to mine all rules satisfying these two constraints by one-step, i.e., we will put these two constraints in the process of mining CARs. The lattice is also used to fast generate CARs. Experimental results show that our approach is more efficient than mining CARs using two steps.

Keywords: Data mining · Class association rules · Left constraint · Right constraint · Synthesis constraints

1 Introduction

Classification plays an important role in knowledge discovery and data mining techniques [3]. Classification rule mining aims to discover a small set of rules in the dataset that forms an accurate classifier. A class association rule is a special case of association rule. In association rule mining, the target of discovery is not pre-determined, while class association rule has only a pre-determined target. Mining class association rules (CARs) is the most common method in data mining. In recent years, many methods have been proposed to mine CARs based on these approaches to generate rules: (i) Apriori approach like CBA (Classification Based on Associations) [8] which scans the dataset many times, uses heuristics to build the classifier and uses one rule to

© Springer International Publishing AG 2017
N.T. Nguyen et al. (Eds.): ACIIDS 2017, Part I, LNAI 10191, pp. 556–565, 2017.
DOI: 10.1007/978-3-319-54472-4_52

predict a new case. MAC (Multiclass Associative Classification) [1] is also based on Apriori and intersection between Obidsets (a set of object identifiers in the dataset). Therefore, it scans the dataset only once and uses multiple rules to predict a new case; (ii) FP-tree approach needs to scan the dataset twice like CMAR (Classification based on Multiple Association Rules) [7] which uses the database coverage to build the classifier and uses multiple rules to predict a new case; (iii) IT-tree approach just only scan the dataset once like ECR-CARM (Equivalence Class Rule tree) [19] which is based on ECR-tree and Obidset to mine all rules, CAR-Miner [14], CAR-Miner-Diff [13]. Mining CARs in data streams [6]. In 2011, Zhang et al. proposed a method to filter and rank class association rules to enhance the accuracy [20].

Lazy-based approaches using class association rules have also been proposed in [16–18].

The above algorithms can mine all the rules that satisfy the minimum support threshold (*minSup*) and the minimum confidence threshold (*minConf*). In many cases, the users never used all of rules, therefore, many rules become redundant. Mining all these rules will take more time and more memory to store.

Hence, the problem is how to mine rules in an efficient way. For example, users give some conditions on their queries, the algorithms have to mine rules satisfying these conditions. Recently, Nguyen et al. [9] proposed a method to mine rules that satisfy the itemset constraints. After that, they also proposed another method to mine rules satisfying class constraints [10]. The problem is that if we want to mine rules that satisfy these two constraints, we can apply both previous two methods one-by-one to get the results. This method consequently consumes more time than the single step method. Additionally, temporal memory for storing rules is also bigger than using one-step for mining rules.

In this paper, we propose a method to mine all rules that satisfy these two kinds of constraints. As discussed in the above example, we investigate the single step to generate the rules that satisfy all the conditions given by the users. The proposed algorithm uses left constraint and right constraint (synthesis constraints) the same time so we save time and memory usage to mine and store all rules.

The rest of the paper is organized as follows. Section 2 outlines related works on mining CARs and mining CARs with constraints. The main contributions of this work are described in Sect. 3. Section 4 presents the experimental results of the proposed method and the conclusions and future works are discussed in Sect. 5.

2 Related Works

2.1 Mining Class Association Rules

There are many methods for mining CARs. The first method is CBA [8], this method is divided into two parts: (i) rule generator (called CBA-RG) is based on the Apriori algorithm [2] for mining CARs and (ii) classifier builder (called CBA-CB). CBA generates *1-ruleitems* that satisfy *minSup* and *minConf*. From the *1-ruleitems*, it generates candidates of *2-ruleitems* and computes their support to generate *2-ruleitems*. The same steps are performed for 3, 4, ..., *k-ruleitems* until there is no candidate generation. The main purpose of CBA-RG is to find rules in the dataset whose supports satisfy *minSup* and confidences satisfy *minConf*. After mining all the set of rules, they can check each rule based on the dataset to eliminate the redundant rules.

The second method is CMAR [7] which is based on FP-tree [4] for mining CARs and uses multiple rules for prediction. Given a new case for prediction, CMAR selects a small set of high confidence, highly related rules and analyzes the correlation among those rules. To avoid the bias, the authors develop a new measure called weighted χ^2, which is used to evaluate the strength of rules under both conditional support and class distribution. An extensive performance study shows that CMAR in general had higher accuracy than those of CBA [8] and C4.5 [15]. To improve the efficiency, CMAR employs a novel data structure, CR-tree, to compactly store and efficiently retrieve a large number of rules for classification. CR-tree is a prefix tree structure that exploits sharing among rules, making it compact. CR-tree itself is also an index structure for rules, allowing efficient rule retrieval.

Another method is ECR-CARM [19] which is based on a tree structure, named ECR-tree, to mine all CARs. This approach scans the dataset only once and uses Obidset (a set of object identifiers in the dataset) for computing the support of itemsets quickly. Therefore, its runtime is fast. Although it saves more time than previous methods but all itemsets with the same attributes are arranged into one group and put them in one node on the tree. It leads to consume a lot of memory to store Obidsets and takes a lot of time to compute the intersection of Obidsets. Besides, it spends time to generate-and-test many candidates.

Nguyen et al. [13, 14] modify ECR-tree into MECR-tree and proposed efficient algorithms, named CAR-Miner and CAR-Miner-Diff, for mining CARs. CAR-Miner used some pruning techniques to fast remove nodes that cannot generate rules from the tree.

2.2   Mining CARs with Constraints

Above methods aim to mine all CARs that satisfy the *minSup* and *minConf*. Recently, constraint-based approaches for mining have been proposed [9, 10].

For class constraints, Nguyen et al. proposed a method for mining class association rules with class constraints [10]. The proposed algorithm only generates rules from nodes that satisfy class constraints so the search space is smaller than that of mining all CARs.

For itemset constraint(s), Nguyen and Vo proposed SC-CAR-Miner algorithm for mining CARs with the itemset constraints [12]. SC-CAR-Miner considers only the single itemset constraint. Then, Nguyen et al. [11] proposed CCAR, an improved version of SC-CAR-Miner, for mining CARs with multiple itemset constraints. In 2016, Nguyen et al. proposed a method for mining CARs with itemset constraints [9]. They use the lattice structure to store all frequent itemsets. After that, they mine all CARs satisfying the itemset constraints from built lattice. Diffset strategy is also used to save memory usage. By using lattice to fast generate rules and determine the duplication, LD-CARM-IC [9] is more efficient than CCAR [11] in both the runtime and memory usage.

3 Mining CARs with Synthesis Constraints

3.1 Definitions and Problem Statement

Let D be the set of training data with n attributes $A_1, A_2, \ldots, A_n$ and $|D|$ objects (records) where each record has an object identifier (*OID*). Let $C = \{c_1, c_2, \ldots, c_k\}$ be a list of class labels (k is the number of classes). A specific value of an attribute A_i and class C are denoted by the lower-case letters a and c, respectively [16].

Definition 1. An itemset is a set of some pairs of attributes and a specific value, denoted as $\{(A_{i1}, a_{i1}), (A_{i2}, a_{i2}), \ldots, (A_{im}, a_{im})\}$.

Definition 2. A class-association rule r is of the form $\{(A_{i1}, a_{i1}), \ldots, (A_{im}, a_{im})\} \rightarrow c$, where $\{(A_{i1}, a_{i1}), \ldots, (A_{im}, a_{im})\}$ is an itemset, called left hand side of the rule, and $c \in C$ is a class label, called right hand side of the rule.

Definition 3. The actual occurrence $ActOcc(r)$ of a rule r in D is the number of rows of D that match $r's$ condition.

Definition 4. The support of a rule r, denoted $Sup(r)$, is the number of rows that match $r's$ condition and belong to $r's$ class.

Definition 5. The confidence of a rule r, denoted by $Conf(r)$, is defined as:

$$Conf(r) = \frac{Supp(r)}{ActOccr(r)}$$

Example 1. Table 1 shows a sample dataset which contains eight objects, three attributes (A, B, and C), and three classes (1 and 2). For example, consider rule $r = \{< (A, a1)> \rightarrow 1\}$. We have $ActOccr(r) = 3$ and $Sup(r) = 2$ because there are three objects with $A = a1$, in that two objects have the same class 1. We also have

$$Conf(r) = \frac{Supp(r)}{ActOccr(r)} = \frac{2}{3}$$

Table 1. An example of training dataset.

ID	A	B	C	Class
1	a1	b1	c1	1
2	a1	b2	c1	2
3	a2	b2	c1	2
4	a3	b3	c1	1
5	a3	b1	c2	2
6	a3	b3	c1	1
7	a1	b3	c2	1
8	a2	b2	c2	2

Definition 6 (Itemset constraints) [8]. Let $\alpha = \{X_1, X_2, ..., X_k\}$ be a set of itemsets (each itemset is called an *itemset constraint*). A rule $X \rightarrow c_j$ satisfies the itemset constraints α iff $\exists X_i \in \alpha: X_i \subseteq X$.

Definition 7 (Class constraints). Let $\beta = \{c_{i1}, c_{i2}, ..., c_{im}\}$ be a set of class labels (each value is called a *class constraint*). A rule $X \rightarrow c_j$ satisfies the class constraints β iff $c_j \in \beta$.

Example 2. Given dataset in Table 1, $\alpha = \{<(A, a3), (C, c1)>; <B, b3>\}$, and $\beta = \{1\}$, i.e., we want to mine all rules that their left hand side must contain $<(A, a3), (C, c1)>$ or contain $<B, b3>$ and the right hand side (class label) is 1. Rule $r_1 = \{<(A, a1), (C, c2) > \rightarrow 1\}$ satisfies β constraint but it does not satisfy α constraints; and rule $r_2 = \{<(A, a3), (B, b2), (C, c1)> \rightarrow 1\}$ satisfies both α and β because in α exists an itemset $<(A, a3), (C, c1)>$ such that it is a subset of $<(A, a3), (B, b2), (C, c1)>$ and $1 \in \beta$.

Problem statement: Given α is itemset constraints and β is class constraints, a *minSup*, and a *minConf*, the problem of mining CARs satisfying these constraints is to discover a set of CARs *RS* such that $RS = \{r \mid Sup(r) \geq minSup \wedge Conf(r) \geq minConf \wedge r$ satisfies $\alpha \wedge r$ satisfies $\beta\}$.

3.2   Algorithm

Theorem 1. Node X and its child nodes cannot generate rules satisfying β constraints if $|X \cdot Obidset_i| < minSup(\forall i \in \beta)$.

Proof. Please refer [9].

In [9], the authors show that using lattice to mine CARs with itemset constraints is more efficient than that of using CCAR [12] with post-processing approach to remove duplication rules.

In GCSC algorithm (Fig. 1), we also use lattice to mine CARs with synthetic constraints. First, procedure BUILD-LATTICE [9] is called to build a lattice L from a given dataset Line 2 in the algorithm GENERATE-CARs-with-SYNTHETIC-CON-STARINTS (GCSC). Then, for each itemset in α, we find it in the lattice L (Lines 4 and 5) to generate CARs that satisfy α and β constraints (Line 7). In our algorithm, we use a *temp* variant to store nodes that contain itemset constraints to fast reset them (Lines 1, 6, 8, and 9) (In [9], the authors must traverse the lattice L again to reset them).

Consider procedure TRAVERSE-LATTICE: Line 10 is used to avoid the duplication as CCAR [12]. In case of node l does not traverse, Theorem 1 is checked (Line 11), if the condition is satisfied, we need not to traverse all its child nodes (Line 12 return this procedure). In case of the condition of Theorem 1 does not satisfy, it calls GENERATE-RULE procedure to generate rule that satisfies at least on class in the β constraints (Lines 19–21). l is set traversed (Line 14) and Lines 15 to 18 are used for traversing all child nodes of l to generate CARs.

Input: Dataset D and *minSup, minConf,* α, β
Output: *RS* contains all CARs that satisfy α and β constraints
GCSC (*minSup, minConf,* α, β)
1.   temp = $\varnothing$;
2. L = BUILD-LATTICE(D, *minSup*); // from [9]
3. $RS = \varnothing$;
4.   **for each** *itemset* $\in \alpha$ **do**
5.   find node l which contains *itemset* in lattice L;
6.   temp = temp $\cup l$;
7.   TRAVERSE-LATTICE(l, *minSup, minConf,* β);
8.   **for each** $l \in$ temp **do**
9.   RESET-LATTICE(l); // from [9]

TRAVERSE-LATTICE(l, *minSup, minConf,* β)
10. **if** $l.traverse = false$ **then**
11.   **if** $\{\forall t \in \beta: l.Obidset_t < minSup\}$ **then** // satisfy Theorem 1
12.   return; // l and all its child nodes cannot generate rules
13.   GENERATE-RULE(l, *minConf,* β);
14. $l.traverse = true$;
15.   **for each** child node X in $l.childrenEC$ **do**
16.   TRAVERSE-LATTICE(X, *minSup, minConf,* β);
17.   **for each** child node Y in $l.childrenL$ **do**
18.   TRAVERSE-LATTICE(Y, *minSup, minConf,* β);

GENERATE-RULE(l, *minConf,* β)
19. conf = $|l.Obidset_{l.pos}| / l.total$;
20. **if** conf $\geq minConf$ and $l.pos \in \beta$ **then**
21. $RS = RS \cup \{l.itemset \rightarrow c_{pos} (|l.Obidset_{l.pos}|, conf)\}$;

Fig. 1. The GCSC algorithm

3.3   An Illustrative Example

In this section, we use the dataset in Table 1 to illustrate the process of GCSC with *minSup* = 20% and *minConf* = 60%, and synthesis constraints are $\alpha = \{<(A, a3), (C, c1)>; <B, b3>\}$, and $\beta = \{1\}$.

The lattice for the dataset in Table 1 is shown in Fig. 2.

Assume that we want to generate rules containing itemset $\{<(A, a3), (C, c1)>; <B, b3>\}$ or $\{<(5 \times a3c1)>; <(2 \times b3)>\}$ and $\beta = \{1\}$ from the lattice.

The TRAVERSE-LATTICE first find nodes contain the α in the lattice.
$RS = \varnothing$;

Consider node $l = \dfrac{5 \times a3c1\,(\emptyset, 5)}{(2, 0)}$: Traverse this node to generate rules with the

$\beta = \{1\}$, we have $RS = \{<(A, a3), (C, c1)> \rightarrow 1\ (2,1)$;
    $<(A, a3), (B, b3), (C, c1)> \rightarrow 1\ (2,1)\}$

Consider node $l = \dfrac{2 \times b3\,(\underline{467}, \emptyset)}{(3, 0)}$: Traverse this node to generate rules with the

$\beta = \{1\}$, we have the final results as:

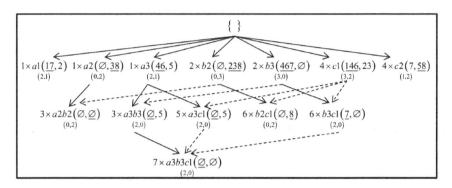

Fig. 2. The lattice structure for the dataset in Table 1 [9]

$$RS = \{ <(A, a3), (C, c1) > \;\rightarrow 1(2, 1);$$
$$<(A, a3), (B, b3), (C, c1) > \;\rightarrow 1(2, 1);$$
$$<(B, b3) > \;\rightarrow 1(3, 1);$$
$$<(B, b3), (C, c1) > \;\rightarrow 1(2, 1);$$
$$<(A, a3), (B, b3) > \;\rightarrow 1(2, 1); \}$$

4 Experiments

The algorithms used in the experiments were coded with C# 2012, and run on a laptop with Windows 8.1 OS, CPU i5-4200U, 1.60 GHz, and 4 GBs RAM.

Experimental datasets were downloaded from UCI Machine Learning Repository (http://mlearn.ics.uci.edu) and their characteristics are showed in Table 2.

Table 2. Characteristics of experimental datasets.

Dataset	#attributes	#classes	#records
Breast	12	2	699
German	21	2	1000
Glass	10	6	214
Chess	37	2	3196

To show the efficient of proposed algorithm, with each dataset, we use the selectivity of a left hand side constraint as the ratio of the number of items selected to be the constraint against the total number of items as used in [8]. For right hand side constraints, we choose two first classes of each dataset. Table 3 shows the comparison of the method using two-step (i.e., mining CARs with the left hand side constraints first, after that filtering CARs satisfied the right hand side constraints).

Table 3. Runtime of two methods.

Dataset	Selectivity (%)	Two-step		GCSC	
		Class = 0	Class = 1	Class = 0	Class = 1
Breast *minSup* = 0.10%	10	0.057	0.048	0.021	0.031
	20	0.094	0.14	0.041	0.09
	30	0.175	0.591	0.041	0.094
	40	0.198	0.187	0.054	0.136
	50	0.285	0.274	0.076	0.189
	60	0.341	0.327	0.081	0.21
	70	0.366	0.368	0.148	0.286
	80	0.386	0.416	0.148	0.435
	90	1.813	1.456	0.5	0.949
Total time		3.715	3.807	1.11	2.42
German *minSup* = 1%	10	1.243	1.255	0.35	1.199
	20	1.286	1.285	0.35	1.177
	30	1.248	1.301	0.369	1.209
	40	1.265	1.253	0.35	1.192
	50	1.241	1.334	0.348	1.236
	60	1.277	1.282	0.353	1.208
	70	1.282	1.341	0.35	1.182
	80	1.904	1.266	0.348	1.806
	90	1.333	1.955	0.384	1.273
Total time		10.836	11.017	2.852	10.283
Glass *minSup* = 0.10%	10	0.028	0.024	0.006	0.006
	20	0.062	0.106	0.005	0.011
	30	0.084	0.104	0.008	0.018
	40	0.19	0.11	0.01	0.023
	50	0.138	0.15	0.014	0.03
	60	0.196	0.189	0.016	0.038
	70	0.21	0.217	0.025	0.038
	80	0.211	0.236	0.017	0.108
	90	0.274	0.238	0.019	0.047
Total time		1.393	1.374	0.12	0.319
Chess *minSup* = 45%	10	0.462	0.54	0.01	0.433
	20	0.628	0.612	0	0.607
	30	0.645	0.701	0	0.616
	40	1.064	1.163	0	1.018
	50	1.62	1.529	0	1.635
	60	1.949	1.978	0	1.984
	70	2.526	2.537	0	2.458
	80	3.227	3.201	0	3.108
	90	4.273	4.36	0	4.094
Total time		16.394	16.621	0.01	15.953

Results from Table 3 shows that when we change class constraint, the runtime of two-step-based method does not change more. However, GCSC sometimes change more when we change the class constraint. For example, consider Chess dataset with class = 0, the runtime is almost zero but with class = 1, the runtime is the same two-step method. This is because in the rules set does not contain any rule with class = 0, it means all rules have class = 1.

5   Conclusions and Future Work

This paper proposes a method for mining class association rules that satisfy synthesis constraints using a lattice-based approach. The contributions are as follows:

1. We define the problem of mining class association rules with synthetic constraints.
2. We develop an algorithm for fast mining class association rules that satisfy these constraints in one-step.

The experimental results show that our algorithm is faster than filtering rules that satisfy right hand side constraint after mining rules using left hand side constraint.

In the future, we will continue to study how to put more constraints in the mining process. Besides, we will expand our method for mining class association rules with constraints on the process of mining non-redundant class association rules. We will also apply our approach in intrusion detection [5].

Acknowledgments. This work was carried out during the tenure of an ERCIM 'Alain Bensoussan' Fellowship Programme.

This research is funded by NTTU Foundation for Science and Technology Development.

References

1. Abdelhamid, N., Ayesh, A., Thabtah, F., Ahmadi, S., Hadi, W.: MAC: a multiclass associative classification algorithm. J. Inf. Knowl. Manag. **11**, 1–10 (2012)
2. Agrawal, R., Srikant, R.: Fast algorithms for mining association rules in large databases. In: Proceedings of the 20th International Conference on Very Large Data Bases, San Francisco, CA, USA, pp. 487–499 (1994)
3. Do, T.N.: Parallel multiclass stochastic gradient descent algorithms for classifying million images with very-high-dimensional signatures into thousands classes. Vietnam J. Comput. Sci. **1**(2), 107–115 (2014)
4. Han, J., Pei, J., Yin, Y.: Mining frequent patterns without candidate generation. In: Proceedings of SIGMODKDD 2000, Boston, MA, USA, pp. 1–12 (2000)
5. Kolaczek, G., Pieczynska-Kuchtiak, A., Juszczyszyn, K., Grzech, A., Katarzyniak, R.P., Nguyen, N.T.: A mobile agent approach to intrusion detection in network systems. In: Khosla, R., Howlett, Robert, J., Jain, Lakhmi, C. (eds.) KES 2005. LNCS (LNAI), vol. 3682, pp. 514–519. Springer, Heidelberg (2005). doi:10.1007/11552451_69
6. Kompalli, P.L.: Effcient mining of data streams using associative classification approach. Int. J. Softw. Eng. Knowl. Eng. **25**(3), 605–631 (2015)

7. Li, W., Han, J., Pei, J.: CMAR: accurate and efficient classification based on multiple class-association rules. In: Proceedings of the 1st IEEE International Conference on Data Mining, San Jose, California, USA, pp. 369–376 (2001)

8. Liu, B., Hsu, W., Ma, Y.: Integrating classification and association rule mining. In: Proceedings of the 4th International Conference on Knowledge Discovery and Data Mining, New York, USA, pp. 80–86 (1998)

9. Nguyen, D., Nguyen, L.T.T., Vo, B., Pedrycz, W.: Efficient mining of class association rules with the itemset constraints. Knowl. Based Syst. **103**, 73–88 (2016)

10. Nguyen, D., Nguyen, L.T.T., Vo, B., Hong, T.P.: A novel method for constrained class-association rule mining. Inf. Sci. **320**, 107–125 (2015)

11. Nguyen, D., Vo, B., Le, B.: CCAR: an efficient method for mining class association rules with itemset constraints. Eng. Appl. Artif. Intell. **37**, 115–124 (2015)

12. Nguyen, D., Vo, B.: Mining class-association rules with constraints. In: KSE, pp. 307–318 (2013)

13. Nguyen, L.T.T., Nguyen, N.T.: An improved algorithm for mining class association rules using the difference of obidsets. Expert Syst. Appl. **42**(9), 4361–4369 (2015)

14. Nguyen, L.T.T., Vo, B., Hong, T.P., Thanh, H.C.: CAR-miner: an efficient algorithm for mining class-association rules. Expert Syst. Appl. **40**(6), 2305–2311 (2013)

15. Quinlan, J.R.: C4.5: Program for Machine Learning. Morgan Kaufmann, USA (1992)

16. Veloso, A., Meira Jr., W., Zaki, M.J.: Lazy associative classification. In: The 2006 IEEE International Conference on Data Mining (ICDM 2006), Hong Kong, China, pp. 645–654 (2006)

17. Veloso, A., Meira Jr., W., Goncalves, M., Almeida, H.M., Zaki, M.J.: Multi-label lazy associative classification. In: The 11th European Conference on Principles of Data Mining and Knowledge Discovery, Warsaw, Poland, pp. 605–612 (2007)

18. Veloso, A., Meira Jr., W., Goncalves, M., Almeida, H.M., Zaki, M.J.: Calibrated lazy associative classification. Inf. Sci. **181**(13), 2656–2670 (2011)

19. Vo, B., Le, B.: A novel classification algorithm based on association rules mining. In: Richards, D., Kang, B.-H. (eds.) PKAW 2008. LNCS (LNAI), vol. 5465, pp. 61–75. Springer, Heidelberg (2009). doi:10.1007/978-3-642-01715-5_6

20. Zhang, X., Chen, G., Wei, Q.: Building a highly-compact and accurate associative classifier. Appl. Intell. **34**(1), 74–86 (2011)

Towards Auto-structuring Harmony Transcription

Marek Kopel[(⊠)]

Wroclaw University of Science and Technology,
Wybrzeze Wyspianskiego 27, 50-370 Wroclaw, Poland
marek.kopel@pwr.edu.pl
http://www.ii.pwr.wroc.pl/~kopel

Abstract. In recent years one can observe a significant progress in transcribing harmony from songs. New methods and applications had made it easy to automatically retrieve harmony, i.e. the chord progression for any song one can find. The shortcoming of these applications however is presentation of the transcription. Even though in most cases verses, choruses or other parts of a song share the same harmony, theyre never presented in a compact form. Even when two chords are repeated throughout the entire song one has to look at the entire transcription - from the first to the last occurrence of a chord - to realize that. This paper researches approaches to structuring the transcription, like using repetition notation (e.g. "x2") or finding the shortest commonly repeated chord progression, which may be a riff.

Keywords: Music transcription · Pattern recognition · Song harmony · Chord progression

1 Introduction

It is actually not uncommon that today pop songs are using just a short chord progression pattern repeated over and over throughout the song. Sometimes they may use one pattern for a verse and another pattern for a chorus. And the patterns are usually 3–6 chords long.

But looking at the transcription from automatic recognition one will just see a long sequence of chords - which usually needs scrolling - so it is hard for human to extract the pattern. And only seeing the pattern would make the harmony and the chord progression easy to remember. And even if one extracts the 3 chords pattern repeated throughout the song - she may never be sure that it wouldn't change towards the end of the song.

So the goal for this work is to find the harmony structure, extract the chord progression patterns and make the notation of the chord progression as compact as possible and thus easy to remember. But the problem with achieving this goal is that the automatic transcriptions are not consistent. First of all, the transcriptions never use repetition notation. And that is because the repetitions

© Springer International Publishing AG 2017
N.T. Nguyen et al. (Eds.): ACIIDS 2017, Part I, LNAI 10191, pp. 566–574, 2017.
DOI: 10.1007/978-3-319-54472-4_53

of a pattern are transcribed in various ways. Sometimes changing the chords in the same places of the pattern and sometimes putting in other chords between them. So a machine won't be able to discover the repetition pattern automatically. So this work first should focus on comparing repetitions (first be it verses or choruses) within a song to see how much their harmony notation would differ.

1.1   Song Structure

First thing that comes to mind - while thinking about structure of a song - is that they usually use the same building blocks repeated in a sequence. The most common building blocks (or sections) are: verse. chorus, bridge and solo. And probably the most popular sequence is: (verse, verse, chorus, verse, solo, chorus). Many of early Beatles songs follow this pattern hence the pattern is usually called The Beatles song structure. It may come in many variations, e.g. having a bridge instead of a solo. But all these structures and its variations origin from the AABA song form. This form is a most common theme throughout popular songs.

The Beatles often used AABA form for their songs. Not only as young songwriters ("From Me To You", "Yesterday"), but also as already experienced authors ("Lady Madonna", "Hey Jude", "Something" "Oh Darling" just to name a few most obvious examples). The structure becomes even more clear when connected with title or main hook repeated in the lyrics for the following sections [1] give these examples for "first line of the verse": "A Hard Day's Night", "Free As A Bird", "The Long And" Winding Road or "last line in the verse": "And I Love Her", "Back In The U.S.S.R." and so on. Of course not every song must follow the pattern. There are also songs with bizarre structures, e.g. "Happiness Is A Warm Gun". But they are in a minority.

1.2   Related Works

The problem of finding a songs structure has been researched for many years now and with different approaches. Mainly the extraction of the structure methods work directly with audio signal. In works like [6,10,11] authors use self-similarity of signal spectrum to find some meaningful structure.

Authors of [2,4] use TPSD (Tonal Pitch Step Distance) as a similarity measure for harmony and chord progressions. Explicit Semantic Analysis (ESA) is used for finding similarity between original and song cover in [5].

In [9] A structural segmentation algorithm is a pre-processing step for the chord extraction. The algorithm is used to identify repeated sections at the verse-chorus level. The structure of the song can help in recognising chord sequences because of sections repetitions.

In [7] author mentions chord progression comparison as a possible future work, while dealing with low level information on harmony extracted from audio.

In this paper however the research deals not with audio signal, but with already extracted high abstraction metadata like beat, chords and segments.

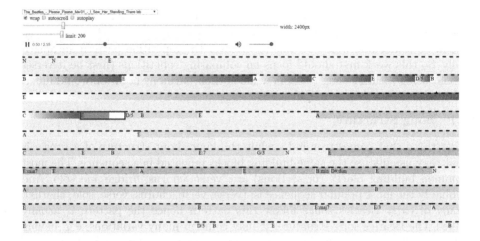

Fig. 1. First iteration of a service for interactive following songs chord progression and finding harmony structure. The progression is showing the following chords using flexbox, and adjusting each chord box width to its duration. The duration is yet based on direct time provided by metadata (difference between ending and starting millisecond) and not the number of bars. A dashed top border (added for comparison purpose only) shows that boxes of chord that last the same number of bars are not of the same width. Current chord (estimated on audio player time) is displayed in a solid frame, which fills left to right with speed appropriate to chords duration (concept known from karaoke systems). Audio can be navigated with a standard player rendered with HTML <audio> tag or by clicking a chord box. The synchronization works both ways: choosing chord skips forward the audio track and sliding on audio player timeline moves solid frame to corresponding chord box. The longest repeating chord progressions are highlighted only to show that found repeatings usually consist of an ending of a verse and a beginning of a chorus. And this knowledge is useless without knowing actual segments boundaries.

A problem to be solved by this research is the problem that services like Chordify (first described in [3]) and RiffStation already experienced. The problem is to display song harmony as a chord progression in a compact way. The way it is used in songbooks. The way it would fit on a single screen. Not as a window to an endless stream of unstructured chord progressions.

2   Problems and Possible Solutions

First trial within the research was to find repeating chord progressions. But this soon prove to be more complex than finding the longest repeating fragment of the sequence (see Fig. 1). So the work pivoted towards finding how similar the sequenced within segments - that are supposed to be identical - really are. In other words first we need to check the coherency of the chord progression notation in repeated verses and choruses. And this shall be measured by comparing them with one another.

Since the work deals with extracted metadata and not raw data, its nature is more textual than numerical. Notation of a chord progression is of string type. Thus comparing the sequences may be done using string similarity metrics, like Levenshtein distance. To embrace also the temporal aspect of a progression (each chord lasts some fixed time/number of bars), notation string may be enhanced with whitespace characters. A number of that characters - proportional to the time a chord is played - may be inserted after the considered chord. This approaches shortcoming, however, is the fact, that different chords may be notated with different number of characters, e.g. "C:min" vs "Eb:sus4(b7)".

To eliminate this unwanted feature each chord may be represented as a chroma vector. Chroma vector is a binary vector of fixed length. The length is 12, as the number of pitch notes in musical scale. Positive value at a certain position informs that a corresponding pitch is part of the represented chord. This makes comparing chords more accurate from the harmony perspective. Comparing "C" and "C:maj7" based on notation would show low similarity. But their corresponding chroma vector would only differ on one of 12 positions which resonates with actual harmony similarity. Additionally it may eliminate the problem of different notations of the same harmony, e.g. "C:6" and "A:min/b3" shall share the same chroma vector. And simple operation of vector shift makes it easy to compare chord progressions independent of songs keys.

But in practice it turns out, that within a single metadata set chord notation is consistent and using chroma vectors is not as critical as moving chord durations from time domain (see Fig. 1 description) to number of bars. Even with the same notation of the same chord progression, misalignment of chord changes by one beat may change the similarity of two analogous segments dramatically (see Fig. 2).

3   Dataset

The dataset used in this research is part of the metadata set provided by [8] extracted from audio of 12 Beatles studio albums (178 songs total). The metadata describing high level features is called annotations. As dataset description puts it: "The annotations [...] fall into four categories: chords, keys, structural segmentations, and beats/bars. The aim is to allow more musically driven Music Information Retrieval by combining several features that are intrinsically linked."

Each annotation is a series of events in the time domain. Each event has a starting and ending point (or start and duration) on timeline corresponding with audio recording timeline. In order to find harmony structure a combination of 3 categories is used: beats, segmentations and chords. The 3 event streams allow to transition chord progressions to beat/bars domain like demonstrated on Fig. 2. Although, as the description explains, the alignment is not always correct. But the transition is essential for comparing chord sequences in repeated segments. When the sequences are identical or very similar - harmony notation may become more compact. And as shown in Fig. 3 number of repeating segments in each song seems to give a huge potential for that.

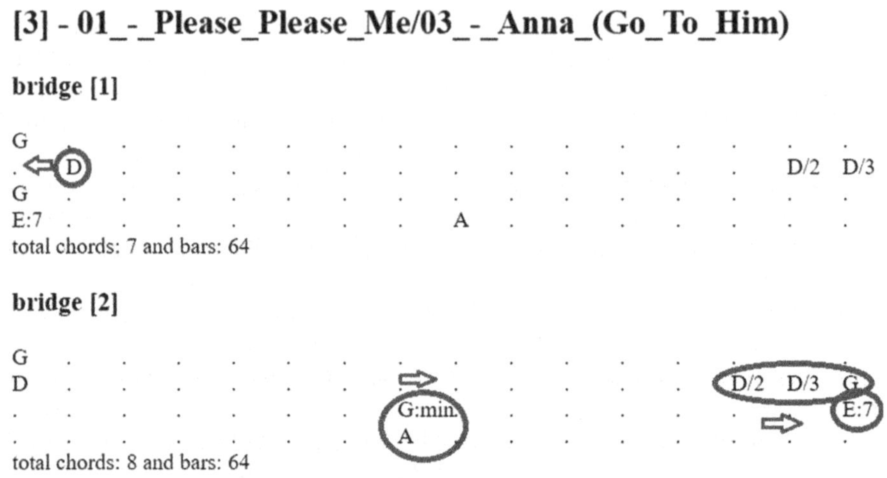

[3] - 01_-_Please_Please_Me/03_-_Anna_(Go_To_Him)

bridge [1]

total chords: 7 and bars: 64

bridge [2]

total chords: 8 and bars: 64

Fig. 2. The main problem of converting chord durations from milliseconds to number of bars is that beat and thus bar length differs throughout a song. In used dataset the beat durations (quarters of a bar) differ from $+/-$ 0.03 s in faster songs up to $+/-$ 0.05 s in slower ones. This is without intended tempo changes, like downbeat intros or slowing down towards songs end. This figure shows an output of a diagnostic tool, which maps chord progression to beat. In this edge case circles mark misaligned chords (by one beat) and arrows show shift direction for fixing the misalignment. Though easy to spot by human this kind of misalignment in general may be a challenging AI problem.

4   Experiment

As mentioned in introduction first experiment aims at comparing chord progressions. This is done by calculating difference according to algorithm in Fig. 4. As pseudocode follows, difference calculation is done for all songs with repeating segments. And each segment is compared with another of the same type (lines 1–2). When two compared segments are of the same size, the difference is the number of beats at which those sections differ (line 4). Otherwise, there are 2 cases: the longer segment has additional progression either at the beginning (e.g. intro included in a verse) or at the end (e.g. outro or pre-chorus). Both cases are checked (lines 6–7) and better similarity (lower difference) is chosen (line 8). A normalized difference is also calculated (line 9). After comparing all sections of a song an aggregation at song level is calculated: sum of all song differences (line 10) and average of the normalized differences (line 11). Both aggregation values for the dataset are presented in Fig. 5.

5   Findings

First problem discovered during the experiment is the alignment of the chord progression to the beat. This is probably the consequence of feature extraction

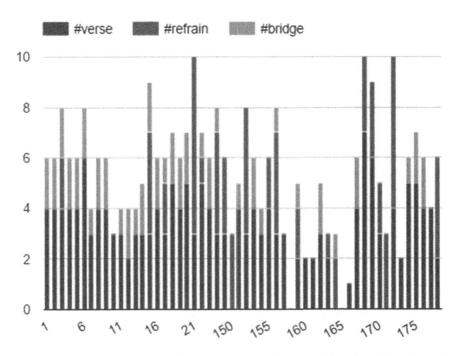

Fig. 3. Number of 3 most popular segment in each song of first 2 and last 2 studio albums by The Beatles (readability purpose). Axis y show the actual cumulative number of segments of type: verse, refrain and bridge for each song labeled at axis x. Songs 1–14: "Please Please Me" (1963); 15–28: "With the Beatles" (1963); 150–166: "Abbey Road" (1969); 167–178: "Let It Be" (1970). This diagram shows also how similar, strict structure have early Beatles compositions, using mostly the same amount of verses and bridges (solos, intros and other segments usually occurring at most once per song are omitted in the diagram); and how the structure becomes more loose and organic in their late recordings.

```
1 For each song
2   For each pair of segments s1 and s2 of the same type
3     If length(s1) == length(s2)
4       d =  difference(s1,s2)
5     Else
6       dl =  difference(s1,substring(s2, left) // assuming s2 is longer
7       dr =  difference(s1,substring(s2, right)
8       d = min(dl, dr)
9     norm_d = d / min( length(s1), length(s2) )
10    sum_d = sum (d)
11    avg_norm_d = avg(norm_d) *100
```

Fig. 4. Pseudocode of algorithm for calculating difference in chord progression of repeating segments of a song

inaccuracy and maybe also from the human aspect of the musicians performing a song (unwitting tempo change).

Another problem encountered while comparing chord progressions is often different length of the corresponding segments. To quickly overcome this obstacle a heuristic was proposed: align compared segments to the left (leave the ending of the longer segment out of comparison) and to the right (leave out the beginning), then choose higher similarity. This intuitive heuristic seems to work well in most cases.

Since the comparisons are made within a single song, which is supposed to have a consistent chord notation (the same extraction method must have been used for the entire song) the considered transition to chroma vectors turns out not needed at this stage.

Chord misalignment problem is counter-intuitively more common in songs with simple harmony. E.g. song 151 "Something" having quite complex harmony features almost perfect chords alignment in repetition segments. On the other hand, quite straightforward harmony in song 3 "Anna" shows little alignment.

Songs from 2 last albums (150–178) feature no repeating segments or their chord alignment is almost perfect. Notable exceptions from these are 4 songs:152 "Maxwell's Silver Hammer", 157 "Because", 173 "Maggie Mae" and 176 "The Long and Winding Road". Note: songs with no repetitions or 0 differences were removed from Fig. 5.

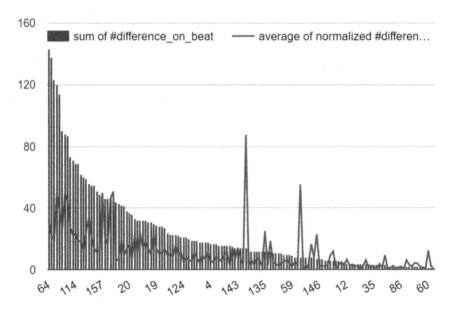

Fig. 5. Sum of beats at which repeating segments features different chords and average of normalized version of the discordant beats measure. The 2 values are presented for each song (axis x). Axis y represents the actual number for the sum or percent in case of the average.

Some clear problems with the dataset also got discovered. E.g. song 53 "I Don't Want to Spoil the Party" sections are mislabeled, i.e. 4th verse and 2nd bridge labels are switched. This highly impacted the song position in Fig. 5 (close to the left) - it is the second peak of the average difference from the left = 50%. Right next to it is song 67 "Yesterday" with 44% of average difference. This score is mostly influenced by comparisons of verse 1 with next 3. Verse 1 if the only one accompanied by guitar only (then enter strings). This probably was the reason that extraction algorithm assigned to it quite different chord progression.

First peak of the average difference from the left is 33 "And I Love Her" which has a key change before 4th verse. This means that all the chords are different, so the comparison of verse 4 with 1, 2 and 3 shows 100% difference, which highly influenced the songs position.

55% is the second highest peak of average. It is song 47 "Mr. Moonlight" and its high difference score is cause by the fact, that it has 2 quite harmonically different verses: verse_(initial) and verse_(variant). And as these verses occur a few times the calculated difference is high.

The same thing happens with song 173 "Maggie Mae", which 87% score is the highest peak of average difference in the whole result set. The song has 2 harmonically completely different verses, but this time they are not even labeled differently.

6 Conclusions

Research described in this paper takes first steps toward solving a problem already shared by a few automatic harmony transcription tools. The problem is to find song harmony structure and based on repeating segments propose a compact notation of chord progression. In order to be able to compress the transcription the repeating segments should actually have the chord progression. Described there experiment conducted on an authoritative dataset proved that quite often this is not the case.

Further steps on the research shall be taken. Proposed comparison based on chroma vector in a few cases would allow finding higher similarity (e.g. for song 100 "Getting Better"). It may also allow comparing progression in different keys, which would help in mentioned case of song 33 "And I Love Her". Using key independent comparison would also allow for extension of the dataset to e.g. covers of the original dataset songs. These covers may be additional source of chord progression for a particular segment, allowing to reinforce original findings and ultimately lead to finding the compact notation.

References

1. Crew, S.: A guide to song forms (2015). AABA song form. http://www.songstuff.com/song-writing/article/aaba-song-form/
2. De Haas, B., Veltkamp, R.C., Wiering, F.: Tonal pitch step distance: a similarity measure for chord progressions. In: ISMIR, pp. 51–56 (2008)

3. De Haas, W.B., Magalhaes, J.P., Ten Heggeler, D., Bekenkamp, G., Ruizendaal, T.: Chordify: chord transcription for the masses. In: Demonstration presented at the International Society for Music Information Retrieval Conference, pp. 8–12 (2012)

4. Haas, W.B., Robine, M., Hanna, P., Veltkamp, R.C., Wiering, F.: Comparing approaches to the similarity of musical chord sequences. In: Ystad, S., Aramaki, M., Kronland-Martinet, R., Jensen, K. (eds.) CMMR 2010. LNCS, vol. 6684, pp. 242–258. Springer, Heidelberg (2011). doi:10.1007/978-3-642-23126-1_16

5. Englmeier, D., Hubig, N., Goebl, S., Böhm, C.: Musical similarity analysis based on chroma features and text retrieval methods. In: BTW Workshops, pp. 183–192 (2015)

6. Grohganz, H., Clausen, M., Jiang, N., Müller, M.: Converting path structures into block structures using eigenvalue decompositions of self-similarity matrices. In: ISMIR, pp. 209–214 (2013)

7. Harte, C.: Towards automatic extraction of harmony information from music signals. Ph.D. thesis, Department of Electronic Engineering, Queen Mary, University of London (2010)

8. Mauch, M., Cannam, C., Davies, M., Dixon, S., Harte, C., Kolozali, S., Tidhar, D., Sandler, M.: OMRAS2 metadata project 2009. In: Late-breaking Session at the 10th International Conference on Music Information Retrieval, Kobe, Japan (2009)

9. Mauch, M., Noland, K., Dixon, S.: Mirex submissions for audio chord detection (no training) and structural segmentation. MIREX Submission Abstracts (2009)

10. McFee, B., Ellis, D.: Analyzing song structure with spectral clustering. In: ISMIR, pp. 405–410 (2014)

11. Serra, J., Müller, M., Grosche, P., Arcos, J.L.: Unsupervised music structure annotation by time series structure features and segment similarity. IEEE Trans. Multimedia **16**(5), 1229–1240 (2014)

Computer Vision Techniques

Boosting Detection Results of HOG-Based Algorithms Through Non-linear Metrics and ROI Fusion

Darius Malysiak[1(✉)], Anna-Katharina Römhild[2], Christoph Nieß[1], and Uwe Handmann[1]

[1] Computer Science Institute, Hochschule Ruhr West, Mülheim, Germany
darius.malysiak@hs-ruhrwest.de
[2] Hochschule Bochum, Bochum, Germany

Abstract. Practical application of object detection systems, in research or industry, favors highly optimized black box solutions. We show how such a highly optimized system can be further augmented in terms of its reliability with only a minimal increase of computation times, i.e. preserving realtime boundaries. Our solution leaves the initial (HOG-based) detector unchanged and introduces novel concepts of non-linear metrics and fusion of ROIs. In this context we also introduce a novel way of combining feature vectors for mean-shift grouping. We evaluate our approach on a standarized image database with a HOG detector, which is representative for practical applications. Our results show that the amount of false-positive detections can be reduced by a factor of 4 with a negligable complexity increase. Although introduced and applied to a HOG-based system, our approach can easily be adapted for different detectors.

Keywords: Augmentation · Object detection · GPGPU · High performance computing · Histogram of oriented gradients · HOG · OpenCL · CUDA · Meanshift grouping · SVM

1 Introduction and Previous Work

Histograms of oriented gradients [4] are a fundamental building block for many object detection systems. Even with the advent of deep-learning, several of today's state-of-the-art systems, e.g. [9] or [6]. [5] still continue to use HOGs as supplementary information in order to boost their performance. In other cases, the benefits of HOGs, e.g. less required training data, outweigh those of other systems such as a slightly better recognition rate with much more training data. One can see these effects in the comparison of MultiFtr+CSS and HOG in [2].

Yet, the practical application of object detection systems, in research or industry, favors highly optimized black box solutions. Complex systems require an intrinsic understanding of their parameters in order to boost their performance, due to time constraints it is often unfeasible to study and retrofit an

© Springer International Publishing AG 2017
N.T. Nguyen et al. (Eds.): ACIIDS 2017, Part I, LNAI 10191, pp. 577–588, 2017.
DOI: 10.1007/978-3-319-54472-4_54

existing system. For industrial applications it is a costly task to train an SVM classifier and optimize the involved parameters, a similar thought regarding time holds for research applications in which the detector results are merely used as supplementary feature elements. In this paper we show how such a highly optimized system can be further augmented in terms of its reliability with only a minimal increase of computation times, i.e. increasing the detection quality while preserving realtime boundaries. Our solution leaves the initial detector unchanged. Although introduced and applied to a HOG-based system, our approach can easily be applied for other detectors as well.

Section 2 briefly explains the challenge of grouping and selecting detections within the HOG algorithm. Furthermore it introduces the key elements of our approach; metric scaling of SVM weights and ROI fusion. Section 3 introduces a processing pipeline which applies the mentioned elements in order to boost the detectors performance. We conclude this paper with Sects. 4, 5 and 6 which describe our setup, present our results on a standard image database and give an outlook for additional research, respectively.

2   Boosting Results Through ROI Fusion and Non-linear Metrics

Let us assume an already trained HOG detector, i.e. all HOG parameters and the involved SVM training have been optimized for some training/verification set of images. Even in case of a huge training set, the pure HOG detection will yield a large amount of false-positive detections. This is often addressed by discarding all detections y_i whose SVM score ω_i lies below a certain theshold t. This can reduce the e.g. ≈ 10000 positively classified patches down to ≈ 50, which usually removes many false positives yet keeps many adjacent scales and positions for correct classifications. In order to reduce these detection groups down to an (ideally) single representant one applies clustering methods such as the mean shift approach. Yet as the mean-shift algorithm incorporates the SVM scores, it is also possible to loose true-positives (in case of true positives with small SVM scores). Finetuning the threshold t yields only marginal improvements and results in an increased amount of detections, which in turn can significantly slow down clustering algorithms.

2.1   ROI Fusion

The following approach is motivated by the results of [4], who utilized a weighted variant of mean-shift clustering for the grouping of multiple detections in (x, y, s) space. Let D_1, D_2 be existing HOG detectors which are trained to find distinct parts of an object, e.g. an upper-body and a head detector respectively. Just as with classical mean shift algorithms, e.g. [3], in which one iteratively estimates the modes y_m (of n points y_i) of a distribution by

$$y_m = H_h(y_m) \sum_{i=1}^{n} \overline{\omega}_i(y_m) H_i^{-1} y_i \tag{1}$$

with

$$\overline{\omega}_i(y_m) = \frac{|H_i|^{-1/2} \exp(-D^2[y_m, y_i, H_i]/2)}{\sum_{j=1}^{n} |H_j|^{-1/2} \exp(-D^2[y_m, y_j, H_j]/2)} \tag{2}$$

Let $y_i = (x, y, s) \in \mathbb{R}^3$ be the elements of the sampled data (i.e. the windows obtained by a complete multiscale HOG run, x, y, s denoting the position of the window center and scale respectively), $H_i = diag(\sigma_x, \sigma_y, \sigma_s)$ the diagonal uncertainty matrix and

$$D^2[y_m, y_j, H_j] := (y_m - y_i)^T H_i^{-1} (y_m - y_i) \tag{3}$$
$$= \sigma_x((y_m)_1 - (y_i)_1)^2 + \sigma_y((y_m)_2 - (y_i)_2)^2 + \tag{4}$$
$$\sigma_s((y_m)_3 - (y_i)_3)^2 \tag{5}$$

the Mahalanabois distance between y_m and y_i ($(y)_i$ indicates the i-th vector element). We propose the following weighted extension, let y^1, y^2 denote the resulting windows from D_1, D_2 respectively and ω^1, ω^2 the corresponding SVM scores/weights. First one has to create feasible 5-dimensional features

$$\tilde{y}_k := ((y_i^1)_1, (y_i^1)_2, (y_j^2)_1, (y_j^2)_2, (y_i^1)_3), \quad \tilde{\omega}_k := \omega_i^1 \omega_j^2 \tag{6}$$

by grouping all feasible D_2 windows y_j^2 for a single D_1 window y_i^1. The selection criteria for this combination, which must be fulfilled, are as follows

1. $\alpha_1(y_i^1)_3 \le (y_j^2)_3 \le \alpha_2(y_i^1)_3, \quad \alpha_1, \alpha_2 \in (0, 1], \alpha_1 < \alpha_2$
2. Let w, h denote the width and height of y^1:
   $\beta_1 w \le (y_j^2)_1 - (y_i^1)_1 \le \beta_2 w, \quad \beta_1 < \beta_2, \beta_1, \beta_2 \in (0, 1]$
3. $\beta_3 h \le (y_j^2)_2 - (y_i^1)_2 \le \beta_4 h, \quad \beta_3 < \beta_4, \beta_3, \beta_4 \in (0, 1]$

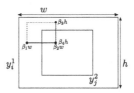

Fig. 1. The fusion of detections, each detection y_i^1 of detector D^1 is combined with all detections y_j^2 from detector D^2 if its left upper corner lies in the dashed rectangle. The same applies for the scale.

These rules represent position restrictions which combine windows y_j^2 only if they lie in a certain boundary relative to y_i^1 (see Fig. 1). A practical example would be to consider only head windows which lie completely within the upper-body window. This grouping lifts the upper-body windows into a 5-dimensional space and adds the position variance of each fitting head window y_i^2 to it. The scale remains unchanged since a scale equivalent is defined with criteria 1. The mean-shift clustering was changed to a weighted variant

$$\bar{\omega}_i(y_m) = \frac{|H_i|^{-1/2}\tilde{\omega}_i \exp(-D^2[y_m, y_i, H_i]/2)}{\sum_{j=1}^n |H_j|^{-1/2}\tilde{\omega}_j \exp(-D^2[y_m, y_j, H_j]/2)} \tag{7}$$

with $H_i = diag(\sigma_x^1, \sigma_y^1, \sigma_x^2, \sigma_y^2, \sigma_s)$. The uncertainty values σ_x^2, σ_y^2 should be set to a smaller value than σ_x^1, σ_y^1, since might be reasonable to put more certainty into D_2 so that it might stabilize the detection windows for the upper body. Since the amount of D_1 detections is increased we refer to this combination approach as sample spreading. The evaluation in Sect. 5 shows that this strategy can yield a significant improvement in detection quality compared to D_1 alone. It should be pointed out that k can (in theory) reach values up to $|\{y_i^1\}| \cdot |\{y_j^2\}|$, which can even slow down an efficient implementation.

2.2   A Nonlinear Metric for SVM Weights

In order to reduce the computation time and increase the detection quality in the context of HOG applications, one usually first filters out all result windows with a SVM weight below a given threshold t_1. This strategy can be applied to accommodate the problem of too many items for the mean shift clustering. Yet this simple method can also remove a large amount of correct detections. The reason for this lies in the HOG algorithm, for large objects it will scale the corresponding image area down to the detection window size, this removes a large quantity of high-res image information. Such windows will exhibit a smaller SVM weight compared to smaller regions. Filtering according to t_1, which obviously will be chosen according to the higher SVM values (and thus the smaller windows), will remove many if not all candidates for large objects. Our approach addresses this problem under the assumption that not all large windows have been filtered out. We developed a simple strategy by rescaling the SVM weights (after filtering with t_1) according to

$$\omega'(\omega_i) := f(\omega_i)\omega_i \tag{8}$$

with for example

$$f(\omega_i) := \begin{cases} \tau(-\theta((I_h - (y_i)_2)/I_h))(y_i)_3 & (y_i)_3 \geq \rho \\ 1 & else \end{cases} \tag{9}$$

with I_h being the image height and θ a constant. The general effect of this transformation should be that SVM weights of large windows will be increased to rival with those of smaller windows during the mode estimation. Not only can this approach retain large windows but also remove infeasible small windows in an area of large windows. The scaling function f must be chosen to accommodate this goal. In the example above the scaling function f exhibits a linear behaviour, yet one might also use an nonlinear tesselated transformation for more complex scenes. Large objects usually appear in the lower part of the image while small objects inhibit the upper portion. Thus SVM weights of window candidates in the lower region should be scaled up, while the scaling vanishes linear in the

Fig. 2. Weighting of near field windows: The left image shows the use of a single upper-body detector; the person in the lower part is not detected due to small amounts of candidate windows. Using transformed SVM weights one can see on the right image that the same detector now finds the person in the lower part and suppresses the small alse detection.

upper image area. Furthermore only the weights of windows above a certain scale will be transformed, this prevents small windows in the lower area to be transformed as well. An example for this can be seen in Fig. 2.

3   A Detection-Pipeline for Boosting the Detection Quality

In order to further motivate the techniques from Sect. 2 we conducted a thorough evaluation by embedding them in a detection pipeline. This section shows that an efficient implementation of the previously described algorithms and ideas can boost HOG-based systems in terms of their detection quality while still maintaining previous time constraints.

The detection pipeline is depicted in Fig. 3. The first step may consist of any form of image preprocessing, the output is directed into the HOG detector, which performs the initial detection (of at least one feature) without any form of detection grouping. All detections are forwarded into a metric-based selector, which consists of two steps; a thresholded reduction of detections and a metric scaling of SVM weights. After this point the results are forwarded into two parallel grouping branches, each consisting of two steps; detection grouping and a sanity check, which may use any available scene information in order to remove detections with impossible positions. The calculated detections from both branches are finally fused with a mean-shift grouping, these detections are forwarded into a so called *streaking* block. The streaking block is utilized in video streams and applies a simple heuristic (which is described in Algorithm 1) in order to predict detections and eliminate short term detection gaps. The algorithm leaves out the details of how to find corresponding detections, i.e. it does not specify the form of feature vectors. In the following evaluation normalized intensity histograms have been used, additionally the distance between detections has been checked.

More precisely

- The feature vector f_i consists of $n \cdot 256$ real numbers, 256 for each color channel.
- Feature vectors are compared by calculating the euclidean distance, which can not exceed a threshold t_f
- The distance between the upper left corner for two detections can not exceed a pixels along the x-axis and b pixels along the vertical.

Although simple in its design, this approach yields significant improvements compared to the canonical HOG algorithm. Yet, depending on the situation, e.g. image quality, a color histogram might become unfeasible since it is susceptible to image noise effects. This histogram based metric can be exchanged for more complex descriptors and similarity measures, e.g. a gradient based descriptor with a weight based metric, thus one can adapt the described pipeline for such a scenario. The streaking results will be grouped by a last mean-shift step after having been filtered by a final sanity check.

Note that the classic HOG algorithm can be obtained by setting less restrictive parameters for the metric selection, defining the sanity check of the fusion branch to filter all results, setting the streaking history size $s = 1$, adapting the parameters of the last two grouping steps and the last sanity check.

Algorithm 1. Streaking

Require: Image I, detections $\mathcal{D}$, history size s, detection history $\mathcal{H} = \{(d_1, f_1, x_1, c_1), ..., (d_k, f_k, x_k, c_k)\}$ with shift buffer x_i of size s, feature vector f_i, match counter c_i and mismatch threshold t

$b = 0$;
for each $d \in \mathcal{D}$ **do**
    for each $e_i \in \mathcal{H}$ **do**
        compare the feature vector e_i and that of d (include additional sanity checks)
        if If the vectors match **then**
            Add position x of d to x_i;
            Update histogram of f_i with data at the position of d;
            Set $c_i = 0$;
            $b = 1$;
            break;
        else
            Set $c_i = c_i + 1$;
        end if
        if If $c_i \geq t$ **then**
            remove e_i from $\mathcal{H}$;
        end if
    end for
    if $b == 0$ **then**
        Add new entry for d to $\mathcal{H}$;
    end if
    $b = 0$;
end for

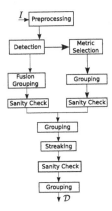

Fig. 3. A detection pipeline with ROI fusion and metric scaling of SVM weights. The image I will be preprocessed before being forwarded into the HOG-based detector without grouping. The third block removes all detections below a certain threshold and rescales the weights according to a scene specific metric. The initial and remaining detections are sent into two parallel grouping branches; a fusion grouping and a canonical grouping, respectively. The grouped results of both branches will be merged afterwards. In case of video streams from a static scene the streaking block can be utilized for reduction of detection gaps. All sanity checks are scene specific heuristics which remove detections at unfeasible places. The pipelines output consists of a grouped detection set $\mathcal{D}$.

4   Evaluation

The motivation behind the design of the pipeline was to demonstrate the applicability of the developed algorithmic concepts. Since the work in this thesis mainly targets the improvement of efficiency while preserving scalability we will show the works potential by improving the detection rate of an existing HOG implementation while preserving the previous time constraints.

Two HOG detector were trained on the INRIA training set, one for the detection of entire human bodies (H_B) and one for detecting upper bodies (H_{UB}), the training was done according to the original protocol by [4]. Each corresponding SVM was obtained by a decadic grid search over $C \in [10^{-5}, 10^3]$ with a 10-fold cross-validation at every step. Since the INRIA database only provides labels for complete bodies, all upper body labels were extracted by using the upper third of each rectangular label ROI. Both HOG detectors have been utilized in the pipeline's detection block. Table 1 states the HOG parameters in more detail. The pipeline was evaluated on the CAVIAR [1] database since it provides targets for all blocks in the pipeline;

1. The metric scaling of SVM weights becomes applicable due to strong size differences between objects in near and far field.
2. The ROI fusion is trivially applicable.
3. The streaking is applicable since the images are extracted from a continuous video stream.

Futhermore, this database represents a typical field for many parallel video streams; surveillance. In order to get comparable results between the classic standalone HOG detectors and the pipeline results, the mean-shift grouping parameters were kept identical for all grouping steps, $\sigma_x = 16, \sigma_y = 8, \sigma_s = 1.05, \epsilon = 1.0$. The fusion grouping was done with $\alpha_1 = \alpha_2 = 0.05, \beta_1 = \beta_2 = 0.1, \beta_3 = \beta_4 = 0.1$ and $\sigma_x^1 = \sigma_x^2 = \sigma_x, \sigma_y^1 = \sigma_y^2 = \sigma_y$. The iteration count was limited to a maximum of 100.

A detection d is considered to be a true positive if it can be associated with a ground truth date d_{gt} such that

$$\frac{|d \cap d_{gt}|}{|d \cup d_{gt}|} \geq 0.7 \tag{10}$$

The parameters for the metric scaling were set to $\tau = \beta = 1.0$ and $\rho_{Body} = -0.2, \rho_{UpperBody} = -0.1$, this represents an entirely linear scaling over the complete image. One should note that ρ represents an individual thresholding for each detector. All sanity checks consist of checking if a detection lies in an unfeasible region, which are defined through a polygon set $P = \{p_1, ..., p_n\}$.

Since all images within the CAVIAR dataset have been recorded with a rather low resolution of 640×480 the preprocessing consists of a GPU accelerated upscaling to 1600×1200. This step is reasonable in the sense of applicability, the already trained detector should not need to be retrained for each different image format. Furthermore one would need to shrink the training images even more for smaller window sizes, this would introduce further information loss and a reduction in detection quality.

Two image sets of a specific scene were chosen, *WalkByShop1Cor* and *Three-PastShop1Cor*, the scene and the corresponding set P is visualized in Fig. 4. Besides the parameters for the metric weight scaling and the defined polygons no additional scene specific optimizations have been utilized.

Fig. 4. A scene from the CAVIAR image database, the surveillance camera's position induces a significant size difference between objects in near and far field. Three polygons (P_1, P_2, P_3) have been defined and enclose image regions which should not contain pedestrians.

Table 1. Parameters for both detectors, i.e. body and upper body HOG detectors

Set	Cell size	Block size	Window size	Block stride	Window stride	Bin count	Scale	σ
H_B	8×8	16×16	64×128	(8,8)	(8,8)	9	1.05	1.0
H_B	8×8	16×16	96×88	(8,8)	(8,8)	9	1.05	1.0

5   Results

The plots in Fig. 5 illustrate the differences between the classic HOG algorithm and elements of the pipeline. Let $TP_i = (x_0, x_1, ..., x_k, ...)$ be the sequence of image-wise true positive counts obtained with algorithm i for each image, analogously let FP_i be the sequence of false positives. Both plots in the first row of Fig. 5 depict the difference $\delta^{TP}_{HOG,Metric} := TP_{Metric} - TP_{HOG}$ and $\delta^{FP}_{HOG,Metric} := FP_{Metric} - FP_{HOG}$, respectively. If $\delta^{TP}_{HOG,Metric}(k) > 0$ for some image index k then more true positives were obtained by using the HOG algorithm than with metric scaling of the weights. The same holds for the false positive count. It becomes obvious that less true positives were obtained with metric scaling, yet one has to accept significantly more false positives. This indicates a stabilizing effect onto the canonical HOG approach, which is also visible in the second row of plots, i.e. the results of comparing the HOG against the pipeline's fusion branch. The last row in Fig. 5 depicts the comparison between the HOG and the entire pipeline. The amount of false positives is significantly reduced while keeping the amount of false positives close to that of the classic

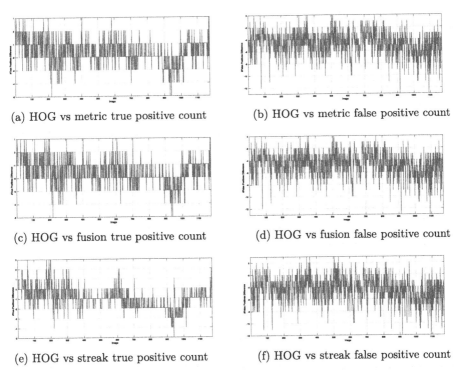

(a) HOG vs metric true positive count   (b) HOG vs metric false positive count

(c) HOG vs fusion true positive count   (d) HOG vs fusion false positive count

(e) HOG vs streak true positive count   (f) HOG vs streak false positive count

Fig. 5. Comparison of recall statistics for the canonical HOG body detector and pipeline segments on the *ThreePastShop1Cor* image set. Image (a) shows the difference: #true positives fusion branch - #true positives classic HOG, image (b) the corresponding false positive difference.

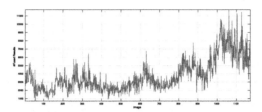

Fig. 6. Statistics for the multidimensional mean-shift grouping on the *ThreePast-Shop1Cor* image set. The graph depicts the amount of computed detection combinations for the multidimensional mean-shift grouping.

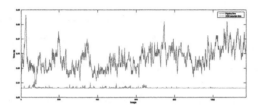

Fig. 7. Complete processing time (blue line) of the pipeline for the *ThreePastShop1Cor* image set. The red line depicts the detection time, note that these time values are invariant to the amount of detections since the grouping has been left out. (Color figure online)

HOG algorithm. One obtains a mean value of 4.3643 less false positives per image and 1.02 less true negatives per image.

These results illustrate the potential of the constructed pipeline, by using the detections of an existing HOG detector one obtain significantly less false positives while preserving the amount of true positives. As Fig. 6 shows, it turned out that a significant amount of fused detections was constructed within the fusion branch. This in turn posed a bottleneck for the pipeline's applicability, since grouping times would reach values up ≈ 200 s (right plots in Fig. 6). Yet, by using the concept for massively parallelized mean shift computation [7] this time could be reduced to a maximum of ≈ 10 ms, which in turn led to the pipelines complete processing times as depicted in Fig. 7. The pipelines maximal detection time was ≈ 88 ms, which still enables one to process ≈ 11 frames per second. An additional speed-up could be achieved by using the tile image approach from [8] since the actual SVM-based detection process makes up about 1/3 of the total processing time (see the red line in Fig. 7).

One has to note that the detection quality is determined to a large extend by the initial HOG detector. An improvement of the underlying detections would further boost the pipelines results, such an improvement may include scene specific SVM training or boosting approaches from machine learning. Very similar results were obtained on the *ThreePastShop1Cor* image set.

6   Conclusion

A very common application for object detection is that of video surveillance, in which a system must process many parallel video streams. Section 3 shows how an existing HOG based system can be augmented for this application; using the described pipeline it becomes possible to increase the systems reliability, the induced complexity increase is negligable and can easily be compensated by techniques from [7,8]. The pipeline incorporates two developed concepts; metric scaling of SVM weights and ROI fusion, both being introduced in Sects. 2.2 and 2.1, respectively. The results indicate a stabilizing effect onto the initial HOG detector, i.e. the amount of false positive detections can be reduced while retaining the amount of true positives. As shown in Sect. 5 one can expect, considering an adequate choice of parameters, a reduction of false positives by a factor of ≈ 4 while reducing the amount of true positives only marginally with a factor of ≈ 0.2. Increasing the detectors initial reliability will most likely eliminate the reduction of false positives, yet this remains a question for future research. The developed concepts can be applied to any object detector, yet the resulting gain in detection quality might be different, e.g. it may have the same stabilizing effect or even provide an increase of true positives. Furthermore it should be studied how the boosting parameters can be infered from the camera perspective itself, this may provide a simple applicable "black box" boost solution for existing detectors, no redesign or retraining is required.

References

1. Caviar: Context aware vision using image-based active recognition. http://homepages.inf.ed.ac.uk/rbf/CAVIAR/. Accessed: 1 Mar 2016
2. Benenson, R., Omran, M., Hosang, J., Schiele, B.: Ten years of pedestrian detection, what have we learned? In: Agapito, L., Bronstein, M.M., Rother, C. (eds.) ECCV 2014. LNCS, vol. 8926, pp. 613–627. Springer, Heidelberg (2015). doi:10.1007/978-3-319-16181-5_47
3. Comaniciu, D., Meer, P.: Mean shift: a robust approach toward feature space analysis. IEEE Trans. Pattern Anal. Mach. Intell. **24**(5), 603–619 (2002)
4. Dalal, N., Triggs, B.: Histograms of oriented gradients for human detection. In: IEEE Computer Society Conference on Computer Vision and Pattern Recognition, CVPR 2005, vol. 1, pp. 886–893. IEEE (2005)
5. Fukui, H., Yamashita, T., Yamauchi, Y., Fujiyoshi, H., Murase, H.: Pedestrian detection based on deep convolutional neural network with ensemble inference network. In: Intelligent Vehicles Symposium (IV), 2015 IEEE, pp. 223–228 (2015)
6. Luo, P., Tian, Y., Wang, X., Tang, X.: Switchable deep network for pedestrian detection. In: Proceedings of the IEEE Conference on Computer Vision and Pattern Recognition, pp. 899–906 (2014)
7. Malysiak, D., Handmann, U.: An algorithmic skeleton for massively parallelized mean shift computation with applications to gpu architectures. In: 2014 IEEE 15th International Symposium on Computational Intelligence and Informatics (CINTI), pp. 109–116. IEEE (2014)

8. Malysiak, D., Markard, M.: Increasing the efficiency of GPU-based HOG algorithms through tile-images. In: Nguyen, N.T., Trawiński, B., Fujita, H., Hong, T.-P. (eds.) ACIIDS 2016. LNCS (LNAI), vol. 9621, pp. 708–720. Springer, Heidelberg (2016). doi:10.1007/978-3-662-49381-6_68
9. Zou, W.Y., Wang, X., Sun, M., Lin, Y.: Generic object detection with dense neural patterns and regionlets. arXiv preprint arXiv:1404.4316 (2014)

Automatic Interactive Video Authoring Method via Object Recognition

Ui-Nyoung Yoon, Myung-Duk Hong, and Geun-Sik Jo[✉]

Department of Information Engineering, Inha University, Incheon, South Korea
{entymos13,hmdgo}@eslab.inha.ac.kr, gsjo@inha.ac.kr

Abstract. Interactive video is a type of video which provides interactions for obtaining video related information or participating in video content. However, authors of interactive video need to spend much time to create the interactive video content. Many researchers have presented methods and features to solve the time-consuming problem. However, the methods are still too complicated to use and need to be automated. In this paper, we suggest an automatic interactive video authoring method via object recognition. Our proposed method uses deep learning based object recognition and an NLP-based keyword extraction method to annotate objects. To evaluate the method, we manually annotated the objects in the selected video clips, and we compared proposed method and manual method. The method achieved an accuracy rate of 43.16% for the whole process. This method allows authors to create interactive videos easily.

Keywords: Interactive video · Object recognition · Keyword extraction

1 Introduction

Interactive video is a type of video which provides interactions for obtaining video-related information or participating in video content [1, 2]. A viewer can obtain information related to a current scene while watching interactive video by clicking a hotspot or object [3]. These interactions are created and annotated by author in the authoring time. Viewers can reduce the time and effort needed to search for information on the internet and obtain interesting information instantly and easily if it is already annotated. However, a critical problem of interactive video is that authors need to spend much time to create interactive video content, which involves the following [4].

1. Choosing objects in a video to search for related information
2. Describing the object metadata which includes the start time, end time, position, and size for showing the information on the video
3. Searching for object-related information on the web and choosing interesting information
4. Summarize the information and save into the object metadata.

Many researchers and interactive video services have presented methods and features to solve this problem.

© Springer International Publishing AG 2017
N.T. Nguyen et al. (Eds.): ACIIDS 2017, Part I, LNAI 10191, pp. 589–598, 2017.
DOI: 10.1007/978-3-319-54472-4_55

For example, Wirewax [2] provides face detection and object recognition features to help authors who want to choose interesting objects and describe object metadata. Several interactive video services [2, 5] provide templates for displaying information, which sometimes removes the need to summarize the information. However, these methods are still too complicated to use, they are inefficiently and they are not fully automatic.

In this paper, we propose an automatic interactive video authoring method via object recognition to create interactive videos easily. We used the deep learning based object recognition technique for choosing objects to obtain more information and describe object metadata automatically. We also present a method for keyword extraction method from the object related documents on the web to obtain the exact name of the object and information associated with it.

The rest of this paper is organized as follows. The next section describes background knowledge and related work. Section 3 discusses the proposed method. Section 4 presents experimental results, and Sect. 5 concludes the paper and discusses future works.

2   Background Knowledge and Related Work

2.1   Background Knowledge

More people watch and generate videos, research has been devoted to developing methods such as video retrieval and video browsing to help people [6, 7]. In order to retrieve and browse video, the video should have metadata for describing video content and related information, and various video annotation techniques have been studied for this reason. Video annotation techniques can be classified into three categories: manual video annotation, semi-automatic video annotation, and automatic video annotation [8]. Recently, automatic video annotation methods have been actively researched to reduce the effort for annotators [9]. Automatic video annotation tools usually use approaches based on text, voice recognition, and image processing [1, 9].

Object recognition is an important feature for annotating video and creating interactive video automatically. In the last decade, research for recognizing objects have used methods based on SIFT features, SURF features, Haar-like features or HOG [8]. However, deep-learning-based techniques have precise results, and research using these techniques for object recognition has been increasing [10].

Recognizing objects can be divided into two separate tasks: class-level object recognition and instance-level object recognition. Class-level object recognition methods based on deep learning now have high accuracy in real time. For example, Faster R-CNN is a state-of-the-art method for recognizing objects in real time based on the Pascal VOC dataset [10]. However, it is still difficult to recognize the instance of an object, such as the name of a person or a car.

Therefore, we used Google image search to recognize instance-level keywords of an object after that we got the class of the object using Faster R-CNN.

2.2 Related Work

To reduce the time to annotate a clickable object in a video, Yoon et al. proposed a method based on face detection [3]. An annotator annotates objects in the relative position of an actor's face in a selected keyframe. The same objects are annotated automatically until the end of the shot sequence in which the face area has been detected. Oh et al. proposed a text-based semantic video annotation method using closed captions. Closed captions were used to obtain highly related cooking information. Interactive objects are automatically generated by using this information [4]. Park et al. introduced a method for creating freeform areas as triggers to obtain more information. To create a freeform area, they provide a sketching interface and use an object tracking method to track the area [3].

Goldman et al. presented a system for annotating moving video objects. To detect a moving object, they used particle tracking and grouping in an off-line preprocess. Based on the object tracking technique, they showed various applications for creating interactive video content [11]. Wirewax is a well-known interactive video online authoring tool [2]. It uses face and object detection to reduce the time needed to choose an object in a he video. If user annotates information to the automatically detected object or manually selected object by using bounding box, then the object with annotated information is automatically tracked in the frame sequence.

These interactive video authoring methods allow users to annotate objects easily and efficiently. However, the role of the user for annotation is still important because only some parts of the whole process are carried out automatically to create an interactive video content.

3 Automatic Interactive Video Authoring Method via Object Recognition

To create interactive video content, a video and object metadata describing the video content are needed. To reduce the cost of creating content, we present the following three main steps:

1. Shot detection and object recognition (class-level object recognition)
2. Object-related information search and keyword extraction (Instance-level object recognition)
3. Ontology matching

Figure 1 shows the whole process of the proposed method. First, we extract keyframes to reduce the cost of the object recognition process, which are representative frames for each shot in the video. We recognize objects shown in the keyframes and save the object information as object metadata, such as the start time, end time, location, size, object image, and object class. To obtain the object-related documents, we crawl the documents using the object image in the object metadata. We then extract keywords from the text which that results from the previous step. Finally, we match the entities with the Uniform Resource Identifier (URI) on the Linked Open Data (DBPedia [15]) related to the object.

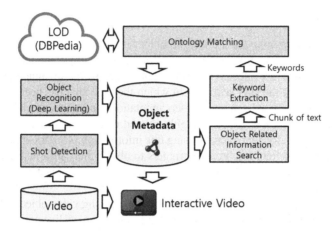

Fig. 1. Proposed method

3.1 Shot Detection and Object Recognition

Shot Detection. General video is structured as scenes, shots, frames. Frames are consecutive images and a fundamental element of video structure [12]. A shot is a sequence of continuing frames captured by a single camera. In other words, the histogram between two successive frames in a shot are very similar. we use the shot detection method [13] to detect shots in the video considering various video editing effects such as hard cuts (abrupt transitions), fade effects (gradual transitions). After all the shots are detected, we select the first frame of the shot after video editing effects as a keyframe. Objects in the first frame are usually in most of other frames in the same shot and the same position, so we can reduce the frame count for the object recognition process.

Object Recognition. Faster R-CNN is an efficient and accurate object recognition method based on deep learning [10]. This algorithm is suitable for interactive video because it can detect the exact position and size of an object in an image with the class label of the object in real time. We used a pre-trained model the based on the Pascal VOC dataset, with which we can recognize more than 20 classes of objects.

However, Faster R-CNN still cannot recognize all objects in the real world. To overcome this limitation, we present a simple architecture for instance-level object recognition, as shown in Fig. 2. In the first step, we obtain the class name and boundary of an object using Faster R-CNN with the pre-trained model. In the second step, we use a classification method based on Google Image Search [14] using an instance-level dataset on the web for each class. Finally, we obtain the exact name (label) of the object. For example, if we obtain "car" as the class name of the object in the first step, we can obtain "Lamborghini" as the exact name of the car in the second step.

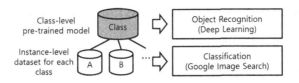

Fig. 2. Object recognition process

Object Metadata. A recognized object from a video has a variety of information, such as position, size, and class name. Figure 3 shows the object metadata ontology that is modeled to create interactive video content and manage the automatically annotated data. Object metadata can be divided into user-generated metadata and auto-generated metadata.

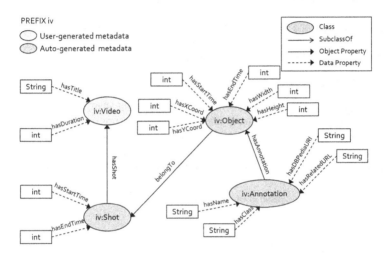

Fig. 3. Object metadata ontology

An example of user-generated metadata is the video title. If a video is uploaded on YouTube, we can obtain more user-generated metadata such as a description of video and replies to the video. This metadata can be used for obtaining more relevant information about the object.

Auto-generated metadata is generated in the shot-detection step and object recognition step. As shown in Fig. 3, the generated data is linked to the video-related class (the shot) and object-related classes such as the object and annotation. Such information is used as auxiliary information to retrieve information associated with an object and used to overlay on the video. Specific data properties such as hasName, hasDBPediaURI [15], and hasRelatedURL are generated by keyword extraction and ontology matching to provide object-related information to the viewer. The hasRelatedURL data property can indicate any URL, and we assign the Wikipedia page of the object to it in this paper.

3.2   Object Related Information Search and Keyword Extraction

Object Related Information Search. Google has many user-generated images and documents. Google generally provides text-based search and image-based search. In text-based search, users use keywords or sentences to search on Google. The results of this search method are documents and images that are relevant to the keywords. In image-based search, Google finds images and documents with similar appearance to an image uploaded by user. Google then responds with a large quantity of images and documents ranked by their relevance to the image [14]. There have been many studies related to images uploaded to Google [16, 17].

In this paper, we use image-based search to find information related to on object based on the object image created by object recognition. The results are a summary of the document, a description, and titles of similar images. We deal with all of this information as a list of sentences. However, many times, image-based search on Google is not enough to provide the information that the user wants because the results can contain similar but not exact information with multiple topics that are not related to the object. Therefore, we filter the search result using natural language processing (NLP).

Keyword Extraction. To obtain the object related information, we extract the keywords from the result of the object-related information search, as shown in Fig. 4. First, we tokenize the sentences from the previous step using pre-defined punctuation such as blanks and hyphens. We then remove the stop words such as prepositions, conjunctions, and adverbs. Next, we use a word-level n-gram model to extend the keyword set to cover a multi-word object name (for example, "Barack Hussein Obama").

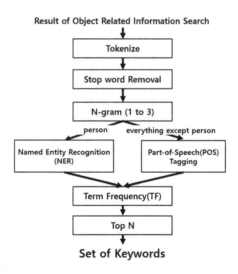

Fig. 4. Keyword extraction process

We made two paths to extract a keyword. The first path is for named entity recognition (NER) to recognize a person. The second path is for part-of-speech (POS) tagging to recognize nouns only (NNS, NNP, MN) [19]. The only there classes classified by the NER are "person," "organization," and "location" [18], so we divided the processing path. Finally, we determine the 5 top ranked keywords by calculating the term frequency to find highly related to the object. The highest ranked keyword is the main keyword, and the other keywords are sub-keywords. For example, the following keywords here extracted in an experiment: (1) Obama, (2) Barack, (3) Barack Obama, (4) BARACK OBAMA, (5) BARACK. We choose "Obama" as the main keyword and the rest as sub-keywords.

3.3 Ontology Matching

To make an interactive video that provides rich information related to an object in the video, it should be connected to informative resources on the web [1]. Therefore, we find semantically related entities of an object in DBPedia to obtain the URI using keywords that are selected as the entity name of the object. To obtain the information, we made a simple query that uses the entity name from the main keyword and the class name of the object using the lookup API of DBPedia.

However, there are often too many results to obtain highly related information and they are not related to the object. Accordingly, we filter the results using the sub-keywords. The following are steps to filter the results: (1) checking that the labels in the results include the main-keyword (substring matching), (2) checking that the class labels in the results include the object class (substring matching), (3) checking that the description in the results include the sub-keywords (substring matching). (4) If the substrings are matched in the previous steps, we evaluate the scores each step. (5) After we sort the results using the scores from each step, then we select the result that obtained the best score. If the scores are the same, we select all the results that have the same score. (6) We obtain the URIs of the results as exact object-related information.

4 Experiments

We tested the proposed method on 6 video clips selected from YouTube, including objects that might be interesting to people. The topics of the selected video clips include the US president election, soccer player, and a car advertisement.

The object recognition process using Faster R-CNN recognized 326 objects in the 5 classes "person", "bottle", "chair", "car", and "TV monitor". We use the objects in the classes "person" and "car" because the recognized objects in the other classes are too noisy to use (the objects are too small or overlapped for example).

To evaluate the method, we manually annotated the exact name of interesting objects in the keyframes of shots in the video clips using a manual authoring tool. We annotated 5 classes and 20 objects that are not duplicated (127 objects were annotated in total).

The experimental results can be divided into three steps, as explained in Sect. 3. We use the methods for step 1 [10, 13]. Especially, in [10], Faster R-CNN achieved 73.2% of mean average precision on the PASCAL VOC 2007. However, we obtained a lower accuracy of about 61% when drawing bounding box, in contrast to recognizing the object class. We also we did experiments for step2 (object related information search and keyword extraction) and step3 (ontology matching).

Figure 5 shows the accuracy rate of step 2. The bottom axis shows the object name and video number. The average accuracy of step 2 is 49.69%. This result is not high, but the proposed method worked fully automatically without user feedback. We analyzed the reason of this accuracy rate includes 0% accuracy of 9 objects. In step 1, Faster R-CNN could not draw the bounding box of the object exactly, and even if the bounding box correctly matched the object, the object could have been seen from a side view, it overlapped other objects, only specific areas of the object were shown, or there was low resolution, as a few examples. In these cases, the accuracy decreased in step 2. Although the results of step 1 were perfect, the accuracy decreased if the object had similar appearance to objects in the same class on the web, or the object-related information is rare on the web.

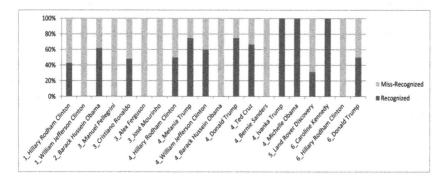

Fig. 5. Accuracy rate of step2

Figure 6 shows the accurate rate of whole process (step 1 to step 3). The average accurate is 43.16%. The reason of decreased accuracy is shown in Fig. 7. The accuracy rate of step 3 is the result of by using objects for which the exact object names are correctly recognized in step 2. The average rate of step 3 is 91.26%. This means that if we recognize the correct name of the object, there is a high probability of finding entities that are highly related to the object. In the case of missed matching, we found the reason to be that the exact name of the object was one of the sub-keywords. To increase the accuracy of step 3, we need to consider the context of the object in the scene and video.

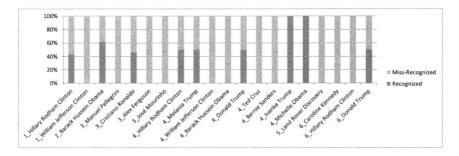

Fig. 6. Accuracy rate of whole process including step 3

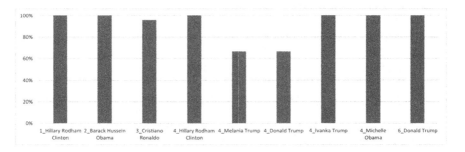

Fig. 7. Accuracy rate of step3 with correctly recognized object

5   Conclusions and Future Work

We proposed a framework for an automatic interactive video authoring method via object recognition. We showed that we can automatically annotate the object-related information on a video using our object metadata without user feedback.

The average accuracy of whole process was 43.16%. Especially, accuracy rate of keyword extraction using recognized objects step was 49.69%. There are two reasons for this accuracy. First, the deep learning based object recognition detected wrong boundary of object. Second reason is that the image-based information search obtained unnecessary and erroneous object-related information in many cases. And we also got an average accuracy of ontology matching is 91.26%, which is pretty good result, but it can be improved.

Future work is that we will improve the accuracy of image based information search by using deep learning based classification method with a large set of images related to the video content. Also, we will improve the ontology matching by using more accurate NLP techniques and semantic analysis techniques.

References

1. Oh, K.J., Hong, M.D., Yoon, U.N., Jo, G.S.: Automatic generation of interactive cooking video with semantic annotation. J. Univ. Comput. Sci. **22**(6), 742–759 (2016)
2. WIREWAX. http://www.wirewax.com/
3. Park, T.J., Kim, J.K., Choy, Y.C.: Creating a clickable TV program by sketching and tracking freeform triggers. Multimedia Tools Appl. **59**(3), 833–850 (2012)
4. Yoon, U.N., Ga, M.H., Jo, G.S.: Annotation method based on face area for efficient interactive video authoring. J. Intell. Inf. Syst. **21**(1), 83–98 (2015)
5. Zentrick. https://www.zentrick.com/
6. Yoon, U.N., Ko, S.H., Oh, K.J., Jo, G.S.: Thumbnail-based interaction method for interactive video in multi-screen environment. In: IEEE International Conference on Consumer Electronics, pp. 3–4 (2016)
7. Zhang, H.J., Wu, J., Zhong, D., Smollar, S.W.: An integrated system for content-based video retrieval and browsing. Pattern Recogn. **30**(3), 643–658 (1997)
8. Bianco, S., Ciocca, G., Napoletano, P., Schettini, R.: An interactive tool for manual, semi-automatic and automatic video annotation. Comput. Vis. Image Underst. **131**, 88–99 (2015)
9. Sun, S.W., Wang, Y.C.F., Hung, Y.L., Chang, C.L., Chen, K.C., Cheng, K.C., Cheng, S.S., Wang, H.M., Liao, H.Y.M.: Automatic annotation of web videos. In: Proceedings of IEEE International Conference on Multimedia and Expo, Barcelona (2011)
10. Ren, S., He, K., Girshick, R., Sun, J.: Faster R-CNN: towards real-time object detection with region proposal networks. In: Advances in Neural Information Processing Systems (2015)
11. Goldman, D.B., Curless, B., Salesin, D., Seitz, S.M.: Interactive video object annotation. In: ACM Computing Surveys (2007)
12. Chasanis, V.T., Likas, A.C., Galatsanos, N.P.: Scene detection in videos using shot clustering and sequence alignment. Trans. Multimedia **11**(1), 89–100 (2009)
13. Apostolidis, E., Mezaris, V.: Fast shot segmentation combining global and local visual descriptors. In: Conference of Acoustics, Speech and Signal Processing, Italy (2014)
14. Google Image Search engine. http://images.google.com/
15. DBPedia. http://wiki.dbpedia.org/
16. Wang, X.J., Zhang, L., Li, X., Ma, W.Y.: Annotating images by mining image search results. IEEE Trans. Pattern Anal. Mach. Intell. **30**(11), 1919–2008 (2008)
17. Jing, Y., Rowley, H., Wang, J., Tsai, D., Rosenberg, C., Covell, M.: Google image swirl a large-scale content-based image visualization system. In: Proceedings of the International Conference on World Wide Web, France, pp. 539–540 (2012)
18. Stanford Named Entity Recognizer. http://nlp.stanford.edu/software/CRF-NER.shtml
19. Stanford Log-linear Part-Of-Speech Tagger. http://nlp.stanford.edu/software/tagger.shtml

Target Object Tracking-Based 3D Object Reconstruction in a Multiple Camera Environment in Real Time

Jinjoo Song, Heeryon Cho, and Sang Min Yoon[✉]

HCI Laboratory, School of Computer Science, Kookmin University,
77 Jeongneung-ro, Sungbuk-gu, Seoul 02707, Korea
smyoon@kookmin.ac.kr

Abstract. The visualization of a three-dimensional target object reconstruction from multiple cameras is an important issue in high-dimensional data representations with application for medical uses, sports scene analysis, and event creation for film. In this paper, we propose an efficient 3D reconstruction methodology to voxelize and carve the 3D scene in focus on 3D tracking of the object in a large environment. We applied sparse representation-based target object tracking to efficiently trace the movement of the target object in a background clutter and reconstruct the object based on the estimated 3D position captured from multiple images. The voxelized area is optimized to the target by tracking the 3D position and then effectively reduce the process time while keeping the details of the target. We demonstrate the experiments by carving the voxels within the 3D tracked area of the target object.

1 Introduction

The visualization of three-dimensional object captured from sparse images has been received numerous concerns in computer vision and computer graphics. Most of previously proposed 3D visualization methodologies have focused on the rendering of an arbitrary viewpoint of the 3D shapes of deformable objects using meshes or point-sets. Considering the techniques that recover the shapes of target object, classical approaches commonly used for motion analysis rely on 3D scanners [1–4]. Other techniques aim to recover the shape of the object using image-based approaches [5–8].

In particular, techniques of image-based 3D scene reconstruction and understanding have typically adopted intensity-based matching and carving methods or feature-based methods, multiple camera calibration, and reconstruction of the 3D shape by triangulation and surface voxel carving. However, traditional 3D reconstruction algorithms start from the assumption that views are close together convexly and corresponding features must be maintained over many views in a static camera environment.

We present an efficient image-based 3D reconstruction technique by estimating the 3D position of the object using multiple cameras to overcome the limitations of previous

---

J. Song and H. Cho are equally contributed.

© Springer International Publishing AG 2017
N.T. Nguyen et al. (Eds.): ACIIDS 2017, Part I, LNAI 10191, pp. 599–608, 2017.
DOI: 10.1007/978-3-319-54472-4_56

approaches. A moving object captured by multiple cameras around the scene is tracked in real-time as shown in Fig. 1. The area to be voxelized within scene is determined by the 3D position of the target object to only focus on volume carving procedure to the ROI (Region of Interest) of the target object. We show that our proposed method succeeds in reconstructing the target object in a large environment by tracking its position. Moreover, the method is very robust and effective in reconstructing the details of the target object in 3D comparing to previous approaches.

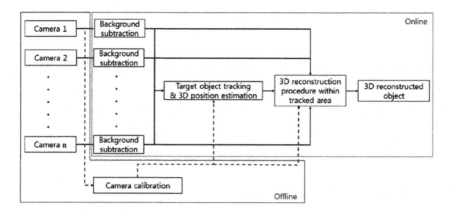

Fig. 1. Multiple image-based 3D reconstruction procedure by tracking the target object to efficiently voxelize and carve the object in a 3D environment.

The paper proceeds as follows: In Sect. 2, we briefly survey the remarkable previous approaches. We explain the technical details how to track the target object in real-time and estimate the area to be voxelized in a given environment by intersecting the 3D cones in Sect. 3. The 3D reconstruction results including comparison between our proposed approach and previous approaches will be presented in Sect. 4. In Sect. 5, we summarize the paper and suggest future research on the tracking-based 3D scene reconstruction in real-time.

2   Related Work

Generally, 3D volume data reconstruction of the target object starts from the assumption that target object can be reconstructed from a bounded area in a given 3D scene. Among the previous remarkable image-based 3D reconstruction algorithms, voxel coloring [7] and Image Based Visual Hull [8] provided an effective 3D reconstruction technique. The Visual Hull based voxel carving algorithms [2, 8] compute a 3D coarse shape estimation of the object with the use of only few 2D projected images because it depends on the following factors: the number of reference images used, the position of the view considered, the deformation of the object's shape, and multiple camera calibration data. The voxel coloring method focused on reconstruction of the color at the 3D volume data by projecting to each images, which requires a great deal of processing time in dividing the 3D scene in detail. Even though numerous approaches on voxel carving use color consistency to carve the bulky voxels, the performance of the 3D reconstruction is very dependent on the initial bounding volume of the given scene [3]. In particular, the static camera-based voxel carving methods typically falls into trouble when applied to recon-struct a moving target object in a large scene.

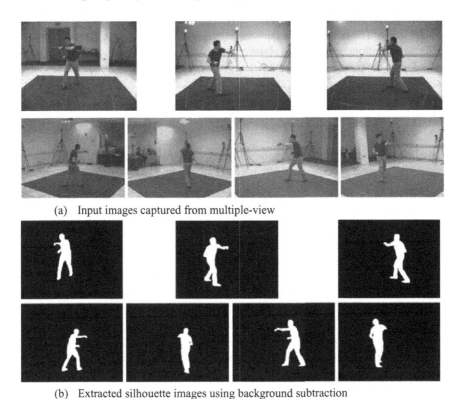

(a)   Input images captured from multiple-view

(b)   Extracted silhouette images using background subtraction

Fig. 2. Silhouette extraction using background subtraction from HumanEva dataset.

3   3D Reconstruction by Tracking the Target Object

The photo-realistic 3D reconstruction procedure of a target object using multiple images can be largely separated into two subsections; one is camera calibration and the extrac-tion of silhouette images via background subtraction; the other is 3D reconstruction by tracking the target's position in the 3D space. We will discuss the projection of the 3D rays onto reference images and the lift of the 2D points back-projected to the 3D scene.

3.1   3D Position Estimation from Calibrated Multiple Camera Environment

The extraction of intrinsic and extrinsic parameters from 2D images via camera calibration is a required off-line job to accurately predict the pixel axis of a 3D voxel in the task of 3D object reconstruction. The input includes a large number of points with their 3D coordinates, as well as the pixel coordinates.

Figure 2 represents the overview of 3D volume by estimating the object and its rays to the 3D scene. The point C is the camera center, and M is an inverse projection matrix transforming homogeneous image plane $x = [\, u \; v \; 1 \,]^T$ into rays in the 3D scene, where u and v denote the pixel position in the image plane. The pair $[M|C]$ indicates the camera location in the 3D space. The computation of a ray in the 3D space is solved by the following equation:

$$X(t) = C + tMx \tag{1}$$

The point a within an image plane and the projection of the 3D point A onto the image of the camera specified by $[M|C]$ are calculated by the following equation:

$$a = M^{-1}(A - C) \tag{2}$$

We compute the 3D ray lifting by back-projection of its two points onto the reference image and determine the 3D line penetrating the points.

To efficiently extract and segment the targets from static cameras with as low false alarm rate as possible, the objects are detected in the 3D environment typically via background subtraction, which is achieved by subtracting a new frame to a model of the scene background. On the other hand, we adopt a non-parametric method based on estimating the kernel density of the probability density function (PDF) of every pixel's intensity within each image [9]. The purpose of background modeling based on estimating the kernel density is finding and saving the up-to-date information about every new consecutive frame of image sequence. This makes it possible to recognize and extract fast changes in the background even when the pixel's intensity value changes rapidly. This, consequently, enables us to estimate the density function of the pixel's distribution at any moment even if only very recent information is available, so very sensitive detection is achievable. With the recent information of the pixels, the PDF of each pixel will have intensity value $I(u, v)$ at time t and it is represented as the following non-parametric estimation using the kernel, K:

$$pdf(I_t) = \frac{1}{N} \sum_{i=1}^{N} K(I_t - I_i) \tag{3}$$

where N is the recent intensity value of a certain pixel for comparing the pixel value of the current image. The density function can be expressed as below when Gaussian kernel is applied:

$$pdf(I_t) = \frac{1}{N} \sum_{i=1}^{N} \prod_{j=1}^{3} \frac{1}{\sqrt{2\pi\sigma_j^2}} e^{-\frac{1(I_{tj} - I_{ij})^2}{2\sigma_j^2}} \tag{4}$$

where j means the number of the color channel (R, G, and B) and σ denotes the standard deviation of Gaussian kernel. If an adequate threshold which comes from Eq. (4) is set, the foreground region is successfully segmented. Figure 2 shows the extracted silhouette images which are captured from multiple cameras.

The 3D rays which come from Eq. (2) intersect within the given environment and determine the object position in the 3D scene by the following steps. Firstly, the centers of gravity $g_1, g_2, \ldots, g_n$ of the silhouettes are extracted with respect to each image from the centers of the target object, which are calculated using background subtraction of each camera. Secondly, G, the position of the target object in the 3D space, is estimated by intersecting the n 3D rays passing through the centers of gravity.

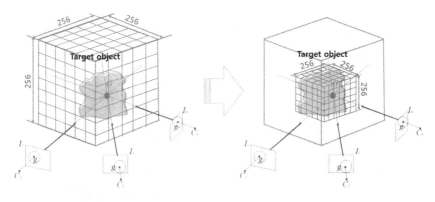

(a)   3D reconstruction without tracking   (b) 3D reconstruction with tracking

Fig. 3. Comparison of the tracking-based 3D object reconstruction and previous approach within $256 \times 256 \times 256$ volume size.

Figure 3 shows the difference between our 3D reconstruction procedure with tracking and the traditional volume carving procedure in the given scene. Even though the whole scene is voxelized with $256 \times 256 \times 256$ 3D lattices, the performance of the target object is lower than the given 3D lattice, but we can represent the details of the target by fitting the 3D lattice to the tracked 3D position of the target object.

3.2   3D Carving Methodology from Tracked 3D Region of the Target Object

We focus on a single 3D scene obtained by multiple calibrated static cameras. To discretize the 3D scene, we choose one voxel in position V from the 3D lattice and pay attention to its state. Now, we model the way the image formation is affected by knowing the occupancy state of voxel V with the assumption that a static appearance model is built with background subtraction based on estimating the kernel density. For simultaneous estimating the position of the object in the 3D space, we applied the sparse representation-based object tracking methodology, which is efficiently and robustly separate the target object from background even in existing background clutter, partial occlusion, and illumination change. The target object, which is separated from complex using kernel-based background subtraction [9], is used as the training sample and the target template is represented in terms of a linear combination of a set of all training samples.

We can find the combination coefficients by solving for L1-minimization [10] to model the appearance of an object as a sparse linear representation framework of the union of subspaces in a binary library.

As shown in Fig. 3, we constantly follow the gravity center $g_1, g_2, \ldots, g_n$ of the silhouette images from each reference image and the point G is produced in the 3D scene by intersecting multiple 3D rays. We set the visual rays in the 3D space according to all silhouette images to form a generalized cone within the rays with respect to the same target object. The integration of all object images lets us extract the 3D lattice, which is determined by the intersection of the cones, to be used for reconstruction with the assistance of the camera calibration information. When the 3D volume is determined, the photo-consistency measure is performed to decide which voxels belong to the candidate region for 3D reconstruction [11, [12].

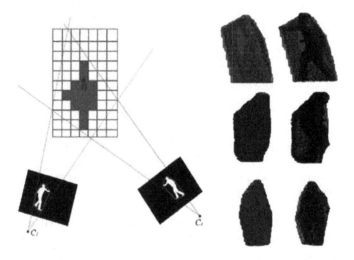

Fig. 4. Volume carving procedure in a given candidate region by estimating the position of the target in the 3D space.

Fig. 5. Candidates of 3D volume by estimating the position of the target in the 3D space and its carved 3D object from HumanEva dataset.

Our proposed target object tracking-based 3D volume carving algorithm deals with only a subset of the input cameras for each voxel extracted by intersecting the 3D rays of the silhouettes. To ensure that all cameras are considered, we repeatedly perform the color consistency by projecting the object to the reference images through the 3D volume, corresponding to the six principle direction. The complete algorithm is as follows:

```
Step 1:  Extract silhouette from input image using
         background subtraction based on estimating the
         kernel density
Step 2:  Track the position of the object in the 3D
         scene using L1-minimization
Step 3:  Initialize V from the intersected 3D rays of
         the convex hull silhouette images
Step 4:  For each voxel v on V
```
   a. let $C_1, C_2, \ldots, C_n$ be the camera centers that participate in the consistency check for each voxel v, project to an unmarked pixel on each image plane;
   b. determine the photo-consistency of v;
   c. if v is not consistent, then set V=V-{v}, otherwise paint the pixels to each v projects

Figure 4 shows the 3D lattice constructed by intersecting multiple visual cones via tracking the 3D gravity center of the object. We visualize the result of the 3D object reconstruction task within the 3D lattice. Within the 3D lattice, we carve the bulky volume data by examining the color consistency among multiple images. By fitting the 3D candidate region of the object based on the tracking position, the bulky 3D volume data can be adapted to the target object. Figure 5 also shows the reconstructed 3D object in various viewpoint.

4   Experiment Results

4.1   Experiment Setup

We evaluated the experimental results based on moving target object in a large scene and static object to compute the comparison between our proposed algorithm and remarkable previous approaches like voxel coloring. The techniques are performed on a desktop computer with an Intel Core i5 3.2 GHz CUP and 4 GB memory.

(a) Multiple images and its camera calibration data from HumanEva dataset.

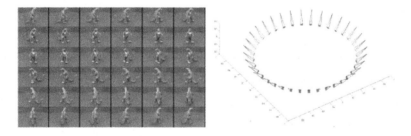

(b) Dinosaur images and its camera calibration data from Oxford Univ.

Fig. 6. Input images and its camera calibration for photo-realistic image-based 3D object reconstruction

For the evaluation of the moving object, we used the HumanEva (http://humaneva.is.tue.mpg.de/) dataset which provides the 7 color/gray images in a large scene including camera calibration data. HumanEva dataset is very suitable for our experiments to reconstruct the moving target object in a large environment. We also used 37 images of dinosaur which are captured in a constraint environment for fair comparison (http://www.robots.ox.ac.uk/~vgg/data/data-mview.html). The input images of the dataset were resized to 320 × 240 pixels for fast evaluation. Figure 6 shows the input images and its camera calibration information which is used in our experiments.

4.2   Experiment Evaluation

To qualitatively evaluate the performance of our proposed tracking-based 3D volume carving methodology, we compare the generic 3D reconstruction method using voxel coloring [6] in a fixed volume size as 256 × 256 × 256. We only voxelize the 3D scene the candidate region of the object instead of whole scene. By focusing to the candidate region to be carved by color-consistency check, we can improve the quality of the reconstructed object within a constrained memory and processing time.

We also compared the accuracy of the 3D reconstruction of our approach to the original voxel coloring approach. Figure 7 shows the 3D reconstructed model comparison between our approach (Fig. 7(a)) and the original voxel coloring approach [6] (Fig. 7(b)) in 256 × 256 × 256 dimensions. Our proposed method provides more detail in the arms and legs of the deformable object because we discretize the 3D scene at the intersection of the convex cone of each camera, while the original voxel coloring method discretize the whole 3D scene. In particular, our proposed method is more accurate in representing the characteristics of the arms and legs of 3D reconstructed human body models.

(a) 3D reconstruction using our approach   (b) 3D reconstruction using voxel coloring

Fig. 7. Comparison between our approach and the approach with voxel coloring

5   Conclusion

In this paper, we proposed the 3D reconstruction technique that provides more accurate details of the object captured from multiple images. By continuously estimating the 3D position, we only need to focus on the candidate region of the target object. We will continuously update our moving camera calibration and segmentation algorithm to reconstruct the target object within a multiple moving camera environment.

Acknowledgment. This research was supported by the Institute for Information & Communications Technology Promotion (IITP) grant funded by the Korean government (MSIP) (No. R0115-15-1009, An Interaction Platform for Learning Foreign Languages with Tangible Reality Based on Experiential Awareness). It was also supported by the National Research Foundation of Korea (NRF) funded by the Ministry of Science, ICT & Future Planning (NRF-2016R1D1A1B04932889, 3D reconstruction of high-resolution optical microscopy software development).

References

1. Allen, B., Curless, B., Popvic, Z.: The space of human body shapes: reconstruction and parameterization. ACM Trans. Graphics **22**(3), 587–593 (2003)
2. de Bonet, J.S., Viola, P.: Roxels: responsibility weighted 3D volume reconstruction. In: Proceeding of International Conference on Computer Vision (2001)
3. Mulayim, A.Y., Yilmaz, U., Atalay, V.: Silhouette-based 3D model reconstruction from multiple images. IEEE Trans. Syst. Man Cybern. Part B **33**(4), 582–591 (2001)
4. Dyer, C.R.: Volumetric scene reconstruction from multiple views. In: Davis, L.S. (ed.) Foundations of Image Understanding, pp. 469–489. Springer, Heidelberg (2001)
5. Angueloy, D., Sriniysan, P., Koller, D., Thrun, S., Rodgers, J., Davis, J.: Scape: shape completion and animation of people. ACM Trans. Graph. **24**(3), 408–416 (2005)
6. Seutz, S.M., Dyer, C.R.L.: Photorealistic scene reconstruction by voxel coloring. In: IEEE Conference on Computer Vision and Pattern Recognition, pp. 1067–1073 (1997)
7. Szeliski, R.: Rapid octree construction from image sequences. CVGIP Image Underst. **58**(1), 1067–1073 (1997)
8. Matusik, W., Buehler, C., Raskar, R., Gortler, S.J., McMillan, L.: Image based visual hulls. In: Proceeding of SIGGRAPH, pp. 369–374 (2000)
9. Han, B., Comaniciu, D., Davis, L.S.: Sequential kernel density approximation through mode propagation: applications to background modeling. In Proceeding of ACCV (2004)
10. Xie, Y., Zhang, W., Li, C., Lin, S., Qu, Y., Zhang, Y.: Discriminative object tracking via sparse representation and online dictionary learning. IEEE Trans. Cybern. **44**(4), 539–553 (2014)
11. Grau, O.: 3D sequence generation from multiple cameras. In: IEEE 6th Workshop on Multimedia Signal Processing, pp. 391–394 (2004)
12. Agrawal, M., Davis, L.S.: A probabilistic framework for surface reconstruction from multiple images. In: Proceeding of International Conference on Computer Vision and Pattern Recognition (2001)

Boosting Discriminative Models for Activity Detection Using Local Feature Descriptors

Van-Huy Pham[1], My-Ha Le[2], and Van-Dung Hoang[3(✉)]

[1] Ton Duc Thang University, Ho Chi Minh City, Vietnam
phamvanhuy@tdt.edu.vn
[2] Ho Chi Minh City University of Technology and Education, Ho Chi Minh City, Vietnam
halm@hcmute.edu.vn
[3] Quang Binh University, Dong Hoi City, Quang Binh, Vietnam
zunghv@gmail.com

Abstract. This paper presents a method for daily living activity prediction based on boosting discriminative models. The system consists of several steps. First, local feature descriptors are extracted from multiple scales of the sequent images. In this experiment, the basic feature descriptors based on HOG, HOF, MBH are considered to process. Second, local features based BoW descriptors are studied to construct feature vectors, which are then fed to classification machine. The BoW feature extraction is a pre-processing step, which is utilized to avoid strong correlation data, and to distinguish feature properties for uniform data for classification machine. Third, a discriminative model is constructed using the BoW features, which is based on the individual local descriptor. Sequentially, final decision of action classes is done by the classifier using boosting discriminative models. Different to previous contributions, the sequent-overlap frames are considered to convolute and infer action classes instead of an individual set of frames is used for prediction. An advantage of boosting is that it supports to construct a strong classifier based on a set of weak classifiers associated with appropriate weights to obtain results in high performance. The method is successfully tested on some standard databases.

Keywords: Action recognition · Boosting discriminative models · Histograms of oriented gradients · Motion boundary · Optical flow

1 Introduction

Recently, activities prediction based on vision has become an important task in a variety of survey systems, intelligent transportation, living assistance systems and robotics [1–7]. Prediction of human activity aims to identify some kinds of action class of human concerning to realistic contextual conditions of daily living. The literature briefly evaluated the state of the art methods applying in activity recognition field. Some advantages and limitations of them were also pointed out in that literature. Besides that, some available online datasets are used as a benchmark for competition and testing and evaluation, such as HMDB51 [8], UCF50, UCF101 [9]. There are many activity prediction methods [10, 11] used the local feature descriptors, such as HOG (histogram of orientated gradient) [12], HOF

© Springer International Publishing AG 2017
N.T. Nguyen et al. (Eds.): ACIIDS 2017, Part I, LNAI 10191, pp. 609–618, 2017.
DOI: 10.1007/978-3-319-54472-4_57

(histogram of optical flow) and MBH (motion boundary histogram) [13]. Wang *et al.* in [12] presented a method to transform the gradient-based features into spatiotemporal concatenation with Gaussian convolution for extracting spatial location features with expected to improve the precision rate. Some other groups focus on global feature descriptors for improving the result, such as [14, 15]. Authors in [14] proposed a method based on multiview skeletal for action recognition using a weighted averaging fusion to merge skeletal data from multiple views. Other while, authors in [15] investigated a method hybrid classification model using SVM (Support Vector Machine) and *k*-NN (*k*-Nearest Neighbor) using human silhouette and grids for modeling feature vectors. A graph model based method in [3] was proposed for multiple instance learning. Authors used a graph for presenting the local information, which expected faster than the previous subspace learning methods with computational complexity. Global feature descriptors usually produce an acceptable result with complex computation.

In generally, one versus all technique is used to classify the set of action classes. In that, two or several kinds of activities are focused on recognizing from a realistic daily living of human activities. On another hand, human activities usually switch each other while human does actions. In this situation, confusion of actions is not only artificial intelligent systems but also for a human when to predict action classes.

This paper focuses on a boosting multiple features descriptors in order to solve the problems of variety and ambiguity of human appearance. The activity motion and contextual information are analyzed to discover activity relationship for distinguishing features of similar human actions. Similarly to [11], dense features based on HOG, HOF, MBH trajectory disparity are extracted for constructing feature vectors. Some research groups focus on the use of BoW for activity recognition [16]. Additionally, a motion filter is proposed for discarding noise, which results in dense feature extraction and dynamic views. The filter method is utilized for significant improving action prediction results. A lot of activities are confused for clarifying such as jogging, running, walking. Most of the difficulties of action recognition are that they are quite similar each other. In this study, we expect to improve an action recognition result by using local feature descriptors with boosting multiple classifiers.

2   Overview of System

There are several major steps of activity prediction as briefly presented as follows. First, local feature descriptors are extracted based on multiple scales of sequent images. In this experiment, the basic feature descriptors based on HOG, HOG, MBH and displacement of trajectory are considering to process. Second, taking into account advantages of BoW for machine learning in order to emphasize the feature representation, it is used to solve the problem of difference of the feature number. Furthermore, The BoW descriptor is a preprocessing task, which avoids strong correlation of data, and distinguishes feature properties for improving the accuracy of the prediction system. The BoW is considered to apply for clustering feature descriptors and then fed to classification machine. In this experiment, each kind of descriptors using single individual BoW dictionary, e.g. BoW dictionary of each HOG, HOF, MBH feature descriptor. Third, learning a discriminative model for classification based on individual descriptor based BoW. In this step, discriminative models with respect to BoW dictionaries are trained. These classification models based individual descriptor use as a weak classifier for pre-predicting activity class. Finally, a boosting technique is used to construct a strong classifier. The feature descriptors are extracted in sequent frames within the period L frames for each action prediction. Different to previous contributions, an activity class is predicted in each interval of L frames and inferred based on the set of previously predicted classes. This task is necessary due to a human can switch his action while doing action, for example from running to walking or jumping and vice versa. On the other hand, it is difficult to identify that where are the starting and the ending point of the cycle of each action. The order of object appearance in sequent frames is a very important criterion to recognize activity class. Overall schedule of feature collection and action recognition is described in Fig. 1.

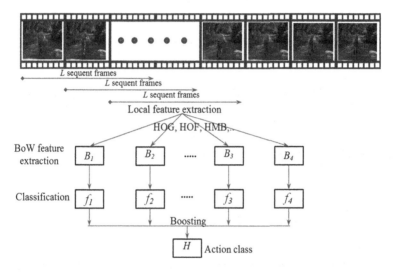

Fig. 1. Overview architecture of fusion feature descriptors for action recognition

3   Local Feature Descriptors

There are some advantages of low-level feature descriptors, e.g. it is robustness to geometry properties, complex background, independent machine learning and tractable filter. There are several major steps to extract features. First, it is necessary to determine feature locations, where to consist of rich structure for robust tracking. Some kinds of feature descriptors are extracted at detected locations. The feature locations are tracking within an interval of the action cycle. At end of each action cycle, spatiotemporal features are extracted based on consecutive frames, as represented in Fig. 2. They are fed to next step of action prediction processing. Similar to previous contributions [11], this study focuses on some kind of feature descriptors such as HOG, HOF, MBH, motion trajectory.

Fig. 2. Trajectories of the local features are extracted and tracked from actions: (a) running, (b) jogging, (c) waking.

A dense sample feature extraction is generated by cell splitting in multiple scale levels. The input image is divided into the cells with a cell size (c_w, c_h), where c_w, c_h are width and height of the cell, respectively. Points of interesting (POI) for extracting features called feature location. Homogeneous regions with the poor structure of content are suggested to discard in order to extract robust feature locations for tracking them in sequent frames. In practical, POI tracking results in low accuracy inhomogeneous region because it is poor texture, no more benchmarks for tracking. Then all robust feature locations are achieved based on the gradient of intensities, dense optical flow. Dense optical flows are extracted using the Farneback algorithm in [17]. It is applicable for real-time processing in the dense optical flow. The optical flow of two sequent images is computed based on polynomial expansion.

In general, it is difficult to determine the time or the number of frames, which human require for full doing an action period. For example, an action happens in a half of second with 24 f/s frame rate that means there are 12 frames required for each action cycle. Furthermore, it depends on the kind of action. However, it should be covered action interacting. In the experiment, the trial and error examination method is used to determine the number of frames, which is defined as the happening interval of action cycle. Therefore, we are tracking feature location within an interval of L frames for extracting spatiotemporal features.

Consequently, many POIs of background also sampled and tracking, which are a negative influence on action prediction result, as mentioned in above. Therefore, they are necessary removed. A post-processing is proposed to do this task. Abnormal feature locations of object parts or backgrounds are detected and discarded. The second column shows the result of dense POIs, which are detected and tracked. The third column presented results of post-processing for avoiding abnormal POIs.

4 BoW Extraction

Local descriptors are usually high dimension and a strong correlation of data representation. The problems are challenges in image processing. Therefore, a framework of BoW is discussed and proposed for application in order to reduce data correlation and distinguish feature properties. There are several steps for constructing the dictionary BoW. Firstly, local descriptors are extracted from video, as illustrated in Fig. 2 and discussed in the previous subsection. Sequent frames are matching and new resampling local features at every frame. The feature positions are filtered out for any robust tracking within interval L frames. In general, some kinds of feature descriptors can be concatenated for constructing codebook. However, it causes of huge dimensional feature and expensive computation cost. Therefore, an individual descriptor is used for generating a BoW dictionary, such as BoW of HOG, BoW of HOF. They are individually processed. There are some advantages, as follows. First, the low dimension of descriptors is faster in clustering for constructing a dictionary of BoW. Second, each kind of descriptor is comprised different physic and geometry properties, for example, HOG presents a gradient of illumination and context of the object while HOF present of optical flow properties. Individual feature descriptor based classifiers are solved by boosting technique to infer results for activity prediction.

In BoW feature extraction task, there are many approaches for matching and computing BoW feature, such as linear searching, greedy method, Brute-Force matching, SSD, RANSAC, and so on. The amount of them, the matching algorithm of fast approximate nearest neighbors (FANN [18]) method is a significant solution to feature extraction. In that contribution, authors have investigated many different algorithms for approximate nearest neighbor search on the set of big data with high dimension. In this study, we used the FANN method to extract BoW feature. The objective is optimization problem of finding closest word in the dictionary as follows

$$\underset{w_j \in X}{\arg\min} \sum \left\| x - w_j \right\|^2 \tag{1}$$

where x is input sample as a feature vector in n dimension, w_j is the closest element to the input sample x.

5 Training and Prediction

This section presents a boosting discriminative model of SVM for constructing an activity detection system. The main advantage of this boosting technique is its ability to boost weak classifiers for constituting strong. The number of weak classifiers inside a strong classifier is a total number of descriptors. The SVM is widely applied successfully in many fields but is limited by its high computational costs, which are affected by the data dimensions and support vectors. Therefore, a boosting technique should be introduced to reduce the time consumed while still maintain detection accuracy. The details of basic SVM will not be reviewed here for it is standard nowadays and successfully implemented [19]. In order to adaptive with practical data, the weighting of each weak classifier is automatically decided by a statistic algorithm. An advantage of the cascade classification structure is its usefulness for reducing the time required for the classification process.

There are three steps for training the activity prediction system. BoW dictionary BoW training, this task processes on the individual local descriptor. That means each kind of local descriptor of HOG, HOF, MBH, trajectory displacement is trained for getting a model of it. As mention in the previous section, unsupervised training method of k-means is considered to apply due to its advantages. Each word of the dictionary is the clustering center, which represents a strong correlation features group. The dimension number n depends on the kind of descriptors. For example, 96 elements concerning HOG feature, which consists 3 intervals of the spatiotemporal descriptor, which formed by 32 elements for each POI. Spatiotemporal feature extracted in a period of σ_t frames. The volume feature is defined as $\sigma_x \times \sigma_y \times (L/\sigma_t)$ elements. In the experiment, we used 500,000 feature vectors of each action class/each descriptor for training data.

The discriminative model using SVM is applied to classify activity class using individual local descriptor based BoW features. The details of the SVM are referenced to in [19, 20]. The training of SVM is solved the primal optimization issue of maximum margin of hyperplane, as the following

$$\min_{w,b,\zeta} \frac{1}{2}\|w\|^2 + C\sum_{i=1}^{n} \zeta_i \tag{2}$$
$$s.t. \ y_i(w^T\phi(x_i) + b) \geq 1 - \zeta_i, \ \zeta_i \geq 0, i - 1, \ldots, n$$

where $\phi()$ is BoW function with respect to a local feature descriptor x_i in high dimensional space. Coefficient C is the non-negative parameter, which assigns as a penalty bias to errors.

As mentioned in the previous section, the HOG, HOF, MBH based BoW descriptors the SVM- based partial detectors are used as weak classifiers for boosting strong detector. For boosting classifiers, the probability of the SVM is used for boosting instead of labels of activity classes.

$$H(x) = \max_{\omega \in R} \sum_{\forall i \in D} \omega_i f_i(\phi(x)) \tag{3}$$

where D is the number of local descriptor domain, ω is a coefficient of boosting component, and f is classifier corresponding to a local descriptor.

6   Experiment

Three benchmark datasets are considered for evaluation the method, which consists of the KTH [21], HMDB51 [8], UCF50 [9]. First, KTH dataset was obtained by recording 25 people performed six activity classes such as Boxing, Handclapping, Hand waving, Jogging, Running, Walking, as depicted in the first row of Fig. 3. Dataset collected under four different scenarios, such as static and homogenous background, scale variations, apparent differences, and lighting variations. This dataset includes 100 videos for each class. 600 videos were recorded with each situation includes 25 videos per a class. The configuration of the video is with 25 fps and 160 × 120 pixels resolution. Second, HMDB51 dataset includes 51 activity classes of realistic videos, which recorded from many kinds of activities such as do exercises, sports, music instruments, and other daily living activities. Totally, there are 6,766 videos of 51 action categories. Videos were normalised to the frame rate of 25 fps and video resolution of 320 × 240 pixels. Third, UCF-50 dataset includes 50 action classes from realistic action videos, which retrieved from YouTube. The dataset of 50 action categories consists of 6,681 videos. The set of videos is with a diversity object appearance and pose, object scales, viewpoint, cluttered background, illumination conditions, camera motion and so on. Videos were made a uniform with the frame rate of 25 fps and video resolution of 320 × 240 pixels. We selected six classes such as eye makeup, lipstick, haircut, head massage to show in the second row of Fig. 3. Among of that, KTH dataset is quite uniform of human style, background and selected scenarios and environment and acting performance, while HMDB51 and UCF101 datasets are realistic actions, which were collected from films, YouTube.

Fig. 3. Some kind of action classes using for evaluation from KTH, UCF-101.

Figure 4 shows evaluation results of the prediction system using an individual local descriptor, concatenate all descriptors and our approach based on boosting detectors. For simple explanation, feature descriptor concerns to the BoW descriptor using a local descriptor of HOG, HOF and so on. Experimental results showed that most of the distinguishing activity classes are similar in accuracy between a concatenation of all descriptors approach and boosting detectors based on all descriptors. However, the presented approach outperformer than the concatenating all descriptors method.

Table 1 shows some prediction results on KTH dataset. Each video consists of frequent actions. Each interval of 15 frames is used for extracting features, BoW processing, and input to classification. Crosscheck results show that accuracy of the system can reach to 94.7% precision rate. In some confusion activities, the miss prediction rate is high such as Jogging versus Running.

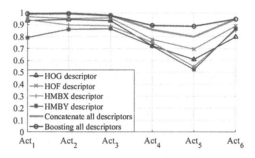

Fig. 4. Evaluation results on accuracy criterion using individual and all descriptors.

Table 1. Confusion matrix of prediction results on KTH dataset.

	Boxing	Clapping	Waving	Jogging	Running	Walking
Boxing	**99.18**	0.29	0.08	0.01	0.00	0.44
Clapping	0.55	**99.03**	0.36	0.00	0.00	0.06
Waving	0.22	2.13	**97.65**	0.00	0.00	0.00
Jogging	0.11	0.02	0.02	**89.17**	7.28	3.40
Running	0.01	0.01	0.01	10.06	**88.62**	1.29
Walking	0.02	0.01	0.00	3.56	1.79	**94.62**

Table 2 presents evaluation results on three datasets. The prediction values are show in mean of all classes of each dataset. The result shows that the presented approach is outperformer that single detector based on concatenating descriptors. This approach is appropriate for application in general condition of realistic activities.

Table 2. The mean of prediction activity class on the whole dataset.

	KTH	HMDB51	UCF50
HOG	82.3%	52.5%	84.2%
HOF	87.2%	47.6%	85.6%
MBHX	81.3%	43.5%	73.3%
MBHY	77.1%	43.2%	73.6%
Concatenation	92.9%	54.6%	89.6%
Our approach	94.7%	61.1%	90.4%

Table 3 shows the evaluation results on three datasets. The results of [11, 22, 23] in Table 3 are provided from the original papers of [11, 23]. The results of [23] were evaluated on two scenarios of automatic human detection (HD) processing and without detection task (non HD). The experimental evaluation showed that in a simple dataset of KTH, our approach are outperformers in comparison to other approaches. On some complex datasets with more action classes such as UCF50 and HMDB51, our approach also results in acceptable accuracy.

Table 3. Comparison of the our approach to the state of the art methods.

	KTH	HMDB51	UCF50
Kovashka [22]	94.53%	(*)	(*)
Wang [11]	94.2%	(*)	(*)
Wang [23] (non HD)	(*)	55.9%	90.5%
Wang [23] (HD)	(*)	57.2%	**91.2%**
Our approach	**94.7%**	**61.1%**	90.4%

(*) non data

7   Conclusion

This paper introduces a solution for activity prediction. Usually, traditional methods use concatenation of all local descriptors for classification machines. Our approach using boosting detectors based on individual descriptors produces good results in both situations of distinguished and confused activities. Furthermore, different to previous contributions, the feature locations are tracked in dense and sharing sequent frames for an inferring activity sequence. The BoW descriptor is utilized to reduce the feature space and correlation. Naturally, multiple activities are predicted by continuously frequent action processing based on inferring technique. The method is implemented for testing on some standard datasets. The experiment results proved that the investigated method outperforms the standard approaches. The most inhibit of the action prediction using vision sensor is more expensive computation in local feature extraction task. In future work, we are focusing on speedup local feature extraction for reducing computational cost for real-time applications in the limited resource systems.

Acknowledgment. This research is funded by Vietnam National Foundation for Science and Technology Development (NAFOSTED) under grant number 102.05-2015.09

References

1. Geronimo, D., Lopez, A.M., Sappa, A.D., Graf, T.: Survey of pedestrian detection for advanced driver assistance systems. IEEE Trans. Pattern Anal. Mach. Intell. **32**(7), 1239–1258 (2010)
2. Sharma, A.: A combined static and dynamic feature extraction technique to recognize handwritten digits. Vietnam J. Comput. Sci. **2**(3), 133–142 (2015)
3. Yi, Y., Lin, M.: Human action recognition with graph-based multiple-instance learning. Pattern Recogn. **53**, 148–162 (2016)
4. Pham, V., Lee, B.: An image segmentation approach for fruit defect detection using k-means clustering and graph-based algorithm. Vietnam J. Comput. Sci. **2**(1), 25–33 (2015)
5. Ziaeefard, M., Bergevin, R.: Semantic human activity recognition: a literature review. Pattern Recogn. **48**(8), 2329–2345 (2015)
6. Stefic, D., Patras, I.: Action recognition using saliency learned from recorded human gaze. Image Vis. Comput. **52**, 195–205 (2016)

7. Hoang, V.-D., Jo, K.-H.: Path planning for autonomous vehicle based on heuristic searching using online images. Vietnam J. Comput. Sci. **2**(2), 109–120 (2015)
8. Kuehne, H., Jhuang, H., Garrote, E., Poggio, T., Serre, T.: HMDB: a large video database for human motion recognition. In: 2011 International Conference on Computer Vision, pp. 2556–2563 (2011)
9. Soomro, K., Zamir, A.R., Shah, M.: UCF101: a dataset of 101 human actions classes from videos in the wild. In: CRCV-TR-12-01 (2012)
10. González, S., Sedano, J., Villar, J.R., Corchado, E., Herrero, Á., Baruque, B.: Features and models for human activity recognition. Neurocomputing **167**, 52–60 (2015)
11. Wang, H., Kläser, A., Schmid, C., Liu, C.-L.: Action recognition by dense trajectories. In: IEEE Conference on Computer Vision and Pattern Recognition (CVPR), pp. 3169–3176 (2011)
12. Dalal, N., Triggs, B.: Histograms of oriented gradients for human detection. In: Conference on Computer Vision and Pattern Recognition, pp. 886–893 (2005)
13. Dalal, N., Triggs, B., Schmid, C.: Human detection using oriented histograms of flow and appearance. In: 9th European Conference on Computer Vision, pp. 428–441 (2006)
14. Azis, N.A., Jeong, Y.S., Choi, H.J., Iraqi, Y.: Weighted averaging fusion for multi-view skeletal data and its application in action recognition. IET Comput. Vis. **10**(2), 134–142 (2016)
15. Vishwakarma, D.K., Kapoor, R.: Hybrid classifier based human activity recognition using the silhouette and cells. Expert Syst. Appl. **42**(20), 6957–6965 (2015)
16. Peng, X., Wang, L., Wang, X., Qiao, Y.: Bag of visual words and fusion methods for action recognition: comprehensive study and good practice. Comput. Vis. Image Underst. **150**, 109–125 (2016)
17. Farnebäck, G.: Two-frame motion estimation based on polynomial expansion. In: Bigun, J., Gustavsson, T. (eds.) SCIA 2003. LNCS, vol. 2749, pp. 363–370. Springer, Heidelberg (2003). doi:10.1007/3-540-45103-X_50
18. Muja, M., Lowe, D.G.: Fast approximate nearest neighbors with automatic algorithm configuration. In: VISAPP (1), vol. 2, pp. 331–340 (2009)
19. Chih-Chung, C., Chih-Jen, L.: LIBSVM: a library for support vector machines. ACM Trans. Intell. Syst. Technol. **2**(3), 1–27 (2011)
20. Maji, S., Berg, A.C., Malik, J.: Efficient classification for additive kernel SVMs. IEEE Trans. Pattern Anal. Mach. Intell. **35**(1), 66–77 (2013)
21. Schuldt, C., Laptev, I., Caputo, B.: Recognizing human actions: a local SVM approach. In: Proceedings of the 17th International Conference on Pattern Recognition, pp. 32–36 (2004)
22. Kovashka, A., Grauman, K.: Learning a hierarchy of discriminative space-time neighborhood features for human action recognition. In: IEEE Conference on Computer Vision and Pattern Recognition (CVPR 2010), pp. 2046–2053 (2010)
23. Wang, H., Schmid, C.: Action recognition with improved trajectories. In: Proceedings of the IEEE International Conference on Computer Vision, pp. 3551–3558 (2013)

Highlights Extraction in Sports Videos Based on Automatic Posture and Gesture Recognition

Kazimierz Choroś[✉]

Faculty of Computer Science and Management, Wrocław University of Science and Technology,
Wybrzeże Wyspiańskiego 27, 50-370 Wrocław, Poland
kazimierz.choros@pwr.edu.pl

Abstract. Content-based indexing of sports videos is usually based on the automatic detection of video highlights. Highlights can be detected on the basis of players' or referees' gestures and postures. Some gestures and postures of players are very typical for special sports events. These special gestures and postures can be recognized mainly in close-up and medium close view shots. The effective view classification method should be first applied. In the paper sports video shots favorable to detect gesture and posture of players are characterized and then experimental results of the tests with video shot categorization based on gesture recognition are presented. Then important and interesting moments in soccer games are detected when referees hold the penalty card above the head and look towards the player that has committed a serious offense. This recognition process is based only on visual information of sports videos and does not use any sensors.

Keywords: Content-based video indexing · Sports videos · Highlights detection · Video shot categorization · Player posture recognition · Gesture recognition · View type recognition

1 Introduction

The automatic methods of text indexing are developed over many years. These methods unable us to define a list of key words for a given text and then to use it in retrieval process. There are also methods enabling us to extract the most informative and the most important short parts of a text to generate a text summary in a automatic way. It would be desirable to apply similar approaches to index sports videos. Instead of – for example – whole long soccer matches we could admire only the most interesting moments, such as goals scored, free kicks or penalties, corners, fouls, etc. These most attractive parts of videos, the most watched by soccer fans are called video highlights. The automatic detection of highlights is a key problem in automatic video summarization process.

The automatic methods of highlight detection based on only visual feature analyses use different information extracted from the video such as information on detected player fields, dominant colors, colors of player sports wears, detected sports objects, player faces, etc. and also sports actions as well as player postures and gestures. The detection of specific postures or gestures in sports videos is not easy because in many video frames the size of humans is very small, players are in rapid movements and, therefore, they

© Springer International Publishing AG 2017
N.T. Nguyen et al. (Eds.): ACIIDS 2017, Part I, LNAI 10191, pp. 619–628, 2017.
DOI: 10.1007/978-3-319-54472-4_58

are usually blurred, and then the players are often in the group mainly in team games, so the single figures are not easily identified.

Nevertheless, the opportunity to detect some of this special situations characterized by typical postures and typical gestures of players is very attractive leading to the detection of the most interesting parts of sports videos. One of such interesting situations in soccer games is the moment when the referee shows the yellow card a player. Yellow and red cards are used in many sports categories as a means of warning, reprimanding or penalizing a player, coach or team official. A penalty card is shown by a sports referee or sports umpire to indicate that a player has committed an offense. In such a case according to the sporting rules the referee is obliged to hold the card above his head while looking or pointing towards the player that has committed the offense. Usually the referee is holding the yellow or red card during several seconds. The detection of these cases, very specific for certain sports categories, can be also useful in shot categorization of sports videos.

In this paper a method of the detection of the referee holding the yellow penalty card is presented. Some experimental results are also discussed. The paper is structured as follows. The next section describes related work on automatic sports video summarizing based on highlights as well as on posture and gesture recognition. The approach applied to detect and classify posture and gesture in soccer videos is described in the third section. The fourth section discusses the cases in sports videos of successful and unsuccessful detections of the referee showing the yellow card a player. The final conclusions are discussed in the last section.

2   Related Work

Sports highlights are probably the most frequently viewed on the Internet and the most frequently shown in the TV sports news. They are always used as a summary of sports events and usually broadcasted instead of the whole sports events. The most interesting and attractive parts of sports videos are [1]: specific events like the scoring of a goal in soccer videos, actions occurring around the goal area in a soccer game, reactions from the live audience such as applause and cheering, excited speech of a sports commentator, the cutting rate, score caption, and replay sequence, repeated footages in news, video parts with high motion and colorful, or those viewed most often by the user, but also segments rare and unusual, or presenting exciting people.

The best sports highlights are usually included in headlines of TV news or TV sports news [2]. Headlines can be automatically detected in sports news [3] but highlights for headlines are manually chosen by editors of TV shows, news, reports, etc. It would be desirable to generate highlights for a given sports event also in automatic way.

Human posture and gesture recognition methods are mainly applied in human computer interactions. The thorough survey of present research related to gesture recognition in human-computer interaction and the comparative analysis of the methods using different key parameters are presented in [4]. The human posture and gesture detection, gait recognition, and event recognition in digital videos are also main processes in human activity recognition. The overview of research on human activity recognition, including objectives, challenges, representations, classifiers, and datasets is presented in a recently edited book on human activity recognition and prediction [5]. But posture and gesture recognition can be useful not only in human-computer interaction but also in the analyses of archive sports videos and the recognition and categorization of shots, scenes, and events.

In the past as well as nowadays much research has been conducted on analyzing soccer games as one of the most popular sports category in the world. The most exciting parts of soccer games happen near the goal post, hence the method proposed in [6] concentrates near goal post. These significant and most exciting segments near goal post were extracted by extracting and analyzing key-frames. First the key-frames are categorized according to a view type, into far-view, mid-view and out-view. Further analyzes are performed with mid-view frame and goal post is determined using the Hough transform. Then the soccer highlights are chosen along the frames containing goal post. Finally, events are classified as significant and insignificant using fuzzy classifier.

The detection of players can be also applied to reduce undetected cuts [7]. If two (or even more) consecutive shots present two different sports events but of the same category they can be detected as one shot. It often happens for soccer events reported in the sports news. The strong similarity mainly of green color of soccer fields makes it difficult to detect a cut. The differences in numbers of player detected in the field and mainly the differences of colors of their sports wears are strong signals of shot changes.

However, it has been already observed [8] that in many sports views, the camera usually covers a large part of the playing fields, and as a result the resolution of players is very low making the recognition of player postures and gestures a challenging task. The task becomes more difficult if in addition a camera is rapidly moved. In soccer videos, the players often have low resolutions because the executive producer knows that the TV audience prefer to watch a wide view of the player field in order to understand the tactic of a game and relative position of players in the field. This is the reason why for soccer fans the worst seat in the audience is better than the very comfortable armchair in front of a TV set. The wide views are also preferred in surveillance video where fixed cameras are usually used in practice. The consequence is that the tests of the methods for posture and gesture recognition in sports videos have been performed more and more frequently but with tennis, basketball, and other sports categories.

In the early research some solutions have been proposed to automatically recognize the behavior of players or umpires for example in cricket games [9]. The hierarchical hidden Markov model was applied in conjunction with a filler model for segmenting and classifying gestures at differing levels of detail. However, in this approach described accelerometer sensors in the form of wrist bands were used. Sports referees were obliged to wear the sensors during the game and therefore their movements were recorded and analyzed throughout the game.

In [10] the tennis games have been analyzed. The method was based on the approach combining non-dominant color extraction and edge detection filters to effectively separate players from background and to obtain the raw player figure. Furthermore, these figures were then refined using a modern shadow removal scheme. To locate the player in the first frame of a video a window called an initial search window was used. Once the players were detected, their locations were applied in the subsequent frames using a much smaller search window. It has been proposed to adapt this search window to efficiently follow the changing player figure since – as it is observed in sports videos – a player is moving and the player posture is varying from frame to frame. Such a adaptation approach better detected the complete player figure and also reduced the processing time and significantly reduced the noise. The proposed method has been found very efficient for different colors of courts and different lighting conditions.

Whereas in the method of player detection described in [11] the largest object found in the frame was found and filled by removing the small one.

Also basketball videos have been often analyzed. Detecting and tracking every player in a basketball video seems to be still a challenging task because basketball players are frequently and quickly moving in broadcast sports videos. On the other hand due to the co-occurrence between player actions and camera motions action recognition process can be efficient [12]. It could be even improved by separating the player motion from the background motion.

In order to detect players in basketball videos it has been proposed in [13] to use a model based on mixture of non-oriented pictorial structures and information on locations. Players including their body parts are detected on the basketball court. However, during the player detection process it happened that people from the audience and the referees have been detected as players. Some solutions based on the detection of court boundaries and the area below the basket have been also proposed to remove fans or referees. Thereby, only the positions of players were precisely determined.

In [14] the method of the movement prediction of basketball referees has been proposed using a multilayered perceptron neural network. Such a network is working on the basis of a ball movement during a play action.

In the system presented in [15] not only the players were detected, their positions were determined, but also the players were identified. Player identification is not an easy process in sports videos because of blurry faces due to fast camera motion as well as relatively low-resolution of broadcast videos. Moreover, jersey numbers which could be very useful in player identification are invisible or even when visible are usually deformed because of player movements.

Also sports videos of other categories were examined to find the solutions to automatically detect players and to analyze the sports actions, these are for example American football [16], handball [17], or badminton [18].

A completely different and interesting approach for automatic identification of highlights in currently displayed sports videos has been presented in [19]. Highlights are detected basing on the observations, tracking, and recording by a video camera of viewer facial expressions and heart rate signals.

However, in our research only visual information of archives sports videos are tested as input data. Only digital frames of a sports video are analyzed without using any sensors, any additional camera, or any other additional equipment.

3 Posture and Gesture Recognition in Sports Videos

The process of posture and gesture recognition has been implemented in the AVI Indexer [20]. It is performed in several steps:

– selection of a single frame from the sports video,
– shot view classification,
– detection of objects,
– feature extraction,
– object classification.

At the beginning single frames are extracted from an analyzed soccer sports video. First of all for a given frame its view should be determined. A method of view classification has been based on the approach presented in [21] and then applied many times in different experiments and implementations [22]. Three view types are usually defined for soccer games: long views, medium views, and out of field or close-up views. For further analyses only shots of medium views are considered. Medium view shots are such shots that a whole player body is usually visible. It has been observed that although the occurrence of a single medium view shot between long view shots often corresponds to a play, whereas a sequence of medium view shots usually indicates a break in the soccer game. The situations of showing a player penalty cards is such a moment of a break in the game. Also replays, replays of important parts of the game, are shown as a medium view shot more frequently than other two shot types.

The view type detection methods use the share of the green color in the frame, i.e. dominant color of soccer pitch. The efficiency of this method can be improved by removing upper parts of the analyzed frame. The frame is first divided according to the ratio 3 to 5 to 3 (which is close to the golden ratio) in both directions. Then the upper segments of the frame which are very probable out of the field are removed.

Next, the background of the frame is determined. And all other parts of the analyzed frame are treated as objects neglecting their colors (Fig. 1).

Fig. 1. The separation of objects and green background of the soccer pitch. (Color figure online)

All adjacent pixels forming groups of pixels are supposed to be objects if the number of pixels of a single group is greater than a certain threshold. All very small groups of pixels are eliminated from further analyses.

The detected objects are then classified. Two well-known algorithms have been implemented [23] in the AVI Indexer: naive Bayes classifier and k-nearest neighbors classification algorithm. The objects are classified into following nine classes:

- standing player – player usually slowly moving with hands near the body,
- running player – player usually moving, the hands and feet at a certain small distance from the body,
- sitting or lying player – player sitting or lying,
- player with hand up – player with hand raised (the most important cases for our further tests),
- player with hand to the side – player with one hand pointing to the side,
- player with hands to the side – player with both hands pointing to the side,
- player group – a group of players as one object,
- player in a group of players – a group of players but one of the players makes a gesture with his hand,
- unidentified object – small object, it might be a part of a player body or any other object like banner advertising, audience, or table with results.

In the training set there were 450 images of postures with gestures, 50 images of each class. In the analyzed soccer game 1801 postures and gestures have been recognized. The defined postures do not occur with the same frequency. The group class is most frequently observed, but it is not surprising because the soccer game is a team game in which players act in a close proximity to each other. In some cases even for humans it is not easy to decide to which class a given object should be classified. In the case of more than one possible decision, for example player running with hand to the side, it has been assumed that the gesture has higher priority than the posture.

Table 1 presents the efficiency of these two classification algorithms applied for a soccer game recorded off the TV in the resolution 720×576, broadcasted during the World Cup 2014.

Table 1. Precision and recall of the classification process applying two implemented methods and received for the recorded soccer game of the World Cup 2014.

		Naive Bayes classifier		k-nearest neighbors	
Player position with a given gesture	Number of objects	Precision	Recall	Precision	Recall
Standing player	413	0.34	0.58	0.40	0.49
Running player	257	0.26	0.35	0.26	0.39
Sitting or lying player	110	0.48	0.18	0.35	0.23
Player with hand up	23	0.08	0.48	0.06	0.58
Player with hand to the side	137	0.35	0.29	0.37	0.39
Player with hands to the side	25	0.16	0.28	0.17	0.31
Player group	480	0.51	0.11	0.50	0.19
Player in a group of players	269	0.33	0.30	0.36	0.24
Unidentified object	87	0.72	0.49	0.67	0.49

The results are not satisfactory. The most interesting results are obtained for the group class – high precision and for the player with hand up class – high recall. This last case is typical for a very important referee action during a soccer game. When a player has committed an offense, the referee or umpire shows a player a penalty card, yellow or red. It usually indicates a caution given to a player or a coach regarding his or her conduct, it may also indicate a temporary suspension of a player. These parts of soccer games are often presented in headlines of sports news or in game summary. So, they are one of the types of soccer highlights. It would be desirable to automatically detect such a referee action.

4   Detection of Referees Showing Penalty Card

During the Soccer World Cup 2014 many yellow cards have been shown players and our purpose was to verify if the methods of posture and gesture recognition could be useful to automatically find these parts of soccer videos. Even if the efficiency of posture and gesture recognition is not impressive it is interesting to know whether they are sufficient for the detection of a very specific part of soccer match – the referee with the yellow card.

Fig. 2. Analyzed frames from the sports video shots with referees showing a player the yellow card during soccer games. (Color figure online)

First, the shots of medium view type have been selected using the methods of automatic playing field detection, dominant color extraction, and estimation of the share of playing field color in the frame [24]. The frames from these shots have been examined. It enabled us to identify the reasons of unsuccessful as well as successful results of automatic detection. The examples presented in Fig. 2 show that the reasons of unsuccessful detection are mainly due to the fact that even if the referee with yellow card is recorded in a medium view shot and the referee is usually detected as an object however the posture is not always correctly classified, with required gesture, i.e. hand up and moreover not always the yellow card can be detected in his hand.

Unfortunately, the positions of referees in the soccer video shots with the situations when the referee holds the yellow card are not always favorable to detect a human posture. In the first case the referee with yellow card is shown rather close-up (Fig. 2a), his posture is partially cut and the algorithm was not able to detect him as a human with hand up. Although this view has been classified as medium, however it was rather very close-up. In the second case the referee held up a hand but unfortunately his palm is recorded on the border of the image on the background of the banner advertising and in consequence yellow card has not been found in this shot (Fig. 2b). The third frame presents the referee in a green shirt (Fig. 2c). Although the posture could be detected, unfortunately the green shirt has been included into the player field area and in consequence the referee has not been detected at all. Also in the fourth case the referee with the yellow card has not been detected. In this case the referee was in a group of players and his posture was difficult to recognize (Fig. 2d), although his hand with yellow card was easily observed.

Only the fifth case (Fig. 3) has been correctly identified. The posture of the referee is well seen, his hand is held up, the yellow card is also distinguishable against the green grass on the soccer pitch.

Fig. 3. Successful case of soccer shot with the referee showing a player the yellow card during soccer games. (Color figure online)

These examples prove that the efficiency of the detection of the referee showing a player the yellow card depends not only on the techniques of human posture and gesture recognition but also on some other conditions, such as positions of referees, view types, and even the color of referee clothing.

5   Conclusions

It seems that the automatic methods of highlights detection in soccer videos should successfully apply currently developed methods of posture and gesture recognition. However, these methods are not yet sufficiently effective, especially when they are applied for analyses of relatively low resolution videos like sports games recorded off the TV. Furthermore, in soccer games long views are preferred, so the players are really very small objects and their gesture are difficult to recognize.

In the AVI Indexer two algorithms were implemented: naive Bayes classifier and k-nearest neighbors classification algorithm. The objects detected were classified into nine classes: standing player, running player, sitting or lying player, player with hand up, player with hand to the side, player with hands to the side, player group, player in a group of players, and unidentified object. The promising result – high recall was obtained for the player with hand up class. Such a posture is observed when the referee shows a player the yellow card. The opportunity to automatically detect the referee holding the yellow penalty card is examined in the paper. It would be useful for extracting highlights from soccer videos.

The tests proved that the efficiency of the detection of the referee showing a player the yellow card depends not only on the techniques of human posture and gesture recognition. Other features, such as positions of referees, view types, and even the color of referee clothing have also a great influence.

However, in a shot of medium view type where the whole posture of a referee is observed such a situation during soccer games can be automatically detected and can be used in the process of highlight extraction.

References

1. Truong, B.T., Venkatesh, S.: Video abstraction: a systematic review and classification. ACM Trans. Multimedia Comput. Commun. Appl. (TOMM) **3**(1), 1–37 (2007). Article 3
2. Choroś, K.: Video structure analysis for content-based indexing and categorisation of TV sports news. Int. J. Intell. Inf. Database Syst. **6**(5), 451–465 (2012)
3. Choroś, K.: Automatic detection of headlines in temporally aggregated TV sports news videos. In: Proceedings of the 8th International Symposium on Image and Signal Processing and Analysis (ISPA 2013), pp. 147–152. IEEE (2013)
4. Rautaray, S.S., Agrawal, A.: Vision based hand gesture recognition for human computer interaction: a survey. Artif. Intell. Rev. **43**(1), 1–54 (2015)
5. Fu, Y.: Human Activity Recognition and Prediction. Springer, Switzerland (2016)
6. Naik, V., Rathod, G.: An algorithm for retrieval of significant events near goal post from soccer videos using fuzzy systems. Int. J. Emerg. Technol. Adv. Eng. **3**(3), 808–814 (2013)
7. Choroś, K.: Improved video scene detection using player detection methods in temporally aggregated TV sports news. In: Hwang, D., Jung, J.J., Nguyen, N.-T. (eds.) ICCCI 2014. LNCS (LNAI), vol. 8733, pp. 633–643. Springer, Heidelberg (2014). doi: 10.1007/978-3-319-11289-3_64
8. Roh, M.-C., Christmas, B., Kittler, J., Lee, S.-W.: Robust player gesture spotting and recognition in low-resolution sports video. In: Leonardis, A., Bischof, H., Pinz, A. (eds.) ECCV 2006. LNCS, vol. 3954, pp. 347–358. Springer, Heidelberg (2006). doi: 10.1007/11744085_27
9. Chambers, G.S., Venkatesh, S., West, G.A.: Automatic labeling of sports video using umpire gesture recognition. In: Fred, A., Caelli, T.M., Duin, R.P.W., Campilho, A.C., de Ridder, D. (eds.) SSPR/SPR 2004. LNCS, vol. 3138, pp. 859–867. Springer, Heidelberg (2004). doi: 10.1007/978-3-540-27868-9_94
10. Hsieh, C.H., Huang, C.P., Jiang, Y.C.: Player detection, tracking and segmentation in broadcast tennis video. J. CCIT **43**(1), 25–40 (2014)

11. Archana, M., Geetha, M.K.: An efficient ball and player detection in broadcast tennis video. In: Berretti, S., Thampi, S.M., Srivastava, P.R. (eds.). AISC, vol. 384, pp. 427–436. Springer, Heidelberg (2016). doi:10.1007/978-3-319-23036-8_37

12. Takahashi, M., Naemura, M., Fujii, M., Little, J.J.: Recognition of action in broadcast basketball videos on the basis of global and local pairwise representation. In: Proceedings of the IEEE International Symposium on Multimedia (ISM), pp. 147–154. IEEE (2013)

13. Ivankovic, Z., Rackovic, M., Ivkovic, M.: Automatic player position detection in basketball games. Multimedia Tools Appl. **72**(3), 2741–2767 (2014)

14. Pecev, P., Racković, M., Ivković, M.: A system for deductive prediction and analysis of movement of basketball referees. Multimedia Tools Appl. **75**(23), 16389–16416 (2016)

15. Lu, W.L., Ting, J.A., Murphy, K.P., Little, J.J.: Identifying players in broadcast sports videos using conditional random fields. In: Proceedings of the IEEE Conference on Computer Vision and Pattern Recognition (CVPR), pp. 3249–3256. IEEE (2011)

16. Atmosukarto, I., Ghanem, B., Saadalla, M., Ahuja, N.: Recognizing team formation in American football. In: Computer Vision in Sports, pp. 271–291. Springer International Publishing, Switzerland (2014)

17. Santiago, C.B., Sousa, A., Reis, L.P.: Vision system for tracking handball players using fuzzy color processing. Mach. Vis. Appl. **24**(5), 1055–1074 (2013)

18. Peng, Y., Ma, X., Gao, X., Zhou, F.: Background estimation and player detection in badminton video clips using histogram of pixel values along temporal dimension. In: Proceedings of the Sixth International Conference on Electronics and Information Engineering, p. 979409. International Society for Optics and Photonics (2015)

19. Chakraborty, P.R., Tjondronegoro, D., Zhang, L., Chandran, V.: Automatic identification of sports video highlights using viewer interest features. In: Proceedings of the International Conference on Multimedia Retrieval, pp. 55–62. ACM (2016)

20. Choroś, K.: Video structure analysis and content-based indexing in the automatic video indexer AVI. In: Nguyen, N.T., Zgrzywa, A., Czyżewski, A. (eds.) Advances in Multimedia and Network Information System Technologies. AISC, vol. 80, pp. 79–90. Springer, Heidelberg (2010). doi:10.1007/978-3-642-14989-4_8

21. Ekin, A., Tekalp, A.M., Mehrotra, R.: Automatic soccer video analysis and summarization. IEEE Trans. Image Process. **12**(7), 796–807 (2003)

22. Li, L., Zhang, X., Hu, W., Li, W., Zhu, P.: soccer video shot classification based on color characterization using dominant sets clustering. In: Muneesawang, P., Wu, F., Kumazawa, I., Roeksabutr, A., Liao, M., Tang, X. (eds.) PCM 2009. LNCS, vol. 5879, pp. 923–929. Springer, Heidelberg (2009). doi:10.1007/978-3-642-10467-1_83

23. Cholewa B.: Detection and categorization of objects on the basis of player gestures and postures in sports videos. Unpublished Master's Thesis (in Polish), Wrocław University of Science and Technology (2015)

24. Choroś, K.: Automatic playing field detection and dominant color extraction in sports video shots of different view types. In: Zgrzywa, A., Choroś, K., Siemiński, A. (eds.). AISC, vol. 506, pp. 39–48. Springer, Heidelberg (2017). doi:10.1007/978-3-319-43982-2_4

Advanced Data Mining Techniques and Applications

A High-Performance Algorithm for Mining Repeating Patterns

Ja-Hwung Su[1,2(✉)], Tzung-Pei Hong[3,4], Chu-Yu Chin[5,6],
Zhi-Feng Liao[4], and Shyr-Yuan Cheng[6]

[1] Department of Information Management, Cheng Shiu University, Kaohsiung, Taiwan
bb0820@ms22.hinet.net
[2] Department of Information Management, Kainan University, Taoyuan, Taiwan
[3] Department of Computer Science and Information Engineering,
National University of Kaohsiung, Kaohsiung, Taiwan
[4] Department of Computer Science and Engineering, National Sun Yat-sen University,
Kaohsiung, Taiwan
[5] Computer Science and Information Engineering, National Cheng Kung University,
Tainan, Taiwan
[6] Telecommunication Laboratories, Chunghwa Telecom Co. Ltd., Taoyuan, Taiwan

Abstract. A repeating pattern is a sequence composed of identical elements, repeating in a regular manner. In real life, there are lots of applications such as musical and medical sequences containing valuable repeating patterns. Because the repeating patterns hidden in sequences might contain implicit knowledge, how to retrieve the repeating patterns effectively and efficiently has been a challenging issue in recent years. Although a number of past studies were proposed to deal with this issue, the performance cannot still earn users' satisfactions especially for large datasets. To aim at this issue, in this paper, we propose an efficient algorithm named *Fast Mining of Repeating Patterns (FMRP)*, which achieves high performance for finding repeating patterns by a novel index called *Quick-Pattern-Index (QPI)*. This index can provide the proposed *FMRP* algorithm with an effective support due to its information of pattern positions. Without scanning a given sequence iteratively, the repeating patterns can be discovered by only one scan of the sequence. The experimental results reveal that our proposed algorithm performs better than the compared methods in terms of execution time.

Keywords: Repeating pattern · Fast mining of repeating patterns · Quick-pattern-index · Data mining · Knowledge discovery

1 Introduction

Knowledge discovery has been a hot topic in recent years. Actually patterns hidden in the real data can be viewed as knowledge and therefore how to mine the valuable patterns from the real data also has been a challenging issue. To deal with such issue, a number of mining algorithms are proposed nowadays. In recent studies, the patterns are categorized into several categories, including association patterns, sequential patterns, cyclic patterns, repeating patterns and so on. Different patterns are useful to different fields of

© Springer International Publishing AG 2017
N.T. Nguyen et al. (Eds.): ACIIDS 2017, Part I, LNAI 10191, pp. 631–640, 2017.
DOI: 10.1007/978-3-319-54472-4_59

knowledge engineering. Among these categories, repeating patterns are popular because they can be regarded as a set of representative patterns that facilitates object recognition. Typically, a repeating pattern contains a set of sequential elements, which can also be viewed as a sequence repeats in a regular form. For example, a string {1, 2, 3, 1, 2, 4, 1, 2} contains a repeating substring {1, 2} which is identified as a repeating pattern. In fact, lots of repeating patterns appear in our life, such as musical and medical data.

For musical data, a music piece can be viewed as a sequence of bytes. The semantics of a music piece is abridged where it is regarded as an incomprehensible large object. Lots of meaningful feeling is comprehended by human beings as they listen to a music object. However, it is very difficult to automatically extract the feeling from the raw data of the audio object [1, 3, 10]. Ghias, *et al.* proposed an approach to transform a music query example into a string that represents a higher note (denoted by, *"U"*), lower note (denoted by, *"D"*), or the same note (denoted by, *"S"*) [4]. Furthermore, some technique [5] was also proposed for finding the repeating patterns from a sequence of musical chords and melodies. Therefore, the musical repeating patterns can be retrieved to support semantic music retrieval. In the field of medical biology, repeating subsequences are a kind of repeating patterns. Actually, the repeating patterns in biological cell DNA occur in multiple copies throughout a genome. The functions and descriptions of these subsequences are currently being characterized by scientists. Tandem repeat is a kind of repeating sequence occurring in DNA when a pattern of one or more nucleotides repeats and the repetitions are directly adjacent to each other. Several protein domains also form tandem repeats within their amino acid primary structures. In practical, tandem repeats are very helpful in the field of bioinformatics.

Although a set of previous work was proposed to discover repeating patterns, the performance is not satisfactory. To accelerate the retrieval of repeating patterns, in this paper, we propose an algorithm named *Fast Mining of Repeating Patterns (FMRP)* to achieve high performance of mining repeating patterns by the proposed *Quick-Pattern-Index (QPI)*. With this index, the occurrences and positions of the patterns are kept to reduce the cost of searching the repeating patterns. Hence, without scanning the sequence iteratively, the repeating patterns can be discovered by only one scan of the input sequence. The experimental results reveal our proposed algorithm performs better than the compared methods in terms of execution time. The rest of this paper is organized as follows. In Sect. 2, a brief review of related work is shown. In Sect. 3, we demonstrate our proposed method for mining of repeating patterns in great detail. An example is given in Sect. 4. The experimental evaluations of the proposed algorithm are depicted in Sect. 5. Finally, conclusions and future work are shown in Sect. 6.

2   Related Work

In real applications, the repeating patterns hidden in object sequences might be viewed as implicit knowledge for identification of the objects. Hence, how to retrieve valuable repeating patterns from sequences has been a hot issue in the last decades. For this issue, a number of past studies were conducted. However, the high performance is not easy to achieve and the brief review of these predecessors is shown in the followings.

I. Algorithms of mining of repeating patterns. Hsu *et al.* [5] proposed a method to generate the repeating patterns. In this method, a string-join operation and a data structure called RP-tree are proposed to achieve high performance of mining of repeating patterns. The basic idea of this approach is to iteratively join two short repeating patterns into a long one. To speed up the join procedure, a tree structure named RP-tree was proposed. Although the tree structure can reduce the time complexity of join operations, the checking cost is so high that the generation of repeating patterns is inefficient. In addition to RP-tree, another tree structure for generating the repeating patterns is suffix tree. Basically, the suffix tree is the compressed tree for the nonempty suffixes of a string Q. Since a suffix tree is a compressed tree, it is an important idea that, the procedure of mining of repeating patterns highly refers to its sub-trees. Once constructed, several operations can be performed quickly, for instance, locating a substring in S, locating matches for a regular expression pattern etc. Suffix trees also provide the first linear-time solutions for the longest common substring problem [8, 9]. Unfortunately, the construction of such a tree for the string Q takes much time and space.

II. Applications of mining of repeating patterns. Repeating patterns are usually supported for the matching of local image features. They can be modeled as a set of sparse repeated features in which the crystallographic group theory. P. Muller *et al.* proposed an approach to detect symmetric structures and to reconstruct a 3D geometric model [2]. Liu *et al.* proposed a new method for detection of repeated patterns following a *Kronecker Product formulation*. They handle problems of pose variation and varying brightness by employing the low-rank part of the rearranged input facade image [6, 12]. *Automatic video summarization* presented an effective way to accelerate the video browsing and retrieval. The video structure was first analyzed by spatial-temporal analysis. Then, they extracted video nontrivial repeating patterns to remove the visual-content redundancy among videos [7, 11].

3 Proposed Method

3.1 Overview of the Proposed Method

In recent years, several schemes have been proposed to discover repeating patterns. However, the execution performance is limited in the search strategy, which is not satisfactory. To improve the execution performance, in this paper, we propose an efficient algorithm that utilizes a novel data structure to facilitate the discovery of repeating patterns. In detail, without scanning a given sequence iteratively, the repeating patterns can be discovered by only one scan of the sequence. As shown in Fig. 1, the mining process can be divided into two primary phases, namely construction of the index and generation of repeating patterns.

I. Construction of the index. This is the base of the proposed method and the major goal of this phase is to construct an informative index to reduce the cost of mining of repeating patterns. Basically, this index called *Quick-Pattern-Index (QPI)* contains two of pattern information, including positions and occurrences.

II. Generation of repeating patterns. In this phase, the main goal is to search the repeating patterns using the constructed index. To this end, a novel mining method called Fast Mining of Repeating Patterns (FMRP) is proposed in this work. Through the prefix search strategy, the repeating subsequences (also called repeating patterns) are mined successfully.

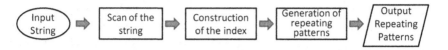

Fig. 1. Workflow of the proposed method.

Before describing our proposed method, the repeating patterns are defined as follows.

Definition 1. For a string B, if a substring A appears more than once, its length is larger than one and no any substring contains A, we call A is a repeating pattern of B. For example, the repeating pattern of a string $\{1, 2, 3, 1, 2, 3\}$ is $\{1, 2, 3\}$.

3.2   Construction of the Index

Figure 2 shows the process of constructing the index. In this process, two indexes are constructed, including an array storing all patterns and an array storing the pattern positions in the input string. Through these two indexes, the next process called generation of repeating patterns can be performed efficiently.

Input: A string $ST=\{s_1, s_2, ..., s_k\}$ containing g unique symbols (also called pattern);
Output: A set QPI;
Process: Construction of the index
1.   initialize the string array $S[i]$ as a null array;
2. $i=0$;
3.   **for** each pattern s_i in ST **do**
4. $i++$;
5.   insert s_i into $S[i]$;
6. $p_i{\rightarrow}count++$; // $p_i{\rightarrow}count$ denotes the occurrence of pattern s_i, where $s_i = p_i$;
7. $p_i{\rightarrow}pos[p_i{\rightarrow}count]=i$;// $p_i{\rightarrow}pos[p_i{\rightarrow}count]$ denotes the position of pattern s_i;
8. $P= P{\cup}p_i$; //P is the set containing patterns, occurrences and positions;
9.   **end for**
10.   **return** $S[i]$ and P as QPI;

Fig. 2. Process of construction of the index.

3.3   Generation of Repeating Patterns

Figure 3 shows the algorithm of generating the repeating patterns. Basically, it can be regarded as a prefix search paradigm that mines the repeating patterns prefixed by a pattern. Hence, as shown in Line 1, the whole procedure starts with one of the distinct patterns and mines the repeating patterns one by one prefixed patterns. Under the prefixed pattern p_i, Lines 2–7 show the patterns in the next positions of p_i are grouped into the set *NextStep* and they are linked into the paths where the root is p_i. Next, Lines

9–23 show that, for each pattern p_x in *NextStep*, all patterns in the next positions of p_x are further grouped into the set $p_x.NextTwoStep$.

```
Input: A set QPI;
Output: A set of repeating patterns RPS;
Process: Generation of the repeating patterns;
1.    for i=1 to g do
2.      for j=1 to p_i→count do  //where p_i denotes the i^th pattern in S
3.        let p_x=S[p_i→pos[j]+1];
4.        p_x→pos= p_i→pos[j]+1;
5.        insert the pattern p_x into the set NextStep;
6.        link p_i and p_x as a path pк_x^i ; //where p_i is the root of pк_x^i ;

7.      endfor
8.      cnt=|NextStep|;
9.      for each p_x∈NextStep do
10.     temp=φ;
11.     if p_x→count=1 then
12.     let all patterns in path pк_x^i be the set temp;

13.         if temp∉ RPS then
14.         insert all patterns in path pк_x^i into the resulting set RPS;

15.     else
16.         for j=1 to p_x→count do
17.           let p_y=S[p_x→pos[j]+1];
18.           p_y→pos=p_x→pos[j]+1;
19.           insert the pattern p_y in the next position of p_x into the set p_x.NextTwoStep;
20.           insert p_y into the path pк_x^i ; //where p_i is the root of pк_x^i ;

21.     endfor
22.     endif
23.     endfor
24.     while cnt>0 do
25.     cnt=0;
26.       for each p_x∈NextStep do
27.         for each p_y∈p_x.NextTwoStep do
28.     temp=φ;
29.       if p_y→coun=1 then
30.     let all patterns in path pк_x^i be the set temp;

31.     if temp∉ RPS then
32.             insert all patterns in path pк_x^i into the resulting set RPS;

33.     else
34.             for j=1 to p_y→count do
35.               let p_z=S[p_y→pos[j]+1];
36.           p_z→pos=p_y→pos[j]+1;
37.               insert the pattern p_z in the next position of p_y into the set  p_y.NextTwoStep;
38.               insert p_z into TMP;
39.               insert p_z into the path pк_x^i ; //where p_i is the root of pк_x^i ;

40.     end for
41.         end if
42.     endfor
43.       end for
44.       set NextStep = NULL;
45.       set NextStep = TMP;
46.       set TMP = NULL;
47.     cnt=|NextStep|;
48.     endwhile
49.     endfor
50.   return RPS;
```

Fig. 3. Algorithm of generation of repeating patterns.

In this process, if the number of patterns in the next positions of p_x is less than 2, check whether the patterns in the referred path exist in the results or not. If they have not existed in the results, the patterns in the referred path will be returned as the resulting patterns. Until this process, the whole search is divided into a number of search paths prefixed by p_x where the root is p_i. In the following process of the algorithm, Lines 21–42 show that the prefixed searches are performed to mine all repeating patterns.

4   Example

For more detail, an example for mining process including *construction of the index* and *generation of repeating patterns* is given in this section.

4.1   Construction of the Index

In this example, we give a string $ST = \{1\ 3\ 5\ 2\ 1\ 3\ 7\ 4\ 1\ 3\ 5\ 6\ 1\ 3\ 7\ 8\}$ which contains sixteen symbols (called patterns in the example) and the kind of unique patterns is 8. In this process, first, the string array $S[]$ is initialized as a null array. Second, ST is loaded into the array $S[]$ as shown in Fig. 4, where $S[0]$ keeps the length of ST. Finally, a table containing occurrences and positions of patterns is conducted as shown in Table 1.

Fig. 4.   Example of the resulting array S.

Table 1.   Example of the resulting set P.

Pattern	Occurrence	Position
p_1	4	1, 5, 9, 13
p_2	1	4
p_3	4	2, 6, 10, 14
p_4	1	8
p_5	2	3, 11
p_6	1	12
p_7	2	7, 15
p_8	1	16

4.2   Generation of Repeating Patterns

By referring to Fig. 3, the exampling process is shown step by step as follows.

Step 1: Considering Table 1, a pattern is selected from Table 1.
Step 1.1: The selected pattern p_i is conducted as the root and the next positions of p_i are grouped into the set *NextStep*. Here, the example pattern p_i is p_1 and its *NextStep* is P_3.
Step 1.2: For each p_i, find each pattern p_x in p_i.*NextStep*. If the occurrences of p_x are more than 1, link p_x and p_i as the paths where the root is p_i. Then, all patterns in the next positions of p_x are further grouped into the set p_x.*NextTwoStep*. If the occurrences of p_x are more than 1, link p_x and all patterns in the next positions of p_x. In this example shown in Fig. 5, the green grid is defined as {pattern#, (next positions), (parents)}. For example, $\{p_3, (3, 7, 11, 15), (p_1)\}$ denotes that, the pattern number is 3 where the next position set is $\{3, 7, 11, 15\}$ and the parent is 1. Next, because the occurrences of all

patterns $\{p_5, p_7\}$ in the p_3.*NextTwoStep* are more than 1, $\{p_3\}$ and $\{p_5, p_7\}$ are linked. After generating p_x.*NextTwoStep*, for each pattern p_x in *NextStep*, let $p_x = p_i$ as shown in Fig. 6. If the occurrences of all patterns $\{p_5, p_7\}$ in the p_3.*NextTwoStep* are less than 2, check whether the patterns in the path of the root generated in Step 1 have existed in the result set *RPS* or not. If the patterns in the path of the root generated in Step 1 have not existed in the result set *RPS*, insert them into *RPS*. In this example, this step does not stop until p_x.*NextTwoStep* is null. Because the occurrences of all patterns $\{p_5, p_7\}$ in the p_3.*NextTwoStep* are less than 2, $\{p_1, p_3, p_5\}$ and $\{p_1, p_3, p_7\}$ are inserted into *RPS* as shown in Fig. 7.

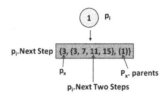

Fig. 5. Example of {pattern#, (next positions), (parents)}

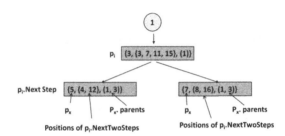

Fig. 6. Example for generating new *NextStep* whose p_x is 5 and 7.

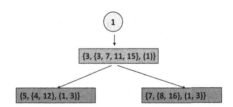

Fig. 7. Example of the resulting *RPS* are $\{P_1, P_3, P_5\}$ and $\{P_1, P_3, P_7\}$.

Step 2: Repeat Step 1 until all repeating patterns are generated.

5   Empirical Study

5.1   Experimental Settings

All of the experiments were implemented in C++ and executed on a PC with Intel Core i5-4590 @ 3.3 GHz of CPU and 16 GB of memory. The parameter settings for yielding the experimental data are shown in Table 2. The compared methods with our proposed method are RP-tree [5] and Suffix tree [8].

Table 2. Parameter settings

	Minimum	Maximum	Interval
#Transactions	500	1000	100
#Transaction length	500	1000	100
#Pattern categories	10	200	50

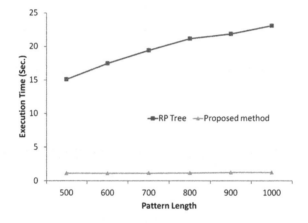

Fig. 8. Execution time of the compared methods under different transaction lengths.

5.2   Comparisons of the Proposed Method and Other Methods in Terms of Execution Time

The first evaluation we want to show is the performance by varying the transaction length as the data size is 1000 and the number of pattern categories is 200. The related experimental results are shown in Fig. 8 that depicts several points. First, the longer the transaction length, the larger the execution time. Second, RP-tree performs much worse than the proposed method. Third, our proposed method performs stably while the transaction length increases. That is, the execution time increases slightly for the proposed method. In summary, even facing the larger transaction length, our proposed index can deal with issue of searching the repeating patterns.

The second experiment is to evaluate the proposed method under different numbers of pattern categories. This evaluation was conducted as the data size is 1000 and the transaction length is 1000. Figure 9 depicts the experimental results which delivers some aspects. First, similar to Fig. 8, our proposed method performs better than RP-tree. Second, the number of pattern categories increases, the execution time decreases. Note that, this result is limited in the same data size and transaction length. This is because the occurrences of patterns reduce significantly so that the number of repeating patterns thereby reduces. Hence, few patterns decrease the execution time.

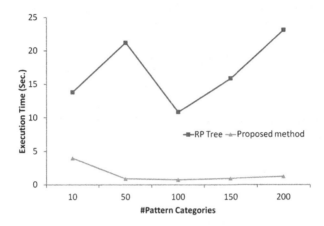

Fig. 9. Execution time of the compared methods under different numbers of pattern categories.

The final empirical evaluation is to show the impact of the data size. This experiment is made as the transaction length is 1000 and the number of pattern categories is 200. Figure 10 shows the experimental results that, the execution time of RP-tree increases significantly in contrast to the our proposed method. It says that, RP-tree needs much effort to identify the repeating patterns for a large amount of data. To sum up, if the parameter value is large, the execution time of our proposed method is much smaller than that of RP-tree. On the other hand, our proposed method performs more stably than RP-tree if enlarging the data parameter values. In overall, whatever the data is, our proposed method performs better than the compared methods RP-tree.

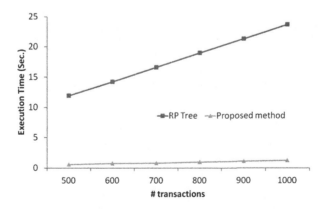

Fig. 10. Execution time of the compared methods under different numbers of transactions.

6   Conclusion and Future Work

In the last decades, how to retrieve valuable repeating patterns from sequences has been a hot issue. Although a number of past studies were proposed to discover repeating patterns, the high performances are not easy to achieve. In this paper, we propose an algorithm to accelerate the retrieval of repeating patterns. First, we construct an informative index called *QPI* containing positions and occurrences of patterns. It can effectively reduce the cost of mining of repeating patterns. By *QPI*, an algorithm named *FMRP* performs efficiently on mining repeating patterns. The experimental results reveal our proposed method is more efficient than the compared methods for different data. In the future, the main idea will be applied to the real applications.

Acknowledgements. This research was supported by Ministry of Science and Technology, Taiwan, R.O.C. under grant no. MOST 105-2221-E-230-011-MY2 and MOST 105-2632-S-424-001.

References

1. Cooper, M., Foote, J.: Automatic music summarization via similarity analysis. In: The 2002 IEEE International Conference on Music Information Retrieval, pp. 81–85 (2002)
2. Friedman, S., Stamos, I.: Online detection of repeated structures in point clouds of urban scenes for compression and registration. Int. J. Comput. Vis. **102**(1–3), 112–128 (2013)
3. Gool, L.V., Zeng, G., Wonka, P., Muller, P.: Image-based procedural modeling of facades. In: The ACM SIGGRAPH Conference on Computer Graphics, pp. 63–130 (2007)
4. Han, B.J., Hwang, E., Rho, S.: An efficient voice transcription scheme for music retrieval. In: The 2007 IEEE International Conference on Multimedia and Ubiquitous Engineering, pp. 28–26 (2007)
5. Hsu, J.L., Liu, C.C., Chen, L.P.: Discovering non-trivial repeating patterns in music data. IEEE Trans. Multimedia **3**(3), 311–325 (2001)
6. Liu, J., Psarakis, E., Stamos, I.: Automatic kronecker product model based detection of repeated patterns in 2D Urban Images. In: The 2013 IEEE International Conference on Computer Vision, pp. 401–408 (2013)
7. Ma, Y.F., Lu, L., Zhang, H.J., Li, M.J.: A user attention model for video summarization. In: The Tenth ACM International Conference on Multimedia, pp. 533–542 (2002)
8. Singh, A.: Ukkonen's suffix tree construction (2014). http://www.geeksforgeeks.org/ukkonens-suffix-tree-construction-part-6/
9. Ukkonen, E.: On-line construction of suffix tree. Algorithmica **14**(3), 249–260 (1995)
10. Wang, M., Lu, L., Zhang, H.H.: Repeating pattern discovery from acoustic musical signals. In: The 2004 IEEE International Conference on Multimedia and Expo, vol. 3, pp. 2019–2022 (2004)
11. Xiao, R.G., Wang, Y.Y., Pan, H., Wu, F.: Automatic video summarization by spatio-temporal analysis and non-trivial repeating pattern detection. In: The 2008 IEEE Congress on Image and Signal Processing, pp. 555–559 (2008)
12. Zhao, P., Fang, T., Xiao, J., Zhang, H., Zhao, Q., Quan, L.: Rectilinear parsing of architecture in Urban environment. In: The 2010 IEEE Computer Vision and Pattern Recognition, pp. 342–349 (2010)

Recognition of Empathy Seeking Questions in One of the Largest Woman CQA in Japan

Tatsuro Shimada[1]([✉]) and Akito Sakurai[2]

[1] Connehito Inc., Tokyo, Japan
shimada@connehito.com
[2] Keio University, Kanagawa, Japan
sakurai@ae.keio.ac.jp

Abstract. Many questions are posted on community websites in the world. Some of these questions are actually asked in order to receive empathy for the feelings of questioners, instead of getting specific answers to the questions asked. However, it is difficult to receive answers for these questions compared with questions that are asked for seeking responses other than for empathy. If such questions that are asked for the purpose of receiving empathy can get responses, it serves as an important factor to increase satisfaction of users. This paper reports on our attempt to improve response rate to the questions by classifying those questions that are asked for seeking empathy and those that are not by using machine learning and showing the questions classified as the ones seeking empathy to the prospective respondents who have been answered to these questions with higher rate.

Keywords: CQA · Community · Machine learning · NLP

1 Introduction

In Japan, a Q&A site called mamariQ [1] is the largest Community Question Answering (abbreviated as CQA hereinafter) website for women to find solutions to their questions about pregnancy, childbirth, and child raising. On CQA websites, users post their questions, and other users respond to them. A typical example is YAHOO ANSWERS. In light of the characteristics of such CQA websites, in this paper, we define questions as both questions that are asked for seeking empathy, focused by this paper, and those that are asked for seeking responses other than for empathy. If such questions that are asked for the purpose of receiving empathy can get responses, it serves as an important factor to increase satisfaction of users.

As of the end of 2015, more than 100,000 people registered with mamariQ and entered their children's information. When calculated based on the Vital Statistics conducted by the Ministry of Health, Labour and Welfare [2], 1 out of 10 people whose estimated date of birth is in 2015. Therefore, mamariQ is one of the largest CQA services in Japan. Amid such a circumstance, along with

© Springer International Publishing AG 2017
N.T. Nguyen et al. (Eds.): ACIIDS 2017, Part I, LNAI 10191, pp. 641–650, 2017.
DOI: 10.1007/978-3-319-54472-4_60

an increase in the number of mamariQ users, users started to post questions by which they try to receive empathy for their feelings rather than getting specific answers to the questions they ask. For example, such questions include the one shown below.

"Because of being tired from caring for my children, I take it out on others... I've never raised a hand against them though; I find myself being irritated and yelling at them. I think I've become frustrated and nervous since we had moved because of my husband's work transfer. I hate myself, and it gives me pain."

This kind of question that is asked for seeking empathy accounts for about 20% of all the questions. However, such a question actually does not explicitly ask questions. Therefore, it is difficult for other users to give responses.

We picked up 300 questions manually from both question types, those asked for seeking empathy and those asked for getting responses other than empathy, from mamariQ. They are listed on Tables 1 and 2. Based on these selected questions, we compared the number of unanswered questions and the number of unique responders. Section 3 will describe the details of question types for manually sorting out questions asked for seeking empathy and those asked to get responses other than for empathy. Table 1 indicates that 138 questions asked for seeking empathy are unanswered out of all the 300 questions, while only 1 question is unanswered out of questions asked for getting responses for other than empathy. This shows that such questions asked for seeking empathy tend to be unanswered when compared to those questions asked for getting responses other than for empathy. The number of unique responders between questions asked for seeking empathy and questions asked to get responses other than for empathy was compared. Table 2 indicates that for the 300 questions, 309 users responded to those questions asked for seeking empathy, while 847 users responded. The number of responders outnumbered the questions. This is because sometimes a single question has multiple responders. Comparison of these two values indicates a difference of about 2.7 times. This shows that a fewer number of unique responders respond to questions asked for seeking empathy when compared to those asked to get responses other than for empathy.

However, from five-star reviews of mamariQ posted on Apple's App Store, one of the app markets where mamariQ is provided, we randomly selected 100 reviews. Of these 100 reviews, in 58 reviews, reviewers said that they were happy to have certain responses to the questions they posted. We can see the fact that being able to receive responses to questions posted is one of the important factors to increase satisfaction of users. With that, we set the final goal of this paper to improve the response rate for questions asked for seeking empathy and to the unanswered rate.

Table 1. The number of questions unanswered

Question type	Answered	Unanswered
Empathy	162	138
Other	299	1

Table 2. The number of unique responders

User type	Empathy	Other
Have birth experience	255	565
Don't have birth experience	54	282
Total	309	847

Table 3. Percentage of the responders

User type	Empathy	Other
Have birth experience	83	67
Don't have birth experience	17	33

When the responders of the 300 questions mentioned above are divided into those who have birth experience and those who do not, Table 3 indicates that the percentage of the responders differs between questions asked for seeking empathy and those asked to get responses other than for empathy. As for questions asked for seeking empathy, 83% of responders had the birth experience. On the other hand, 67% of the responders of questions asked to receive responses other than for empathy had the birth experience. This fact shows that responders to questions asked to receive empathy have a higher percentage of having the birth experience.

Based on above, we considered that encouraging those having the birth experience to respond to questions asked for seeking empathy to increase the rate of response to such questions. As one of the user interface (UI) for community services including mamariQ on smartphones, there exist a format where questions posted by users flow from top down as shown in Fig. 1. Service users click their questions from those flowing down from the top in order to respond to them. On this list of questions screen, controlling the types of questions that are displayed according to user attribute can make it possible to improve the response rate. Based on the purpose of increasing the response rate for questions asked for seeking empathy on CQA services, in this paper, we adopted a method for displaying questions, which are asked for seeking empathy, targeting those responders who have high response rates to questions asked for seeking empathy. To implement this method, by using the features extracted from question sentences and by using machine learning, we classified questions that are asked for seeking empathy and those that are asked to get other responses.

2   Related Work

There are several existing studies that have been made on question classification in CQA services. Kim et al. [3] manually classified 465 questions and the answers

Fig. 1. mamariQ UI

chosen as best answers by questioners in order to examine the reasons why questioners chose them as the best answers. Kim et al. classified the questions into 4 classes: Information (searching specific facts and understanding phenomena), Suggestions (searching advice, recommendation, and feasible solutions), Opinion (examination of opinions and feelings of others, and starting discussion), and Others (those not classified into any of the previous three classes). As a result, the percentage of each of these classes was 35% for Information, 23% for Suggestions, and 39% for Opinion. They also indicated that the distribution of reasons for choosing the best answers could differ according to the question type. As for questions classified as Opinion, their study showed that socio-emotional factors, such as attitudes of responders and their emotional support, could serve as a significant factor for choosing the best answers.

Kuriyama et al. [4] manually divided 500 questions asked on Yahoo! Chiebukuro into three classes. The first type was an information search type for which users can search answers by using Internet search engines or library reference services. The questions of this type ask about facts, authenticities, definitions and descriptions, methods and means, reasons and causes, and effects and consequences. The second type was the social-survey type for which we need to conduct questionnaire surveys targeting specified individuals or groups and for which we cannot obtain objective answers. The questions of this type ask about advices, opinions, implementations, recommendations, and experiences. The third type was the non-question type which is expressed with descriptions for seeking reactions or responses toward the questioner's point of view. The purpose of this type is not to get objective or subjective answers for the question by means of information search or questionnaire survey. This type includes questions which analyzers could not understand the meaning of question. Kuriyama et al. suggested that when service users post their questions, in addition to providing question categories based on topics, showing these question types could serve as support for users.

Qu et al. [5] and Aikawa et al. [6] made studies on classification of questions by using natural language processing and machine learning and by features based on text data of questions and answers. By using text data of questions and answers, Qu et al. categorized questions posted on Yahoo! Answers by using SVM, Naive Bayes, and Maximum Entropy. Aikawa et al. classified questions posted on Yahoo! Chiebukuro based only on features obtained by question text data as to whether these questions were asked to get subjective answers or objective answers by using SVM and Naive Bayes. In addition to using features obtained from text data of questions and answers, Zhou et al. [7] attempted to use system-provided functions, such as Like (an action of users done when they feel the question/answer is beneficial), Vote (an action of users to vote for the best answer), Source (the reference source of the answer), Poll and Survey (questioning by using voting function), and Answer Number (the number of answers given). By using these factors based on the system, they classified questions as to whether questions were subjective or not without using manual procedures.

Many CQA services classify question categories according to the topic as described above, and questions posted whether they seek subjective answers or objective answers. However, they do not classify those questions asked for seeking empathy and other questions. In the present study, while defining the question type that seeks empathy where certain needs exist in the actual society, we attempted to classify questions in an applicable way by classifying them by using only features obtained from question text data. Additionally, although many related studies have used two to three different kinds of classifiers for conducting comparative research, in our study, we used seven classifiers for classification.

3   Definition of Question Classification Tasks

3.1   Definition of Question Types

In the present study, we classified questions into the following two question types.

Questions Asked for Receiving Empathy. In this paper, questions asked for seeking empathy are defined as questions having any of the following factors.

- Replying by a sentence to show empathy, such as "I understand" is sufficient to respond to the question
- There is no need to give specific answers or proposals to solve the question

Questioners who post this type of questions do so with a desire to be agreed with or understood.

Questions Asked for Receiving Responses other than Empathy. Questions that are defined as questions asked for receiving responses other than for empathy in this paper have the following factors.

- Providing a sentence to show empathy, such as "I understand" is insufficient to respond to the question
- There is a need to give specific answers or proposals to solve the question

  Questioners asking this type of questions seek diverse answers or correct facts.

3.2   Specific Example of Each Question Type

A question, "I have bad morning sickness. Is there any way I can feel more comfortable?" is one of the examples of questions to seek for specific answers.

Through this question, the questioner wants to get specific methods to ease her morning sickness. On the other hand, take a look at the following question: "Because of being tired from caring for my children, I take it out on others... I've never raised a hand against children though; I find myself being irritated and yelling at them. I think I've become frustrated and nervous since we had moved because of my husband's work transfer. I hate myself, and it gives me pain."

This is an example of questions asked for seeking empathy. Those users who received a response, such as "I can understand how you feel, it really hurts." answered in an interview, felt less anxious and comfortable by knowing that there were people in the same situation in the world and share the same concern.

3.3   Discussion of Question Classification Tasks

Some questions are difficult to classify. For example, the following question is a typical example.

"My husband totally makes me sick, he doesn't help with the housework!! I can't understand why he doesn't take out even one bag of garbage even though I cook every day!! Don't you agree?"

The last portion of this question is asked in order to receive agreement from readers. When this last portion is interpreted as a question asked to responders, this question could be classified into the question type that seeks diverse answers. However, since this question is determined to express the questioner's feeling of being sick about her husband and as seeking agreement, it is classified into the question type that seeks empathy.

4   Classification Methodology

4.1   Classification Based on Characteristic Words and Phrases

Questions are classified based on specified key phrases and characteristic words and phrases as to whether they are seeking for empathy or not. The classification results will be served as the baseline. Table 4 is the list of example of character-istic words and phrases that are used for the present classification method. The examples are translated from Japanese into English. This list of characteristic words and phrases is created based on patterns extracted by Kuriyama et al. [4] In total, the number of words and phrases is 103. These patterns are expressions (nouns, adjectives, adjective verbs, and sentence end expressions) that were con-sidered to be characteristic in the information search type and the social survey type of questions. As described earlier, many of those questions asked for seeking empathy do not take the form of a question. Therefore, we decided to use this list of characteristic words and phrases. Questions were to be classified into the question asked for seeking empathy where the relevant question does not contain any of the character string in this list.

Table 4. Example of characteristic words and phrases (translated from Japanese into English)

why(なぜ)	who(誰)	where(どこ)	when(何時)	how(どうやって)
can(できますか)	experience(経験)	reason(理由)	name(名前)	place(場所)
way(方法)	how much(いくら)	evaluation(評価)	measure(対策)	etc ...

4.2 Classification Based on Machine Learning

Questions are classified by using machine learning as to whether they are asked for seeking empathy or not. While several classification methods exist in natural language processing, in the present study, we

1. Extract words from the question sentences by using the morphological analy-sis program (MeCab [8]),
2. Select characteristic words from those extracted words,
3. Count the characteristic words of each sentence and create the characteristic vector,
4. Classify the questions by using a classifier.

Based on this process, questions were classified as to whether they are asked for seeking empathy or not. As for part-of-speech, nouns, verbs, adjectives, and symbols were used.

5 Experiment

5.1 Data Set

The data set we used for the experiments is the real question texts posted on mamariQ without any attributes of users. In accordance with the definition described in Sect. 3, questions were manually classified into those asked for seek-ing empathy and those asked to get other responses, and labeled as such. As a whole, 1500 classified questions were prepared composing of 1000 teacher questions and 500 test questions. Of these 1500 questions, 300 questions, about 20%, were those asked for seeking empathy. Of these 300 questions, 200 questions were included in the teacher questions, while 100 questions were included in test questions.

5.2   Experimental Method

Questions were classified based on two classification methods, each of which is described in Sects. 4.1 and 4.2, in order to compare classification precision. As a classifier, we use Logistic Regression, SVM, Gaussian Naive Bayes, K-nearest Neighbors, Random Forest, AdaBoost and Decision Tree. By using scikit-learn [9], two kernels, rbf and linear, were used for SVM. Default values were set on all parameters for other classifiers.

5.3   Definition of Target Values

In order to measure whether questions asked for seeking empathy are correctly classified in a quantitative manner, precision, recall, and F1 were used. Although about 20% of all questions account for those asked for seeking empathy, as indicated by Table 1, many questions ended up unanswered when compared to questions asked to get other responses. However, in order to activate the community to be a place where questioners can be satisfied, these unanswered questions should also be answered. Therefore, as shown in Fig. 1, in order to have questions answered, it is important to show those questions asked for seeking empathy mainly to service users (as shown in Tables 2 and 3) those who have the birth experience and relatively tend to answer questions whenever possible.

However, a limited number of service users actually have the birth experience. In mamariQ, half of the service users have the birth experience. Although all types of questions were allocated to postpartum service users, if the number of questions per one user increased, the frequency and quality of answer could decrease. On the other hand, there is no method for measuring these things. Therefore, it was assumed that the acceptable amount of questions that one user can answer remains the same as the present level. If questions asked for seeking empathy are correctly allocated to service users having the birth experience, the possibility of each question to be answered is heightened while a more proper answer could be obtained. Questions asked for seeking empathy account for 20% of all the questions. Therefore, by the precision value to 0.4 after improving the recall value of the question asked for seeking empathy to 1 as much as possible, questions asked for seeking empathy can be allocated to those postpartum users while the number of questions per one user remains at the same level. These values are the target values for this experiment.

5.4   Experiment Result

As for the experimental result, Table 5 indicates that recall was 0.4 and precision was 0.13 in classification based on characteristic words and phrases. On the other

Table 5. Experiment result

Classifier	Precision	Recall	F1
Characteristic words and phrases	0.13	0.40	0.20
Logistic regression	**0.80**	0.91	**0.85**
SVM (rbf)	0.31	**0.94**	0.47
SVM (linear)	0.75	0.86	0.80
Gaussian Naive Bayes	0.39	0.75	0.52
K-nearest Neighbors	0.35	0.93	0.51
Random forest	0.74	0.71	0.72
AdaBoost	0.66	0.80	0.72
Decision tree	0.65	0.73	0.69

hand, where SVM with the rbf kernel was used as a classifier, recall was 0.94 and precision was 0.31. K-nearest Neighbors classifier's recall is 0.93 and precision was 0.35. These values exceeded the baseline just enough, further they are extremely close to the targeted values.

When we practically implement our attempt in products, questions asked for seeking empathy are allocated to users having the birth experience. If the cost of answering questions asked for seeking empathy is high, users having birth experience have to use more energy than before. In this case, the number of questions which users having birth experience answer should decrease. Logistic Regression was the best in the experiments because its precision was 0.8 while recall rate was kept at 0.91.

6   Conclusion

In this study, with the goal of increasing the response rate for questions that are considered to be difficult to answer because of being asked seeking empathy, we classified questions into those seeking empathy and those seeking responses other than for empathy by using machine learning based on the characteristics extracted from question data. Besides, we compared seven classifiers for classification. As a result, we were able to obtain greatly-improved classification results when compared to those obtained by the classification method based on manually selected characteristic words and phrases. As a consequence, we believe we can make our users satisfied in our community. Our future issue is to practically implement our attempt in products in order to increase the response rate.

References

1. mamariQ. http://qa.mamari.jp/
2. The Vital Statistics conducted by the Ministry of Health, Labour and Welfare. http://www.mhlw.go.jp/toukei/saikin/hw/jinkou/geppo/nengai15/index.html

3. Kim, S., Oh, J., Oh, S.: Best-answer selection criteria in a social Q&A site from the user-oriented relevance perspective. In: Annual Meeting of American Society for Information Science and Techonology (ASIS&T 2007). Milwaukee, Wisconsin (2007)
4. Kuriyama, K., Kando, N.: Analysis of Questions and Answers in Q&A Site, IPSJ SIG Technical report, vol. 2009DBS1, No.19, p.1. UTF20138 2009
5. Bo, Q., Cong, G., Li, C., Sun, A., Chen, H.: An evaluation of classification models for question topic categorization. JASIST **63**(5), 889–903 (2012)
6. Aikawa, N., Sakai, T., Yamana, H.: Community qa question classification: is the asker looking for subjective answers or not? IPSJ Online Trans. **4**, 160–168 (2011)
7. Zhou, T.C., Si, X., Chang, E.Y., King, I., Lyu, M.R.: A datadriven approach to question subjectivity identification in community question answering. In: Proceedings AAAI (2012)
8. Kudo, T., Yamamoto, K., Matsumoto, Y.: Applying Conditional Random Fields to Japanese Morphological Analysis. In: Proceedings of the Conference on Empirical Methods in Natural Language Processing (EMNLP-2004), pp. 230–237 (2004)
9. scikit-learn. http://scikit-learn.org/stable/

A Content-Based Image Retrieval Method Based on the Google Cloud Vision API and WordNet

Shih-Hsin Chen[1,2] and Yi-Hui Chen[1,2(✉)]

[1] Department of Information Management, Cheng Shiu University, No. 840,
Chengcing Road, Niaosong District, Kaohsiung City 83347, Taiwan R.O.C.
shchen@csu.edu.tw
[2] Department of M-Commerce and Multimedia Applications,
Asia University, No. 500, Lioufeng Road, Wufeng,
Taichung 41354, Taiwan R.O.C.
chenyh@asia.edu.tw

Abstract. Content-Based Image Retrieval (CBIR) method analyzes the content of an image and extracts the features to describe images, also called the image annotations (or called image labels). A machine learning (ML) algorithm is commonly used to get the annotations, but it is a time-consuming process. In addition, the semantic gap is another problem in image labeling. To overcome the first difficulty, Google Cloud Vision API is a solution because it can save much computational time. To resolve the second problem, a transformation method is defined for mapping the undefined terms by using the WordNet. In the experiments, a well-known dataset, Pascal VOC 2007, with 4952 testing figures is used and the Cloud Vision API on image labeling implemented by R language, called Cloud Vision API. At most ten labels of each image if the scores are over 50. Moreover, we compare the Cloud Vision API with well-known ML algorithms. This work found this API yield 42.4% mean average precision (mAP) among the 4,952 images. Our proposed approach is better than three well-known ML algorithms. Hence, this work could be extended to test other image datasets and as a benchmark method while evaluating the performances.

Keywords: Content Based Image Retrieval · Image annotation · Google Cloud Vision API · WordNet · Pascal VOC 2007

1 Introduction

Content-Based Image Retrieval (CBIR) [13,22,27] is a typical image retrieval method. The main way of CBIR is to analyze the content of an image and extract the features (e.g. colors, texture, shape, and so on). It makes use of the low-level image features to retrieve similar images from the image dataset according to the similarity of image features. The image annotation [16] is roughly divided into classification-based methods [20] and probabilistic modeling-based methods [5]. In Classification-based methods, a machine learning (ML) algorithm is commonly

© Springer International Publishing AG 2017
N.T. Nguyen et al. (Eds.): ACIIDS 2017, Part I, LNAI 10191, pp. 651–662, 2017.
DOI: 10.1007/978-3-319-54472-4_61

used which extract the low-level features from images to train as patterns by supervised learning methods [2]. An unlabeled image will be classified into the categories according to the results that the features compared to the patterns. Later on, the labels of categories, such as animal, building, etc., are the annotations for the image. That is, the classification-based methods can recognize the unlabeled images to add the corresponding labels according to the trained classifier.

The classification-based approach has great performances, but it is a time-consuming process during training [37]. The cost is even higher if experts are involved to assigned the labels [29]. In addition, once a corresponding category is not trained, the model is not able to assign a correct label for that figure. To overcome these mentioned problems, this work proposes a new approach done by the Google Cloud Vision API[1]. The Vision API applies the deep learning algorithms and convolution neural network [18]. Because the API is with a large number of pictures training and millions of CPU hours of the training time, the researchers or practitioners could be benefited from the pre-trained model.

Even though Google Cloud Vision API could supplies useful labels, there is a semantic gap between the labels returned by Cloud Vision API and the image datasets. For example, when an image is labeled as a airplane by the Vision API, other datasets may name it as aeroplane or plane. It is apparently that these terms are the same. We design a transformation approach to mapping the terms by WordNet. Most important of all, we like to understand what does the Cloud Vision API could do well.

This paper is organized as follows. Section 3 presents the transformation method which could map the image annotations of the Cloud Vision API to the testing datasets. In Sect. 4, we take five well-known datasets and compare the Vision API with some state-of-art algorithms in the literature.

2   Literature Reviews

Content Based Image Retrieval (CBIR) [13,22,27,34] is a typical image retrieval method. The main way of CBIR is to analyze the content of image and extract the features (e.g. colors, texture, shape, and so on). The similarities of the features are used to calculate for collecting similar images from the image dataset. Nevertheless, there is a problem that computer uses a series of numerical value to express an image, which is widely divergent from the languages and words of human being, called semantic gap [3,11,20,23]. To overcome the problem of the semantic gap, image annotation is a way to annotate text information on images. The image annotation [16] is roughly divided into classification-based methods [14,20,31] and probabilistic modeling-based methods [7,16,17]. Classification-based methods extract low-level features from images to train as patterns by supervised learning methods. An un-labeled image will be classified into the categories according to the results that the features compared to the patterns. Later on, the labels of categories, such as human, building, etc., are as the annotations for the image. That is, the classification-based methods can recognize

---

[1] https://cloud.google.com/vision/.

the un-labeled images to add the corresponding labels according to the trained classifier. We show the two important parts in Sects. 2.1 and 2.2, respectively.

2.1 Classification-Based Methods

One of the classification-based methods is based on SIFT features to generate bag-of-visual words (BVW) for object recognition [14,20,26]. However, there are thousands of SIFT keypoint in an image, which takes a lot of time to train classifiers. Also, the accuracy of classification is affected by the noises.

Kesorn and Poslad [14] proposed a method to improve the qualities of visual words. The idea of [14] is to combine the close keypoints and removes the cluster that has high document frequency and small statistical association with all the categories (concept) in the dataset. Lu and Wang [24] developed a semantic regularized matrix factorization based on Laplacian regularization to improve the efficiency during the training process of BVW. In addition to BVW model method, AICMD [35] proposed by Su et al. to create different models to represent the images and all the features are used to train the classifiers by SVM (Support Vector Machine). Instead of SVM, some researches use Hidden Markov Models [1,19].

Classification-based approach has great performances, but it is a time consuming process during training. Nevertheless, it is difficult to recognize the object as an instance class, e.g., Barack Obama, St. Peter's Basilica. In addition, it suffers from ambiguity problem. To resolve time consuming problem, Feng et al. [8] rank tags in the descending order according to their relevance to the given image to reduce the learning space. Also, Zhang et al. and Xia et al. proposed the method about refining and enriching the imprecise tag words in 2013 and 2014 separately to solve the ambiguity problem [36,39]. As for hierarchical concept, Yuan et al. proposed a hierarchical image annotation system to generate the hierarchical tags for images [38]. Moreover, Fang et al. [4]proposed an ontology hierarchically concepts and concept relationships to create the semantic understanding for information retrieval.

2.2 Probabilistic Modeling-Based Method

The probabilistic modeling-based method calculates the joint probability between image content and the corresponding annotations. Mori [26] calculates the co-occurrence as the relationship between the sub-images and the corresponding labels. Although it takes shorter time of the process than classification-based method, the accuracy is worse than that one.

Kuric and Bielikov [16] consider both the local and global features of the images. Zhang et al. [39] propose ObjectPatchNet, which calculates the co-occurrence among each cluster and also considers the probability between image patches and labels to present the relationship. The limits of probabilistic modeling-based method are (1) the low-level feature is lack of semantic, and (2) the low-level features of the same individual objects, but different orientation are judged as dissimilar. To resolve the semantic gap, Hong et al. [10] creates the relationship between semantic concepts based on the data from image commercial engine.

The methods mentioned above are based on image recognition, but lots of abstract concepts (e.g., location, cannot be defined according to the image features). It is difficult to get accurate results because the images are not with abstract concepts. To meet the requirements, it desires that the labels for image annotation are with semantic meaning by using ontology theory to identify various definitions, attributes and the relationships between individuals [12,28,30,33].

3   Methods

This paper employs the Google Cloud Vision APIs to annotate the labels of each figures. We could verify the labels which meet to the original name. Table 1 demonstrates a figure taken from ImageNet [15]. We also show the three returned labeling results in JSON format. In each JSON record, the mid, description, and the score are the message ID, image annotation, and the score of this annotation, respectively. The score value represents the confidence of this annotation. From this example, the image annotation with the highest score (i.e., 98.55%) is the polar bear. The second one is the mammal whereas the corresponding score is 95.87% and the third label is the animal. The number of available labels is different for each figure. In this polar bear example, we could obtain more than ten labels. For example, the 12th label is the biology together with the score 60.71%. On the other hand, the Cloud Vision API may not return any labels because their score values are too low in some figures.

By using this example, Cloud Vision API might be capable of extracting annotations from images. However, human may not use the same term to identify an object. For example, the terms of the polar bear could be ice bear, thalarctos maritimus, or ursus maritimus according to WordNet [25]. As a result, after we validate the labels generated by Google Cloud Vision API does not fit the instance label, we make a further comparison by using WordNet. The whole research framework is shown in Fig. 1.

Table 1. Image labels of a polar bear photo taken from ImageNet

{"mid": "/m/0633h",
"description": "polar bear",
"score": 0.985539},
{"mid": "/m/04rky",
"description": "mammal",
"score": 0.9587503},
{"mid": "/m/0jbk",
"description": "animal",
"score": 0.9344022}

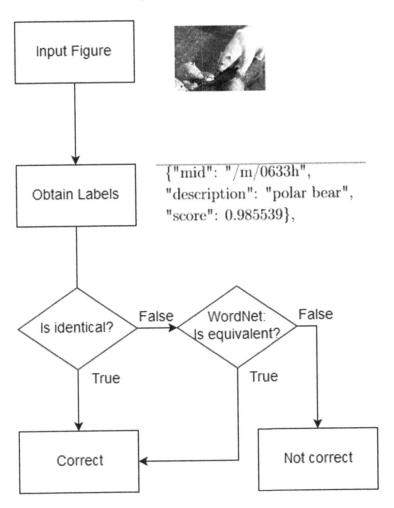

{"mid": "/m/0633h",
"description": "polar bear",
"score": 0.985539},

Fig. 1. Research procedure

We code our programs in R language which parse the image and then obtain the labels. The labels of each image are stored into a database. It is useful for the researchers to do further studies. The codes are available on the Github [2]. The detail information is given in Sect. 3.1.

Because the annotations given by Cloud Vision API might be different from the original category names of the selected image datasets, we design a transformation approach to mapping the terms. Our approach actually apply the method given by WordNet. We already finish the work of this mapping method in R language. We explain the code in Sect. 3.2.

---

[2] https://github.com/worldstar/GoogleCloudVisionAPI_RLanguage.

3.1   The Core Programs of Calling the Vision API

The core functionality of this paper is to access Google Cloud Vision API which parses the contents of these images. In order to get the image of the label content, we follow the standard process according to the API. Our code is shown on GitHub[2] already, which is coded by R language. The corresponding testing results is shown in Table 1. The primary method, called parseFigure, is to read the binary content into the Base64 string, and then parse JSON data. Later on, JSON data is returned with the parsed tag content by using function parseFigure.

3.2   Comparing the Meaning of the Labels by Using WordNet

WordNet has defined a set of nouns, verbs, adjectives and adverbs which are grouped into sets of cognitive synonyms [25]. By using their synonyms, we can verify the words given by the names given by the image datasets and the Google Cloud Vision API. We carry out this work by using the R lanaguage as follows. A function, called checkWords, is defined to specify the word from dataset in wordnet and the Google Cloud Image API. Take the Pascal 2007 dataset for instance, they use aeroplane while the Google Cloud Vision API applies the airplane. Even though the words are different, we understand both of them refer to the same thing. Hence, by writing the above code, we are able to map the words of the image datasets and the names given by Vision APIs. Via this approach, we expect we could map the terms of the image datasets and the labels returned by Google Cloud Vision API. Then, we would calculate the accuracy of the Google Cloud Vision API and then compared it with existing machine learning algorithms.

```
 1  library("wordnet")
 2
 3  checkWords <- function(dataSetName, myWord){
 4    if(!initDict()) return(null)
 5
 6    wordBags = synonyms(dataSetName, "NOUN")
 7    isEqual = FALSE
 8
 9    for(i in wordBags){
10      if(i == myWord){
11        isEqual = TRUE
12        break
13      }
14    }
15
16    return(isEqual)
17  }
18
19  isEqual = checkWords("aeroplane", "airplane")
20  isEqual
```

4   Experiment Results

To verify the effectiveness of the Google Cloud Vision, we employ a well-known image dataset, which is the Pascal VOC 2007. There are 4952 images and 20 categories of these images. We input all the images as the testing ones and then obtain their own labels by Google Cloud Vision APIs. After that, we validate the correctness given by Google Cloud Vision APIs, which is named the Method 1. Then, because there is a meaning gap of the labels, we further use WordNet to fill up the gap, named as the Method 2. We show the equivalent names obtained from WordNet in Table 2.

From Table 2, we read the synonyms of the aeroplane are airplane and plane. The synonyms of the car are auto, automobile, an so on. However, we found the terms given by WordNet of chair is not the one we need because the WordNet shows chairman, chairperson, chairwoman, electric chair, hot seat, and so on. The synonym of motorbike does not include the motorcycle. In addition, the categories include the diningtable, pottedplant, sheep, and tvmonitor do not yield any synonym. Hence, we expect the number of corrects could be improved by WordNet. Nonetheless, there are still semantic gaps between the category names defined in the dataset and the synonyms given by WordNet.

The detail results of each category are shown in Table 3 of the Method 1 and Method 2. The mean average percentage of Method 1 and Method 2 are 32.1% and 42.4% indicating the improvement by WordNet. There is 10.3% improvement between the two approaches. The major reason is due to the average precision of aeroplane, car and sofa are all 0% in the beginning. After we apply the WordNet, they are 78.5%, 77.4%, and 19.1%, respectively. Hence, it might be necessary for a researcher or a practitioner to apply the WordNet if we want to improve the accuracy.

On the other hand, the performances of diningtable, motorbike, pottedplant, and tvmonitor are all zero. We explained that the WordNet is not able to supply adequate terms. Once this problem is solved, we expect the accuracy of the proposed method could be further improved.

Later on, we further compare our approaches with the results in the literature, including the R-CNN FT fc_7 BB [9], DPM v5 [6], DPM HSC [32], and DPM HSC [32]. Their accuracy represents the mean average precision (mAP). These four algorithms remain use 4952 testing images from the Pascal VOC 2007 dataset. We found the R-CNN FT fc_7 BB is the best algorithm. The Method 2 proposed in this paper comes to the second place which also outperforms the DPM v5, DPM HSC, and DPM HSC. This results shows the Method 2 yileds at least 8% relative improvement compared to the rest of the algorithms (Table 4).

By running the Pascal VOC 2007 image dataset, the overall results show the Google Cloud Vision API plus the WordNet could yield an acceptable result

Table 2. The corresponding synonym(s) given by WordNet

Category	Synonym(s)
Aeroplane	Airplane, plane
Bicycle	Bike, cycle, wheel
Bird	Birdie, boo, Bronx cheer, chick, dame, doll, fowl, hiss, hoot, raspberry, razz, razzing, shuttle, shuttlecock, skirt, nort, wench
Boat	Gravy boat, gravy holder, sauceboat
Bottle	Bottleful, feeding bottle, nursing bottle
Bus	Autobus, busbar, bus topology, charabanc, coach, double decker, heap, jalopy, jitney, motorbus, motorcoach, omnibus, passenger vehicle
Car	Auto, automobile, cable car, elevator car, gondola, machine, motorcar, railcar, railroad car, railway car
Cat	African tea, Arabian tea, big cat, bozo, CAT, Caterpillar, cat-o'-nine-tails, computed axial tomography, computed tomography, computerized axial tomography, computerized tomography, CT, guy, hombre, kat, khat, qat, quat, true cat
Chair	Chairman, chairperson, chairwoman, death chair, electric chair, hot seat, president, professorship
cow	Moo-cow
Diningtable	-
Dog	Andiron, blackguard, bounder, cad, Canis familiaris, click, detent, dog-iron, domestic dog, firedog, frank, frankfurter, Frump, heel, hotdog, hot dog, hound, pawl, weenie, wiener, wienerwurst
Horse	Buck, cavalry, Equus caballus, gymnastic horse, horse cavalry, knight, sawbuck, sawhorse
Motorbike	Minibike
Person	Individual, mortal, somebody, someone, soul
Pottedplant	-
Sheep	-
Sofa	Couch, lounge
Train	Caravan, gear, gearing, geartrain, power train, railroad train, string, wagon train
Tvmonitor	-

even though we do not train their own model. Without using a prior training step, it saves a lot of computational time. Hence, even though the CPU time of compared algorithms is not reported, our proposed method would be faster than these algorithms in the literature.

Table 3. The detail results of each category

Category	Total	Method 1		Method 2	
		Corrects	Avg (%)	Corrects	Avg (%)
Aeroplane	200	0	0.0	157	78.5
Bicycle	189	128	67.7	131	69.3
Bird	275	196	71.3	197	71.6
Boat	162	104	64.2	104	64.2
Bottle	130	17	13.1	17	13.1
Bus	156	86	55.1	129	82.7
Car	580	0	0.0	449	77.4
Cat	305	213	69.8	213	69.8
Chair	238	42	17.6	42	17.6
Cow	123	0	0.0	0	0.0
Diningtable	109	0	0.0	0	0.0
Dog	384	271	70.6	274	71.4
Horse	212	131	61.8	131	61.8
Motorbike	181	0	0.0	0	0.0
Person	845	194	23.0	194	23.0
Pottedplant	119	0	0.0	0	0.0
Sheep	93	56	60.2	56	60.2
Sofa	215	0	0.0	41	19.1
Train	250	171	68.4	171	68.4
Tvmonitor	192	0	0.0	0	0.0
mAP			32.1		42.4

Table 4. The comparison results

Methods	Corrects	mAP
Method 1 (Raw results)	1609	32.1%
Method 2 (Using WordNet)	2306	42.4%
R-CNN FT fc$_7$ BB [9]	2897	58.5%
DPM v5 [6]	1669	33.7%
DPM ST [21]	1441	29.1%
DPM HSC [32]	1699	34.3%

5   Conclusions

This paper is the first work which distinguishes the effectiveness of the Google Cloud Vision API compared to some efficient algorithms in the literature. When Google Cloud Vision API works with the WordNet, it could provide a good precision compared to prior algorithms even though some category names are not available. The reason is the WordNet fill up the semantic gap between the labels generated by Google Cloud Vision API and the image dataset. Most important of all, the computational effort is reduced because we do not make any training for the proposed methods. As a result, the proposed approach might show a good direction for researchers or practitioners. The researchers could employ our proposed algorithm into their own research framework. In industry, they could directly apply this framework to deal with their daily routine which needs an image recognition procedure. For the future research, we will modify the category named fit to the labels given by Google Cloud Vision APIs. Moreover, we evaluate the proposed scheme to test more well-known datasets.

References

1. Bashir, F.I., Khokhar, A.A., Schonfeld, D.: Object trajectory-based activity classification and recognition using hidden markov models. IEEE Trans. Image Process. **16**(7), 1912–1919 (2007)
2. Chang, S.-F., Ma, W.-Y., Smeulders, A.: Recent advances and challenges of semantic image/video search. In: IEEE International Conference on Acoustics, Speech and Signal Processing, ICASSP, vol. 4, pp. IV-1205. IEEE (2007)
3. Dorai, C., Venkatesh, S.: Bridging the semantic gap with computational media aesthetics. IEEE Multimed. **10**(2), 15–17 (2003)
4. Fang, Q., Xu, C., Sang, J., Hossain, M., Ghoneim, A.: Folksonomy-based visual ontology construction and its applications. IEEE Trans. Multimed. **18**(4), 702–713 (2016)
5. Fei-Fei, L., Fergus, R., Perona, P.: Learning generative visual models from few training examples: an incremental bayesian approach tested on 101 object categories. Comput. Vis. Image Underst. **106**(1), 59–70 (2007)
6. Felzenszwalb, P.F., Girshick, R.B., McAllester, D., Ramanan, D.: Object detection with discriminatively trained part based models. IEEE Trans. Pattern Anal. Mach. Intell. **32**(9), 1627–1645 (2010)
7. Feng, S., Manmatha, R., Lavrenko, V.: Multiple bernoulli relevance models for image and video annotation. In: Proceedings of the IEEE Computer Society Conference on Computer Vision and Pattern Recognition, CVPR, vol. 2, pp. II-1002. IEEE (2004)
8. Feng, S., Feng, Z., Jin, R.: Learning to rank image tags with limited training examples. IEEE Trans. Image Process. **24**(4), 1223–1234 (2015)
9. Girshick, R., Donahue, J., Darrell, T., Malik, J.: Rich feature hierarchies for accurate object detection and semantic segmentation. In: Proceedings of the IEEE Conference on Computer Vision and Pattern Recognition, pp. 580–587 (2014)
10. Hong, R., Yang, Y., Wang, M., Hua, X.-S.: Learning visual semantic relationships for efficient visual retrieval. IEEE Trans. Big Data **1**(4), 152–161 (2015)

11. Hu, X., Li, K., Han, J., Hua, X., Guo, L., Liu, T.: Bridging the semantic gap via functional brain imaging. IEEE Trans. Multimed. **14**(2), 314–325 (2012)
12. Im, D.-H., Park, G.-D.: Linked tag: image annotation using semantic relationships between image tags. Multimed. Tools Appl. **74**(7), 2273–2287 (2015)
13. Kekre, H., Sarode, T.K., Thepade, S.D., Vaishali, V.: Improved texture feature based image retrieval using kekres fast codebook generation algorithm. In: Pise, S.J. (ed.) Thinkquest~ 2010, 143–149. Springer, Heidelberg (2011)
14. Kesorn, K., Poslad, S.: An enhanced bag-of-visual word vector space model to represent visual content in athletics images. IEEE Trans. Multimed. **14**(1), 211–222 (2012)
15. Krizhevsky, A., Sutskever, I., Hinton, G.E.: Imagenet classification with deep convolutional neural networks. In: Pereira, F., Burges, C.J.C., Bottou, L., Weinberger, K.Q. (eds.) Advances in Neural Information Processing Systems, vol. 25, pp. 1097–1105. Curran Associates Inc. (2012). http://papers.nips.cc/paper/4824-imagenet-classification-with-deep-convolutional-neural-networks.pdf
16. Kuric, E., Bielikova, M.: ANNOR: efficient image annotation based on combining local and global features. Comput. Graph. **47**, 1–15 (2015)
17. Lavrenko, V., Manmatha, R., Jeon, J.: A model for learning the semantics of pictures. In: Advances in Neural Information Processing Systems (2003). p. None
18. LeCun, Y., Bengio, Y., Hinton, G.: Deep learning. Nature **521**(7553), 436–444 (2015)
19. Li, J., Wang, J.Z.: Automatic linguistic indexing of pictures by a statistical modeling approach. IEEE Trans. Pattern Anal. Mach. Intell. **25**(9), 1075–1088 (2003)
20. Li, L.-J., Wang, C., Lim, Y., Blei, D.M., Fei-Fei, L.: Building and using a semantivisual image hierarchy. In: IEEE Conference on Computer Vision and Pattern Recognition (CVPR), pp. 3336–3343. IEEE (2010)
21. Lim, J.J., Zitnick, C.L., Dollár, P.: Sketch tokens: a learned mid-level representation for contour and object detection. In: Proceedings of the IEEE Conference on Computer Vision and Pattern Recognition, pp. 3158–3165 (2013)
22. Liu, G.-H., Yang, J.-Y.: Content-based image retrieval using color difference histogram. Pattern Recogn. **46**(1), 188–198 (2013)
23. Liu, Y., Zhang, D., Lu, G., Ma, W.-Y.: A survey of content-based image retrieval with high-level semantics. Pattern Recogn. **40**(1), 262–282 (2007)
24. Lu, Z., Wang, L.: Learning descriptive visual representation for image classification and annotation. Pattern Recogn. **48**(2), 498–508 (2015)
25. Miller, G.A.: Wordnet: a lexical database for english. Commun. ACM **38**(11), 39–41 (1995)
26. Mori, Y., Takahashi, H., Oka, R.: Image-to-word transformation based on dividing and vector quantizing images with words. In: First International Workshop on Multimedia Intelligent Storage and Retrieval Management, pp. 1–9. Citeseer (1999)
27. Murala, S., Maheshwari, R., Balasubramanian, R.: Local tetra patterns: a new feature descriptor for content-based image retrieval. IEEE Trans. Image Process. **21**(5), 2874–2886 (2012)
28. Osman, T., Thakker, D., Schaefer, G.: Utilising semantic technologies for intelligent indexing and retrieval of digital images. Computing **96**(7), 651–668 (2014)
29. Pan, Y., Yao, T., Mei, T., Li, H., Ngo, C.-W., Rui, Y.: Click-through-based cross-view learning for image search. In: Proceedings of the 37th International ACM SIGIR Conference on Research & Development in Information Retrieval, pp. 717–726. ACM (2014)

30. Pesquita, C., Ferreira, J.D., Couto, F.M., Silva, M.J.: The epidemiology ontology: an ontology for the semantic annotation of epidemiological resources. J. Biomed. Semant. **5**, 4 (2014)
31. Poslad, S., Kesorn, K.: A multi-modal incompleteness ontology model (mmio) to enhance information fusion for image retrieval. Inf. Fusion **20**, 225–241 (2014)
32. Ren, X., Ramanan, D.: Histograms of sparse codes for object detection. In: Proceedings of the IEEE Conference on Computer Vision and Pattern Recognition, pp. 3246–3253 (2013)
33. Rodríguez-García, M.Á., Valencia-García, R., García-Sánchez, F., Samper-Zapater, J.J.: Ontology-based annotation and retrieval of services in the cloud. Knowl. Based Syst. **56**, 15–25 (2014)
34. Sarker, I.H., Iqbal, S.: Content-based image retrieval using haar wavelet transform and color moment. SmartCR **3**(3), 155–165 (2013)
35. Su, J.-H., Chou, C.-L., Lin, C.-Y., Tseng, V.S.: Effective semantic annotation by image-to-concept distribution model. IEEE Trans. Multimed. **13**(3), 530–538 (2011)
36. Xia, Z., Peng, J., Feng, X., Fan, J.: Automatic abstract tag detection for social image tag refinement and enrichment. J. Signal Process. Syst. **74**(1), 5–18 (2014)
37. Yang, J., Yu, K., Gong, Y., Huang, T.: Linear spatial pyramid matching using sparse coding for image classification. In: IEEE Conference on Computer Vision and Pattern Recognition, CVPR 2009, pp. 1794–1801. IEEE (2009)
38. Yuan, Z., Xu, C., Sang, J., Yan, S., Hossain, M.S.: Learning feature hierarchies: a layer-wise tag-embedded approach. IEEE Trans. Multimed. **17**(6), 816–827 (2015)
39. Zhang, S., Tian, Q., Hua, G., Huang, Q., Gao, W.: Objectpatchnet: towards scalable and semantic image annotation and retrieval. Comput. Vis. Image Underst. **118**, 16–29 (2014)

A Personalized Recommendation Method Considering Local and Global Influences

Hendry[1], Rung-Ching Chen[1(✉)], and Lijuan Liu[1,2]

[1] Department of Information Management, Chaoyang University of Technology, Taichung, Taiwan
{s10314905,crching}@cyut.edu.tw
[2] College of Computer and Information Engineering, Xiamen University of Technology, No. 600, Ligong Road, Jimei District, Xiamen, 361024 Fujian, China
ljliu@xmut.edu.cn

Abstract. Social Media is one of the largest media data storage in the website. Many researchers utilize this to do some research about user interest and recommendation system. This data is like a treasure vault waiting to be utilized to develop the recommendation systems. Social common interest is one of the methodologies to implement the recommendation system among users. It performs well in community with similar interest. The drawback of it ignores the outside influence from other communities. In this paper, a methodology to calculate the global influence from outside community and to implement the recommendation system is proposed. The results could be utilized to make the recommendation system not only in local communities but also notice the outside influence of item in social media.

Keywords: Social media analysis · Recommendation system · Social influence · Global influence

1 Introduction

As social media becomes popular, many users put their information to social media application such as Facebook, Tweeter, Google Plus, etc. The development of the web 2.0 also encourages user to have more participation in content management, such as share, create and update content in website [1]. Social media content usually consists of users interest fields [2] such as videos [3], music [4], and user's interests [5]. The data their share also could be "comments," "likes" or "follows" another users. Circle of social media also has factors and attributes about what information and recommendation that user will choose, because this factor influences the user similarity preference based on their friend's interest [6]. In social media sites like Facebook, each user provides their personal information such as name, gender, birthdates, hobby, etc. Users also could make a page for their favorite things, such as movies, music or books.

In social media, relationship is described by post and share in the wall page. Every post and like in user's wall page will make an influence to other users. Users with the same interests could form a cluster and we call it user's community. Zhong Li, et al.,

© Springer International Publishing AG 2017
N.T. Nguyen et al. (Eds.): ACIIDS 2017, Part I, LNAI 10191, pp. 663–672, 2017.
DOI: 10.1007/978-3-319-54472-4_62

stated that social similarity in local activity is ignored in different levels. More activity from outside community also could influence to the local community [7]. In this paper, we propose to find how the global item could influence the local community have the local influence among users inside. Figure 1 depicted the illustration about local communities are created and the intersections of the area are outside influence from the global item that affects the local community.

Social interests between users in local community are proved as one of the factors to provide the personal recommendation system. Boley et al., proposed the Tag-based social interest discovery [8]. This method could enable users to do the tag about things their interested by giving specific ID discovery system or ISID. This methodology works well in local community where users have similar interest. Chen et al. [9], also proposed a personal recommendation system by combining user similarity and familiarity between users in local community. The recommendation results are implemented to three categories of movie, music, and book. Social common interests are clearly defined when a user joins a specific user's group whose members share a common interest. On the other hand, sometimes a different condition occurs while users, who do not know each other and do not join a certain group, have similar interest. In this paper, we propose an algorithm to discover the global influence affects the local community. The personal recommendation system could be enhanced to recommend items to the user with the cold start problem. The cold start users don't have the user interest data, and no record in community could help to calculate the recommendation results for them because the lack of data. We try to enhance our personal recommendation system based on what item is popular outside the community that could be interested to local community, and recommend the cold start problem users with better recommendation items. Challenge to do the research are (1) Data sampling in Facebook need user permission to collect the data. Not all the respondent willing to share their profile and data in the process. (2) Some users didn't fill all their profiles, interests, also some event still left in blank. This make a preprocessing process need to be done.

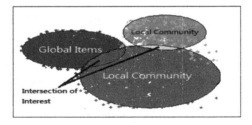

Fig. 1. Local communities with intersection area that shows the global influence to each community

The rest of the paper is organized as follows. Section 2 is literature review. Section 3 is our proposed methodology, we proposed a global influence factor to calculate the recommendation items to solve cold start problem. Section 4 is our experimental results and discussion. Section 5 is conclusions of our work.

2 Literature Review

User Interest is one of the factors that influence the decision of the user makes the decision to what item she/he want to choose [10]. Many researchers use this factor to calculate the similarity items between users. User interest could be affected by both individual preferences and interpersonal influences [11]. Individual preference is chose by users; sometimes could be affected by influences of their social circle such as their friends, family or closest persons between them. Interpersonal influences are what their communities influence other users; this influence could be affected by user relation in social media.

The finding of social common interest in a social network helps to connect users with similar interest to find out what current trending topic is in the social media, and even recommend content that fits to the user's interest [12]. For business owners, social common interest is useful to identify the type of product or service to be introduced to the market. Kwona et al. [13] described Acceptance Rate (AR) to calculate the user preferences from each user. The data are collected then classified into each genre of each category. These data will be used to calculate the similarity rank between users.

There are many researches about similarity search in social media, such as Euclidean distance [14], cosine similarity, Hamming feature distance, weighted tree similarity, and semantic similarity. We used weighted tree similarity algorithm in our previous work to calculate the similarity between users. This tree represents all interests from all users in system. The weight or nodes of this tree could changes depends on the size of users. In other hand, individual tree is a tree which represents personal interest from each user. In similarity tree measure, there are some issues to tackles regarding the general shape of the tree, their recursive nature, and their arbitrary size. Weighted Similarity Tree is an algorithm which could solve all the issues, so it will be used to compute the similarity between two trees.

Similarity rank also could be calculated by other method such as XOR method [9]. This method simplifies the calculation with normalization process. After calculating all the normalized user preferences AR, the system will obtain the candidate user with the same interest values. The similarity values will be sorted from the highest point. The smaller distance between two users will be the closest users to the candidate user's preferences. Personalized recommendation system becomes popular to many researchers. With the development of the social media, this system becomes more and more popular. There are many methods to calculate the recommendation item, such as using user interest between users in social circle. The system uses the similarity between user interests to recommend items. Personalized recommendation system also recommends items based on the familiarity between users with the same topic in community. This method considers how users have the same like or comments of a topic. More similar between users choice about page they like and comments means this user have same interest, and could be have same recommendation items.

In this paper, we will generate a personal recommendation system that collects data from Facebook. The system collects the data using crawler engine. This engine has scheduler to automatically crawl the Facebook user data and save it to our database.

Facebook user data are collected, such as basic profile, friends list, pages their follow, etc., will be sent into our web server to be processed and construct the relationship.

3 Research Method

The section describes the operation of our proposed method. We describe the architecture of our system, personalized recommendation system, calculation of similarity and familiarity and global influence item that affect the decision of local community.

3.1 The System Operations

Every change in the user data implies that the crawler needs to read the data again and modify the archive to make it identical with the most recent data. This may occurs because the system contains a scheduler engine that always detects Facebook user data, and is notified if there are changes made by the user. We also perform pre-processing to repair and modify missing data or data not relevant to our research. User Affecting Index (UAI) is a value how much a user could affect others user to choose the same items with them. UAI consist of two factors, they are the similarity of user and familiarity. UAI value is the degree of influence user to other users. The value range is from 1 to 0, which is high influence low influence. The UAI calculation is start from user preference AR and Similarity Rank. Similarity Rank then combines with the value of familiarity. We calculate AR using Facebook user's preferences; it consists of FB users likes, comments and post of movies, books and music pages in Facebook. The dynamic rules are generated to give recommendations have the highest UAI rank. The algorithm of User Preference calculation is show in Fig. 2.

```
Input: User Preference
Output: Similarity Count
Similarity (User, Member)
Begin
  User_Preference ← Select from User
  Member_Preference ← Select from Member
  Normalize_User ← Normalization (User)
  Normalize_Member ← Normalization (Member)
  For i in Every Member m
      Similar_Count←∑XOR(Normalize_User_{i,m}, Normaliz
      e_Member_{i,m})
  Return Similar_Count
End.
```

Fig. 2. Similarity rank algorithm

The system factors are similarity between users, and familiarity. UAI is user affecting influence which combines similarity rank and familiarity. Equation (1) describes the formulation of UAI.

$$UAI = \left(r_{i,j} \left(\alpha \, SR_{i,j} + \beta \, FR_{i,j} \right) \right) \tag{1}$$

Where $r_{i,j}$ is posts, likes or comments from user i and j to movies, music and book pages. SR is similarity rank between users, where α is a coefficient value. FR is familiarity factor where β is a coefficient value. The range of values of α and β are ranged from 0 to 1. The maximum summation value of α and β is 1. The calculation of similarity SR consists of AR. AR is the threshold used to calculate the normalization of user preferences. Local community is formed by nearest distance of user preferences. When the AR value of user is equal 1 with NP the user will classify as the same community or local community, and others value will be classify as global community. The system collects all the user preferences from database. The user preferences are counting how many likes or comments user from movie titles. The data will be classified into each genre and category. The genre attribute is chosen because this attribute appears in all categories. We use 3 categories from dataset those are movie, music and book category. Genre also makes dataset less sparse, and some missing data for genre are easy to collect the information through the web. After all user preferences are collected, the next step is to normalize all of the user preferences. Equation (2) shows the formulation of our normalization formula.

$$NP = \begin{cases} 0, & Normalization(AR) < \theta \\ 1, & Normalization(AR) \geq \theta \end{cases} \tag{2}$$

```
Input: User Preference
Output: Candidate Decision List
Candidate_Decision
Begin
  User←rand (member_list)
  Remove (user, member_list)
  While (member_list!=NULL)
    Member_?←Rand (member_list)
    UAI←Similarity (User, Member_?)
          +Familiarity (User, Member_?)
    If (UAI=0)
        First_candidate_list ← Member_?
    Else If (UAI=1)
        Second_candidate_list← Member_?
    Else
        Third_candidate_list ← Member_?
  Return First_candidate_list,
    Second_candidate_list,
    Third_candidate_list
  End.
```

Fig. 3. Candidate decision list algorithm

Where θ is the threshold of our normalization value. All normalization values that are below the threshold will be 0 and greater or equal than the threshold will be 1. After the system calculates all the normalization user preferences, the next step is to calculate the similarity rank. The first step to calculate the similarity rank is to find the candidate user with the same interests.

$$SR_{i,j} = \sum_{c=1}^{C} XOR(NP_{ic}, NP_{jc}) \qquad (3)$$

The systems use XOR operator to search all candidates' users, the similarity formula is shown by Eq. (3), where C is number of classes, and c is index of the class. NP is the normalization process value for the compared users. The smaller distance value SR between users indicates users have greater similarity and closer candidate user preferences. After the calculation, the system will sort the similarity from the closest to the farthest. Figure 2 shows the calculation of similarity rank and Fig. 3 shows the candidate decisions to classify community members into candidate rank based on the similarity.

Familiarity also one of the factors affect the UAI. Familiarity is the degree of closeness between users. Familiarity is calculated based on how many likes and comments users make to other users wall page. Equation (4) shows the formula of familiarity.

$$FR_{i,j} = \begin{cases} 1 + Log(U(i,j)) \, r_{i,j} \in R \\ 0, otherwise \end{cases} \qquad (4)$$

Where U is social activity on Facebook, such as user's post, likes and comments to other user's wall. Social activity between users indicates how well the user know other users. The higher the value of social activity, more familiar this user with other users.

3.2 Local and Global Influences

User Influence could derive from local community and from outside community. UAI recommendation model only pay attention from the local influence inside their community. It is based on the closest user interest between users form their local community. Figure 4 depicted the global influence from local community to other community. Global items is what the popular items are been discussed.

$$LG - UAI = \left(r_{i,j} \left(\alpha \ SR_{i,j} + \beta \ FR_{i,j} + \gamma \ G_{i,j}\right)\right) \qquad (5)$$

Equation (5) is our proposed formula to enhance the UAI with global influence. We called the system name is Local-Global User Affecting Index (LG-UAI). We try to solve the cold start problem for users with more accurate recommendation items. If the user is new to the community is not with history and records of their interests, the system will get the values from function G as one of the factors to generate the recommendation items and γ is coefficient value.

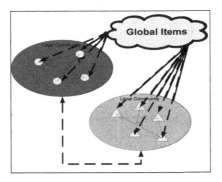

Fig. 4. Global influences affect local community

Function G is based on whether the global items will be recommended to the user or not. If the value of the global influences popularity, and most of them are popular in the local community, the items will be recommended to the user and the value of G is 1, otherwise the value is 0. We define the threshold as the value to determine whether need to recommend items or not. GP is global items the system generated from outside community. Equation (6) shows the formula.

$$G_{i,j} = \begin{cases} 1, & Normalization(GP) \geq \theta \\ 0, & Normalization(GP) < \theta \end{cases} \tag{6}$$

4 Experiments

This subsection we will discuss the results of our experiments. We discover at there are differences genre popularity within our local community with global community.

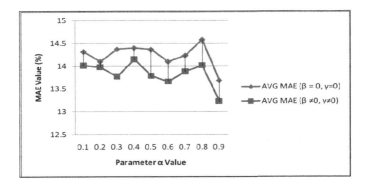

Fig. 5. Performance comparison

The system calculates the average from each category into single MAE value for each parameter. Figure 5 shows that considering both factors of familiarity and similarity factor could return better result than only considering similarity factor. The best accuracy for recommendation is α = 0.9 and β = 0.1 with MAE value is 13.22.

We get top 10 movie genres for local community as depicted in Fig. 6. Top 3 of local community popular genres are Drama, Drama/Romance and Animation. In other hands, we also get the top 10 movie genres throughout our data as depicted in Fig. 8.

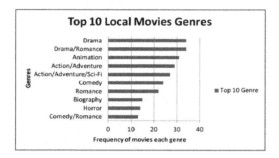

Fig. 6. Top 10 local movie genres

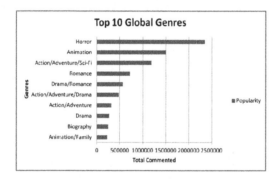

Fig. 7. Top 10 local community movie genres

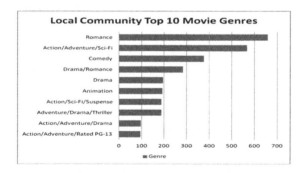

Fig. 8. Global movie genres

The top 3 of the genres for global items are Horror, Animation and Action/Adventure/Sci-fi. In the top 3 of the genres we found our local community movie popularity is different with global popularity. Event animation genre at least has the same popularity. From these results we assume genre animation has the bigger chance to be recommended to cold start user as they recommendation items.

Fig. 9. Percentage global items influence to local community

In Fig. 7 depicted the local community comments or likes for movie pages in Facebook. We found top 3 genres are Romance, Action/Adventure/Sci-Fi and comedy. From this top 3 also have same genre with global popular genre. We also calculate the percentage for global popular items affect the local community. In Fig. 9, we get 33% the recommendation items based on global influence items. For our future work, we would like to examine this value could be used as threshold for determining whether global influence will affect the user better, if Eq. (6) threshold will be set to $\theta = 0.33$ based on these experiments.

5   Conclusions

In this paper, we proposed personalized recommendation system with considering global influence item to solve the cold start of user problems. The system calculates the recommendation items based on user influence combine similarity, familiarity and global influence. We found that the 33% of global items affect the decision in our local community decision of recommendation items. In the future, we aim to calculate the accuracy of the system, with the threshold value we will calculate the function to discover whether the accuracy of the personalized recommendation system could be increased or not. We also aim to solve the cold start problem users with no record of user interest.

Acknowledgment. This paper is supported by Ministry of Science and Technology, Taiwan, with project number: MOST-103-2221-E-324-028 and MOST-104-2221-E-324-019-MY2.

References

1. Kaplan, A.M., Haenlein, M.: Users of the world, unite! The challenges and opportunities of social media. Bus. Horiz. **53**(1), 59–68 (2010)
2. Feng, H., Qian, X.: Mining user-contributed photos for personalized product recommendation. Neurocomputing **129**, 409–420 (2014)
3. Shen, J., Cheng, Z.: Personalized video similarity measure. Multimedia Syst. **17**(5), 421–433 (2011)
4. Shen, J., Pang, H., Wang, M., Yan, S.: Modeling concept dynamics for large scale music search. In: Proceedings of the 35th International ACM SIGIR Conference on Research Development Information Retrieval, New York, NY, USA, pp. 455–464 (2012)
5. Qian, X., Liu, X., Zheng, C., Du, Y., Hou, X.: Tagging photos using users vocabularies. Neurocomputing **111**, 144–153 (2013)
6. Qian, X., Feng, H., Zhao, G., Mei, T.: Personalized recommendation combining user interest and social circle. IEEE Trans. Knowl. Data Eng. **26**(7), 1763–1777 (2014)
7. Li, Z., Wang, C., Yang, S., Jiang, C., Li, X.: LASS: local-activity and social-similarity based data forwarding in mobile social networks. IEEE Trans. Parallel Distrib. Syst. **26**(1), 174–184 (2015)
8. Boley, H., Bhavsar, V.C., Singh, D.A.: Agent matcher search in weighted, tree structured learning object metadata. In: Learning Objects Summit, pp. 29–30 (2004)
9. Rung-Ching, C., Hendry: A domain ontology in social networks for identifying user interest for personalized recommendations. J. Univ. Comput. Sci. **22**(3) (2016)
10. Bond, R., Smith, P.B.: Culture and conformity: a meta-analysis of studies using Asch's (1952b, 1956) line judgement task. Pshycol. Bull. **119**(1), 111–137 (1996)
11. Jiang, M.: Social contextual recommendation. In: Proceedings of the 21st ACM International on CIKM, New York, NY, USA, pp. 1267–1275 (2012)
12. Li, X., Guo, L., Zhao, Y.E.: Tag-based social interest discover. In: Proceedings of the 17th International Conference on World Wide Web, pp. 675–684. ACM (2008)
13. Kwona, O., Leea, Y., Sarangibc, D.: A Galois lattice approach to a context aware privacy negotiation service. Expert Syst. Appl. **38**(10), 12619–12629 (2011)
14. Nguyen, N.P., Dinh, T.N., Tokala, S., Thai, M.T.: Overlapping communities in dynamic networks: their detection and mobile applications. In: Proceedings of the ACM 17th Annual International Conference on Mobile Computing and Networking, pp. 85–96 (2011)

Virtual Balancing of Decision Classes

Marzena Kryszkiewicz[(✉)]

Institute of Computer Science, Warsaw University of Technology,
Nowowiejska 15/19, 00-665 Warsaw, Poland
mkr@ii.pw.edu.pl

Abstract. It has been observed in the literature and practice that the quality of classifying based on confidences of decision rules is poor when a decision table consists of decision classes which significantly differ in the number of objects. A typical approach to overcome negative consequences of the occurrence of this phenomenon is to apply oversampling of minority decision classes and/or undersampling of majority decision classes. In this paper, we introduce a notion of a virtual balancing of decision classes, which does not require any replication of data, but produces the same results as a physical balancing of decision classes. Also, we derive a number of properties of selected evaluation measures (coverage, confidence, lift and growth) of decision rules and relations among them w.r.t. virtually (and by this, physically) balanced decision classes. In particular, we show how to determine threshold values for confidence, lift, growth and coverage so that resulting sets of decision rules were identical.

1 Introduction

It has been observed in the literature and practice that the quality of classifying based on confidences of decision rules is poor when a decision table consists of decision classes which significantly differ in the number of objects. A typical approach to overcome negative consequences of the occurrence of this phenomenon is to apply oversampling of minority decision classes and/or undersampling of majority decision classes [1, 2, 4, 6, 8]. In this paper, we introduce a notion of virtual balancing of decision classes, which does not require any replication of data, but produces the same results as a physical balancing of decision classes. Also, we derive a number of properties of selected evaluation measures (coverage, confidence, lift and growth) of decision rules w.r.t. virtually (and by this, physically) balanced decision classes. In particular, we show how to determine threshold values for confidence, lift, growth and coverage so that resulting sets of decision rules were identical. Our contribution enables derivation of relations among rule evaluation measures in (virtually or physically) balanced decision tables on a theoretical basis.

The layout of the paper is as follows. In Sect. 2, we recall basic notions and properties useful in the paper. Our new contribution is presented in Sect. 3. In Subsect. 3.1, we introduce the notion of a physical balancing and discuss its consequences. Then, in Subsect. 3.2, we introduce the notion of a virtual balancing, which, unlike physical balancing, does not require replication of objects in data, but enables deriving

© Springer International Publishing AG 2017
N.T. Nguyen et al. (Eds.): ACIIDS 2017, Part I, LNAI 10191, pp. 673–684, 2017.
DOI: 10.1007/978-3-319-54472-4_63

the same decision rules and of the same power as in the case of a physical balancing of a data set. Section 4 summarizes our work.

2 Basic Notions and Properties

2.1 Decision Tables and Descriptors

A *decision table* is defined as a pair $DT = (O, AT \cup \{d\})$, where O is a non-empty set of objects, $d \notin AT$ is a distinguished attribute called the *decision attribute*, and the elements of AT are called *conditional attributes*. The set of all values of an attribute $a \in AT \cup \{d\}$ is denoted by V_a.

In the paper, an attribute-value pair is called an *atomic descriptor*. In addition, we distinguish the following types of *descriptors*: a *simple positive descriptor*, a *simple negative descriptor* and a *mixed descriptor*. A conjunction of atomic descriptors $\bigwedge(a = v)$, where $a \in A \subseteq AT \cup \{d\}$ and $v \in V_a$, is called a *conjunctive descriptor*. A formula $\bar{t}$, where t is a conjunctive descriptor, is called a *negated conjunctive descriptor*. A formula $p \wedge \bar{t}$, where p and t are conjunctive descriptors, is called a *mixed descriptor*.

The set of objects satisfying a descriptor t will be denoted by $O(t)$. Clearly, $O(\bar{t}) = O \backslash O(t)$, while $O(t \wedge p) = O(t) \cap O(p)$. The cardinality of the object set $O(t)$ will be called *support* of t and will be denoted by $sup(t)$. The probability that an object satisfies descriptor t will be denoted by $P(t)$ and is equal to $sup(t) / |O|$.

A descriptor t is called a *conditional descriptor* if it refers only to conditional attributes. A descriptor s is called a *decision descriptor* if it refers only to the decision attribute d.

2.2 Decision Rules and Their Example Evaluation Measures

In the paper, we consider *decision rules* (shortly, *rules*) of the form: $t \to s$ and $t \to \bar{s}$, where t is a conditional conjunctive descriptor $\bigwedge(a = v)$, where $a \in A \subseteq AT$, and s is an atomic decision descriptor $(d = w)$, $w \in V_d$. The left-hand side (t) of a rule is called its *antecedent*, while the right-hand side (s or $\bar{s}$, respectively) is called its *consequent*.

In the remainder of the paper, without loss of generality, we assume that the domain of the decision attribute $V_d = \{1, ..., r\}$. In addition, an atomic descriptor $(d = l)$, which refers to the decision attribute d, will be denoted by s_l, where $l \in \{1, ..., r\}$. $O(s_l)$, where $l \in \{1, ..., r\}$, will be called *l-th decision class*. Clearly, $O(\bar{s}_l)$ is the complement of the decision class $O(s_l)$: $O(\bar{s}_l) = O \backslash O(s_l)$; that is, $O(\bar{s}_l)$ is the union of all decision classes different from $O(s_l)$ $(O(\bar{s}_l) = \bigcup_{i \in \{1,...,r\} \backslash \{l\}} O(s_i))$.

Example measures used to evaluate decision rules are support, confidence, coverage, lift and growth. Let us recall their definitions.

The *support* of rule $t \to s_l$ is denoted by $sup(t \to s_l)$ and is defined as the number of objects in O satisfying both t and s_l; that is, $sup(t \to s_l) = sup(t \wedge s_l)$.

The *confidence* of rule $t \to s_l$ is denoted by $conf(t \to s_l)$ and is defined as the fraction of objects in O satisfying t that satisfy also s_l; that is: $conf(t \to s_l) = \frac{sup(t \wedge s_l)}{sup(t)}$.

After dividing the numerator and denominator of the above expression by the number of all objects $|\,O\,|$, we obtain: $conf(t \rightarrow s_l) = \frac{P(t \land s_l)}{P(t)}$, which allows us to interpret $conf$ $(t \rightarrow s_l)$ as the conditional probability that s_l is satisfied by an object given t is satisfied by the object.

An orthogonal measure to confidence is *coverage*: The *coverage* of rule $t \rightarrow s_l$ is denoted by $cov(t \rightarrow s_l)$ and is defined as the fraction of objects in O satisfying s_l that satisfy also t; that is: $cov(t \rightarrow s_l) = \frac{sup(t \land s_l)}{sup(s_l)} = \frac{P(t \land s_l)}{P(s_l)}$. $cov(t \rightarrow s_l)$ can be perceived as the probability that an object from the l-th decision class satisfies descriptor t.

Table 1. Decision table DT

O	a	d
1	2	1
2	2	1
3	2	1
4	2	1
5	2	1
6	2	1
7	2	1
8	1	1
9	1	2
10	2	2

Example 1. Table 1 presents the decision table $DT =$ $(O, AT \cup \{d\})$, where $AT = \{a\}$. There are two decision descriptors: $s_1 = (d, 1)$, $s_2 = (d, 2)$. O consists of 8 objects in the decision class $O(s_1)$ and 2 objects in $O(s_2)$. We note that $conf((a, 1) \rightarrow s_1) = conf((a, 1) \rightarrow s_2) = \frac{1}{2}$, which contradicts the intuition that $(a, 1)$ is more typical for $O(s_2)$ rather than for $O(s_1)$ as suggested by the coverage measure: cov $((a, 1) \rightarrow s_1) = 1/8$, while $cov((a, 1) \rightarrow s_2) = 1/2$. □

Among popular measures of rules is also *lift* [3]. The *lift* of rule $t \rightarrow s_l$ is denoted by $lift(t \rightarrow s_l)$ and is defined as the ratio of the conditional probability that s_l is satisfied by an object given t is satisfied by the object to the probability that s_l is satisfied by the object; that is, $lift(t \rightarrow s_l) = \frac{conf(t \rightarrow s_l)}{P(s_l)}$. The lift measure can be also defined equivalently as shown in Property 1.

Property 1. $lift(t \rightarrow s_l) = \frac{P(t \land s_l)}{P(t) \times P(s_l)} = \frac{cov(t \rightarrow s_l)}{P(t)}$.

Property 1 enables perceiving *lift* as an indicator of a statistical dependence between consequent and antecedent of a decision rule; namely, if $lift(t \rightarrow s_l) = 1$, then $P(t \land s_l) = P(t) \times P(s_l)$, so s_l is independent of t. Otherwise, s_l is regarded as dependent on X. The *dependence* of s_l on t regarded as *positive* if $lift(t \rightarrow s_l) > 1$, and as *negative* if $lift(t \rightarrow s_l) < 1$. It follows from Property 1 that *lift* is a symmetric measure with respect to antecedent and consequent of a rule: namely, $lift(t \rightarrow s_l) = lift(s_l \rightarrow t)$.

It has been shown in [5] that decision rules with high values of growth measure (at least greater than 1, but preferably much higher) are particularly useful for classification purposes. The *growth* of rule $t \rightarrow s_l$ is denoted by $growth(t \rightarrow s_l)$ and is defined as the ratio $\frac{cov(t \rightarrow s_l)}{cov(t \rightarrow \bar{s_l})}$.

Property 2. $growth\ (t \rightarrow s_l) = \frac{P(t \land s_l)}{P(s_l)} / \frac{P(t \land \bar{s_l})}{P(\bar{s_l})} = \frac{lift(t \rightarrow s_l)}{lift(t \rightarrow \bar{s_l})}$.

Proof. $growth\ (t \rightarrow s_l) = \frac{cov(t \rightarrow s_l)}{cov(t \rightarrow \bar{s_l})} = \frac{P(t \land s_l)}{P(s_l)} / \frac{P(t \land \bar{s_l})}{P(\bar{s_l})} = \frac{P(t \land s_l)}{P(t) \times P(s_l)} / \frac{P(t \land \bar{s_l})}{P(t) \times P(\bar{s_l})} = /*$ by Property 1*/ $= \frac{lift(t \rightarrow s_l)}{lift(t \rightarrow \bar{s_l})}$. □

2.3 Underlying Probabilities for *ACBC*-Measures of Rules

As observed in [7], a large number of measures of a rule, say, $t \rightarrow s_l$ is definable at most in terms of the probability $P(t)$ of the *antecedent*, the probability $P(s_l)$ of the *consequent*, the probability $P(t \wedge s_l)$ of the conjunction of the antecedent and consequent (or *base*) and constants. Such measures are called *ACBC-measures* there. The three probabilities $P(t)$, $P(s_l)$, $P(t \wedge s_l)$ will be called *underlying* for *ACBC*-measures of a rule $t \rightarrow s_l$.

In fact, *ACBC*-measures include also measures typical definitions of which refer to marginal probabilities $P(\bar{t})$ and $P(\bar{s}_l)$ as well as joint probabilities $P(t \wedge s_l)$, $P(\bar{t} \wedge s_l)$, $P(t \wedge \bar{s}_l)$ and $P(\bar{t} \wedge \bar{s}_l)$. This comes from the fact that the latter probabilities are derivable from $P(t)$, $P(s_l)$, $P(t \wedge s_l)$, as shown in Property 3.

Property 3. The marginal probabilities $P(\bar{t})$, $P(\bar{s}_l)$ and joint probabilities $P(t \wedge s_l)$, $P(\bar{t} \wedge s_l)$, $P(t \wedge \bar{s}_l)$ and $P(\bar{t} \wedge \bar{s}_l)$ are derivable from $P(t)$, $P(s_l)$, $P(t \wedge s_l)$ as follows:

(a) $P(\bar{t}) = 1 - P(t)$
(b) $P(\bar{s}_l) = 1 - P(s_l)$
(c) $P(\bar{t} \wedge s_l) = P(s_l) - P(t \wedge s_l)$
(d) $P(t \wedge \bar{s}_l) = P(t) - P(t \wedge s_l)$
(e) $P(\bar{t} \wedge \bar{s}_l) = P(\bar{t}) - P(\bar{t} \wedge s_l) = 1 - P(t) - P(s_l) + P(t \wedge s_l)$

Please see Table 2 for definitions of *conf*, *cov*, *lift* and *growth* as example *ACBC*-measures of rules expressed in terms of underlying probabilities.

Table 2. Example *ACBC*-measures of a rule $t \rightarrow s_l$ expressed in terms of underlying probabilities $P(t)$, $P(s_l)$, $P(t \wedge s_l)$

ACBC-measure	Definition		
$conf(t \rightarrow s_l)$	$\frac{P(t \wedge s_l)}{P(t)}$		
$cov(t \rightarrow s_l)$	$\frac{P(t \wedge s_l)}{P(s_l)}$		
$lift(t \rightarrow s_l)$	$\frac{conf(t \rightarrow s_l)}{P(s_l)} = \frac{P(t \wedge s_l)}{P(t) \times P(s_l)}$		
$growth(t \rightarrow s_l)$	$\frac{cov(t \rightarrow s_l)}{cov(t \rightarrow \bar{s}_l)} = \frac{P(t \wedge s_l)}{P(s_l)} / \frac{P(t \wedge \bar{s}_l)}{P(\bar{s}_l)}$	$= \frac{P(t \wedge s_l)}{P(s_l)} / \frac{P(t) - P(t \wedge s_l)}{1 - P(s_l)}$	

In the remainder of the paper, we will use the fact that the probability of a conditional descriptor t is derivable from the joint probabilities $P(t \wedge s_i)$, where each s_i is a decision descriptor, $i \in \{1, \ldots, r\}$.

Property 4. $P(t) = \sum_{i \in \{1, \ldots, r\}} P(t \wedge s_i)$.

Proposition 1. The knowledge about probabilities: $P(s_1)$, ..., $P(s_r)$, $P(t \wedge s_1)$,..., $P(t \wedge s_r)$ is equivalent to the knowledge of the underlying probabilities: $P(t)$, $P(s_i)$ and $P(t \wedge s_i)$ for all decision descriptors s_i, where $i \in \{1, \ldots, r\}$.

Proof. By Property 4.   □

3 Balancing of Decision Classes

It has been observed that the quality of classifying based on confidences of decision rules is poor when a decision table consists of decision classes which significantly differ in the number of objects (see Example 1 for an illustration of the problem with the confidence measure in the presence of significantly imbalanced decision classes). In Subsect. 3.1, we consider a naïve approach to solving this problem by physical balancing of decision classes, which might result in creating a balanced decision table with unacceptably large number of objects. We treat this naïve approach as a reference one and do not really expect it to be applied in practice. Then, in Subsect. 3.2, we define and examine a virtual balancing of decision classes, which will yield the same values of any *ACBC*-measure (confidence or another) without any augmentation of the original decision table as in the case of applying physically balanced decision table.

3.1 Physical Balancing of Decision Classes

Let K be a sequence $<k_1, ..., k_r>$ of natural numbers. Let DT^K be a decision table obtained from DT by replicating each object in each decision class $O(s_i)$ k_i times, $i = \{1, ..., r\}$. In the remainder of the paper, we will use replication numbers' sequence K in superscript whenever referring to augmented in this way decision table DT^K or when determining values of measures based on DT^K. Clearly, the cardinality of the decision class $O^K(s_i)$ will be k_i times larger than the cardinality of the original decision class $O(s_i)$, $i = \{1, ..., r\}$.

DT^K is called a *physically balanced decision table* if the cardinalities of all its decision classes are identical; that is, if $k_i \times sup(s_i) = k_j \times sup(s_j)$ for $i, j \in \{1, ..., r\}$.

Table 3. DT^K, $K = <1, 4>$

O	a	d
1	2	1
2	2	1
3	2	1
4	2	1
5	2	1
6	2	1
7	2	1
8	1	1
9	1	2
9a	1	2
9b	1	2
9c	1	2
10	2	2
10a	2	2
10b	2	2
10c	2	2

Example 2. Table 3 shows a physically balanced decision table DT^K obtained from DT from Table 1 for sequence $K = <1, 4>$ of replication numbers. O^K consists of 8 objects in the decision class $O^K(s_1)$ and 8 objects in $O^K(s_2)$. $O^K(s_1)$ stores 1 instance of each object from $O(s_1)$, while $O^K(s_2)$ stores 4 instances for each object from $O(s_2)$ (i.e. 4 instances of object 9 and 4 instances of object 10). Values of *conf* and *cov* calculated in DT^K are as follows: $conf^K((a,1) \to s_1) = 1/5$, $conf^K((a,1) \to s_2) = 4/5$, $cov^K((a,1) \to s_1) = 1/8$, $cov^K((a,1) \to s_2) = 4/8$. Note that after applying the balancing, both *cov* and *conf* indicate that $(a,1)$ is more typical for the decision class $O^K(s_2)$ rather than $O^K(s_1)$, which is consistent with the intuition.☐

Proposition 2. Let DT^K be a physically balanced decision table, $l \in \{1, ..., r\}$ and t be a descriptor.

(a) $sup^K(s_l) = sup(s_l) \times k_l$
(b) $sup^K(s_i) = sup^K(s_l)$ for all $i \in \{1, ..., r\}$

(c) $\left| O^K \right| = \sum_{i \in \{1,\dots,r\}} sup^K(s_i) = r \times sup^K(s_l) = r \times sup(s_l) \times k_l$

(d) $sup^K(\bar{s}_l) = \sum_{i \in \{1,\dots,r\} \setminus \{l\}} sup^K(s_i) = (r-1) \times sup^K(s_l)$

(e) $sup^K(t \wedge s_l) = sup(t \wedge s_l) \times k_l$

(f) $sup^K(t) = \sum_{i \in \{1,\dots,r\}} sup^K(t \wedge s_i) = \sum_{i \in \{1,\dots,r\}} (sup(t \wedge s_i) \times k_i)$

(g) $sup^K(t \wedge \bar{s}_l) = \sum_{i \in \{1,\dots,r\} \setminus \{l\}} sup^K(t \wedge s_i) = \sum_{i \in \{1,\dots,r\} \setminus \{l\}} (sup(t \wedge s_i) \times k_i)$

Proof. Ad (a) By definition of DT^K.

Ad (b) By definition of a physically balanced DT^K.

Ad (c) $\left| O^K \right| = \sum_{i \in \{1,\dots,r\}} sup^K(s_i)$ comes from the fact that all r decision classes are mutually exclusive and cover all objects in DT^K. $\sum_{i \in \{1,\dots,r\}} sup^K(s_i) = r \times sup^K(s_l)$ comes from the fact that all r decision classes in DT^K have identical cardinalities.

Ad (d) Comes from the fact that the negative descriptor $\bar{s}_l$ is satisfied by the objects from r-1 decision classes in DT^K (different from the l-th decision class), which, nevertheless, have identical cardinalities as the l-th decision class.

Ad (e) Comes from the fact that each object in the l-th decision class of DT (in particular, an object that satisfies descriptor t) is replicated k_l times in DT^K.

Ad (f) Follows from Property 4 and Proposition 2e.

Ad (g) $sup^K(t \wedge \bar{s}_l) = \sum_{i \in \{1,\dots,r\}} sup^K(t \wedge s_i)$ comes from the fact that $t \wedge \bar{s}_l$ is satisfied by all objects from all decision classes except for the l-th decision class in DT^K that satisfy descriptor t. $\qquad \square$

Theorem 1. Let DT^K be a physically balanced decision table and t be a descriptor. Then, for all $l \in \{1, \dots, r\}$:

(a) $P^K(s_l) = \frac{1}{r}$

(b) $P^K(t \wedge s_l) = \frac{cov(t \to s_l)}{r}$

Proof. Ad (a) All r decision classes in the physically balanced decision table DT^K have identical cardinalities. Hence, the probability that an object from DT^K belongs to one of these decision classes is the same and equals $\frac{1}{r}$.

Ad   (b) $P^K(t \wedge s_l) = \frac{sup^K(t \wedge s_l)}{|O^K|} = / *$ by Propositions 2c and 2e $* / =$

$\frac{sup(t \wedge s_l) \times k_l}{r \times sup(s_l) \times k_l} = \frac{sup(t \wedge s_l)}{r \times sup(s_l)} = \frac{cov(t \to s_l)}{r}$. $\qquad \square$

Theorem 2. Let DT^K be a physically balanced decision table and t be a descriptor. Then, for all $l \in \{1, \dots, r\}$:

$$P^K(t) = \sum_{i \in \{1,\dots,r\}} P^K(t \wedge s_i) = \sum_{i \in \{1,\dots,r\}} \frac{cov(t \to s_i)}{r}.$$

Proof.

$$P^K(t) = /* \text{ by Property } 4 \; */ = \sum_{i \in \{1,\ldots,r\}} P^K(t \wedge s_i) = /* \text{ by Theorem 1b } */$$
$$= \sum_{i \in \{1,\ldots,r\}} \frac{cov(t \to s_i)}{r}. \qquad \Box$$

As follows from Theorem 2, the probability of t in a physically balanced decision table DT^K equals the average coverage of the decision classes by the objects satisfying descriptor t in the original decision table DT.

By Theorem 1, for each physically balanced decision table DT^K obtained from decision table DT, the learnt probabilities of decision descriptors are identical $(P^K(s_1) = \ldots = P^K(s_r) = \frac{1}{r})$ and depend only on the number of decision classes in DT, while the joint probabilities $P^K(t \wedge s_1) = \frac{cov(t \to s_1)}{r}, \ldots, P^K(t \wedge s_r) = \frac{cov(t \to s_r)}{r})$ depend only on the number and the coverages of the decision classes in original DT. Hence, the corresponding probabilities learnt in any two physically balanced decision tables DT^K and DT^M will be the same and will be independent of replication numbers' sequences K and M. In addition, the probabilities $P^K(s_1) = P^M(s_1), \ldots, P^K(s_r) = P^M(s_r), P^K(t \wedge s_1) = P^M(t \wedge s_1), \ldots, P^K(t \wedge s_r) = P^M(t \wedge s_r)$ can be learnt directly from original DT.

For that reason, in the next section, instead of applying a physical balancing, we propose to apply a *virtual balancing of DT*, which does not require replication of objects, but produces the same results as any physical balancing of DT does.

3.2 Virtual Balancing of Decision Classes

Let us start with introducing notions of *virtually balanced probabilities*.

A *virtually balanced probability of a decision descriptor* s_i, $l \in \{1, \ldots, r\}$, is denoted by $P^V(s_i)$ and is defined as follows:

$$P^V(s_i) = \frac{1}{r}.$$

A *virtually balanced probability of a conjunction of a conditional descriptor t and a decision descriptor* s_i, $i \in \{1, \ldots, r\}$, is denoted by $P^V(t \wedge s_i)$ and defined as follows:

$$P^V(t \wedge s_i) = \frac{cov(t \to s_i)}{r}.$$

Please note that values of virtually balanced probabilities $P^V(s_i)$ and $P^V(t \wedge s_i)$ are identical as corresponding probabilities $P^K(s_i)$ and $P^K(t \wedge s_i)$ from Theorem 1, which were calculated in a physically balanced DT^K.

Virtually balanced probabilities of t, $\bar{t}$, $\bar{s}_l$, $t \wedge s_l$, $\bar{t} \wedge s_l$, $t \wedge \bar{s}_l$ and $\bar{t} \wedge \bar{s}_l$, where $l = 1, \ldots, r$, are denoted by $P^V(t)$, $P^V(\bar{t})$, $P^V(\bar{s}_l)$, $P^V(t \wedge s_l)$, $P^V(\bar{t} \wedge s_l)$, $P^V(t \wedge \bar{s}_l)$, $P^V(\bar{t} \wedge \bar{s}_l)$, respectively, and are calculated according to Properties 4 and 3 based on virtually balanced probabilities of $s_1, \ldots, s_r, t \wedge s_1, \ldots, t \wedge s_r$. An *ACBC*-measure μ of a rule $t \to s_l$ calculated based on virtually balanced probabilities will be denoted as $\mu^V(t \to s_l)$ and will be called a *virtually balanced ACBC-measure*.

Theorem 3. Let $l \in \{1, \ldots, r\}$ and t be a descriptor. Then:

$$P^V(t) = \sum_{i \in \{1,\ldots,r\}} P^V(t \wedge s_i) = \sum_{i \in \{1,\ldots,r\}} \frac{cov(t \to s_i)}{r}.$$

Proof. $P^V(t) = / *$ by Property 4 $* / = \sum_{i \in \{1,\ldots,r\}} P^V(t \wedge s_i) = / *$ by definition of a virtually balanced $P^V(t \wedge s_i) * / = \sum_{i \in \{1,\ldots,r\}} \frac{cov(t \to s_i)}{r}$.   □

As follows from Theorem 3, the virtually balanced probability that an object satisfies descriptor t in *DT* equals the average probability that descriptor t is satisfied by an object from a decision class in *DT*. Note that this virtually balanced probability is the same as calculated in a physically balanced decision table (see Theorems 2–3).

Corollary 1. Values of underlying probabilities $P^V(t)$, $P^V(s_i)$, $P^V(t \wedge s_i)$ and, by this, also of each *ACBC*-measure μ^V of rule $t \to s_l$ determined based on *DT* are the same as those determined in each physically balanced decision table DT^K, irrespective of replications numbers in sequence K.

We will derive now formulae for the evaluation measures of decision rules calculated w.r.t. virtually balanced probabilities, and, by this, holding in each physically balanced decision table.

Lemma 1. Let $l \in \{1, \ldots, r\}$ and t be a descriptor. Then:

(a) $P^V(\bar{s}_l) = 1 - \frac{1}{r} = \frac{r-1}{r}$

(b) $P^V(t \wedge \bar{s}_l) = \sum_{i \in \{1,\ldots,r\}\setminus\{l\}} \frac{cov(t \to s_i)}{r}$

Proof. Ad (a). $P^V(\bar{s}_l) = 1 - P^V(s_l) = 1 - \frac{1}{r} = \frac{r-1}{r}$.

Ad (b) $P^V(t \wedge \bar{s}_l) = P^V(t) - P^V(t \wedge s_l) = / *$ by Theorem 3 $* / = \sum_{i \in \{1,\ldots,r\}\setminus\{l\}}$
$P^V(t \wedge s_i) = \sum_{i \in \{1,\ldots,r\}\setminus\{l\}} \frac{cov(t \to s_i)}{r}$.   □

Theorem 4. Let $l \in \{1, \ldots, r\}$ and t be a descriptor. Then:

(a) $cov^V(t \to s_l) = cov(t \to s_l)$

(b) $cov^V(t \to \bar{s}_l) = \frac{\sum_{i \in \{1,\ldots,r\}\setminus\{l\}} cov(t \to s_i)}{r-1}$

(c) $conf^V(t \to s_l) = \frac{cov(t \to s_l)}{\sum_{i \in \{1,\ldots,r\}} cov(t \to s_i)}$

(d) $conf^V(t \to \bar{s}_l) = \frac{\sum_{i \in \{1,\ldots,r\}\setminus\{l\}} cov(t \to s_i)}{\sum_{i \in \{1,\ldots,r\}} cov(t \to s_i)}$

(e) $lift^V(t \to s_l) = r \times conf^V(t \to s_l) = \frac{cov(t \to s_l)}{\left(\frac{\sum_{i \in \{1,\ldots,r\}} cov(t \to s_i)}{r}\right)}$

(f) $lift^V(t \to \bar{s}_l) = \frac{r}{r-1} \times conf^V(t \to \bar{s}_l) = \dfrac{\left(\frac{\sum_{i\in\{1,...,r\}\setminus\{l\}} cov(t \to s_i)}{r-1}\right)}{\left(\frac{\sum_{i\in\{1,...,r\}} cov(t \to s_i)}{r}\right)}$

(g) $growth^V(t \to s_l) = \frac{cov^V(t \to s_l)}{cov^V(t \to \bar{s}_l)} = \dfrac{cov(t \to s_l)}{\left(\frac{\sum_{i\in\{1,...,r\}\setminus\{l\}} cov(t \to s_i)}{r-1}\right)} = \dfrac{lift(t \to s_l)}{\left(\frac{\sum_{i\in\{1,...,r\}\setminus\{l\}} lift(t \to s_i)}{r-1}\right)}$

(h) $growth^V(t \to s_l) = \frac{lift^V(t \to s_l)}{lift^V(t \to \bar{s}_l)}$

(i) $growth^V(t \to s_l) = \frac{conf^V(t \to s_l)}{\left(\frac{conf^V(t \to \bar{s}_l)}{r-1}\right)}$

Proof. Ad (a) $cov^V(t \to s_l) = \frac{P^V(t \wedge s_l)}{P^V(s_l)} = /*$ by definition, $P^V(t \wedge s_l) = \frac{cov(t \to s_l)}{r}$ and $P^V(s_l) = \frac{1}{r} */ = cov(t \to s_l)$.

Ad (b) $cov^V(t \to \bar{s}_l) = \frac{P^V(t \wedge \bar{s}_l)}{P^V(\bar{s}_l)} = /*$ by Lemmas 1a and 1b $*/ = \frac{\sum_{i\in\{1,...,r\}\setminus\{l\}} cov(t \to s_i)}{r-1}$.

Ad (c) $conf^V(t \to s_l) = \frac{P^V(t \wedge s_l)}{P^V(t)} = /*$ by the fact that $P^V(t \wedge s_l) = \frac{cov(t \to s_l)}{r}$ and by Theorem 3 $*/ = \frac{cov(t \to s_l)}{\sum_{i\in\{1,...,r\}} cov(t \to s_i)}$.

Ad (d) $conf^V(t \to \bar{s}_l) = \frac{P^V(t \wedge \bar{s}_l)}{P^V(t)} = /*$ by Lemma 1b and Theorem 3 $*/ = \frac{\sum_{i\in\{1,...,r\}\setminus\{l\}} cov(t \to s_i)}{\sum_{i\in\{1,...,r\}} cov(t \to s_i)}$.

Ad (e) $lift^V(t \to s_l) = \frac{conf^V(t \to s_l)}{P^V(s_l)} = /*$ by definition, $P^V(s_l) = \frac{1}{r} */ = r \times conf^V(t \to s_l) = /*$ by Theorem 4c $*/ = \frac{cov(t \to s_l)}{\left(\frac{\sum_{i\in\{1,...,r\}} cov(t \to s_i)}{r}\right)}$.

Ad (f) $lift^V(t \to \bar{s}_l) = \frac{conf^V(t \to \bar{s}_l)}{P^V(\bar{s}_l)} = /*$ by Lemma 1a $*/ = \frac{r}{r-1} \times conf^V(t \to \bar{s}_l) = /*$ by Theorem 4d $*/ = \dfrac{\left(\frac{\sum_{i\in\{1,...,r\}\setminus\{l\}} cov(t \to s_i)}{r-1}\right)}{\left(\frac{\sum_{i\in\{1,...,r\}} cov(t \to s_i)}{r}\right)}$.

Ad (g) By definition, $growth^V(t \to s_l) = \frac{cov^V(t \to s_l)}{cov^V(t \to \bar{s}_l)} = /*$ by Theorems 4a and b $*/ = \frac{cov(t \to s_l)}{\left(\frac{\sum_{i\in\{1,...,r\}\setminus\{l\}} cov(t \to s_i)}{r-1}\right)} = \frac{cov(t \to s_l)/P(t)}{\left(\frac{\sum_{i\in\{1,...,r\}\setminus\{l\}} cov(t \to s_i)/P(t)}{r-1}\right)} = \frac{lift(t \to s_l)}{\left(\frac{\sum_{i\in\{1,...,r\}\setminus\{l\}} lift(t \to s_i)}{r-1}\right)}$.

Ad (h) $\frac{lift^V(t \to s_l)}{lift^V(t \to \bar{s}_l)} = /*$ by Theorems 4e and f $*/ = \frac{cov(t \to s_l)}{\left(\frac{\sum_{i\in\{1,...,r\}\setminus\{l\}} cov(t \to s_i)}{r-1}\right)} = /*$ by Theorem 4g $*/ = growth^V(t \to s_l)$.

Ad (i) $\frac{conf^V(t \to s_l)}{\left(\frac{conf^V(t \to \bar{s}_l)}{r-1}\right)} = /*$ by Theorems 4c and d $*/ = \frac{cov(t \to s_l)}{\left(\frac{\sum_{i\in\{1,...,r\}\setminus\{l\}} cov(t \to s_i)}{r-1}\right)} =$

$/*$ by Theorems 4a and b $*/ = \frac{cov^V(t \to s_l)}{cov^V(t \to \bar{s}_l)} = growth^V(t \to s_l)$. $\square$

Corollary 2. Let $l \in \{1, ..., r\}$ and t be a descriptor. Then:

(a) $conf^V(t \rightarrow s_l)$ reaches the greatest value, which equals 1, if t is satisfied only by objects in the l-th decision class.
(b) The greatest value of $lift^V(t \rightarrow s_l)$ is equal to the number r of decision classes.

Proof. Ad (a) Follows Theorem 4c.

  Ad (b) Follows from Corollary 2a and Theorems 4c and e.   □

Theorem 4 allows us to infer that a decision rule with a given antecedent that has the highest value of *cov* in the original *DT*, will have the highest values of *cov*, *conf*, *lift* and *growth* after applying the balancing (see Corollary 3). Also, Theorem 4 enables deriving corresponding threshold values for these measures application of which make it possible to derive identical sets of decision rules after applying balancing (see Theorem 5 and Corollary 4).

Corollary 3. Let $l \in \{1, ..., r\}$ and t be a descriptor. Then the following statements are equivalent:

(a) $cov(t \rightarrow s_l) = max\{cov(t \rightarrow s_i)| \ i \in \{1, ..., r\}\}$
(b) $cov^V(t \rightarrow s_l) = max\{cov^V(t \rightarrow s_i)| \ i \in \{1, ..., r\}\}$
(c) $conf^V(t \rightarrow s_l) = max\{conf^V(t \rightarrow s_i)| \ i \in \{1, ..., r\}\}$
(d) $lift^V(t \rightarrow s_l) = max\{lift^V(t \rightarrow s_i)| \ i \in \{1, ..., r\}\}$
(e) $growth^V(t \rightarrow s_l) = max\{growth^V(t \rightarrow s_i)| \ i \in \{1, ..., r\}\}$

Theorem 5. Let $l \in \{1, ..., r\}$, t be a descriptor and $\varepsilon \in [0, 1]$. Then the following statements are equivalent:

(a) $conf^V(t \rightarrow s_l) > \varepsilon$
(b) $lift^V(t \rightarrow s_l) > \varepsilon \times r$
(c) $growth^V(t \rightarrow s_l) > \frac{r-1}{\frac{1}{\varepsilon}-1}$
(d) $cov(t \rightarrow s_l) > \varepsilon \times \sum_{i \in \{1,...,r\}} cov(t \rightarrow s_i)$
(e) $cov^V(t \rightarrow s_l) > \varepsilon \times \sum_{i \in \{1,...,r\}} cov(t \rightarrow s_i)$

Corollary 4. Let $l \in \{1, ..., r\}$, t be a descriptor, $\varepsilon \in [0, 1]$ and DT^K be a physically balanced decision table. Then the following sets of decision rules are identical.

(a) the set of all decision rules with virtually balanced confidence greater than ε
(b) the set of all decision rules with virtually balanced lift greater than $\varepsilon \times r$
(c) the set of all decision rules with virtually balanced growth greater than $\frac{r-1}{\frac{1}{\varepsilon}-1}$
(d) the set of all decision rules with virtually balanced coverage greater than $\frac{\sum_{i \in \{1,...,r\}} cov(t \rightarrow s_i)}{r}$
(e) the set of all decision rules with confidence in DT^K greater than ε
(f) the set of all decision rules with lift in DT^K greater than $\varepsilon \times r$
(g) the set of all decision rules with growth in DT^K greater than $\frac{r-1}{\frac{1}{\varepsilon}-1}$

(h) the set of all decision rules with coverage in DT^K greater than $\dfrac{\sum_{i\in\{1,...,r\}} cov(t\to s_i)}{r}$

(i) the set of all decision rules with coverage in DT greater than $\dfrac{\sum_{i\in\{1,...,r\}} cov(t\to s_i)}{r}$

Corollary 5. Let $l \in \{1, ..., r\}$ and t be a condition descriptor. Then:

$$conf^V(t \to s_l) > \frac{1}{r} \Leftrightarrow cov(t \to s_l) > \frac{\sum_{i\in\{1,...,r\}} cov(t\to s_i)}{r} \Leftrightarrow cov^V(t \to s_l) >$$

$$\frac{\sum_{i\in\{1,...,r\}} cov(t\to s_i)}{r} \Leftrightarrow growth^V(t \to s_l) > 1 \Leftrightarrow lift^V(t \to s_l) > 1 \Leftrightarrow \text{the decision descriptor } s_l$$

is positively dependent on conditional descriptor t in each physically balanced decision table DT^K.

Proof. Follows from Theorem 5.   □

4 Summary

In the paper, we first introduced the notion of a physical balancing and discussed its consequences. Then, we introduced the notion of a virtual balancing of decision classes, which does not require any replication of data, but produces the same results as physical balancing. Also, we derived a number of properties of selected evaluation measures (coverage, confidence, lift and growth) of decision rules w.r.t. virtually (and by this, physically) balanced decision classes. In particular, we showed how to determine threshold values for confidence, lift, growth and coverage so that resulting sets of decision rules were identical after applying (virtual or physical) balancing. Our contribution enabled derivation of relations among rule evaluation measures in (virtually or physically) balanced decision tables on a theoretical basis.

References

1. Ali, A., Shamsuddin, S.M., Ralescu, A.L.: Classification with class imbalance problem: a review. Int. J. Adv. Soft Comp. Appl. **7**(3), 176–204 (2015)
2. Batista, G., Prati, R., Monard, R.: A study of the behavior of several methods for balancing machine learning training data. ACM SIGKDD Explor. Newslett. **6**(1), 20–29 (2004)
3. Brin, S., Motwani, R., Ullman, J.D., Tsur, S.: Dynamic itemset counting and implication rules for market basket data. In: ACM SIGMOD 1997, pp. 255–264. ICMD (1997)
4. Chawla, N.V.: Data mining for imbalanced datasets: an overview. In: Maimon, O., Rokach, L. (eds.) Data Mining and Knowledge Discovery Handbook, pp. 853–867. Springer, Heidelberg (2005)
5. Dong, G., Zhang, X., Wong, L., Li, J.: CAEP: classification by aggregating emerging patterns. In: Arikawa, S., Furukawa, K. (eds.) DS 1999. LNCS (LNAI), vol. 1721, pp. 30–42. Springer, Heidelberg (1999). doi:10.1007/3-540-46846-3_4
6. Galar, M., Fernandez, A., Barrenechea, E., Bustince, H., Herrera, F.: A review on ensembles for the class imbalance problem: bagging-, boosting-, and hybrid-based approaches. IEEE Trans. Syst. Man Cybern. Part C Appl. Rev. **99**, 1–22 (2011)

7. Kryszkiewicz, M.: A lossless representation for association rules satisfying multiple evaluation criteria. In: Nguyen, N.T., Trawiński, B., Fujita, H., Hong, T.-P. (eds.) ACIIDS 2016. LNCS (LNAI), vol. 9622, pp. 147–158. Springer, Heidelberg (2016). doi:10.1007/978-3-662-49390-8_14

8. Stefanowski, J.: Dealing with data difficulty factors while learning from imbalanced data. In: Matwin, S., Mielniczuk, J. (eds.) Challenges in Computational Statistics and Data Mining. SCI, vol. 605, pp. 333–363. Springer, Heidelberg (2016). doi:10.1007/978-3-319-18781-5_17

Evaluation of Speech Perturbation Features for Measuring Authenticity in Stress Expressions

Branimir Dropuljić, Leo Mršić$^{(\boxtimes)}$, Robert Kopal, Sandro Skansi, and Andrijana Brkić

IN2data Data Science Company Ltd., Zagreb, Croatia
{branimir.dropuljic,leo.mrsic,robert.kopal,
sandro.skansi,andrijana.brkic}@in2data.hr

Abstract. Expressions can vary by the authenticity level, i.e. the real amount of emotion present within the person when expressing it. They are often sincere, and thus authentic and natural; the person expresses what he/she feels. But play-acted expressions are also present in our lives in a form of deception, movies, theater, etc. It was shown in the literature that those two type of expressions are often hard to distinguish. While some studies concluded that play-acted expressions are more intense, exaggerated or stereotypical than the natural ones, other authors failed to detect such a behavior. The goal of our analysis is to investigate whether speech perturbation features, i.e. jitter, shimmer, variance and features of disturbances in laryngeal muscle coordination, can be used as a robust measure for the analysis of the stress expression authenticity. Two subsets of the SUSAS database (Speech Under Simulated and Actual Stress) – the Roller-coaster subset and the Talking Styles Domain – are used for this purpose. It was shown that perturbation features in general show statistically significant difference between realistic and acted expressions, only the jitter features generally failed to discriminate these two type of expressions. The rising trend of perturbation feature values is observed from acted- to real-stress expressions.

Keywords: Speech perturbation features · Authenticity analysis · Emotional stress · Speech under stress

1 Introduction

The purpose of this paper is to evaluate speech perturbation features in the context of stress (or emotional) expression authenticity; also to discover whether such features can be used for distinguishing authentic (realistic) expressions from those that are acted. Of course, acted expressions are not necessarily nonauthentic. "Although natural expressions are partly staged, acted expressions are also partly natural." [1] Actors' portrayals can be influenced by subjective feelings, especially when produced via techniques based on emotional imagination or memory [2,3]. It was therefore argued that authentic expressions and play-acted ones are sometimes very difficult to distinguish [3].

© Springer International Publishing AG 2017
N.T. Nguyen et al. (Eds.): ACIIDS 2017, Part I, LNAI 10191, pp. 685–694, 2017.
DOI: 10.1007/978-3-319-54472-4_64

Only a few studies compare these two types of expressions [4–8] and most of them concluded that play-acted expressions are more intense, exaggerated or stereotypical [7,9–12]. E.g. it was shown in [13] that speech fundamental frequency (F_0) contour varies more and is generally higher in play-acted expressions than in authentic ones. Additionally, differences between professional and non-professional actors' expressions were analyzed and compared to the authentic expressions. It was shown that in terms of the acoustic characteristics, vocal expressions delivered by professional actors were not more similar to authentic expressions than the ones by non-actors [14]. Furthermore, the results do not support the view that play-acted expressions are necessarily stereotyped caricatures of authentic expressions. The similar findings were reported in [5,15,16], in which authors failed to detect an exaggerating behavior in the case of acting expressions.

The perturbation features (i.e. jitter and shimmer) were included in the authentic vs. play-acted expression analysis in [17] and the results show that there is no statistically significant difference between these two type of expressions for jitter measure, and that there is a significant falling trend from acted to authentic expressions for shimmer measure in the case of vowel "e" (the vowel "a" did not result in any difference). The authors used acted utterances produced by professional actors under four emotional states: 'fear', 'anger', 'joy' and 'sadness', and the same emotions were extracted as authentic set from various emotional reportages about situations in the past. The authors expect higher arousal for acted expressions, which is related with the exaggerating behavior, and some of the "arousal features" show such a behavior, and some of them don't. E.g. shimmer and F_0 contour variability decrease from acted to authentic emotions, which is an indicator of arousal decrease, while several measures like formant bandwidths and peak frequencies indicate that arousal was increased.

There are many vocal components of emotional (arousal) expression. Some of them are under voluntary control and some of them are autonomic [18]. As real arousal is not present, or is minimally present, within the person in the case of acted expressions, we expect that the autonomic vocal features of arousal will have smaller values than in the case of authentic (realistic) expressions. On the other hand, features that are affected by the voluntary control (e.g. F_0 variations, energy/intensity of the utterance, etc.) can have smaller or larger values, depending on the style of the expression (e.g. larger when a person is exaggerating in the expression). As speech perturbations increase as the arousal (and thus emotional stress) increases [19,20], perturbation features should show the rising trend from acted to real-stress expressions if include only of the autonomic component of perturbations, i.e. if they are not under voluntary control.

The goal of our analysis is to investigate how speech perturbation features behave in acted- and real-stress conditions, compared to neutral state. We also investigate whether the features can be used as a robust measure for the analysis of the expression authenticity. An expanded set of perturbation features compared to [17] is proposed for this purpose, as will be described in the next section. Results presented in [21], indicates that some perturbation features are less affected by the voluntary control and, on the other hand, more related to autonomic disturbances in laryngeal muscle coordination. Such features should therefore not depend on the style of acted expression; they should be relatively good at distinguishing between acted and realistic expressions (show a rising trend) regardless of whether acted expressions are exaggerated or moderate.

2   Speech Perturbation Features

Four groups of speech perturbation features are used for the analyses purpose:

1. *peSNS* features – speech features of disturbances in laryngeal muscle coordination; 8 features proposed in [21]: *var* (variance of the perturbation contour), *mean abs* (mean absolute value of the perturbation contour), *max abs* (maximal absolute value of the perturbation contour), *dur per sec (v)* (total duration of the perturbation intervals per second of a voiced speech), *dur per sec* (total duration of the perturbation intervals per second), *mean dur* (mean duration of the perturbation intervals), *max dur* (maximal duration of the perturbation intervals) and *quan* (perturbation quantity: the product of *mean abs* and *dur per sec (v)* features)
2. **jitter features** – measures of period-to-period fluctuations in glottal-cycle durations; 4 features calculated using the Praat functions [22]: *local, local (abs), rap* and *ppq5*
3. **shimmer features** – measures of the period-to-period variability of the speech amplitude value; 5 features calculated using the Praat functions [22]: *local, local (dB), apq3, apq5* and *apq11*
4. **variation features** – 2 features: *var diff* (F_0) (variance of the first differential of F_0 contour) and *std diff* (F_0) (standard deviation of the first differential of F_0 contour)

Jitter, shimmer and variation features are state-of-the-art speech perturbation features for emotion and stress recognition [23]. *peSNS* features are proposed in [21] by the research group of which the authors of this paper are members, and are derived from F0 contour decomposition described in [24].

3   Analyses

For the purpose of our analyses, we combined two subsets of the SUSAS database – the Actual Speech Under Stress Domain (Roller-coaster subset) and the Talking Styles Domain [25].

The Roller-coaster subset of the database is defined as actual high level stress and contains utterances of seven speakers (four male and three female) pronouncing 35 keywords like "break", "change", "gain", etc., typical for communication between pilots and air traffic controllers. They were recorded at the ground (neutral samples) as well as during the ride on the roller-coaster (stress samples). A total of 414 stress samples and 701 neutral samples are available from this subset, not equally distributed between people. The Talking Styles Domain is defined as simulated stress. It consists of utterances of nine people (actors), all male, talking in seven different styles (acted samples): slow, fast, angry, question, soft, loud, clear, as well as talking normally thus producing a neutral speech (neutral samples). For each person and each style (plus neutral speech) there are 70 samples, which is in total $9 \times 8 \times 70 = 5040$ samples. The same 35 words as in the Roller-coaster subset are used.

3.1 1st Analysis: *Neural – Acted – Stress*

We first combined the described data sets into one data set with three levels – *neutral* (composed of both neutral samples from the first and the second domain; t-test show that there is no statistically significant difference between these two sets at α level of 0.05), *acted* (a portion of the Talking Styles Domain) and *stress* (a portion of the stress samples from the Roller-coaster subset). The combined data set was equally sampled over different people and consists of 1242 utterances (414 neutral, 414 acted, 414 stress).

We assume that perturbations are higher in stressful situations than in neutral situations or while acting and not actually experiencing stress (or an emotion). It was shown in [21] that perturbation features (in general) show a rising trend from neutral to real-stress samples. The relative position of acted-stress samples compared to neutral and real-stress samples is the focus of this paper. A separate 'neutral vs. acted' and 'acted vs. stress' analyses were therefore performed. The aim is also to evaluate speech perturbation features in order to explore which of them are relevant for the authenticity analysis in stress expressions. All features were thus analyzed individually for all samples. The left- and right-tailed t-tests, as well as the support vector machines (SVM) method are used. We have chosen the SVM classifier because it generally outperforms other classifiers in emotional speech recognition tasks.

The trend is calculated separately for 'neutral' to 'acted' changes in perturbation level and 'acted' to 'stress' changes. A moderate rising trend ($\nearrow$) is declared if left-tailed t-test show statistically significant difference at α level of 0.05 and a moderate falling trend ($\searrow$) is declared if right-tailed t-test show statistically significant difference at α level of 0.05. Additionally, strong rising ($\nearrow\nearrow$) and falling ($\searrow\searrow$) trends are defined in cases where p values of left- and right-tailed t-tests, respectively, are smaller than 0.0001. If neither left- nor right-tailed t-tests show statistically significant difference, then no trend ($=$) is declared.

A two-class[1] SVM model is trained for each type of comparison ('neutral vs. acted'; 'neutral vs. stress'; and 'acted vs. stress') and for each feature separately, i.e. with the observation vector consisting of only one feature. LIBSVM implementation of the SVM is used [26]. The following parameters were applied:

---

[1] The reason we use three two-class classifications instead of a multiclass classification is that we want the classifier to perform as close to random as possible when discriminating the classes 'neutral' and 'acted', while giving a good discrimination of 'stress' at the same time.

a 10-fold cross-validation (CV) process ($k = 10$) was selected; the radial basis function (RBF) was used as a kernel function; γ and C were set to 1; the threshold ε was set to 0.001.

The results are presented in Table 1. As the main goal of the analysis is to investigate a feature potential for separating 'acted' and 'stress' classes, the best features from each perturbation category are marked gray in the table, defined on the basis of 'acted vs. stress' SVM accuracy. The best feature from the entire feature set (*peSNS*: the maximal duration of the perturbation intervals) is marked with a dark gray color in the table. The box plot of the feature is presented in Fig. 1.

Table 1. *Neutral-acted-stress* analysis (N = neutral, A = acted and S = stress)

Feature	Trend $N{\to}A$	$A{\to}S$	N vs. A [%]	N vs. S [%]	A vs. S [%]
peSNS features					
var	=	↗↗	53.98	78.20	74.97
mean abs	=	↗↗	51.20	76.24	73.24
max abs	↗	↗↗	51.20	80.42	76.83
dur per sec (v)	↗	↗↗	55.37	86.03	81.09
dur per sec	↗↗	↗↗	60.68	88.38	78.96
mean dur	↗	↗↗	52.34	84.60	80.03
max dur	↗	↗↗	53.48	87.47	82.42
quan	↗	↗↗	54.74	85.77	80.69
jitter features					
local	↗	=	50.95	52.61	51.66
local (abs)	=	↘	50.95	52.61	51.66
rap	↘	↗	49.30	52.61	51.66
ppq5	=	=	48.29	52.61	51.66
shimmer features					
local	=	↗↗	50.95	70.63	70.71
local (dB)	=	↗↗	55.88	69.06	68.18
apq3	↘	↗↗	50.95	61.88	68.18
apq5	=	↗↗	48.93	73.50	74.03
apq11	=	↗↗	50.70	61.10	57.26
variation features					
var diff (F_0)	↗	=	60.05	75.07	63.38
std diff (F_0)	↗↗	↗	53.86	72.45	65.91

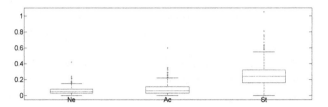

Fig. 1. Box plot of *peSNS max dur* feature distribution per three classes: Ne = neutral, Ac = acted and St = stress

It can be seen in the table that the SVM generally fails to discriminate jitter features. The deviation from the expected 50% is due to several NaN values in the features. Other feature groups are relatively good at separating acted-stress samples from the real-stress samples, while some of them also show significant rising trend from neutral to acted-stress samples. The possible explanation of such trend is that the talking style samples are partly realistic, but the possibility that perturbation features are partly affected by the voluntary control must not be disregarded. E.g. there is a possibility that some perturbations occur in voice when a person shout, even if there is no real anger behind. Results on a synthetic case, presented in [21], show that *peSNS* features have a potential for isolating autonomic perturbations from the voluntary components in speech, but further research must be done in this field.

One more parameter that must be taken into account is the level of stress, which a person tries to express during acting. The best way to compare the perturbations of acted- and real-stress samples is to ensure that the samples of these two data sets consist of the same level of stress expressions (not the real stress level as it is assumed that there is no real stress – or is minimal – in the case of acted expressions). The 'loud' talking style samples are thus chosen as acted-stress samples for the next analysis, as this style (according to our opinion) best suits the type and the level of stress within the roller-coaster samples.

3.2   2nd Analysis: *Neural – Loud – Stress*

Utterances from the 'loud' talking style were used as acted-stress samples for the purpose of this analysis. A total of 414 'loud' utterances were randomly sampled from this subset, which form, together with 414 neutral samples and 414 real-stress (roller-coaster) samples, a new data set for the analysis. The 'loud' subset is used because the style and intensity of expression is very similar to the roller-coaster stress expression. It can be seen in Fig. 2 that the root-mean-square energy of a speech signal and a standard deviation of F_0 contour are rather similar for the samples in these two classes when compared with the neutral state.

The same SVM parameters and the same rules for trend calculation were used as for previous analysis. Results are presented in Table 2 and the best features from each feature group are again marked as gray. The 'acted vs. stress' SVM accuracy is used as a measure for this. The best feature in this analysis (shimmer: the five-point amplitude perturbation quotient – apq5) is marked with a dark gray color in the table and the box plot is presented in Fig. 3.

It can be seen in the table that jitter features again results with the lowest accuracies of 'acted vs. stress' SVM classifies. Shimmer features performed better even with smaller perturbations in the 'acted' class compared with the 'neutral' class. Most of features show a statistically significant difference between 'acted' and 'stress' classes with the rising trend between these two classes. Again, the rising trend between the neutral and acted-stress samples (in this case the loud samples) is present and is even stronger in the case of *peSNS* features. A more detailed analysis per each speaking style is presented in the next section.

Table 2. *Neutral-loud-stress* analysis (N = neutral, A = acted and S = stress)

	Trend		Accuracy		
Feature	$N \to A$	$A \to S$	N vs. A [%]	N vs. S [%]	A vs. S [%]
peSNS features					
var	=	↗↗	60.69	78.20	72.46
mean abs	↗↗	↗↗	59.46	76.24	69.63
max abs	↗↗	↗↗	63.88	80.42	75.93
dur per sec (v)	↗↗	↗↗	65.85	86.03	74.39
dur per sec	↗↗	↗↗	72.60	88.38	68.08
mean dur	↗↗	↗↗	62.04	84.60	73.10
max dur	↗↗	↗↗	64.99	87.47	78.25
quan	↗↗	↗↗	62.53	85.77	76.96
jitter features					
local	=	=	48.53	52.61	53.28
local (abs)	↘	=	50.86	52.61	53.28
rap	↘	↗↗	50.86	52.61	53.28
ppq5	↘	↗↗	50.86	52.61	53.28
shimmer features					
local	↘	↗↗	60.44	70.63	79.28
local (dB)	↘	↗↗	67.08	69.06	79.28
apq3	↘	↗↗	47.91	61.88	76.58
apq5	↘	↗↗	48.89	73.50	81.60
apq11	↘	↗↗	54.42	61.10	73.49
variation features					
var diff (F_0)	=	↗	68.43	75.07	56.89
std diff (F_0)	↗	↗↗	70.15	72.45	61.13

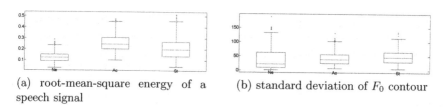

(a) root-mean-square energy of a speech signal

(b) standard deviation of F_0 contour

Fig. 2. Box plots of feature distributions per three classes: Ne = neutral, Ac = acted and St = stress

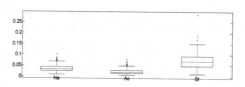

Fig. 3. Box plot of *shimmer apq5* feature distribution per three classes: Ne = neutral, Ac = acted and St = stress

3.3 3rd Analysis: *Neural – 7 Speaking Styles – Stress*

All seven speaking styles are compared with neutral and roller-coaster stress samples within this analysis. A total of 414 randomly chosen samples were selected for each speaking style class, i.e. for 'angry', 'clear', 'fast', 'loud', 'question', 'slow' and 'soft' type of stress expression. A data set is thus created with the total of $9 \times 414 = 3726$ samples.

It can be seen in Fig. 4 how samples are distributed for two best features from the previous analyses. As already mentioned, some perturbations exist in speaking styles and its level differs per styles and per features. The largest *peSNS max dur* perturbations are present for speaking style 'angry', while the largest *shimmer apq5* perturbations are in speaking style 'soft'.

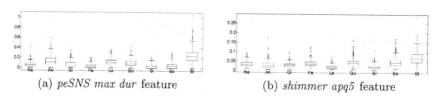

(a) *peSNS max dur* feature

(b) *shimmer apq5* feature

Fig. 4. Box plots of feature distributions per nine classes: Ne = neutral, An = angry, Cl = clear, Fa = fast, Lo = loud, Qu = question, Sl = slow, So = soft and St = stress

4   Conclusion

We can conclude that there is a statistically significant difference between various speaking style (acted) stress samples and realistic roller-coaster stress samples for most of perturbation features. The rising trend of perturbations is achieved from acted to realistic samples. Only the jitter features generally failed to discriminate these two classes. When all talking-style samples were included in the same 'acted' class and compared with the real-stress samples, the maximal duration of *peSNS* perturbation intervals turns to be the most discriminative feature resulting with the highest 'acted vs. stress' SVM accuracy. The best feature when comparing these two classes (with only 'loud' speech samples included in 'acted' class) is the shimmer: five-point amplitude perturbation quotient (apq5). The main goal is thus achieved. We can say that *peSNS* and shimmer perturbation features can be used for distinguishing authentic (realistic) expressions from the acted ones. However, a significant rising trend in perturbation level is observed between neutral and acted samples, which varies per speaking style and per features. Such trend can appear as a result of two potential effects. First, maybe a significant amount of the expression authenticity was present within the speakers when pronouncing speaking style utterances and the perturbation features actually measure the authenticity level – the real stress behind the expressions. Second, a robustness of perturbation features for voluntary components in speech is only initially proved on a synthetic case in [21], so there is still a possibility that the features are partly affected by the voluntary control, i.e. by speaking loudly etc. In order to eliminate the voluntary component completely, a future research must be undertaken to analyze the specific type of perturbations from various neurological sources and also on the isolation of only autonomic, emotionally related, perturbations from *peSNS* intervals. In this way it will be possible to claim that the perturbation features measure pure authenticity of emotional of stressful expression.

References

1. Banse, R., Scherer, K.R.: Acoustic profiles in vocal emotion expression. J. Personal. Soc. Psych. **70**(3), 614–636 (1996)
2. Gosselin, P., Kirouac, G., Dore, F.Y.: Components and recognition of facial expression in the communication of emotion by actors. J. Personal. Soc. Psych. **68**, 83–96 (1995)
3. Scherer, K.R., Banziger, T.: On the use of actor portrayals in research on the emotional expression. In: Scherer, K.R., Bänziger, T., Roesch, E. (eds.) A Blueprint for an Affectively Competent Agent: Cross-Fertilization Between Emotion Psychology, Affective Neuroscience, and Affective Computing, pp. 166–176. Oxford University Press, Oxford (2010)
4. Auberge, V., Audibert, N., Rilliard, A.: E-Wiz: a trapper protocol for hunting the expressive speech corpora in Lab. In: Proceedings of the 4th LREC, Lisbon, Portugal (2004)
5. Drolet, M., Schubotz, R.I., Fischer, J.: Authenticity affects the recognition of emotions in speech: behavioral and fMRI evidence. Cogn. Affect. Behav. Neurosci. **12**, 140–150 (2012)
6. Greasley, P., Sherrard, C., Waterman, M.: Emotion in language and speech: methodological issues in naturalistic approaches. Lang. Speech **43**, 355–375 (2000)

7. Laukka, P., Audibert, N., Auberge, V.: Exploring the determinants of the graded structure of vocal emotion expressions. Cogn. Emot. **26**, 710–719 (2012)

8. Williams, C.E., Stevens, K.N.: Emotions and speech: some acoustical correlates. J. Acoust. Soc. Am. **52**, 1238–1250 (1972)

9. Wilting, J., Krahmer, E., Swerts, M.: Real vs. acted emotional speech. In: INTER-SPEECH 2006, Pittsburgh PA, USA (2006)

10. Barrett, L.F.: Was Darwin wrong about emotional expressions? Curr. Dir. Psychol. Sci. **20**, 400–406 (2011)

11. Batliner, A., Fischer, K., Huber, R., Spilker, J., Noth, E.: Desperately seeking emotions or: actors, wizards and human beings. In: ISCA Workshop on Speech and Emotion, Newcastle, Northern Ireland (2000)

12. Douglas-Cowie, E., Campbell, N., Cowie, R., Roach, P.: Emotional speech: towards a new generation of databases. Speech Commun. **40**, 33–60 (2003)

13. Drolet, M., Schubotz, R.I., Fischer, J.: Recognizing the authenticity of emotional expressions: F_0 contour matters when you need to know. Front. Hum. Neurosci. **8**, 1–11 (2014)

14. Jurgens, R., Grass, A., Drolet, M., Fischer, J.: Effect of acting experience on emotion expression and recognition in voice: non-actors provide better stimuli than expected. J. Nonverbal Behav. **39**, 195–214 (2015)

15. Jurgens, R., Drolet, M., Pirow, R., Scheiner, E., Fischer, J.: Encoding conditions affect recognition of vocally expressed emotions across cultures. Front. Psychol. **4** (2013)

16. Scherer, K.R.: Vocal markers of emotion: comparing induction and acting elicitation. Comput. Speech Lang. **27**, 40–58 (2013)

17. Jurgens, R., Hammerschmidt, K., Fischer, J.: Authentic and play-acted vocal emotion expressions reveal acoustic differences. Front. Psychol. **2**, 1–11 (2011)

18. Scherer, K.R.: Vocal correlates of emotional arousal and affective disturbance. In: Wagner, H., Manstead, A. (eds.) Handbook of Psychophysiology: Emotion and Social Behavior, pp. 165–197. Wiley, London (1989)

19. Scherer, K.R.: Nonlinguistic vocal indicators of emotion and psychopathology. In: Izard, C.E. (ed.) Emotions in Personality and Psychopathology, pp. 495–529. Plenum Press, New York (1979)

20. Scherer, K.R.: Vocal affect expression: a review and a model for future research. Psychol. Bull. **99**, 143–165 (1986)

21. Dropuljic, B., Petrinovic, D., Cosic, K.: Novel speech features of disturbances in laryngeal muscle coordination. In: 7th IEEE International Conference on Cognitive Infocommunications, pp. 175–180 (2016)

22. Praat. http://www.fon.hum.uva.nl/praat/

23. Schuller, B., et al.: Recognising realistic emotions and affect in speech: state of the art and lessons learnt from the first challenge. Speech Commun. **53**(9), 1062–1087 (2011)

24. Dropuljic, B.: Emotional state estimation based on data mining of acoustic speech features. Ph.D. Thesis (in Croatian). University of Zagreb, Croatia (2014)

25. Hansen, J.H.L., et al.: The impact of speech under stress on military speech technology. Nato Project 4 (2000)

26. LibSVM. http://www.csie.ntu.edu.tw/~cjlin/libsvm/

Intelligent and Context Systems

Context Injection as a Tool for Measuring Context Usage in Machine Learning

Maciej Huk[(✉)]

Department of Information Systems, Wroclaw University of Science and Technology,
Wroclaw, Poland
maciej.huk@pwr.edu.pl

Abstract. Machine learning (ML) methods used to train computational models are one of the most valuable elements of the modern artificial intelligence. Thus preparing tools to evaluate ML training algorithms abilities to find inside the training data information (the context) crucial to build successful models is still an important topic. Within this text we introduce a new method of quantitative estimation of effectiveness of context usage by the ML training algorithms based on injection of predefined context to the training data sets. The results indicate that the proposed solution can be used as a general method of analyzing differences in context processing between ML training methods.

Keywords: Context analysis · Context injection · Effectiveness of context usage · Gradient boosting · Deep neural networks · Classification

1 Introduction

Machine learning methods used to train computational models are one of the most valuable elements of the modern artificial intelligence. At the same time, various computational systems are using data described as the "context". Measuring the context related properties of ML training methods is an important task, if we want to analyze their context-wise abilities not only qualitatively but also with quantitative methods.

The meaning of the term "context" is described typically as the set of circumstances or facts that surround a particular event, the interrelated conditions in which something exists or occurs (after Merriam Webster dictionary) or as the parts of data (e.g. of a written or spoken statement) usually influencing its meaning [1]. From this, one can conclude, that without context, information can be misinterpreted and that with context, information can be understood. Analogously without context, the problem can be hard to solve or not solvable (having no solution or too many solutions). Thus we can define context more precisely as the data that is needed to solve the given problem. Especially in the case of otherwise unsolvable problem the context creates additional dimension that allows solving it.

The above definition of the context covers all cases, where context includes objects, states/properties of objects, relations between objects and/or goals and their priorities (a goal is an object, state or relation that needs to be created or persisted). Thus it catches contextual character of data that is crucial for functioning of many systems - such as

© Springer International Publishing AG 2017
N.T. Nguyen et al. (Eds.): ACIIDS 2017, Part I, LNAI 10191, pp. 697–708, 2017.
DOI: 10.1007/978-3-319-54472-4_65

geo-location, user preferences, actual battery level, etc. [2–4]. It also covers importance of more complex contextual data: system knowledge about procedures and relations [5], communication channels and protocols of the data processing [6] and priorities of goals (see the Laws of Robotics initially proposed by Isaac Asimov) [7, 8].

What connects all of the above examples of contextual data is, that in such cases, context is typically introduced to the system by predefined source – system creator, user, sensor [9], data base [10] or communication channel [6]. But one can also notice that contrary situation is often found in ML – when the training algorithm is trying to find within the training data the context that is needed to solve the problem given by the data (discriminative features, relation between attributes or between data vectors) [11–13]. Thus in such situation, the training algorithm searches for an unknown context, which can be fully available within the search domain, available partially or not existing.

Within presented view, the trained model is a notation of the context found by the training algorithm, joined with potential noise included in the data and context introduced into the training algorithm by its author. Thus evaluating the performance of the trained model, in terms of such estimators as e.g. classification accuracy, generalization, AUC [14], Wilcoxon-Mann-Whitney statistic [15], correlation coefficient, root mean squared error, entropy gain [16, 17], cross-entropy [18] or confusion entropy [19], do not give clear information about how effective the given training algorithm is in searching for the context within the data.

Especially mentioned estimators can't give precise information about the effectiveness of the training algorithm when the evaluation is performed with use of the benchmark or real life data about which it is not known if and where the context can be found within the data. In such situation one can't say, which portion of the classifier inaccuracy is caused by the possible incompleteness of the context inside the data and which is the result of inability of the training algorithm to find and use of the available context. In special cases it can also happen, that context built-in into the training algorithm can mask the incompleteness of the context inside the data.

Also when the algorithm (or the model) explicitly performs features selection using statistical relations between features and data vector labels (calculated with such metrics as e.g. entropy, information gain [20], Fisher Score [21]), the situation is not much better. Even if it is possible to check which attributes or features are used by the resulting model, it is hard to check their relation to the context. This is because the context can be located in different subsets of attributes for different data vectors. Even if the selected subset of features is the same for every data vector, then it can be both: too wide or too narrow. And in both cases there is no guarantee that some parts of the context are not covered by the analyzed attributes or features.

In the above situations, the estimations which are used to measure properties of the machine learning models (e.g. classifiers) or of their attributes, can give only coarse evaluation of the training algorithms and their ability to locate and use the context. For that reason training algorithms are tested typically with many benchmark problems, e.g. from the UCI ML repository [22]. Still, such approach allows only qualitative evaluation of the ML algorithms. Because of that it is an actual problem to search for appropriate estimators that could quantitatively describe effectiveness of context location and its usage by machine learning training algorithms.

Within this paper we propose a method of estimation of effectiveness of context usage by machine learning training algorithms. It is based on analyzing changes of properties of classifiers prepared for data set without and with predefined context. For the clearness of the analysis we initially prepare context-free data sets which then are base for injecting context in four different forms. After defining the estimation method and presentation of the data sets creation process, results are presented for two boolean functions, six UCI ML repository benchmark problems and four selected training algorithms: gradient boosting machine, deep learning of neural network, random forest and C4.5. The results indicate that the proposed method can be useful to quantitatively analyze context related properties of ML training algorithms.

2 Measuring the Effectiveness of Context Usage

The context is a crucial element that is needed to successfully use machine learning methods. This is because even for otherwise unsolvable problem the context creates additional dimension that is needed to find a solution. This leads to the conclusion that to precisely measure context usage by machine learning methods one should exactly know which parts of the processed data belong to the context and if the context is full. Such requirements can be met by preparation of dedicated data sets. Thus the general idea of proposed method for measurement of effectiveness of context usage is composed of the following steps:

1. prepare hard to solve or unsolvable problem (e.g. classification),
2. extend input vectors of the original data with the complete context (it indicates correct output values for all the input vectors),
3. test if the system will find and how will use the context for both data sets - train a model for both data sets, calculate their accuracies and the effectiveness of context usage.

If the initial problem is very hard to solve or unsolvable, and the injected context is enough to calculate correct output values, the best solution for the classifier from the second step would be using only the context and ignoring the rest of the data, fully or at least for the problematic original vectors. If the context only adds missing information, second model should use it for all the ambiguous vectors. In the effect, comparison of the accuracies of both models allows checking if the second model is effectively using additional information injected with the context.

What is important, by changing representation of the context and relation of its properties to the original data one can change the level of difficulty of this test. This can be easily observed for the AND-CF context-free problem and its extensions after injections of different forms of the context (Fig. 1). In the most complicated cases the context can be not complete (no proper context for selected input vectors), dispersed over many inputs or hidden – processed with some algorithm (e.g. coded differently for different sub-problems, superimposed with the original data or even ciphered with unknown key).

The proposed solution can be viewed as an analogy to typical construction of IQ tests. We are hiding the needed clues among almost useless data and the training system is trying to locate and understand those clues as precisely as possible. If the training system will succeed, how well the context was hidden informs about the abilities of that system to become aware of and to use the given context. In such setup both the resulting model accuracy and the format of the context indicate properties of the training algorithm.

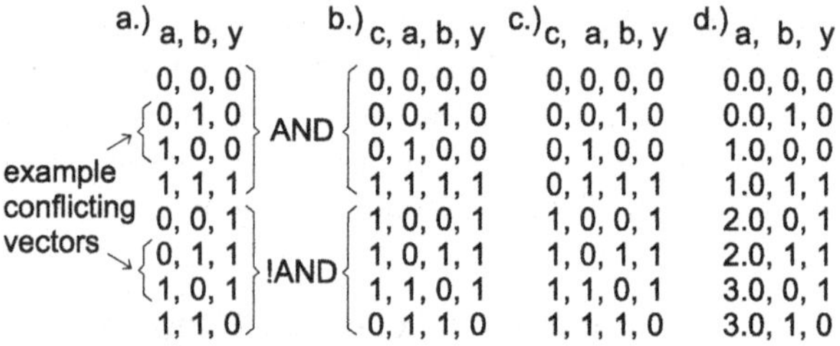

a.) a, b, y	b.) c, a, b, y	c.) c, a, b, y	d.) a, b, y
0, 0, 0	0, 0, 0, 0	0, 0, 0, 0	0.0, 0, 0
0, 1, 0	0, 0, 1, 0	0, 0, 1, 0	0.0, 1, 0
1, 0, 0	0, 1, 0, 0	0, 1, 0, 0	1.0, 0, 0
1, 1, 1	1, 1, 1, 1	0, 1, 1, 1	1.0, 1, 1
0, 0, 1	1, 0, 0, 1	1, 0, 0, 1	2.0, 0, 1
0, 1, 1	1, 0, 1, 1	1, 0, 1, 1	2.0, 1, 1
1, 0, 1	1, 1, 0, 1	1, 1, 0, 1	3.0, 0, 1
1, 1, 0	0, 1, 1, 0	1, 1, 1, 0	3.0, 1, 0

example conflicting vectors — AND (top four rows), !AND (bottom four rows)

Fig. 1. Context-free classification problem construction from the AND logical function (a), with examples of data sets after injection of different types of context: direct (b), indirect (c), hidden (d). Designations of columns - inputs: a, b, output: y, context: c.

Creation of the hard-to-solve problem for further context injection, can be done e.g. by joining many sub-problems. But to construct a context-free data set one need to make sure that for each data vector exists at least one conflicting data vector. Conflicting vectors have the same values of input attributes but different output classes. In turn, the context can be prepared in different forms and at least three main types of the context can be used:

- direct context, where for each data vector the context includes direct indication of the output class,
- indirect context, where the context does not indicate directly the output class, but allows for distinguishing conflicting vectors (it can be done e.g. by assigning different numbers to different sub-problems that make conflicting vectors),
- hidden context, where the context is superimposed with the original data.

Finally, the context can be injected to the data in the form of one or more additional attributes or merged with the values of existing attributes (see Fig. 1).

2.1 Estimation of Context Usage Effectiveness

Having context-depleted and context-rich data sets we can use the analyzed training algorithm to build related models and calculate accuracies of those models. Then for the accuracies of the models built for data sets without and with injected context, noted further respectively as γ_1 and γ_2, one can construct an estimator of training algorithm context usage effectiveness given by the formula:

$$Ecu(\gamma_1, \gamma_2) = Min\left(\gamma_2, \frac{100(\gamma_2 - \gamma_1)}{101 - \gamma_2}\right), \quad \gamma_1, \gamma_2 \in [0 \ldots 100]. \tag{1}$$

The above formula spans its values from 100 to almost −100. Its positive values indicate that injected context was successfully used by the training algorithm to increase the accuracy of the constructed classifier. Zero and near zero values indicate that the context was not found or not used to increase the accuracy. And the negative values show that injected context made the data harder to analyze for the training algorithm what resulted in lowering the accuracy of the classifier. The proposed relation between effectiveness of context usage and initial and resulting accuracies is presented on the Fig. 2.

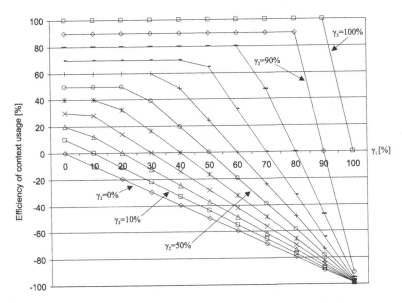

Fig. 2. The change of the percentage of correctly classified vectors from initial level γ_1 to the target level γ_2 shows how efficiently the system uses the context.

The general character of the proposed method makes it usable for ML algorithms dealing with different kinds of problems like classification, clustering, text analysis, etc. This is because the only requirements to use it are: possibility to measure model accuracy and to prepare data sets with predefined forms of context information. To limit the scope of the experiments, in this work we are focusing on the classification problems. But the potential positive results can indicate that the context injection technique and proposed formula for estimation of effectiveness of context usage can be valuable also in the case of problems other than classification.

3   Benchmark Datasets for Context Usage Analysis

To perform proposed analysis of context usage of machine learning algorithms we construct a set of 36 reference data sets. Eight of them are context-free data sets from which the other sets are derived by injection of different forms and types of the context.

The data sets are created on the basis of logical "AND" and "XOR" functions as well as modified versions of selected benchmark problems from the UCI ML repository: Breast-Cancer-Wisconsin, CRX, Heart-disease, Iris, Sonar and Votes.

3.1 Context-Free Data Sets

The context-free data sets are created by the following procedure. For the initial data set with N data vectors and C classes, we create and join its $C!$ different copies having not changed input data and all possible permutations of class names (without repetition) as the output. The resulting data set has $N * C!$ data vectors and for each original data vector inside the resulting data a set of conflicting data vectors exist with all the C-1 classes. In the effect, such data set consists of $C!$ different sub-problems of the same classification difficulty level. Together those conflicting sub-problems form a set of vectors that can't be all classified correctly without additional context.

The basic example of a context-free data set for the two-class AND problem is presented on the Fig. 1a. The more complex example for the three-class Iris problem is shown on the Fig. 3a. In this case one can observe how classes assignment change in selected vectors of each shown consecutive sub-problem.

a.)	b.)
6.30,3.30,4.70,1.60,Iris-versicolor	6.30,3.30,4.70,1.60,Iris-versicolor
5.80,4.00,1.20,0.20,Iris-setosa	5.80,4.00,1.20,0.20,Iris-setosa
6.70,3.30,5.70,2.10,Iris-virginica	6.70,3.30,5.70,2.10,Iris-virginica
(...)	(...)
6.30,3.30,4.70,1.60,Iris-virginica	6.30,3.30,4.70,4.60,Iris-virginica
5.80,4.00,1.20,0.20,Iris-setosa	5.80,4.00,1.20,3.20,Iris-setosa
6.70,3.30,5.70,2.10,Iris-versicolor	6.70,3.30,5.70,5.10,Iris-versicolor
(...)	(...)
6.30,3.30,4.70,1.60,Iris-setosa	6.30,3.30,4.70,7.60,Iris-setosa
5.80,4.00,1.20,0.20,Iris-versicolor	5.80,4.00,1.20,6.20,Iris-versicolor
6.70,3.30,5.70,2.10,Iris-virginica	6.70,3.30,5.70,8.10,Iris-virginica
(...)	(...)
6.30,3.30,4.70,1.60,Iris-virginica	6.30,3.30,4.70,10.60,Iris-virginica
5.80,4.00,1.20,0.20,Iris-versicolor	5.80,4.00,1.20,9.20,Iris-versicolor
6.70,3.30,5.70,2.10,Iris-setosa	6.70,3.30,5.70,11.10,Iris-setosa
(...) (...) (...)	(...) (...) (...)
6.10,2.80,4.70,1.20,Iris-versicolor	6.10,2.80,4.70,16.20,Iris-versicolor
5.90,3.20,4.80,1.80,Iris-versicolor	5.90,3.20,4.80,16.80,Iris-versicolor

Fig. 3. Selected vectors of the Iris-CF context-free data set (a) and their form after hidden context injection in the Iris-HCxR data set (b). Context is injected within the rightmost continuous column. The beginnings of four from six included sub-problems are shown.

The basic properties of the constructed context-free data sets are presented in the Table 1. They have the same number of attributes and classes as the original data sets but increased number of data vectors. The table includes also the results of measurements of accuracy achieved by the classifiers trained with selected machine learning algorithms.

Table 1. Characteristics of proposed context-free data sets ($Ecu = 0$) with average accuracy of classification for models built with selected ML algorithms: Gradient Boosting Machine, training of Deep Neural Network, Random Forest and C4.5. All data sets represent related two-class problems except the Iris three-class problem.

Data set name	# of Instances	# of Attributes	GBM [%]	DNN [%]	RF [%]	C4.5 [%]
AND-CF	8	2	50	50	50	50
XOR-CF	8	2	50	50	50	50
Breast cancer-CF	1398	10	50	50	50	50
CRX-CF	1380	15	50	50	50	50
Heart disease-CF	606	19	50	50	50	50
Iris-CF	900	4	33.3	33.3	33.3	33.3
Sonar-CF	416	60	50	50	50	50
Votes-CF	870	16	50	50	50	50

Presented results are fully consistent with the expectations – due to specific construction of the context-free data sets none of the machine learning algorithms can construct models with classification accuracy better than $1/C$. This confirms that the constructed context-free data set does not include context needed to distinguish conflicting vectors. Thus they can be easily used for further controlled context injection.

3.2   Data Sets Created with Context Injection

Having the context-free data sets we next use them to create data sets with dedicated context. The context is injected to the data in two forms: as indirect and hidden context. In both cases we code the context with nominal (categorical) values and with real (continuous) values. The indirect context is presented as a single additional data column of values that allows identify the conflicting sub-problems. For C class problem this extra attribute uses exactly $C!$ different values. The example of indirect context coded with nominal (categorical) values is presented on the Fig. 1c. For such data sets we add "ICxC" suffix in their name. By analogy, for distinguishing data sets with indirect context in the form of continuous values we add "ICxR" suffix.

The different method is used to inject the hidden context. No additional attribute is added and one of the existing data columns is used to present both its original values and the context. To do that, for the selected real-valued attribute of the initial data set, we check its range of values. Then we calculate the difference of its maximal and minimal values. In the next step we increase calculated difference by adding selected constant (e.g. 1) and for each sub-problem of the context-free data set we multiply this difference by the index of the sub-problem (from 0 to $C! - 1$). Finally, such offset is added to the values of the real-valued attribute for data vectors of the given sub-problem. In the effect, the values of the selected attribute for different sub-problems are in disjunctive ranges. In the case of categorical attribute, the hidden context is injected similarly. For that attribute for each sub-problem we use modified values that are constructed as concatenation of the original name of the value and the unique designation of the sub-problem.

Both above solutions create hidden context that allows differentiation of sub-problems within the data set. Finally, the constructed data sets with hidden context in the form of continuous values we name with the "HCxR" suffix, and data sets with context hidden within categorical attribute have the suffix "HCxC". The examples of data sets with hidden context coded within continuous attribute are presented on the Fig. 1d and on the Fig. 3b. All the proposed benchmark data sets are available for research use and can be downloaded from ICxS Contextual Data Repository [23].

4   Results of Experiments

After defining the data sets described in the previous section we have used Eq. 1 to estimate effectiveness of context usage (*Ecu*) of four machine learning algorithms: Gradient Boosting Machine[1], training of Deep Neural Network[2], Random Forest[3] and C4.5[4]. All results are averages of measurements repeated 10 times. Cross-validation was not used because it could falsify the classification accuracy by allowing training methods to favor the majority class of the training sets.

The values of measured efficiency of context usage for indirect context in categorical form ("ICxC" data sets) and in continuous form ("ICxR" data sets) are collected within Tables 2 and 3, respectively. Further, the results for hidden context in categorical form ("HCxC" data sets) and in continuous form ("HCxR" data sets) are presented within Tables 4 and 5. The lower number of data sets with hidden than with indirect context is the result of the fact that not all of the analyzed problems are described both with categorical and continuous attributes.

Table 2. Efficiency of context usage by selected ML algorithms for data sets with indirect context injected as categorical attribute.

Data set name	GBM [%]	DNN [%]	RF [%]	C4.5 [%]
AND-ICxC	75	100	100	75
XOR-ICxC	0	100	100	0
Breast cancer-ICxC	99.4	99.7	99.9	98.9
CRX-ICxC	95.5	99.0	99.7	94.9
Heart disease-ICxC	97.4	99.8	100	94.7
Iris-ICxC	0.51	92.6	0	98
Sonar-ICxC	100	100	100	98.1
Votes-ICxC	100	100	100	97.7

---

[1] Gradient Boosting Machine training parameters: 50 trees, maximum tree depth = 5, learning rate = 0.1, implementation: H2O Flow 3.10.0.8.

[2] Deep Neural Network training parameters: 100×100 hidden neurons, activation function: rectifier, max number of training epochs = 300, implementation: H2O Flow 3.10.0.8.

[3] Random Forest classifier training parameters: bag size = 100, number of iterations = 100, unlimited tree size, implementation: Weka 3.8.0.

[4] C4.5 tree (not pruned) implementation: C4.5 v8 by R. Quinlan.

Table 3. Efficiency of context usage by selected ML algorithms for data sets with indirect context injected as continuous attribute.

Data set name	GBM [%]	DNN [%]	RF [%]	C4.5 [%]
AND-ICxR	75	100	100	75
XOR-ICxR	0	100	100	0
Breast cancer-ICxR	99,4	99,9	99,9	98,9
CRX-ICxR	95,4	98,6	99,7	94,9
Heart disease-ICxR	100	100	100	94,7
Iris-ICxR	0.51	92,4	0	0
Sonar-ICxR	100	99,8	100	98,1
Votes-ICxR	100	100	100	97,7

The results show that effectiveness of context usage can differ considerably between different data sets and training algorithms. For Breast cancer, CRX and Heart disease based data sets *Ecu* is high for all context types and training methods. But this is not the case for XOR, Iris and Sonar based data sets. For Iris-ICxR data only the DNN has high *Ecu* value and for Iris-HCxR all the training algorithms are efficient with finding the hidden context. Unlike is for Sonar datasets – all the training methods have high *Ecu* both for Sonar-ICxC and Sonar-ICxR, but for Sonar-HCxR and C4.5 algorithm surprisingly the *Ecu* is zero. Reverse anomaly we find for the XOR related data: C4.5 has high *Ecu* only for the XOR-HCxR. It is also interesting that for XOR based data, all analyzed types of injected context generate high *Ecu* for DNN and RF methods and *Ecu* = 0 for the GBM.

At the current stage of experiments it is hard to explain the detailed reasons of the above anomalies and *Ecu* variability. But we can try to get more general results by averaging values from Tables 2, 3, 4 and 5 by context type, its format and used machine learning algorithm. Such aggregated results are presented in the Table 6.

Table 4. Efficiency of context usage by selected ML algorithms for data sets with hidden context injected to one of existing continuous attribute.

Data set name	GBM [%]	DNN [%]	RF [%]	C4.5 [%]
AND-HCxR	75	100	100	75
XOR-HCxR	0	75	100	100
Breast cancer-HCxR	99,1	99.7	100	98.5
CRX-HCxR	93,0	96.3	99.0	93.8
Heart disease-HCxR	97,7	100	100	95.2
Iris-HCxR	97,9	92.6	100	98
Sonar-HCxR	100	100	100	0

The data aggregated in the Table 6 indicate that for the analyzed data sets training of the Deep Neural Network and building of the Random Forest classifier use the context information considerably more effectively than Gradient Boosting and C4.5 algorithms. This relation is valid both for data sets with indirect and hidden context as well as for both used context formats.

Table 5. Efficiency of context usage by selected ML algorithms for data sets with hidden context injected to one of existing categorical attribute.

Data set name	GBM [%]	DNN [%]	RF [%]	C4.5 [%]
AND-HCxC	75	87.5	100	0
XOR-HCxC	0	87,5	100	0
CRX-HCxC	96,0	98,8	99,0	93,6
Heart disease-HCxC	96,5	95,1	100	97,0
Votes-HCxC	99,4	100	100	98,4

Table 6. Average efficiency of context usage by selected ML algorithms for analyzed reference data sets groupped by context format and context injection method.

Subgroup of analyzed data sets	GBM [%]	DNN [%]	RF [%]	C4.5 [%]
All data sets	73,9	96,9	92,8	73,8
Indirect context data sets	71,1	98,9	87,5	76,0
Hidden context data sets	77,5	94,4	99,8	70,8
Categorical context data sets	71,9	96,9	92,2	72,8
Continuous context data sets	75,5	97,0	93,2	74,7

On the other hand, all analyzed training methods a little more effectively use the context given in continuous attributes than when it is presented with categorical values (on average 1.6 percentage point better). And what is interesting, algorithms building single classifiers (DNN, C4.5) have on average the Ecu higher by 4.5 percentage point for indirect than for hidden context, and algorithms constructing ensembles of classifiers (GBM, RF) are more effective when the context is hidden (on average 9.3 percentage points more).

5   Conclusions

In this paper, we have presented a new method of quantitative estimation of effectiveness of context usage by the ML training algorithms. It is based on comparison of the accuracy of model built for data set without the context and for its version after injection of predefined context. The method allowed identification of differences among abilities of typical ML training methods to find and use the two defined types of the context (indirect and hidden). With the proposed solution it was also shown that learning of deep neural network and the training of random decision forest use all types of context more effectively than gradient boosting and C4.5 algorithms.

To evaluate proposed method we have defined a set of dedicated benchmark data sets that include context-free and their context-enriched versions of classification UCI ML problems. It would be interesting to perform analogous experiments for wider set of benchmark data sets (especially with number of classes greater than two) and for greater number of ML methods. But the proposed method can be used also for algorithms solving problems other than classification, what can be a valuable direction of further research.

Planning the continuation of the above work it is also worth to remember, that the ability of the system to find and use a context is what some of the intelligence tests are concentrating on – e.g. Raven's Progressive Matrices test. Such methods often use hidden and dispersed context information, visual or abstract in nature, to detect intelligence. Thus it is also needed to check how the ML training algorithms use the context represented in different dispersed forms – e.g. coded within more than one attribute, with attributes of high entropy, etc. Especially it is planned to check if the latter form of the context would allow for smooth regulation of the difficulty of use of the injected context.

References

1. Chen, P., Xu, B., Yang, M., Li, S.: Clause sentiment identification based on convolutional neural network with context embedding. In: 12th International Conference on Natural Computation, Fuzzy Systems and Knowledge Discovery (ICNC-FSKD), pp. 1532–1538. IEEE Press (2016)
2. Tang, K., Paluri, M., Fei-Fei, L., Fergus, R., Bourdev, L.: Improving image classification with location context. In: 2015 IEEE International Conference on Computer Vision (ICCV), pp. 1008–1016. IEEE Press (2015)
3. Kapitsaki, G.M.: Reflecting user privacy preferences in context-aware web services. In: 2013 IEEE 20th International Conference on Web Services (ICWS), pp. 123–130. IEEE Press (2013)
4. Datta, S.K., Bonnet, C., Nikaein, N.: Self-adaptive battery and context aware mobile application development. In: 2014 International Wireless Communications and Mobile Computing Conference (IWCMC), pp. 761–766 (2014)
5. Klingelschmitt, S., Eggert, J.: Using context information and probabilistic classification for making extended long-term trajectory predictions. In: 2015 IEEE 18th International Conference on Intelligent Transportation Systems, pp. 705–711 (2015)
6. Spaulding, J., Krauss, A., Srinivasan, A.: Exploring an open WiFi detection vulnerability as a malware attack vector on iOS devices. In: 2012 7th International Conference on Malicious and Unwanted Software (MALWARE), pp. 87–93 (2012)
7. Nguyen, T.C., Nguyen, X.H., Nguyen, V.K.: Hybrid priority schemes for the message scheduling for CAN-based Networked Control Systems. In: 2014 IEEE Fifth International Conference on Communications and Electronics (ICCE), pp. 264–269 (2014)
8. Murphy, R., Woods, D.D.: Beyond Asimov: the three laws of responsible robotics. IEEE Intell. Syst. **24**, 14–20 (2009)
9. Wang, J., Qiu, M., Guo, B., Shen, Y., Li, Q.: Low-power sensor polling for context-aware services on smartphones. In: 2015 IEEE 12th International Conference on Embedded Software and Systems (ICESS), 2015 IEEE 7th International Symposium on Cyberspace Safety and Security (CSS), pp. 617–622 (2015)
10. Pallotta, G., Jousselme, A.L.: Data-driven detection and context-based classification of maritime anomalies. In: 2015 18th International Conference on Information Fusion (Fusion), pp. 1152–1159 (2015)
11. Duma, D., Sutton, C., Klein, E.: Context matters: towards extracting a citation's context using linguistic features. In: 2016 IEEE/ACM Joint Conference on Digital Libraries (JCDL), pp. 201–202 (2016)
12. Kang, S., Kim, D., Cho, S.: Efficient feature selection-based on random forward search for virtual metrology modeling. IEEE Trans. Semicond. Manuf. **29**, 391–398 (2016)

13. Chakraborty, G., Horie, S., Yokoha, H., Kokosiński, Z.: Minimizing sensors for system monitoring - a case study with EEG signals. In: 2015 IEEE 2nd International Conference on Cybernetics (CYBCONF), pp. 206–211. IEEE Press (2015)
14. Fan, X., Tang, K.: Enhanced maximum AUC linear classifier. In: 2010 7th International Conference on Fuzzy Systems and Knowledge Discovery (FSKD), pp. 1540–1544 (2010)
15. Yan, L., Dodier, R., Mozer, M.C., Wolniewicz, R.: Optimizing classifier performance via the Wilcoxon-Mann-Whitney statistic. In: 20th International Conference on Machine Learning (ICML-03), pp. 848–855. American Association for Artificial Intelligence (2003)
16. Trigg, L.: An entropy gain measure of numeric prediction performance. Working paper 98/11, Department of Computer Science, University of Waikato (1998)
17. Patil, L.H., Atique, M.: A novel feature selection based on information gain using WordNet. In: Science and Information Conference (SAI), pp. 625–629 (2013)
18. Wu, G., Wang, L., Zhao, N., Lin, H.: Improved expected cross entropy method for text feature selection. In: 2015 International Conference on Computer Science and Mechanical Automation (CSMA), pp. 49–54 (2015)
19. Wang, X.N., Wei, J.M., Jin, H., Yu, G., Zhang, H.W.: Probabilistic confusion entropy for evaluating classifiers. Entropy **15**, 4969–4992 (2013)
20. Sofeikov, K.I., Tyukin, I.Y., Gorban, A.N., Mirkes, E.M., Prokhorov, D.V., Romanenko, I.V.: Learning optimization for decision tree classification of non-categorical data with information gain impurity criterion. In: 2014 International Joint Conference on Neural Networks (IJCNN), pp. 3548–3555 (2014)
21. Bhasin, V., Bedi, P., Singhal, A.: Feature selection for steganalysis based on modified Stochastic Diffusion Search using Fisher score. In: 2014 International Conference on Advances in Computing, Communications and Informatics (ICACCI), pp. 2323–2330 (2014)
22. UCI Machine Learning Repository. http://archive.ics.uci.edu/ml
23. ICxS Contextual Data Repository. http://www.icxs.pwr.edu.pl/cx

An Approach for Multi-Relational Data Context in Recommender Systems

Nguyen Thai-Nghe[✉], Mai Nhut-Tu, and Huu-Hoa Nguyen

College of Information and Communication Technology, Can Tho University,
3/2 Street, Can Tho City, Vietnam
{ntnghe,nhhoa}@ctu.edu.vn, mntu.it@gmail.com

Abstract. Matrix factorization technique has been successfully used in recommender systems. Currently, many variations are developed using this technique, e.g., biased matrix factorization, non-negative matrix factorization, multi-relational matrix factorization, etc. In the context of multi-relational data, this paper proposes another multi-relational approach for recommender systems by including all of the information from latent factor matrices to the prediction functions so that the models have more data to learn. To validate the proposed approach, experiments are conducted on standard datasets in recommender systems. Experimental results show that the proposed approach is promising.

Keywords: Multi-relational data · Matrix factorization · Recommender Systems

1 Introduction

Recommender Systems (RS) have been used in many different areas of information systems. The RS helps to solve problems of information overload and to choose items quickly by presenting appropriate contents to individual user. For efficiently providing recommendations, the RS needs a prediction model, which can predict the unseen items based on the past data then providing suggestions for the user. Therefore, using accurate algorithms in RS is very important.

In the RS, many algorithms are proposed. They can be grouped into three main groups [1, 2]:

- Content-based Filtering approach: Based on description of the items and profile of the users. These algorithms try to recommend items that are similar to those that the users liked/selected in the past using attributes of the items or profile of the users.
- Collaborative Filtering approach: Algorithms in this group are neighborhood-based models, which use historical data of the similar users (user-based approach) or use historical data of the similar items (item-based approach). Another approach is model-based, which builds the prediction models based on data collected in the past.
- The third group is a combination of the two above approaches.

In the collaborative filtering approach, matrix factorization (MF) is one of the most successful methods (the state-of-the-art) in rating prediction [3, 4]. However, the MF

© Springer International Publishing AG 2017
N.T. Nguyen et al. (Eds.): ACIIDS 2017, Part I, LNAI 10191, pp. 709–720, 2017.
DOI: 10.1007/978-3-319-54472-4_66

algorithm focuses on exploiting information in a single relationship between the user
and the item (e.g., the relation "Rates" between "User" and "Movie" entities).
Therefore, they have not used all relevant information from the other relationships
between the users and the items. To utilize more information, multi-relational matrix
factorization (MRMF) approach was proposed [5, 6]. However, in these researches, the
prediction formula does not include all of the information from latent factor matrices.

In this work, for the context of having multi-relational data, we propose a
multi-relational matrix factorization approach, which allows utilizing information from
different relationships between the users and the items for building the prediction
models. For validating the proposed approach, we have experimented on the standard
data sets in both recommender systems (entertainment area) and Intelligent Tutoring
Systems (education area). Experimental results show that the proposed approach can
improve the accuracy of the prediction models.

2 Matrix Factorization Approaches

First, we briefly summarize the matrix factorization (MF) [4] on a single relationship
and the multi-relational matrix factorization (MRMF) [5–8]. Based on these approaches, we then propose a new multi-relational matrix factorization approach for incorporating all of available information to the prediction models.

2.1 Matrix Factorization (MF)

Matrix factorization is the task of approximating a matrix $\mathbf{R}$ by the product of two
smaller matrices $\mathbf{W_1}$ and $\mathbf{W_2}$, i.e. $\mathbf{R} \approx \mathbf{W_1}\mathbf{W_2^T}$ as illustrated in Fig. 1 [4]. In these
notations, $\mathbf{W_1} \in \mathbb{R}^{|U| \times K}$ is a matrix where each row u is a vector containing K latent
factors describing the user u and $\mathbf{W_2} \in \mathbb{R}^{|I| \times K}$ is a matrix where each row i is a vector
containing the K latent factors describing the item i. Let $w_{1_{uk}}$ and $w_{2_{ik}}$ be the elements
and $\mathbf{w}_{1_u}$ and $\mathbf{w}_{2_i}$ be the vectors of $\mathbf{W}_1$ and $\mathbf{W}_2$ respectively, then the rating r given by
user u to item i is predicted by:

$$\hat{r}_{ui} = \sum_{k=1}^{k} \mathbf{w}_{1_{uk}}\mathbf{w}_{2_{ik}} = \mathbf{w}_{1_u}\mathbf{w}_{2_i}^T \tag{1}$$

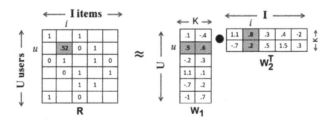

Fig. 1. An example of matrix factorization

$\mathbf{W_1}$ and $\mathbf{W_2}$ are the model parameters (the so-called latent factor matrices) which can be learnt by optimizing objective function (2) (e.g., using stochastic gradient descent)

$$O^{MF} = \sum\nolimits_{(u,i)\in\mathbf{R}} \left(\mathbf{R}_{ui} - \mathbf{w}_{1_u}\mathbf{w}_{2_i}^T\right)^2 + \lambda\left(\|\mathbf{W}_1\|_F^2 + \|\mathbf{W}_2\|_F^2\right) \qquad (2)$$

where $\|\cdot\|_F^2$ is a Frobenius norm and λ $(0 \le \lambda < 1)$ is a regularization term which is used to prevent over-fitting.

2.2 Multi-Relational Matrix Factorization (MRMF)

In previous section, we have briefly described the MF which uses only one relation type between two entity types (e.g., the relation "rates" between "user" and "movie" in Fig. 2). In the MRMF [5, 6], we can include more than one relationship and more than two entity types to the models (Figs. 2 and 3).

Let $\{\mathbf{E}_1, \mathbf{E}_2,..., \mathbf{E}_N\}$ be a set of N entity types, $\{\mathbf{R}_1, \mathbf{R}_2,..., \mathbf{R}_M\}$ be a set of M binary relation types and $\mathbf{R}_r = \{(E_{1_r}; E_{2_r})\}$ $(r = 1..M)$, then the objective function of the MRMF is presented by [5, 6]:

$$O^{MRMF} = \sum\nolimits_{r=1}^{M} \sum\nolimits_{(u,i)\in\mathbf{R}_r} \left((\mathbf{R}_r)_{ui} - \mathbf{w}_{r_1u}\mathbf{w}_{r_2i}^T\right)^2 + \lambda\left(\sum\nolimits_{j=1}^{N} \|\mathbf{W}_j\|_F^2\right) \qquad (3)$$

where M is the number of relation types and $\{\mathbf{W}_j\}_{j=1...N}$ are the latent factor matrices of N entity types. The objective function (3) is optimized by using stochastic gradient descent. For learning process, the MRMF updates its latent factors using Eqs. (4) and (5):

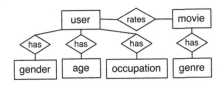

Fig. 2. Example on ER-diagram of MovieLens dataset

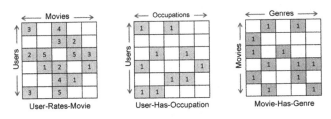

Fig. 3. An example of matrix representations for the ERD in Fig. 2

$$\mathbf{w}_{r_1 u}^{new} = \mathbf{w}_{r_1 u}^{old} - \beta \left(\frac{\partial O^{MRMF}}{\partial \mathbf{w}_{r_1 u}^{old}} \right) \tag{4}$$

$$\mathbf{w}_{r_2 i}^{new} = \mathbf{w}_{r_2 i}^{old} - \beta \left(\frac{\partial O^{MRMF}}{\partial \mathbf{w}_{r_2 i}^{old}} \right) \tag{5}$$

where β ($0 < \beta < 1$) is a learning rate; and the gradients $\frac{\partial O^{MRMF}}{\partial \mathbf{w}_{r_1 u}}$ and $\frac{\partial O^{MRMF}}{\partial \mathbf{w}_{r_2 i}}$ are determined by:

$$\frac{\partial O^{MRMF}}{\partial \mathbf{w}_{r_1 u}} = -2 \left((\mathbf{R}_r)_{ui} - \mathbf{w}_{r_1 u} \mathbf{w}_{r_2 i}^{T} \right) \mathbf{w}_{r_2 i} + \lambda \mathbf{w}_{r_1 u} \tag{6}$$

$$\frac{\partial O^{MRMF}}{\partial \mathbf{w}_{r_2 i}} = -2 \left((\mathbf{R}_r)_{ui} - \mathbf{w}_{r_1 u} \mathbf{w}_{r_2 i}^{T} \right) \mathbf{w}_{r_1 u} + \lambda \mathbf{w}_{r_2 i} \tag{7}$$

Besides, [6] has applied MRMF to predict the performance of the students. The authors have used scalability of MRMF which can utilize many relationships between many entities to take full advantage of student information and the task information that the students must solve, thus, making the prediction model with high accuracy. Moreover, the authors also introduced a variant of MRMF that is Weighted MRMF (WMRMF). This technique is similar to MRMF but it allows assigning the weights (Θ_r) to each relation for setting important levels of the relations. With WMRMF, the objective function in Eq. (3) now becomes:

$$O^{WMRMF} = \sum_{r=1}^{M} \Theta_r \sum_{(u,i) \in \mathbf{R}_r} \left((\mathbf{R}_r)_{ui} - \mathbf{w}_{r_1 u} \mathbf{w}_{r_2 i}^{T} \right)^2 + \lambda \left(\sum_{j=1}^{N} \| \mathbf{W}_j \|_F^2 \right) \tag{8}$$

Although previous works (MRMF/WMRF) can utilize more information than the single MF, their prediction functions still use the formula (1) for generating the prediction scores. Therefore, MRMF/WMRMF has not used all the information from latent factor matrices in the prediction.

In this work, we propose a different approach that try to employ all of relevant information to the model so that the model have more data to learn, thus, it would get more prediction accuracy. This is an extended work from [6, 9].

3 Proposed Method

We propose a multi-relational factorization approach that can integrate all information from the latent factor matrices. Therefore, the number of model parameters of the new approach are different from the MRMF. The proposed methods are named **MRMF++** (Multi-Relational Matrix Factorization Plus Plus) and **WMRMF++** (Weighted MRMF++).

3.1   Multi-Relational Matrix Factorization Plus Plus (MRMF++)

Figure 4 presents differences about the number of model parameters between the MRMF and the MRMF++. Let $\{\mathbf{E}_1, \mathbf{E}_2,..., \mathbf{E}_N\}$ be a set of N entity types, $\{\mathbf{R}_1, \mathbf{R}_2,..., \mathbf{R}_M\}$ be a set of M binary relation types and $\mathbf{R}_r = \{(E_{1_r}; E_{2_r})\}$ $(r = 1..M)$, the MRMF will have N latent factor matrices and the MRMF++ will have $2M$ latent factor matrices. Based on the idea of using all information from the latent factor matrices for prediction, we present the prediction formula of the MRMF++ as the following

$$\hat{r}_{ui} = \left(\sum_{x=1}^{P} w_{1_x}\right)_u \left(\sum_{y=1}^{Q} w_{2_y}\right)_i^T = \sum_{k=1}^{K}\left(\left(\sum_{x=1}^{P} w_{1_x}\right)_{uk}\left(\sum_{y=1}^{Q} w_{2_y}\right)_{ik}\right) \quad (9)$$

where P and Q are the number of latent factor matrices of u and i respectively; w_{1_x} is matrix at index x in P matrices of u; w_{2_y} is matrix at index y in Q matrices of i. To reduce the length of the formula, we set

$$X = \sum_{x=1}^{P} w_{1_x} \ and \ Y = \sum_{y=1}^{Q} w_{2_y}$$

Then the formula (9) is rewritten as following

$$\hat{r}_{ui} = X_u Y_i^T \quad (10)$$

Similar to the MRMF, model parameters of the MRMF++ can be learned by optimizing objective function (11) given a criterion, e.g., root mean squared error (RMSE), using stochastic gradient descent

$$O^{MRMF++} = \sum_{r=1}^{M} \sum_{(u,i)\in R_r}\left((\mathbf{R}_r)_{ui}-X_{ru}Y_{ri}^T\right)^2 + \lambda\left(\sum_{j=1}^{2M}\|W_j\|_F^2\right) \quad (11)$$

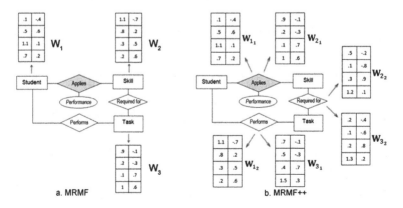

Fig. 4. Comparison of model parameters between MRMF and MRMF++

where M is the number of relation types and $\{\mathbf{W}_j\}_{j=1...2M}$ are the latent factor matrices of M relation types, $(\mathbf{R}_r)_{ui}$ is real value of relation r, $X_{ru}Y_{ri}^T$ is the predicted value of relation r, $\lambda\left(\sum_{j=1}^{2M}\|\mathbf{W}_j\|_F^2\right)$ is a regularization term. For learning process, the MRMF++ updates its latent factors for each relation at iteration n via Eqs. (12) and (13):

$$X_{ru}^n = X_{ru}^{n-1} - \beta\left(\frac{\partial O^{MRMF++}}{\partial X_{ru}^{n-1}}\right) \qquad (12)$$

$$Y_{ri}^n = Y_{ri}^{n-1} - \beta\left(\frac{\partial O^{MRMF++}}{\partial Y_{ri}^{n-1}}\right) \qquad (13)$$

where β is a learning rate; and the gradients $\frac{\partial O^{MRMF++}}{\partial X_{ru}}$ and $\frac{\partial O^{MRMF++}}{\partial Y_{ri}}$ are determined by:

$$\left(\frac{\partial O^{MRMF++}}{\partial X_{ru}}\right) = \lambda X_{ru} - 2\left((\mathbf{R}_r)_{ui} - X_{ru}Y_{ri}^T\right)Y_{ri} \qquad (14)$$

$$\left(\frac{\partial O^{MRMF++}}{\partial Y_{ri}}\right) = \lambda Y_{ri} - 2\left((\mathbf{R}_r)_{ui} - X_{ru}Y_{ri}^T\right)X_{ru} \qquad (15)$$

The MRMF++'s learning process is summarized in a **LearnMRMF++** procedure. We initialize the latent factor matrices from the normal distribution $N(\mu, \sigma2)$, e.g., mean $\mu = 0$ and standard deviation $\sigma2 = 0.01$, and initialize the weight value for each relation types. While the stopping condition is not met, e.g. reaching the maximum number of iterations or converging, the latent factors are updated iteratively (converging: $O_{Iter_{(n-1)}}^{MRMF++} - O_{Iter_n}^{MRMF++} < \in$).

After the learning process, the model parameters $\{\mathbf{W}_j\}_{j=1...2}M$ are obtained, then we can generate the prediction for any relation using the Eq. (9).

1: procedure **LearnMRMF++**($\mathbf{E}_1, \ldots, \mathbf{E}_N$: Entity types; $\mathbf{R}_1, \ldots, \mathbf{R}_M$: Relation types; λ: Regularization term; β: Learning rate; K: #Latent factors; Stopping criterion)

2:   **for** $j \leftarrow 1 \ldots 2M$ **do**

3:      $\mathbf{W}_j \leftarrow$ Draw randomly from $\mathcal{N}(\mu, \sigma^2)$

4:   **end for**

5:   **while** (Stopping criterion is NOT met) **do**

6:      **for** each relation $\mathbf{R}_r = \{(E_{1_r}; E_{2_r})\}$ in $\{\mathbf{R}_1, \ldots, \mathbf{R}_M\}$ **do**

10:        **for** $l \leftarrow 1 \ldots |\mathbf{R}_r|$ **do**

11:          Draw randomly (u, i) in $\mathbf{R}_r$

12:          $X_{ru} \leftarrow X_{ru} - \beta\left(\frac{\partial O^{MRMF++}}{\partial X_{ru}}\right)$

13:          $Y_{ri} \leftarrow Y_{ri} - \beta\left(\frac{\partial O^{MRMF++}}{\partial Y_{ri}}\right)$

14:        **end for**

15:      **end for**

16: **end while**

17: **return** $\{\mathbf{W}_j\}_{j=1...2M}$

18: **end procedure**

3.2 Weighted Multi-Relational Matrix Factorization Plus Plus (WMRMF++)

Using the MRMF++, we can utilize many relationships between many entities for rating prediction. However, this method treats the important role of all relations equally. Clearly, we can see that the main relation which contains the target variable (e.g. "User-Rates-Movie" in Fig. 2) is more important than the other supplement relations, thus it should have higher weight. Based on [6], we propose the WMRMF++ to take into account the importance of the main relation. The objective function in Eq. (11) now becomes:

$$O^{WMRMF++} = \sum_{r=1}^{M} \Theta_r \sum_{(u,i)\in R_r} \left((R_r)_{ui} - X_{ru}Y_{ri}^T\right)^2 + \lambda\left(\sum_{j=1}^{2M} \|W_j\|_F^2\right) \quad (16)$$

where Θ_r is a weight function, for example, it sets the weight to maximum for the main relation and reduces the weight for the rest, as in Eq. (17). However, other choices could also be considered.

$$\Theta_r = \begin{cases} 1, & \text{if } r \text{ is the main relation} \\ \theta, & \text{else } (0 < \theta \leq 1) \end{cases} \quad (17)$$

where θ is a hyperparameter which can be determined from the training data. Another important property of the WMRMF++ is that in an extreme case ($\theta = 1$), the WMRMF++ is still equivalent to the MRMF++.

For learning process, the WMRMF++ updates its latent factors for each relation using Eqs. (18) and (19):

$$X_{ru}^n = X_{ru}^{n-1} - \beta\left(\frac{\partial O^{WMRMF++}}{\partial X_{ru}^{n-1}}\right) \quad (18)$$

$$Y_{ri}^n = Y_{ri}^{n-1} - \beta\left(\frac{\partial O^{WMRMF++}}{\partial Y_{ri}^{n-1}}\right) \quad (19)$$

where the gradients $\frac{\partial O^{WMRMF++}}{\partial X_{ru}}$ and $\frac{\partial O^{WMRMF++}}{\partial Y_{ri}}$ are determined by:

$$\left(\frac{\partial O^{WMRMF++}}{\partial X_{ru}}\right) = \lambda X_{ru} - 2\Theta_r\left((R_r)_{ui} - X_{ru}Y_{ri}^T\right)Y_{ri}$$

$$\left(\frac{\partial O^{WMRMF++}}{\partial Y_{ri}}\right) = \lambda Y_{ri} - 2\Theta_r\left((R_r)_{ui} - X_{ru}Y_{ri}^T\right)X_{ru}$$

The WMRMF++'s learning process is summarized in **LearnWMRMF++** algorithm below. We initialize the latent factor matrices from the normal distribution $N(\mu, \sigma 2)$, e.g., mean $\mu = 0$ and standard deviation $\sigma 2 = 0.01$, and initialize the weight value for each relation types. While the stopping condition is not met, e.g., reaching the maximum number of iterations or converging (converging: $O_{Iter_{(n-1)}}^{MRMF++} - O_{Iter_n}^{MRMF++}$ $< \in$), the latent factors are updated iteratively.

1: **procedure** LEARNWMRMF++($\mathbf{E}_1, \ldots, \mathbf{E}_N$: Entity types;
$\mathbf{R}_1, \ldots, \mathbf{R}_M$: Relation types; λ: Regularization term; β: Learning
rate; K: #Latent factors; ψ: Weight value; Stopping criterion)
2:   **for** $j \leftarrow 1 \ldots 2M$ **do**
3:      $\mathbf{W}_j \leftarrow$ Draw randomly from $\mathcal{N}(\mu, \sigma^2)$
4:   **end for**

2:   **for** $r \leftarrow 1 \ldots M$ **do**

3:      Initialize Θ_r using equation (17)

4:   **end for**
5:   **while** (Stopping criterion is NOT met) **do**
6:      **for** each relation $R_r = \{(E_{1_r}; E_{2_r})\}$ in $\{\mathbf{R}_1, \ldots, \mathbf{R}_M\}$ **do**
10:        **for** $l \leftarrow 1 \ldots |\mathbf{R}_r|$ **do**
11:          Draw randomly (u, i) in $\mathbf{R}_r$
12:          $X_{ru} \leftarrow X_{ru} - \beta \left(\dfrac{\partial O^{WMRMF++}}{\partial X_{ru}} \right)$
13:          $Y_{ri} \leftarrow Y_{ri} - \beta \left(\dfrac{\partial O^{WMRMF++}}{\partial Y_{ri}} \right)$
14:        **end for**
15:      **end for**
16:   **end while**
17:   **return** $\{\mathbf{W}_j\}_{j=1\ldots2M}$
18: **end procedure**

After the learning process, the model parameters $\{\mathbf{W}_j\}_{j=1\ldots2}M$ are obtained, we also generate the prediction for any relation using the Eq. (9).

4 Experiments

4.1 Datasets

For experiments, we have used datasets from two different fields which are entertainment and education. Movielens 100 k dataset is collected by GroupLens (www.grouplens.org). This data was extracted from a movie recommender system. It has 100,000 rating, 943 users and 1,682 movies. This data set contains user information, e.g., age, gender, occupation and movie information, e.g., title, release date, genre, etc. The second dataset is Assistments-2009–2010 (Assistments) which was extracted from ASSISTments system (teacherwiki.assistment.org). This dataset represents the log files of interactions between students and the tutoring systems. While students solve the problems in the tutoring system, their activities, success and progress indicators are logged as individual rows in the data set. This data can be mapped to the concepts of recommender systems as student $\rightarrow$ user; task $\rightarrow$ item; and performance (CFA) $\rightarrow$ rating. *Clearly, in these datasets, there are several relationships between data attributes that we can exploit.* Information about the number of users, items, and ratings on these datasets are summarized in Table 1.

Table 1. Information of datasets

Dataset	User	Item	Rating
Movielens 100 k	943	1,682	100,000
Assistments	8,519	35,798	1,011,079

4.2 Entity Relationship Diagram (ERD)

To use the MRMF, MRMF++ and WMRMF++, we need to provide a list of entities and relations which are input parameters, therefore the datasets need to be preprocessed. Parts of ERDs are presented in Fig. 5 for Movielens and Assistments dataset.

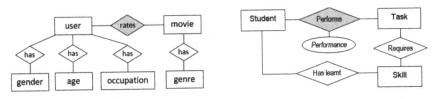

Fig. 5. ERD for Movielens and Assistments Data sets (the filled gray color relation is the main relation)

4.3 Experimental Setting

4.3.1 Baseline

The proposed methods are compared with several methods such as *global average, user average, item average, user-kNN,* and *item-kNN*. Please refer to [8] for details about these methods. Moreover, we also compare the proposed approach with *matrix factorization* (MF) and *multi-relational matrix factorization* (MRMF). The prediction functions of these methods are presented in the following:

- Global average: The rating r of the user u on the item i is predicted by

$$\hat{r}_{ui} = \mu = \frac{\sum_{(u,i,r) \in \mathcal{D}^{train}} r}{|\mathcal{D}^{train}|}$$

- User average: The rating of the user u on the item i is predicted by

$$\hat{r}_{ui} = \frac{\sum_{(u',i,r) \in \mathcal{D}^{train}|u'=u} r}{|\{(u',i,r) \in \mathcal{D}^{train}|u' = u\}|}$$

- Item average: The rating of the user u on the item i is predicted by

$$\hat{r}_{ui} = \frac{\sum_{(u,i',r) \in \mathcal{D}^{train}|i'=i} r}{|\{(u,i',r) \in \mathcal{D}^{train}|i' = i\}|}$$

- User-kNN: The rating of the user u on the item i is predicted by

$$\hat{r}_{ui} = \bar{r}_u + \frac{\sum_{u' \in K_u} sim(u, u') \cdot (r_{u'i} - \bar{r}_{u'})}{\sum_{u' \in K_u} |sim(u, u')|}$$

where K_u is a set of K nearest neighbors of user u; $\bar{r}_u$ and $\bar{r}_{u'}$ are average rating over all the items of user u and u' respectively; $sim(u, u')$ is the similarity between user u and user u', computed by using Cosine similarity:

$$sim_{cosine}(u, u') = \frac{\sum_{i \in I_{uu'}} r_{ui} \cdot r_{u'i}}{\sqrt{\sum_{i \in I_{uu'}} r_{ui}^2} \sqrt{\sum_{i \in I_{uu'}} r_{u'i}^2}}$$

where $I_{uu'}$ is a set of items rated by both user u and user u'.

- Item-kNN: The rating of the user u on the item i is predicted by

$$\hat{r}_{ui} = \bar{r}_i + \frac{\sum_{i' \in k_i} sim(i, i') \cdot (r_{ui} - \bar{r}_{i'})}{\sum_{i' \in K_i} |sim(i, i')|}$$

where K_i is a set of K nearest neighbors of item i; $\bar{r}_i$ and $\bar{r}_{i'}$ are average rating over all users of item i and i' respectively; $sim(i, i')$ is the similarity between item i and item i'.

4.3.2 Evaluation Measure

To compare among the methods, we use the standard measure in recommender systems, which is root mean squared error (RMSE).

$$RMSE = \sqrt{\frac{1}{|D^{test}|} \sum_{u,i,r \in D^{test}} (r_{ui} - \hat{r}_{ui})^2}$$

4.3.3 Hyperparameter Setting

The hyperparameter search is applied for the MF, MRMF, MRMF++ and WMRMF++ to search the best hyperparameters such as the number of iterations (Iter), the number of latent factors K, learning rate β, regularization term λ.

4.4 Experimental Results

Figures 6 and 7 present the RMSE results on Movielens and Assistment datasets, respectively. The experimental results show that the proposed MRMF++ and WMRMF++, which take into account multiple relationships between entities, have improvements compared to the others. These results show that (W) MRMF++ is a feasible approach for the multi-relational data.

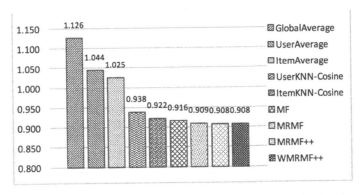

Fig. 6. RMSE on Movielens data set. The lower the better

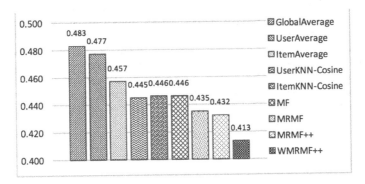

Fig. 7. RMSE on Assistments data set. The lower the better

5 Conclusion

In many real systems, we can adopt several relationships among data (e.g. product attributes, user attributes, etc.), thus we have introduced a new multi-relational approach for recommender systems (MRMF++ and WMRMF++) in the context of having multi-relational data. With this approach, the prediction model can use all the information from the latent factor matrices in prediction, so it has improvements compared to the others. The experimental results show that the proposed methods work well on both entertainment and education data. In future work, we will test on other datasets to get more validations on the proposed approach.

References

1. Ricci, F., Rokach, L., Shapira, B., Kantor, P.B. (eds.): Recommender Systems Handbook. Springer, US (2011)
2. Su, X., Khoshgoftaar, T.M.: A survey of collaborative filtering techniques. Adv. Artif. Intell. **2009**, 4:1–4:19 (2009)
3. Bell, R.M., Koren, Y.: Scalable collaborative filtering with jointly derived neighborhood interpolation weights. In: Proceedings of the 7th IEEE International Conference on Data Mining (ICDM 2007), Washington, USA, pp. 43–52. IEEE CS (2007)
4. Koren, Y., Bell, R., Volinsky, C.: Matrix factorization techniques for recommender systems. J. Comput. **42**(8), 30–37 (2009). IEEE Computer Society Press
5. Lippert, C., Weber, S.H., Huang, Y., Tresp, V., Schubert, M., Kriegel, H.P.: Relation prediction in multi-relational domains using matrix factorization. In: Proceedings of the NIPS 2008 Workshop: Structured Input-Structured Output, Vancouver, Canada, December, 2008
6. Thai-Nghe, N., Schmidt-Thieme, L.: Multi-relational factorization models for student modeling in intelligent tutoring systems. In: 2015 Seventh International Conference on Knowledge and Systems Engineering (KSE), pp. 61–66. IEEE (2015)
7. Singh, A.P., Gordon, G.J.: Relational learning via collective matrix factorization. In: Proceeding of the 14th ACM SIGKDD International Conference on Knowledge Discovery and Data Mining (KDD 2008). KDD 2008, pp. 650–658. ACM, New York (2008)
8. Drumond, L., Diaz-Aviles, E., Schmidt-Thieme, L., Nejdl, W.: Optimizing multi-relational factorization models for multipletarget relations. In: Proceedings of the 23rd ACM International Conference on Information and Knowledge Management (CIKM 2014) (2014)
9. Thai-Nghe, N.: Predicting Student Performance in an Intelligent Tutoring System. Ph.D. thesis. University of Hildesheim, Germany (2012). hildok.bsz-bw.de

A Hybrid Feature Selection Method Based on Symmetrical Uncertainty and Support Vector Machine for High-Dimensional Data Classification

Yongjun Piao and Keun Ho Ryu[✉]

Database/Bioinformatics Laboratory, School of Electrical and Computer Engineering,
Chungbuk National University, Cheongju, South Korea
{pyz,khryu}@dblab.chungbuk.ac.kr

Abstract. MicroRNA (miRNA) is a small, endogenous, and non-coding RNA that plays a critical regulatory role in various biological processes. Recently, researches based on microRNA expression profiles showed a new aspect of multiclass cancer classification. Due to the high dimensionality, however, classification of miRNA expression data contains several computational challenges. In this paper, we proposed a hybrid feature selection method for accurately classification of various cancer types based on miRNA expression data. Symmetrical uncertainty was employed as a filter part and support vector machine with best first search were used as a wrapper part. To validate the efficiency of the proposed method, we conducted several experiments on a real bead-based miRNA expression datasets and the results showed that our method can significantly improve the classification accuracy and outperformed the existing feature selection methods.

Keywords: MicroRNA · Feature selection · Support vector machine · Symmetrical uncertainty · Classification

1 Introduction

MicroRNA (miRNA) is a small, endogenous, and non-coding RNA that plays a critical regulatory role in various biological processes [1, 2]. Before early 2000s, miRNAs were not considered as an independent group of regulation factors [3–5]. In the past decade, miRNA has proven to be a highly tissue related markers and a lot evidences have been shown that dysregulation of miRNAs is highly associated with various human diseases including cancer [6–9]. Several studies have shown that miRNA expression data have significant advantage in cancer classification compared with mRNA expression [10, 11], i.e., the expression pattern of miRNAs is more close to the practical level of the gene. Due to the high dimensionality, however, classification of miRNA expression data contains several computational challenges. Dimension reduction techniques that discover a reduced set of miRNAs are needed to achieve better classification result. From the point of data mining, selecting important miRNAs related

© Springer International Publishing AG 2017
N.T. Nguyen et al. (Eds.): ACIIDS 2017, Part I, LNAI 10191, pp. 721–727, 2017.
DOI: 10.1007/978-3-319-54472-4_67

to cancers can be seen as a feature selection problem that aims to find most relevant subset of features for classification.

Feature selection is an essential pre-processing procedure in data mining for identifying relevant subset for classification. High dimensionally of the data may cause various problems such as increasing the complexity and reducing the accuracy, i.e., curse of dimensionality. The goal of feature selection is to provide faster construction of prediction models with a better performance [12]. Feature selection approaches can be broadly grouped into three categories: filter, wrapper, and hybrid [13]. The main difference of filter and wrapper method is in whether they adopt a machine learning algorithm to guide the feature selection or not. In general, filters employ independent evaluation measures thus are fast but can generate local-optimal result. In contrast, wrapper methods adopt a searching algorithm to iteratively generate several subsets, evaluate them based on the classification algorithm, and finally choose the subset with best classification performance. Wrappers usually can produce better results than filters but they are computationally expensive. Hybrid methods combine the advantages of filter and wrapper techniques to achieve better learning performance with a similar computational cost of filters [14].

In this paper, we propose a hybrid feature selection technique for accurately classification of various cancer types based on miRNA expression data. Symmetrical uncertainty is employed to serve as a filter part and support vector machine with best first search to be the wrapper part. To validate the efficiency of the proposed method, we conducted several experiments on a real bead-based expression datasets and the results showed that our method can significantly improve the classification accuracy and outperformed existing feature selection methods.

2   Method

2.1   Symmetrical Uncertainty

Various studies have demonstrated that symmetrical uncertainty (SU) was a good measure for selecting relevant features [15, 16]. SU is a correlation measure of a feature that is a normalized form of mutual information. The SU between a feature and class can be calculated as follows:

$$IG(F|C) = H(F) - H(F|C) \tag{1}$$

$$SU(F, C) = 2 * IG(F|C)/(H(F) + H(C)) \tag{2}$$

Where $IG(F|C)$ is the information gain of a feature F after observing class C. $H(F)$ and $H(C)$ are the entropy of feature F and class C, respectively.

2.2   Support Vector Machine

In the past few years, a lot of studies have reported that SVM has good performance on high-dimensional data to avoid curse of dimensionality. The major idea of SVM is to find the decision boundary with a maximal margin [17]. In order to explain how SVM works, consider the following two-class classification problem as Eq. 3:

$$(x_i, y_i), x_i \in R^n, y_i \in +1, -1, i = 1, 2, \ldots, n \tag{3}$$

Where x_i is the feature set of the ith vector and y_i denotes the class label. Based on the parameters w and b of the decision boundary, two parallel hyperplanes h_{i1}, h_{i2} and the margin d can be expressed as following Eqs. 4, 5 and 6:

$$h_{i1}: w \cdot x + b = +1 \tag{4}$$

$$h_{i2}: w \cdot x + b = -1 \tag{5}$$

$$d = \frac{2}{||w||} \tag{6}$$

Consequently, learning the SVM model can be considered as the constrained optimization problem, which can be formulized as the following Eq. 7:

$$L_p = \frac{1}{2}||w||^2 - \sum_{i=1}^{N} \sigma_i(y_i(w \cdot x_i + b) - 1) \tag{7}$$

Where the parameters σ_i refers to the Lagrange. Based on the Lagrange multipliers, the classification function f(x) can be written as the following Eq. 8:

$$f(x) = sgn(\sum_{i=1}^{n} \sigma_i y_i (x_i, x) + b) \tag{8}$$

For non-linear classification problem, the input instances are projected into a high dimensional space via a mapping function K, which can be written as the following Eq. 9:

$$f(x) = sgn(\sum_{i=1}^{n} \sigma_i y_i K(x_i, x) + b) \tag{9}$$

2.3 Hybrid Feature Selection

Here, we describe our proposed hybrid feature selection method. The overall framework of generating optimal feature subset is illustrated in Fig. 1. In the proposed method, symmetrical uncertainty is used as the filter part and the best first search with support vector machine is adopted as the wrapper part. For each feature in the original space, the symmetrical uncertainty value is calculated and the features that have smaller uncertainty value than the pre-defined threshold are removed. Then the remaining features are sorted in the descending order of their uncertainty value. Starting from the most relevant one (feature has largest uncertainty value), adding one feature at one time, the subset is evaluated by the support vector machine until the best feature subset that has highest accuracy is selected.

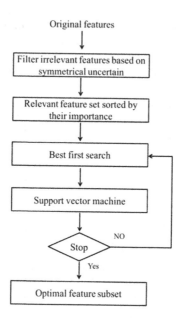

Fig. 1. Framework of proposed hybrid feature selection

3   Results

3.1   Dataset

The miRNA expression dataset used in the experiment was from Lu et al. [10]. The expression levels were quantified using bead-based approach resulted in expression profiles of 217 mammalian miRNAs from 73 samples including 5 cancer types. The details of the cancer types and the number of samples in each class are shown in Table 1.

Table 1. The number of samples in each cancer type

Cancer	Number of samples
Colon	10
Pancreas	9
Uterus	10
B Cell ALL	26
T Cell ALL	10
Total	73

3.2   Performance Evaluation

To achieve reliable results, we conducted 10-fold cross validation during all the experiments. In 10-fold cross validation, 9 parts were used to train the model and the remaining one was used to test the classification model. Moreover, we adopted several evaluation measures such as classification accuracy, sensitivity, specificity, and area under ROC (AUC) to test the performance of classification.

During the filter part, we set the symmetrical uncertainty cutoff to 0.2 so that the features that have smaller symmetrical uncertainty value than 0.2 were removed from the analysis. As a result, 149 features were selected as a relevant subset. After wrapper part, 9 features, denoted as $F_1, F_2, F_3, F_4, F_5, F_6, F_7, F_8, F_9$, were finally selected as a best feature subset for classification. The list of the miRNAs with their symmetrical values was shown in Table 2. Table 3 shows the sensitivity, specificity, and AUC of each feature subset during wrapper construction.

Table 2. Seletected features and their symmetrical uncertainty values in the filter part. The cutoff value of the symmetrical uncertainty was set to 0.2

Feature No.	SU value	miRNA annotation
F_1	0.628	mmu-miR-10b
F_2	0.601	mmu-miR-337
F_3	0.553	rno-miR-151
F_4	0.502	hsa-miR-194
F_5	0.477	hsa-miR-99a
F_6	0.469	mmu-miR297
F_7	0.445	hsa-miR-142-5p
F_8	0.259	hsa-miR-106a
F_9	0.222	mmu-miR-7b

Table 3. Classification performance of differnt feature subsets during wrapper construction

No.	Feature subset	Sen	Spe	AUC
1	F_1	0.507	0.781	0.741
2	F_1, F_2	0.534	0.813	0.805
3	F_1, F_2, F_3	0.781	0.955	0.940
4	F_1, F_2, F_3, F_4	0.836	0.970	0.967
5	F_1, F_2, F_3, F_4, F_5	0.904	0.978	0.975
6	$F_1, F_2, F_3, F_4, F_5, F_6$	0.918	0.975	0.973
7	$F_1, F_2, F_3, F_4, F_5, F_6, F_7$	0.945	0.987	0.984
8	$F_1, F_2, F_3, F_4, F_5, F_6, F_7, F_8$	0.945	0.987	0.984
9	$F_1, F_2, F_3, F_4, F_5, F_6, F_7, F_8, F_9$	0.959	0.989	0.987

Then we compared the performance of classification with that of using original feature space and several commonly used feature selection methods such as gain ratio, information gain, ReliefF, and Pearson correlation. To make a fair comparison, the cutoff values of feature selection were set to 0.2 for all the methods. Figure 2 shows the classification accuracy of the 5 feature selection method and 1 without applying any feature selection. From the figure, it was clear that our proposed method was found to be result in the best classification accuracy, which was 0.959. The other methods were found to be 0.904, 0.877, 0.877, 0.904, and 0.890. One interesting observation was that the traditional feature selection methods could not improve the miRNA expression data classification accuracy since the results of the existing feature selection methods were even worse than that of without dimensionality reduction except ReliefF.

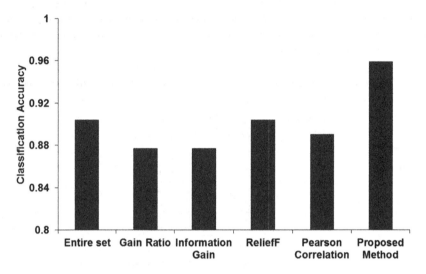

Fig. 2. Performance comparison with other feature selection methods in terms of classification accuracy.

4   Conclusion

In this paper, we proposed a hybrid feature selection technique based on symmetrical uncertainty and support vector machine for classification of high-dimensional miRNA expression data. To validate the proposed method, we conducted several experiments and compared the performance with existing feature selection approaches on a real bead-based miRNA expression data. During the experiments, we have shown that the existing methods were difficult to improve the classification accuracy. Moreover, the proposed method yielded good results and significantly outperformed the other methods. We believe that the excellent performance of the proposed method can accurately identify functionally significant miRNAs related cancers to facilitate the understanding of the roles of miRNA in caners.

Acknowledgement. This research was supported by Basic Science Research Program through the National Research Foundation of Korea (NRF) funded by the Ministry of Science, ICT & Future Planning (No. 2013R1A2A2A01068923) and the MSIP (Ministry of Science, ICT and Future Planning), Korea, under the ITRC (Information Technology Research Center) support program (IITP-2016-H8501-16-1013) supervised by the IITP(Institute for Information & communication Technology Promotion).

References

1. Calin, G.A., Croce, C.M.: MicroRNA signatures in human cancers. Nat. Rev. Canc. **6**, 857–866 (2006)
2. Croce, C.M., Calin, G.A.: miRNAs, cancer, and stem cell division. Cell **122**, 6–7 (2005)
3. Lagos-Quintana, M., Rauhut, R., Lendeckel, W., Tuschl, T.: Identification of novel genes coding for small expressed RNAs. Science **26**, 853–858 (2001)
4. Lau, N.C., Lim, L.P., Weinstein, E.G., Bartel, D.P.: An abundant class of tiny RNAs with probable regulatory roles in Caenorhabditis elegans. Science **294**, 858–862 (2001)
5. Lee, R.C., Ambros, V.: An extensive class of small RNAs in Caenorhabditis elegans. Science **294**, 862–864 (2001)
6. Mencía, A., Modamio-Høybjør, S., Redshaw, N., Morín, M., Mayo-Merino, F., Olavarrieta, L., Aguirre, L.A., del Castillo, I., Steel, K.P., Dalmay, T., Moreno, F., Moreno-Pelayo, M.A.: Mutations in the seed region of human miR-96 are responsible for nonsyndromic progressive hearing loss. Nat. Genet. **41**, 609–613 (2009)
7. Hughes, A.E., Bradley, D.T., Campbell, M., Lechner, J., Dash, D.P., Simpson, D.A., Willoughby, C.E.: Mutation altering the miR-184 seed region causes familial keratoconus with cataract. Am. J. Hum. Genet. **89**, 628–633 (2011)
8. Musilova, K., Mraz, M.: MicroRNAs in B cell lymphomas: how a complex biology gets more complex. Leukemia **5**, 1004–1017 (2015)
9. Malumbres, M.: miRNAs and cancer: an epigenetics view. Mol. Aspects Med. **34**, 863–874 (2013)
10. Lu, J., Getz, G., Miska, E.A., Alvarez-Saavedra, E., Lamb, J., Peck, D., Sweet-Cordero, A., Ebert, B.L., Mak, R.H., Ferrando, A.A., Downing, J.R., Jacks, T., Horvitz, H.R., Golub, T.R.: MicroRNA expression profiles classify human cancers. Nature **435**, 834–838 (2005)
11. He, L., Thomson, J.M., Hemann, M.T., Hernando-Monge, E., Mu, D., Goodson, S., Powers, S., Cordon-Cardo, C., Lowe, S.W., Hannon, G.J., Hammond, S.M.: A microRNA polycistron as a potential human oncogene. Nature **435**, 828–833 (2005)
12. Piao, Y., Piao, M., Park, K., Ryu, K.H.: An ensemble correlation-based gene selection algorithm for cancer classification with gene expression data. Bioinformatics **28**, 3306–3315 (2012)
13. Hsu, H.H., Hsieh, C.W., Lu, M.D.: Hybrid feature selection by combining filters and wrappers. Expert Syst. Appl. **38**, 8144–8150 (2011)
14. Xie, J., Wang, C.: Using support vector machines with a novel hybrid feature selection method for diagnosis of erythemato-squamous diseases. Expert Syst. Appl. **38**, 5809–5815 (2011)
15. Zeng, Z., Zhang, H., Zhang, R., Yin, C.: A novel feature selection method considering feature interaction. Pattern Recogn. **48**, 2656–2666 (2015)
16. Kannan, S.S., Ramaraj, N.: A novel hybrid feature selection via symmetrical uncertainty ranking based local memetic search algorithm. Knowl. Based Syst. **23**, 580–585 (2010)
17. Akay, M.F.: Support vector machines combined with feature selection for breast cancer diagnosis. Expert Syst. Appl. **36**, 3240–3247 (2009)

Selecting Important Features Related to Efficacy of Mobile Advertisements

Goutam Chakraborty[1(✉)], L.C. Cheng[1], L.S. Chen[2], and Cedric Bornand[3]

[1] Faculty of Software and Information Science,
Iwate Prefectural University, Takizawa 020-0693, Japan
`goutam@iwate-pu.ac.jp`
[2] Department of Information Management,
Chaoyang University of Technology, Wufeng, Taichung, Taiwan
[3] Department of Industrial Technologies,
University of Applied Sciences (HES-SO), Yverdon-les-Bains, Switzerland

Abstract. With growing use of mobile devices, mobile advertisement is playing increasingly important role. It can reach potential customers at any time and place based on individual's real-time needs. Factors for success of mobile advertisements are different from similar media like Television or large screen monitors. We investigated the important factors to enhance click through rate (CTR) for a mobile Ad. As CTR is directly related to revenue, it is used to measure success of a mobile Ad. To identify important factors that determine CTR, we took two approaches - one directly asking subjects, and the other, from analyzing their selective attention. Subjects were asked to respond to questionnaire. From the responses important features were selected using Least Absolute Shrinkage and Selection Operator (LASSO). For the other approach, selective attention was inferred from subjects' eye-tracking data. When results from two approaches were compared, the findings were similar. Those features will be helpful for designing Ads favored by users, as well as could earn more revenues.

Keywords: Mobile advertisement · Selective attention · Eye-tracking · LASSO

1 Introduction

In this work, we have two motivations - (1) to find which features of a mobile Ad are important for high CTR; and at the same time (2) to find whether there is a correlation between our explicit knowledge that we express verbally and natural selective attention (visual in this case).

1.1 Background

With growing use of mobile devices, mobile advertisement is playing a more important role. People now spend a lot of time with their smart phones, while

© Springer International Publishing AG 2017
N.T. Nguyen et al. (Eds.): ACIIDS 2017, Part I, LNAI 10191, pp. 728–737, 2017.
DOI: 10.1007/978-3-319-54472-4_68

the device is connected with the internet. Mobile Ads pop-up to attract attention. If it is clicked, it would earn revenue. We investigate features that increase the CTR.

Interactive Advertising Bureau (IAB) [8], reported that in 2015 the internet advertising revenue is totaled to $59.6 billion, up 20.4% from $49.5 billion reported in 2014. Of that, only mobile advertisement revenue in the United States totaled $20.7 billion during fiscal year 2015, a 66% increase from the previous year's total of $12.5 billion. The market is expected to grow up to $41.9 billion by 2017 [7]. There is no doubt that from the business point of view, a design which attracts more CTR, is important.

1.2 Motivation

Click through rate is our performance measure, to judge whether a mobile Ad is effective or not. A product launched by Medialets' called Servo Total Attribution helps advertisers track Ads that appear on display but not clicked. Using that data, Ads could be filtered. Some big advertisers, like Facebook and Twitter, do not use it because they keep tight control of data within their systems.

What would attract attention of a prospective viewer is the important question - whether it is an attractive picture, or whether some rebate/promotion is offered or not, etc. Several works pointed different features as relevant. But, their degree of relevance never experimented. From the works of [1], we know that visual attention has a strong relation to the decision to click an Ad. In our study, we did experiments with two approaches. We asked subjects to rank features they think are important to induce clicking. The collected data was then used for feature selection, to identify important features for high CTR. Subsequently, we use eye-tracker to detect users' visual attention, and analyze the gaze point data to identify locations that attract attention and corresponding features. While correlating them, we found that similar features were identified.

The rest of the paper is organized as follows. In Sect. 2, we talk about proposed methods. Section 2.1 is related works where 20 possible features were referred as important. In Sect. 3, we explain how the experiments were setup, data collected and analyzed. Section 4 describes the experimental results. Section 5 is the conclusion and plan for future extension.

2   Proposed Method

2.1   Related Work

The American Marketing Association defines advertisement as "the placement of announcements and persuasive messages in time or space purchased in any of the mass media by business firms, non-profit organizations, government agencies, and individuals who **seek to inform and/or persuade members of a particular target market or audience about their products, services, organizations, or ideas**" [2]. From this definition, we can say that advertisement is a message from enterprise to customer; a one way communication of information. As it is one way, it is difficult to know how it reacts. In addition, unlike traditional advertising media such as newspapers, magazines, television and radio, for which we have experience, mobile advertising is new. It offers advantages like reaching the consumers at the right time and place [3], which the business needs to exploit.

2.2 Questionnaire and Response Data

In our first approach, we asked a set of questions to the subjects who respond by ranking the influence of different factors. The set of potential factors influencing mobile advertisement are summarized from the works as listed in Table 1

Different authors proposed features they considered to be important. In Table. 1 we listed 20 potential factors which helped us to design the questionnaire we used to get feed-back from real users to identify which potential factors significantly influence click through rate.

Table 1. Main references of mobile advertisement and corresponding factors

Serial number	Factors of mobile ad	References							
		A	B	C	D	E	F	G	H
1	Involvement	O							
2	Language	O							
3	Type of Website	O							
4	Information Privacy		O						
5	Entertainment			O	O	O			
6	Irritation				O	O			
7	Perceived Usefulness				O	O			
8	Perceived ease of use					O			
9	Credibility			O		O			
10	Price						O		
11	Preference						O		
12	Promotion						O		
13	Interest						O		
14	Brand Name						O		
16	Mobile Device						O		
17	Informativeness				O				
18	Incentives				O				
19	Social Media								O
20	Rich Media							O	

A: William et al. (2014) [14]   E: Yang et al. (2013) [15]
B: Ufuoma and Ayesha 2015 [13]   F: Chen and Hsieh (2012) [2]
C: Kim and Han (2014) [10]   G: Erik (2014) [6]
D: Jos et al. (2013) [9]   H: Lauren(2013) [11]

Feature Selection. In machine learning, feature selection is to find the set of minimum features that will deliver maximum classification precision. Lowering the number of features/factors is important, because it facilitates efficient computation, understanding the nature of data, and improves classification results by eliminating irrelevant factors which act as noise. Feature selection is different from dimensionality reduction, where the original data is projected on a lower dimension space by using statistical tools like, Principal Component Analysis (PCA), Singular value decomposition (SVD) or Sammon's mapping (a non-linear Artificial Neural Network based approach).

Feature selection methods are classified into two - Filter method and Wrapper method. In a filter method, an individual feature is evaluated using some statistical methods like Chi squared test, information gain or correlation coefficient score. Features are selected according to their scores. In wrapper method a model is used, and a subset of features is evaluated using the model. The model could be anything, like a regression model, K-nearest neighbor, or a neural network. Searching for optimum subset of features could be heuristic, stochastic or forward-backward to add and remove features. We used wrapper method with regression as model.

On the set of data collected as questionnaire response, we used least absolute shrinkage and selection operator (LASSO) as feature selection method. Among existing feature selection algorithms, LASSO has been demonstrated as the most practical one because of its robustness and high precision [4,16]. As it uses linear regression as model, it is very fast. The number of selected features could be tuned giving it flexibility of use. It is known that variable selection and parameter estimation via LASSO are more stable than other subset selection procedure and produce better prediction accuracy compared to other methods of similar efficiency.

2.3   Visual Attention

In this experiment, we collected 6 Ads of the same category. Subjects watch them for a brief period of time. Their eye-tracking data are collected from which we conclude which features of the Ad attract visual attention. Here, visual attention is used as a mean to determine important features.

Eye-tracking data consists of alternate fixations and saccades. Fixations are the point of interest. Fixation points are dense around the location which attracts attention. In this work, we cluster the fixation points, and find the duration of time the subject spends at that location. If the duration crosses some threshold, we identify that as a point of attention. Finally, we correlate a feature corresponding to that location's image in the Ad.

Clustering. As fixation points are many with short durations, it is difficult to consider them individually. Clustering analysis is a very important and useful technique to find the pattern from lots of gaze points. Clustering algorithms, like DBSCAN [12], K-means [5], are efficient to organize enormous amounts of data.

The K-means method is a widely used clustering technique that seeks to minimize the average squared distance between points in the same cluster. It is by far the most popular clustering algorithm because of its simplicity and efficiency. The first k-means algorithm is proposed by Lloyd, also called Lloyd's algorithm which is an iterative, partition clustering algorithm that assigns n observations to one of k clusters defined by centroids chosen before the algorithm starts. The result is not robust, and depends on initialization. After about 25 years, Arthur and Vassilvitskii [5] proposed a new algorithm called K-means++ which improves the run time of Lloyd's algorithm, and the quality of the final solution. In this study, we use K-mean++ to find the pattern of attention from data collected by an eye-tracker.

3   Experimental Set-Up and Data Analysis

As mentioned in Sect. 2, we proposed two methods to find important features for mobile Ads. The detail of data collection is explained in the following sections.

3.1   Data Analysis from Subjects' Questionnaire Responses

In this approach, data collection and analysis is done in the following steps.

- **Step 1 Collect data:** The subjects are young adults, between the age of 20–24, who frequently use mobile devices and have prior experience of using mobile advertisements. The questionnaire included 20 potential factors listed in Table 1. For every factor there are two aspects, whether this feature is well represented on the Ad or not. Subjects are to score both aspects, from 1 to 5. Lower score means less important. For example, the two aspects of the price factor are as follows.
  Question1: Mobile advertisement provides price information of product/service clearly.
  Question2: Mobile advertisement does not provide price information, or it is vague/confused.
  The subject has to fill in a score from a five-rating scale ranging from very unimportant to very important, representing a factor's importance level. All the listed factors need to be responded.
- **Step 2 Data preprocessing:** We normalized the data to the range [0, 1].
- **Step 3 Feature selection:** We used 2 feature selection methods - correlation and LASSO. LASSO is a wrapper method where linear regression is used as classifier model. Regularization parameter λ is used to tune the number of selected features [4]. High value of λ leads to selection of less number of features.
- **Step 4 SVM classifier:** Once features are selected, we use libsvm classifier to find the accuracy of 2-class classification (CT or not). Five fold cross validation was used. We used five evaluation criterion namely: accuracy, precision, recall, F1-score, time taken for feature selection and classification.

– **Step 5 Comparison:** To sum up, we find the important factors of mobile advertisements, and find their performance using the supervised data available from questionaire-response. Finally, we evaluated the selected factors by classification accuracy using SVM, and compare results when all 20 features are used, features selected by correlation, and by LASSO.

3.2   Data Analysis from Subjects' Eye-Tracking

In this approach, we collect information about subjects' selective attention using eye-tracking data. Eye tracking data logs the position coordinates of fovea, the central location of vision. Data collection and analysis is done in the following steps:

– **Step 1 Data collection:** We used three categories of advertisements. Every category has six different advertisements. Respondents will see every set for 30 s. The mobile Ads were displayed on a Desktop monitor (27 in.) for higher resolution of eye-tracking data. The viewing distance between eye-tracker and respondent is 50 to 70 cm. We had 5 subjects, four male and 1 female. All subjects were between age 22 to 24.
   Figure 1, we have shown the health-supplement category of advertisement, of the three categories we used. Six Ads of the same category are grouped together. One category of Ad is shown to the subject at a time. Subjects watch it for 30 s. The eye-tracking data is logged.
– **Step 2 Clustering:** Our sample points are the fixation points of vision. We need to find a region, where the subject stared for long. We did clustering of fixation points, where the distance metric was naturally Euclidean distance.

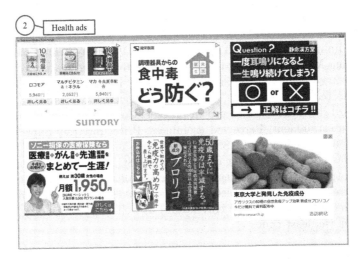

Fig. 1. Six ads from dietary supplement category

We used K-means++ clustering for its robustness. We started with a large k equal of 100, so that every cluster covers only a small area of the display, because we want to pinpoint image features of the advertisement. If k is less, area covered by a cluster will be too large, from which identifying feature is not possible.

- **Step 3 Extracting features:** As mentioned, k is set to 100. Every cluster is ranked by the duration of time the subject stared in that cluster location. Finally, only top 10 clusters are acknowledged as important. These 10 important clusters' locations are mapped on the image, and the features relevant to that part of image is manually identified.

Finally, we compare the features selected from questionnaire responses, and from gaze data. A striking similarity confirms that our natural instinct has good correlation with impression verbally expressed as questionnaire response.

4   Experimental Results

The results from the two approaches will be discussed in the next two sections, after which we will give a comprehensive summary of the results from both approaches.

4.1   Results from Questionnaire Responses

We collected data from about 500 subjects, out of which 482 were valid records. 64% of the subjects were male and the rest female. They were all young adults, spending a few hours on smart phone every day. They also have experience in on-line purchase using their mobile devices.

We use LASSO with 5-fold cross validation to select the features. The mean squared prediction error (MSE) on validation data (new data not used for training) for the linear regression model was plotted against λ. MSE at $\lambda = 0.16$ was 0.88. Minimum MSE was 0.85 at $\lambda = 0.06$. When we relax MSE to 0.88, corresponding $\lambda = 0.16$, leading to a small set of 6 selected features.

In Table 2, we compared classification accuracy using SVM when all features, features selected by correlation, and features selected by LASSO were used. The standard deviations are shown in bracket. Correlation selected 7 features, namely - Entertainment, Perceived usefulness, Price, Interest, Informativeness, Rich media, and Credibility. LASSO selected the same features, except Credibility which was not selected. We can see that LASSO performs best compared to using all features or features selected by correlation. Classification performance is much better when feature selection is done. The total execution time, feature selection (in case of LASSO and correlation) as well as SVM training, is shown. LASSO took the least time.

Table 2. Comparison of classification results - all 20 features, features selected by correlation, features selected by LASSO

Evaluation criterion	Using all 21 features	7 features selected by correlation	6 features selected by LASSO
Accuracy (%)	67.69 (4.33)	69.17 (4.40)	71.67 (6.14)
Precision (%)	66.18 (5.05)	67.90 (5.24)	71.46 (7.37)
Recall (%)	67.72 (4.34)	69.18 (4.41)	71.66 (6.15)
F1-score (%)	65.96 (4.45)	67.72 (4.90)	70.30 (5.55)
Time (s)	24.20	19.56	17.17

4.2   Results from Eye-Tracking Data

The six advertisements, in a group, were chosen such that they contain different features, to the extent possible. We run k-means to group them into 100 clusters. We need clusters covering small areas, so that we can pin point the feature at fixation location. For 100 clusters, we calculate the time spent in watching every cluster. Highest 10 were selected.

Features on locations of those 10 clusters are manually identified. For health products, those features with corresponding images, identified by 10 clusters are shown in Fig. 2. Images and corresponding feature names in text are included.

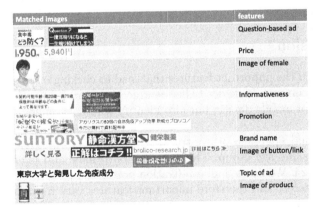

Fig. 2. Features extracted from eye-tracking data while subjects were watching health supplement advertisements

4.3   Summarizing the Two Results

We compiled the features selected by LASSO from subjects' questionnaire responses and from eye-tracking data in Table 3. In the first approach, there was no particular type of Ads the subjects were responding. In addition, all subjects are aged between 20–24. The type of things they purchase online are different from some category of items used in second experiment, like dietary supplement. Still, a striking similarity in features could be found. We put together features that are identical or very similar. In the first approach, "Entertainment" was a feature, which could not be identified from visual image (from eye-tracking data). In addition, "Question based Ad" in dietary supplement Ad was discovered as an important feature, which was not included in the first approach in the list of 20 features. Those two are put at the last row of the table. Price, clear information about the product, lots of nice pictures and brand name are features in which prospective customers are interested.

Table 3. Summarizing features selected from two different approaches

Type of advertisement			
Electronic gadgets	Health supplement	Tourism	No specific product
Price	Price Promotion	Promotion	Price
Informativeness	Informativeness	Informativeness	Informativeness
Image of Product	Image of Product	Image of Product	Rich Media
Image of female	Image of female	Image of monkey	
Image of button	Image of button	Image of button	
Brand name	Brand name	Brand name	Perceived usefulness
Topic of Ad	Topic of Ad	Topic of Ad	Interest
	Question Based Ad		Entertainment

5   Conclusion

We investigated the important features that lead to clicking of an advertisement on a small screen mobile device. The identification could not only lead to successful design of mobile device, but also identify what customers are looking for from an Ad.

In the second approach of eye-tracking, the number of subjects was small, and from the same age group. This included both male and female, all of age between 22 to 24. We merged all eye-fixation points and performed clustering. It is true that interest and therefore important features vary from male to female and for different age groups. In future, we plan to get data using subsets from different sex and age groups, so that we would be able to identify the change of attention for different sets of subjects.

Acknowledgment. This project was partially supported by research grant from Iwate Prefectural University, iMOS research center

References

1. Bridget, K.B., Mikyeung, B., Patricia, T.H., Lynnell, S.: The effect of involvement on visual attention and product choice. J. Retail. Consum. Serv. **24**, 1021 (2015)

2. Chen, P.T., Hsieh, H.P.: Personalized mobile advertising: its key attributes, trends, and social impact. Technol. Forecast. Soc. Chang. **79**, 543–557 (2012)

3. Wong, C.H., Tan, W.H., Tan, B.I., Ooi, K.B.: Mobile advertising: the changing landscape of the advertising industry. Telematics Inform. **32**, 720–734 (2015)

4. Tibshirani, R.J.: Regression shrinkage and selection via the LASSO. J. Roy. Stat. Soc.: Ser. B (Methodol.) **58**, 267–288 (1996)

5. Arthur, D., Vassilvitskii, S.: K-means++: the advantages of careful seeding. In: SODA 2007, Proceedings of the Eighteenth Annual ACM-SIAM Symposium on Discrete Algorithms, pp. 1027–1035 (2007)

6. Erik, S.: 5 mobile ad trends to watch in 2014 (2013). http://www.inc.com/erik-sherman/make-your-online-ads-move-into-mobile.html

7. Gartner Inc., Gartner Says Mobile Advertising Spending Will Reach $18 Billion in 2014, newsroom of Gartner (2014). http://www.gartner.com/newsroom/id/2653121

8. Interactive Advertising Bureau (IAB), IAB internet advertising revenue report, PwC Advisory Services LLC (PwC), April 2016. https://www.iab.com/wp-content/uploads/2016/04/IAB-Internet-Advertising-Revenue-Report-FY-2015.pdf

9. Jos, M.P., Silvia, S.B., Carla, R.M., Joaquin, A.M.: Key factors of teenagers mobile advertising acceptance. Ind. Manage. Data Syst. **113**(5), 732–749 (2013)

10. Kim, Y.J., Han, J.: Why smartphone advertising attracts customers: a model of Web advertising, flow, and personalization. Comput. Hum. Behav. **33**, 256–269 (2014)

11. Lauren, J.: Mobile marketer reported: top 10 mobile advertising campaigns of 2013, Mobile Marketer (2013). http://www.mobilemarketer.com/cms/news/advertising/16847.html

12. Mahesh, K.K., Reddy, A.R.M.: A fast DBSCAN clustering algorithm by accelerating neighbor searching using groups method. Pattern Recogn. **58**, 39–48 (2016)

13. Ufuoma, A., Ayesha, B.D.: Mobile advertisements and information privacy perception amongst South African generation of students. Telematics Inform. **32**, 1–10 (2015)

14. William, F., Chen, J.C., William, H.R.: The effect of variations in banner ad, type of product, website context, and language of advertising on internet users attitudes. Comput. Hum. Behav. **31**, 37–47 (2014)

15. Yang, B., Kim, Y., Yoo, C.: The integrated mobile advertising model: the effects of technology- and emotion-based evaluations. J. Bus. Res. **66**, 1345–1352 (2013)

16. Zhou, Q., Song, S., Huang, G., Wu, C.: Efficient lasso training from a geometrical perspective. Neurocomputing **168**, 234–239 (2015)

17. Chakraborty, G., Kozma, R., Murata, T., Zhao, Q.: Awareness in brain, society and beyond, a bridge connecting raw data to perception and cognition. IEEE Syst. Man Cybern. Mag. **1**, 9–16 (2015)

A Comparative Study of Evolutionary Algorithms with a New Penalty Based Fitness Function for Feature Subset Selection

Atsushi Kawamura and Basabi Chakraborty[✉]

Faculty of Software and Information Science, Iwate Prefectural University,
152-52 Sugo, Takizawa 020-0693, Japan
basabi@iwate-pu.ac.jp

Abstract. Feature subset selection is an important task for knowledge extraction from high dimensional huge data which reduces dimension of data, accelerates processing of data and improves classification accuracy. For mining of knowledge from huge data, feature subset selection acts as extraction of the context for classification process. Feature subset selection is basically an optimization problem in which a search technique is used to find out the best possible feature subset from all possible subsets of a large feature set with the use of a feature evaluation function. Evolutionary techniques are well known for their efficiency as search algorithms and are used in feature subset selection problem. In this work, a comparative study of well known EC algorithms (GA, PSO) and not so known algorithms (CS, GSA, FireFly and BAT) used for feature subset selection has been done with classification accuracy of a SVM classifier as the wrapper fitness function. A new fitness function with an added penalty term based on two objectives of improving classification accuracy while reducing dimension is proposed and its efficiency over classification accuracy alone is examined by simulation experiments with bench mark data sets. The simulation results show that the new fitness function is more effective than classification accuracy based fitness function. It also produces better results in reducing dimension and improving classification accuracy than using popular multi-objective search algorithm NSGA II for feature selection.

1 Introduction

With the increasing generation of information and data due to advancement of internet and communication technologies, the analysis and summarization of data for knowledge extraction from huge data becoming a challange day by day. For efficient categorization, mining or classification, the data need to be preprocessed to contain the characteristic or discriminatory information while being free from redundant and irrelevant information. Feature selection aims to solving this problem and it is an important preprocessing task in the area of pattern recognition or data mining [1,2] prior to classification or clustering. A sample data or pattern in the paradigm of pattern recognition or machine

© Springer International Publishing AG 2017
N.T. Nguyen et al. (Eds.): ACIIDS 2017, Part I, LNAI 10191, pp. 738–747, 2017.
DOI: 10.1007/978-3-319-54472-4_69

learning is represented by a n-dimensional vector or a point in a n-dimensional space where individual dimension represents individual feature. Feature subset selection refers to the processing of selecting a subset of d features from the set of n features by discarding irrelevant features and retaining discriminatory informative features. Reduction of features facilitates speedy processing of data and improves classification accuracy. Feature extraction also reduces dimensionality by projecting original high dimensional feature set to a lower dimensional set in which the new features are created instead of retaining a subset of original features. This paper focuses on feature selection paradigm.

Basically feature subset selection process needs to define two things: an evaluation function to evaluate the goodness of a feature or a feature subset and a search algorithm to find the best feature subset from all possible feature subsets according to the evaluation function. Depending on the nature of the evaluation function, the algorithms of feature selection are of two types, filter and wrapper. Filter algorithms evaluate the data set without reference to a particular classifier while wrapper algorithms use classifier accuracy as the evaluation function. The history of pattern recognition is long and early reasearches on feature selection evolved from statistical community. A lot of statistical feature selection algorithms have been proposed so far [3]. However, real world problems are often characterised by vagueness rather than randomness and are difficult to be modelled by rigid framework of mathematics or statistics. Soft computing technologies emerged to bridge this gap and lots of algorithms based on neural computation, fuzzy logic, rough set theory, evolutionary algorithms have been proposed for feature selection and classification in the area of pattern recognition and data mining [4–6]. Evolutionary computational(EC) algorithms are well known tools for solving optimization problems and have been efficiently used for search stage in feature subset selection problem. Among EC based algorithms, Genetic Algorithms (GA) [7], Particle Swarm Optimization (PSO) [8] and Ant Colony Optimization (ACO) [9] are widely used for feature selection. Other less commonly used algorithms are Cuckoo Search (CS) [10], Gravitational Search Algorithm (GSA) [11], Firefly Algorithm (FA) [12], Bat Algorithm(BA) [13] etc.

In this work, a comparative study of evolutionary computation (EC) based feature subset selection algorithms with classification accuracy as a wrapper fitness function (default) has been done. A new penalty based fitness function has been proposed for EC based feature selection. The efficiency of the new fitness function over the default function has been studied by simulation experiments with a number of bench mark data sets from UCI repository. The next section represents a brief overview of the algorithms used for study in this work. The following section describes comparative study of different EC algorithms and the proposal of a new penalty based fitness function followed by the simulation experiments and results in the next section. The final section contains discussion and conclusion.

2   EC Based Feature Subset Selection

Evolutionary algorithms are now becoming popular for solving feature subset selection problem. A brief presentation of the algorithms used for feature subset selection in this paper are done in the following subsections.

2.1   Genetic Algorithm

Genetic algorithm(GA), a randomized heuristic and adaptive search technique based on the principal of natural selection and the most popular evolutionary approach is a good candidate for solving optimization problems where the search space is large [14,15]. Several research works for solving feature subset selection problem with GA have been reported. In genetic algorithm, a population of possible solutions i,e the possible candidate feature subsets from a feature set of n features, encoded as a binary string of n bits, are maintained through several generations. In each generation, genetic operators such as crossover and mutation are used to generate new population from the most elite pairs of the current generation and the good ones are retained after evaluation by a fitness function. Through the generations, the population is led to the better solution space and finally produces the near optimal solution in the final generation. GA requires no domain knowledge and quite robust than other random or local search methods.

2.2   Particle Swarm Optimization

Recently particle swarm optimization (PSO), specially binary particle swarm optimization (BPSO) [16] have been also become popular for feature subset selection [17–19]. Particle Swarm Optimization [8] is a population based evolutionary algorithm. The conventional PSO algorithm begins by initializing a random swarm of m particles in d dimensional space charaterizing candidate solution like genetic algorithm. However PSO is motivated by simulation of social behaviour instead of survival of fittest and each particle is associated with a velocity. The particles fly through the search space, constantly adjusting their velocity according to corresponding particle's experience and the particle's neighbours' experience. Each particle X_i makes use of its individual memory and knowledge gained by the swarm as a whole to find the best solution. At each iteration, the fitness of each particle is evaluated by an appropriate fitness function and the algorithm progressively stores and replaces two best values, called *pbest* and *gbest*. $pbest_i, (i = 1, 2, \cdots, m)$ denotes the best position associoated with the best fitness value achieved so far for each individual and *gbest* denotes the position corresponding to global best value. Binary Particle Swarm optimization (BPSO) algorithm also proposed by Kennedy and Eberhart [16] is an extension of PSO to solve optimization problems with discrete valued parameters. Here each particle (candidate solution) represents a position in a binary multidimensional space, i,e components of X_i can take only binary values instead of continuous values. The velocity vector associated with each particle is real valued.

2.3   Gravitational Search Algorithm

Gravitational Search algorithm (GSA) is a nature inspired heuristic optimization algorithm based on the law of gravity and mass interactions. The algorithm is comprised of collection of agents which interact with each other through the gravity force. The agents are considered as objects and their performances are measured in terms of their masses. The gravity force causes a global movement where all objects move toward other objects with heavier masses. In GSA, the agent has four parameters which are position, inertial mass, active gravitational mass and passive gravitational mass. The position represent the solution of the problem. The gravitational and inertial masses are determined by fitness function. The algorithm is navigated by adjusting gravitational and inertial mass. Finally the position of the heaviest mass presents the optimum solution. The details are found in [11]. A binary version of GSA, known as BGSA is found in [20]

2.4   Cuckoo Search

Cuckoo search is an optimization algorithm belonging to the class of swarm intelligence (SI) based algorithms like particle swarm optimization. It is inspired by the obligate interspecific brood parasitism of some cuckoo species that lay their eggs in the nests of other host birds. In this behavior of reproduction, there are two possible cases for a cuckoo egg dumped into a host bird nest including: the host bird does not recognize the cuckoo egg and the cuckoo egg will hatch and carry over to the next generation or the host bird identifies the cuckoo egg and either throw it away or abandon its nest to build a new one. The two mentioned phenomena have been inspired in the CSA method for two phases of new solution generation including the exploration phase via Levy flights (the first phenomenon) and the exploitation phase via replacement of a fraction of eggs (the second phenomenon). The detail algorithm is presented in [10]. A binary version of the algorithm is presented in [21].

2.5   Firefly Algorithm

Firefly algorithm (FFA) is also another SI based optimization algorithm inspired by the flashing pattern of tropical fireflies. It is based on three rules: (1) the fireflies are unisex and one is attracted by other irrespective of sex (2) attractiveness is proportional to brightness, less brighter firefly moves to more brighter one, brightness decreases as their distance inreases (3) brightness is determined by landscape of the objective function. The objective function of a given optimization problem is based on differences of light intensity. The fireflies are characterized by light inensity which helps to change their position iteratively to more attracting position in order to obtain optimal solution. The details are in [12]. A binary version of the algorithm BFFA is proposed in [22].

2.6 Bat Algorithm

Bat algorithm (BA) is a newly proposed SI based metaheuristic optimization algorithm based on echolocation behaviour of bats. Microbats, small bats, use extensive echolocation. They use a type of sonar, to detect prey and to avoid obstacles and locate their resting crevices in the dark. These bats emit a very loud sound pulse and listen for the echo that bounces back from the surrounding objects. Bat algorithm is a modification of particle swarm optimization in which the position and the velocity of virtual microbats are updated based on frequency of their emitted pulses and loudness. The pseudocode of the algorithm and the details can be found in [13]. A binary version of bat algorithm BBA is proposed in [23].

3   Comparative Study of Different Algorithms

In this paper a comparative study of different EC algorithms for feature subset problem has been done. The solution space is considered as a binary multidimensional space and represented by a binary string or a binary vector, a point in binary multidimensional space. Binary versions of the algorithms BGA, BPSO, BGSA, BCS, BFFA and BBA are used in this study. The fitness function of the EC algorithm is the classification accuracy of a SVM classifier as a default wrapper fitness function which is defined as:

Fitness function S_1 = No. of test samples correctly classified (T_c)/Total no. of test samples (T).

3.1   Proposal of a New Fitness Function with Penalty

The objective of feature subset selection is two fold: to reduce the dimensionality to lower computational cost as well as to increase the classification accuracy to make the performance higher. But it seems that this two objectives are somewhat contradictory. Reduction of features leads to lower classification accuracy, so use of classification accuracy as the evaluation function of the optimization algorithm is not sufficient for obtaining optimally reduced feature set. So a new fitness function is proposed with the addition of a penalty term in S_1. The new fitness function S_2 is given by

$$S_2 = S_1 - \alpha \times \frac{D}{N} \qquad (1)$$

Where D and N represent the number of features in the selected feature subset and total number of features respectively, whereas α is a control parameter used to adjust the weight of the penalty term in the fitness function. The above fitness function is used in conjunction with various evolutionary algorithms to find the feature subset. The performance of the new fitness function S_2 is compared to the performance of the default fitness function S_1 in terms of the final reduction of feature set and classification accuracy. The performance of the new fitness function with various EC algorithms (single objective) is also compared with the performance of NSGA II, a popular multiobjective genetic algorithm [24] using classification accuracy and reduction of features in the feature set as two seperate objectives.

4 Simulation Experiments and Results

Simulation experiments are done with several benchmark data sets from UCI machine learning repository [25]. The EC algorithms used in the simulation experiments are BGA, BPSO, BGSA, BCS, BFFA abd BBA, The parameters of different algorithms are set by trial and error experiments so that maximum number of comparison in the search algorithms are 10,000. Table 1 represents the parameters used. For BGA, two point crossover and rank based selection is used. P_c and P_m represent probability of crossover and mutation respectively. For other algorithms, relevant parameter values (details are omitted here due to lack of space, can be found in the references) are noted in Table 1.

Table 1. Parameters of EC algorithms

	Population size	Maximum epoch	Parameter values
BGA	8	1250	$P_c = 0.1, P_m = 0.05$
BPSO	20	500	$c_1, c_2 = 1, w = 0.5$
BCS	20	500	$\alpha = 0.1, \beta = 1.5, p = 0.25$
BGSA	20	500	$R_p = E_c = 1, min - flag = 0$
BFFA	20	25	$\alpha = 0.25, \beta = 0.2, \gamma = 1$
BBA	20	500	$loudness = 0.25, r = 0.1$
NSGA	20	250	same as BGA

Wine, Cancer and Sonar data sets with number of features 13, 30 and 60, number of classes 3, 2 and 2 and number of data samples 178, 569 and 208 respectively are used for experiment with default fitness function and penalty based fitness function with control parameter $\alpha = 0.05$ to 0.4 for feature subset selection by different EC algorithms. Finally SVM is used for measuring classification accuracy with the final reduced subset. Different training -test ratio of samples are used for experiments. The evaluation of the fitness function is done by the final number of features in the reduced feature subset and classification accuracy with the final selected subset.

Tables 2, 3 and 4 represent the effeciveness of the proposed fitness function with penalty (for $\alpha = 0.15$) for wine, cancer and sonar data set respectively in terms of average classification accuracy and average number of features in the final selected feature subset. For all the data sets we found that control parameter $\alpha = 0.15$ produces the best result.

Table 2. Simulation results for Wine data

Algorithm	Default fitness		Fitness with penalty	
	Accuracy	Av. no. of features	Accuracy	Av. no. of features
BGA	0.95224	7.87	0.93374	3.84
BPSO	0.96009	8.72	0.93915	3.61
BCS	0.94514	7.89	0.9286	4.74
BFFA	0.94561	6.24	0.93444	3.01
BBA	0.95999	8.42	0.93813	3.95
BGSA	0.95421	8.43	0.92869	4.69

Table 3. Simulation results for Cancer data

Algorithm	Default fitness		Fitness with penalty	
	Accuracy	Av. no. of features	Accuracy	Av. no. of features
BGA	0.94933	14.77	0.94476	7.55
BPSO	0.96632	13.45	0.96059	4.16
BCS	0.94965	13.05	0.94004	8.82
BFFA	0.96061	8.37	0.95193	3.71
BBA	0.96488	13.27	0.95658	4.43
BGSA	0.95173	13.51	0.94412	8.64

Table 4. Simulation results for Sonar data

Algorithm	Default fitness		Fitness with penalty	
	Accuracy	Av. no. of features	Accuracy	Av. no. of features
BGA	0.85392	30.16	0.94476	24.65
BPSO	0.88224	30.51	0.96059	17.87
BCS	0.82344	30.48	0.94004	21.23
BFFA	0.83592	18.82	0.95193	10.56
BBA	0.88808	29.88	0.95658	16.11
BGSA	0.84312	31.62	0.94412	22.0

In all the cases, new fitness function produce final feature subset with lesser number of features without much degradation in classification accuracy. Also it is found that BFFA produced the highest reduction in feature set. For comparison with multiobjective algorithm, we used NSGA II with two objective functions classification accuracy and number of features in the selected feature subset. The classification accuracy and number of features in the final selected subset for wine, cancer and sonar data set came out to be 0.95, 0.95, 0.85 and 8.07, 14.65 and 29.5 respectively. It seems that our proposed penalty based single objective fitness function is better in efficiency than NSGAII in terms of reducing feature number in the final feature subset without having much degradation in classification accuracy.

We repeated the experiment with three other high dimensional datasets, Hill, Gas and Madelon from UCI repository with number of features 100, 128 and 500 respectively. Tables 5, 6 and 7 represent the simulation results. Here we used only three EC algorithms GA, BFFA and BBA as those are found to be the effective algorithms for feature subset selection.

Table 5. Simulation results for Hill data

Algorithm	Default fitness		Fitness with penalty	
	Accuracy	Av. no. of features	Accuracy	Av. no. of features
GA	0.95987	54.18	0.95234	45.7
BFFA	0.93545	47.7	0.92666	34.66
BBA	0.97677	55.36	0.97059	39.5

Table 6. Simulation results for Gas data

Algorithm	Default fitness		Fitness with penalty	
	Accuracy	Av. no. of features	Accuracy	Av. no. of features
GA	0.97148	68.7	0.96869	58.16
BFFA	0.98916	18.8	0.98697	12.4
BBA	0.98243	68.06	0.97894	44.26

Table 7. Simulation results for Madelon data

Algorithm	Default fitness		Fitness with penalty	
	Accuracy	Av. no. of features	Accuracy	Av. no. of features
GA	0.65150	250.76	0.59433	246.3
BFFA	0.62129	125.32	0.59141	73.31
BBA	0.67824	249.72	0.62487	242.38

For high dimensional data also, it is found that the penalty based fitness function works better than the default fitness function in reducing the number of features in optimal feature subset, though the classification accuracy falls drastically for too high dimensional data after reducing the dimension to 50% (Madelon data set). Here also we found that BFFA is the most effective EC algorithm. NSGA II is also used for the high dimensional data sets and it is found that the result is similar to single objective GA and worse than using penalty based fitness function.

5   Conclusion

Optimal feature subset selection is an extremely important preprocessing step for any pattern recognition or machine learning problem. The successful elimination of redundant and irrelevant information increases the performance of the classifier while retaining important and informative feature is highly needed for improved performance. For high dimensional data, the judicious selection of feature from available features becomes more important as reduction of feature having relevance to the class reduces classification accuracy while retaining all the features heavily increases the computational cost. So feature evaluation function should be carefully designed. In search algorithm based optimal feature subset selection, feature evaluation function is used as the fitness function of the search algorithm. For wrapper method, classification accuracy itself is generally used as the default fitness function of the search algorithm.

In this work various evolutionary algorithm has been used for searching the optimal feature subset from a set of features with classification accuracy as the default fitness function and a newly proposed fitness function in which a penalty term for high number of features in the selected subset is added to classification accuracy. The new penalty function seems to be very effective in reducing the number of features in the selected subset without much degradation in classification accuracy. The proposed fitness function also seems to be effective compared to multiobjective genetic algorithm. Among the EC algorithms, BFFA produced the best result in terms of reduction of number of features. Simulation experiments with high dimensional data also show that the algoriths are quite effective for data sets of dimension 100 to 200. Further experiments are now carried on for modifications of the fitness function for filter type to be used with high dimensional data having dimension in the range of 1000.

References

1. Devijver, P.A., Kittler, J.: Pattern Recognition: A Statistical Approach. Prentice Hall International, London (1982)
2. Liu, H., Motoda, H.: Feature Selection for Knowledge Discovery and Data Mining. Springer, Heidelberg (1998)
3. Duda, R.O., Hart, P.E., Stork, D.G.: Pattern Classification. Wiley-Interscience, New York (2001)
4. Verikas, A., Bacauskiene, M.: Feature selection with neural networks. Pattern Recogn. Lett. **23**(11), 1323–1335 (2002)
5. Oh, I., Lee, J., Moon, B.: Hybrid genetic akgorithm for feature selection. IEEE Trans. PAMI **26**(11), 1424–1437 (2004)
6. Chakraborty, B.: Genetic algorithm with fuzzy operators for feature subset selection. IEICE Trans. Fundam. Electron. Commun. Comput. Sci. **E85–A**(9), 2089–2092 (2002)
7. Goldberg, D.E.: Genetic Algorithms in Search, Optimization and Machine Learning. Addison-Wesley, Boston (1989)
8. Kennedy, J., Eberhart, R.C.: Particle swarm optimization. In: Proceedings of IEEE International Conference On Neural Network, vol. 4, pp. 1942–1948 (1995)

9. Dorigo, M., Stuzle, T.: Ant Colony Optimization. The MIT Press, Cambridge (2004)
10. Yang, X.S., Deb, S.: Cuckoo search via levy flights. In: Proceedings of NaBIC 2009, pp. 210–214 (2009)
11. Reshedi, E., Nezamabadi-pour, H., Saryazdi, S.: GSA: a gravitational search algorithm. Inf. Sci. **179**(13), 2232–2248 (2009)
12. Yang, X.S.: Nature Inspired Metaheuristic Algorithms. Luniver Press, Bristol (2008)
13. Yang, X.S.: A new metaheuristic bat-inspired algorithm. In: Gonzalez, J.R. (ed.) NISCO 2010. SCI, vol. 284, pp. 65–74. Springer, Heidelberg (2010)
14. Chakraborty, B.: Evolutionary computational approaches to feature subset selection. Int. J. Soft Comput. Bioinform. **1**(2), 59–65 (2010)
15. Chakraborty, B.: Genetic algorithm with fuzzy fitness function. In: Proceedings of IEEE International Symposium on Industrial Electronics (IEEE-ISIE 2002), pp. 315–319 (2002)
16. Kennedy, J., Eberhart, R.C.: A discrete binary version of the particle swarm algorithm. In: Proceedings of IEEE International Conference on Computational Cybernetics and Simulation, vol. 5, pp. 4104–4108 (1997)
17. Chakraborty, B.: Feature subset selection by particle swarm optimization with fuzzy fitness function. In: Proceedings of ISKE 2008, 3rd International Conference on Intelligent Systems and Knowledge Engineering, pp. 1038–1042 (2008)
18. Chakraborty, B.: Binary particle swarm optimization based algorithm for feature subset selection. In: Proceedings of ICAPR 2009, pp. 145–148 (2009)
19. Chakraborty, B.: Feature subset selection using hybrid particle swarm optimization. ICIC Express Lett. **6**(5), 1161–1168 (2012)
20. Reshedi, E., Nezamabadi-pour, H., Saryazdi, S.: BGSA: binary gravitational search algorithm. Nat. Comput. **9**(3), 727–745 (2009)
21. Rodrigues, D., Pereira, L.A.M., Souza, N., Ramos, C.C.O., Yang, X.S.: BCS: a Binary Cuckoo Search algorithm for feature selection. In: 2013 IEEE International Symposium on Circuits and Systems, pp. 465–468 (2013)
22. Crawford, B., et al.: A binary coded firefly algorithm that solves the set covering problem. Rom. J. Inform. Sci. Technol. **17**(3), 252–264 (2014)
23. Nakamura, Y.M., Pereira, L.A.M., Costa, K.A., Rodrigues, D., Papa, J.P., Yang, X.S.: Binary bat algorithm. Neural Comput. Appl. **25**(3), 663–668 (2014)
24. Deb, K., Pratap, A., Agarwal, S., Meyarivan, T.: A fast and elitist multiobjective genetic algorithm: NSGA-II. IEEE Trans. Evolu. Comput. **2**(3), 182–197 (2002)
25. UCI Machine learning Repository. http://archive.ics.uci.edu/ml/

A Context-Aware Fitness Function Based on Feature Selection for Evolutionary Learning of Characteristic Graph Patterns

Fumiya Tokuhara[1(✉)], Tetsuhiro Miyahara[1], Tetsuji Kuboyama[2],
Yusuke Suzuki[1], and Tomoyuki Uchida[1]

[1] Graduate School of Information Sciences,
Hiroshima City University, Hiroshima 731-3194, Japan
mb67015@e.hiroshima-cu.ac.jp,
{miyares17,y-suzuki,uchida}@info.hiroshima-cu.ac.jp
[2] Computer Centre, Gakushuin University, Tokyo 171-8588, Japan
ori-icxs17@tk.cc.gakushuin.ac.jp

Abstract. We propose a context-aware fitness function based on feature selection for evolutionary learning of characteristic graph patterns. The proposed fitness function estimates the fitness of a set of correlated individuals rather than the sum of fitness of the individuals, and specifies the fitness of an individual as its contribution degree in the context of the set. We apply the proposed fitness function to our evolutionary learning, based on Genetic Programming, for obtaining characteristic graph patterns from positive and negative graph data. We report some experimental results on our evolutionary learning of characteristic graph patterns, using the context-aware fitness function and a previous fitness function ignoring context.

1 Introduction

Evolutionary learning methods such as Genetic Algorithm (GA) and Genetic Programming (GP) [4,12] are widely used as probabilistic methods for solving computationally hard learning problems. The fitness, i.e. the evaluation value, of individuals plays a central role in controlling the process of evolutionary learning. Usual fitness of individuals depends on the performance of individuals only.

In this paper, we consider a new fitness function aware of context in the sense that the fitness of an individual depends on not the individual only but a set of individuals relevant to it. We propose a context-aware fitness function based on a feature selection method [16] for evolutionary learning of characteristic graph patterns. The proposed fitness function estimates the fitness of a set of correlated individuals rather than the sum of fitness of the individuals, and defines the fitness of an individual as its contribution degree in the context of the set of correlated individuals.

The problem we address is to find a general pattern that covers positive examples as much as possible and does not cover negative examples as much

© Springer International Publishing AG 2017
N.T. Nguyen et al. (Eds.): ACIIDS 2017, Part I, LNAI 10191, pp. 748–757, 2017.
DOI: 10.1007/978-3-319-54472-4_70

as possible. Therefore, in the course of evolutionary computation, it is necessary to diversify patterns to cover all the positive examples as much as possible. To ensure this diversity with a small number of individuals, we introduce a consistency-based feature selection algorithm, super-CWC [16]. This algorithm extracts a minimal set of features by which positive examples are consistently discriminated from negative examples as much as possible.

As the amount of graph structured data has increased, graph classification [13], whose main task is to classify graphs with class labels in a graph database into two or more classes, has received wide attention. Then we apply the context-aware fitness function to our evolutionary learning, based on Genetic Programming, for obtaining characteristic graph patterns which classify given positive and negative graph data. Outerplanar graphs are known to be suited to model many chemical compounds [1]. Block preserving outerplanar graph patterns (bpo-graph patterns, for short) [14,19] are graph structured patterns with structured variables and can represent characteristic graph structures of outerplanar graphs.

In our previous work [11,18], we have proposed evolutionary learning methods, which are based on Genetic Programming, using a fitness function ignoring context, for acquiring characteristic graph structures, represented as bpo-graph patterns, from positive and negative graph structured data, represented as outerplanar graphs.

We discuss related work. Using context awareness in Genetic Programming is considered in different setting, in which context-aware crossover operator is proposed [5]. Genetic Network Programming (GNP) [3] and Graph Structured Program Evolution (GRAPE) [17] are proposed as evolutionary learning methods from graph structured data. We proposed GP-based learning of characteristic tree or graph patterns other than bpo-graph patterns from both positive and negative tree or graph structured data [7–10]. Mining frequent subgraphs in outerplanar graphs [1], finding minimally generalized bpo-graph patterns and enumerating frequent bpo-graph patterns from positive outerplanar graphs [14,19] are known. These approaches [1,14,19] are different from our evolutionary learning method in that our method learns from positive and negative graph structured data.

This paper is organized as follows. In Sect. 2, we summarize our evolutionary learning framework for acquiring characteristic graph patterns. In Sect. 3, we propose a GP-based learning method for acquiring characteristic bpo-graph patterns from positive and negative outerplanar graph data, by incorporating a context-aware fitness function which is based on the feature selection method [16]. In Sect. 4, we report some experimental results on our evolutionary learning of characteristic graph patterns, using the context-aware fitness function and a previous fitness function ignoring context. Finally, in Sect. 5, we present the conclusion and future work.

2   Preliminaries

2.1   Block Preserving Outerplanar Graph Patterns and Their Tree Representations

In this subsection, we overview block preserving outerplanar graph patterns, which are graph representations of characteristic features of graph structured data, and block tree patterns, which are tree representations of block preserving outerplanar graph patterns, according to [11,14,18,19].

A cut-vertex of a connected graph is a vertex whose removal makes the graph disconnected. A bridge of a connected graph is an edge whose removal makes the graph disconnected. We remark that both endpoints of a bridge are cut-vertices. A block of a connected graph is a maximal connected subgraph which has at least 3 vertices but has no cut-vertex. An outerplanar graph is a planar graph which can be embedded in the plane such that all vertices are on the outer boundary. By an outerplanar graph we mean a connected outerplanar graph with vertex labels and edge labels.

We consider graph patterns with *structured variables* (*variables*, for short) in order to represent characteristic graph structures. There are two types of variables, (1) a *terminal variable* which is a list of just one vertex, and (2) a *bridge variable* which is a list of two vertices. A *block preserving outerplanar graph pattern* (*bpo-graph pattern*) is a graph pattern which is obtained from an outerplanar graph by replacing some bridges with bridge variables and adding terminal variables to vertices.

We can replace a variable in a bpo-graph pattern with an arbitrary outerplanar graph. Each variable is labeled with a variable name. All variables in a bpo-graph pattern is assumed to have unique labels. In a bpo-graph pattern, let us consider a terminal variable (v_1), or a bridge variable (v_1, v_2). Let (u_1) or (u_1, u_2) be a list of vertices in an outerplanar graph g such that v_i and u_i have the same label for any $i \in \{1, 2\}$. We can replace the variable (v_1) or (v_1, v_2) with the outerplanar graph g by identifying the vertices (v_1) or (v_1, v_2) with the vertices (u_1) or (u_1, u_2), respectively. A bpo-graph pattern p is said to *match* an outerplanar graph G if G is obtained from p by replacing all the variables with certain outerplanar graphs. We use, block tree patterns $t(p)$ [14,19], which are tree representations of bpo-graph patterns p and have the structure of unrooted and unordered trees, as individuals in Genetic Programming, which is an evolutionary learning method dealing with tree structures.

As examples we give a bpo-graph pattern p and the block tree patterns $t(p)$ of p, and outerplanar graphs g_1, g_2, g_3 and G in Fig. 1. The bpo-graph pattern p has 2 bridge variables, (v_1, v_2) labeled with X and (v_5, v_7) labeled with Y, and a terminal variable, (v_4) labeled with Z. We obtain the outerplanar graph G from the bpo-graph pattern p by replacing the variables (v_1, v_2), (v_5, v_7) and (v_4) with outerplanar graphs g_1, g_2 and g_3, respectively, as follows. First, we replace the variable (v_1, v_2) with the outerplanar graph g_1 by identifying the vertices v_1 and v_2 in the variable labeled with X in p with the vertices u_1 and u_2 in g_1. Second, we replace the variable (v_2, v_3) with the outerplanar graph g_2 by identifying the

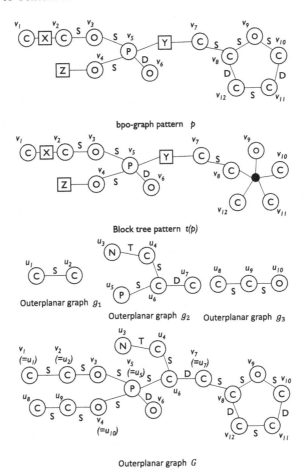

Fig. 1. A bpo-graph pattern p, the block tree pattern $t(p)$ of p. Outerplanar graphs g_1, g_2, g_3 and G. A box with lines to its elements represents a variable. A variable label is inside a box. The bpo-graph pattern p matches the outerplanar graph G.

vertices v_5 and v_7 in the variable labeled with Y in p with the vertices u_5 and u_7 in g_2. Third, we replace the variable (v_4) with the outerplanar graph g_3 by identifying the vertex v_4 in the variable labeled with Z in p with the vertices u_{10} in g_3. Thus we see that p matches G.

2.2 Obtaining Outerplanar Graph Patterns Using Genetic Programming

In this subsection, we explain a genetic programming-based learning method [18] for obtaining characteristic bpo-graph patterns from positive and negative outerplanar graph data. The problem we address is the following [11].

Problem of Acquisition of Characteristic Block Preserving Outerplanar Graph Patterns:

Input: A finite set D of positive and negative outerplanar graph data.

Problem: Find a bpo-graph pattern having high fitness with respect to D.

A bpo-graph pattern as an individual is a binary classifier of outerplanar graph data. The *fitness* of a bpo-graph pattern p with respect to D, denoted by $fitness_D(p)$, is defined to be a kind of accuracy of classifying positive and negative graph data with respect to D. We therefore consider a bpo-graph pattern having high fitness a characteristic bpo-graph pattern, which matches many positive and few negative data represented as outerplanar graphs. We give a context-aware fitness function, which is defined by an individual ranking method based on feature selection in Sect. 3, for our evolutionary learning method.

In applying a GP operator to a block tree pattern, we set a root in the block tree pattern in order to specify the affected portion of the block tree pattern by the GP operator. For example, applying crossover operator to block tree patterns is illustrated in Fig. 2. Our learning method [18] generates block tree patterns such that the corresponding bpo-graph patterns satisfy valence relations, by using label information of connecting vertices and edges from positive examples of outerplanar graphs.

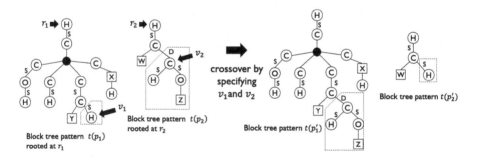

Fig. 2. A genetic operator **crossover** applied to block tree patterns $t(p_1)$ and $t(p_2)$.

3   A New Fitness Function Based on Feature Selection

In this section, we give a GP-based learning method for acquiring characteristic bpo-graph patterns. In this method, we propose a new context-aware fitness function CWCFITNESS. The fitness function CWCFITNESS employs a state-of-the art feature selection algorithm, super-CWC [16] (CWC, for short). While conventional fitness functions evaluate individuals in population, CWCFITNESS evaluates a set of individuals. This aspect is very important to find a general pattern to cover the whole set of positive examples because even if each selected pattern has relatively high coverage of positive examples, each coverage does not necessarily imply the high coverage of the whole positive examples by the selected individuals.

3.1   Consistency-Based Feature Selection CWC

Supervised feature selection is the problem of finding a small set of features for discriminating class labels. CWC is a very fast and accurate consistency-based feature selection algorithm of filter approach proposed by Shin et $al.$ [16]. For example, consider the feature selection of the data consisting of four instances $\{d_1, \ldots, d_4\}$ shown in Table 1. Each feature is denoted by $F_i \in \{F_1, \ldots, F_4\}$, and the variable of class labels is denoted by C. In a conventional way,

Table 1. An example data

ID	F_1	F_2	F_3	F_4	C
d_1	1	0	0	1	+
d_2	1	0	1	0	+
d_3	1	0	0	0	−
d_4	0	1	1	1	−

features are sorted by a correlation measure, e.g. mutual information, between the values of a single feature and class labels. The mutual information between F_i and C is denoted by $I(F_i; C)$. Now we have $I(F_3; C) = 0.0$ and $I(F_4; C) = 0.0$. Hence, F_3 and F_4 are useless to determine the class label C at all if F_3 or F_4 is used alone. However, the set of features $\{F_3, F_4\}$ can uniquely determine class labels since $I(\{F_3, F_4\}; C) = 1.0$. On the other hand, even if F_2 is selected in addition to F_1, it does not add any information to determine class labels since $I(F_1; C) = I(F_2; C) = I(\{F_1, F_2\}; C) \approx 0.3$. This example suggests that not only a single feature but a group of features should be also considered to evaluate the effect of selected features relevant to class labels. CWC is a consistency-based feature selection method. Consistency is the property of a feature set by which all the class labels are uniquely determined. CWC explores a minimal consistent subset of given features based on a backward-deletion strategy with very high accuracy.

3.2   Fitness Function Based on CWC-Ranking

Let p be a bpo-graph pattern and D a finite set of positive and negative graph structured data with respect to a specific phenomenon. If a bpo-graph pattern p matches an example graph G, then we classify G as positive according to p. Otherwise we classify G as negative according to p. Let TP denote the number of true positive examples in D according to p, TN the number of true negative examples, FP the number of false positive examples and FN the number of false negative examples. The balanced accuracy of p with respect to D, denoted by $balanced\_accuracy(p)$, is a basic measure used in our previous work [7,18] and defined as follows.

$$balanced\_accuracy(p) = \frac{1}{2} \times \left(\frac{TP}{TP + FN} + \frac{TN}{TN + FP} \right).$$

Furthermore, we use Matthews correlation coefficient (MCC), which is an established measure in biology and is suited to data of imbalanced numbers of positive and negative data. The original definition of MCC of p with respect to D, denoted by $MCC(p)$, is as follows [6].

$$MCC(p) = \frac{TP \times TN - FP \times FN}{\sqrt{(TP + FP)(TP + FN)(TN + FP)(TN + FN)}}.$$

If the denominator of $\mathrm{MCC}(p)$ is zero, we set $\mathrm{MCC}(p) = 0$. Examples of such cases where the denominator of $\mathrm{MCC}(p)$ is zero would be when a bpo-graph pattern p matches all data or matches no data. On such cases, the correlation $\mathrm{MCC}(p)$ should be set to zero [15]. Then we define $\mathrm{MCC}^*(p)$ according to [7] as follows. $\mathrm{MCC}^*(p) = (\mathrm{MCC}(p) + 1)/2$.

We define a context-aware fitness function by using ranked patterns obtained from a CWC output. The input of CWC is a dataset described by a set of features $\{F_1, \ldots, F_N\}$ with the class C. The output of CWC is a minimal subset $S \subseteq \{F_1, \ldots, F_N\}$ such that S has the binary consistency, i.e. S uniquely determines class labels, where each feature $F_i \in S$ is assigned evaluation value. We call such the set S of features *the selected minimal set* by CWC.

In this paper, we consider a bpo-graph pattern p as a feature, a set of bpo-graph patterns in a generation as a set of features, and a positive or negative of an example as a class label $c = 1$ or 0, respectively. If a pattern p matches an example, then the value of feature p is set as 1, otherwise 0. As the correlation measure between two patterns p and q, we use the *phi coefficient* $\phi(p, q)$. Also we introduce two parameters ℓ and u which determine the scope of bpo-graph patterns affected by the selected minimal set by CWC. Let (ℓ, u) be $(*, *)$ or a pair of real numbers ℓ and u with $0 \leq \ell \leq u \leq 1$. We define the predicate $\mathrm{CWCRANK}(p, (\ell, u))$ for a bpo-graph pattern p and parameters ℓ and u as follows. Let S be the selected minimal set by CWC. The predicate $\mathrm{CWCRANK}(p, (*, *))$ is true if p is in S, and false otherwise. The predicate $\mathrm{CWCRANK}(p, (\ell, u))$ is true if (1) p is in S, or (2) p is not in S, and there exists a pattern q in S such that $\ell \leq \phi(p, q) \leq u$ holds. Otherwise, the predicate $\mathrm{CWCRANK}(p, (\ell, u))$ is false.

The *CWC-fitness* of a bpo-graph pattern p and parameters ℓ and u with respect to D, denoted by $\mathrm{CWCFITNESS}(p, (\ell, u))$, is defined as follows.

$$\mathrm{CWCFITNESS}(p, (\ell, u)) = \begin{cases} \mathrm{MCC}^*(p) & \text{if } \mathrm{CWCRANK}(p, (\ell, u)) \text{ is true} \\ balanced\_accuracy(p) & \text{otherwise} \end{cases}$$

The value of the above defined two fitness functions $balanced\_accuracy(p)$ and $\mathrm{CWCFITNESS}(p, (\ell, u))$ of a bpo-graph pattern p is between 0 and 1, and a bpo-graph pattern having high fitness is a good classifier of graph structured data.

4   Experimental Results

We implemented our GP-based learning method for acquiring characteristic bpo-graph patterns from positive and negative outerplanar graph data, by using a CWC-ranking method in Sect. 3. Let p be a bpo-graph pattern and D a finite set of positive and negative graph structured data with respect to a specific phenomenon. We report some experimental results on our evolutionary learning of characteristic graph patterns, using the context-aware fitness function $\mathrm{CWCFITNESS}(p, (\ell, u))$ with parameters ℓ and u which is proposed in Sect. 3 and a previous fitness function $balanced\_accuracy(p)$ ignoring context which is used in [18].

Our method records canonical representations of block tree patterns and their fitness, using canonical representations as keys and fitness values as values in a HashMap table [18]. By using the matching algorithm [14,19] for bpo-graph patterns and outerplanar graphs, we calculate the fitness of bpo-graph patterns represented by block tree patterns as individuals. We implemented our method in Java and Scala on Windows 10 (64-bit), and compared it with our previous method. GP parameters we use in the experiments are as follows: population size: 50, reproduction probability: 0.05, crossover probability: 0.50, mutation probability: 0.45, selection method: roulette wheel selection, tournament selection (size 2), elite selection (size 3), maximum number of generations: 200. As the experimental data we use 88 positive data and 88 negative data which have the structure of outerplanar graph. The experimental data are extracted from the file "CAD2DA99.sdz" in the NCI database [2].

For each of the proposed method using $\mathrm{CWCFITNESS}(p, (\ell, u))$ as fitness and the previous method using $balanced\_accuracy(p)$ as fitness, we performed 10 GP runs for acquiring characteristic block preserving outerplanar graph patterns from positive and negative outerplanar graph data, by using the same GP parameters and experimental data above mentioned. For the proposed method we set two parameters ℓ and u as $(\ell, u) = (*, *), (1.0, 1.0), (0.8, 1.0), (0.8, 0.9)$.

The best individuals by the proposed method mean the best individuals over all generations, and the best individuals by the previous method mean the best individuals at the final generation. Some of the best individuals by the proposed method are isomorphic to best individuals by the previous method. The average values means the average valued over 10 GP runs.

Table 2 shows the average fitness of best individuals of the proposed method and the previous method. Table 2 shows the average number of vertices and the average specificness of individuals of final populations for the both methods. The specificness of a bpo-graph pattern p is defined as $EN/(EN + VN)$, where EN and VN denote the number of edges and the number of variables of the corresponding block tree pattern $t(p)$ of p. A bpo-graph pattern which has many vertices or high specificness is considered to be a specific graph pattern.

Table 2. The average fitness of best individuals, the average number of vertices and the average specificness of individuals of final populations for the proposed and previous methods over 10 GP runs

	Proposed method				Previous method
Parameters (ℓ, u)	$(*, *)$	$(1.0, 1.0)$	$(0.8, 1.0)$	$(0.8, 0.9)$	
Average fitness of best individuals	0.781	0.777	0.777	0.781	0.756
Average number of vertices	4.95	4.25	5.14	4.60	3.99
Average specificness	0.140	0.188	0.218	0.157	0.124
Average run-time (sec)	409	373	453	416	276

The both average values of the proposed method with any parameters (ℓ, u) are higher than those of the previous method. Table 2 shows average run-time for the both methods. The proposed method, however, required much run-time than the previous method.

Tables 3 and 4 show the fitness used in the proposed method with $(\ell, u) = (0.8, 0.9)$ of best individuals obtained by the previous method, and vice versa. The fitness of the proposed method is slightly higher than that of the previous method. From these experimental results, we can say that the proposed method obtained characteristic bpo-graph patterns which have slightly higher fitness and are more specific than the obtained bpo-graph patterns of the previous method.

Table 3. The fitness of best individuals of the previous method and their fitness used in the proposed method with $(\ell, u) = (0.8, 0.9)$ over 10 GP runs

	1	2	3	4	5	6	7	8	9	10	Avg.
Previous method	0.756	0.756	0.756	0.756	0.756	0.756	0.756	0.756	0.756	0.756	0.756
Proposed method	0.756	0.767	0.756	0.756	0.756	0.767	0.767	0.756	0.767	0.756	0.760

Table 4. The fitness of best individuals of the proposed method with $(\ell, u) = (0.8, 0.9)$ and their fitness used in the previous method over 10 GP runs

	1	2	3	4	5	6	7	8	9	10	Avg.
Proposed method	0.782	0.782	0.781	0.781	0.782	0.782	0.782	0.782	0.781	0.781	0.781
Previous method	0.716	0.756	0.756	0.750	0.750	0.756	0.756	0.756	0.756	0.750	0.750

5   Conclusion

In this paper we have proposed a context-aware fitness function based on feature selection for evolutionary learning of characteristic block preserving outerplanar graph patterns from positive and negative outerplanar graph data. We have reported some experimental results on our evolutionary learning of characteristic graph patterns, using the context-aware fitness function and a previous fitness function ignoring context. The proposed method obtained characteristic bpo-graph patterns which have slightly higher fitness and are more specific than the obtained bpo-graph patterns of the previous method. We plan to apply our method to a large data set of chemical compounds or other types of graph data.

Acknowledgments. We would like to thank the anonymous referees for their helpful comments. This work was partially supported by JSPS KAKENHI Grant Numbers JP15K00312 and JP26280090.

References

1. Horvath, T., Ramon, J., Wrobel, S.: Frequent subgraph mining in outerplanar graphs. Data Min. Knowl. Disc. **21**, 472–508 (2010)
2. National Cancer Institute. The NCI Open Database. Release 1 Files (1999)
3. Katagiri, H., Hirasawa, K., Hu, J.: Genetic network programming - application to intelligent agents. In: Proceedings of IEEE SMC 2000, pp. 3829–3834 (2000)
4. Koza, J.R.: Genetic Programming: On the Programming of Computers by Means of Natural Selection. MIT Press, Cambridge (1992)
5. Majeed, H., Ryan, C.: Using context-aware crossover to improve the performance of GP. In: Proceedings of GECCO 2006, pp. 847–854 (2006)
6. Matthews, B.W.: Comparison of the predicted and observed secondary structure of T4 phage lysozyme. Biochim. Biophys. Acta **405**, 442–451 (1975)
7. Miyahara, T., Kuboyama, T.: Learning of glycan motifs using genetic programming and various fitness functions. J. Adv. Comput. Intell. Intell. Inf. (JACIII) **18**(3), 401–408 (2014)
8. Nagai, S., Miyahara, T., Suzuki, Y., Uchida, T.: Acquisition of characteristic ttsp graph patterns by genetic programming. In: Proceedings of IIAI AAI 2012, pp. 340–344 (2012)
9. Nagamine, M., Miyahara, T., Kuboyama, T., Ueda, H., Takahashi, K.: A genetic programming approach to extraction of glycan motifs using tree structured patterns. In: Orgun, M.A., Thornton, J. (eds.) AI 2007. LNCS (LNAI), vol. 4830, pp. 150–159. Springer, Heidelberg (2007). doi:10.1007/978-3-540-76928-6_17
10. Nakai, S., Miyahara, T., Kuboyama, T., Uchida, T., Suzuki, Y.: Acquisition of characteristic tree patterns with vldc's by genetic programming and edit distance. In: Proceedings of IIAI AAI 2013, pp. 147–151 (2013)
11. Ouchiyama, Y., Miyahara, T., Suzuki, Y., Uchida, T., Kuboyama, T., Tokuhara, F.: Acquisition of characteristic block preserving outerplanar graph patterns from positive and negative data using genetic programming and tree representation of graph patterns. In: Proceedings of IEEE IWCIA 2015, pp. 95–101 (2015)
12. Poli, R., Langdon, W., McPhee, N.: A Field Guide to Genetic Programming. Lulu Press, Raleigh (2008)
13. Rehman, S.U., Khan, A.U., Fong, S.: Graph mining: a survey of graph mining techniques. In: Proceedings of ICDIM 2012, pp. 88–92 (2012)
14. Sasaki, Y., Yamasaki, H., Shoudai, T., Uchida, T.: Mining of frequent block preserving outerplanar graph structured patterns. In: Blockeel, H., Ramon, J., Shavlik, J., Tadepalli, P. (eds.) ILP 2007. LNCS (LNAI), vol. 4894, pp. 239–253. Springer, Heidelberg (2008). doi:10.1007/978-3-540-78469-2_24
15. Seehuus, R., Tveit, A., Edsberg, O.: Discovering biological motifs with genetic programming. In: Proceedings of GECCO 2005, pp. 401–408 (2005)
16. Shin, K., Kuboyama, T., Hashimoto, T., Shepard, D.: Super-cwc and super-lcc: super fast feature selection. In: Proceedings of IEEE Big Data 2015, pp. 61–67 (2015)
17. Shirakawa, S., Ogino, S., Nagao, T.: Graph structured program evolution. In: Proceedings of GECCO 2007, pp. 1686–1693 (2007)
18. Tokuhara, F., Miyahara, T., Suzuki, Y., Uchida, T., Kuboyama, T.: Acquisition of characteristic block preserving outerplanar graph patterns by genetic programming using label information. In: Proceedings of IIAI AAI 2016, pp. 203–210 (2016)
19. Yamasaki, H., Sasaki, Y., Shoudai, T., Uchida, T., Suzuki, Y.: Learning block-preserving graph patterns and its application to data mining. Mach. Learn. **76**, 137–173 (2009)

Multiple Model Approach to Machine Learning

Authenticating ANN-NAR and ANN-NARMA Models Utilizing Bootstrap Techniques

Nor Azura Md. Ghani[1(✉)], Saadi bin Ahmad Kamaruddin[2],
Norazan Mohamed Ramli[1], and Ali Selamat[3,4]

[1] Faculty of Computer and Mathematical Sciences, Center for Statistical Studies and Decision
Sciences, Universiti Teknologi MARA, Selangor Darul Ehsan, Malaysia
{azura,norazan}@tmsk.uitm.edu.my
[2] Department of Computational and Theoretical Sciences, Kulliyyah of Science,
International Islamic University Malaysia, Pahang Darul Makmur, Malaysia
saadiak@iium.edu.my
[3] Faculty of Computing, UTM-IRDA Digital Media Centre and Software Engineering
Department, Universiti Teknologi Malaysia, 81310 Skudai, Johor Darul Takzim, Malaysia
aselamat@utm.my
[4] FIM, Center for Basic and Applied Research, University of Hradec Kralove,
Rokitanskeho, 62, 500 03 Hradec Kralove, Czech Republic

Abstract. Neural system procedures have a colossal reputation in the space of
gauging. In any case, there is yet to be a sure strategy that can well accept the last
model of the neural system time arrangement demonstrating. Thus, this paper
propose a way to deal with accepting the said displaying utilizing time arrangement square bootstrap. This straightforward technique is different compared to
the traditional piece bootstrap of time-arrangement based, where it was composed
by making utilization of every information set in the information apportioning
procedure of neural system demonstrating; preparing set, testing set and approval
set. At this point, every information set was separated into two little squares,
called the odd and even pieces (non-covering pieces). At that point, from every
piece, an arbitrary inspecting with substitution in a rising structure was made, and
these duplicated tests can be named as odd-even square bootstrap tests. In time,
the examples were executed in the neural system preparing for last voted expectation yield. The proposed strategy was forced on both manufactured neural system
time arrangement models, which were nonlinear autoregressive (NAR) and
nonlinear autoregressive moving normal (NARMA). In this study, three changing
genuine modern month to month information of Malaysian development materials value records from January 1980 to December 2012 were utilized. It was
found that the suggested bootstrapped neural system time arrangement models
beat the first neural system time arrangement models.

Keywords: Non-overlapping block bootstrap · Time series prediction · Neural
networks · Malaysian construction material price indices

© Springer International Publishing AG 2017
N.T. Nguyen et al. (Eds.): ACIIDS 2017, Part I, LNAI 10191, pp. 761–771, 2017.
DOI: 10.1007/978-3-319-54472-4_71

1   Introduction

Nonlinear modelling has had its speak to numerous analysts around the world, and a standout amongst the most famous displaying is the manufactured neural systems. One of the vital strategies in nonlinear displaying is contrasting the models with achieving the best one for further usage. Therefore, a model selection based on good approximation of their generalization errors among all models developed [1] is essential.

In the zone of nonlinear displaying, time arrangement expectation is a conspicuous movement among numerous analysts for a long time now, particularly in the regions of computational knowledge and the budgetary fields [6]. This is because of the inborn trait of the time arrangement information themselves, in particular deficient specimen size, incredible clamor, awesome number of anomalies, missing qualities, non-stationary and non-linearity [7]. In this manner, in this paper our proposition is found in a bootstrap strategy in view of the neural system approach, in particular odd-even time arrangement piece bootstrap and we look to contrast the execution and the current standard bootstrap and the moving square time arrangement bootstrap by [6]. In Malaysia, the Tenth Malaysia Plan (RMK-10) earns the trust of making more grounded the private budgetary activities (PFI) ventures and in addition development material costs as they introduce what's to come. The material cost increment is not another issue to every single financial part. Effective undertaking administration furthermore all around evaluated development material costs may diminish the likelihood of the material value change, and in the meantime, empower the development venture to be liable to appropriate execution.

Next, in Sect. 2, the related literatures will be discussed, and the data background of this research will be presented in Sect. 3. In Sect. 4, the technique review will be highlighted. The results and discussions will be presented in Sect. 5. Finally, in Sect. 6 will include the conclusion and a recommendation for future studies.

2   Related Literature

There are a few effectively settled and broadly utilized techniques as a part of the model choice errand. They are [1];

(i)   The wait (HO)
(ii)   The Monte-Carlo cross-validation
(iii) The k-fold cross-validation
(iv)   Leave-one-out (LOO)
(v)   The bootstrap

In HO, information is taken out from the learning set to an approval set. In cross-acceptance, a few approval sets from HO are drawn indiscriminately. At that point, in f-fold cross-validation, the learning set is divided into k equal parts, and each part functions as a validation set. In addition, LOO is a k-fold cross-acceptance however the approval size is 1. The bootstrap includes information replication from the first specimen under center whether with or without substitution [4, 5]. Bootstrap 632 and 632+ are the advanced versions of the ordinary bootstrap [1], where further details can be referred to in [3]. There is adequate writing on enhancing relapse execution utilizing various ensembling systems. [6] in their paper reports that bootstrap packing [8] and boosting [9, 10] are the most prominent techniques among the ensembling systems [6]. The experimental study on the assortment of troupe techniques can be alluded to in [11, 12].

Where the predictive modeling is concerned, neural systems have been appeared to beat the traditional prescient models, for example, autoregressive coordinated moving normal (ARIMA) models [6]. As a rule, there are three routinely utilized neural system models as a part of time arrangement estimating, and they are conventional backpropagation (BP), scaled conjugate slope (SCG) and Bayesian regularization (BR) as said in [13, 14].

3 Data Background

This segment talks about the information foundation. It was settled that cement's controlled cost had begun to be made incapable by the Malaysian government, on 5 June 2008 [2]. From that point forward, there has been a sensational increment of the concrete cost in June 2008 which is by 23.3% in Peninsula Malaysia [2]. The expansion is likewise relevant to a great deal of other construction materials. Such an emotional addition to the costs of development materials has been known not an effect to the suppliers, subcontractors, contractual workers and proprietors [4] or pertinent gatherings that would not have the scarcest thought what they were getting included in and the sort of monetary battles that they would need to confront.

The information were assembled from three unique sources which are, Unit Kerjasama Awam Swasta (UKAS) of Prime Minister's Department, Construction Industry Development Board (CIDB) and Malaysian Statistics Department from the Central area of Peninsular Malaysia which comprises of three states specifically Kuala Lumpur Federal Territory, Selangor, Negeri Sembilan and Melaka. Government Territory, Selangor, Negeri Sembilan and Melaka. Month to month information from the year 1980 to 2012 (1980 = 100) of three distinctive development material value lists was utilized; aggregate, sand, steel reinforcement.

4 Overview of the Methodology

Figure 1 shows the genuine month to month time arrangement information of total from January 1980 to December 2012. Taking into account Fig. 1, the time arrangement information were compartmentalized into three subsets; preparing set (70%), testing set (15%) and approval set (15%) [15]. The issue identified with information apportioning can allude in [2]. In this examination, we propose an option of utilizing odd-notwithstanding nonoverlapping square bootstrap for information resampling or replication. Moreover, the execution of the proposed bootstrap with correlation with the covering moving normal [2] and standard piece bootstrap techniques utilizing MSE, MAD, RMSE, and MAPE, and also R-squared qualities are likewise given in this paper.

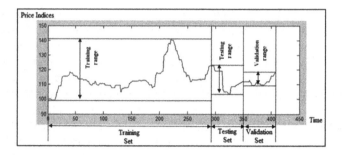

Fig. 1. Time series of Malaysian aggregate data

In general, there are two types of block bootstrap which are non-overlapping and overlapping as shown in Figs. 2 and 3.

(a) Nonoverlapping block bootstrap.

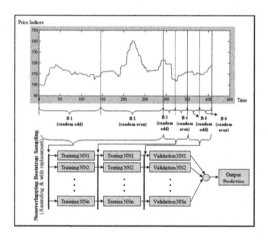

Fig. 2. Nonoverlapping Odd-Even block bootstrap

Given a time series of R = {Xi, i = 1, 2, ... n}. The series can be divided into three subsets and each subset can be divided into two different subsets (odd and even numbers) and altogether forming 6 different non-overlapping blocks. For fixed length 1: BTrain1 = {Xj, j = 1, 2, ... 1}, BTrain2 = {Xl + j, j = 1, 2, ... l); for fixed length k: BTest3 = {Xj, j = 1, 2, ... k}, BTest4 = {Xk + j, j = 1, 2, ... k); for fixed length m: BValid5 = {Xj, j = 1, 2, ... m}, BValid6 = {Xm + j, j = 1, 2, ... m). Then the blocked bootstrap is created through the sampling done in ascending form (with replacement) of block in the form of {B1, B2, B3, B4, B5, B6}.

(b) Overlapping block bootstrap.

Figure 3 displays the moving block bootstrap with the replacement for nonlinear neural network time series as suggested by [2]. This type of bootstrap is overlapping in ascending form whereby for the fixed length l: BTrain = {Xl − 1 + j, j = 1, 2, ... l}; for fixed length k: BTest = {Xk − 1 + j, j = 1, 2, ... k}; for fixed length m: BValid = {Xm − 1 + j, j = 1, 2, ... m}.

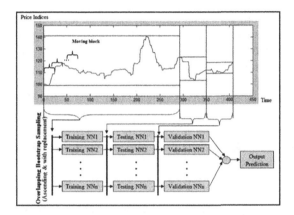

Fig. 3. Overlapping Moving block bootstrap

Moreover, Fig. 4 shows the ordinary bootstrap with replacement. This type of bootstrap is overlapping, also in ascending form where for the fixed length l: BTrain = {Xj, j = 1, 2, ... l}; for fixed length k: BTest = {Xj, j = 1, 2, ... k}; for fixed length m: BValid = {Xj, j = 1, 2, ... m}.

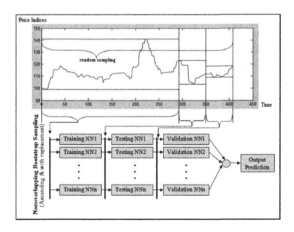

Fig. 4. Ordinary bootstrap

For each block bootstrap sample, a two-layer neural system was prepared with sigmoid move capacity in the concealed layer and direct move capacity in the yielding layer. Here, the repeated information was conveyed as inputs to the system, and their yields had converged to get a definitive expectation yield.

5   Results and Discussions

Table 1 displays the outline measurements of the variables of interest. The aggregate N = 408 (12 months × 34 years) from January 1980 to 2013 (base 1980 = 100). The mean of sand is the most elevated (198.6969), trailed by rooftop materials (131.6038) and total (113.7731). Indisputably, the cost of sand is the most excessive when contrasted with total and rooftop materials. Sand likewise demonstrates the most astounding standard deviation which is 68.4966, contrasted with total (7.63405) and rooftop materials (8.21297). Both total and sand are emphatically skewed which are 1.409 and 0.143 individually. Then again, rooftop materials are contrarily skewed which is –0.321. Notwithstanding, taking into account the Jarque-Bera test for ordinariness, each of the three variables is profoundly critical at 99% certainty interim; total (J-B = 0.873, p = 0.000), sand (J-B = 0.828, p = 0.000), and rooftop materials (J-B = 0.786, p = 0.000).

Table 1. Summary statistics of the construction materials price indices data of the study

No.	Variable	Notation	N	Mean	Std.Dev.	Max	Min	Skewness	Kurtosis	J-B
1.	Agg	Aggregate	408	113.7731	7.63405	140.63	99.2	1.409	2.803	0.873**
2.	Sand	Sand	408	198.6969	68.4966	287.88	100	0.143	–1.730	0.828**
3.	RM	Roof Materials	408	131.6038	8.21297	150.04	100	–0.321	3.508	0.786**

Note: * and ** indicate significance at the 5% and 1% levels respectively

Table 2 demonstrates the consequences of nonlinear autoregressive and nonlinear autoregressive moving normal of counterfeit neural system models regarding the variables of enthusiasm utilizing moving piece bootstrap. Also, Table 3 demonstrates the consequences of nonlinear autoregressive and nonlinear autoregressive moving normal of counterfeit neural system models as for the variables of enthusiasm utilizing conventional square bootstrap. Likewise, Table 4 demonstrates the aftereffects of nonlinear autoregressive and nonlinear autoregressive moving normal of manufactured neural system models as for the variables of enthusiasm utilizing odd-even piece bootstrap.

Table 2. Results of ANN-NAR and ANN-NARMA models with Moving Block bootstrap

No.	Variable	Bootstrap type	Train Algo	NAR						NARMA					
				MSE	MAD	RMSE	MAPE	PEARSON	R²	MSE	MAD	RMSE	MAPE	PEARSON	R²
1.	Agg		Traincg f	0.003	0.029	0.052	10.204	0.992	0.984	0.000	0.013	0.020	4.415	0.998	0.998
			Traincg p	0.010	0.061	0.102	20.161	0.970	0.940	0.007	0.049	0.083	16.687	0.980	0.961
			Traincg b	0.003	0.032	0.055	11.145	0.991	0.982	0.000	0.012	0.020	4.189	0.999	0.998
			Trainscg	0.010	0.062	0.098	21.611	0.974	0.945	0.006	0.058	0.080	21.924	0.984	0.963
			Trainbfg	0.002	0.029	0.052	9.971	0.992	0.984	0.001	0.015	0.024	6.258	0.998	0.997
			Trainoss	0.004	0.040	0.063	14.866	0.989	0.977	0.002	0.026	0.040	10.886	0.996	0.991
2.	Sand	Moving Block	Traincg f	0.003	0.034	0.060	65.535	0.994	0.989	0.001	0.023	0.039	65.535	0.998	0.995
			Traincg p	0.002	0.024	0.046	65.535	0.996	0.993	0.001	0.021	0.034	65.535	0.998	0.996
			Traincg b	0.006	0.048	0.078	65.535	0.991	0.980	0.004	0.039	0.067	65.535	0.994	0.986
			Trainsc g	0.009	0.065	0.098	65.535	0.986	0.969	0.006	0.050	0.080	65.535	0.990	0.979
			Trainbfg	0.002	0.025	0.049	65.535	0.996	0.992	0.001	0.015	0.027	65.535	0.999	0.998
			Trainoss	0.002	0.027	0.051	65.535	0.996	0.992	0.001	0.017	0.029	65.535	0.999	0.997
3.	RM		Traincg f	0.023	0.092	0.151	38.363	0.930	0.856	0.008	0.053	0.087	22.827	0.980	0.952
			Traincg p	0.023	0.091	0.152	35.408	0.925	0.855	0.008	0.056	0.089	22.453	0.977	0.950
			Traincg b	0.025	0.087	0.156	35.444	0.919	0.849	0.010	0.062	0.094	22.243	0.970	0.948
			Trainscg	0.023	0.091	0.152	35.408	0.925	0.855	0.008	0.056	0.089	22.453	0.977	0.950
			Trainbfg	0.027	0.072	0.159	35.532	0.910	0.833	0.016	0.070	0.100	22.196	0.966	0.943
			Trainoss	0.023	0.091	0.152	35.408	0.925	0.855	0.008	0.056	0.089	22.453	0.977	0.950

Table 3. Results of ANN-NAR and ANN-NARMA models with ordinary bootstrap

No.	Variable	Bootstrap type	Train Algo	NAR						NARMA					
				MSE	MAD	RMSE	MAPE	PEARSON	R^2	MSE	MAD	RMSE	MAPE	PEARSON	R^2
1.	Agg		Traincgf	0.003	0.031	0.055	10.027	0.993	0.985	0.001	0.020	0.033	6.794	0.998	0.995
			Traincgp	0.006	0.048	0.077	16.441	0.986	0.971	0.002	0.034	0.049	10.335	0.994	0.988
			Traincgb	0.006	0.046	0.076	15.804	0.986	0.973	0.004	0.040	0.061	12.681	0.992	0.982
			Trainscg	0.005	0.041	0.069	14.171	0.989	0.977	0.003	0.044	0.058	13.518	0.994	0.984
			Trainbfg	0.003	0.032	0.052	10.166	0.994	0.987	0.001	0.023	0.033	7.201	0.997	0.995
			Trainoss	0.027	0.072	0.159	35.532	0.910	0.833	0.016	0.070	0.100	22.196	0.966	0.943
2.	Sand	Ordinary	Traincgf	0.005	0.034	0.070	65.535	0.993	0.986	0.005	0.036	0.072	65.535	0.993	0.985
			Traincgp	0.018	0.085	0.133	65.535	0.975	0.950	0.009	0.067	0.097	65.535	0.987	0.974
			Traincgb	0.005	0.037	0.074	65.535	0.992	0.985	0.003	0.024	0.055	65.535	0.996	0.992
			Trainscg	0.012	0.070	0.110	65.535	0.984	0.966	0.008	0.055	0.091	65.535	0.989	0.977
			Trainbfg	0.023	0.091	0.152	65.535	0.925	0.855	0.008	0.056	0.089	65.535	0.977	0.950
			Trainoss	0.031	0.077	0.175	65.535	0.941	0.885	0.015	0.070	0.123	65.535	0.978	0.943
3.	RM		Traincgf	0.001	0.012	0.026	9.781	0.975	0.949	0.001	0.010	0.024	8.293	0.979	0.957
			Traincgp	0.002	0.025	0.041	16.341	0.947	0.878	0.002	0.024	0.039	15.811	0.950	0.886
			Traincgb	0.001	0.020	0.035	13.544	0.960	0.907	0.001	0.018	0.033	12.349	0.964	0.920
			Trainscg	0.002	0.022	0.043	15.396	0.931	0.861	0.002	0.021	0.042	14.809	0.935	0.870
			Trainbfg	0.001	0.021	0.038	16.599	0.957	0.891	0.001	0.020	0.038	16.340	0.957	0.891
			Trainoss	0.023	0.091	0.152	35.408	0.925	0.855	0.008	0.056	0.089	22.453	0.977	0.950

Table 4. Results of ANN-NAR and ANN-NARMA models with Odd-Even bootstrap

No.	Variable	Bootstrap type	Train Algo	NAR						NARMA					
				MSE	MAD	RMSE	MAPE	PEARSON	R²	MSE	MAD	RMSE	MAPE	PEARSON	R²
1.	Agg	Odd-Even	Traincgf	0.008	0.051	0.090	19.769	0.974	0.949	0.001	0.023	0.038	9.727	0.997	0.991
			Traincgp	0.024	0.098	0.155	37.399	0.923	0.848	0.013	0.079	0.112	28.481	0.969	0.920
			Traincgb	0.008	0.053	0.091	21.385	0.973	0.947	0.002	0.026	0.044	10.771	0.995	0.988
			Trainscg	0.023	0.091	0.152	35.408	0.925	0.855	0.008	0.056	0.089	22.453	0.977	0.950
			Trainbfg	0.008	0.051	0.091	20.058	0.974	0.948	0.002	0.032	0.048	13.097	0.994	0.985
			Trainoss	0.023	0.092	0.151	38.363	0.930	0.856	0.008	0.053	0.087	22.827	0.980	0.952
2.	Sand		Traincgf	0.031	0.089	0.176	65.535	0.942	0.883	0.010	0.061	0.100	65.535	0.984	0.962
			Traincgp	0.031	0.077	0.175	65.535	0.941	0.885	0.015	0.070	0.123	65.535	0.978	0.943
			Traincgb	0.028	0.071	0.166	65.535	0.947	0.897	0.004	0.032	0.061	65.535	0.993	0.986
			Trainscg	0.103	0.193	0.321	65.535	0.822	0.614	0.045	0.130	0.213	65.535	0.920	0.830
			Trainbfg	0.028	0.090	0.167	65.535	0.950	0.896	0.007	0.064	0.086	65.535	0.992	0.972
			Trainoss	0.031	0.077	0.175	65.535	0.941	0.885	0.015	0.070	0.123	65.535	0.978	0.943
3.	RM		Traincgf	0.007	0.021	0.082	11.102	0.871	0.757	0.000	0.009	0.020	6.615	0.993	0.986
			Traincgp	0.023	0.091	0.152	35.408	0.925	0.855	0.008	0.056	0.089	22.453	0.977	0.950
			Traincgb	0.007	0.024	0.085	13.030	0.861	0.739	0.001	0.013	0.029	8.984	0.985	0.970
			Trainscg	0.027	0.072	0.159	35.532	0.910	0.833	0.016	0.070	0.100	22.196	0.966	0.943
			Trainbfg	0.007	0.022	0.084	11.214	0.866	0.749	0.001	0.014	0.027	9.418	0.987	0.974
			Trainoss	0.025	0.087	0.156	35.444	0.919	0.849	0.010	0.062	0.094	22.243	0.970	0.948

In Table 2, for moving piece bootstrap on the Aggregate variable, the Fletcher-Reeves conjugate slope calculation (traincgf) with setup 11-10-1 plays out the best result in the underlying model, NAR with R-squared quality 0.984, and in the possible model, NARMA with R-squared equivalents to 0.998. The outcome beat the best ANN-NAR and ANN-NARMA models utilizing normal and odd-even square bootstrap on Aggregate information (allude to Table 4) of trainbfg (R^2 ANN-NAR = 0.987; R^2 ANN-NARMA = 0.995), and traincgf (allude to Table 5) (R^2 ANN-NAR = 0.949; R^2 ANN-NARMA = 0.991) separately. In this way, it can be presumed that the moving square time arrangement bootstrap in the two-layer neural system with 11-10-1 design of traincgf calculation delivered the best model to figure Malaysian Aggregate value list (RMSE ANN-NAR = 0.052; RMSE ANN-NARMA = 0.020).

Again in Table 3, moving piece bootstrap at the end of the day beat the other square bootstrap strategies on the Sand variable, where the BFGS semi Newton calculation (traincgf) with arrangement 11-10-1 plays out the best result in the underlying model, NAR with R-squared quality 0.992, and in the consequent model, NARMA with R-squared equivalents to 0.998. The outcome beat the best ANN-NAR and ANN-NARMA models utilizing normal and odd-even square bootstrap on Sand information (allude to Table 4) of traincgb (R^2 ANN-NAR = 0.985; R^2 ANN-NARMA = 0.992), likewise traincgb (R^2 ANN-NAR = 0.897; R^2 ANN-NARMA = 0.986) individually. In this manner, it can be presumed that the moving piece bootstrap with a two-layer neural system with 11-10-1 setup of trainbfg calculation is the best model to estimate Malaysian Sand value list (RMSE ANN-NAR = 0.049; RMSE ANN-NARMA = 0.027).

Be that as it may, distinctive results were found for the rooftop materials information (allude to Table 4), where the odd-even piece bootstrapping was found to beat the other square bootstrapping strategies. Here, the Fletcher-Reeves conjugate slope calculation (traincgf) with design 11-10-1 plays out the best result in the underlying model, NAR with R-squared worth 0.757, and in the inevitable model, NARMA with R-squared equivalents to 0.986. The outcome outflanked the best ANN-NAR and ANN-NARMA models utilizing moving and standard piece bootstrapping procedures on Sand information (allude to Table 3) of traincgf (R2ANN-NAR = 0.856; R2ANN-NARMA = 0.952), and traincgf (allude to Table 4) (R2ANN-NAR = 0.949; R2ANN-NARMA = 0.957) separately. In this way, it can be inferred that the moving piece time arrangement bootstrap in the two-layer neural system with 11-10-1 setup of traincgf calculation delivered the best model to conjecture Malaysian Roof Materials value file (RMSEANN-NAR = 0.082; RMSEANN-NARMA = 0.020).

6 Conclusions

In this specific study, it can be reasoned that the moving piece bootstrap in the nonlinear autoregressive moving normal model is the most appropriate to conjecture both Malaysian Aggregate and Sand data. Then again, odd-even piece bootstrap in nonlinear neural system model is the most reasonable to foresee the future value list of Malaysian Roof Materials. This implies the execution of various piece bootstrapping time arrangement strategies may rely on upon the information themselves, particularly with regards to genuine information in light of the fact that diverse sort of information carry on distinctively and comprise of various characteristic issues.

Later in the attempt, the comparative square bootstrapping methods should be performed in other diverse Malaysian development materials value records information, and also other genuine mechanical information.

Acknowledgment. We might want to commit our gratefulness and appreciation to Unit Kerjasama Awam Swasta (UKAS) of Prime Minister's Department, Construction Industry Development Board (CIDB) and Malaysian Statistics Department. Extraordinary thanks additionally go to Universiti Teknologi MARA and Malaysian Ministry of Higher Education (MOHE) for supporting this exploration under the Research Grant No. 600-RMI/DANA 5/3/CIFI (65/2013) and No. 600-RMI/FRGS 5/3 (137/2014). Additionally, appreciation to International Islamic University Malaysia and MOHE for the research grant awarded to this project, RIGS 16-092-0256.

References

1. Lendasse, A., Wertz, V., Verleysen, M.: Model selection with cross-validations and bootstraps — application to time series prediction with RBFN models. In: Kaynak, O., Alpaydin, E., Oja, E., Xu, L. (eds.) ICANN/ICONIP -2003. LNCS, vol. 2714, pp. 573–580. Springer, Heidelberg (2003). doi:10.1007/3-540-44989-2_68
2. He, H., Shen, X.: Bootstrap methods for foreign currency exchange rates prediction. In: International Joint Conference Neural Networks 2007, pp. 1272–1277, August 2007
3. Kamruzzaman, J., Sarker, R.A., Ahmad, I.: SVM based models for predicting foreign currency exchange rates. In: Third IEEE International Conference on Data Mining, pp. 557–560, November 2003
4. Singh, K., Xie, M.: Bootstrap: A Statistical Method, Rutgers University, The State University of New Jersey, Department of Statistics (2008)
5. Dixon, P.M.: The bootstrap. Iowa, USA: Iowa State University, Department of Statistics (2001)
6. Arlot, S., Calisse, A.: A survey of cross-validation procedures for model selection. Stat. Surv. **4**, 40–79 (2010)
7. Breiman, L.: Bagging predictors. Mach. Learn. **24**(2), 123–140 (1996)
8. Freund, Y., Schapire, R.E.: Experiments with a new boosting algorithm. In: International Conference on Machine Learning, vol. 96, pp. 148–156, July 1996
9. Schapire, R.E.: The strength of weak learnability. Mach. Learn. **5**(2), 197–227 (1990)
10. Zhou, Z.H., Wu, J., Tang, W.: Ensembling neural networks: many could be better than all. Artif. Intell. **137**(1), 239–263 (2002)
11. Maclin, R., Opitz, D.: Popular ensemble methods: an empirical study. J. Artif. Intell. Res. **1**, 169–198 (1999)
12. Kamruzzaman, J., Sarker, R.A.: ANN-based forecasting of foreign currency exchange rates. Neural Inf. Process. Lett. Rev. **3**(2), 49–58 (2004)
13. Kamruzzaman, J., Sarker, R.A.: Forecasting of currency exchange rates using ANN: a case study. In: Proceedings of the 2003 IEEE International Conference on Neural Networks and Signal Processing, vol. 1, pp. 793–797, December 2003
14. Foad, H.M., Mulup, A.: Harga siling simen dimansuh 5 Jun, Utusan, Putrajaya, 2 June 2008
15. Law, K.C.: Analysts mixed on cement price outlook this year, The Star, 10 January 2009

Method for Aspect-Based Sentiment Annotation Using Rhetorical Analysis

Łukasz Augustyniak[1]([✉]), Krzysztof Rajda[2], and Tomasz Kajdanowicz[1]

[1] Department of Computational Intelligence,
Wroclaw University of Technology, Wroclaw, Poland
{lukasz.augustyniak,tomasz.kajdanowicz}@pwr.edu.pl
[2] Kenaz Technologies, Leszno, Poland
kenaz.technologies@gmail.com

Abstract. This paper fills a gap in aspect-based sentiment analysis and aims to present a new method for preparing and analysing texts concerning opinion and generating user-friendly descriptive reports in natural language. We present a comprehensive set of techniques derived from Rhetorical Structure Theory and sentiment analysis to extract aspects from textual opinions and then build an abstractive summary of a set of opinions. Moreover, we propose aspect-aspect graphs to evaluate the importance of aspects and to filter out unimportant ones from the summary. Additionally, the paper presents a prototype solution of data flow with interesting and valuable results. The proposed method's results proved the high accuracy of aspect detection when applied to the gold standard dataset.

Keywords: Sentiment analysis · Opinion mining · Aspect-based sentiment analysis · Rhetorical analysis · Rhetorical Structure Theory

1 Introduction

Modern society is an information society bombarded from all sides by an increasing number of different pieces of information. The 21st century has brought us the rapid development of media, especially in the internet ecosystem. This change has caused the transfer of many areas of our lives to virtual reality. New forms of communication have been established. Their development has created the need for analysis of related data. Nowadays, unstructured information is available in digital form, but how can we analyse and summarise billions of newly created texts that appear daily on the internet? Natural language analysis techniques, statistics and machine learning have emerged as tools to help us. In recent years, particular attention has focused on sentiment analysis. This area is defined as the study of opinions expressed by people as well as attitudes and emotions about a particular topic, product, event, or person. Sentiment analysis determines the polarisation of the text. It answers the question as to whether a particular text is a positive, negative, or neutral one.

© Springer International Publishing AG 2017
N.T. Nguyen et al. (Eds.): ACIIDS 2017, Part I, LNAI 10191, pp. 772–781, 2017.
DOI: 10.1007/978-3-319-54472-4_72

Our goal is to build a comprehensive set of techniques for preparing and analysing texts containing opinions and generating user-friendly descriptive reports in natural language - Fig. 1. In this paper, we describe briefly the whole workflow and present a prototype implementation. Currently, existing solutions for sentiment annotation offer mostly analysis on the level of entire documents, and if you go deeper to the level of individual product features, they are only superficial and poorly prepared for the analysis of large volumes of data. This can especially be seen in scientific articles where the analysis is carried out on a few hundred reviews only. It is worth mentioning that this task is extremely problematic because of the huge diversity of languages and the difficulty of building a single solution that can cover all the languages used in the world. Natural language analysis often requires additional pre-processing steps, especially at the stage of preparing the data for analysis, and steps specific for each language. Large differences can be seen in the analysis of the Polish language (a highly inflected language) and English (a grammatically simpler one). We propose a solution that will cover several languages, however in this prototype implementation we focused on English texts only.

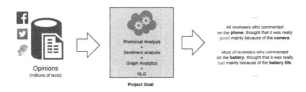

Fig. 1. The workflow for Rhetorical and Sentiment Analysis.

In this paper, we present analysis and workflow inspired by the work of Joty, Carenini and Ng [7]. We experimented with several methods in order to validate aspect-based sentiment analysis approaches and in the next steps we want to customize our implementation for the Polish language.

The paper presents in Sect. 1 an introduction to sentiment analysis and its importance in business, then in Sect. 2 - related work from rhetorical and sentiment analysis areas is presented. Section 3 covers description of our method. Implementation and the dataset are described in Sect. 4. Section 5 refers to the results. The last Sect. 6 consists of conclusions and future work.

2 Related Work

2.1 Rhetorical Analysis

Rhetorical analysis seeks to uncover the coherence structure underneath the text, which has been shown to be beneficial for many Natural Language Processing (NLP) applications including text summarization and compression [12], machine translation evaluation [5], sentiment analysis [9], and others. Different formal theories of discourse analysis have been proposed. Martin [15] proposed discourse relations based on discourse connectives (e.g., because, but) expressed in the text. Danlos [2] extended sentence grammar and formalize discourse structure. Rhetorical Structure Theory or RST - used in our experiments - was proposed by Mann and Thompson [13]. The method proposed by them is perhaps the most influential theory of discourse in computational linguistics. Moreover, it was initially intended to be used in text generation tasks, but it became popular for parsing the structure of a text [18]. Rhetorical Structure Theory represents texts by hierarchical structures with labels. This is a tree structure, which comprises Discourse Trees (DTs). Presented at Fig. 2 this Discourse Tree is a representation of the following text:

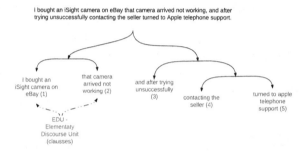

Fig. 2. An exemplary Discourse Tree based on Rhetorical Structure Theory.

2.2 Sentiment Analysis

A sentiment analysis can be made at the level of (1) the whole document, (2) the individual sentences, or (what is currently seen as the most attractive approach) (3) at the level of individual fragments of text. Regarding document level analysis [1,10] - the task at this level is to classify whether a full opinion expresses a positive, negative or neutral attitude. For example, given a product review, the model determines whether the text shows an overall positive, negative or neutral opinion about the product. The biggest disadvantage of document level analysis is an assumption that each document expresses views on a single entity. Thus, it is not applicable to documents which evaluate or compare multiple objects. As for sentence level analysis [6] - The task at this level relates to sentences and determines whether each sentence expressed a positive, negative, or neutral opinion. This level of analysis is closely related to subjectivity classification which distinguishes sentences (called objective sentences) that express factual information from sentences (called subjective sentences) that express subjective views and opinions. However, we should note that subjectivity is not equivalent to sentiment as many objective sentences can imply opinions. With feature/aspect level analysis [20] - both the document level and the sentence level analyses do not discover what exactly people liked and did not like. A finer-grained analysis can be performed at aspect level. Aspect level was earlier called feature/aspect level. Instead of looking at language constructs (documents, paragraphs, sentences, clauses or phrases), aspect level directly looks at the opinion itself. It is based on the idea that an opinion consists of a sentiment (positive or negative) and a target (of opinion). As a result, we can aggregate the opinions. For example, the phone display gathers positive feedback, but the battery is often rated negatively. The aspect-based level of analysis is much more complex since it requires more advanced knowledge representation than at the level of entire documents only. Also, the documents often consist of multiple sentences, so saying that the document is positive provides only partial information. In the literature, there exists some initial work related to aspects. There exist initial solutions that use SVM-based algorithms [19] or conditional random field classifiers [3] with manually engineered features. There also exist some solutions based on deep neural networks, such as connecting sentiments with the corresponding aspects based on the constituency parse tree [20].

3   Method for Aspect-Based Sentiment Analysis

The proposed Rhetorical and Sentiment Analysis flow is divided into four main tasks:

1. Rhetorical analysis with sentiment detection.
2. Aspect detection in textual data.
3. Methods, techniques, and graph analytics of aspect inter-relations.
4. Abstractive summary generation in natural language (not included in prototype workflow yet).

The overall characteristics and flow organisation can be seen in Fig. 3. Each of the mentioned steps of the proposed method is described in the following subsections.

3.1   Rhetorical Analysis

The goal of discourse analysis in our method is the segmentation of the text for the basic units of discourse structures EDU (Elementary Discourse Units) and connecting them to determine semantic relations. The analysis is performed separately for each source document, and as the output we get Discourse Trees (DT) such as in Fig. 2. At this stage, existing discourse parsers will model the structure and the labels of a DT separately. They do not take into account the sequential dependencies between the DT constituents. Then existing discourse parsers will apply greedy and sub-optimal parsing algorithms and build a Discourse Tree. During this stage, and to cope with the mentioned limitation The inferred (posterior) probabilities can be used from CRF parsing models in a probabilistic CKY-like bottom-up parsing algorithm [8] which is non-greedy and optimal. Finally, discourse parsers do not discriminate between intra-sentential parsing (i.e., building the DTs for individual sentences) and multi-sentential parsing (i.e., building a DT for the whole document) [7]. Hence, this part of the analysis will extract for us distributed information about the relationship between different EDUs from parsed texts. Then we assign sentiment orientation to each EDU.

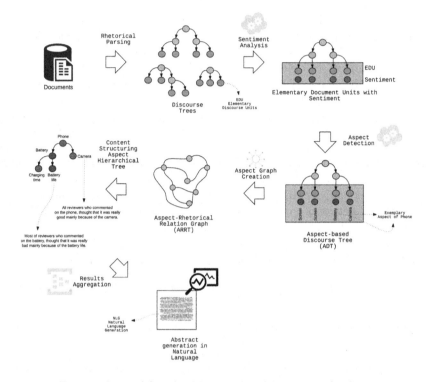

Fig. 3. The workflow for Rhetorical and Sentiment Analysis.

3.2 Aspect Detection in Textual Data

The second step covers aspect extraction and creation of aspect-based discourse trees ADT - see Fig. 3. Aspect detection from textual data is based commonly on detection of names or noun-phrases [14] and we used exactly this approach.

3.3 Analysis of Aspect Inter-relations

The third step consists of an Aspect-Rhetorical Relation Graph (ARRG) and content Structuring Aspect Hierarchical Tree (see Fig. 3). Discourse Trees of individual documents are processed (the order of EDU is not changed) to form association rules. Then, an Aspect-Rhetorical Relation Graph based on a set of these rules is created. Each node represents an aspect and each edge is one of the relations between the EDU's aspects. A graph will be created for all documents used in the experiment. The graph can be represented with weighted edges (association rules confidence, a number of such relations in the whole graph etc.), but there is a need to check and compare different types of graph representations. Then, it is possible to characterize the whole graph and each node (aspect) with graph metrics (PageRank [17], degree, betweenness or other metrics). These metrics will be used for estimating the cut threshold – removing uninformative or redundant aspects. Hence, we will end up with only the most important aspects derived from analysed corpora. Then the graph will be transformed into an Aspect Hierarchical Tree. This represents the correlation between aspects and enables us to generate natural language-based descriptions.

3.4   Abstractive Summary Generation in Natural Language

The last step covers summary (abstract) generation in natural language. Natural language generation models use parameterized templates (very limited and dependent on the size of the rule-based system responsible for the completions of the text), or deep neural networks [21].

4   Experimental Scenario

For the Rhetorical Parsing part of our experiment, we used a special library implemented for such purposes [4]. As a sentiment analysis model, we used the Bag of Word vectorization method with a Logistic Regression classifier trained on 1.2 million (1, 3 and 5-star rating only) of Electronic reviews from SNAP Amazon Dataset [16]. The BoW vectorization method built a vocabulary that considers the top 50,000 terms only ordered by their frequency across the corpus, similarly to supervised learning examples presented in our previous works in [1]. We used a noun and noun phrases extractor according to part-of-speech tagger from the Spacy Python library[1]. In order to create an Aspect-Rhetorical Relation Graph we used breadth-first search (BFS) algorithm for each Discourse Tree.

4.1   Dataset

We used Bing Liu's dataset [11] for evaluation. It contains three review datasets of three domains: computers, wireless routers, and speakers as in Table 1. Aspects in these review datasets were annotated manually.

Table 1. Bing Liu's dataset [11] statistics.

Dataset	# of documents	# of distinct aspects
Computer	531	354
Wireless router	879	307
Speaker	689	440

---

[1] https://spacy.io.

4.2   Experimental Setup

We implemented our framework in Python. The first computational step was to load the dataset and parse it into individual documents. Next, each document was processed through the Discourse Parser [4] and transformed into a Discourse Tree (DT). Then we extracted Elementary Discourse Units (EDUs) from the DT and each EDU was processed through the Logistic Regression sentiment algorithm. All neutral EDUs were taken off from consideration to ensure that the discovered aspects are correlated with authors' emotions. The remaining EDUs were processed through part-of-speech tagger to extract nouns and noun phrases which we decided to treat as potential aspects. The result of this step was a set of Aspect-based Discourse Trees (ADTs). Then, from each ADT relations between aspects were extracted using breadth-first search, and an Aspect-Rhetorical Relation Graph (ARRG) was created by using aspects and relations such as nodes and edges respectively. Next, we evaluated the importance of aspects using a PageRank algorithm. Our approach resulted in complete list of aspects sorted by PageRank score. We applied a user-selected importance threshold to filter trivial aspects.

5   Results

In Table 2 there are presented some examples of the results of our approach compared with the annotated data from Bing Liu's dataset. In the first sentence, the results of the analysis differ because we decided to treat only nouns or noun phrases as aspects, while annotators also accepted verbs. In some cases, such as sentences 2 or 4, our approach generated more valuable aspects than the annotators found, but in some cases, like sentence 5, we found fewer. This is

Table 2. Examples of proposed analysis results

No.	Input content	Annotated aspect : sentiment	Detected aspect : sentiment
1	I have this connected to my late 2008 MacBook Pro, and it works flawlessly	Works : positive	Macbook pro : positive
2	We are well pleased with the monitor and the company	Monitor : positive	Monitor : positive Company : positive
3	The changing colors help to tell, with a quick glance	Colors : positive	Colors : positive
4	The screen is a very pleasing matte, and the colors are great	Colors : positive	Screen : negative Colors : positive
5	I would not recommend this or any Acer product to anyone except perhaps my ex	Acer product : negative	– : negative
6	I purchased this as a Christmas gift	– : –	– : negative

possibly the result of our method of filtering valuable aspects - if some aspects were not frequent enough in the dataset, we can treat them as void. In cases where there is neither aspect nor sentiment in the dataset, such as sentence 6, we measure sentiment as well, as one of our analysis steps.

Figure 4 shows the agreement between our aspects and that of the dataset. We assumed two aspects as equal when they were textually the same. We made some experiments using text distance metrics, such as the Jaro-Winkler distance, but the results did not differ significantly from an exact matching. We fitted the importance factor value (on the X axis) so as to enrich final aspects set: a higher factor resulted in a larger aspects set and a higher value of precision metric, with slowly decreasing recall. First results (blue line on charts) were not satisfactory, so we removed a sentiment filtering step of analysis (orange line on chart), which doubled the precision value, with nearly the same value of recall. The level of precision for whole dataset (computer, router, and speaker) was most of the time at the same level. However, the recall of router was significantly worse than speaker and computer sets.

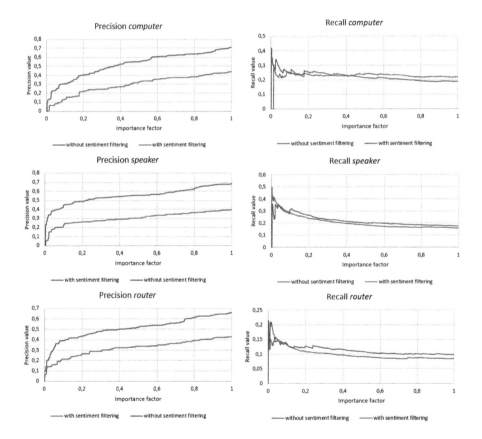

Fig. 4. Agreement between found aspects and gold standard with (blue line) and without sentiment filtering (orange line) (color figure online)

6   Conclusions and Future Work

We have proposed a comprehensive flow of analysing aspects and assigning sentiment orientation to them. The advantages of such an analysis are that: it is a grammatically-based and coherent solution, it shows opinion distribution, it doesn't need any aspect ontology, it is not limited to the number of aspects and really important, it doesn't need training data (unsupervised method). The method proved it has a big potential in generating summary overviews for aspect and sentiment distribution across analysed documents. In our next steps, we want to improve the aspect extraction phase, probably using neural network approaches. Moreover, we want to expand the analysis of the Polish language.

Acknowledgment. The work was partially supported by the National Science Centre grants DEC-2016/21/N/ST6/02366 and DEC-2016/21/D/ST6/02948, and from the European Unions Horizon 2020 research and innovation programme under the Marie Skłodowska-Curie grant agreement No. 691152 (RENOIR project).

References

1. Augustyniak, Ł., Szymański, P., Kajdanowicz, T., Tuligłowicz, W.: Comprehensive study on Lexicon-based ensemble classification sentiment analysis. Entropy **18**(1), 4 (2015)
2. Danlos, L.: D-STAG: a formalism for discourse analysis based on SDRT and using synchronous TAG. In: Groote, P., Egg, M., Kallmeyer, L. (eds.) FG 2009. LNCS (LNAI), vol. 5591, pp. 64–84. Springer, Heidelberg (2011). doi:10.1007/978-3-642-20169-1_5
3. De Clercq, O., Van de Kauter, M., Lefever, E., Hoste, V.: LT3: applying hybrid terminology extraction to aspect-based sentiment analysis. In: Proceedings of the 9th International Workshop on Semantic Evaluation - SemEval 2015, vol. 1997, pp. 719–724 (2015)
4. Feng, V.W., Hirst, G.: Two-pass discourse segmentation with pairing and global features. ArXiv e-prints 1407.8215, July 2014
5. Guzmán, F., Joty, S., Màrquez, L., Nakov, P.: Using discourse structure improves machine translation evaluation. In: Proceedings of the 52nd Annual Meeting of the Association for Computational Linguistics (Volume 1: Long Papers), pp. 687–698 (2014)
6. Joty, S., Carenini, G., Ng, R.T.: A novel discriminative framework for sentence-level discourse analysis (2012)
7. Joty, S., Carenini, G., Ng, R.T.: CODRA: a novel discriminative framework for rhetorical analysis. Comput. Linguist. **41**(3), 1–50 (2015)
8. Jurafsky, D., Martin, J.H.: Speech and Language Processing: An Introduction to Natural Language Processing, Computational Linguistics, and Speech Recognition, pp. 21:0–21:934 (2009)
9. Lazaridou, A., Titov, I., Sporleder, C.: A Bayesian model for joint unsupervised induction of sentiment, aspect, discourse representations. In: Proceedings of the 51st Annual Meeting of the Association for Computational Linguistics (Volume 1: Long Papers), pp. 1630–1639 (2013)
10. Liu, B.: Sentiment analysis and subjectivity (2010)

11. Liu, Q., Gao, Z., Liu, B., Zhang, Y.: Automated rule selection for aspect extraction in opinion mining. In: International Joint Conference on Artificial Intelligence, IJCAI 2015, pp. 1291–1297 (2015)

12. Louis, A., Joshi, A.K., Nenkova, A., Louis, C., Joshi, A.: Discourse indicators for content selection in summaization, pp. 147–156 (2010)

13. Mann, W.C., Thompson, S.A.: Rhetorical Structure Theory: toward a functional theory of text organization (1988)

14. Pontiki, M., Galanis, D., Papageorgiou, H., Manandhar, S., Androutsopoulos, I.: SemEval-2015 Task 12: aspect based sentiment analysis. In: Proceedings of the 9th International Workshop on Semantic Evaluation (SemEval 2015), Denver, Colorado, pp. 486–495 (2015)

15. Martin, J.R.: English Text. John Benjamins Publishing Company, Amsterdam (1992)

16. McAuley, J., Leskovec, J.: Hidden factors, hidden topics: understanding rating dimensions with review text. In: Proceedings of the 7th ACM conference on Recommender systems - RecSys 2013, pp. 165–172 (2013)

17. Page, L., Brin, S., Motwani, R., Winograd, T.: The PageRank citation ranking: bringing order to the web. World Wide Web Internet Web Inf. Syst. **54**(1999–66), 1–17 (1998)

18. Taboada, M.: Discourse markers as signals (or not) of rhetorical relations. J. Pragmatics **38**(4), 567–592 (2006)

19. Wagner, J., Arora, P., Cortes, S., Barman, U., Bogdanova, D., Foster, J., Tounsi, L.: DCU: aspect-based polarity classification for SemEval task 4. In: Proceedings of the 8th International Workshop on Semantic Evaluation - SemEval 2014, pp. 223–229 (2014)

20. Wang, B., Liu, M.: Deep learning for aspect-based sentiment analysis, pp. 1–9 (2015)

21. Wen, T-H., Gasic, M., Mrksic, N., Su, P-H., Vandyke, D., Young, S.: Semantically conditioned LSTM-based natural language generation for spoken dialogue systems. In: Proceedings of the 2015 Conference on Empirical Methods in Natural Language Processing, pp. 1711–1721, September 2015

On Quality Assesement in Wikipedia Articles Based on Markov Random Fields

Rajmund Kleminski[✉], Tomasz Kajdanowicz, Roman Bartusiak,
and Przemyslaw Kazienko

Faculty of Computer Science and Management,
Wroclaw University of Science and Technology, Wrocław, Poland
{rajmund.kleminski,tomasz.kajdanowicz,roman.bartusiak,
przemyslaw.kazienko}@pwr.edu.pl

Abstract. This article investigates the possibility of accurate quality prediction of resources generated by communities based on the crowd-generated content. We use data from Wikipedia, the prime example of community-run site, as our object of study. We define the quality as a distribution of user-assigned grades across a predefined range of possible scores and present a measure of distribution similarity to quantify the accuracy of a prediction. The proposed method of quality prediction is based on Markov Random Field and its Loopy Belief Propagation implementation. Based on our results, we highlight key problems in the approach as presented, as well as trade-offs caused by relying solely on network structure and characteristics, excluding metadata. The overall results of content quality prediction are promising in homophilic networks.

Keywords: Wikipedia · Quality prediction · Iterative classification

1 Introduction

Over the years, we have observed the growth and expansion of multiple internet communities characterized by heavy reliance on the efforts of their members to generate content. While these services often attempt to include self-policing and self-moderating mechanisms, the question of created content's quality and objectivity is an open one.

The sheer volume of the user generated content makes it challenging for the community-sourced quality assessment to work accurately. A viable method should possess an ability to work on the basis of relatively small community input and with the inclusion of community-specific guidelines.

We propose a new approach to the task of quality assessment in crowd generated resources on the example of Wikipedia. The method is characterized by its fully network-based operations, which allow the exclusion of language processing techniques (thus providing a potential for cross-language applications). Additionally, predictive mechanisms are included in an attempt to offset the impact of

© Springer International Publishing AG 2017
N.T. Nguyen et al. (Eds.): ACIIDS 2017, Part I, LNAI 10191, pp. 782–791, 2017.
DOI: 10.1007/978-3-319-54472-4_73

data incompleteness, natural for community-sourced quality assessments. Furthermore, predictive output is generalized to the form of a distribution of possible 'scores', aiming to provide a greater insight into community's view on the item, as opposed to assigning simply an aggregate number. It is important to note that this approach is very general in its nature; methods developed specifically for narrow tasks are very likely to outperform it.

The article is organized as follows: in Sect. 2, we present the work related to our research topic. Section 3 introduces the idea of a homophily and details our approach. Section 4 details the experimental setup and discusses the results, while Sect. 5 presents the conclusions and directions for the future work.

2 Related Work

Past approaches towards the topic of quality measurement have been focused on metrics mined directly from the articles' text and edit history.

The most straightforward method of assessing quality is presented in an article by Dalip et al. [2], where a multitude of metrics were used in a linear regression models. The features used for this task ranged from ones based on length, style and structure (text-based features) to more complex ones, calculated on the basis of review and readability scores, as well as typical network features for a network representing the articles and their connections. The research highlighted the most useful pairs and groups of features, suggesting that simple, textual metrics can be used to the best effect.

Hu's et al. [3] proposed method used the concept of authority in models built on the basis of the peer-review process, with the leading assumption that the article is of higher quality when the total authority of authors (or authors and reviewers) is higher. In this case, a 'reviewer' was a contributor adding to the article at a later point. Their work also showed that user interaction alone is not sufficient of a metric, warranting the use of article-focused metrics as well.

In Calzada and Dekhtyar's [1] work, a machine learning technique was applied to a body of articles grouped into categories dependent on their 'type'. This division included both an actual type of an article (like a list or a disambiguation page) as well as that article's state of evolution (stabilized, controversial, or stub). The quality used in this research came from a study which asked people to rate the articles. This research confirmed that different 'categories' of articles can't be accurately rated with the same model.

A research on lifecycle-based metrics [12] analyzed the usefulness of metrics related to time and contributions. By dividing contributions into persistent and transient, it was possible to build a model of quality prediction inferring from the survival rates of results of the editing process. These two metrics, when aggregated, allow for the modeling of the lifecycles to be used in evaluation effort. Based on their relation to the quality of an article, further metrics were constructed and used in quality assessment, showing promising results.

Liu and Ram [5] researched a relation between quality and user interactions shown by a pattern of collaboration. In this method, edit-based metrics are used to categorize an editor into one of six editor types, which are later on used to infer about the quality of an article. This approach effectively combined the data mined from edits and user activity, contributing a way of understanding the collaborative processes based on deeply implicit features.

The relationship between contributions and the contributors was taken further in [9], where a series of self-correcting calculations was implemented to weigh on the quality of both. Mutual evaluation was achieved by performing initial calculations and the re-evaluating those by including the initial value of edit quality into the initial editor's quality and vice versa. The proposed method of calculating these metrics is resistant to vandalism, reducing the noise from the data.

3   The Method for Wikipedia Page Quality Assessment Based on Homophily

This section contains brief introduction about the underlying phenomenon called homophily that has been adopted from social sciences [7].

3.1   Homophily

Homophily is the similarity of entities within the network; two nodes are homophilic if they are demonstrably similar in regards to specific qualities that describe them within the data. For our purposes, two articles are similar if their score distribution within the Article Feedback Tool (explained in Sect. 4.1) can be identified as being of a similar shape. This requirement necessitated a search for a properly discriminating measure of similarity between the distributions.

A measure known as Earth's Mover Distance has been selected for the task, based on its discriminating properties; it was determined that, from among the examined measures (including Hellinger's, Bhattacharaya's and chi-squared's distance as well as Jensen-Shannon's divergence), its reaction to differences in the actual shape of a distribution is the sharpest.

3.2   The New Method Based on Loopy Belief Propagation

The basis of our approach was the use of the Loopy Belief Propagation algorithm, an iterative approach to a collective classification problem. The defining characteristic of this method is the global objective function, defined as its optimization goal. This function is based on the idea of a pairwise Markov Random Field, as described in [10] and sucessfully applied in [4].

LBP works on the message-passing's principle, transferring the label of a node v_i among the nodes connected to it (i.e. v_j), where $(v_i, v_j \in V)$ and $(v_i, v_j) \in E$. This action can be understood as spreading the *belief* of what the label of v_j should be from v_i's point of view. A message to be propagated across the neighbours is calculated according to Eq. 1:

$$m_{i \to j}(l_j) = \alpha \sum_{l_i \in L} \Psi_{ij}(l_i, l_j)\phi(l_i) \prod_{v_k \in V^{UK} \setminus v_j} m_{k \to i}(l_i) \tag{1}$$

where $m_i \rightarrow_j (l_j)$ denotes a message of the label l_j to be sent from node v_i to v_j. A normalization constant of α is introduced, guaranteeing that the sum of messages will be equal to one. Ψ and ϕ are clique potentials as explained in [8].

Our modified method is no longer a labelling algorithm, however. Instead, it uses the score distribution of an article, stored as a vector depicting how many of each of the grades it scored, and attempts to propagate these vectors across the network. This qualifies our approach as a predictive method, as opposed to a classification method, the validity of which is supported by the network-wide homophily; the more similarity we can observe between the neighbours in a network, the higher the expected accuracy such prediction should achieve.

Algorithm 1. Loopy Belief Propagation based estimator

---

1 function loopyBeliefPropagationEstimator $(graph, iterations)$;
   Input   : Graph $graph$ where each vertex data $vertex.data$ contains AFT score
               or is empty
   Output: Graph $outputGraph$ which has same structure as input $graph$ but
               each empty vertex contains estimated AFT score
2 **for** 0 *to iterations* **do**
3 $\quad$ $messages = List()$;   // List of pairs (vertex, messageToVertex)
4 $\quad$ **for** $edge(fromVertex, toVertex) \in graph$ **do**
5 $\quad\quad$ **if** $isVertexToAproximate(toVertex)$; // Check if vertex is known
6 $\quad\quad$ **then**
7 $\quad\quad\quad$ $|\quad messages = messages + +(toVertex, fromVertex.data)$
8 $\quad\quad$ **end**
9 $\quad$ **end**
10 $\quad$ $dataToProcess = messages.groupByTargetVertex();$ // Group messages
       by recipient
11 $\quad$ **for** $(vertex, dataList) \in dataToProcess$ **do**
12 $\quad\quad$ $newVertexData =$
       $dataList.sumAFTScores().divideAFTScores(dataList.size)$;
       // Calculate average value of each AFT score
13 $\quad\quad$ $graph.setVertexData(vertex, newVertexData)$
14 $\quad$ **end**
15 **end**

---

4   Experiments and Results

This section describes the dataset in detail (Sect. 4.1), and provides information about the experimental setup (Sect. 4.2) and details about the experimental results (Sect. 4.3).

4.1   Wikipedia Dataset

Wikipedia Articles. The experiments were conducted on a dataset containing the entirety of English Wikipedia, represented as a network encompassing all of the articles and their relations within the Wikipedia. Each article was treated as a single node in the network, identified by its unique ID.

The edges within the network were based on the in-Wikipedia hyperlinks; whenever an article linked to another article in any section, it was considered as a basis for forming an actual connection within the network used for the experimental purposes. As such, differences between different kinds of in-site links (related articles, articles within the same category and so on) were not represented within the resulting graph.

Article Feedback Tool. AFT was Wikimedia Foundation's initiative designed to gather reader's feedback about an individual article's quality. It has been since discontinued.

Quality assessment through the AFT was a community-driven process, utilising a previously untapped source of feedback. Traditionally, the quality of an article was judged by selected members of Wikipedia's community, while AFT was open to every user of the site.

The Tool offered readers the ability to express their opinion on an article through a four-dimensional, five-point scale. The qualities (dimensions) that one had to rate were: Trustworthy, Objective, Complete and Well-written. For each of these, the user was to assign a grade ranging from 1 to 5, to express how well the article was realising these qualities.

For the purposes of our research, we assigned each node in the network a full set of one dimension's grades as their attribute, in the form of a vector. Each article is thus described with 5 numbers, corresponding to the count of specific grades (1 s, 2 s, ...) it scored in the aforementioned dimension over the entire period of AFT being in use.

The reasoning behind using only one of the four possible dimensions is that, if we expect the inter-connected articles to be of similar quality, the similarity should be apparent in each of the qualities. As such, analysing one of these qualities should provide sufficient insight about other dimensions' behaviour across the network. The quality of Trustworthiness was selected for the experiments.

Homophily Distribution. Prior to our experiments, we have analysed the homophily distribution within the dataset. This was done to offer an early intuition as to whether the proposed approach should be valid on the data used for the research.

Based on the topological structure of the network, we have measured the EMD (Earth's Mover Distance) between the grade distributions of each pair of connected nodes in our network. Given that EMD is a measure and not a classifier, its output is a numeric value dependent on the distributions in question and on additional distance matrix that serves as a calibrating tool. The amount of "work" needed to shift one distribution into another, the measure offered by the EMD, is thus not an output that points out at similarity or lack of it directly.

For this reason, additional analysis was conducted, aimed at determining the sufficient cut-off point for the specific case of our research. This included

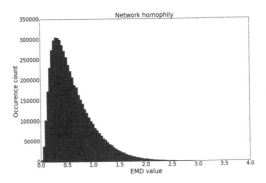

Fig. 1. The distribution of EMD values for each pair of connected nodes in the network

analysing heatmaps and plots of actual grade distributions in an effort to find an EMD value which denoted a big enough difference between the actual shapes of five-point distributions to be considered a satisfying discriminating point.

The value we have settled on was 0.74, as it offered partial leniency towards small differences in distribution shape, while properly filtering out obviously dissimilar cases. Knowing that value, it can be observed on the Fig. 1 that the majority of the network does fit the criteria of homophily defined as having the EMD value of a link between the given article and their neighbour not higher than the aforementioned 0.74. Based on this observation, we have concluded that the application of our proposed method had a chance to yield valuable results.

Per-Score Homophily. In addition to measuring the overall homophily of the network, we analysed the data to gain possible insight into the similarity of the articles on a more granular level. To achieve that, we created heatmaps illustrating the density of connections between the nodes, characterised by having

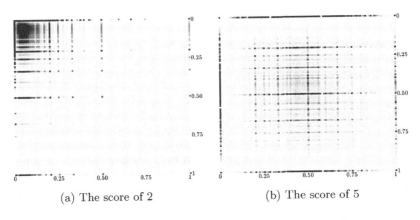

(a) The score of 2   (b) The score of 5

Fig. 2. The heatmap of score correlations for specific grades only

a specific proportion of a given score against all of the grades received. These proportions are marked on both the Y and X axis, for the source and target nodes of an edge, respectively, and range from 0 (an article hasn't received the specified grade) to 1 (all of the grades received by the article were the one we are focusing on). The figures below provide an example of these per-score distributions (Fig. 2).

4.2   Experimental Setup

The Wikipedia article network, as specified above, was the object of our research on which the validity of the predictive method was evaluated. For the purposes of our experiment, we treated all of the articles that received no scores as unknown. Additionally, over the course of several repetitions of the experiment, the articles that were graded less than 1, 10, 50 and 100 times were also treated as if their AFT scores were unknown. Finally, a varying percentage (between 5% and 95%, in 5 point increments) of articles that exceeded these numbers were also treated as unknown. Our method used the known articles as its operational basis to predict the grade distributions for the articles considered unknown. Prediction accuracy was verified on the unknown articles that received at least one grade in AFT.

Additionally, we verified the prediction not only for the full range of grades (from 1 to 5) but also partial prediction on all the applicable sub-ranges (from 1 to 2, from 1 to 3 etc.).

All presented results are mean numbers, achieved from 100 independently repeated realizations of the experiment for a given set of parameters (minimal number of grades, amount of known nodes, and sub-range of grades) (Table 1).

Table 1. Article count versus grade count

Grade count	≥ 0	≥ 1	≥ 10	≥ 50	≥ 100
Articles	1 576 178	1 430 643	174 085	27 045	10 252

The implementation of the algorithm presented in Sect. 3.2 was used to perform the prediction of article quality, understood as the distribution of AFT scores on all nodes marked as 'unknown'. The validity of the prediction process was measured on those of the 'unknown' articles which, received actual grading from Wikipedia users. A prediction was assumed to be valid if the EMD measure for the factual score distribution and the predicted score distribution was not exceeding the value of 0.74, for the reasons explained earlier in the article.

Additionally, the results of our prediction were compared to the results of a simple benchmark method. This baseline assumed that, if we were to follow the idea of a mean field to an extreme degree, the more elaborate predictive algorithm could be substituted with a method that assigns every article a quality vector representing mean quality over an entire dataset. More specifically, the amount of grades of 1 in such a vector would be the mean amount of 1 s received by the articles in the entire network; similar was done for 2 and the following grades. The accuracy of the benchmark method was measured in the same way as the accuracy of our proposed algorithm.

Implementation. Method was implemented using modern technologies following latest trends in software development. For our experimental environment, we had a distributed computing cluster available for our use. Because of that, we were able to conduct fast, reliable and reproducible experiments. In order to do those, we implemented the method using Apache Spark framework, with the use of GraphX [13]. The availability of Pregel [6] operator gave us a perfectly fitting base for our method. As can be seen in the pseudo-code (Algorithm 1), the algorithm is perfectly designed for systems that use the Bulk-Synchronous-Parallel [11] communication paradigm. That design makes our method highly parallelizable and distributable. Thanks to that, it can be easily applied to large datasets without any problems.

4.3 Results

Experimental results show low accuracy of prediction on a full range of grades (Fig. 3a); for the base of known articles consisting of those with at least 10 grades, the accuracy does not exceed 30%. The result is marginally better than the accuracy of a benchmark method, but follows a similar general trend. When the non-homophilic grade of 5 is not considered in the experiment (Fig. 3b), however, the differences in accuracy grow considerably, allowing our method to significantly outperform the naive prediction based on a network-wide mean.

In both of the presented setups, we can observe an interesting anomaly related to the percentage of the known nodes. As the proportion of known to unknown nodes in the network grows in favour of the known nodes, the prediction accuracy falls for both the benchmark and our method. Notably, the falloff is higher for the mean-based benchmark method, as the set of nodes the verification is performed

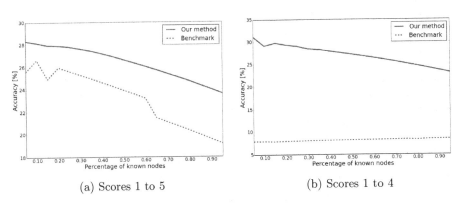

(a) Scores 1 to 5   (b) Scores 1 to 4

Fig. 3. Experimental results for two sub-ranges of grades, for known articles graded at least 10 times

on grows smaller, increasing the chance of the vast majority of it containing deprecated articles (graded very sparsely in AFT). Similar reasoning can be applied to our method's results, however its stability is much higher.

Figure 3b illustrates the highest grade's impact on the predictions. When removed, the mean-based benchmark's accuracy is reduced by roughly two-thirds, while our method remains on a similar level of accuracy as it was before. This strongly suggests that while the quality of the data does affect our proposed approach to a degree, the extreme nature of a specific grade's distribution over the network is virtually meaningless. In this sense, our method is robust.

5   Conclusions and Future Work

While the presented method failed to achieve high accuracy scores, the results obtained point towards its future potential and applicability on a richer, more balanced data. Its low accuracy was caused mainly by a low average number of grades assigned to the articles.

As presented in Sect. 4.1, there are significant differences in the homophily of articles when individual scores are considered. Extreme grades (i.e. 5), are assigned to articles that do not neighbour with other highly-rated articles. This can be a result of unwillingness of the reviewers to assign high scores, or showcases that the articles which are heavily worked on are picked from among various topics, instead of being localized in specific themes. This directly impacts the proposed method focused on the topology of the network.

During the research, all of the possible in-Wikipedia links were treated equally for the purpose of creating the network. This might have lead to the emergence of connections that either don't introduce useful data or create outright noise. This possible interaction warrants further investigation in potential future work.

Future work will be focused on application of a more sophisticated relational learning models to provide more accurate and less variable results. Progressing from a single-layer network towards the multi-layer approach, to build more nuanced relations between the items, is also a viable direction for the research. Additionally, the method could be adapted for different uses, for example as a supportive tool in search engines.

Acknowledgement. The authors express their gratitude towards R. Nielek and A. Wierzbicki from the Polish-Japanese Academy of Information Technology for the dataset. The work was supported by Wroclaw Centre for Networking and Supercomputing as well as Faculty of Computer Science and Management's statutory funds.

References

1. De la Calzada, G., Dekhtyar, A.: On measuring the quality of wikipedia articles. In: Proceedings of the 4th Workshop on Information Credibility, WICOW 2010, pp. 11–18. ACM (2010)

2. Dalip, D.H., Gonçalves, M.A., Cristo, M., Calado, P.: Automatic quality assessment of content created collaboratively by web communities: a case study of wikipedia. In: Proceedings of the 9th ACM/IEEE-CS Joint Conference on Digital Libraries, JCDL 2009, NY, USA, pp. 295–304. ACM, New York (2009)

3. Hu, M., Lim, E.P., Sun, A., Lauw, H.W., Vuong, B.Q.: Measuring article quality in wikipedia: models and evaluation. In: Proceedings of the Sixteenth ACM Conference on Conference on Information and Knowledge Management, CIKM 2007, pp. 243–252. ACM (2007)

4. Kazienko, P., Kajdanowicz, T.: Label-dependent node classification in the network. Neurocomputing **75**(1), 199–209 (2012)

5. Liu, J., Ram, S.: Who does what: collaboration patterns in the wikipedia and their impact on article quality. ACM Trans. Manage. Inf. Syst. **2**(2), 11:1–11:23 (2011)

6. Malewicz, G., Austern, M.H., Bik, A.J., Dehnert, J.C., Horn, I., Leiser, N., Czajkowski, G.: Pregel: a system for large-scale graph processing. In: Proceedings of the 2010 ACM SIGMOD International Conference on Management of data, pp. 135–146. ACM (2010)

7. McPherson, M., Smith-Lovin, L., Cook, J.M.: Birds of a feather: homophily in social networks. Ann. Rev. Sociol. **27**(1), 415–444

8. Sen, P., Namata, G.M., Bilgic, M., Getoor, L., Gallagher, B., Eliassi-Rad, T.: Collective classification in network data. AI Mag. **29**(3), 93–106 (2008)

9. Suzuki, Y., Yoshikawa, M.: Mutual evaluation of editors and texts for assessing quality of wikipedia articles. In: Proceedings of the Eighth Annual International Symposium on Wikis and Open Collaboration, WikiSym 2012, NY, USA, pp. 18:1–18:10. ACM, New York (2012)

10. Taskar, B., Abbeel, P., Koller, D.: Discriminative probabilistic models for relational data. In: Proceedings of the Eighteenth Conference on Uncertainty in Artificial Intelligence, UAI 2002, pp. 485–492. Morgan Kaufmann Publishers Inc., San Francisco (2002)

11. Valiant, L.G.: A bridging model for parallel computation. Commun. ACM **33**(8), 103–111 (1990)

12. Wohner, T., Peters, R.: Assessing the quality of wikipedia articles with lifecycle based metrics. In: Proceedings of the 5th International Symposium on Wikis and Open Collaboration, WikiSym 2009, NY, USA, pp. 16:1–16:10. ACM, New York (2009)

13. Xin, R.S., Gonzalez, J.E., Franklin, M.J., Stoica, I.: Graphx: a resilient distributed graph system on spark. In: First International Workshop on Graph Data Management Experiences and Systems, p. 2. ACM (2013)

Is a Data-Driven Approach Still Better Than Random Choice with Naive Bayes Classifiers?

Piotr Szymański$^{(\boxtimes)}$ and Tomasz Kajdanowicz

Department of Computational Intelligence, Wrocław University of Technology,
Wybrzeże Stanisława Wyspiańskiego 27, 50-370 Wrocław, Poland
piotr.szymanski@pwr.edu.pl

Abstract. We study the performance of data-driven, a priori and random approaches to label space partitioning for multi-label classification with a Gaussian Naive Bayes classifier. Experiments were performed on 12 benchmark data sets and evaluated on 5 established measures of classification quality: micro and macro averaged F1 score, subset accuracy and Hamming loss. Data-driven methods are significantly better than an average run of the random baseline. In case of F1 scores and Subset Accuracy - data driven approaches were more likely to perform better than random approaches than otherwise in the worst case. There always exists a method that performs better than a priori methods in the worst case. The advantage of data-driven methods against a priori methods with a weak classifier is lesser than when tree classifiers are used.

Keywords: Multi-label classification · Label space clustering · Data-driven classification

1 Introduction

In our recent work [11] we proposed a data-driven community detection approach to partition the label space for the multi-label classification as an alternative to random partitioning into equal subsets as performed by the random k-label sets method proposed by Tsoumakas et al. [13]. The data-driven approach works as follows: we construct a label co-occurrence graph (both weighted and unweighted versions) based on training data and perform community detection to partition the label set. Then, each partition constitutes a label space for separate multi-label classification sub-problems. As a result, we obtain an ensemble of multi-label classifiers that jointly covers the whole label space. We consider a variety of approaches: modularity-maximizing techniques approximated by fast greedy and leading eigenvector methods, infomap, walktrap and label propagation algorithms. For comparison purposes we evaluate the binary relevance (BR) and label powerset (LP) - which we call a priori methods, as they a priori assume a total partitioning of the label space into singletons (BR) and lack of any partitioning (LP).

© Springer International Publishing AG 2017
N.T. Nguyen et al. (Eds.): ACIIDS 2017, Part I, LNAI 10191, pp. 792–801, 2017.
DOI: 10.1007/978-3-319-54472-4_74

The variant of RAkEL evaluated in this paper is an approach in which the label space is either partitioned into equal-sized subsets of labels. This approach is called RAkELd - RAkEL distinct as the label sets are non-overlapping. RAkELd takes one parameter - the number of label sets to partition into k. We assumed that all partitions are equally probable and that the remainder of the label set smaller than k becomes the last element of the otherwise equally sized partition family.

In [11] we compared community detection methods to label space divisions against RAkELd and a priori methods on 12 benchmark datasets (*bibtex* [6], *delicious* [14], *tmc2007* [14], *enron* [7], *medical* [9], *scene* [1], *birds* [2], *Corel5k* [4], *Mediamill* [10], *emotions* [12], *yeast* [5], *genbase* [3]) over five evaluation measures with Classifier and Regression Trees (CART) as base classifiers. We discovered that data-driven approaches are more efficient and more likely to outperform RAkELd than binary relevance or label powerset is, in every evaluated measure. For all measures, apart from Hamming loss, data-driven approaches are significantly better than RAkELd ($\alpha = 0.05$), and at least one data-driven approach is more likely to outperform RAkELd than a priori methods in the case of RAkELd's best performance. This has been the largest RAkELd evaluation published to date with 250 samplings per value for 10 values of RAkELd parameter k on 12 datasets published to date.

In this paper we extend our result and evaluate whether the same results hold if instead of using tree-based methods, we employ a weak and Gaussian Naive Bayesian classifier from the scikit-learn python package [8]. The experimental setup remains identical to the one presented in tree-based scheme, except for the change of base classifier. Bayesian classifiers remain of interest in many applications due to their low computational requirements.

We thus repeat the research questions we have asked in the case of tree-based classifiers, this time for Naive Bayes based classifiers:

RH1: Data-driven approach is significantly better than random ($\alpha = 0.05$)

RH2: Data-driven approach is more likely to outperform RAkELd than a priori methods

RH3: Data-driven approach is more likely to outperform RAkELd than a priori methods in the worst case

RH4: Data-driven approach is more likely to perform better than RAkELd in the worst case, than otherwise

2   Results

Micro-averaged F1 Score. While a priori methods such as Binary Relevance and Label Powerset exhibit a higher median likelihood of outperforming RAkELd - we note that the highest mean likelihood is obtained by label propagation data-driven label space division on an unweighted label co-occurrence graph. Unweighted label propagation is also most likely to outperform RAkELd in the worst case.

Table 1. P-values of data-driven methods performing better than an average run of RA*k*EL*d* for each measure tested using non-parametric Friedman test with Rom's post-hoc test. Only methods with p-values greater than $\alpha = 0.05$ are presented. All approaches not listed explicitly were significantly better than RA*k*EL*d* in all measures.

	FG	FGW	LE	LEW	WTW
Macro-averaged F1	0.068	0.37	0.054	0.37	0.37
Micro-averaged F1	**0.011**	0.071	**0.003**	**0.011**	**0.043**
Jaccard Score	**0.026**	0.07	**0.008**	**0.026**	0.070

Table 2. Likelihood of performing better than RA*k*EL*d* in Micro-averaged F1 score of every method for each data set

	Minimum	Median	Mean	Std
BR	0.369565	**1.000000**	0.796143	0.283117
LP	0.369565	0.999076	0.789076	0.294146
fastgreedy	0.263556	0.781778	0.737634	0.232243
fastgreedy-weighted	0.322667	0.601848	0.633698	0.160196
infomap	0.448000	0.869778	0.817113	0.194957
infomap-weighted	0.091556	0.797333	0.705199	0.299230
label_propagation	**0.529778**	0.908000	**0.843744**	0.172125
label_propagation-weighted	0.317778	0.662356	0.703653	0.243097
leading_eigenvector	0.302667	0.829778	0.748593	0.250929
leading_eigenvector-weighted	0.341778	0.632063	0.684237	0.185325
walktrap	0.321333	0.717391	0.719968	0.246686
walktrap-weighted	0.239556	0.600889	0.632683	0.221396

Thus we reject **RH2** and accept **RH3** and **RH4**. The best performing and recommended community detection method for micro-averaged F1 score - unweighted label propagation - is better than average performance of RA*k*EL*d* with statistical significance, we thus accept **RH1**.

Macro-averaged F1 Score. In case of macro averaged F1 score Label Powerset is the most likely to outperform RA*k*EL*d* both in median and mean cases, while underperforms in the worst case. Label propagation data-driven label space division on an unweighted label co-occurrence graph is the most likely data-driven approach to outperform RA*k*EL*d* - although other approaches also yield good results. Unweighted label propagation is also most likely to outperform RA*k*EL*d* in the worst case. It is also better than an average run of RA*k*EL*d* with statistical significance. Thus we accept **RH1**, reject **RH2** and accept **RH3** and **RH4**.

Subset Accuracy. In case of Subset Accuracy label propagation performed on an unweighted graph approach to dividing the labels space is the most resilient

Table 3. Likelihood of performing better than RAkELd in Micro-averaged F1 score of every method for each data set. BR - Binary Relevance, LP - Label Powerset, FG - fastgreedy, FGW - fastgreedy weighted, IN - infomap, INW - infomap weighted, LPG - label propagation, LPGW - label propagation weighted, LE - leading eigenvector, LEW - leading eigenvector weighted, WT - walktrap, WTW - walktrap weighted.

	BR	LP	FG	FGW	IN	INW	LPG	LPGW	LE	LEW	WT	WTW
Corel5k	0.39	0.37	0.85	0.79	0.87	0.09	0.99	0.32	0.9	0.91	0.43	0.68
bibtex	0	0	0.26	0.32	0.45	0.3	0.53	0.34	0.30	0.34	0.32	0.24
birds	0	0.999	0.62	0.6	0.66	0.66	0.66	0.66	0.79	0.62	0.95	0.34
delicious	0	0	0.78	0.59	0.87	0.59	0.87	0.62	0.83	0.72	0.54	0.58
emotions	0.43	0.37	0	0.52	0	0	0	0.57	0	0.52	0	0.57
enron	0.98	0.98	0.94	0.88	0.93	0.93	0.93	0.93	0	0.99	0.79	0.997
mediamill	0	0	0.55	0.65	0.91	0.8	0.91	0.91	0.45	0.69	0.68	0.6
medical	0	0	0.51	0.58	0.51	0.60	0.60	0.60	0.41	0.59	0.51	0.60
scene	0.37	0.37	0.72	0.63	0.80	0.80	0.80	0.80	0.72	0.63	0.72	0.63
tmc2007-500	0	0	0.89	0.55	0	0	0	0	0.85	0.63	0	0.89
yeast	0.58	0.59	0.99	0.85	0.99	0.99	0.99	0.99	0.99	0.88	0.99	0.83

Table 4. Likelihood of performing better than RAkELd in Macro-averaged F1 score of every method for each data set

	Minimum	Median	Mean	Std
BR	0.456522	**1.000000**	**0.868708**	0.222246
LP	0.434783	**1.000000**	0.850310	0.227355
fastgreedy	0.376444	0.836000	0.799503	0.210402
fastgreedy-weighted	0.378222	0.753333	0.679727	0.175535
infomap	**0.519630**	0.806861	0.810572	0.164820
infomap-weighted	0.188444	0.739130	0.728628	0.247947
label_propagation	**0.519630**	0.878667	0.841961	0.163304
label_propagation-weighted	0.500000	0.739130	0.751203	0.186984
leading_eigenvector	0.367111	0.806861	0.746465	0.232450
leading_eigenvector-weighted	0.358667	0.832457	0.722748	0.215705
walktrap	0.253778	0.877333	0.789586	0.225409
walktrap-weighted	0.302222	0.800444	0.745813	0.235022

approach both in the worst case and in the average (mean/median) likelihood. The weighted version performers equally well in the worst case, so does unweighted infomap. As the worst case performance of three data-driven methods is greater than 0.5 we accept **RH4** for Subset Accuracy. While Label Powerset performs better than label propagation in case of the median/mean likelihood of being better than RAkELd - it performs worse by 12 pp. in the worst case. Thus while rejecting **RH2** and accepting **RH3** we still recommend using

Table 5. Likelihood of performing better than RA*k*EL*d* in Macro-averaged F1 score of every method for each data set. BR - Binary Relevance, LP - Label Powerset, FG - fastgreedy, FGW - fastgreedy weighted, IN - infomap, INW - infomap weighted, LPG - label propagation, LPGW - label propagation weighted, LE - leading eigenvector, LEW - leading eigenvector weighted, WT - walktrap, WTW - walktrap weighted.

	BR	LP	FG	FGW	IN	INW	LPG	LPGW	LE	LEW	WT	WTW
Corel5k	0.94	0.78	0.37	0.37	0.89	0.18	0.997	0.76	0.36	0.36	0.25	0.3
bibtex	1.0	1.0	0.53	0.57	0.67	0.52	0.88	0.55	0.52	0.61	0.6	0.47
birds	1.0	1.0	0.98	0.84	0.52	0.52	0.52	0.52	0.99	0.96	0.97	0.97
delicious	1.0	1.0	1.0	0.79	1.0	1.0	1.0	1.0	1.0	0.85	0.88	0.97
emotions	0.46	0.46	0.93	0.52	0.93	0.93	0.93	0.5	0.93	0.52	0.93	0.5
enron	1.0	0.998	0.986	0.89	0.66	0.66	0.66	0.66	0.88	0.89	0.99	0.91
mediamill	1.0	1.0	0.84	0.75	0.99	0.91	0.99	0.99	0.76	0.84	0.89	0.8
medical	1.0	1.0	0.7	0.45	0.7	0.74	0.74	0.74	0.39	0.45	0.70	0.74
scene	0.46	0.43	0.65	0.65	0.74	0.74	0.74	0.74	0.65	0.65	0.65	0.65
tmc2007-500	1.0	1.0	0.99	0.8	1.0	1.0	1.0	1.0	0.92	0.83	1.0	0.98
yeast	0.7	0.68	0.8	0.83	0.81	0.81	0.81	0.81	0.81	1.0	0.81	0.91

Table 6. Likelihood of performing better than RA*k*EL*d* in Subset Accuracy of every method for each data set

	Minimum	Median	Mean	Std
BR	0.217391	0.886667	0.777640	0.285316
LP	0.413043	**1.000000**	**0.924946**	0.174772
fastgreedy	0.028637	0.585333	0.621030	0.304067
fastgreedy-weighted	0.007852	0.586728	0.512003	0.225171
infomap	0.429000	0.978500	0.887924	0.203588
infomap-weighted	**0.533487**	0.934783	0.831424	0.195409
label_propagation	**0.533487**	0.998222	0.912394	0.165066
label_propagation-weighted	**0.533487**	0.934783	0.834437	0.180916
leading_eigenvector	0.000000	0.644000	0.604389	0.355451
leading_eigenvector-weighted	0.000000	0.568988	0.499787	0.304284
walktrap	0.133487	0.600000	0.625201	0.295569
walktrap-weighted	0.000000	0.608696	0.499824	0.331589

data-driven label propagation approach instead of Label Powerset. Label propagation performs better than RA*k*EL*d* with statistical significance - we accept **RH1** (Tables 1, 2, 3, 4, 5, 6, 7, 8, 9, 10 and 11).

Jaccard Score. Among data-driven methods the label propagation performed on an unweighted graph approach to dividing the labels space is the most resilient approach both in the worst case and in the average (mean/median) likelihood.

Table 7. Likelihood of performing better than RA*k*EL*d* in Subset Accuracy of every method for each data set. BR - Binary Relevance, LP - Label Powerset, FG - fastgreedy, FGW - fastgreedy weighted, IN - infomap, INW - infomap weighted, LPG - label propagation, LPGW - label propagation weighted, LE - leading eigenvector, LEW - leading eigenvector weighted, WT - walktrap, WTW - walktrap weighted.

	BR	LP	FG	FGW	IN	INW	LPG	LPGW	LE	LEW	WT	WTW
Corel5k	0.34	0.87	0.59	0.68	0.99	0.59	0.998	0.83	0.0	0.0	0.34	0.0
bibtex	0.89	1.0	0.37	0.69	0.96	0.61	0.998	0.7	0.64	0.73	0.37	0.31
birds	0.996	0.997	0.029	0.007	0.53	0.53	0.53	0.53	0.09	0.0	0.13	0.03
delicious	1.0	1.0	0.997	0.63	1.0	0.999	1.0	1.0	1.0	0.79	0.79	0.92
emotions	0.21	0.41	1.0	0.48	1.0	1.0	1.0	0.61	1.0	0.48	1.0	0.61
enron	0.86	1.0	0.58	0.57	0.98	0.98	0.98	0.98	0.79	0.67	0.6	0.65
mediamill	1.0	1.0	0.45	0.29	0.96	0.87	0.96	0.96	0.41	0.38	0.57	0.28
medical	1.0	1.0	0.43	0.64	0.43	0.64	0.64	0.64	0.33	0.65	0.43	0.64
scene	0.63	0.93	0.63	0.3	0.93	0.93	0.93	0.93	0.63	0.3	0.63	0.3
tmc2007-500	1.0	1.0	0.75	0.59	1.0	1.0	1.0	1.0	0.75	0.57	1.0	0.87
yeast	0.62	0.96	0.999	0.76	0.999	0.999	0.999	0.999	0.999	0.92	0.999	0.88

Table 8. Likelihood of performing better than RA*k*EL*d* in Jaccard Similarity of every method for each data set

	Minimum	Median	Mean	Std
BR	0.326087	**1.000000**	0.784597	0.303331
LP	0.369565	**1.000000**	0.847350	0.240611
fastgreedy	0.183372	0.756000	0.674557	0.274675
fastgreedy-weighted	0.177367	0.586957	0.591697	0.194144
infomap	**0.411085**	0.925333	0.831944	0.218665
infomap-weighted	0.053778	0.804889	0.686328	0.327207
label_propagation	**0.411085**	0.974500	**0.86552**	0.203504
label_propagation-weighted	0.239111	0.630435	0.689132	0.281967
leading_eigenvector	0.308000	0.777333	0.693396	0.272005
leading_eigenvector-weighted	0.116859	0.653745	0.624674	0.222935
walktrap	0.359556	0.696444	0.668188	0.252658
walktrap-weighted	0.080370	0.586957	0.580375	0.244502

It is followed by infomap. While a priori methods are perform better in case of the median likelihood by 3 pp., they perform worse than data-driven methods in the mean and worst case. We thus confirm **RH2** and **RH3**. The worst case likelihood of data-driven methods outperforming RA*k*EL*d* is not grater than 0.5 we thus reject **RH4**. Unweighted infomap performs better than the average run of RA*k*EL*d* with statistical significance - we thus accept **RH1**.

Table 9. Likelihood of performing better than RA*k*EL*d* in Jaccard Similarity of every method for each data set

	BR	LP	FG	FGW	IN	INW	LPG	LPGW	LE	LEW	WT	WTW
Corel5k	0.35	0.47	0.76	0.8	0.9	0.05	0.996	0.24	0.78	0.83	0.43	0.57
bibtex	1.0	1.0	0.31	0.42	0.86	0.40	0.99	0.45	0.42	0.45	0.36	0.28
birds	1.0	0.999	0.18	0.18	0.41	0.41	0.41	0.41	0.32	0.12	0.45	0.08
delicious	1.0	1.0	0.7	0.53	0.77	0.44	0.77	0.47	0.74	0.7	0.49	0.49
emotions	0.33	0.37	0.98	0.52	0.98	0.98	0.98	0.63	0.98	0.52	0.98	0.63
enron	0.993	1.0	0.84	0.82	0.97	0.97	0.97	0.97	0.999	0.87	0.74	0.88
mediamill	1.0	1.0	0.54	0.65	0.93	0.80	0.93	0.93	0.44	0.68	0.7	0.6
medical	1.0	1.0	0.41	0.55	0.41	0.55	0.55	0.55	0.31	0.56	0.41	0.55
scene	0.39	0.85	0.80	0.59	0.93	0.93	0.93	0.93	0.8	0.59	0.8	0.59
tmc2007-500	1.0	1.0	0.89	0.6	1.0	1.0	1.0	1.0	0.85	0.65	1.0	0.9
yeast	0.57	0.63	0.994	0.85	0.994	0.994	0.994	0.994	0.994	0.91	0.994	0.82

Table 10. Likelihood of performing better than RA*k*EL*d* in Hamming Loss of every method for each data set

	Minimum	Median	Mean	Std
BR	0.110667	0.558538	0.579872	0.376954
LP	0.080889	0.652174	0.592830	0.379345
fastgreedy	**0.208889**	0.418222	0.513625	0.276367
fastgreedy-weighted	0.111111	0.260870	0.302981	0.223065
infomap	0.112889	0.735111	0.684758	0.292563
infomap-weighted	**0.204889**	**0.847826**	**0.727799**	0.291282
label_propagation	0.111111	0.735111	0.684971	0.312656
label_propagation-weighted	0.237778	0.735111	0.714660	0.237049
leading_eigenvector	0.121333	0.498029	0.552381	0.315482
leading_eigenvector-weighted	0.111111	0.260870	0.337735	0.227415
walktrap	0.111111	0.418667	0.541611	0.331449
walktrap-weighted	0.094226	0.328113	0.387505	0.228658

Hamming Loss. The data-driven methods that are most likely to outperform RA*k*EL*d* are infomap and label propagation performed on a weighted label co-occurence graph. We recommend using weighted infomap which is also most resilient in the worst case, although much less resilient than the desired 0.5 likelihood of outperforming RA*k*EL*d* in the worst case. As a result the case of Hamming Loss we confirm **RH2** and **RH3** but reject **RH4**. Weighted infomap perform significantly better than an average run of RA*k*EL*d* - we accept **RH1**.

Table 11. Likelihood of performing better than RA*k*EL*d* in Hamming Loss of every method for each data set. BR - Binary Relevance, LP - Label Powerset, FG - fastgreedy, FGW - fastgreedy weighted, IN - infomap, INW - infomap weighted, LPG - label propagation, LPGW - label propagation weighted, LE - leading eigenvector, LEW - leading eigenvector weighted, WT - walktrap, WTW - walktrap weighted.

	BR	LP	FG	FGW	IN	INW	LPG	LPGW	LE	LEW	WT	WTW
Corel5k	0.11	0.15	0.42	0.35	0.33	0.96	0.23	0.86	0.21	0.22	0.42	0.42
bibtex	0.11	0.08	0.31	0.24	0.11	0.20	0.11	0.24	0.27	0.24	0.17	0.26
birds	1.0	0.99	0.27	0.16	0.7	0.7	0.7	0.7	0.37	0.2	0.41	0.09
delicious	0.11	0.11	0.34	0.11	0.36	0.24	0.36	0.39	0.41	0.11	0.11	0.2
emotions	0.43	0.30	1.0	0.28	1.0	1.0	1.0	0.54	1.0	0.28	1.0	0.54
enron	0.40	0.69	0.31	0.3	0.73	0.73	0.73	0.73	0.94	0.57	0.27	0.43
mediamill	0.998	0.999	0.21	0.23	0.74	0.51	0.74	0.74	0.12	0.31	0.35	0.22
medical	1.0	1.0	0.77	0.94	0.77	0.88	0.88	0.88	0.79	0.93	0.77	0.88
scene	0.65	0.65	0.52	0.26	0.85	0.85	0.85	0.85	0.52	0.26	0.52	0.26
tmc2007-500	1.0	1.0	0.57	0.17	1.0	1.0	1.0	1.0	0.5	0.26	1.0	0.64
yeast	0.56	0.55	0.94	0.3	0.94	0.94	0.94	0.94	0.94	0.33	0.94	0.33

Table 12. The summary of evaluated hypotheses and proposed recommendations of this paper

	Micro-averaged F1	Macro-averaged F1	Subset accuracy	Jaccard similarity	Hamming loss
RH1	Yes	Yes	Yes	Yes	Yes
RH2	Undecided	No	No	Undecided	Yes
RH3	Yes	Yes	Yes	Yes	Yes
RH4	Yes	Yes	Yes	No	No
Recommended data-driven approach	Unweighted label propagation	Unweighted label propagation	Unweighted label propagation	Unweighted label propagation	Weighted infomap

3 Conclusion and Outlook

We have examined the performance of data-driven, a priori and random approaches to label space partitioning for multi-label classification with a Gaussian Naive Bayes classifier. Experiments were performed on 12 benchmark data sets and evaluated on 5 established measures of classification quality. Table 12 summarizes out findings. Data-driven methods are significantly better than an average RA*k*EL*d* run that had not undergone parameter estimation - i.e. when results are compared against the mean result of all evaluated RA*k*EL*d* paramater values. When compared against the likelihood of outperforming a RA*k*EL*d* in the evaluated parameter space - in case of F1 scores and Subset Accuracy - data driven approaches were more likely to perform better than RA*k*EL*d* than otherwise in the worst case. There always exists a method that performs better than a priori methods in the worst case.

Data driven methods perform better than a priori methods in the mean likelihood but worse in median when it comes to micro-averaged F1 and Subset Accuracy. This can be attributed to differences in how likelihoods per data set distribute - data-driven methods perform better in worst case, but are also less likely to be always better than RA*k*EL*d* as opposed to a priori methods. The advantage of data-driven methods against a priori methods with a weak classifier is lesser than when tree classifiers are used. The authors acknowledge support from the National Science Centre research projects decision no. 2016/21/N/ST6/02382 and 2016/21/D/ST6/02948.

References

1. Boutell, M.R., Luo, J., Shen, X., Brown, C.M.: Learning multi-label scene classification. Pattern Recogn. **37**(9), 1757–1771 (2004). http://www.sciencedirect.com/science/article/pii/S0031320304001074
2. Briggs, F., Lakshminarayanan, B., Neal, L., Fern, X.Z., Raich, R., Hadley, S.J.K., Hadley, A.S., Betts, M.G.: Acoustic classification of multiple simultaneous bird species: a multi-instance multi-label approach. J. Acoust. Soc. Am. **131**(6), 4640–4650 (2012). http://scitation.aip.org/content/asa/journal/jasa/131/6/10.1121/1.4707424
3. Diplaris, S., Tsoumakas, G., Mitkas, P.A., Vlahavas, I.: Protein classification with multiple algorithms. IEEE Trans. Pattern Anal. Mach. Intel., 448–456 (2005). http://www.springerlink.com/index/P662542G78792762.pdf
4. Duygulu, P., Barnard, K., Freitas, J.F.G., Forsyth, D.A.: Object recognition as machine translation: learning a lexicon for a fixed image vocabulary. In: Heyden, A., Sparr, G., Nielsen, M., Johansen, P. (eds.) ECCV 2002. LNCS, vol. 2353, pp. 97–112. Springer, Heidelberg (2002). doi:10.1007/3-540-47979-1_7
5. Elisseeff, A., Weston, J.: A kernel method for multi-labelled classification. In: Advances in Neural Information Processing Systems, vol. 14, pp. 681–687. MIT Press (2001)
6. Katakis, I., Tsoumakas, G., Vlahavas, I.: Multilabel text classification for automated tag suggestion. In: Proceedings of the ECML/PKDD-08 Workshop on Discovery Challenge (2008)
7. Klimt, B., Yang, Y.: The Enron corpus: a new dataset for email classification research. In: Boulicaut, J.-F., Esposito, F., Giannotti, F., Pedreschi, D. (eds.) ECML 2004. LNCS (LNAI), vol. 3201, pp. 217–226. Springer, Heidelberg (2004). doi:10.1007/978-3-540-30115-8_22
8. Pedregosa, F., Varoquaux, G., Gramfort, A., Michel, V., Thirion, B., Grisel, O., Blondel, M., Prettenhofer, P., Weiss, R., Dubourg, V., Vanderplas, J., Passos, A., Cournapeau, D., Brucher, M., Perrot, M., Duchesnay, E.: Scikit-learn: machine learning in Python. J. Mach. Learn. Res. **12**, 2825–2830 (2011)
9. Read, J., Pfahringer, B., Holmes, G., Frank, E.: Classifier chains for multi-label classification. Mach. Learn. **85**(3), 333–359 (2011). http://link.springer.com/article/10.1007/s10994-011-5256-5
10. Snoek, C.G.M., Worring, M., Gemert, J.C.V., Geusebroek, J.M., Smeulders, A.W.M.: The challenge problem for automated detection of 101 semantic concepts in multimedia. In: Proceedings of the ACM International Conference on Multimedia, pp. 421–430. ACM Press (2006)

11. Szymanski, P., Kajdanowicz, T., Kersting, K.: How is a data-driven approach better than random choice in label space division for multi-label classification? Entropy **18**(8), 282 (2016). http://dx.doi.org/10.3390/e18080282

12. Trohidis, K., Tsoumakas, G., Kalliris, G., Vlahavas, I.P.: Multi-label classification of music into emotions. ISMIR **8**, 325–330 (2008)

13. Tsoumakas, G., Vlahavas, I.: Random k-Labelsets: an ensemble method for multilabel classification. In: Kok, J.N., Koronacki, J., Mantaras, R.L., Matwin, S., Mladenič, D., Skowron, A. (eds.) ECML 2007. LNCS (LNAI), vol. 4701, pp. 406–417. Springer, Heidelberg (2007). doi:10.1007/978-3-540-74958-5_38

14. Tsoumakas, G., Katakis, I., Vlahavas, I.: Effective and efficient multilabel classification in domains with large number of labels. In: Proceedings of ECML/PKDD 2008 Workshop on Mining Multidimensional Data (MMD 2008), pp. 30–44 (2008)

An Expert System to Assist with Early Detection of Schizophrenia

Sonya Rapinta Manalu[1,2], Bahtiar Saleh Abbas[1], Ford Lumban Gaol[1],
Lukas[1,3], and Bogdan Trawiński[4(✉)]

[1] Computer Science Doctoral Study Program, Bina Nusantara University, Jakarta, Indonesia
{smanalu,Bahtiars,fgaol}@binus.edu, lukas@atmajaya.ac.id
[2] School of Computer Science, Bina Nusantara University, Jakarta, Indonesia
[3] Cognitive Engineering Research Group (CERG), Faculty of Engineering,
Universitas Katolik Indonesia Atma Jaya, Jakarta, Indonesia
[4] Department of Information Systems, Faculty of Computer Science and Management,
Wrocław University of Science and Technology, Wrocław, Poland
bogdan.trawinski@pwr.edu.pl

Abstract. Schizophrenia is one of neurobiological disorders whose symptoms appear in young age. Psychiatrists use client-centered therapy to recognize five symptoms of schizophrenia. They comprise: delusions, hallucinations, negative symptoms, grossly disorganized or abnormal motor behavior, and disorganized thinking and speech. Patients experiencing their acute psychotic episode must be quarantined to prevent their unsocial behavior. An early detection of the schizophrenia symptoms is necessary to avoid acute psychotic episodes. An expert system to aid in diagnosing early schizophrenia symptoms is presented in the paper. The system represents psychiatrist knowledge in the form of rules, facts and events and uses them to assess whether a patient suffers from schizophrenia. The knowledge-base is displayed in the form of questions to the patient. The expert system uses forward chaining while gathering the answers from patient. All answers are transformed into facts and processed using Boolean reasoning to generate the diagnosis. The diagnosis states whether the patient has schizophrenia, and if so, it indicates also the type of schizophrenia.which

Keywords: Schizophrenia · Symptoms · Expert system · Boolean reasoning

1 Introduction

Schizophrenia is one of mental health disorders whose symptoms appear in young age. Psychiatrist, as an assessor, recognizes if a patient suffers from schizophrenia trying to diagnose its five symptoms: delusions, hallucinations, negative symptoms, grossly disorganized or abnormal motor behavior, and disorganized thinking and speech [1, 2], as shown in Fig. 1. Delusion is a firm belief in something untrue or not based on reality; patients misinterpret events and their significance. Hallucination is a false perception of five senses: vision, hearing, smell, taste, or touch; patients hear voices that no one else can hear or see, smell or feel things that others do not. Negative symptom is a condition in which patients experience a decrease in quality of life. Sadner [3] lists three negative

© Springer International Publishing AG 2017
N.T. Nguyen et al. (Eds.): ACIIDS 2017, Part I, LNAI 10191, pp. 802–812, 2017.
DOI: 10.1007/978-3-319-54472-4_75

symptoms: affective flattening, alogia and avolition. Grossly disorganized or abnormal motor behavior including catatonia is characterized by childish behavior or emotional outbursts. Disorganized thinking or speech is a condition when patients have trouble in concentrating and organizing their thoughts logically making effective communication impaired. When patients experience acute psychotic episodes they must be quarantined to prevent their unsocial behavior. The community often stigmatizes persons with schizophrenia labelling them as the demented. According to Depkes RI (Departemen Kesehatan Republik Indonesia) 1.7 per thousand of Indonesia population experience schizophrenia. Assistance of psychiatrists is needed to support the patients and educate them how to cope with their illness [4].

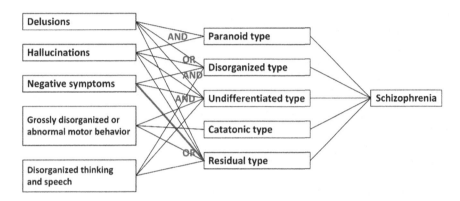

Fig. 1. Schizophrenia symptoms and types (*source* [1, 2])

Psychiatrists apply client-centered therapy to assess whether a patient suffers from schizophrenia [5, 6]. The client-centered therapy uses a question and answer approach to respond statements according to a number of keywords [7]. This therapy is conducted by face-to-face sessions and all the answers are written down. Each schizophrenia symptom has its own set of keywords [7]. For example, delusions have ten keywords and each of them has its own standard questions to indicate whether the patient exhibits symptoms of delusions. These standard questions are used during the client-centered therapy as the basic questions to be answered by the patient. The psychiatrist analyzes the answers and evaluates the patient for a mental disorder. Client-centered therapy is used to assess delusions, hallucinations and disorganized thinking and speech. During the session the psychiatrist evaluates the patient for grossly disorganized or abnormal motor behavior, including catatonia, by analyzing his/her face and body gestures. The psychiatrist also interprets changes in emotions or unbridled emotions to identify the negative symptoms [1, 2]. At the end of a session the psychiatrist gives a diagnosis whether a person has schizophrenia and determines the type of the illness. There are five types of schizophrenia as recognized by the DSM-5 (Diagnostic & Statistical Manual of Mental Disorders) [1] and shown in Fig. 1. The first one is paranoid schizophrenia, if the person experiences delusions and hallucinations. Disorganized schizophrenia occurs if the patient manifests negative symptoms and disorganized thinking and speaking and also shows delusion or hallucination. Undifferentiated schizophrenia arises when the person reveals general symptoms of schizophrenia without any dominant symptom. The catatonic type of schizophrenia takes place when the patient displays grossly disorganized or abnormal motor behavior including catatonia. And the last type is residual schizophrenia which occurs when the patient suffers from all symptoms with a negative symptom as the dominant one. Acute psychotic episodes can be prevented provided the symptoms are detected early and psychiatrist administers treatment according to symptoms.

The psychiatrist can be assisted during the therapeutic session by a computer program like ELIZA and PARRY [8–10]. Both ELIZA and PARRY simulated the conversation between the psychiatrist and patient. ELIZA was developed by Joseph Weizenbaum using pattern matching techniques and played as a psychiatrist to response the patient. In turn, PARRY, created by Kenneth Colby, attempted to simulate a person exhibiting paranoid schizophrenia. Both of ELIZA and PARRY could simulate a client-centered therapy, where computer acted as a psychiatrist to gather the patient data during a therapeutic session. However, they were early natural language processing computer programs and their usefulness was limited.

Nowadays, expert systems (ES) are the class of software capable of helping psychiatrists in diagnosing mental disorders [11–13]. Some ES are designed specifically to support detecting schizophrenia symptoms [14–16]. Much attention has been also paid to clinical decision support in diagnosing schizophrenia based on the analysis of EEG signals of schizophrenic patients [17–19]. The other computer-aided diagnosis methods for schizophrenia encompass the analysis of abnormalities in brain activation patterns using functional magnetic resonance imaging (fMRI) [20–22].

An expert system designed to assist the psychiatrist with diagnosing early schizophrenia symptoms is presented in the paper. The system represents psychiatrist's knowledge in the form of rules, facts and events. The system holds a dialogue with a patient and while gathering the answers from the patient it uses the forward chaining method to evaluate the rules. Next, all answers are transformed into facts and processed using Boolean reasoning to generate the diagnosis.

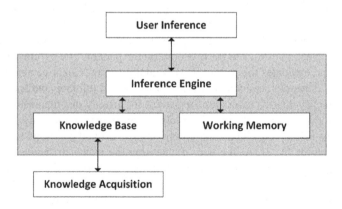

Fig. 2. General architecture of the proposed expert system

2   Structure of the Proposed Expert System

Expert systems (ES) are part artificial intelligence and constitute a class of computer programs which simulate human reasoning to resolve complicated decision-making problems [23, 24]. ES performs reasoning over a representation of human knowledge rather than through conventional procedural code. The aim of ES is to replace or aid a human expert in accomplishing complex tasks, such as diagnosing diseases or equipment faults, making financial or weather forecasts, and scheduling vehicle routes or robotic actions.

General architecture of proposed ES to aid in schizophrenia diagnosis is depicted in Fig. 2. The system comprises five main components: knowledge acquisition, knowledge base, user interface, working memory, and inference engine. The knowledge acquisition consists in acquiring information on schizophrenia symptoms, diagnosis and therapy from medical books and journals as well as the psychiatrist expertise and transforming it into rules which are stored in the knowledge base. The knowledge base is composed of 48 rules which are converted into questions presented to the user during a session. The user interface controls the dialog between ES and the user. It displays a series of questions and reads the user's responses into the working memory. The working memory is utilized as a storage of data entered by the user and results provided by the inference engine. The inference engine imitates human reasoning and based on the user's responses and knowledge base generates the conclusion whether the patient suffers from schizophrenia. Moreover, it determines the type of schizophrenia if this is the case. The engine employs the forward chaining technique to deduce its diagnosis [15]. Boolean reasoning is used while mapping rules and facts. The inference engine divides rules into five main categories as shown in Fig. 1, namely delusion, hallucination, negative symptoms, disorganized thinking and speech, and grossly disorganized or abnormal motor behavior. Next, the engine checks the patient's symptoms to generate the type of schizophrenia. There are five types of schizophrenia, namely paranoid, disorganized, undifferentiated, catatonic, and residual schizophrenia, as illustrated in Fig. 1. ES generates and displays its diagnosis to the end user through the user interface after all rules are examined.

ES applies the client-centered therapy in the form of a question-answer session. There are no dominant rules in this research. Inference engine employs a nominal scale for each patient's answer, i.e. each YES answer is denoted by 1 and NO answer is marked by 0. ES concludes that the patient suffers from a given symptom of schizophrenia if he/she answers at least one YES for symptom's questions. But according to schizophrenia diagnostic criteria the patient has one or all schizophrenia symptoms if he/she exhibits signs of schizophrenia for at least six months and this happens frequently. Therefore, ES asks also five supportive questions for each rule. The questions are as follows:

1. How long does he/she experience the symptom?
2. When did the symptom start?
3. When did the symptom last appear?
4. How often does the symptom occur?
5. How and what does he/she feel while exhibiting the symptoms?

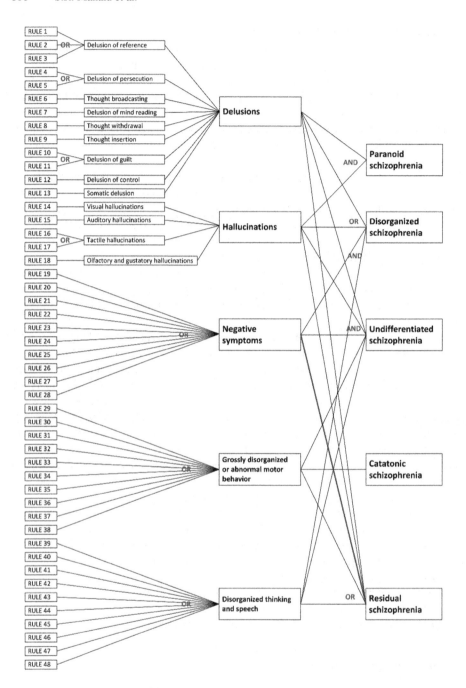

Fig. 3. Map of expert system rules

The ES rules are divided into five categories corresponding to the five main symptoms of schizophrenia, as shown in Fig. 3. The first category, delusion, has nine subcategories. Each subcategory of delusion is evaluated as depicted in Fig. 4. Delusion of reference contains three rules: RULE 1, RULE 2, and RULE 3. If at least one of the three rules is fired then the signs of delusion of reference are proved. Delusion of persecution comprises two rules: RULE 4 and RULE 5. At least one of the two rules should be satisfied to prove the symptoms of delusion of persecution. Thought broadcasting, delusion of mind reading, thought withdrawal, and thought insertion consist of RULE 6, RULE 7, RULE 8, and RULE 9, respectively. In turn, delusion of guilt embraces two rules: RULE 10 and RULE 11. At least one of the two rules should be fired to prove the symptoms of delusion of guilt. And finally, delusion of control and somatic delusion include RULE 12 and RULE 13, respectively. A patient suffers from delusion, if he/she satisfies at least one of the aforementioned rules.

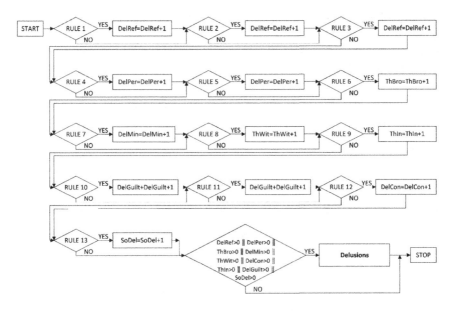

Fig. 4. Evaluation of the delusion symptoms

The evaluation of the second category, hallucination symptoms, is illustrated in Fig. 5. This category has five main rules and four subcategories. Visual hallucinations and auditory hallucinations contain RULE 14 and RULE 15, respectively. In turn, tactile hallucinations comprise two rules: RULE 16 and RULE 17. If at least one of the two rules is fired then the signs of tactile hallucinations are proved. And finally, olfactory and gustatory hallucinations consist of RULE 18. Appraisal of this hallucination symptoms is almost the same as in the case of delusion. A patient suffers from hallucination, if he/she satisfies at least one of the aforementioned rules.

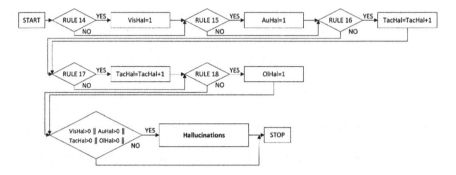

Fig. 5. Evaluation of the hallucination symptoms

The assessment of the third category, grossly disorganized or abnormal motor behavior including catatonia, is presented in Fig. 6. This category comprises ten rules that need to be verified together with the five supportive questions. If all the rules and supportive questions are satisfied, the patient is identified to exhibit grossly disorganized or abnormal motor behavior.

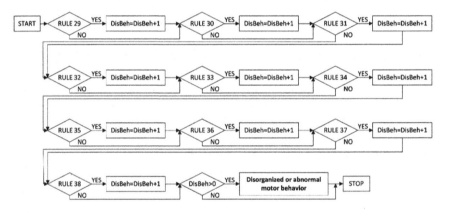

Fig. 6. Evaluation of the grossly disorganized or abnormal motor behavior

The fourth category, disorganized thinking and speech, embraces ten rules. Its evaluation procedure is shown in Fig. 7. Similarly to the three preceding symptoms, this category also contains supportive questions for each rule to verify whether the patient shows the symptom. The last category, negative symptoms, also has ten rules. Evaluation of this category is almost the same as in the case of the other symptoms (see Fig. 8).

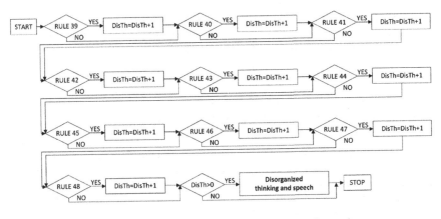

Fig. 7. Evaluation of disorganized thinking and speech

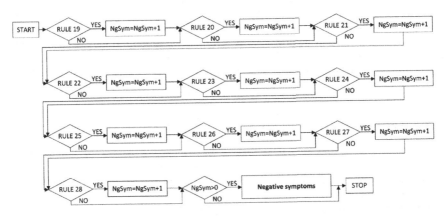

Fig. 8. Evaluation of negative symptoms

3   Results and Discussion

All rules embedded in the knowledge base have equal weights. Boolean reasoning was used to generate the result during this research. Each question connected with a given rule must be answered either YES or NO. Five additional questions about the symptoms have to be answered to support each rule with YES response. These questions ask how long a given symptom recurs, how often the symptom occurs, when it starts, when the symptom last appeared, and how he/she feels when the symptoms appears. The questions strengthen the diagnosis because the patient is recognized to suffer from schizophrenia if he/she has the symptoms at least six months and the symptoms appear frequently during this period. If the patient satisfies a given rule and the criteria expressed by supportive questions are met, then ES adds to the conclusions that he/she exhibits the rule's symptom. ES generates the result through following steps:

1. A rule with YES answer is saved in the working memory.
2. Check whether the answer for supportive questions meets the criteria.
3. Determine a sub symptom according to (1) and (2).
4. Determine the symptom according to (3).
5. Determine the schizophrenia type according to (4).

ES shows the type of schizophrenia with supported facts in conclusion. The first one is the paranoid type of schizophrenia. This type is diagnosed when at least one symptom of delusions or hallucinations is recognized. The second type, disorganized schizophrenia, is determined if at least one sign of disorganized thinking and speech and negative symptoms is distinguished, but also if at least one of delusion or hallucination rules is satisfied. The third type, catatonic schizophrenia, is indicated if at least one rule of disorganized or abnormal motor behavior is fulfilled. The forth type, undifferentiated schizophrenia, is diagnosed if at least one rule for each symptom is met, but there is no dominant symptom. The last type, residual schizophrenia, is identified if at least one negative symptom is dominant and at least one positive sign (of delusions or hallucinations) is exhibited.

ES diagnoses whether a patient suffers from schizophrenia if he/she meets criteria depicted in Figs. 3, 4, 5, 6, 7 and 8, where each symptom has at least six month episode's period. ES processes the patient's answers and maps them into five schizophrenia types (Figs. 1 and 3). For example:

1. Patient answered YES to the questions connected with the 6th and 15th rules.
2. ES analyzed the answers to supportive questions.
3. ES determined that the patient exhibits the delusion of percussion and auditory hallucination sub symptoms.
4. From these sub symptoms ES concludes that the patient has delusion and hallucination symptoms.
5. ES performs mapping as in Figs. 3, 4 and 5 and formulated the diagnosis that the patient suffers from the paranoid type of schizophrenia.

Due to limitations of this ES it is necessary that the parents or guardians also answer a number of questions. They should answer all questions connected with disorganized behavior and negative symptoms. They should answer to a number of questions to verify some patient's responses particularly those related with the period of an episode, date of the last episode and frequency pattern.

ES shows only the diagnosis if the patient suffers from schizophrenia and what type of schizophrenia he/she has. The system does not handle the action to be taken, the psychiatrist may use results provided by ES as hints about therapy. According to [1, 2, 26] there are several actions to be taken. Antipsychotic medications are used for positive symptoms, antidepressant are applied for depression symptoms, glycine supplements are administered for negative symptoms, omega-3 fatty acids help to reduce negative and positive symptoms, as well as antioxidants. Patients in severe conditions need to be quarantined to prevent their unwanted actions. Psychotherapy can be used to educate patients how they can deal with their conditions and socialize.

4   Conclusions and Future Work

The expert system presented in the paper can diagnose schizophrenia according to the medical diagnostic criteria and the psychiatrist's expertise. The expert system helps the psychiatrist in early detection of schizophrenia. The system generates the diagnosis based on the patient's responses to a series of questions representing the knowledge-base. It shows supported facts but does not propose the action to be done. The psychiatrist should administer the treatment according to the result provided by the expert system.

Future research is needed to introduce dominant rules and evaluate them to optimize the diagnosis process. The user interface should be designed based on Human Computer Interaction (HCI) principles adapted especially to schizophrenia needs. The new interface could optimize the process of gathering data from patients as well as improve the usability of the system and increase user's satisfaction. Voice recognition is also needed to automate the client-centered therapy especially in voice analysis. The noise emitted by the patient while mumbling or talking to himself cannot be heard by humans. According to psychiatrist expertise such a noise might be the symptom of delusion or hallucination. The psychiatrists also suggest to implement gesture recognition to analyze mood, true answers, and disorganized behavior during the client-centered therapy. The user interface can be also equipped with natural language processing to analyze the answers to the five supportive questions. Machine learning methods could be also employed to evaluate the stage of schizophrenia to help the psychiatrist to take the right treatment.

References

1. Diagnostic and Statistical Manual of Mental Disorders: DSM-5 (5th edn.). American Psychiatric Association, Arlington (2013)
2. Sue, D., Sue, D.W., Sue, D.M., Sue, S.: Understanding Abnormal Behavior, 11th edn. Wadsworth Publishing, Belmont (2015)
3. Sadner, G.: Schizophrenia: a cognitive science viewpoint. In: Almann, K.V. (ed.) Schizophrenia Research Trends, pp. 55–93. Nova Science Publishers, New York (2008)
4. Depkes RI. Lighting the Hope for Schizoprenia Warnai Peringatan Hari Kesehatan Jiwa tahun 2014. http://www.depkes.go.id/article/view/201410270010/lighting-the-hope-forschizoprenia-warnai-peringatan-hari-kesehatan-jiwa-tahun-2014.html. Accessed 28 Oct 2016
5. Gendlin, E.T.: Client-centered developments and work with schizophrenics. J. Couns. Psychol. 9(3), 205–212 (1962)
6. van Blarikom, J.: A person-centered approach to schizophrenia. Person-Centered and Experiential Psychotherapies 5(4), 155–173 (2006)
7. Rogers, C.R.: Significant aspects of client-centered therapy. Am. Psychol. 1(10), 415–422 (1946)
8. Weizenbaum, J.: ELIZA - a computer program for the study of natural language communication between man and machine. Commun. ACM 9(1), 36–45 (1966)
9. Shah, H., Warwick, K., Vallverdú, J., Wu, D.: Can machines talk? Comparison of Eliza with modern dialogue systems. Comput. Hum. Behav. 58, 278–295 (2016)

10. Güzeldere, G., Franchi, S.: Dialogues with colorful personalities of early AI. Stanford Humanities Review, SEHR 4:2, Constructions of the Mind (1995) http://web.stanford.edu/group/SHR/4-2/text/dialogues.html
11. Yap, R.H., Clarke, D.M.: An expert system for psychiatric diagnosis using the DSM-III-R, DSM-IV and ICD-10 classifications. In: Proceedings of the AMIA Annual Fall Symposium, pp. 229–233 (1996)
12. Yankovskaya, A., Kitler, S.: Mental disorder diagnostic system based on logical-combinatorial methods of pattern recognition. Comput. Sci. J. Moldova **21**(3), 391–400 (2013)
13. Singh, S.: A fuzzy rule based expert system to diagnostic the mental illness (MIDExS). Int. J. Innovative Res. Comput. Commun. Eng. **3**(9), 8759–8764 (2015)
14. Razzouk, D., Mari, J.J., Shirakawa, I., Wainer, J., Sigulem, D.: Decision support system for the diagnosis of schizophrenia disorders. Braz. J. Med. Biol. Res. **39**(1), 119–128 (2006)
15. Singh, P.K., Sarkar, R.: A simple and effective expert system for schizophrenia detection. Int. J. Intell. Syst. Technol. Appl. **14**(1), 27–49 (2015)
16. Nunes, L.C., Pinheiro, P.R., Cavalcante, T.P., Pinheiro, M.C.D.: Handling diagnosis of schizophrenia by a hybrid method. Comput. Math. Methods Med. Article Id 987298 (2015) http://doi.org/10.1155/2015/987298
17. Sabeti, M., Katebi, S.D., Boostani, R., Price, G.W.: A new approach for EEG signal classification of schizophrenic and control participants. Expert Syst. Appl. **38**(3), 2063–2071 (2011)
18. Hiesh, M.H., et al.: Classification of schizophrenia using genetic algorithm-support vector machine (ga-svm). In: 5th Annual International Conference of the IEEE Engineering in Medicine and Biology Society (EMBC). IEEE, pp. 6047–6050 (2013). doi:10.1109/EMBC.2013.6610931
19. Elgohary, M.I., Alzohairy, T.A., Eissa, A.M., Eldeghaidy, S., Hussein, M.: An intelligent system for diagnosing schizophrenia and bipolar disorder based on MLNN and RBF. Int. J. Sci. Res. Sci. Eng. Technol. **2**(4), 117–123 (2016)
20. Rashid, B., Damaraju, E., Pearlson, G.D., Calhoun, V.D.: Dynamic connectivity states estimated from resting fMRI identify differences among schizophrenia, bipolar disorder, and healthy control subjects. Front. Human Neurosci. **8**, 897 (2014). doi:10.3389/fnhum.2014.00897
21. Juneja, A., Rana, B., Agrawal, R.K.: A combination of singular value decomposition and multivariate feature selection method for diagnosis of schizophrenia using fMRI. Biomed. Signal Process. Control **27**, 122–133 (2016)
22. Varshney, A., Prakash, C., Mittal, N., Singh, P.: A multimodel approach for schizophrenia diagnosis using fMRI and sMRI dataset. Intelligent Systems Technologies and Applications 2016. AISC, vol. 530, pp. 869–877. Springer, Heidelberg (2016). doi:10.1007/978-3-319-47952-1_69
23. Lucas, P., Van Der Gaag, L.: Principles of Expert Systems (International Computer Science Series). Addison-Wesley, Boston (1991)
24. Giarratano, J.C., Riley, G.D.: Expert Systems: Principles and Programming, 4th edn. Course Technology, Boston (2004)
25. Gupta, S., Singhal, R.: Fundamentals and characteristics of an expert system. Int. J. Recent Innov. Trends Comput. Commun. **1**(3), 110–113 (2013)
26. Lehman, A.F., et al.: Practice guideline for the treatment of patients with schizophrenia. Am. J. Psychiatry **161**, 1–56 (2004). Second edition

Author Index

Printed in the United States
By Bookmasters